Handbook of
Experimental Pharmacology

Volume 175

Editor-in-Chief
K. Starke, Freiburg i. Br.

Editorial Board
G.V.R. Born, London
S.P. Duckles, Irvine, CA
M. Eichelbaum, Stuttgart
D. Ganten, Berlin
F. Hofmann, München
W. Rosenthal, Berlin
G. Rubanyi, Richmond, CA

Neurotransmitter Transporters

Contributors

G. Ahnert-Hilger, W. Armsen, M. H. Bazalakova, H. Betz,
R. D. Blakely, C. Blex, H. Bönisch, I. Brunk, M. Brüss†,
M. A. Cervinski, N. Chen, L. J. DeFelice, R. H. Edwards,
V. Eulenburg, H. Farhan, J. D. Foster, M. Freissmuth, A. Galli,
K. Gerstbrein, U. Gether, J. Gomeza, B. K. Gorentla,
D. Gründemann, M. Höltje, B. von Jagow, A. Lazar, K.-P. Lesch,
Z. Lin, B. K. Madras, M. S. Mazei-Robison, R. Mössner,
K. Nørgaard-Nielsen, I. Pahner, D. Piston, M. W. Quick,
M. E. A. Reith, M. B. Robinson, J. D. Rothstein, G. Rudnick,
R. Sattler, E. Schömig, J. W. Schwartz, R. P. Seal, H. H. Sitte,
J. Sternberg, R. J. Vandenberg, R. A. Vaughan, J. M. Williams,
S. Winter, J. Zhen

Editors
Harald H. Sitte and Michael Freissmuth

Prof. Dr.
Harald H. Sitte
Medical University Vienna
Institute of Pharmacology
Center for Biomolecular Medicine
and Pharmacology
Waehringer Str. 13a
A-1090 Wien
Austria
harald.sitte@meduniwien.ac.at

Prof. Dr.
Michael Freissmuth
Medical University Vienna
Institute of Pharmacology
Center for Biomolecular Medicine
and Pharmacology
Waehringer Str. 13a
A-1090 Wien
Austria
michael.freissmuth@meduniwien.ac.at

With 67 Figures and 29 Tables

ISSN 0171-2004

ISBN-10 3-540-29783-9 Springer Berlin Heidelberg New York

ISBN-13 978-3-540-29783-3 Springer Berlin Heidelberg New York

This work is subject to copyright. All rights reserved, whether the whole or part of the material is concerned, specifically the rights of translation, reprinting, reuse of illustrations, recitation, broadcasting, reproduction on microfilm or in any other way, and storage in data banks. Duplication of this publication or parts thereof is permitted only under the provisions of the German Copyright Law of September 9, 1965, in its current version, and permission for use must always be obtained from Springer. Violations are liable for prosecution under the German Copyright Law.

Springer is a part of Springer Science+Business Media
springer.com

© Springer-Verlag Berlin Heidelberg 2006
Printed in Germany

The use of general descriptive names, registered names, trademarks, etc. in this publication does not imply, even in the absence of a specific statement, that such names are exempt from the relevant protective laws and regulations and therefore free for general use.

Product liability: The publishers cannot guarantee the accuracy of any information about dosage and application contained in this book. In every individual case the user must check such information by consulting the relevant literature.

Editor: Simon Rallison, Heidelberg
Desk Editor: Susanne Dathe, Heidelberg
Cover design: *design & production* GmbH, Heidelberg, Germany
Typesetting and production: LE-TEX Jelonek, Schmidt & Vöckler GbR, Leipzig, Germany
Printed on acid-free paper   27/3100-YL - 5 4 3 2 1 0

Preface

Neurotransmitter transporters play a pivotal role in synaptic transmission: They function as a reset button and thus define the shape and duration of signal transfer across the synaptic cleft. This statement nowadays appears to be a self-evident truism, but the concept is fairly young: It builds on landmark studies on "reuptake sites" that were done by Georg Hertting and Julius Axelrod about four decades ago. It is, in this context, a special pleasure that Georg Hertting will celebrate his 80th birthday while this volume goes to press. Initially, the pharmacological and biochemical characterization relied on classical organ and tissue preparation techniques as well as on brain homogenates and platelets. This allowed for the characterization of the substrate specificities of individual transport processes and the identification of specific inhibitors, and it spurred conjecture about the co-transported ions and the stoichiometric coupling of ion influx and substrate translocation. The impact on clinical medicine of these efforts is best documented by the continuous proliferation of improved—that is, more selective—antidepressant drugs. A major breakthrough was achieved by the purification of a prototypical transporter and by the more or less concomitant isolation of cDNAs that encoded various transporters some 15 years ago. This marked the advent of a new era: What had been an assorted mix of functional activities that was best defined by pharmacological criteria was now attributable and accounted for by defined molecular species. Researchers were given the opportunity to address the structural and functional mysteries of individual transporter proteins (which rapidly proliferated). Transporter proteins displayed unexpected new functional features, e.g. channel-like properties with various types of currents. Transporters had originally been conceived as static clamps; the glory of regulation (desensitization, endocytotic recycling, resensitization) went to the receptors. However, it is now clear that transporter proteins also integrate input from kinases that they subject to endocytotic internalization and recycling. Similarly, in neurons, transporters must be delivered to a very specific area, namely the perisynaptic zone, and retained at these sites, presumably by interacting with accessory proteins. Thus, transporters serve as interesting models to understand the process underlying sorting and targeting.

Because the gene loci of individual transporters were assigned, it was possible to search for links to diseases and to search for genetic polymorphisms in non-coding regions, which were of pathophysiological significance in disease states. Work on the serotonin transporter has been particularly inspiring, because it not only had repercussions on clinical medicine but also on the debate that pitted the advocates of nurture against those of nature (genetic versus social determinism). With the dawn of the new millennium, the crystal era has also reached the field of neurotransmitter transport. The acme was reached by contributions that visualized the structures of several bacterial transporters in atomic detail. Because some of these proteins are reasonably closely related to the major mammalian neurotransmitter transporter families, the available structures shed light on the structural constraints that have to be imposed on the models developed from mutagenesis. One is tempted to speculate that this additional knowledge may substantially refine drug development; there is, for instance, an unmet need to develop a true SNRI (selective serotonin and norepinephrine reuptake inhibitor), because the currently available compounds either block one of the two transporters poorly (e.g. venlafaxine) or with low selectivity (tricyclic antidepressants, which block many other receptors).

If we look back to the early 1960s, it is evident that our understanding has increased dramatically. However, there are still many open questions. As a matter of fact, these still include the very mechanism of substrate translocation and the actions of antidepressant drugs that inhibit some transporters and drugs of abuse that induce transport reversal.

The present volume on neurotransmitter transporters for the Springer *Handbook of Experimental Pharmacology* aims at providing an overview of insights that were generated in the past 5 years. If the volume serves as both a useful compendium of current concepts and an inspiring starting point for future research, it will have fulfilled its mission and will be a source for students interested in this emerging field as well as for experienced scientists looking for an update.

This volume is the brainchild of the series editor Klaus Starke—awe-inspiring to all pharmacologists of younger generations—who motivated us by his very interest in our work. Our special thanks go to all contributors; they found the time to cope with our requests, to meet deadlines and to share their ideas and insights to make this book possible. Last but not least, we wish to express our gratitude to Susanne Dathe, Springer Heidelberg, for her help in finalizing this exciting project.

As mentioned above, this volume goes to print as one of the fathers of the field of neurotransmitter transport celebrates his 80th birthday; we therefore dedicate this volume to Georg Hertting, whose unabashed self-irony, witty comments and uncompromising honesty many of us have enjoyed over time.

Vienna, Austria,	Michael Freissmuth
Vienna, Austria,	Harald H. Sitte
January 2006	

List of Contents

Zn^{2+} Modulation of Neurotransmitter Transporters 1
   K. Nørgaard-Nielsen, U. Gether

Molecular Microfluorometry: Converting Arbitrary Fluorescence Units
into Absolute Molecular Concentrations to Study Binding Kinetics
and Stoichiometry in Transporters 23
   J. W. Schwartz, D. Piston, L. J. DeFelice

Structure/Function Relationships in Serotonin Transporter:
New Insights from the Structure of a Bacterial Transporter 59
   G. Rudnick

The Importance of Company: Na^+ and Cl^- Influence Substrate
Interaction with SLC6 Transporters and Other Proteins 75
   M. E. A. Reith, J. Zhen, N. Chen

Currents in Neurotransmitter Transporters 95
   K. Gerstbrein, H. H. Sitte

Mutational Analysis of Glutamate Transporters 113
   R. J. Vandenberg

The Diverse Roles of Vesicular Glutamate Transporter 3 137
   R. P. Seal, R. H. Edwards

Extraneuronal Monoamine Transporter and Organic Cation
Transporters 1 and 2: A Review of Transport Efficiency 151
   E. Schömig, A. Lazar, D. Gründemann

The Role of SNARE Proteins in Trafficking and Function
of Neurotransmitter Transporters 181
   M. W. Quick

Regulation of the Dopamine Transporter by Phosphorylation 197
   J. D. Foster, M. A. Cervinski, B. K. Gorentla, R. A. Vaughan

The Dopamine Transporter: A Vigilant Border Control
for Psychostimulant Action 215
   J. M. Williams, A. Galli

Oligomerization of Neurotransmitter Transporters:
A Ticket from the Endoplasmic Reticulum to the Plasma Membrane . . 233
   H. Farhan, M. Freissmuth, H. H. Sitte

Acute Regulation of Sodium-Dependent Glutamate Transporters:
A Focus on Constitutiveand Regulated Trafficking 251
   M. B. Robinson

Regulation and Dysregulation of Glutamate Transporters 277
   R. Sattler, J. D. Rothstein

Regulation of Vesicular Monoamine and Glutamate Transporters
by Vesicle-Associated Trimeric G Proteins: New Jobs for Long-Known
Signal Transduction Molecules 305
   I. Brunk, M. Höltje, B. von Jagow, S. Winter, J. Sternberg, C. Blex,
   I. Pahner, G. Ahnert-Hilger

Human Genetics and Pharmacology of Neurotransmitter Transporters 327
   Z. Lin, B. K. Madras

ADHD and the Dopamine Transporter:
Are There Reasons to Pay Attention? 373
   M. S. Mazei-Robison, R. D. Blakely

Inactivation of 5HT Transport in Mice: Modeling Altered
5HT Homeostasis Implicated in Emotional Dysfunction,
Affective Disorders, and Somatic Syndromes 417
   K. P. Lesch, R. Mössner

Lessons from the Knocked-Out Glycine Transporters 457
   J. Gomeza, W. Armsen, H. Betz, V. Eulenburg

The Norepinephrine Transporter in Physiology and Disease 485
   H. Bönisch, M. Brüss

The High-Affinity Choline Transporter:
A Critical Protein for Sustaining Cholinergic Signaling
as Revealed in Studies of Genetically Altered Mice 525
   M. H. Bazalakova, R. D. Blakely

Subject Index 545

List of Contributors

Addresses given at the beginning of respective chapters

Ahnert-Hilger, G., 305
Armsen, W., 457

Bazalakova, M. H., 525
Betz, H., 457
Blakely, R. D., 373, 525
Blex, C., 305
Bönisch, H., 485
Brunk, I., 305
Brüss, M., 485

Cervinski, M. A., 197
Chen, N., 75

DeFelice, L. J., 23

Edwards, R. H., 137
Eulenburg, V., 457

Farhan, H., 233
Foster, J. D., 197
Freissmuth, M., 233

Galli, A., 215
Gerstbrein, K., 95
Gether, U., 1
Gomeza, J., 457
Gorentla, B. K., 197
Gründemann, D., 151

Höltje, M., 305

Jagow, B. von, 305

Lazar, A., 151
Lesch, K. P., 417
Lin, Z., 327

Madras, B. K., 327
Mazei-Robison, M. S., 373
Mössner, R., 417

Nørgaard-Nielsen, K., 1

Pahner, I., 305
Piston, D., 23

Quick, M. W., 181

Reith, M. E. A., 75
Robinson, M. B., 251
Rothstein, J. D., 277
Rudnick, G., 59

Sattler, R., 277
Schömig, E., 151
Schwartz, J. W., 23
Seal, R. P., 137
Sitte, H. H., 95, 233
Sternberg, J., 305

Vandenberg, R. J., 113
Vaughan, R. A., 197

Williams, J. M., 215
Winter, S., 305

Zhen, J., 75

Zn^{2+} Modulation of Neurotransmitter Transporters

K. Nørgaard-Nielsen · U. Gether (✉)

Molecular Neuropharmacology Group, Department of Pharmacology,
The Panum Institute, University of Copenhagen, 2200 Copenhagen N, Denmark
gether@neuropharm.ku.dk

1	Introduction	2
2	Synaptic Zn^{2+} Is a Potential Modulator of Several Neurotransmitter Systems	5
3	Endogenous Zn^{2+}-Binding Sites in Na^+/Cl^--Dependent Neurotransmitter Transporters	7
4	Endogenous Zn^{2+}-Binding Sites in the Excitatory Amino Acid Transporters	9
5	Zn^{2+} Modulation of DAT Function	11
6	Reversal of Zn^{2+} Sensitivity by Intracellular Mutations in the DAT	13
7	Zn^{2+} Is Found in Presynaptic Vesicles and is Released Upon Neuronal Stimulation	16
	References	18

Abstract Neurotransmitter transporters located at the presynaptic or glial cell membrane are responsible for the stringent and rapid clearance of the transmitter from the synapse, and hence they terminate signaling and control the duration of synaptic inputs in the brain. Two distinct families of neurotransmitter transporters have been identified based on sequence homology: (1) the neurotransmitter sodium symporter family (NSS), which includes the Na^+/Cl^--dependent transporters for dopamine, norepinephrine, and serotonin; and (2) the dicarboxylate/amino acid cation symporter family (DAACS), which includes the Na^+-dependent glutamate transporters (excitatory amino acid transporters; EAAT). In this chapter, we describe how the identification of endogenous Zn^{2+}-binding sites, as well as engineering of artificial Zn^{2+}-binding sites both in the Na^+/Cl^--dependent transporters and in the EAATs, have proved to be an important tool for studying the molecular function of these proteins. We also interpret the current available data on Zn^{2+}-binding sites in the context of the recently published crystal structures. Moreover, we review how the identification of endogenous Zn^{2+}-binding sites has indirectly suggested the possibility that several of the transporters are modulated by Zn^{2+} in vivo, and thus that Zn^{2+} can play a role as a neuromodulator by affecting the function of neurotransmitter transporters.

Keywords Neurotransmitter transporters · Synaptic zinc · Modulation · Conformational changes

1
Introduction

Synaptic neurotransmission mediated by neurotransmitters is a crucial component of neuronal communication in the central nervous system. Upon arrival of an action potential, the neurotransmitters are released from presynaptic vesicles into the synaptic cleft where the chemical substance can exert its action by binding to distinct pre- and postsynaptic receptors and hence promote downstream signaling events (Amara and Arriza 1993). Integral membrane neurotransmitter transporters located either in the presynaptic or glial cell membrane are responsible for the stringent and rapid clearance of the transmitter from the synapse and thus for termination of signaling and thereby the duration of the synaptic input (Iversen 1971). Importantly, disturbances in the homeostasis of the major neurotransmitters play roles in various pathological conditions including epilepsy, cerebral ischemia, Alzheimer's disease, Parkinson's disease, major depression, and schizophrenia (Amara and Fontana 2002; Dauer and Przedborski 2003; Fava and Kendler 2000; Lewis and Lieberman 2000).

Based on sequence homology, two distinct gene families of plasma membrane neurotransmitter transporters have been identified: (1) the neurotransmitter sodium symporter family (NSS), which includes the Na^+/Cl^--dependent transporters for dopamine, norepinephrine, serotonin, γ-aminobutyric acid (GABA) and glycine; and (2) the dicarboxylate/amino acid cation symporter family (DAACS), which includes the Na^+-dependent glutamate transporters (excitatory amino acid transporters, EAAT) (Sonders et al. 2005). The transporters for dopamine, norepinephrine, and serotonin are the targets for several psychostimulants including amphetamine and cocaine as well many therapeutic antidepressive agents (Fava and Kendler 2000; Rothman and Baumann 2003). The EAATs are not the direct target for any currently available drugs; however, recent evidence suggests that an increase in the expression of GLT-1 (also known as EAAT2) by β-lactam antibiotics can ameliorate the symptoms of neurodegenerative disease in mice (Rothstein et al. 2005).

Both families of neurotransmitter transporters are secondary active carriers that couple the movement of substrate to the cotransport of sodium down its concentration gradient (Sonders and Amara 1996). Whereas substrate uptake by the Na^+/Cl^--dependent transporters are coupled to the influx of both Na^+ and Cl^-, the uptake of amino acids by the EAATs are coupled to Na^+ and H^+ influx and a counter transport of K^+ (Sonders and Amara 1996). By coupling substrate transport to the movement of ions, the transport process is potentially electrogenic and the transporters can be subjected to electrophysiological measurements (Sonders and Amara 1996). Interestingly, such measurements have shown that the transport-associated currents generally exceed the currents expected from the predicted stoichiometrically coupled current (Sonders and Amara 1996). This means that, in addition to the stoichiometrical transport-coupled current, an uncoupled current is also observed in the presence of substrate (Fairman et al. 1995; Galli et al. 1995; Mager et al. 1994; Risso et al. 1996; Sonders et al. 1997; Wadiche et al. 1995). In the EAATs, the uncoupled current is of substantial magnitude and known to be carried by Cl^- (Fairman et al. 1995; Wadiche et al. 1995). The biological significance of the uncoupled conductances is, nevertheless, not yet clear, but it might be important for regulating neuronal excitability (Fairman et al. 1995; Ingram et al. 2002; Wadiche et al. 1995). Apart from the substrate-associated current, several of the transporters also possess a constitutive leak current, which can be blocked by substrate and uptake inhibitors even in the complete absence of Na^+ (Galli et al. 1995; Mager et al. 1993; Mager et al. 1994; Sonders et al. 1997; Sonders and Amara 1996; Vandenberg et al. 1995). The significance of this leak is also debated because it has been observed exclusively in heterologous expression systems. For example, no leak was observed when studying the electrophysiological properties of the dopamine transporter (DAT) in cultivated dopaminergic neurons (Ingram et al. 2002).

Analysis of the primary structures of the Na^+/Cl^--dependent transporters predicts a topology with 12 transmembrane segments connected by alternating extracellular and intracellular loops, and intracellular amino- and carboxy-termini (Fig. 1; Amara and Kuhar 1993; Bruss et al. 1995; Chen et al. 1998; Ferrer and Javitch 1998; Hersch et al. 1997; Melikian et al. 1994; Nirenberg et al. 1996). This overall topology has been confirmed by the successful crystallization of a bacterial homolog of these transporters from *Aquifex aeolicus* ($LeuT_{Aa}$) (Yamashita et al. 2005). The structure of this transporter revealed

Neurotransmitter: sodium symporters (NSS)

Dopamine transporter (DAT), serotonin transporter (SERT), norepinephrine transporter (NET), GABA transporters 1, 2, 3, and 4 (GAT1-4), glycine transporters 1 and 2 (GlyT1 and GlyT2)

Ion dependence: Na^+, (Cl^-), (co-transport)

Structure: 12 TMs, dimers

Dicarboxylate/amino acid: cation symporters (DAACS)

Glutamate transporters (EAAT1 [GLAST], EAAT2 [GLT-1], EAAT3 [EAAC1], EAAT4, and EAAT5)

Ion dependence: Na^+, H^+ (co-transport) K^+ (counter transport)

Structure: 6TMs, trimers

Fig. 1 Simple two-dimensional representations of the neurotransmitter transporter structures. Based on the crystal structures of the homologous bacterial transporters $LeuT_{Aa}$ (Yamashita et al. 2005) and Glt_{Ph} (Yernool et al. 2004)

a dimer with each monomer containing 12 transmembrane helical regions organized in a unique fold. The binding site for the substrate, L-leucine, was buried inside the center of the LeuT$_{Aa}$ with transmembrane segments (TM) 1, 3, 6, and 8 forming the binding pocket (Yamashita et al. 2005). Remarkably, a crystal structure of the bacterial homolog of the mammalian EAATs from *Pyrococcus horikoshii* (Glt$_{Ph}$) has also recently been published (Yernool et al. 2004). In agreement with previous predictions, the structure was very different from that of the Na$^+$/Cl$^-$-dependent transporter and revealed the existence of six regular helices N-terminally, followed by a complex C-terminal half containing the substrate and ion-binding sites (Fig. 1; Yernool et al. 2004). The structure, moreover, showed a trimer that formed a characteristic bowl with an extracellular solvent-filled basin extending halfway through the membrane. At the bottom of the bowl, the structure suggested the existence of a glutamate-binding site within each protomer (Yernool et al. 2004).

The prevailing theory for the function of the transporters is that binding of Na$^+$ and substrate produces a series of conformational changes that alters the transporter from assuming an "outward-facing" to an "inward-facing" conformation, which in turn exposes the substrate-binding site to the cell interior and allows release of substrate and Na$^+$ (Jardetzky 1966; Rudnick 1997). A prerequisite for this alternating access model is the existence of "external" as well as "internal" gates, i.e., protein domains that undergo conformational changes during the transport cycle and are capable of controlling access to the substrate-binding site from either the extracellular or the intracellular environment, respectively. Both Glt$_{Ph}$ and LeuT$_{Aa}$ were crystallized in a conformational state in which the substrate-binding sites were inaccessible both from the extracellular and intracellular environments (Yamashita et al. 2005; Yernool et al. 2004). From the crystal structures, it was accordingly possible to envisage residues that contribute to the formation of these gates; however, the dynamic function of the predicted gating domains and thus the exact mechanics of the transport process remains to be further explored.

In this chapter, we describe how the identification of endogenous Zn^{2+}-binding sites, as well as engineering of artificial Zn^{2+}-binding sites both in the Na$^+$/Cl$^-$-dependent transporters and in the EAATs, have proved to be an important tool for studying the molecular function of these proteins. We also interpret the current available data on Zn^{2+}-binding sites in the context of the recently published crystal structures. Moreover, we review how the identification of endogenous Zn^{2+}-binding sites has indirectly suggested the possibility that several of the transporters are modulated by Zn^{2+} in vivo, and thus that Zn^{2+} can play a role as a neuromodulator by affecting the function of neurotransmitter transporters.

2
Synaptic Zn^{2+} Is a Potential Modulator of Several Neurotransmitter Systems

Zn^{2+} is an important metal ion in many biological systems. Next to iron, Zn^{2+} is the most abundant transition metal in the brain (Huang 1997). Zn^{2+} is known to be required for the function of a large number of soluble proteins, playing a valuable role for enzyme catalysis and structural stability of so-called Zn^{2+}-finger proteins (Alberts et al. 1998). In addition to these well-described Zn^{2+}-binding sites in soluble proteins, increasing numbers of membrane proteins in the brain have been shown to be modulated by Zn^{2+} in low micromolar concentrations. These include:

- Ion channels, such as voltage-activated Ca^{2+} channels (Busselberg et al. 1994) and acid-sensing ion channels (Chu et al. 2004)
- Ligand-gated ion channels, such as ionotropic $GABA_A$ receptor (Draguhn et al. 1990; Horenstein and Akabas 1998), the strychnine-sensitive glycine receptor (Bloomenthal et al. 1994; Laube et al. 1995), and glutamate receptors of the α-amino-3-hydroxy-5-methyl-4-isoxazolepropionic acid (AMPA) (Lin et al. 2001; Rassendren et al. 1990) and N-methyl-D-aspartate (NMDA) subtype (Hollmann et al. 1993; Peters et al. 1987; Westbrook and Mayer 1987)
- G protein-coupled receptors—for instance the melanocortin MC_1 and MC_4 receptors (Holst et al. 2002), the tachychinin NK_3 receptor (Rosenkilde et al. 1998), and the $β_2$-adrenergic receptor (Swaminath et al. 2002, 2003)

Furthermore, Zn^{2+}-binding sites have been identified in several members of the NSS family, namely in the DAT (Norregaard et al. 1998), in the glycine transporter 1b (GLYT1b) (Ju et al. 2004) and in the GABA transporter 4 (GAT4) (Cohen-Kfir et al. 2005; Table 1). Additionally, in the DAACS family, Zn^{2+} was shown to cause inhibition of glutamate uptake by the EAAT1 (Vandenberg et al. 1998) as well as decreasing the aspartate-induced currents mediated by EAAT4 (Mitrovic et al. 2001; Table 1). Within both transporter families, the Zn^{2+}-mediated inhibition of transporter function appeared to be highly subtype-specific. Thus, monoamine uptake was not affected by Zn^{2+} in the norepinephrine transporter (NET) (Norregaard et al. 1998) and the serotonin transporter (SERT) (Mitchell et al. 2004), glycine uptake was not affected in the GLYT2a (Ju et al. 2004), GABA uptake was not affected in GAT1 (Cohen-Kfir et al. 2005; MacAulay et al. 2001) and GAT3 (Cohen-Kfir et al. 2005), and glutamate uptake was not affected in EAAT2 (Vandenberg et al. 1998). The physiological significance of the identified Zn^{2+}-binding sites is still unknown; however, their specificity together with the fact that they exist only in distinct members within the different protein families indirectly supports a putative physiological role.

Table 1 Neurotransmitter transporters and Zn^{2+}

	Species	Expression system	Effects of Zn^{2+}	Reference
Na^+/Cl^--dependent transporters				
DAT	Rat	Native striatal synaptosomes	Inhibits [^{3}H]DA uptake. Potentiates cocaine inhibition	Richfield 1993
	Human	COS-7	Inhibits [^{3}H]DA uptake ($IC_{50} \approx$ 0.8 µM) non-competitively. Potentiates cocaine inhibition	Norregaard et al. 1998
	Human	HEK-293	Increases V_{max} for efflux of [^{3}H]MPP$^+$	Scholze et al. 2002
	Human	HEK-293	Increases AMPH-induced efflux of [^{3}H]MPP$^+$. Potentiates DA and AMPH-induced inward currents	Pifl et al. 2004
	Human	Xenopus laevis oocytes	Potentiates DA and AMPH-induced inward currents (voltage-dependent). Increases charge-to-flux ratio	Meinild et al. 2004
GLYT1b	Human	Xenopus laevis oocytes	Inhibits [^{3}H]glycine uptake ($IC_{50} \approx$ 11 µM) non-competitively (voltage-independent). Inhibits glycine-induced currents (unchanged charge-to-flux ratio)	Ju et al. 2004
GAT2	Mouse	Xenopus laevis oocytes	Inhibits of [^{3}H]GABA uptake [93%; at 50 µM (Zn^{2+})]	Cohen-Kfir et al. 2005
GAT4	Mouse	Xenopus laevis oocytes	Inhibits of [^{3}H]GABA uptake ($IC_{50} \approx$ 4.5 µM). Inhibits GABA-induced currents (unchanged charge-to-flux ratio)	Cohen-Kfir et al. 2005
Excitatory amino acid transporters				
EAAT1	Human	Xenopus laevis oocytes	Inhibits [^{3}H]glutamate uptake ($IC_{50} \approx$ 10 µM) non-competitively. Inhibits glutamate- and aspartate-induced currents by inhibition of substrate flux	Vandenberg et al. 1998
EAAT4	Human	Xenopus laevis oocytes	Inhibits aspartate-induced currents by: Inhibition of uncoupled anion conductance	Mitrovic et al. 2001

AMPH, amphetamine; DA, dopamine; DAT, dopamine transporter; EAAT, excitatory amino acid transporter; GABA, γ-aminobutyric acid; GAT, GABA transporter; GLYT, glycine transporter; MPP$^+$, 1-methyl-4-phenylpyridinium

3
Endogenous Zn^{2+}-Binding Sites in Na^+/Cl^--Dependent Neurotransmitter Transporters

In the GLYT1b, Zn^{2+} was shown to inhibit glycine transport noncompetitively and in a voltage-independent manner. (Table 1; Ju et al. 2004). Under voltage-clamp conditions, Zn^{2+} did not alter the charge-to-flux ratio, indicating that Zn^{2+} inhibited currents mediated by substrate translocation by means of decreasing the rate of glycine transport (Table 1; Ju et al. 2004). Further depiction of the Zn^{2+}-binding site revealed two coordinating histidines, one in extracellular loop (ECL) 2 [$His243_{4.14}$; using the generic numbering scheme described by Goldberg and co-authors (2003)] and one in ECL4 ($His410_{7.59}$; Ju et al. 2004; Table 2). Another glycine transporter subtype, GLYT2a, was insensitive to Zn^{2+} both with respect to glycine-mediated currents and [^{3}H]glycine uptake (Ju et al. 2004). Somewhat surprisingly, introducing the two coordinating histidines in GLYT1b to the corresponding positions of GLYT2a did not confer Zn^{2+} sensitivity to this mutant transporter despite the relatively high sequence homology of approximately 50% between the two different glycine transporter subtypes (Eulenburg et al. 2005; Ju et al. 2004). The most conceivable explanation for the observation would be that the two histidines are not sufficient for binding of Zn^{2+}, which might require yet another and still-unknown coordinating residue that is unique to GLYT1b. Alternatively, the prevailing conformation of the two transporters must be different.

Isoform-dependent Zn^{2+} inhibition has also recently been shown for the GABA transporters (Cohen-Kfir et al. 2005). It was observed that GAT1 and GAT3 were not significantly affected by the presence of Zn^{2+}, whereas GABA uptake by GAT2 and GAT4 was strongly inhibited by high concentrations of

Table 2 Endogenous Zn^{2+}-binding sites in neurotransmitter transporters

	Species	Zn^{2+}-coordinating residues	References
Na^+/Cl^--dependent transporters			
DAT	Human	$His193_{3.87}$ in ECL2, $His375_{7.60}$ and $Glu396_{8.34}$ in ECL4	Norregaard et al. 1998; Loland et al. 1999
GLYT1b	Human	$His243_{4.14}$ in ECL2 and $His410_{7.59}$ in ECL4	Ju et al. 2004
GAT4	Mouse	Residues in TM6-TM12 (carboxy-terminal half of the transporter)	Cohen-Kfir et al. 2005
Excitatory amino acid transporters			
EAAT1	Human	His146 and His156 in ECL2	Vandenberg et al. 1998
EAAT4	Human	His154 and His164 in ECL2	Mitrovic et al. 2001

DAT, dopamine transporter; EAAT, excitatory amino acid transporter; ECL, extracellular loop; GAT, GABA transporter; GLYT, glycine transporter

Zn^{2+} (Table 1; Cohen-Kfir et al. 2005). Only the interaction between GAT4 and Zn^{2+} was further characterized in this study. Application of Zn^{2+} abolished substrate-mediated currents and caused a reduction in the total number of charge movements across the oocyte membrane (Cohen-Kfir et al. 2005). Nonetheless, the effective charge movement was not altered by Zn^{2+}, suggesting that the effect of Zn^{2+} was merely due to a decline in GABA uptake and, in consequence, in the substrate-elicited currents (Table 1; Cohen-Kfir et al. 2005). Notably, this was also suggested to be the mechanism of Zn^{2+} inhibition of GLYT1b function (Ju et al. 2004). Since Cohen-Kfir et al. (2005) found that Zn^{2+} inhibited GABA uptake in low micromolar concentrations, it is conceivable that at least three coordinating residues were responsible for the interaction (Regan 1995). A chimeric construct of GAT3 and GAT4 comprising the amino-terminal half of GAT3 and the carboxy-terminal half of GAT4 was also sensitive to Zn^{2+}, showing a similar responsiveness as GAT4 itself (Cohen-Kfir et al. 2005). It was suggested that this could be explained by the presence of the Zn^{2+}-binding site in the carboxy-terminal half of GAT4 (Cohen-Kfir et al. 2005); however, it certainly does not rule out the possibility that one of the coordinating residues in the amino-terminal part of the transporters is conserved between GAT3 and GAT4, whereas Zn^{2+}-binding residues in the carboxy-terminal part of GAT4 are unique to this transporter.

In DAT, Zn^{2+} in micromolar concentrations inhibits dopamine uptake in heterologous cell expression systems (Norregaard et al. 1998) as well as in striatal synaptosomes (Table 1; Richfield 1993). We demonstrated previously that this effect was due to the interaction of Zn^{2+} with a tridentate Zn^{2+}-binding site on the extracellular face of the human DAT (hDAT: $His193_{3.87}$ in ECL2, and $His375_{7.60}$ and $Glu396_{8.34}$ in ECL4; Table 2; Fig. 3; Loland et al. 1999; Norregaard et al. 1998). Despite a high sequence identity between DAT and NET, the NET does not possess an endogenous Zn^{2+}-binding site. However, by introducing $His193_{3.87}$ from the hDAT into the corresponding position in the hNET, we were able to confer full Zn^{2+}-sensitivity to the hNET-$K189_{3.87}$H (Norregaard et al. 1998). Additionally, by removing the histidine in hNET-$K189_{3.87}$H corresponding to $His375_{7.60}$ in hDAT, the mutant NET was no longer inhibited by Zn^{2+}, signifying that the coordination of Zn^{2+} in the DAT WT and the NET-$K189_{3.87}$H were due to binding of Zn^{2+} to identical binding sites in the two transporters (Norregaard et al. 1998). In the GAT1 we were also able to introduce an inhibitory Zn^{2+}-binding site by inserting histidines in positions corresponding to $His375_{7.60}$ and $Glu396_{8.34}$ in the hDAT (MacAulay et al. 2001). In this GAT1 mutant, both GABA uptake and GABA-mediated inward currents were inhibited by the presence of Zn^{2+} (MacAulay et al. 2001). Similarly, a corresponding Zn^{2+}-binding site has been introduced into SERT (Mitchell et al. 2004), further underlining the importance of this region for transporter function.

It is also interesting to reconcile these observations with the recent high-resolution structure of $LeuT_{Aa}$ (Yamashita et al. 2005). ECL2 is much smaller in LeuT$_{Aa}$ as compared to mammalian transporters, however, it is clear from the structure that ECL2 is close to ECL4 and thus indeed position 3.87 (His193 in DAT) can be in close proximity to the coordinating residues in ECL4, i.e., His375$_{7.60}$ and Glu396$_{8.34}$. His375$_{7.60}$ is located in the first of two short helices situated just above the helix of TM7, whereas Glu396$_{8.34}$ is positioned in the second of these helices within the loop (Fig. 3; Yamashita et al. 2005). In the conformation crystallized, the positions in LeuT$_{Aa}$ corresponding to His375$_{7.60}$ and Glu396$_{8.34}$ are too far apart to coordinate Zn^{2+}-binding (the distance between the α-carbons being 15–20 Å). It cannot be excluded that this is because of structural differences between the mammalian and bacterial proteins; nonetheless, we find it more likely that ECL4 undergoes major conformational changes during the transport process; consequently, this brings the residues closer together, thereby permitting Zn^{2+} binding. Notably, in the crystallized conformation of LeuT$_{Aa}$, ECL4 forms a lid that dives into the center of the transporter and forms interactions with residues deep in TM3. In this way, ECL4 covers the substrate-binding site from the outside without being in direct contact with the substrate (Yamashita et al. 2005). It is, therefore, highly conceivable that opening of the transporter to the extracellular milieu allowing access of substrate to the binding site will involve a major conformational rearrangement of ECL4. We propose, accordingly, that this rearrangement moves His375$_{7.60}$ and Glu396$_{8.34}$ into close proximity; thus, Zn^{2+} becomes capable of coordinating binding between these two residues. Obviously, the validity of this hypothesis awaits future experimental verification.

4
Endogenous Zn^{2+}-Binding Sites in the Excitatory Amino Acid Transporters

Studies on the EAAT1 expressed in *Xenopus laevis* oocytes showed that Zn^{2+} inhibited glutamate transport noncompetitively in micromolar concentrations (Table 1; Vandenberg et al. 1998). Co-application of glutamate and 100 µM Zn^{2+} reduced the substrate-mediated current in oocytes expressing EAAT1 by approximately 50%, whereas the EC$_{50}$ for glutamate was comparable in the presence and absence of Zn^{2+} (Table 1; Vandenberg et al. 1998). In contrast, Zn^{2+} in high concentrations (up to 300 µM) did not modulate EAAT2 glutamate transport currents considerably. The glutamate transporters carry an uncoupled Cl$^-$ conductance of varying size (Fairman et al. 1995; Wadiche et al. 1995) and transport of D-aspartate by the EAAT1 activates a markedly greater Cl$^-$ conductance than transport of L-glutamate (Wadiche et al. 1995). Upon application of 100 µM Zn^{2+}, the reversal potential of the D-aspartate transport current was shifted towards the reversal potential for Cl$^-$ in *Xenopus laevis* oocytes, suggesting that under conditions where Zn^{2+} is present, the Cl$^-$ ions contribute more to the transport-associated currents than under conditions where Zn^{2+} is absent (Vandenberg et al. 1998). Moreover, at membrane potentials greater than the reversal potential for D-aspartate, the difference between net transport currents in the presence vs absence of Zn^{2+} was diminished, thus illustrating that Zn^{2+} increases the Cl^- conductance relative to the substrate flux of EAAT1, whereas the absolute Cl^- conductance is not altered during the transport process (Vandenberg et al. 1998).

A mutagenesis analysis revealed that substitution of a histidine (His146) in ECL2 to an alanine caused a more than 20-fold reduction in the Zn^{2+} sensitivity of the transporter (Vandenberg et al. 1998). Additionally, alanine substitution of another histidine in ECL2 (His156) also reduced the effect of Zn^{2+} of EAAT1 in a similar fashion (Vandenberg et al. 1998). These data supported the hypothesis that both His146 and His156 contribute to the Zn^{2+}-binding site of EAAT1 (Table 2). The crystal structure of the bacterial homolog Glt_{Ph} revealed that these histidines are located in the large flexible ECL2 (Yernool et al. 2004) in which accommodation of Zn^{2+} binding is quite likely. Sequence alignment of EAAT1 and EAAT2 showed that His146 is conserved between the two transporters, whereas in EAAT2 a glycine is found in the position corresponding to His156 in EAAT1. By substituting this glycine in EAAT2 with a histidine, Vandenberg and coworkers (1998) were able to confer Zn^{2+} sensitivity to the EAAT2 mutant, thus illustrating that the organization of the EAAT1 and EAAT2 proteins might be very similar. In the EAAT4, two histidines are also found in the positions corresponding to His146 and His156 in EAAT1, and, accordingly, it is not surprising that Zn^{2+} is also able to modulate EAAT4 function; however, the mechanism appears to be somewhat different from that described for the EAAT1 (Mitrovic et al. 2001). In oocytes expressing EAAT4, co-application of Zn^{2+} and L-aspartate reduced the substrate-mediated current, whereas the reversal potential was unaffected (Table 1; Mitrovic et al. 2001). Since the observed decrease was more pronounced at positive potentials, where the outward current is mediated by an anion conductance through the transporter, it suggests that Zn^{2+} inhibits this uncoupled anion conductance of EAAT4 (Mitrovic et al. 2001). This was confirmed by the observation that under Cl^--free conditions there was no difference between the L-aspartate-mediated currents measured in the presence and absence of Zn^{2+} (Mitrovic et al. 2001). Single alanine substitutions of the two histidines in ECL2 of EAAT4 (His154 and His164) corresponding to the Zn^{2+}-coordinating residues identified in EAAT1 (Table 2) rendered the EAAT4 mutants unresponsive to Zn^{2+} modulation, possibly due to a lack of anion conductance inhibition (Mitrovic et al. 2001). While Zn^{2+} binds to sites in EAAT1 and EAAT4 that at least in part are made from coordinating residues organized in a structurally comparable manner, the effects of Zn^{2+} on transporter function appears to be somewhat different. Thus in EAAT4, Zn^{2+} possibly inhibits the uncoupled anion conductance with little effect on substrate fluxes, whereas in EAAT1, Zn^{2+} modulates the coupled substrate fluxes while the anion conductance is not affected (Mitrovic et al. 2001).

5
Zn^{2+} Modulation of DAT Function

Although there are exceptions, it is striking that the coordinating residues of the endogenous Zn^{2+}-binding sites identified within the different members of the NSS family—as well as in the DAACS family of neurotransmitter transporters—have been found in the large ECL2 and in ECL4. This certainly shows that these domains are in close proximity in the tertiary transporter structure; as well, it signifies the functional importance of these transporter domains (see also Sect. 4). Nevertheless, the exact mechanism for how Zn^{2+} affects the function of the transporters is not fully understood and might differ substantially between them.

As stated above, binding of Zn^{2+} to GLYT1b (Ju et al. 2004), GAT4 (Cohen-Kfir et al. 2005), and the DAT (Norregaard et al. 1998) results in potent noncompetitive inhibition of substrate uptake (Table 1). This led to the proposal that Zn^{2+}, upon binding to the transporters, imposes a conformational restraint that inhibits substrate translocation across the cellular membrane (Cohen-Kfir et al. 2005; Ju et al. 2004; Norregaard et al. 1998). Recent investigations have suggested, however, a more complex mechanism, at least for Zn^{2+} binding in the wildtype (WT) DAT. First, it was unexpectedly observed that via an interaction with the endogenous site in DAT, Zn^{2+} did not decrease but caused an increase in the current induced by the substrates dopamine and amphetamine in human embryonic kidney (HEK)293 cells (Pifl et al. 2004) and in *Xenopus Laevis* oocytes (Fig. 2; Table 1; Meinild et al. 2004). This was clearly incompatible with a simple conformational restraint imposed by Zn^{2+} to inhibit substrate translocation. It was also in contrast to the observations in GLYT1b and GAT4, as well as in a GAT1 mutant containing a Zn^{2+}-binding site engineered in ECL4, i.e., in these transporters Zn^{2+} decreased the substrate-induced inward current concomitantly with inhibition of uptake (Table 1; Cohen-Kfir et al. 2005; Ju et al. 2004). We observed, furthermore, that dopamine uptake in DAT under voltage-clamp conditions was not inhibited by Zn^{2+}, which again argued against a simple conformational constraint imposed by Zn^{2+} (Meinild et al. 2004). Instead, we observed a substantial increase in the charge/dopamine-coupling ratio under voltage clamped conditions, suggesting that Zn^{2+} facilitates an ion flux through the transporter (Table 1; Meinild et al. 2004). Ion substitution experiments revealed that the uncoupled inward ion flux was mainly carried by Cl^-, and that Zn^{2+} suggestively was capable of facilitating this anion conductance mediated by the DAT (Meinild et al. 2004). Based on these observations, we hypothesized that by promoting depolarization of the membrane, this Cl^- conductance could explain inhibition of dopamine uptake and enhancement of carrier-mediated efflux by Zn^{2+} under non-voltage-clamp conditions. In full agreement with this, we found that Zn^{2+} strongly enhanced the depolarizing effect on the membrane potential elicited by amphetamine and dopamine in oocytes expressing DAT (Fig. 2). This was

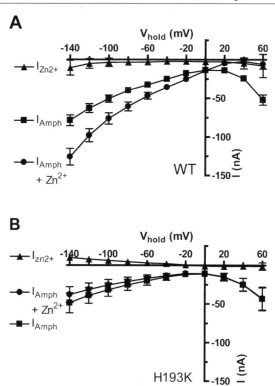

Fig. 2 A,B Zn^{2+} and substrate-mediated currents in *Xenopus laevis* oocytes expressing DAT wildtype (WT) or DAT H193$_{3.87}$K (HI93K). Current-to-voltage relationship of steady-state currents induced by either 10 μM amphetamine (I_{Amph}), 10 μM Zn^{2+} ($I_{Zn^{2+}}$), or 10 μM amphetamine (AMPH) plus 10 μM Zn^{2+} ($I_{Amph+Zn^{2+}}$). **A** Voltage dependence of the substrate-mediated currents in oocytes expressing hDAT WT (WT) ($n = 10$). **B** Voltage dependence of the substrate-mediated currents in oocytes expressing the control mutant hDAT H193$_{3.87}$K (HI93K), which does not bind Zn^{2+} ($n = 5$). Reproduced from Meinild et al. (2004)

predicted to cause a decrease in the driving force for active transport sufficient to account for the 30%–50% uptake inhibition seen in DAT in response to Zn^{2+} application. Importantly, Zn^{2+}-promoted depolarization could also explain the surprising observation that in WT DAT, Zn^{2+} does not slow but rather enhances amphetamine-induced reverse transport of the substrate MPP^+ (1-methyl-4-phenylpyridinium) both in transfected cells and in striatal slices (Scholze et al. 2002). Note that it is well known that reverse transport is augmented upon depolarization of the cell membrane (Khoshbouei et al. 2003). Altogether, the data have defined Zn^{2+} as the only known modulator of a neurotransmitter transporter that differentially modulates inward and outward transport via altering the conductance properties of the transport protein (Table 1; Scholze et al. 2002); however, the physiological significance of the current available data remains to be determined (see Sect. 6).

6
Reversal of Zn^{2+} Sensitivity by Intracellular Mutations in the DAT

The recent crystal structure of LeuT$_{Aa}$ indicates residues that might form the extracellular and intracellular gates that control access to the substrate-binding site from the extracellular and intracellular environment, respectively. The intracellular gate is predicted to be composed of roughly 20 Å of ordered protein structure involving, in particular, the intracellular ends of TM1, 6, and 8 (Yamashita et al. 2005). A key residue, which is conserved throughout evolution in this class of transporters, is a tyrosine (Tyr335$_{6.68}$ in DAT) at the cytoplasmic end of TM6 (Yamashita et al. 2005). This tyrosine is positioned below the substrate-binding site at the cytoplasmic surface of the protein and forms a cation-π interaction with and arginine at the N-terminus just below TM1 (Arg60$_{1.26}$ in DAT) that again forms a salt bridge with an aspartate at the cytoplasmic end of TM8 (Asp436$_{8.74}$ in DAT; Fig. 3). Interestingly, a couple of

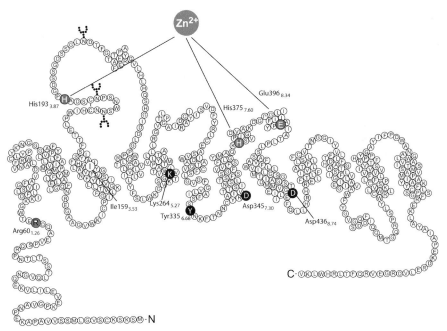

Fig. 3 Two-dimensional representation of the human dopamine transporter. The residues on the extracellular side responsible for coordination of the Zn^{2+} ion are the endogenous Zn^{2+}-binding site of hDAT (*white letters in enlarged gray circles*). The residues highlighted on the intracellular side are residues that upon mutation to alanine revert the Zn^{2+} sensitivity of the transporter (*white letters in enlarged black circles*). Ile159$_{3.53}$ in TM3 that was used for accessibility measurement upon mutation to a cysteine is also highlighted (*black letter in enlarged white circle*). Note that Arg60$_{1.26}$ (*black letter in enlarged gray circle*) forms a salt bridge with Asp436$_{8.74}$ thus contributing to the molecular framework of the internal gate

years ago already we showed that mutation of Tyr335$_{6.68}$ results in a remarkable functional transporter phenotype (Loland et al. 2002; Loland et al. 2004). The V_{max} value for [^{3}H]dopamine uptake was substantially lowered; however, in the presence of Zn^{2+} the [^{3}H]dopamine uptake capacity was potentiated in a dose-dependent manner, with a maximal response at around 10 µM Zn^{2+} (Loland et al. 2002, 2004). Thus, in contrast to the observations in the DAT WT, Zn^{2+} did not cause inhibition of transporter function but restored the ability of the mutant to translocate substrate (Loland et al. 2002, 2004). Another characteristic of the mutant was a dramatic decrease in the apparent affinity for cocaine (~150-fold) and related inhibitors, whereas the apparent affinity for dopamine was increased compared to WT DAT. The presence of 10 µM Zn^{2+} partially restored the apparent affinities, making the mutant more WT-like (Loland et al. 2002, 2004). Importantly, the effects of Zn^{2+} in Y335$_{6.68}$A was due to binding of Zn^{2+} to the same binding site as that in WT DAT, since substitution of either His193$_{3.87}$, His375$_{7.60}$, or Glu396$_{8.34}$ eliminated the ability of Zn^{2+} to potentiate [^{3}H]DA uptake in the Y335$_{6.68}$A (Loland et al. 2002).

Based on these observations, we hypothesized that mutation of Tyr335$_{6.68}$ altered the conformational equilibrium of the transport cycle and likely caused the transporter to assume preferentially an inward instead of an outward facing conformation (Loland et al. 2002). In the presence of Zn^{2+} we predicted that the conformational equilibrium was partially restored allowing transport to occur. Thus, we envisaged—in remarkable agreement with the now-published structure of LeuT$_{Aa}$—that Tyr335$_{6.68}$ is part of a critical molecular network on the cytoplasmic side of the protein responsible for stabilizing the transporter in the outward-facing conformation with the intracellular gate closed. In further agreement with the LeuT$_{Aa}$ structure, we also observed that mutation of Asp436$_{8.74}$ (Fig. 3) results in a phenotype similar to that of Y335$_{6.68}$A (Loland et al. 2004). We have moreover identified two additional residues (K264$_{5.27}$A and D345$_{7.30}$A) that upon mutation to alanines display a phenotype similar to that found for Y335$_{6.68}$A (Loland et al. 2004). A thorough analysis of the LeuT$_{Aa}$ structure in comparison with DAT suggests that these residues might also participate in the network of interactions that controls the function of the intracellular gate and thereby the conformational equilibrium of the transporter (J. Kniazeff, C.J. Loland, and U. Gether, unpublished observation). Strikingly, we were even able to obtain structural evidence for this hypothesis by assessing the accessibility of a cysteine introduced into position 159 (I159$_{3.53}$C) in TM3 of DAT (Fig. 3) to the positively charged sulfhydryl reactive compound MT-SET ([2-(trimethylammonium)ethyl]methanethiosulfonate) in the MTSET-unresponsive background of hDAT E2C (C90$_{1.56}$A/C306$_{6.39}$A; Loland et al. 2004). For the corresponding position in both NET and SERT, the accessibility to MTSET was reported to require an outward-facing transporter conformation (Chen and Rudnick 2000). We found, as predicted, that in the absence of Zn^{2+}, MTSET did not affect [^{3}H]dopamine uptake in the three mutants (E2C

I159$_{3.53}$C/K264$_{5.27}$A; E2C I159$_{3.53}$C/Y335$_{6.68}$A; E2C I159$_{3.53}$C/D345$_{7.30}$A), suggesting that without Zn^{2+} present the transporter assumes an inward-facing conformation (Loland et al. 2004). In contrast, MTSET strongly inhibited uptake of all three mutants in the presence of 10 μM Zn^{2+}, indicating that binding of Zn^{2+} to the DAT increases in the number of transporter molecules assuming an outward-facing conformation, and hence exposes position 159 to the exterior (Loland et al. 2004). Experiments aimed at clarifying the precise role of K264$_{5.27}$ and D345$_{7.30}$ are currently being performed in our laboratory.

The Y335$_{6.68}$A mutant has additionally been subject to electrophysiological analysis using the two-electrode voltage-clamp technique on *Xenopus laevis* oocytes (Meinild et al. 2004). Application of amphetamine to oocytes expressing hDAT Y335$_{6.68}$A under voltage-clamp conditions did not produce any detectable currents (Fig. 4); however, in the presence of Zn^{2+} alone, a cocaine-sensitive voltage-dependent current that reversed at −20 mV was observed (Fig. 4; Meinild et al. 2004). This current was concentration-dependent with the most effective concentrations of Zn^{2+} being 10–30 μM. Moreover, co-application of Zn^{2+} and amphetamine did not enhance the Zn^{2+}-induced current any further (Fig. 4), which is in contrast to the effect of Zn^{2+} on substrate-mediated currents seen in the WT DAT (Fig. 2). Thus, Zn^{2+} itself produces a current in the DAT Y335$_{6.68}$A, whereas it enhances the substrate-mediated current found in the DAT WT (Meinild et al. 2004). A series of ion substitution experiments indicated that both the Zn^{2+}-promoted current in Y335$_{6.68}$A and the uncoupled current potentiated by Zn^{2+} in the WT DAT were carried mainly by Cl^{-} (Meinild et al. 2004). A major question is the molecular basis for this uncoupled Cl^{-} conductance. It is not readily explained within the framework of a simple alternating access model. In particular, it does not explain why Zn^{2+} alone can activate a Cl^{-} conductance in the Y335$_{6.68}$A mutant

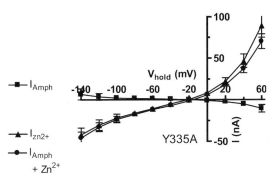

Fig. 4 Zn^{2+} and substrate-mediated currents in *Xenopus laevis* oocytes expressing DAT Y335$_{6.68}$A (Y335A). Current-to-voltage relationship of steady-state currents induced by either 10 μM amphetamine (I_{Amph}), 10 μM Zn^{2+} ($I_{Zn^{2+}}$) or 10 μM AMPH plus 10 μM Zn^{2+} ($I_{Amph+Zn^{2+}}$). Voltage dependence of the substrate-mediated currents in oocytes expressing the hDAT Y335$_{6.68}$A mutant (Y335A) ($n = 8$). Reproduced from Meinild et al. (2004)

and thus why this mutant converts the transporter into a putative "Zn^{2+}-gated ion channel." However, the LeuT$_{Aa}$ structure does not provide any insight into the structural basis for "channel-like" activities of these transporters. The simplest explanation would be that although the transporter primarily might operate via an alternate access scheme, it can also exist in ion-conducting conformational intermediates. An allosteric modulator such as Zn^{2+} could then be able to increase the probability that the transporter enters such states. In a similar fashion, distinct mutants that by themselves are characterized by an altered conformational equilibrium—such as, for example, Y335$_{6.68}$A—might strongly increase the chances that the transporter assumes the ion-conducting mode. Application of Zn^{2+} to such a mutant could then be sufficient to promote the ion-conducting state even when substrate is not present.

7
Zn^{2+} Is Found in Presynaptic Vesicles and Is Released Upon Neuronal Stimulation

The concentration of free extracellular Zn^{2+} is under basal conditions predicted to be low in the brain (1 pM to 10 nM; Frederickson 1989). However, the brain contains substantial amounts of Zn^{2+} and, interestingly, in a specific subset of glutamatergic neurons—mainly in the limbic areas (including the hippocampus and amygdala) as well as in the striatum and thalamus—Zn^{2+} is sequestered in pre-synaptic vesicles together with glutamate (Frederickson 1989; Frederickson and Bush 2001; Slomianka 1992); and in these vesicles the concentration of Zn^{2+} is as high as 3–30 mM (Frederickson and Bush 2001).

Upon neuronal stimulation, Zn^{2+} is released from the presynaptic vesicles in a Ca^{2+}-dependent manner (Li et al. 2001a, b; Ueno et al. 2002). From studies of hippocampal slices, this activity-dependent release of Zn^{2+} leads to a transient rise in the concentration of free synaptic Zn^{2+}, estimated to be in the range of 10 to 20 µM (Li et al. 2001a, b; Ueno et al. 2002; Vogt et al. 2000)—although more optimistic measures suggest that the concentration may even rise to 100–300 µM in the synapse (Assaf and Chung 1984; Vogt et al. 2000). It is thus intriguing to speculate that the amount of free Zn^{2+} transiently present in the synaptic cleft may be sufficient to modulate the function of pre- and postsynaptic receptors as well as neurotransmitter transporters containing endogenous Zn^{2+}-binding sites.

The vesicular accumulation of Zn^{2+} is, nonetheless, dependent on the presence of a specific vesicular Zn^{2+}-transporter (ZnT-3) (Palmiter et al. 1996; Wenzel et al. 1997). Targeted deletion of ZnT-3 has been performed, and in these mice vesicular Zn^{2+} is undetectable and heterozygotes have intermediate levels of vesicular Zn^{2+} (Cole et al. 1999). Surprisingly, an initial "crude" behavioral characterization of the ZnT-3 knockout (KO) mice showed that they had normal learning and memory skills as well as motor coordination compared to control WT mice (Cole et al. 2001). This argues against a role of vesicular Zn^{2+}, but it is important to emphasize that the ZnT-3 KO mice might have compensated for their lack of vesicular Zn^{2+} during embryonic development by altering neuronal architecture, neurotransmitter release, or receptor responsiveness. It must also be emphasized that the high concentration of Zn^{2+} in the synaptic vesicles, and the amount of energy required to maintain this intracellular pool, indirectly supports an important role of vesicular Zn^{2+} in neuronal function. The existence of a strict pattern of glutamate and Zn^{2+} colocalization in the synaptic vesicles also points towards a specific task for Zn^{2+} on excitatory synapses (Vogt et al. 2000). Consistent with this notion, micromolar concentrations of Zn^{2+} are indeed able to modulate the function of the ionotropic NMDA glutamate receptors (Hollmann et al. 1993; Peters et al. 1987; Westbrook and Mayer 1987), the AMPA receptors (Lin et al. 2001; Rassendren et al. 1990), and the glutamate transporters EAAT1 (Vandenberg et al. 1998) and EAAT4 (Mitrovic et al. 2001). Nevertheless, the functional significance of the interaction between the synaptically released Zn^{2+} and the excitatory pathways of the brain has proved difficult to demonstrate (Frederickson et al. 2000).

Dense GAT4 immunoreactivity has been observed in the CA1 and CA3 regions of the rat hippocampus (Cohen-Kfir et al. 2005), regions where Zn^{2+}-containing glutamatergic neurons are found in particularly high numbers (Slomianka 1992). It was, therefore, speculated that during hyperactivation of hippocampal glutamatergic neurons, Zn^{2+} binding to GAT4 and NMDA receptors would result in increased GABAergic inhibitory neurotransmission as well as inhibition of postsynaptic NMDA receptors and, hence, synaptic Zn^{2+} would play an important protective role against glutamate-induced excitotoxicity (Cohen-Kfir et al. 2005).

Interestingly glutamatergic Zn^{2+}-containing neurons and the DAT colocalizes throughout the limbic system and the striatum (Ciliax et al. 1995, 1999; Frederickson et al. 2000). Evidence is growing that there is a close interaction between glutamatergic and dopaminergic neurotransmission. Mice with reduced expression levels of NMDA receptors displayed a phenotype with hyperlocomotion and stereotypy (Mohn et al. 1999), traits that are normally ascribed to increases in dopaminergic tone (Gainetdinov et al. 1999; Giros et al. 1996). This has led to the suggestion that glutamatergic and dopaminergic neurotransmission oppose each other in the control of motor behaviors (Mohn et al. 1999). In accordance with these observations, the locomotor behavior of the DAT KO can be modulated by pharmacological manipulation of either the NMDA or the AMPA receptors (Gainetdinov et al. 2001). Moreover, glutamatergic inputs from the subthalamic nucleus to the substantia nigra has been shown to trigger DAT-mediated dopamine efflux (Falkenburger et al. 2001). Thus, in response to glutamate release in the substantia nigra, the DAT can operate in reverse. The mechanism for this phenomenon remains unclear, but apparently involves metabotropic G protein-coupled glutamate receptors (Falkenburger et al. 2001). It could, therefore, be speculated that the release of glutamate in the substantia nigra and co-release of Zn^{2+} can modify the input to the dopaminergic neurons by involving the DAT. In this way, synaptically released Zn^{2+} may contribute to functional diversity at the molecular level of the neuronal synapses.

References

Alberts IL, Nadassy K, Wodak SJ (1998) Analysis of zinc binding sites in protein crystal structures. Protein Sci 7:1700–1716
Amara SG, Arriza JL (1993) Neurotransmitter transporters: three distinct gene families. Curr Opin Neurobiol 3:337–344
Amara SG, Fontana AC (2002) Excitatory amino acid transporters: keeping up with glutamate. Neurochem Int 41:313–318
Amara SG, Kuhar MJ (1993) Neurotransmitter transporters: recent progress. Annu Rev Neurosci 16:73–93
Assaf SY, Chung SH (1984) Release of endogenous Zn^{2+} from brain tissue during activity. Nature 308:734–736
Bloomenthal AB, Goldwater E, Pritchett DB, Harrison NL (1994) Biphasic modulation of the strychnine-sensitive glycine receptor by Zn^{2+}. Mol Pharmacol 46:1156–1159
Bruss M, Hammermann R, Brimijoin S, Bonisch H (1995) Antipeptide antibodies confirm the topology of the human norepinephrine transporter. J Biol Chem 270:9197–9201
Busselberg D, Platt B, Michael D, Carpenter DO, Haas HL (1994) Mammalian voltage-activated calcium channel currents are blocked by Pb^{2+}, Zn^{2+}, and Al^{3+}. J Neurophysiol 71:1491–1497
Chen JG, Rudnick G (2000) Permeation and gating residues in serotonin transporter. Proc Natl Acad Sci U S A 97:1044–1049
Chen JG, Liu-Chen S, Rudnick G (1998) Determination of external loop topology in the serotonin transporter by site-directed chemical labeling. J Biol Chem 273:12675–12681
Chu XP, Wemmie JA, Wang WZ, Zhu XM, Saugstad JA, Price MP, Simon RP, Xiong ZG (2004) Subunit-dependent high-affinity zinc inhibition of acid-sensing ion channels. J Neurosci 24:8678–8689
Ciliax BJ, Heilman C, Demchyshyn LL, Pristupa ZB, Ince E, Hersch SM, Niznik HB, Levey AI (1995) The dopamine transporter: immunochemical characterization and localization in brain. J Neurosci 15:1714–1723
Ciliax BJ, Drash GW, Staley JK, Haber S, Mobley CJ, Miller GW, Mufson EJ, Mash DC, Levey AI (1999) Immunocytochemical localization of the dopamine transporter in human brain. J Comp Neurol 409:38–56
Cohen-Kfir E, Lee W, Eskandari S, Nelson N (2005) Zinc inhibition of γ-aminobutyric acid transporter 4 (GAT4) reveals a link between excitatory and inhibitory neurotransmission. Proc Natl Acad Sci U S A 102:6154–6159
Cole TB, Wenzel HJ, Kafer KE, Schwartzkroin PA, Palmiter RD (1999) Elimination of zinc from synaptic vesicles in the intact mouse brain by disruption of the ZnT3 gene. Proc Natl Acad Sci U S A 96:1716–1721
Cole TB, Martyanova A, Palmiter RD (2001) Removing zinc from synaptic vesicles does not impair spatial learning, memory, or sensorimotor functions in the mouse. Brain Res 891:253–265
Dauer W, Przedborski S (2003) Parkinson's disease: mechanisms and models. Neuron 39:889–909

Draguhn A, Verdorn TA, Ewert M, Seeburg PH, Sakmann B (1990) Functional and molecular distinction between recombinant rat GABAA receptor subtypes by Zn^{2+}. Neuron 5:781–788

Eulenburg V, Armsen W, Betz H, Gomeza J (2005) Glycine transporters: essential regulators of neurotransmission. Trends Biochem Sci 30:325–333

Fairman WA, Vandenberg RJ, Arriza JL, Kavanaugh MP, Amara SG (1995) An excitatory amino-acid transporter with properties of a ligand-gated chloride channel. Nature 375:599–603

Falkenburger BH, Barstow KL, Mintz IM (2001) Dendrodendritic inhibition through reversal of dopamine transport. Science 293:2465–2470

Fava M, Kendler KS (2000) Major depressive disorder. Neuron 28:335–341

Ferrer JV, Javitch JA (1998) Cocaine alters the accessibility of endogenous cysteines in putative extracellular and intracellular loops of the human dopamine transporter. Proc Natl Acad Sci U S A 95:9238–9243

Frederickson CJ (1989) Neurobiology of zinc and zinc-containing neurons. Int Rev Neurobiol 31:145–238

Frederickson CJ, Bush AI (2001) Synaptically released zinc: physiological functions and pathological effects. Biometals 14:353–366

Frederickson CJ, Suh SW, Silva D, Frederickson CJ, Thompson RB (2000) Importance of zinc in the central nervous system: the zinc-containing neuron. J Nutr 130:1471S–1483S

Gainetdinov RR, Wetsel WC, Jones SR, Levin ED, Jaber M, Caron MG (1999) Role of serotonin in the paradoxical calming effect of psychostimulants on hyperactivity. Science 283:397–401

Gainetdinov RR, Mohn AR, Bohn LM, Caron MG (2001) Glutamatergic modulation of hyperactivity in mice lacking the dopamine transporter. Proc Natl Acad Sci U S A 98:11047–11054

Galli A, DeFelice LJ, Duke BJ, Moore KR, Blakely RD (1995) Sodium-dependent norepinephrine-induced currents in norepinephrine-transporter-transfected HEK-293 cells blocked by cocaine and antidepressants. J Exp Biol 198:2197–2212

Giros B, Jaber M, Jones SR, Wightman RM, Caron MG (1996) Hyperlocomotion and indifference to cocaine and amphetamine in mice lacking the dopamine transporter. Nature 379:606–612

Hersch SM, Yi H, Heilman CJ, Edwards RH, Levey AI (1997) Subcellular localization and molecular topology of the dopamine transporter in the striatum and substantia nigra. J Comp Neurol 388:211–227

Hollmann M, Boulter J, Maron C, Beasley L, Sullivan J, Pecht G, Heinemann S (1993) Zinc potentiates agonist-induced currents at certain splice variants of the NMDA receptor. Neuron 10:943–954

Holst B, Elling CE, Schwartz TW (2002) Metal ion-mediated agonism and agonist enhancement in melanocortin MC1 and MC4 receptors. J Biol Chem 277:47662–47670

Horenstein J, Akabas MH (1998) Location of a high affinity Zn^{2+} binding site in the channel of alpha1beta1 gamma-aminobutyric acid A receptors. Mol Pharmacol 53:870–877

Huang EP (1997) Metal ions and synaptic transmission: think zinc. Proc Natl Acad Sci U S A 94:13386–13387

Ingram SL, Prasad BM, Amara SG (2002) Dopamine transporter-mediated conductances increase excitability of midbrain dopamine neurons. Nat Neurosci 5:971–978

Iversen LL (1971) Role of transmitter uptake mechanisms in synaptic neurotransmission. Br J Pharmacol 41:571–591

Jardetzky O (1966) Simple allosteric model for membrane pumps. Nature 211:969–970

Ju P, Aubrey KR, Vandenberg RJ (2004) Zn^{2+} inhibits glycine transport by glycine transporter subtype 1b. J Biol Chem 279:22983–22991

Khoshbouei H, Wang H, Lechleiter JD, Javitch JA, Galli A (2003) Amphetamine-induced dopamine efflux. A voltage-sensitive and intracellular Na+-dependent mechanism. J Biol Chem 278:12070–12077

Laube B, Kuhse J, Rundstrom N, Kirsch J, Schmieden V, Betz H (1995) Modulation by zinc ions of native rat and recombinant human inhibitory glycine receptors. J Physiol 483:613–619

Lewis DA, Lieberman JA (2000) Catching up on schizophrenia: natural history and neurobiology. Neuron 28:325–334

Li Y, Hough CJ, Frederickson CJ, Sarvey JM (2001a) Induction of mossy fiber→CA3 long-term potentiation requires translocation of synaptically released Zn^{2+}. J Neurosci 21:8015–8025

Li Y, Hough CJ, Suh SW, Sarvey JM, Frederickson CJ (2001b) Rapid Translocation of Zn^{2+} from presynaptic terminals into postsynaptic hippocampal neurons after physiological stimulation. J Neurophysiol 86:2597–2604

Lin DD, Cohen AS, Coulter DA (2001) Zinc-induced augmentation of excitatory synaptic currents and glutamate receptor responses in hippocampal CA3 neurons. J Neurophysiol 85:1185–1196

Loland CJ, Norregaard L, Gether U (1999) Defining proximity relationships in the tertiary structure of the dopamine transporter. Identification of a conserved glutamic acid as a third coordinate in the endogenous Zn(2+)-binding site. J Biol Chem 274:36928–36934

Loland CJ, Norregaard L, Litman T, Gether U (2002) Generation of an activating Zn(2+) switch in the dopamine transporter: mutation of an intracellular tyrosine constitutively alters the conformational equilibrium of the transport cycle. Proc Natl Acad Sci U S A 99:1683–1688

Loland CJ, Granas C, Javitch JA, Gether U (2004) Identification of intracellular residues in the dopamine transporter critical for regulation of transporter conformation and cocaine binding. J Biol Chem 279:3228–3238

MacAulay N, Bendahan A, Loland CJ, Zeuthen T, Kanner BI, Gether U (2001) Engineered Zn(2+) switches in the gamma-aminobutyric acid (GABA) transporter-1. Differential effects on GABA uptake and currents. J Biol Chem 276:40476–40485

Mager S, Naeve J, Quick M, Labarca C, Davidson N, Lester HA (1993) Steady states, charge movements, and rates for a cloned GABA transporter expressed in Xenopus oocytes. Neuron 10:177–188

Mager S, Min C, Henry DJ, Chavkin C, Hoffman BJ, Davidson N, Lester HA (1994) Conducting states of a mammalian serotonin transporter. Neuron 12:845–859

Meinild AK, Sitte HH, Gether U (2004) Zinc potentiates an uncoupled anion conductance associated with the dopamine transporter. J Biol Chem 279:49671–49679

Melikian HE, McDonald JK, Gu H, Rudnick G, Moore KR, Blakely RD (1994) Human norepinephrine transporter. Biosynthetic studies using a site-directed polyclonal antibody. J Biol Chem 269:12290–12297

Mitchell SM, Lee E, Garcia ML, Stephan MM (2004) Structure and function of extracellular loop 4 of the serotonin transporter as revealed by cysteine-scanning mutagenesis. J Biol Chem 279:24089–24099

Mitrovic AD, Plesko F, Vandenberg RJ (2001) Zn^{2+} inhibits the anion conductance of the glutamate transporter EAAT4. J Biol Chem 276:26071–26076

Mohn AR, Gainetdinov RR, Caron MG, Koller BH (1999) Mice with reduced NMDA receptor expression display behaviors related to schizophrenia. Cell 98:427–436

Nirenberg MJ, Vaughan RA, Uhl GR, Kuhar MJ, Pickel VM (1996) The dopamine transporter is localized to dendritic and axonal plasma membranes of nigrostriatal dopaminergic neurons. J Neurosci 16:436–447

Norregaard L, Frederiksen D, Nielsen EO, Gether U (1998) Delineation of an endogenous zinc-binding site in the human dopamine transporter. EMBO J 17:4266–4273

Palmiter RD, Cole TB, Quaife CJ, Findley SD (1996) ZnT-3, a putative transporter of zinc into synaptic vesicles. Proc Natl Acad Sci U S A 93:14934–14939

Peters S, Koh J, Choi DW (1987) Zinc selectively blocks the action of N-methyl-D-aspartate on cortical neurons. Science 236:589–593

Pifl C, Rebernik P, Kattinger A, Reither H (2004) Zn^{2+} modulates currents generated by the dopamine transporter: parallel effects on amphetamine-induced charge transfer and release. Neuropharmacology 46:223–231

Rassendren FA, Lory P, Pin JP, Nargeot J (1990) Zinc has opposite effects on NMDA and non-NMDA receptors expressed in Xenopus oocytes. Neuron 4:733–740

Regan L (1995) Protein design: novel metal-binding sites. Trends Biochem Sci 20:280–285

Richfield EK (1993) Zinc modulation of drug binding, cocaine affinity states, and dopamine uptake on the dopamine uptake complex. Mol Pharmacol 43:100–108

Risso S, DeFelice LJ, Blakely RD (1996) Sodium-dependent GABA-induced currents in GAT1-transfected HeLa cells. J Physiol 490:691–702

Rosenkilde MM, Lucibello M, Holst B, Schwartz TW (1998) Natural agonist enhancing bis-His zinc-site in transmembrane segment V of the tachykinin NK3 receptor. FEBS Lett 439:35–40

Rothman RB, Baumann MH (2003) Monoamine transporters and psychostimulant drugs. Eur J Pharmacol 479:23–40

Rothstein JD, Patel S, Regan MR, Haenggeli C, Huang YH, Bergles DE, Jin L, Dykes Hoberg M, Vidensky S, Chung DS, Toan SV, Bruijn LI, Su Zz, Gupta P, Fisher PB (2005) β-Lactam antibiotics offer neuroprotection by increasing glutamate transporter expression. Nature 433:73–77

Rudnick G (1997) Mechanisms of biogenic amine neurotransporters. In: Reith M (ed) Neurotransmitter transporter: structure, function and regulation. Humana Press, Totowa, pp 73–100

Scholze P, Norregaard L, Singer EA, Freissmuth M, Gether U, Sitte HH (2002) The role of zinc ions in reverse transport mediated by monoamine transporters. J Biol Chem 277:21505–21513

Slomianka L (1992) Neurons of origin of zinc-containing pathways and the distribution of zinc-containing boutons in the hippocampal region of the rat. Neuroscience 48:325–352

Sonders MS, Amara SG (1996) Channels in transporters. Curr Opin Neurobiol 6:294–302

Sonders MS, Zhu SJ, Zahniser NR, Kavanaugh MP, Amara SG (1997) Multiple ionic conductances of the human dopamine transporter: the actions of dopamine and psychostimulants. J Neurosci 17:960–974

Sonders MS, Quick M, Javitch JA (2005) How did the neurotransmitter cross the bilayer? A closer view. Curr Opin Neurobiol 15:296–304

Swaminath G, Steenhuis J, Kobilka B, Lee TW (2002) Allosteric modulation of beta 2-adrenergic receptor by Zn^{2+}. Mol Pharmacol 61:65–72

Swaminath G, Lee TW, Kobilka B (2003) Identification of an allosteric binding site for Zn^{2+} on the beta 2 adrenergic receptor. J Biol Chem 278:352–356

Ueno S, Tsukamoto M, Hirano T, Kikuchi K, Yamada MK, Nishiyama N, Nagano T, Matsuki N, Ikegaya Y (2002) Mossy fiber Zn^{2+} spillover modulates heterosynaptic N-methyl-D-aspartate receptor activity in hippocampal CA3 circuits. J Cell Biol 158:215–220

Vandenberg RJ, Arriza JL, Amara SG, Kavanaugh MP (1995) Constitutive ion fluxes and substrate binding domains of human glutamate transporters. J Biol Chem 270:17668–17671

Vandenberg RJ, Mitrovic AD, Johnston GA (1998) Molecular basis for differential inhibition of glutamate transporter subtypes by zinc ions. Mol Pharmacol 54:189–196

Vogt K, Mellor J, Tong G, Nicoll R (2000) The actions of synaptically released zinc at hippocampal mossy fiber synapses. Neuron 26:187–196

Wadiche JI, Amara SG, Kavanaugh MP (1995) Ion fluxes associated with excitatory amino acid transport. Neuron 15:721–728

Wenzel HJ, Cole TB, Born DE, Schwartzkroin PA, Palmiter RD (1997) Ultrastructural localization of zinc transporter-3 (ZnT-3) to synaptic vesicle membranes within mossy fiber boutons in the hippocampus of mouse and monkey. Proc Natl Acad Sci U S A 94:12676–12681

Westbrook GL, Mayer ML (1987) Micromolar concentrations of Zn^{2+} antagonize NMDA and GABA responses of hippocampal neurons. Nature 328:640–643

Yamashita A, Singh SK, Kawate T, Jin Y, Gouaux E (2005) Crystal structure of a bacterial homologue of Na(+)/Cl(−)-dependent neurotransmitter transporters. Nature 437:215–223

Yernool D, Boudker O, Jin Y, Gouaux E (2004) Structure of a glutamate transporter homologue from Pyrococcus horikoshii. Nature 431:811–818

Molecular Microfluorometry: Converting Arbitrary Fluorescence Units into Absolute Molecular Concentrations to Study Binding Kinetics and Stoichiometry in Transporters

J. W. Schwartz[1] · D. Piston[2] · L. J. DeFelice[3] (✉)

[1]Imaging Center, Stowers Institute for Medical Research, 1000 E 50th St., Kansas City MO, 64110, USA

[2]Department of Molecular Physiology and Biophysics, Department of Molecular Physiology and Biophysics, Vanderbilt University Medical Center, Nashville TN, 37232-8548, USA

[3]Department of Pharmacology, Center for Molecular Neuroscience, Vanderbilt University Medical Center, Nashville TN, 37232-8548, USA
lou.defelice@vanderbilt.edu

1	Introduction	24
1.1	Neurotransmitter Transporters Conduct Transmitters Across Membranes	24
1.2	Methods to Study Transporter Function	27
1.2.1	Radiometric Assay	27
1.2.2	Electrophysiology	29
1.2.3	Amperometry and Cyclic Voltammetry	31
1.2.4	Quantitative Fluorescence Microscopy	32
1.2.5	Confocal Microscopy and Two-Photon Excitation	33
1.2.6	TIRF Microscopy	40
1.2.7	Fluorescence Lifetime Imaging Microscopy	42
1.2.8	Fluorescence Correlation Spectroscopy	45
1.2.9	Fluorescence Recovery After Photobleaching	48
1.2.10	Fluorescence Plate Reader	50
2	Summary	51
	References	52

Abstract Cotransporters use energy stored in Na^+ or H^+ gradients to transport neurotransmitters or other substrates against their own gradient. Cotransport is rapid and efficient, and at synapses it helps terminate signaling. Cotransport in norepinephrine (NET), epinephrine (EpiT), dopamine (DAT), and serotonin (SERT) transporters couples downhill Na^+ flux to uphill transmitter flux. NETs, for example, attenuate signaling at adrenergic synapses by efficiently clearing NE from the synaptic cleft, thus preparing the synapse for the next signal. Transport inhibition with tricyclic antidepressants prolongs neurotransmitter presence in the synaptic cleft, potentially alleviating symptoms of depression. Transport inhibition with cocaine or amphetamine, which respectively block or replace normal transport, may result in hyperactivity. Little is known about the kinetic interactions of substrates or drugs with transporters, largely because the techniques that have been successful in discovering transporter agonists and antagonists do not yield detailed kinetic information. Mechanistic data are for the most part restricted to global parameters, such as K_m and V_{max}, measured from large populations of transporter molecules averaged over thousands of cells. Three relatively new techniques used in transporter research are electrophysiology, amperometry, and microfluorometry. This review focuses on fluorescence-based methodologies, which—unlike any other technique—permit the simultaneous measurement of binding and transport. Microfluorometry provides unique insights into binding kinetics and transport mechanisms from a quantitative analysis of fluorescence data. Here we demonstrate how to quantify the number of bound substrate molecules, the number of transported substrate molecules, and the kinetics of substrate binding to individual transporters. Although we describe experiments on a specific neurotransmitter transporter, these methods are applicable to other membrane proteins.

Keywords Cotransporters · Neurotransmitters · Synapses · Membranes · Norepinephrine · Dopamine · Serotonin · Antidepressants · Cocaine · Amphetamine · Kinetics · Electrophysiology · Amperometry · Microfluorometry · Binding · Fluorescence · Stoichiometry

1
Introduction

1.1
Neurotransmitter Transporters Conduct Transmitters Across Membranes

Central nervous system (CNS) signaling relies on chemical transmission between adjacent neurons. After a stimulus-evoked response, the presynaptic neuron releases neurotransmitter (NT) into the synaptic cleft, activating pre- and postsynaptic ligand-gated channels or secondary messenger cascades (Fig. 1A and B). Chemical transmission is attenuated by NT degradation, diffusion, and clearance (Fig. 1C). The majority of transmitter molecules are sequestered in the presynaptic neuron via NT transporters (Axelrod and Kopin 1969; Graefe et al. 1978; Iversen et al. 1967). For example, the norepinephrine transporter (NET) removes 90% of released NE in peripheral noradrenergic neurons (Blakely 2001), and genetic NET ablation induces a significant increase in extracellular NE longevity (Moron et al. 2002; Wang et al. 1999; Xu et al. 2000). Serotonin (5-HT) and dopamine (DA) have similar transporters, SERT and DAT, respectively. Noradrenergic signaling disruptions have profound physiological effects because they influence attention, learning, memory, emotion, and pain reception (Foote and Aston-Jones 1995; Foote et al. 1980; Valentino et al. 1983). Furthermore, adrenergic dysfunction is associated with mood disorders such as depression (Clark and Russo 1998; Schildkraut et al. 1965) and posttraumatic stress disorder (Maes et al. 1999), hypertension, diabetes cardiomyopathy, and heart failure (Backs et al. 2001; Merlet et al. 1999).

The monoamine transporters, NET, SERT, and DAT are biological targets for drugs of abuse, such as cocaine and amphetamines, and tricyclic antidepressants (Ritz et al. 1990; Sacchetti et al. 1999; Tatsumi et al. 1997). NT transporter

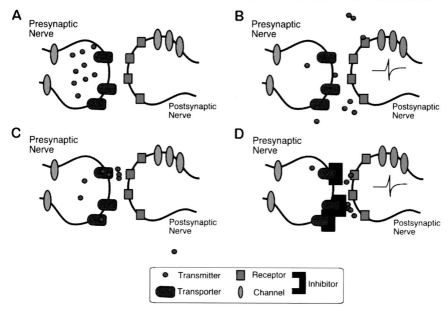

Fig. 1 A–D Neurotransmitter transporters regulate synaptic transmitter concentrations. Resting neurons have a reserved pool of docked synaptic vesicles primed for release (**A**). Electrical stimulation triggers transmitter release (**B**), which is subsequently sequestered into the presynaptic neuron (**C**). Transporter inhibition increases the synaptic cleft neurotransmitter longevity

inhibition delays NT clearance and thus increasing synaptic activity and altering the encoded information (Fig. 1D). In depressed patients, administration of antidepressants, such as Prozac (fluoxetine), alleviates the symptoms of depression via SERT inhibition. Drugs of abuse, such as cocaine, demonstrate more profound effects, which may be attributed to their nonselective nature. NET, SERT, and DAT are all inhibited by cocaine with similar potencies (Torres et al. 2003).

A single gene on chromosomes 5, 16, and 17 encodes DAT, NET, and SERT, respectively (Hahn and Blakely 2002). NET and SERT demonstrate alternative splicing, producing two distinct messages encoding NET and SERT (Bauman and Blakely 2002; Bradley and Blakely 1997; Kitayama et al. 1999, 2001). Both SERT messenger RNA (mRNA) messages maintain expression and function (Bradley and Blakely 1997). However, only one NET mRNA message is expressed and functions (Kitayama et al. 1999, 2001). The coexpression of the nonfunctional mutant with the functional NET protein dominantly inhibits NET activity, suggesting communication between gene products. To date, no DAT splice variants have been described. Due to the biological importance of transporter encoded by single genes, several genetics studies have focused on analysis of the coding sequence for single nucleotide polymorphisms. Patients with complex disorders were examined for monoamine transporter polymorphisms resulting in functional loss. For example, coding single nucleotide polymorphism (cSNPs) in human (h)NET are linked to orthostatic intolerance (OI). Patients suffering from OI demonstrate an increase in standing heart rate (<30 bpm) not accompanied by hypotension (Robertson et al. 2001; Shannon et al. 2000). A subset of OI patients with NE spillover and decreased NE clearance were examined for alterations in the NET gene. These studies reveal an OI-linked heterozygote G to C substitution at nucleotide 247 resulting in a proline substitution for alanine (Robertson et al. 2001; Shannon et al. 2000). In heterologous expression systems, this mutation retards expression and maturation and exhibits a dominant-negative NET inhibition (Hahn et al. 2003). DAT and SERT polymorphism linkage studies correlate with substance abuse/dependence, bipolar disorder, major depressive disorder, attention deficit/hyperactivity disorder and anxiety, unipolar depression, and suicidal patients. Genetic complex linkages are discussed in detail by Hahn and Blakely (2002).

NET, SERT, and DAT belong to the GAT1/NET Na/Cl-dependent NT transporter gene family (Blakely 1992). GAT1 transports the inhibitory NT, GABA (γ-aminobutyric acid). The Na/Cl-dependent, NT-transporter gene family is functionally defined by the millimolar Na^+- and Cl^--concentration dependence for substrate symport (Blakely 1992). The nomenclature for individual transporters is based on the specific neuronal subtype (noradrenergic, dopaminergic, etc.); thus, individual transporters are defined by the endogenous substrate. For example, NET accumulates NE in presynaptic noradrenergic neurons (Schroeter et al. 2000). Although the nomenclature implies selectivity, transporters demonstrate promiscuous substrate selection. 1-methyl-4-phenylpyridine (MPP^+) is a structurally distinct exogenous substrate that is a substrate for NET, SERT, and DAT (Scholze et al. 2002; Sitte et al. 1998, 2000, 2001). MPP^+ is the neurotoxic metabolite of 1-Methyl-4-phenyl-1,2,3,6-tetrahydropyridine, which after intravenous administration induces a DAT-dependent parkinsonian state within 2 weeks. Although monoamine transporters demonstrate selective substrate and inhibitor profiles, separate genes encode NET, SERT, and DAT. The predicted topology describes a 12 transmembrane domain (TMD) protein with an intracellular amino and carboxy termini, and this topology has become a hallmark of the Na^+/Cl^--dependent NT transporter gene family (Blakely 1992). NET and DAT demonstrate 80% amino acid conservation, and NET and SERT demonstrate 60% amino acid conservation (Blakely et al. 1991).

In order to study ionic dependence, radiometric substrate accumulation is traditionally monitored in media in which Na^+ and Cl^- are replaced: Na^+ is typically replaced with *N*-methyl-D-glucamine ($NMDG^+$) or Li^+ at equivalent osmolarity; however, these counter ions do not support substrate accumulation. A similar approach was employed to evaluate Cl^--dependent substrate accumulation. Unlike Na^+, Cl^- removal reduces but does not completely eliminate substrate accumulation. Br^- and SCN^- (thiocyanate) substitution for Cl^- sustains a significant 5-HT SERT-mediated transport activity (~80%; Nelson and Rudnick 1981). Similar NET studies demonstrate that NET-mediated NE accumulation is Na^+ dependent, while Cl^- removal inhibits, but does not abolish, NE accumulation (Bonisch and Harder 1986; Corey et al. 1994; Harder and Bonisch 1985). These studies do not specifically address the ionic-dependent substrate binding. To evaluate the Na and Cl requirements for substrate binding, low-affinity substrate binding prohibits radiometric substrate binding measurements; thus, high-affinity antagonist displacement is used to approximate substrate affinity. Humphreys and coworkers measured [^{3}H]-imipramine and [^{125}I]-β-carbomethoxy-3β-(4-iodophenyl)tropane ([^{125}I]-β-CIT) in the Na-free and Cl-free experimentation; NaCl was replaced with LiCl and Na-isethionate, respectively. According to these data, [^{3}H]-imipramine binding was both Na^+ and Cl^- dependent, while [^{125}I]-β-CIT binding did not depend on Na^+ or Cl^-. 5-HT displaces bound [^{125}I]-β-CIT in Na^+-free and Cl^--free media, which indicates 5-HT interacts with SERT in the absence of Na^+ and Cl^-. Although Humphreys and coworkers argue Li^+ was inert, a recent study describes Li^+-induced conformation changes specific to Li^+ and no other alkali metals (Ni et al. 2001). Li^+ also produces an enhanced current in the absence of substrate (Ramsey and DeFelice 2002).

1.2
Methods to Study Transporter Function

Radiometric substrate accumulation, electrophysiology, and amperometry/cyclic voltammetry measure NET activity in different ways. Each method attempts to enhance one or more of the following criteria: (1) specificity, (2) rapid sampling, (3) dynamic substrate analysis, (4) amenability to single cells, and (5) segregation of substrate binding from substrate accumulation.

1.2.1
Radiometric Assay

In the classic radiometric assay, NET-expressing cells or resealed membrane vesicles are incubated with radiolabeled NE for approximately 5–10 min (shortest assay time is often about 30 s). The applied substrate is removed, and the cells or resealed vesicles are solubilized with detergent; the radioactive native substrate is measured by scintillation. The radiometric approach specifically monitors radiolabeled substrate; thus, it provides a high signal-to-noise ratio. The total number of cells is determined by a hemocytometer; thus, radiometric values provide a mean accumulation rate. The hemocytometer values, however, include dead cells, which reduce the reported accumulation values. Under these conditions, the accumulation rates are an underestimate. The radiometric approach has poor time resolution (>30 s) and cannot dynamically monitor substrate accumulation; thus parallel samples are analyzed at increasing time points to measure transport rates. The maximal transport velocity (V_{max}) is determined by measuring transport rates with increasing substrate concentration until saturation. The theoretical components of maximal transport velocity are:

$$V_{max} = N\upsilon v \quad (1)$$

where N is the number (#) of functional transporters, υ is the number (#) of substrate molecules transported per transport cycle (per turnover), and v is the turnover rate (1/s) of an individual transporter. However, the radiometric method does not directly measure individual transporter kinetics and depends on macroscopic measurements that are extrapolated to underlying molecular properties. The radiometric approach is also too insensitive to monitor single mammalian cells; for example HEK-293 cells—a typical cell line used in experiments—are approximately 10 µm in diameter with an approximate volume of 1.5 nl. These constraints render single-mammalian-cell radiometric assays infeasible; thus, radiometric assays typically employ thousands of cells and the condition of individual cells remains unknown. To get around this limitation, single-cell accumulation measurements have been made in *Xenopus* oocytes, which are approximately 1 mm in diameter and have a volume of 1 µl. Frog oocytes are also amenable to controlled conditions. However, some mRNA transcripts express poorly in oocytes; in particular, NET is a difficult transcript from which to recover active protein.

The classic radiometric approach measures net accumulated substrate, not substrate binding. A radiometric-binding assay exposes monoamine transporter-expressing cells or resealed vesicles to a radiolabeled ligand followed by vacuum filtration to remove excess ligand. Under these conditions, the substrate must remain bound long enough to filter the membranes, thus removing the unbound ligand. The corresponding ligand must remain bound for approximately 1 s. Monoamine transporter substrates do not remain bound for a sufficient time to separate unbound ligand by filtration methods. In order to estimate substrate binding, indirect methods are employed. High-affinity antagonists such as cocaine remain bound to monoamine transporter long enough to quantify, and agonist displacement of antagonist is used as an indirect measure of substrate potency. A radiometric high-affinity antagonist binding assay is employed to calculate the total number of surface monoamine transporters. Monoamine transporter-expressing cells or resealed vesicles are exposed to high-affinity antagonist concentrations. Eventually, all surface transporters are saturated providing the maximum number of bound antagonists (B_{max}).

$$B = \frac{B_{max}}{K_D^n + [L]^n} \quad (2)$$

B is the number (#) of bound ligands at each ligand concentration, B_{max} is the maximum concentration of bound ligands, K_D is a dissociation constant, L is the concentration of ligand exposed to the transporter, and n is the Hill coefficient. The radiometric approach reports the maximal number of inhibitor binding sites, not the monoamine transporter number, N, and not the number of functional transporters. To extrapolate transporter number N from B_{max}, the relative antagonist stoichiometry is assumed to be one ligand bound per functional transporter. These data are often compared to the maximal transport velocity (V_{max}) to establish an estimate of transport cycle (turnover).

$$V = \frac{V_{max}}{K_m^n + [S]^n} \tag{3}$$

V is the velocity at each substrate concentration, V_{max} is the maximum transport velocity, S is substrate concentration, K_m is the Michaelis-Menten constant, and n is the Hill coefficient.

1.2.2
Electrophysiology

Over 10 years ago, Bruns et al. performed a remarkable experiment. They measured SERT-mediated current in Retzius neurons dissociated from the leech and demonstrated a large (500–1000 pA), whole-cell, SERT-mediated currents that were induced by 5-HT application at a holding potential of −80 mV (Bruns et al. 1993). These presynaptic, SERT-mediated currents were sensitive to Na^+ and were inhibited by zimelidine, a specific SERT blocker; furthermore, they preceded the 5-HT-stimulated postsynaptic currents by several milliseconds. The SERT-mediated currents observed by Bruns and coworkers were later recapitulated in *Xenopus* oocytes expressing SERT and monitored with a two-electrode voltage clamp. The *Xenopus* oocyte expression system permits an assessment of the electric current profile in transfected and nontransfected oocytes; nontransfected oocytes do not accumulate 5-HT, nor do they demonstrate 5-HT-induced currents. Surprisingly, the presumably electroneutral SERT produced a large current carried predominately by Na^+ movement (Mager et al. 1994).

According to the classic fixed stoichiometry model of transport, ionic movement via NET produces the redistribution of a single positive charge, thus making the transporter electrogenic (Rudnick and Nelson 1978). Accordingly, HEK-293 cells stably expressing approximately $n = 10^6$ transporters/cell were expected to produce less than 0.2 pA of current, with all transporters working at the same time. When the experiment was performed, NET-mediated currents were 250-fold larger than the predicted value (∼50 pA, depending on membrane voltage). Thus, there are many more charges moving than substrate molecules. Furthermore, NET permits the movement of ions even in the absence of substrate, and NET inhibitors block this constitutive current.

The so-called "leak current"—charge movement without concurrent substrate movement—is revealed easily by the preemptive antagonist administration (Galli et al. 1995). Under whole-cell patch clamp conditions, cocaine administration to hNET-293 cells reduces the NE-induced whole-cell current. DAT electrophysiological profiles also demonstrate unexpectedly large charge movements and leak currents (Sonders et al. 1997). These experiments also describe the voltage-dependent uptake process in which DA accumulation increases with depolarization.

The unexpected substrate-induced currents prompted the determination of relative Na^+ and substrate contributions to the total charge movement (NE, DA, and 5-HT are monovalent cations). The ratio of charge movement to substrate transport (ρ) is an intrinsic transporter property that describes the channel contribution to transporter activity.

$$\rho = \frac{Number\ of\ Charges}{Number\ of\ Substrates} \quad (4)$$

In this equation, ρ measures the stoichiometry of transport. If $\rho = 1$, for example, that would mean one net charge per one NE-transported molecule through NET [in this case, the postulated NE:Na:Cl stoichiometry (Bonisch et al. 1986)]. Values of ρ greater than one are taken to imply an "uncoupled" current, although coupling per se is rarely measured. NET stoichiometry has been assessed using patch clamp and amperometry, yielding $\rho > 0$ (Galli et al. 1996). SERT and DAT have been assessed by concurrent two-electrode voltage clamp and radiolabeled substrate accumulation in frog oocytes. For DAT and SERT, transport velocity V (Eq. 3) increases linearly with hyperpolarization; however, the dependence of ρ on voltage was nonlinear. The maximal ρ values were obtained at −80 mV, with a minimum at 20 mV. Current traces were subsequently analyzed to determine the unitary charge movements associated with both current and transport. These analyses demonstrate that in *Drosophila* SERT ρ exceeds 100. Furthermore, roughly 500 5-HT molecules are translocated per channel opening (Petersen and DeFelice 1999).

The electrophysiological assays described above significantly improve the temporal resolution or transporter function, and they are amenable to single-cell analysis under voltage-controlled conditions. These experiments also provide new mechanistic information. The electrophysiological information, however, cannot provide the information on the species of the charge-carrying ions, and thus they are not as specific as the radiometric transport assay with regard to substrate identity. To alleviate this constraint, concurrent electrophysiological and amperometric approaches can be performed (Galli et al. 1996). These conditions permit identification of the substrate, single-cell assay, and voltage control, but they do not provide good time resolution and they cannot distinguish binding from transport.

1.2.3
Amperometry and Cyclic Voltammetry

NET does not express well in frog oocytes, thus methods used for DAT and SERT for the measurement of ρ cannot be used for NET. Using a series of voltage steps, NET coupling was assessed by concurrent current and amperometric measurements of NET activity in excised patches placed over a carbon fiber electrode. The potential across the carbon fiber is held constant at −700 mV, to oxidize NE. Under the assay conditions, NE is the only available oxidizable ion; thus, the oxidization is attributed to NE. NE oxidation produces two electrons that are subsequently recorded on an amperometric electrode. Due to the stochastic oxidation process, the current produced is directly proportional to the NE concentration exposed to the carbon fiber. Galli and coworkers have demonstrated ρ is directly proportional to voltage. These experimental data show that channel activity is coordinated to transport activity (Galli et al. 1998). These data show rapid NE movement through a patch (lower limit of 33,000 molecules/s; Galli et al. 1998).

The amperometric approach monitors NE from excised patch providing a similar single-cell measurement. Positioning the carbon fiber electrode under the patch pipette can permit the sampling of up to 95% of the total release transmitter (Galli et al. 1998). A carbon fiber electrode placed adjacent to a cell under whole-cell patch clamp conditions samples a subset of release molecules. The use of amperometry in combination with electrophysiology permits rapid sampling of the endogenous substrate, but this method is technically challenging. This combinational approach requires a researcher to pull an excise patch and move the electrode over the corresponding carbon fiber electrode. During the alignment process, the patch must remain intact, which requires delicate dexterity. By placing the patch electrode over the carbon fiber, the transport signal (amperometry) and current signal (patch electrode) are not synchronized. As transmitter exits the patch electrode, there is a delay prior to association with the carbon fiber. The chemical reaction on the carbon fiber adds another kinetic not accounted for in the analysis. The limitation reduced the time resolution for the concurrent amperometric and electrophysiological approach.

Another method proposed to measure substrate binding is cyclic voltammetry, in which a carbon fiber electrode is used to measure bath depletion of an oxidizable substrate. Mammalian cells or resealed vesicles expressing NET or DAT are incubated in a low bath volume with a sensitive electrode. Substrate oxidation is used to dynamically monitor the bath concentration. By examination, a reduction is the exposed concentration; the method extrapolates the internalized substrate concentration. Specific accumulation is defined in the presence of an inhibitor. In order to measure substrate binding, Schenk and coworkers measured DA depletion in Na^+-free conditions, and under these conditions the authors attribute a decrease in bath concentration to DA association with DAT (Batchelor and Schenk 1998; Earles and Schenk 1999; Povlock and Schenk 1997;

Schenk 2002; Wayment et al. 1998). The authors assume that in a Na-replaced media, the DA is only binding to the transporter without transport.

The radiometric, electrophysiological, and carbon-fiber approaches provide multiple methods to measure NET (and other transporters') activity. Each method has some desirable characteristic, whether it be specificity, rapid sampling, or applicability to a single, well-controlled cell. All of the described methods, however, fail to segregate binding and transport, nor do they permit spatially resolved signals with rapid sampling. Concurrent binding and transport measurements, for example, would allow researchers to distinguish functional transporters from plasma-membrane expression.

1.2.4
Quantitative Fluorescence Microscopy

Quantitative fluorescence microscopy offers a method to distinguish binding and transport, and it provides investigators with rapid sampling on single mammalian cells. The method also permits absolute signal quantification while segregating substrate binding and transport, thus allowing measurements of not only transport kinetic but also of substrate-transporter stoichiometry. Quantitative microscopy provides a single-cell, real-time, sub-micron, and simultaneous substrate binding and transport assay.

The methods to be described here calibrate arbitrary fluorescent units (AFUs) to the absolute number of molecules, and therefore to diffusion constants, enzyme kinetics, and photophysical properties. The following contains a brief description of confocal microscopy, two-photon excitation (TPE) microscopy, fluorescence lifetime imaging microscopy (FLIM), total internal reflection (TIRF) microscopy, fluorescence correlation spectroscopy (FCS), and fluorescence recover after photobleaching (FRAP). These methods have been employed to measure NET activity using a fluorescent surrogate substrate 1-4-(4-dimethylaminostyryl)-N-methylpyridinium (ASP$^+$) and green fluorescent protein (GFP) N-terminal tagged NETs. The fluorescent analog of MPP$^+$, ASP$^+$, is a neurotoxic metabolite of MPTP, and is a known substrate of monoamine transporters. Using ASP$^+$, we can assess binding and transport within 50 ms of application with submicron spatial resolution. For each photometric method, we outline the properties for each type of microscopy, along with its advantages and disadvantages.

These methods are employed to quantify fluorescent molecules. Fluorescence occurs when the ground state electrons are excited into a higher energy state via an interaction with a photon. To excite the ground state electrons, the excitation light energy must be at least equivalent to the energy gap between energy states. The energy dependence provides wavelength-dependent excitation, as energy is inversely proportional to wavelength. The excited-state molecule undergoes a series of internal energy conversions between higher energy states. The stored energy is rapidly dissipated by photon release. The emitted photon has decreased energy, thus is emitted at a longer wavelength. Unfortunately, fluorescent molecules are eventually destroyed (i.e., photobleached) by continued excitation and emission, thus limiting the number of photons produced by a molecule. A typical fluorescent molecule, fluorescein, produces approximately 1 million photons prior to photodegradation (Pawley 1995).

Light, such as a fluorescence emission, can be measured with a photomultiplier tube (PMT). A PMT possesses a photoactive surface producing electrons when exposed to light. The resultant electron is magnified through a series of electron-sensitive surfaces that cascade and amplify each incident photon. The PMT signal is directly proportional, over a linear range, to the number of photons striking the photoactive surface; some PMTS are sensitive enough to measure a single photon. The absorbed photons are thus translated into an electrical signal. Another device used to measure light is a charge-coupled device (CCD), which also uses a photoactivated surface to convert the number of photons into an electrical signal. The photoactive surface is divided into subregions and the photons absorbed within each subregion are independently quantified. A computer creates a digital image by registering each sub region. A CCD is a central component in most digital cameras. Although CCDs are often less sensitive than PMT-based detection, typical biological samples provide enough light to trigger a response on CCDs. In general, PMTs and CCDs provide adequate sensitivity for most biological samples.

Noise in a PMT- or CCD-based detection system comprises thermal noise, also known as dark current, readout noise, and shot noise. Thermal noise results from spontaneous PMT or CCD events that occur in the absence of light. The photoactive surface will spontaneously emit electrons, triggering a nonspecific cascade. Thermal noise is reduced by cooling the light recording device, which is a common method used in cooled-CCD cameras. Readout noise affects CCD cameras and can be reduced by slowing down the speed of the camera output. Shot noise is an intrinsic source of noise associated with the photon flux onto the PMT or CCD. A fluorescent molecule produces a stream of photons that excite the photoactive surface of the PMT. Although the average photon flux is constant, the instantaneous signal fluctuates around the mean. Poisson statistics mandate that the deviations are equivalent to the square root of the average number of photons. For example, a molecule producing an average photon count of 100 photons per unit time is accurate within a range from 90 to 110 photons (number of photons $\pm \sqrt{number\ of\ photons}$). To reduce shot noise, the integration times are increased.

1.2.5
Confocal Microscopy and Two-Photon Excitation

Marvin Minsky developed the theory for confocal microscopy in 1957 (Minsky 1957). Confocal microscopy differs from wide-field (conventional) microscopy, permitting acquisition of optical slices of thick samples. Fluorescent specimens are commonly visualized with epi-illumination, in which filtered excitation light is reflected by a dichroic mirror through an objective lens onto the sample. The emitted light, at a lower wavelength, is collected by the same objective lens and transmitted through the dichroic mirror to the detection surface. A confocal microscope inserts pinholes between both the excitation source and emission detector to reject out-of-focus light. Using the original Minsky design, an image is generated by rastering the stage, providing a complete image (Pawley 1995). The stage raster was a major challenge to producing commercial confocal microscopes. A laser excitation source provides high-intensity monochromatic light that does not require an excitation pinhole. Laser scanning confocal microscopy (LSCM) scans the optical field with xy scanning mirrors that move the laser beam across the field. At each point, the light is collected on a PMT and transformed to a digital image by a computer. Adjusting the focus generates optical slices, and the corresponding optical planes are digitally combined to generate a volume. The confocal point spread function (PSF) is estimated by a three-dimensional Gaussian function (Rigler et al. 1993), and the shape and size depend on the objective numerical aperture, wavelength of light, and size of the excitation beam (beam waist) (Rigler et al. 1993).

$$\overline{PSF}_{Confocal}(x, y, z) = \exp\left[-\frac{2(x^2 + y^2)}{\omega_0} - \frac{2z^2}{z_0^2}\right] \quad (5)$$

The PSF describes the volume encompassed by the single point using confocal microscopy, where ω_0 is the beam waist in the lateral direction and Z_o is the beam waist in the z-axial direction (Rigler et al. 1993). For example, a 1.4-numerical aperture (NA) 63 × oil immersion objective with a pinhole at 1.5 airy units (airy units are a normalized optical units, one airy unit equals the full width $1/e^2$ maximum for the first dark ring for the Airy diffraction pattern) produces an approximate volume of 0.30 fl. A complete description of confocal microscopy, including the calculation of the PSF and airy units, can be found in *Handbook of Biological Confocal Microscopy*, edited by James B. Pawley (Pawley 1995).

As described above, the thin segments generated by a confocal microscope are advantageous, but they are not a solution to all imaging problems. A detector for a confocal microscope is usually a PMT, which is limited by shot noise. The small volume generated under confocal microscopy limits the number of fluorescent molecules being sampled, thus reducing the number of photons being collected. In order to reduce noise, the integration time for each focal volume is increased, which increases the acquisition time. To reduce acquisition time, samples can be excited with more intense light, which expedites the photodegradation of the fluorescent molecule. The balance of acquisition time, pixel integration, and photobleaching can limit the time utility of confocal microscopy.

Confocal microscopy illuminates the entire sample, but acquires light emanating only from a small volume, typically roughly 0.15 fl. Fluorescent molecules outside the focal volume are simultaneously photobleached; therefore, the potential fluorescent signal decreases throughout the entire sample. As a sample is optically sliced, the later slices have a decreased signal due to prolonged exposure, thus limiting the penetration depth of confocal microscopy. TPE provides a solution to photobleaching outside the plane of focus. Using TPE, the excitation volume is limited by the photon density. Normally, a fluorescent molecule is excited by a single photon, but simultaneous absorption of two photons at half the energy (double the wavelength) can produce photoexcitation. TPE requires nearly simultaneous absorption of two photons. The probability of two photons being simultaneously absorbed increases in the focal volume, which has a higher photon density. The limited excitation area reduces out-of-plane photobleach.

In general, the TPE excitation wavelength is double the single-photon excitation wavelength, but the TPE excitation spectra can deviate from the doubling paradigm (Dickinson et al. 2003). The photon density is only sufficient to excite a fluorescent molecule within a small volume (Piston et al. 1995; Williams et al. 1994). The PSF for TPE can be well approximated by a Gaussian–Lorentzian function (Berland et al. 1995), and also depends on objective NA, wavelength, and beam waist.

$$\overline{PSF}_{TPE}(\rho, z) = \frac{4\omega_o^4}{\pi^2 \omega^4(z)} \exp\left[\frac{-4\rho}{\omega^2(z)}\right] \; ; \; \omega^2(z) = \omega_o^2 \left(1 + \frac{z\lambda}{\pi\omega_o^2}\right) \quad (6)$$

The Gaussian–Lorentzian PSF describes the volume encompassed by the single point using TPE microscopy, where ω_0 is the beam waist in the lateral direction (Berland et al. 1995). By scanning the laser beam in the x–y direction and moving the objective focus, TPE generates a three-dimensional representation of the object. The volume difference for the wide-field PSF compared to confocal and TPE is significant, but at matched wavelengths with a small pinhole (<1 airy unit) the PSF for confocal microscopy and TPE (Berland et al. 1995) are similar; therefore, under wavelength-matched conditions, TPE does not enhance z-axial resolution compared to confocal microscopy.

TPE has several advantages over confocal microscopy: (1) reduced total sample photobleaching, (2) increased number of photons detected, and (3) enhanced tissue penetration. Due to the spatially constrained fluorescence, an emission pinhole is not necessary. Detectors can be positioned to collect all the light produced by the fluorescent event, significantly increasing the total number of photons sampled and reducing shot noise. The increase in collected photons reduces the integration time required to obtain photon counts, thus reducing the image acquisition time. The spatial constraints prohibit photobleaching the sample outside the excitation volume; thus, as a thick sample is optically sliced, the later slices have not been photobleached. The sustained signal into later planes enhances penetration depth. TPE uses infrared (IR) radiation, which is not readily absorbed by biological samples. TPE is able to penetrate deeper into tissues, because IR radiation is not readily absorbed by tissue. Confocal microscopy often employs ultraviolet or visible excitation wavelengths, which are absorbed by endogenously expressed compounds within tissue. These same molecules do not absorb IR radiation. The enhanced penetration depth and reduction in out-of-plane photobleaching is advantageous for thicker samples (>200 μm).

Confocal microscopy permits optical isolation of subvolumes within thick samples, which is ideal for measuring ASP^+ binding separate from accumulation. By reducing the sample volume, the relative solution contribution to the total signal is negligible, while the cellular contribution remains high. NTs, amphetamines, and neurotoxins bind before being transported, whereas cocaine and antidepressants bind to block the substrate transport s. Although binding is crucial to transport, few assays separate direct effects on binding from effects on transport after binding, nor do they provide adequate temporal or spatial resolution to describe real-time kinetics or localize sites of active uptake. Monitoring changes in ASP^+ fluorescence, fluorescence microscopy distinguishes substrate binding from substrate transport using single-cell, space-resolved, real-time fluorescence microscopy. ASP^+ has micromolar potency for NET (Schwartz et al. 2003). The plasma membrane is visualized by acquiring concurrent differential interference contrast (DIC) images (Fig. 2A). After ASP^+ exposure, the plasma membrane of cells expressing NET demonstrates an immediate increase in ASP^+ fluorescence (Fig. 2B). Prolonged exposure permits cytosolic ASP^+ accumulation (Fig. 2C). Subsequent desipramine (DS) (a selective NET antagonist) displaces bound ASP^+ and inhibits further accumulation (Fig. 2D). Accumulated ASP^+ remains constant after DS administration; the increase in sequestered ASP^+ is parallel to the increase in mitochondrial ASP^+ accumulation (Schwartz et al. 2003). ASP^+ accumulation is Na^+-, Cl^--, cocaine-, and DS-sensitive, and temperature-dependent, and it competes with NE uptake. Monitoring ASP^+ with confocal microscopy provides a single-cell, rapid-sampling binding and transport assay for NET activity.

To evaluate NET surface distribution, we utilized confocal three-dimensional reconstruction to localize GFP-hNET and ASP^+ molecules. The N-terminal GFP-tagged NET (GFP-hNET) maintained 100% wildtype activity and bind ASP^+ with similar affinity. To evaluate the relative surface expression, we reconstructed the corresponding isosurface for HEK-GFP-hNET cells. Setting a threshold value to one AFU does not permit any gaps within the isosurface; this setting outlines the plasma membrane (Fig. 3a–C). Using a threshold value at the beginning of the dynamic range for the GFP signal (threshold = 64), the corresponding isosurface represents the distribution of colocalized GFP-hNETs and ASP^+ molecules. The GFP-hNET completely colocalizes with ASP^+ molecules. Colocalized GFP-hNET and ASP^+ are represented in yellow, and ASP^+ molecules alone are represented by red (Fig. 3d–F). By removing the

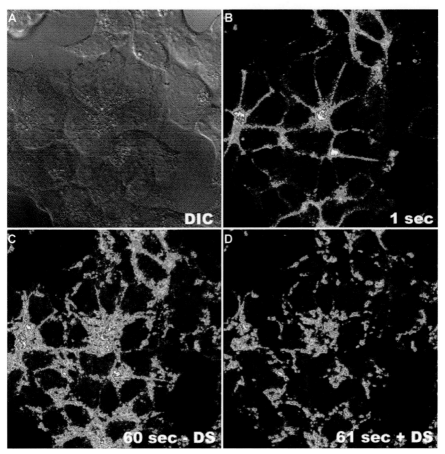

Fig. 2 A–D Desipramine displaces bound ASP^+. A The DIC image for HEK-293 cells expressing NET. A fluorescence image is acquired 1 s after ASP^+ exposure (**B**). The cells continue to incubate for 60 s (an image is recorded every 3 s). The 60-s image (**C**) is recorded prior to the addition of 10 µM desipramine. The next image recorded (3 s later, 63-s image) shows the displacement of bound ASP^+ (**D**)

last three optical sections (3 µm of GFP-hNETs), the underlying mitochondria are exposed (Fig. 3G–I). These data provide transporter localization and demonstrate that transporters are confined to subcellular regions across the cell surface. We further demonstrated that ASP^+ interacts with transporters not only in transfected cells but also in cultured neurons. As seen in Fig. 4, active NET is also localized to subcellular regions along superior cervical ganglia neurons. Because NET transport depends on internal substrate concentration, NET localization to pseudopodia, for example, may generate higher concentrations compared to other regions of the neuron. Thus, cellular localization may profoundly affect the accumulation rate. Indeed, the thin-filament pseudopo-

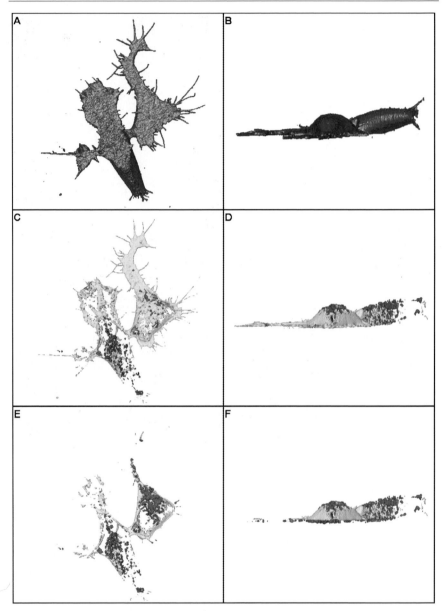

Fig. 3 A–F Isosurface GFP-hNET and ASP$^+$ surface distribution. All isosurface images were generated using Amira image analysis software. Each *column* represents and identical view angle for individual isosurfaces. **A, B** An isosurface of a single GFP-hNET cell. Isosurface was generated without permitting gaps within the isosurface. **C, D** The *yellow* isosurface represents colocalized GFP-hNET and ASP$^+$, while the *red* isosurface represents only ASP$^+$. **E, F** *Yellow* represents the GFP-hNET/ASP$^+$ isosurface, except the bottom three planes have been removed exposing the underlying mitochondria. QuickTime virtual reality (VR) files for each surface are published on the physical biology online version

AFUs to Molecules

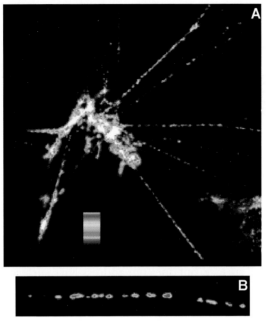

Fig. 4 A,B ASP$^+$ accumulates in superior cervical ganglia (SCG) neurons. **A** ASP$^+$ accumulation in SCG recorded after 60 s. ASP$^+$ accumulation is desipramine-sensitive. **B** A neurite at higher magnification. The *color gradient* represented in *panel a* denotes the color range corresponding to the intensity values

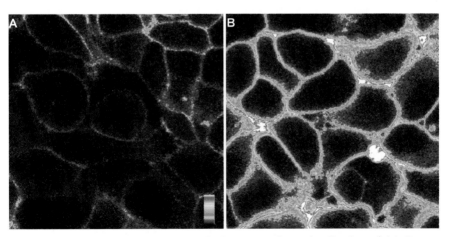

Fig. 5 A–C Cl$^-$ replacement increases cell plasma membrane localized ASP$^+$ binding. **A** hNET-293 cells exposed to 2 µM ASP$^+$ for 1 s in KRH buffer. **B** HNET-293 cells exposed to 2 µM ASP$^+$ for 1 s in KRH buffer with 120 mM NaI replacing NaCl; 100 µM Cl$^-$ present.

dia may represent the endogenous environment. In superior cervical ganglia neurons, NET localizes to small varicosities along the neurites (Fig. 4; see Schwartz et al. 2003). The transporter has a distinct substrate conductance, which is concentration-dependent. The varicosities may represent a diffusion barrier, thus significantly decreasing the volume the transporter is using for a concentration gradient.

Although transport *ex more* depends on Na^+ and Cl^-, binding is actually independent of Na^+ and increases in low Cl^-, while transport per se depends critically on Na^+ but is much less affected by Cl^-. In KRH media containing 120 mM NaCl, ASP^+ addition to HNET-293 cells causes an immediate increase in membrane-localized fluorescence (Fig. 5A, 1 s ASP^+ exposure). In order to evaluate ASP^+-binding ionic dependence, we exposed hNET-293 cells to KRH buffer in which Cl^- was replaced with I^-. Under these conditions, we observe a significant increase in bound ASP^+ (Fig. 5B, 1 s ASP^+). The increase in ASP^+ response is localized to the membrane and represents a twofold increase in plasma membrane ASP^+ fluorescence. These data indicate that more ASP^+ is binding to the cell surface, but does not address NET surface expression.

1.2.6
TIRF Microscopy

To evaluate NET surface expression, we employed TIRF microscopy using specialized optics to generate an evanescent wave with sufficient intensity to excite fluorescent molecules less than 1,000 Å from the coverslip. This thin optical section is ideally suited to measure areas juxtaposed to the coverslip. Using high NA oil immersion lenses (>1.45 NA) or a trapezoidal prism, a laser is adjusted at an angle of incidence greater than the critical angle for total internal reflection (Axelrod 1989, 2001a, b, 2003; Axelrod et al. 1983). At this angle, the change in refractive index between the sample index ($n = 1.33$) and coverslip ($n = 1.52$) completely reflects the incident light. The measured power of the light entering the objective and the corresponding reflected light are equivalent, if no molecules are present to absorb the photons within the evanescent field. TIRF illumination generates an electromagnetic field (called an "evanescent wave") which prorogates from the interface into the sample. Photons tunnel to excite a fluorescent molecule via evanescent field coupling. The field intensity decreases exponentially in the sample, and the intensity is only sufficient to excite molecules for the first 1,000 Å. This optical configuration permits exclusive membrane visualization with a significant enhancement in z-axial resolution compared to confocal or two-photon optical slices. The major disadvantage to TIRF imaging is that the area of interest within the sample must be adjacent to the coverslip. The exponential decay in intensity from the coverslip hinders but does not eliminate the ability to perform quantitative microcopy (Axelrod 2001b, 2003).

As shown above in Fig. 4, ASP$^+$ binding is enhanced twofold by Cl$^-$ substitution with I. This increase may be attributed to a rapid increase in surface expression due to Cl$^-$ removal. To evaluate this hypothesis, we used TIRF microscopy. Under these conditions, we exclusively examined GFP-hNET proteins localized to the membrane in the presence and absence of Cl$^-$. All surface NETs were occupied by ASP$^+$ under normal (120 mM NaCl) conditions. As seen in the TIRF images, the ASP$^+$ and GFP-hNET molecules form a pattern based on cell contact. Previous studies demonstrated that 2 µM ASP$^+$ occupies all surface transporters. According to the TIRF studies (Fig. 6), we observe identical GFP-hNET and ASP$^+$ distributions. Images collected before

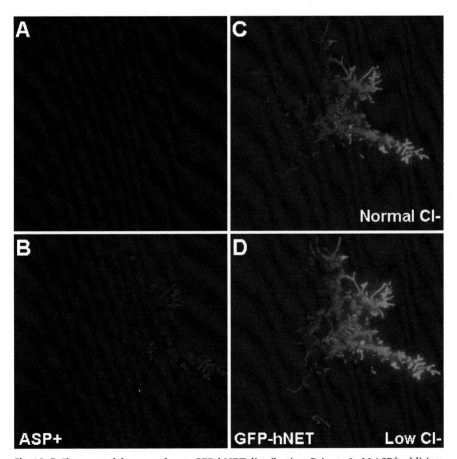

Fig. 6 A–D Cl$^-$ removal does not elevate GFP-hNET distribution. Prior to 2 µM ASP$^+$ addition, GFP fluorescence is not recorded in the ASP$^+$ channel (**A**). Decreasing Cl$^-$ concentrations does not elevate the GFP-hNET membrane distribution in normal, 120 mM, NaCl KRH buffer (**B**). The relative GFP-hNET transporter density does not change upon simultaneous 2 µM ASP$^+$ addition (**C**) in reduced Cl$^-$ media (Cl$^-$<3 mM) (**D**)

ASP⁺ addition indicate that GFP-hNET fluorescence does not appear in the ASP⁺ channel (Fig. 6A, B). After 10 s of 2 µM ASP⁺ exposure in Cl⁻-free media (NaI substation, FC Cl < 3 mM) hNET surface distribution is unaltered. These measurements provided a relative increase in substrate concentration. More extensive experimentation is required to address an absolute substrate concentration.

1.2.7
Fluorescence Lifetime Imaging Microscopy

FLIM provides information about the localized environment, permitting the calibration of confocal images. Fluorescence intensity measurements from a cellular context are not always sufficient to determine the fluorescent molecule concentration. Fluorescence intensity depends on the excitation light intensity (I_o), molar absorptivity (ε), the concentration ($[c]$), volume (L), and the quantum yield. The molar absorptivity, the ability to accept a photon at a defined wavelength, is largely independent on the localized environment (Lakowicz 1999). However, fluorescence is highly sensitive to the local environment, which leads to changes in quantum yield (Lakowicz 1999).

$$F = I_o \varepsilon [c] L Q_f \qquad (7)$$

Fluorescence intensity is equivalent to the product of I_o, the excitation light intensity, ε, molar absorptivity, $[c]$, fluorescent molecule concentration and Q_f, the quantum fluorescence quantum yield (Herman 2001). FLIM provides a concentration-independent measurement to evaluate environmental effects on a fluorescent molecule (Lakowicz 1999). Fluorescence lifetime measures the average time a population of molecules spends in the excited state. After excitation, a fluorescent molecule dissipates stored energy primarily as photons. In the absence of all other energy-releasing processes, the decay of photons from the excited state is defined as the natural decay of the molecule. Fluorescent molecules also dissipate energy via nonradiative mechanisms such as vibrational coupling. The combination of radiative and nonradiative processes results in the measured lifetime (Eq. 8), which can be defined in terms of the number of excited molecules (F_o) and the exponential decay constant.

$$F = F_o e^{-\frac{t}{\tau}} \qquad (8)$$

F_o represents the number of molecules in the excited state and τ is the fluorescence lifetime (Lakowicz 1999). The quantum yield of fluorescence is proportional to the measured fluorescence decay divided by the natural fluorescence decay (Herman 2001).

$$Q_f = \frac{\tau_{measured}}{\tau_{natural}} \qquad (9)$$

Q_f is the quantum yield for a fluorescence molecule. $\tau_{measured}$ is the measured lifetime. $\tau_{natural}$ is the fluorescence lifetime in the absence of all other energy releasing processes (Herman 2001). By substituting Eq. 5 into Eq. 1, fluorescence quantum yield changes between two local environments—such as solution and cellular—can be evaluated. A calibration curve in a solution can thus be calibrated to predict molecular concentration in the cell.

Fluorescence lifetimes are established using either frequency-domain (Fig. 7) or time-domain measurements (Fig. 8). Hanson and coworkers describe in detail the construction of a laser scanning FLIM microscope using frequency domain measurements (Hanson et al. 2002). Briefly, an appropriate TPE source, such as a titanium:sapphire laser (Millinia-pumped Tsunami, Spectra-physics), is coupled to the epi-fluorescence port of an inverted microscope. TPE excitation permits frequencies at 80 MHz; the corresponding TPE frequency must be similar to the fluorescent lifetime. The resultant fluorescence is recorded by a PMT and the sample scanned using an *xy* scanning mirror. A small portion of the excitation beam is diverted and recorded on a reference PMT. The phase and modulation of the high-frequency fluorescence emission are detected relative to the phase and modulation of the high-frequency repetitive light source. The fluorescence lifetime is calculated using the heterodyne frequency modulation method described by Jameson and coworkers (Alcala et al. 1985; Gratton et al. 1984; Jameson et al. 1984). Measuring a sample of known lifetime accounts for the microscope response, and all lifetime changes are additive. Alternatively, the fluorescence lifetime can be established by measuring the fluorescence

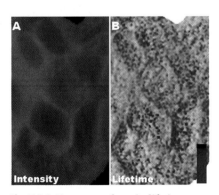
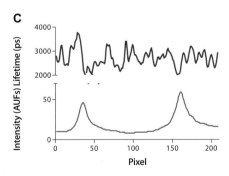

Fig. 7 A,B Frequency domain lifetime measurements. hNET-293 cells are exposed to 2 µM ASP$^+$ for 5 min prior to image acquisition. The changes in fluorescence intensity (**A**) and fluorescence lifetime (**B**) were examined using fluorescence lifetime imaging microscopy (FLIM) (see Sect. 1.2). The *color gradient* in *panel b* indicates the fluorescence lifetime value with *black* as the longest lifetime. Using the identical imaging settings, increasing concentrations of ASP$^+$ in isobutyl alcohol and GFP in water were imaged (*right panel*). The average pixel intensity across an image plan 10 µm from the coverslip was plotted against concentration [average ± standard error of mean (SEM), $n = 3$ ASP$^+$ slope = 26.5 AFU/µM, GFP slope = 15.3 ± 0.41 AFU/µM]. **C** Line scan

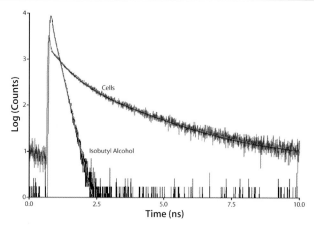

Fig. 8 Time-resolved fluorescence decay measurements establish the fluorescence lifetime. The cellular ASP$^+$ fluorescence lifetime was quantified using time-resolved fluorescence spectroscopy. The ASP$^+$ lifetime in the calibration solvent, isobutyl alcohol, and cellular lifetime was measured in a lifetime spectrometer. Cellular lifetime was measured by preloading hNET-293 cells with 5 μM ASP$^+$ for 10 min prior to harvest and washing. The log value for the photon counts is plotted against time, and the data are fit to the multi-exponential function $\left(Counts = A + Be^{-t/\tau} + Ce^{-t/\tau} + De^{-t/\tau}\ldots\right)$

decay in the time domain. A single focal spot is excited with a femtosecond laser pulse and the resultant fluorescence is recorded on a PMT. The decay is a measure of the fluorescence lifetime. Lifetime spectrometers and some microscopes utilize time-resolved fluorescence lifetime measurements.

FLIM has a few disadvantages. Currently, commercial FLIM systems are not readily available, and FLIM images have a slow acquisition rate. Several data points must be collected for each pixel in order to calculate the corresponding fluorescence lifetime. Multiple data point acquisition significantly delays image acquisition. A lifetime image can require as much as 6 min to obtain a single plane of an HEK-293 cell. To alleviate this constraint, single-pixel measurements at differing cellular locations are often sampled instead of acquiring data over an entire image.

Using FLIM, we measured the fluorescence lifetime of cellular ASP$^+$, which was significantly longer than ASP$^+$ in aqueous buffer. A fluorescence intensity image (Fig. 7A) was acquired concurrently with a fluorescence lifetime image (Fig. 7B). As seen in Fig. 7, for each pixel within the intensity image the corresponding fluorescence lifetime was determined. The solution ASP$^+$ fluorescence lifetime was significantly shorter (<10 ps) compared to cellular ASP$^+$ fluorescence lifetime (~2.5 ns). GFP has the same quantum efficiency as aqueous buffer and the cytosol; thus, no lifetime changes were observed. A solution calibration is directly applicable to cellular GFP. Similar values were determined from corresponding time-domain measurements (Fig. 8). The time-domain measurements demonstrate a multi-exponential decay, which shows that ASP$^+$ decays from several excited states. This is also reflected in frequency domain measurements, as τ_{phase} and τ_{mod} provided different fluorescence lifetime measurements.

In isobutyl alcohol, ASP$^+$ has an average lifetime of 168 ps, compared to 1,200 ps for cellular ASP$^+$. Therefore, a sevenfold increase was used to convert measured cellular fluorescence to absolute ASP$^+$ concentration values. Calibration curves for ASP$^+$ and GFP were examined at identical microscopy configurations. Confocal images of ASP$^+$ in isobutyl alcohol and GFP in water were acquired and the pixel histogram peak was plotted against concentration. GFP fluorescence is relatively independent of the local environment, thus lifetime corrections similar to those for ASP$^+$ were not required for GFP (Patterson et al. 1997). To avoid day-to-day variation, the microscope was calibrated daily. N-terminal GFP tagged hNET (GFP-hNET) maintains full transport activity. We collected dual channel ASP$^+$ and GFP-hNET time series using the calibrated settings. Under the conditions used, crosstalk between the GFP-hNET channel and ASP$^+$ channel was negligible. We converted pixel intensity to molecular values and the corresponding images for ASP$^+$ and GFP-hNET were compared. Panel a shows an overlay image taken at 10 s, and the yellow cell borders signify ASP$^+$ and GFP-hNET colocalization. Dividing a GFP-hNET image by the ASP$^+$ provided the GFP/ASP$^+$ ratio images. In the cell interior, the colocalization ratio approaches zero. Although the GFP-hNET and ASP$^+$ intensity varies along the cell perimeter, the ratio value is approximately constant and time-independent. Measuring the maximum number of GFP-hNET and ASP$^+$ molecules along a four-pixel-wide line scan along the cell circumference established the GFP-hNET to ASP$^+$ ratio. These data are summed over time for each cell ($n = 250$). The distribution ratio peaks at one GFP-hNET molecule per ASP$^+$ molecule; that is, one substrate associates with each hNET protein.

1.2.8
Fluorescence Correlation Spectroscopy

Another means to determine the underlying number of fluorescent molecules is FCS. FCS is a method where fluorescence from a single focal volume is measured over time, and small fluctuations in this signal provide information related to the number of particles, diffusion times, or enzyme kinetics. Over 30 years ago, Madge, Elson, and Webb (Elson et al. 1974; Magde et al. 1972, 1974) developed FCS to examine intercalated fluorescent particles in DNA. In FCS, small changes in the fluorescent signal arise due to an alteration in fluorescent molecule particle concentration from diffusion through the optical volume, enzymatic cleavage, or environmental changes (Fig. 9). In a simple case, as molecules enter the optical volume the fluorescence signal proportionally increases, but as molecules exit the optical volume the fluorescence signal proportionally decreases. Diffusion is only one of many processes that affect fluorescence. For example, after continued excitation, fluorescence molecules enter a dark

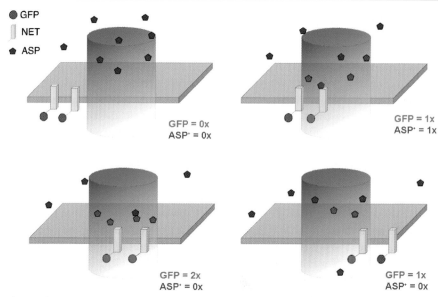

Fig. 9 Fluorescence correlation spectroscopy analyzes the fluorescence fluctuation due to diffusion or enzyme kinetics within a defined optical volume. Molecules outside the optical volume are not observed, thus do not contribute to the fluorescence signal. As fluorescent molecules diffuse into the optical volume, the signal increases. Molecules diffusing out of the optical volume have the opposite effect. The average fluorescence signal is proportional to the particle number, and the fluctuations provide a measure of diffusion time

state in which photoexcitation is not permitted. Under these conditions, the fluorescence intensity decreases although the particle remains in the optical volume. The resulting fluctuations are analyzed for autocorrelation.

$$G(\tau) = \langle F(t) \cdot F(t+\tau) \rangle / \langle F(t)^2 \rangle \qquad (10)$$

The normalized autocorrelation function, $G(\tau)$, is calculated as the time average of the product of the fluctuations of the detected fluorescence [$F(t)$] at every time t, and the fluctuations at the delayed times $t + \tau$, normalized by the average of the fluorescence emission [$F(t)$]. The zero time correlation ($\tau = 0$) is then $G(0) = F(t)^2/F(t)^2$. The temporal variation of $F(t)$ is proportional to $N(t)$, the fluctuations of the number of fluorophores in the probe volume. Since this is a shot-noise-limited measurement, $N(t)$ is sqrt(N), so $G(0)$ is proportional to $N(t)/N(t)2 = 1/N$, where N is the average number of fluorescent molecules in the probe volume.

The general autocorrelation function measures sustained self-similarity. $F(t)$ describes the original time-resolved data. $F(t + \tau)$ describes the translated data by the constant, τ. $F(t)^2$ is the average ensemble average. The resulting fluctuation can be analyzed in many ways, but the most common approach is by autocorrelation. Autocorrelation is a measure for the self-similarity of a time signal. The original fluorescence signal is translated in time by adding a set amount (~10 μs) to each time point, without adjusting the corresponding fluorescence intensity. The resultant translated signal is compared to the original signal for correlation. The original signal is translated by an increased time (20 μs). The translated and original signals are compared for correlation. The process is iterated for increasing time intervals. Deviations in similarity provide characteristic time constants of underlying processes (Bacia and Schwille 2003; Dittrich et al. 2001; Haustein and Schwille 2003; Medina and Schwille 2002; Schwille 2001).

The autocorrelation is fit to a theory derived for the underlying fluctuations, thus experimental condition has an appropriate fit for the autocorrelation function. For example, a fluorescent molecule in solution is fit to a three-dimensional free diffusion model in which the molecule exits and enters the focal volume from any direction. If the molecule undergoes a transition from a dark to the light state, the fluctuations are observed in the autocorrelation. Multiplying the two independent fluctuation theories derives the resultant theory. The experimental condition mandates the autocorrelation theory applied.

$$G(\tau) = \frac{1}{N} \left(\frac{1 - F_B + F_B e^{-\tau/\tau_B}}{1 - F_B} \right) \left(\sum_{i=1}^{n} \frac{f_i}{(1 + \tau/\tau_{Di}) \sqrt{1 + \tau/\omega^2 \tau_{Di}}} \right) \quad (10)$$

The autocorrelation function is derived for a three-dimensional diffusion model for an open system (Bacia and Schwille 2003). N is the average particle number. τ_D is the diffusion time of the particle in microseconds. ω is the axial ratio (ratio of axial to radial dimension of the observed volume). In each case, the observed fluctuations are dependent on the number of particles, N, and the diffusion of those particles in the observation volume. Membrane proteins are often fit to a two-dimensional diffusion model, because the protein movement is restricted to a single plane (Schwille et al. 1999). Schwille and coworkers (Bacia and Schwille 2003; Dittrich et al. 2001; Haustein and Schwille 2003; Medina and Schwille 2002; Schwille 2001) provide a detailed derivation of the autocorrelation function.

Prior to confocal and TPE microscopy, large focal volumes generated by a wide-field microscope required extremely dilute bright samples to minimize particle number. The fluorescent signal from large-particle numbers within the observation volume (Fig. 9) is not dramatically affected by the diffusion of a few particles. FCS sensitivity is inversely related to particle number; thus, low particle concentrations are ideal for FCS analysis. TPE microscopy and ultra-sensitive PMTs dramatically increased sensitivity and reduced particle numbers (Fig. 9), providing enhanced FCS measurements. Using a similar principle, cross-correlation analysis determines the relative similarity between two independent fluorescent signals. These data can provide a measure of association rates and protein–protein interactions or protein–ligand association. Multiple excitation lasers generate two different observation volumes, so dual channel fluorescence cross-correlation spectroscopy (FCCS) requires normalization of these volumes. Calibrations with known diffusion times are used to determine the relative observation volume sizes (Bacia and Schwille 2003).

In summary, FCS provides fast time resolution and the ability to calculate the number of particles independent of local environment. FCS measures events ranging from 10 µs to 1 s, thus providing estimates of chemical kinetics, diffusion, and concentration. A major disadvantage to FCS is the need for low particle numbers. Although the confocal volume can limit the observation volume size, the number of particles must remain low (<50 particles). Also, FCS cannot distinguish single fluorescent molecules from a pair of joined fluorescent molecules. FCS requires a fivefold increase in particle diffusion to segregate distinct particles (Chen et al. 2002; Chen et al. 1999); thus, FCS analysis cannot distinguish a monomeric protein from a dimeric protein.

To investigate the kinetic relationship between ASP^+ and NET, we used FCS to determine the ASP^+ and NET concentration and mobility. Dual-channel FCS simultaneously measures GFP-hNET and ASP^+ fluctuations. Bath ASP^+ demonstrates diminished fluorescence (τ_{water} < 10 ps) compared to cellular ASP^+ (τ_{cells} = 1.2 ns), which permits the identification of bound ASP^+ (versus free ASP^+). Solution ASP^+ did not contribute to the total fluorescence signal. The autocorrelation function of ASP^+ intensity fluctuations reveals an ASP^+ transporter dwell time of 526±25 µs, which is significantly different from the relaxation time for GFP-hNET measured in the same optical volume (~0.15 fl, see Sect. 1.2). Analysis of GFP-tagged transporters gives a relaxation time of 30.2 ± 0.94 ms (D = 0.17 µm^2/s), which represents characteristically slow diffusion for a membrane protein (Meissner and Haberlein 2003; Vrljic et al. 2002). We also observe a 320-µs diffusion time, which correlated to known pH-dependent inter-conversion for GFP (Haupts et al. 1998); thus, FCS measurements are useful as a pH sensor for slow moving GFPs. The τ = 0 intercept predicts the average particle (transporter) number per optical volume (Schwille 2001), thus providing a check on the FLIM calibration. The $G(0)$ autocorrelation values give 26 ± 0.98 (bound) ASP^+ particles and 26 ± 0.5 GFP-hNET (bound and free) particles in the optical volume, and the one-to-one ratio agrees with FLIM calculated ratio. We observe no cross-correlation between ASP^+ and GFP-hNET.

1.2.9
Fluorescence Recovery After Photobleaching

To corroborate the FCS diffusion measurements, we assessed hNET-GFP movement with FRAP. The methodologies presented above attempt to reduce photodamage to acquire a signal with high fidelity; however, FRAP experiments rely on photodegradation to measure fluorescent molecule kinetics. FRAP experimentation is often applied to study the movement of membrane proteins using a GFP-tagged conjugate protein. Under these conditions, a subsection within an image is exposed to a high-intensity laser pulse. Only molecules within the subregion are photodegraded; thus, they appear dark after acquiring an image of the entire field. The recovery of the bleach region is related to the diffusion of the particle of interest. Like FCS, the recorded data are fit to a defined model to evaluate recovery. Siggia and coworkers (Siggia et al. 2000) describe a plausible model for membrane proteins.

GFP-hNET diffusion FRAP measurements provided similar kinetics to the FCS experimentation. Three pre-bleach GFP-hNET cell images were acquired prior to photodegradation of GFP-hNETs in a 4 μm by 18 μm region along the cell surface. Due to the geometry of HEK-293 cells (Fig. 10), we imaged the GFP-hNET cells along the membrane adjacent to the coverslip. HEK-293 cells form flat elongated surfaces adjacent to the membrane, permitting the idealized location for FRAP experiments. GFP-hNET was photodegraded using 100% laser intensity (∼30 mW), followed by minimal exposure (<1 mW) to monitor the recovery. The damaged area recovers after approximately 1 min. These data were subsequently fit to an inhomogeneous diffusion model (Siggia

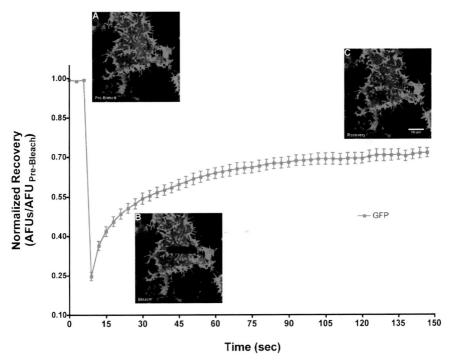

Fig. 10 FRAP measurements predict that GFP-hNET cells diffuse at 0.20 μm²/s. GFP-hNET cells were imaged prior to photobleach (*inset A*) using a 40 × 1.3 NA lens at 4 × digital zoom. A 4-by-18 μm region was bleached from the cell membrane (*inset B*) from the membrane adjacent to the coverslip. After approximately 1 min the bleached region recovered (*inset C*). The time course for GFP-hNET recover provides a diffusion constant of 0.20 μm²/s

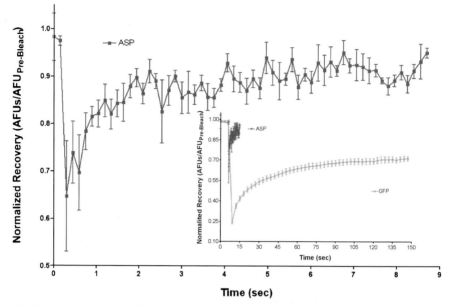

Fig. 11 FRAP experimentation cannot measure ASP$^+$ association rates. hNET-293 cells are exposed to 2 μM ASP$^+$. Several pre-bleach images were acquired prior to high-intensity light exposure. Several images were acquired, and ASP$^+$ recovered within 1 s. The *inset* shows the GFP-hNET and ASP$^+$ recoveries

et al. 2000) that yields an effective diffusion constant of 0.11 ± 0.3 μm^2/s (Fig. 10). These data agree with the FCS-established diffusion times. Similar experimentation was performed using ASP$^+$ photobleach in hNET-293 cells, but the fluorescence recovered faster than the acquisition rate (Fig. 11); thus, these data cannot report an accurate rate measurement. The ASP$^+$ recovery was significantly faster than the GFP-hNET recovery (inset Fig. 11).

1.2.10
Fluorescence Plate Reader

The techniques described above are useful to measure single molecule properties. ASP$^+$ also has utility for large-volume high-information screening. The fluorescence-based assay described is amenable to a high-throughput screen via a fluorescence plate reader. Using a molecular devices flex station, Mason and coworkers monitored ASP$^+$ accumulation in cells expressing NET and DAT. The flex station excites a large volume (compared to confocal microscopy); thus, solution with ASP$^+$ significantly contributes to the total signal. Solution fluorescence is quenched with trypan blue enhancing the signal-to-noise for the automated assay. The plate reader accurately predicted IC$_{50}$ values for inhibitors, such as DS and GBR-1206. These data indicate that this method can be used to screen for other transporter inhibitors. According to the data presented, the fluorescence plate reader cannot distinguish binding from transport, which may be attributed to problem with ASP$^+$ administration.

2
Summary

This review outlines fluorescence methods used to measure absolute molecular concentrations, kinetic interactions of substrates and transporters, and diffusion constants of transporters in the membrane. The advantage of fluorescence methods, as opposed to classical biochemical methods, is that molecules are evaluated within a cellular context. Furthermore, fluorescence-based technologies are amenable to experiments on single molecules using high-throughput screening.

To alleviate the constraints associated with radiometric and other methods, we have used quantitative fluorescence microscopy to monitor dynamically the binding and transport of a fluorescent monoamine transporter substrate. Cells expressing monoamine transporters demonstrated an immediate increase in membrane-localized fluorescence following fluorescent substrate (ASP$^+$) application. After continued ASP$^+$ exposure (>30 s), the cell interior continued to increase in fluorescence. Although HEK-293 cells have non-monoamine transporter-mediated ASP$^+$ accumulation, at concentrations below 5 µM ASP$^+$ does not bind to parental HEK-293 plasma membranes. The nonspecific intracellular ASP$^+$ accumulation is one-tenth that of NET-expressing cells. The immediate ASP$^+$ increase, only seen in transfected cells, was composed of immobile ASP$^+$ molecules, which were displaced by subsequent antagonist (DS) administration. The slower accumulation inside the cell was arrested, but not displaced, by DS. These data indicate that ASP$^+$ initially binds to surface NETs and is subsequently transported. Known blockers or substrates (DS, cocaine, Na$^+$ or Cl$^-$ substitution, and NE) each inhibited substrate binding and accumulation.

Confocal microscopy and TPE define optically isolated subvolumes within a large sample, thus providing "microscope cuvettes." Compartmentalized microscope cuvettes are digitally combined to produce a three-dimensional representation. FLIM provides a concentration-independent measurement to evaluate the local environment within the subvolume. In combination with confocal or TPE microscopy, FLIM establishes molecule concentrations. The established change in quantum yield allows a solution calibration curve that is corrected for environmental changes. FCS measures the diffusion time, concentration, and molecular brightness. In combination, these methods describe a molecular description of subcellular environments and molecular interactions.

Specific subcellular makers colocalized bound ASP$^+$ with the plasma membrane, and accumulated ASP$^+$ is associated with mitochondria. Optical isolation was accomplished by examining plasma membrane localized ASP$^+$ with

TIRF microscopy. The FLIM data showed that ASP^+ is highly sensitive to the local environment. The ASP^+ "fluorescence lifecycle" in our assay is as follows: ASP^+ is quenched by the aqueous solution until bound to NETs. After transport, the aqueous cytosolic environment again quenches ASP^+. ASP^+ subsequently associates with mitochondria, regaining optical activity. Although ASP^+ demonstrates significant environmental quantum yield changes, FLIM images provide the necessary conversion for solution ASP^+ to calibrate cellular ASP^+. ASP^+ is quenched in aqueous environments, such as the cytosol and is optically silent until bound to mitochondria. Calibrated confocal microscopy cannot measure the cytosolic pool, however, which requires complementary methods. Liquid chromatography tandem mass spectroscopy (LC/MS/MS) is employed to measure total ASP^+ accumulation (both dark and bright). These data indicated that mitochondrial ASP comprises only one-fifth the total accumulated ASP^+ at the time point measured. Thus, ASP^+ is less effective as a measure of uptake than it is of binding. This molecular conversion permits the assessment of pharmacological properties, such as surface density, transport velocity, and substrate-to-NET gene product stoichiometry.

FCS and FRAP experimentation provides information about substrate dwell time and transport motion in the plasma membrane. Due to the change in fluorescence lifetime, the fluctuations observed in FCS measurement of ASP^+ estimates the substrate dwell time on NET. Under identical conditions, we observed the relatively slow NET diffusion. The plasma membrane diffusion was substantiated by FRAP experimentation. These data may be substantiated and expanded by the application of image correlation spectroscopy (ICS) and photon-counting histogram (PCH). These methods are complimentary to the presented material, but have not been applied to ASP-bound NETs.

In summary, we have developed a calibrated microfluorometric assay that uniquely visualizes substrate binding and accumulation on single mammalian cells (or groups of cells) in real-time and which is amenable to high-throughput screening. In this review, we have used ASP-bound NETs as an example for multiple quantitative fluorescence microscopes, but these methods are applicable to other biological samples, such as receptor–ligand binding. These new approaches in microscopic fluorometry permit a more complete description and understanding of the underlying membrane protein biology.

Acknowledgements Supported by National Institutes of Health NS-34075.

References

Alcala JR, Gratton E, Jameson DM (1985) A multifrequency phase fluorometer using the harmonic content of a mode-locked laser. Anal Instrum 14:225–250

Axelrod D (1989) Total internal reflection fluorescence microscopy. Methods Cell Biol 30:245–270

Axelrod D (2001a) Selective imaging of surface fluorescence with very high aperture microscope objectives. J Biomed Opt 6:6–13

Axelrod D (2001b) Total internal reflection fluorescence microscopy in cell biology. Traffic 2:764–774

Axelrod D (2003) Total internal reflection fluorescence microscopy in cell biology. Methods Enzymol 361:1–33

Axelrod D, Thompson NL, Burghardt TP (1983) Total internal inflection fluorescent microscopy. J Microsc 129:19–28

Axelrod J, Kopin IJ (1969) The uptake, storage, release and metabolism of noradrenaline in sympathetic nerves. Prog Brain Res 31:21–32

Bacia K, Schwille P (2003) A dynamic view of cellular processes by in vivo fluorescence auto- and cross-correlation spectroscopy. Methods 29:74–85

Backs J, Haunstetter A, Gerber SH, Metz J, Borst MM, Strasser RH, Kubler W, Haass M (2001) The neuronal norepinephrine transporter in experimental heart failure: evidence for a posttranscriptional downregulation. J Mol Cell Cardiol 33:461–472

Batchelor M, Schenk JO (1998) Protein kinase A activity may kinetically upregulate the striatal transporter for dopamine. J Neurosci 18:10304–10309

Bauman PA, Blakely RD (2002) Determinants within the C-terminus of the human norepinephrine transporter dictate transporter trafficking, stability, and activity. Arch Biochem Biophys 404:80–91

Berland KM, So PT, Gratton E (1995) Two-photon fluorescence correlation spectroscopy: method and application to the intracellular environment. Biophys J 68:694–701

Blakely RD (1992) Molecular cloning and characterization of neurotransmitter transporters. NIDA Res Monogr 126:66–83

Blakely RD (2001) Physiological genomics of antidepressant targets: keeping the periphery in mind. J Neurosci 21:8319–8323

Blakely RD, Berson HE, Fremeau RT Jr, Caron MG, Peek MM, Prince HK, Bradley CC (1991) Cloning and expression of a functional serotonin transporter from rat brain. Nature 354:66–70

Bonisch H, Harder R (1986) Binding of 3H-desipramine to the neuronal noradrenaline carrier of rat phaeochromocytoma cells (PC-12 cells). Naunyn Schmiedebergs Arch Pharmacol 334:403–411

Bonisch H, Fuchs G, Graefe KH (1986) Sodium-dependence of the saturability of carrier-mediated noradrenaline efflux from noradrenergic neurones in the rat vas deferens. Naunyn Schmiedebergs Arch Pharmacol 332:131–134

Bradley CC, Blakely RD (1997) Alternative splicing of the human serotonin transporter gene. J Neurochem 69:1356–1367

Bruns D, Engert F, Lux HD (1993) A fast activating presynaptic reuptake current during serotonergic transmission in identified neurons of Hirudo. Neuron 10:559–572

Chen Y, Muller JD, So PT, Gratton E (1999) The photon counting histogram in fluorescence fluctuation spectroscopy. Biophys J 77:553–567

Chen Y, Muller JD, Ruan Q, Gratton E (2002) Molecular brightness characterization of EGFP in vivo by fluorescence fluctuation spectroscopy. Biophys J 82:133–144

Clark MS, Russo AF (1998) Measurement of tryptophan hydroxylase mRNA levels by competitive RT-PCR. Brain Res Brain Res Protoc 2:273–285

Corey JL, Quick MW, Davidson N, Lester HA, Guastella J (1994) A cocaine-sensitive Drosophila serotonin transporter: cloning, expression, and electrophysiological characterization. Proc Natl Acad Sci U S A 91:1188–1192

Dickinson ME, Simbuerger E, Zimmermann B, Waters CW, Fraser SE (2003) Multiphoton excitation spectra in biological samples. J Biomed Opt 8:329–338

Dittrich P, Malvezzi-Campeggi F, Jahnz M, Schwille P (2001) Accessing molecular dynamics in cells by fluorescence correlation spectroscopy. Biol Chem 382:491–494

Earles C, Schenk JO (1999) Multisubtrate mechanism for the inward transport of dopamine by the human dopamine transporter expressed in HEK cells and its inhibition by cocaine. Synapse 33:230–238

Elson EL, Magde D, Webb WW (1974) Fluorescence correlation spectroscopy. II. An experimental realization. Biopolymers 13:1–27

Foote S, Aston-Jones G (1995) Pharmacology and physiology of central noradrenergic systems. In: Bloom FE, Kupfer DJ (eds) Psychopharmacology: the fourth generation of progress. Raven Press, New York, pp 335–345

Foote SL, Aston-Jones G, Bloom FE (1980) Impulse activity of locus coeruleus neurons in awake rats and monkeys is a function of sensory stimulation and arousal. Proc Natl Acad Sci U S A 77:3033–3037

Galli A, DeFelice LJ, Duke BJ, Moore KR, Blakely RD (1995) Sodium-dependent norepinephrine-induced currents in norepinephrine-transporter-transfected HEK-293 cells blocked by cocaine and antidepressants. J Exp Biol 198:2197–2212

Galli A, Blakely RD, DeFelice LJ (1996) Norepinephrine transporters have channel modes of conduction. Proc Natl Acad Sci U S A 93:8671–8676

Galli A, Blakely RD, DeFelice LJ (1998) Patch-clamp and amperometric recordings from norepinephrine transporters: channel activity and voltage-dependent uptake [see comments]. Proc Natl Acad Sci U S A 95:13260–13265

Graefe KH, Bonisch H, Keller B (1978) Saturation kinetics of the adrenergic neurone uptake system in the perfused rabbit heart. A new method for determination of initial rates of amine uptake. Naunyn Schmiedebergs Arch Pharmacol 302:263–273

Gratton E, Jameson DM, Hall RD (1984) Multifrequency phase and modulation fluorometry. Annu Rev Biophys Bioeng 13:105–124

Hahn MK, Blakely RD (2002) Monoamine transporter gene structure and polymorphisms in relation to psychiatric and other complex disorders. Pharmacogenomics J 2:217–235

Hahn MK, Robertson D, Blakely RD (2003) A mutation in the human norepinephrine transporter gene (SLC6A2) associated with orthostatic intolerance disrupts surface expression of mutant and wild-type transporters. J Neurosci 23:4470–4478

Hanson KM, Behne MJ, Barry NP, Mauro TM, Gratton E, Clegg RM (2002) Two-photon fluorescence lifetime imaging of the skin stratum corneum pH gradient. Biophys J 83:1682–1690

Harder R, Bonisch H (1985) Effects of monovalent ions on the transport of noradrenaline across the plasma membrane of neuronal cells (PC-12 cells). J Neurochem 45:1154–1162

Haupts U, Maiti S, Schwille P, Webb WW (1998) Dynamics of fluorescence fluctuations in green fluorescent protein observed by fluorescence correlation spectroscopy. Proc Natl Acad Sci U S A 95:13573–13578

Haustein E, Schwille P (2003) Ultrasensitive investigations of biological systems by fluorescence correlation spectroscopy. Methods 29:153–166

Herman BD (2001) Fluorescence microscopy, 2nd edn. Bios Scientific Publishing, pp 1–170

Iversen LL, de Champlain J, Glowinski J, Axelrod J (1967) Uptake, storage and metabolism of norepinephrine in tissues of the developing rat. J Pharmacol Exp Ther 157:509–516

Jameson DM, Gratton E, Hall RD (1984) The measurement and analysis of heterogeneous emissions by multifrequency phase and modulation fluorometry. Appl Spectrosc Rev 20:55–106

Kitayama S, Ikeda T, Mitsuhata C, Sato T, Morita K, Dohi T (1999) Dominant negative isoform of rat norepinephrine transporter produced by alternative RNA splicing. J Biol Chem 274:10731–10736

Kitayama S, Morita K, Dohi T (2001) Functional characterization of the splicing variants of human norepinephrine transporter. Neurosci Lett 312:108–112

Lakowicz J (1999) Principles of fluorescence spectroscopy, 2nd edn. Kluwer Academic/Plenum Publishers, New York, pp 368–394

Maes M, Lin AH, Verkerk R, Delmeire L, Van Gastel A, Van der PM, Scharpe S (1999) Serotonergic and noradrenergic markers of post-traumatic stress disorder with and without major depression. Neuropsychopharmacology 20:188–197

Magde D, Elson EL, Webb WW (1972) Thermodynamic fluctuations in a reacting system-measurement by fluorescence correlation spectroscopy. Phys Rev Lett 29:705–708

Magde D, Elson EL, Webb WW (1974) Fluorescence correlation spectroscopy. II. An experimental realization. Biopolymers 13:29–61

Mager S, Min C, Henry DJ, Chavkin C, Hoffman BJ, Davidson N, Lester HA (1994) Conducting states of a mammalian serotonin transporter. Neuron 12:845–859

Medina MA, Schwille P (2002) Fluorescence correlation spectroscopy for the detection and study of single molecules in biology. Bioessays 24:758–764

Meissner O, Haberlein H (2003) Lateral mobility and specific binding to GABA(A) receptors on hippocampal neurons monitored by fluorescence correlation spectroscopy. Biochemistry 42:1667–1672

Merlet P, Benvenuti C, Moyse D, Pouillart F, Dubois-Rande JL, Duval AM, Loisance D, Castaigne A, Syrota A (1999) Prognostic value of MIBG imaging in idiopathic dilated cardiomyopathy [see comments]. J Nucl Med 40:917–923

Minsky M (1957) Microscopy apparatus. US Patent No. 3013467

Moron JA, Brockington A, Wise RA, Rocha BA, Hope BT (2002) Dopamine uptake through the norepinephrine transporter in brain regions with low levels of the dopamine transporter: evidence from knock-out mouse lines. J Neurosci 22:389–395

Nelson PJ, Rudnick G (1981) Anion-dependent sodium ion conductance of platelet plasma membranes. Biochemistry 20:4246–4249

Ni YG, Chen JG, Androutsellis-Theotokis A, Huang CJ, Moczydlowski E, Rudnick G (2001) A lithium-induced conformational change in serotonin transporter alters cocaine binding, ion conductance, and reactivity of Cys-109. J Biol Chem 276:30942–30947

Patterson GH, Knobel SM, Sharif WD, Kain SR, Piston DW (1997) Use of the green fluorescent protein and its mutants in quantitative fluorescence microscopy. Biophys J 73:2782–2790

Pawley JB (1995) Handbook of biological confocal microscopy, 2nd edn. Plenum Press, New York, pp 1–632

Petersen CI, DeFelice LJ (1999) Ionic interactions in the Drosophila serotonin transporter identify it as a serotonin channel. Nat Neurosci 2:605–610

Piston DW, Masters BR, Webb WW (1995) Three-dimensionally resolved NAD(P)H cellular metabolic redox imaging of the in situ cornea with two-photon excitation laser scanning microscopy. J Microsc 178:20–27

Povlock SL, Schenk JO (1997) A multisubstrate kinetic mechanism of dopamine transport in the nucleus accumbens and its inhibition by cocaine. J Neurochem 69:1093–1105

Ramsey IS, DeFelice LJ (2002) Serotonin transporter function and pharmacology are sensitive to expression level: evidence for an endogenous regulatory factor. J Biol Chem 277:14475–14482

Rigler R, Mets U, Widengren J, Kask P (1993) Fluorescence correlation spectroscopy with high count rate and low background: analysis of translational diffusion. Eur Biophys J 22:169–175

Ritz MC, Cone EJ, Kuhar MJ (1990) Cocaine inhibition of ligand binding at dopamine, norepinephrine and serotonin transporters: a structure-activity study. Life Sci 46:635–645

Robertson D, Flattem N, Tellioglu T, Carson R, Garland E, Shannon JR, Jordan J, Jacob G, Blakely RD, Biaggioni I (2001) Familial orthostatic tachycardia due to norepinephrine transporter deficiency. Ann N Y Acad Sci 940:527–543

Rudnick G, Nelson PJ (1978) Platelet 5-hydroxytryptamine transport, an electroneutral mechanism coupled to potassium. Biochemistry 17:4739–4742

Sacchetti G, Bernini M, Bianchetti A, Parini S, Invernizzi RW, Samanin R (1999) Studies on the acute and chronic effects of reboxetine on extracellular noradrenaline and other monoamines in the rat brain. Br J Pharmacol 128:1332–1338

Schenk JO (2002) The functioning neuronal transporter for dopamine: kinetic mechanisms and effects of amphetamines, cocaine and methylphenidate. Prog Drug Res 59:111–131

Schildkraut JJ, Gordon EK, Durell J (1965) Catecholamine metabolism in affective disorders. I. Normetanephrine and VMA excretion in depressed patients treated with imipramine. J Psychiatr Res 3:213–228

Scholze P, Norregaard L, Singer EA, Freissmuth M, Gether U, Sitte HH (2002) The role of zinc ions in reverse transport mediated by monoamine transporters. J Biol Chem 277:21505–21513

Schroeter S, Apparsundaram S, Wiley RG, Miner LH, Sesack SR, Blakely RD (2000) Immunolocalization of the cocaine and antidepressant-sensitive l-norepinephrine transporter. J Comp Neurol 420:211–232

Schwartz JW, Blakely RD, DeFelice LJ (2003) Binding and transport in norepinephrine transporters. Real-time, spatially resolved analysis in single cells using a fluorescent substrate. J Biol Chem 278:9768–9777

Schwille P (2001) Fluorescence correlation spectroscopy and its potential for intracellular applications. Cell Biochem Biophys 34:383–408

Schwille P, Korlach J, Webb WW (1999) Fluorescence correlation spectroscopy with single-molecule sensitivity on cell and model membranes. Cytometry 36:176–182

Shannon JR, Flattem NL, Jordan J, Jacob G, Black BK, Biaggioni I, Blakely RD, Robertson D (2000) Orthostatic intolerance and tachycardia associated with norepinephrine-transporter deficiency. N Engl J Med 342:541–549

Siggia ED, Lippincott-Schwartz J, Bekiranov S (2000) Diffusion in inhomogeneous media: theory and simulations applied to whole cell photobleach recovery. Biophys J 79:1761–1770

Sitte HH, Huck S, Reither H, Boehm S, Singer EA, Pifl C (1998) Carrier-mediated release, transport rates, and charge transfer induced by amphetamine, tyramine, and dopamine in mammalian cells transfected with the human dopamine transporter. J Neurochem 71:1289–1297

Sitte HH, Scholze P, Schloss P, Pifl C, Singer EA (2000) Characterization of carrier-mediated efflux in human embryonic kidney 293 cells stably expressing the rat serotonin transporter: a superfusion study. J Neurochem 74:1317–1324

Sitte HH, Hiptmair B, Zwach J, Pifl C, Singer EA, Scholze P (2001) Quantitative analysis of inward and outward transport rates in cells stably expressing the cloned human serotonin transporter: inconsistencies with the hypothesis of facilitated exchange diffusion. Mol Pharmacol 59:1129–1137

Sonders MS, Zhu SJ, Zahniser NR, Kavanaugh MP, Amara SG (1997) Multiple ionic conductances of the human dopamine transporter: the actions of dopamine and psychostimulants. J Neurosci 17:960–974

Tatsumi M, Groshan K, Blakely RD, Richelson E (1997) Pharmacological profile of antidepressants and related compounds at human monoamine transporters. Eur J Pharmacol 340:249–258

Torres GE, Gainetdinov RR, Caron MG (2003) Plasma membrane monoamine transporters: structure, regulation and function. Nat Rev Neurosci 4:13–25

Valentino RJ, Foote SL, Aston-Jones G (1983) Corticotropin-releasing factor activates noradrenergic neurons of the locus coeruleus. Brain Res 270:363–367

Vrljic M, Nishimura SY, Brasselet S, Moerner WE, McConnell HM (2002) Translational diffusion of individual class II MHC membrane proteins in cells. Biophys J 83:2681–2692

Wang YM, Xu F, Gainetdinov RR, Caron MG (1999) Genetic approaches to studying norepinephrine function: knockout of the mouse norepinephrine transporter gene. Biol Psychiatry 46:1124–1130

Wayment H, Meiergerd SM, Schenk JO (1998) Relationships between the catechol substrate binding site and amphetamine, cocaine, and mazindol binding sites in a kinetic model of the striatal transporter of dopamine in vitro. J Neurochem 70:1941–1949

Williams RM, Piston DW, Webb WW (1994) Two-photon molecular excitation provides intrinsic 3-dimensional resolution for laser-based microscopy and microphotochemistry. Faseb J 8:804–813

Xu F, Gainetdinov RR, Wetsel WC, Jones SR, Bohn LM, Miller GW, Wang YM, Caron MG (2000) Mice lacking the norepinephrine transporter are supersensitive to psychostimulants. Nat Neurosci 3:465–471

Structure/Function Relationships in Serotonin Transporter: New Insights from the Structure of a Bacterial Transporter

G. Rudnick

Department of Pharmacology, Yale University School of Medicine, New Haven CT, 06520-8066, USA
gary.rudnick@yale.edu

1	General Background and Significance of SERT	59
2	Mechanism of Transport	61
3	Topology	63
4	The Permeation Pathway	65
5	The Substrate Binding Site	66
6	Conformational Changes	67
7	Future Directions	70
	References	71

Abstract Serotonin transporter (SERT) serves the important function of taking up serotonin (5-HT) released during serotonergic neurotransmission. It is the target for important therapeutic drugs and psychostimulants. SERT catalyzes the influx of 5-HT together with Na^+ and Cl^- in a 1:1:1 stoichiometry. In the same catalytic cycle, there is coupled efflux of one K^+ ion. SERT is one member of a large family of amino acid and amine transporters that is believed to utilize similar mechanisms of transport. A bacterial member of this family was recently crystallized, revealing the structural basis of these transporters. In light of the new structure, previous results with SERT have been re-interpreted, providing new insight into the substrate binding site, the permeation pathway, and the conformational changes that occur during the transport cycle.

Keywords Serotonin · Transporter · Structure · Mechanism · Permeation

1
General Background and Significance of SERT

The neurotransmitter transporters are plasma membrane proteins that take up extracellular neurotransmitters after release and thereby terminate the transmitters' action at extracellular receptor sites. These plasma membrane neurotransmitter transporters represent the first step in the process of transmitter recycling. Subsequent sequestration by synaptic vesicles requires a second transport system in the vesicular membrane. Although the structure and mechanism of the vesicular neurotransmitter transporters are distinct from those of the plasma membrane, the two systems work together to transport extracellular neurotransmitters into the synaptic vesicle, where they are available for release by exocytosis (Rudnick 2002).

The plasma membrane neurotransmitter transporters use transmembrane ion gradients of Na^+, Cl^-, and K^+ and an internal negative membrane potential for transport of their substrate neurotransmitters (Rudnick and Clark 1993; Rudnick 2002). Transporters responsible for reuptake of neurotransmitters across the plasma membrane of neurons and glia fall into two gene families (Amara 1992). Most small neurotransmitters, including glycine, γ-aminobutyric acid (GABA), dopamine (DA), norepinephrine (NE) and serotonin (5-hydroxytryptamine, 5-HT), are transported by proteins belonging to the family designated the neurotransmitter sodium symporter (NSS) family 2.A.22 by Saier (1999). Glutamate, however, is transported by a family of mono- and dicarboxylic amino acid transporters, the DAACS family (Saier 1999). Proteins in both families play important roles in brain function.

Serotonin transporter (SERT) is a member of the NSS family that selectively transports 5-HT into nerve cells together with Na^+ and Cl^- and, in the same reaction, transports a K^+ ion out of the cell. SERT is inhibited by a variety of compounds that are used to treat clinical depression, including fluoxetine (Prozac), sertraline (Zoloft), paroxetine (Paxil), and citalopram (Celexa). These compounds were synthesized as selective serotonin reuptake inhibitors (SSRIs) based on the observation that compounds useful as antidepressants, such as imipramine, inhibited serotonin transport. The widespread use of serotonin reuptake inhibitors makes SERT a molecule of high clinical interest.

In addition to drugs that specifically target SERT, this transporter is also affected by cocaine and amphetamines—psychostimulant drugs that are widely abused. Cocaine acts as a simple inhibitor of SERT and the closely related NSS transporters for NE and DA, NET and DAT, respectively (Gu et al. 1994). Amphetamine and its congeners, however, have a more complex mechanism of action. These compounds are substrates for SERT, NET, and DAT but also diffuse into cells because of their high membrane permeability. This ability to cross membranes in their unprotonated, neutral form allows amphetamines to dissipate the internally acid pH difference (ΔpH) across the synaptic vesicle membrane (Schuldiner et al. 1993). Because this ΔpH is used as an important driving force for accumulation of 5-HT, NE, and DA by synaptic vesicles, collapsing the ΔpH causes release of accumulated neurotransmitter into the cytoplasm. Transport of amphetamines by SERT, NET, or DAT leads to accumulation of Na^+ in the cytoplasm (Khoshbouei et al. 2003), and the combination of high cytoplasmic neurotransmitter and increased cytoplasmic Na^+ leads to reversal of the plasma membrane transporter and appearance of neurotransmitter outside the cell (Rudnick 2002). Thus, both cocaine and amphetamines lead to more neurotransmitter in the synapse, but in the case of amphetamine, cellular stores are actively released, while cocaine raises synaptic transmitter by blocking re-uptake. Among the variety of amphetamine derivatives, MDMA (3,4-methylenedioxymethamphetamine, "ecstasy") is more selective toward releasing 5-HT from serotonergic neurons, and this selectivity is due to MDMA's higher affinity for SERT relative to its affinity for NET or DAT (Wall et al. 1995).

2
Mechanism of Transport

SERT, like other transporters, is believed to function by alternately exposing a substrate binding site to the cytoplasmic and extracellular faces of the plasma membrane. To understand such a mechanism in detail requires knowledge of four key properties of the protein: (1) The nature of the binding site determines how the transporter can selectively transport one substrate and not another. In cases, like SERT, where ions are cotransported with substrate, the relative positioning of substrate and ion binding sites may be critical for coupling. (2) The pathways that the substrate and ions pass through from one side of the membrane to the binding site and then from the binding site to the other side of the membrane need to be tightly coupled to each other so that they are not both open simultaneously, which would lead to uncoupled flux through the transporter. (3) The transporter must undergo conformational changes that close access from one side of the membrane and open access to the other. (4) Occupancy of the binding sites must control conformational changes so that they occur only when the appropriate ligands are bound. Otherwise, the transporter would catalyze uncoupled flux of any solute that occupied its binding site.

According to a mechanism proposed to describe 5-HT transport, SERT binds Na^+, Cl^-, and 5-HT^+ in a 1:1:1 stoichiometry and only then undergoes a conformational change that occludes the binding site from the extracellular medium and exposes it to the cytoplasm (Nelson and Rudnick 1979). After dissociation of Na^+, Cl^-, and 5-HT^+, the transporter returns to its original conformation only after binding a cytoplasmic K^+ ion and releasing it to the extracellular medium. The overall stoichiometry of this process is a 1:1:1:1 electroneutral exchange of K^+ with Na^+, Cl^-, and 5-HT^+ (Rudnick and Nelson 1978; Talvenheimo et al. 1983; Rudnick 1998).

Evidence for this mechanism originally came from studies using platelet plasma membrane vesicles (Rudnick 1977). These studies provided evidence for the stoichiometry and supported the movement of K^+ in a step distinct from the one in which 5-HT was transported (Nelson and Rudnick 1979). SERT is also capable of conducting ionic current that is induced by 5-HT (Mager et al. 1994; Lin et al. 1996; Cao et al. 1997; Cao et al. 1998). Although this would, on the surface, appear to argue against a coupled electroneutral stoichiometry, it has become clear that SERT catalyzes an uncoupled flux in addition to the coupled transport process. An alternative mechanism has been put forward in which 5-HT and Na$^+$ movement are coupled within a channel (Petersen and DeFelice 1999; Adams and DeFelice 2002), but this mechanism does not explain how K$^+$ countertransport could be coupled to 5-HT uptake in an electroneutral process. A recent study demonstrated that interaction with syntaxin 1a could block the uncoupled current, revealing the coupled electroneutral process with the same stoichiometry that was originally proposed (Quick 2003).

For electroneutral 5-HT transport coupled to Na$^+$, Cl$^-$, and K$^+$, the conformational changes that allow these solutes to cross the membrane must serve two functions. First, they must open up the binding site alternately to each side of the membrane to allow binding and dissociation of solutes. Second, they must prevent uncoupled movement of solutes. For example, if the transporter were constantly interconverting between cytoplasmic- and extracellular-facing forms regardless of what solutes were bound, it would catalyze only downhill leakage of 5-HT, Na$^+$, K$^+$, and Cl$^-$. For any transporter to be stoichiometrically coupled, its conformational changes must be linked to the occupancy of the binding site. For SERT, the conformational change should occur when 5-HT, Na$^+$, and Cl$^-$ are bound or when K$^+$ is bound, but not when the binding site is only partly occupied (see Fig. 1).

A major advance in this process resulted from the discovery that the genomes of many prokaryotes (bacteria and archaea) contained genes coding for proteins quite homologous to neurotransmitter transporters. In 2003, the first evidence became available showing that these proteins were actually transporters (Androutsellis-Theotokis et al. 2003). It showed that the TnaT protein of *Symbiobacterium thermophilum* was a Na$^+$-dependent tryptophan transporter with properties similar to those of other NSS transporters. Recently, the laboratory of Eric Gouaux provided a high-resolution structure from another bacterial homolog, LeuT from *Aquifex aeolicus* (Yamashita et al. 2005). Although this structure will certainly differ in details from the structure of the mammalian proteins, it provides a framework for designing further experiments toward a variety of goals, among which are to test the relevance of the structure, to define the particular differences between the bacterial and mammalian transporters, and to understand the molecular motions within the structure that lead to transport.

The structure of the *A. aeolicus* leucine transporter (LeuT$_{Aa}$) provides some surprises, some unique features, and many opportunities to explore mechanistic issues relevant to neurotransmitter transport. An unusual aspect to this structure is that it contains a repeat of two groups of five transmembrane domains in opposite topological orientations. Because of the high resolution of the structure, the Gouaux group was able to identify two Na$^+$ ions bound together with leucine at the active site, thus providing a structural basis for coupling of Na$^+$ and solute fluxes.

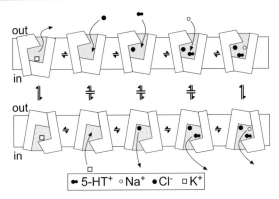

Fig. 1 Possible mechanism of serotonin transport. Transport of 5-HT together with Na$^+$ and Cl$^-$ ions requires binding of each solute to the transporter. These binding events are depicted in the three steps on the *upper right* of the figure. There is no evidence that the binding sequence is strictly ordered. Only after all three solutes are bound is the transporter able to undergo a series of conformational changes that closes off access to the extracellular medium and exposes the binding site to the cytoplasm. This conformational change is depicted on the *right side* of the figure. After dissociation of 5-HT, Na$^+$, and Cl$^-$ on the cytoplasmic side of the plasma membrane, as shown by the three *rightmost* steps on the *lower* part of the figure, a cytoplasmic K$^+$ ion is able to bind (*lower left*). Once K$^+$ has bound, SERT is able to undergo another series of conformational changes that closes off access to the cytoplasm and exposes the binding site to the extracellular medium. Dissociation of K$^+$ to the medium completes the cycle. Note that for effective coupling of 5-HT influx to both influx of Na$^+$ and Cl$^-$ and efflux of K$^+$, transitions between the extracellular-facing and cytoplasmic-facing forms of SERT should occur only when the binding site is occupied with 5-HT, Na$^+$, and Cl$^-$ or with K$^+$

3
Topology

The primary sequence of most NSS family members predicts 12 transmembrane (TM) domains connected by hydrophilic loops. However, several prokaryotic sequences predict only 10 TMs. The fact that the central core of LeuT, including the substrate and Na$^+$ binding sites, is formed by two copies of a 5-TM repeat provides an explanation for the functionality of 10-TM transporters and suggests that TM11 and TM12 are not always required for transport function. In the mammalian transporters, glycosylation sites in the second extracellular loop (EL2, between TM3 and TM4) indicated that EL2 is extracellular (Tate and Blakely 1994). For SERT, many residues predicted by the initial topological predictions to lie in hydrophilic loops were demonstrated to be accessible from the appropriate side of the membrane (Chen et al. 1998; Androutsellis-Theotokis and Rudnick 2002), indicating a 12-TM structure with NH$_2$- and COOH-termini in the cytoplasm. These studies extensively utilized cysteine-scanning mutagenesis of internal and external loops and transmembrane domains. Figure 2 shows a summary of some of results for SERT superimposed on a topology diagram generated from the crystal structure of LeuT and using an alignment of the NSS family. The highlighted positions, where various studies demonstrated chemical reactivity with hydrophilic reagents, indicate many residues in regions exposed to solvent in the structure (Chen et al. 1997a, b, 1998; Chen and Rudnick 2000; Androutsellis-Theotokis et al. 2001; Ni et al. 2001; Androutsellis-Theotokis and Rudnick 2002; Henry et al. 2003; Mitchell et al. 2004; Sato et al. 2004).

In addition to those positions that are clearly in the hydrophilic extracellular or cytoplasmic domains, cysteines were modified at many positions that the LeuT structure predicts to be inaccessible from either face of the membrane. The observation that residues predicted by the structure to be buried were nonetheless accessible in SERT suggests that other conformations of the transporter expose these residues in other conformations of the protein. Thus, the form of LeuT that crystallized is likely to represent only one very restricted conformation of a transporter that must undergo conformational changes to allow substrates to bind and dissociate from both sides of the membrane. In fact, the binding site in the LeuT structure contains the substrate and two Na^+ ions occluded from both faces of the membrane. Conformational changes are required to allow these solutes into and out of the binding site.

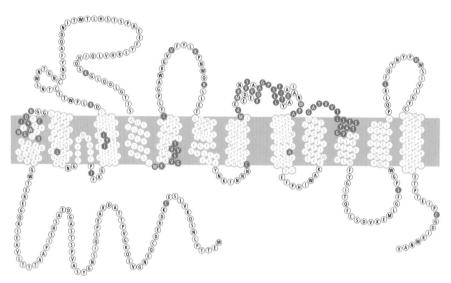

Fig. 2 Topology diagram of SERT. The sequence of SERT was aligned with that of LeuT$_{Aa}$ and presented as a topology diagram, using the beginnings and ends of each helix from the LeuT structure. *Shaded* residues are those where mutation to cysteine or lysine was found to introduce reactivity toward hydrophilic compounds such as MTS reagents. Not all positions were tested, and positions where reactivity was not detected are not marked

4
The Permeation Pathway

Keller et al. (2004) and Zhang et al. (2005) were able to react cysteines in the extracellular half of TM10 and the cytoplasmic half of TM5 that were occluded in the LeuT structure. For example, in the LeuT structure, access of substrate to the extracellular medium is blocked in part by a salt bridge between Arg-30 and Asp-404 (Yamashita et al. 2005). The corresponding residues in SERT are Arg-104 and Glu-493. However, in SERT, access in the EL5 region containing Glu-493 continued up to Pro-499. According to the LeuT model, this position is in TM10, almost in the middle of the membrane (Fig. 3). An example of the same phenomenon from the cytoplasmic side of the membrane is in intracellular loop (IL)2, where positions up to Tyr-289 were found to be accessible to reagents on the cytoplasmic side. In the LeuT structure, Phe-203, which corresponds to SERT Tyr-289 is in TM5 more than halfway across the membrane from IL2 (Fig. 3; Yamashita et al. 2005). It is noteworthy that TM5 and TM10 are related in that each one is the last TM in the two repeated 5-TM structural units that make up the core of the structure. The space between the intracellular end of the exposed region of TM10 and the extracellular end of the exposed region of TM5 contains binding site residues for leucine and Na^+ formed by TMs 1, 3, 6, and 8. Thus, the accessibility of residues in TM5 and TM10 might indicate that these helices form part of the permeation pathway that allows substrates to enter and exit their binding sites.

In the LeuT structure, leucine is occluded, and there is no pathway for it to dissociate either to the periplasmic or cytoplasmic sides of the membrane (Yamashita et al. 2005). Although leucine is occluded in the crystal structure, only a few residues block its exit to the periplasmic (external) side of the membrane. However, there is almost 20 Å of packed protein structure separating leucine

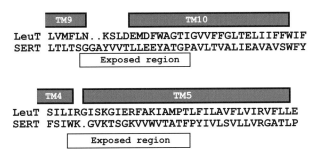

Fig. 3 Exposure of TM domains to aqueous reagents. The sequence of rat SERT was aligned with that of LeuT in the EL5–TM10 region and the IL2–TM5 region. The *shaded* regions indicate the TM helices and the *open boxes* indicate the regions that were accessible to modification with [2-(trimethylammonium)ethyl]methanethiosulfonate (MTSET) (for cysteine replacements in EL5–TM10) or (2-aminoethyl)methanethiosulfonate (MTSEA) (for cysteine replacements in IL2–TM5)

from the cytoplasmic face of the protein. Taken at face value, the pathways from the periplasmic and cytoplasmic faces to the binding site would appear to be quite different. However, another interpretation is that the crystal structure represents an intermediate in which the cytoplasmic pathway is closed and the periplasmic pathway is mostly open. The blockade of the periplasmic permeation pathway (by Tyr-108, Phe-253, Arg-30, and Asp-404) might even be properties of an intermediate that is not on the normal reaction path but that is particularly stable in the crystal. Additional evidence exists that the periplasmic permeation pathway is more condensed in the cytoplasmic-facing form than is suggested by the LeuT structure. Residues in the tip of EL4 (between TMs 7 and 8), Tyr-107, and Ile-108 form parts of the periplasmic pathway in LeuT. In SERT, the corresponding residues are largely protected by 5-HT in the presence of NaCl (Henry et al. 2003; Mitchell et al. 2004), conditions likely to favor the cytoplasmic-facing form. Because the crystal structure may represent a form of LeuT close to the periplasmic-facing form, it is quite possible that—in the cytoplasmic-facing form—the external pathway exists in a more condensed conformation.

5
The Substrate Binding Site

Evidence from many laboratories (Chen et al. 1997b; Barker et al. 1998, 1999; Chen and Rudnick 2000; Adkins et al. 2001; Henry et al. 2003; Melamed and Kanner 2004; Zhou et al. 2004) strongly suggested that the binding site for the substrate and ions is formed, at least in part, by TM1 and TM3. TM2, although it is adjacent to both TM1 and TM3 in the primary sequence, does not contribute directly to the binding site in LeuT, and results from SERT are in agreement that—although mutations in this region have effects on expression and K_M for substrate—most of the positions in TM2 were neither accessible to extracellular reagents nor affected by substrate binding (Sato et al. 2004). However, mutations in TM2 of DAT have strong effects on the affinity of cocaine (Chen et al. 2005; Sen et al. 2005) and may represent part of an inhibitor binding site or it might contribute to the position of TMs 1 and 6. In addition, the LeuT structure revealed that at least one residue, Leu-137—thought to be part of the IL1 loop between TM2 and TM3—is actually part of TM2. L137C reacted with (2-aminoethyl)methanethiosulfonate (MTSEA), and the reaction was inhibited by 5-HT and cocaine (Androutsellis-Theotokis and Rudnick 2002). Similarly, Ala-441 was previously thought to be part of IL4 between TM8 and TM9. The LeuT structure revealed that this position is in TM8 only one helical turn away from binding site residues. When expressed in intact cells, A441C did not react with MTSEA, but in membrane preparations MTSEA inactivated A441C binding activity and this reaction was inhibited by 5-HT and cocaine.

We previously had no indication that other TM domains contributed to the permeation pathway, although analysis of channel proteins strongly suggested that it must consist of more than just TM1 and TM3 (Spencer and Rees 2002). From the LeuT structure, it is apparent that the two corresponding TMs from the second 5-TM repeat, TM6 and TM8, also contribute to the binding site. It will be important to examine residues in TM6 and TM8 to evaluate their contribution to the SERT binding site for 5-HT, Na^+, and drugs such as cocaine, amphetamines, and antidepressants.

The structure of LeuT contains two bound Na^+ ions in proximity to the bound leucine. The residues responsible for coordination of Na^+ are not identical in SERT. The most pronounced difference is that, in LeuT, the leucine carboxyl group coordinates one of the Na^+ ions (Na1). In SERT, as in NET and DAT, the substrate contains no carboxylate group that can fulfill this function, but there is an aspartate residue at position 98 in SERT (Gly-24 in LeuT) that is positioned to take its place. Mutation of this aspartate to glutamate or cysteine was strongly inhibitory for SERT function (Barker et al. 1999; Henry et al. 2003). A second bound Na^+ ion in the LeuT structure (Na2) is coordinated by five residues in the LeuT structure. Three of these residues are identical in SERT. One of the two non-identical residues coordinates with Na2 through its carbonyl oxygen, and the other is a Thr-354 in LeuT and an Asp-437 in SERT. Because Asp-437 could participate in coordinating a Na^+ ion in SERT similar to Na2 in LeuT, it is possible that SERT also binds 2 Na^+ ions. Indeed, previous results comparing the Na^+ dependence of 5-HT transport and imipramine binding by SERT suggested that 2 Na^+ ions were involved in the latter process (Talvenheimo et al. 1983).

However, this poses questions about how Na1 and Na2 are related to Na^+ ions cotransported with substrate. Some transporters in the NSS family, such as GAT-1 and GlyT1b, are known to transport 2 Na^+ ions with each substrate molecule (Keynan and Kanner 1988; Roux and Supplisson 2000). SERT Na^+ stoichiometry has been determined by two methods and found to be 1 Na^+ per 5-HT (Talvenheimo et al. 1983; Quick 2003). The Na^+ binding stoichiometry of LeuT is 2, but the Na^+ stoichiometry for transport is unknown. Many interesting mechanistic questions remain to be answered regarding ion coupling in SERT. Is binding stoichiometry always the same as transport stoichiometry, or is it possible that only 1 of the 2 Na^+ ions in the structure is transported? Similarly, does the single sodium symported with 5-HT by SERT represent a single bound Na^+ ion, or is an additional, non-transported Na^+ bound to SERT? Does either of the Na^+ sites predicted in SERT by the LeuT structure represent the site used for K^+ antiport? How is Cl^- cotransport coupled to 5-HT in SERT?

6
Conformational Changes

Many of the positions indicated by Fig. 2 to react with MTS (methanethiosulfonate) reagents are not close to the binding site for substrate and ions as defined by comparison with the LeuT structure. And yet, for many of these positions, reactivity is sensitive to the presence of 5-HT or cocaine. Because of their distance from the substrate binding site, it is unlikely that these residues are directly occluded by 5-HT or ion binding, but the changes in reactivity are likely to reflect conformational changes in response to occupation of the binding site. Furthermore, many of the changes that were observed required not just 5-HT, but Na^+ and Cl^- as well, suggesting that the change in accessibility represented entry of SERT into the transport cycle.

As discussed in the previous section, two residues in TM3 were identified as being close to the substrate binding site. Part of the evidence for this conclusion was the observation that 5-HT could protect a cysteine at those positions from modification by MTS reagents (Chen et al. 1997b). Further studies demonstrated that 5-HT protected even in the absence of Na^+, indicating that 5-HT binding was not Na^+-dependent (Chen and Rudnick 2000). Thus, it was important to discover that in many SERT cysteine mutants, the effect of 5-HT on reactivity of a cysteine residue required Na^+ (Androutsellis-Theotokis et al. 2001) or both Na^+ and Cl^+ (Mitchell et al. 2004; Sato et al. 2004).

To put these various ion requirements in perspective, it is helpful to consider the likely transport pathway for SERT, as deduced from a variety of approaches (Rudnick 2002). In this pathway, Na^+, Cl^-, and 5-HT all bind from the extracellular side of the membrane to form a quaternary complex with the transporter (see Fig. 1). Only when this complex is formed will the transporter undergo a conformational change to expose the 5-HT binding site to the cytoplasm. After dissociation of Na^+, Cl^-, and 5-HT to the cytoplasm, the binding site can accept a K^+ ion that allows a second conformational change, returning SERT to its original extracellular-facing conformation. After K^+ dissociation to the extracellular medium, the cycle can start over. A key feature of this process is the requirement for a particular set of solutes (either Na^+, Cl^-, and 5-HT or K^+) to be bound before the protein can change conformation. This requirement is responsible for the stoichiometric coupling of 5-HT, Na^+, Cl^-, and K^+ (Rudnick 1998).

Because conformational changes that affect the accessibility of the 5-HT and ion binding sites are such a critical part of the mechanism, it is important to define each of the potential conformations. We assume that there is a form of SERT in which the 5-HT and ion binding sites are directly accessible from the extracellular medium. This conformation is represented by the upper part of Fig. 1 and will be referred to as the "extracellular-facing form." Another conformation (the lower part of Fig. 1) must release 5-HT and ions from their binding sites to the cytoplasm and this will be referred to as the "cytoplasmic-facing form." There may be intermediate forms, such as the form of LeuT in the crystal structure, in which the binding sites are exposed to neither side. These forms will be referred to as "occluded." The pathways from the binding site to the extracellular medium or the cytoplasm that are properties of the extracellular-facing and cytoplasmic-facing forms, respectively, will be referred to as the extracellular and cytoplasmic permeation pathways, respectively.

It would be useful to assign the various changes in cysteine accessibility with different states of the transporter, so as to understand which parts of the protein participate in the conformational changes accompanying binding and transport reactions. It is possible to assign some changes with binding and others with the transport steps. For example, cysteines at some positions (such as I172C and Y176C) were protected by 5-HT or inhibitor binding, did not depend on Na^+, and were protected both at 25 °C and 4 °C (Chen and Rudnick 2000). This can be interpreted as a simple steric occlusion of the reactive residue, which is a property of the outward-facing transporter and does not require conformational changes.

At other positions (such as Cys-357), both 5-HT and inhibitors protected against cysteine modification, but the protection required Na^+ and was observed at 25 °C but not 4 °C (Androutsellis-Theotokis et al. 2001). This behavior suggests an allosteric effect due to a conformational change that occurs when both 5-HT and Na^+ are bound to SERT. It probably does not represent the translocation step, because the protection was seen with both 5-HT and the non-transported inhibitor cocaine. Apparently, binding induced a conformational change that precedes the translocation event.

A third set of residues (such as S404C and Y107C and I108C) were protected only by 5-HT and not cocaine, and the protection required both Na^+ and Cl^- (Henry et al. 2003; Mitchell et al. 2004). This behavior probably represents the conformational change that actually translocates 5-HT across the membrane. The reasoning is as follows: (1) The effect is allosteric, suggesting a conformational change, since some residues were protected (Mitchell et al. 2004) while others were potentiated (they reacted faster with MTS reagents when 5-HT, Na^+, and Cl^- were present) (Sato et al. 2004). (2) Only substrates but not non-transported inhibitors such as cocaine promote the change in reactivity (Mitchell et al. 2004; Sato et al. 2004). (3) Other SERT substrates, such as MDMA, could replace 5-HT (Sato et al. 2004). These data suggest that the presence of 5-HT, Na^+, and Cl^- transforms SERT from a predominantly extracellular-facing conformation to one that is predominantly cytoplasmic-facing, having transported 5-HT and released it on the cytoplasmic side of the membrane.

The ability to manipulate the state of SERT and to determine the effects on accessibility of cysteine residues placed at specific positions will allow the testing of possible transport mechanisms. The goal is to understand how the conformational changes within SERT lead to alternate accessibility of the binding site from the two sides of the membrane. To accomplish this goal, it will be necessary to use biochemical approaches with the functional protein in its native environment. The structure of LeuT provides a framework for these studies, but it does not provide much information about dynamic changes in the transporter structure.

7
Future Directions

Many aspects of SERT structure remain unresolved despite the major advance in our understanding provided by the structure of LeuT. As described above, the difference in ion coupling between SERT and LeuT must be reflected in differences in the structure of the substrate binding site. In addition, there are three regions of SERT where the structure of LeuT provides little or no information. The first of these is EL2, which is much longer in SERT than in LeuT. EL2 is likely to be important for functional expression of SERT because replacing part or all of it with its corresponding sequence from NET led to a protein inactive for transport (Stephan et al. 1997; Smicun et al. 1999). The additional sequence not present in LeuT includes glycosylation sites (Tate and Blakely 1994) and a highly conserved pair of cysteine residues likely to form a disulfide (Chen et al. 1997a). Mutations in these cysteines, or modification of one when the other was mutated, led to severe loss in activity (Chen et al. 1997a). Thus, it is likely that the parts of EL2 that are unique to animal members of the NSS family are functionally important in ways that are not addressed by the LeuT structure.

SERT, like most neurotransmitter transporters, contains much longer NH_2- and COOH-terminal regions than does LeuT. The N-terminal region of SERT has been implicated in the regulation of ion conductance by syntaxin 1a (Quick 2003). It is likely that these domains are important for regulation through interactions with other intracellular pathways. The NH_2- and COOH-terminal regions are also likely targets for agents that control the subcellular localization of SERT, as has been demonstrated for the related norepinephrine and GABA transporters (Perego et al. 1997; Muth et al. 1998; Gu et al. 2001; Farhan et al. 2004). The structure of these domains and their potential interaction with the intracellular face of the central region of SERT are still unknown, and will doubtless be the subject of future study.

Because they show a static structure, the images of LeuT cannot tell us how transporters in this family move substrate and ions from the cell exterior to the binding site and then to the cytoplasm. These movements require conformational changes involving the transmembrane domains and possibly also the hydrophilic loops that connect them. Moreover, these movements are triggered by the binding of appropriate substrates and ions to the transporter (in the case of SERT, Na^+, Cl^-, and 5-HT for the forward reaction and K^+ for the return). Understanding the mechanism by which these binding events allow and control the conformational changes, and comprehending the nature of the conformational changes themselves, are important goals for future research in this area.

References

Adams SV, DeFelice LJ (2002) Flux coupling in the human serotonin transporter. Biophys J 83:3268-3282
Adkins EM, Barker EL, Blakely RD (2001) Interactions of tryptamine derivatives with serotonin transporter species variants implicate transmembrane domain I in substrate recognition. Mol Pharmacol 59:514-523
Amara S (1992) Neurotransmitter transporters—a tale of 2 families. Nature 360:420-421
Androutsellis-Theotokis A, Rudnick G (2002) Accessibility and conformational coupling in serotonin transporter predicted internal domains. J Neurosci 22:8370-8378
Androutsellis-Theotokis A, Ghassemi F, Rudnick G (2001) A conformationally sensitive residue on the cytoplasmic surface of serotonin transporter. J Biol Chem 276:45933-45938
Androutsellis-Theotokis A, Goldberg NR, Ueda K, Beppu T, Beckman ML, Das S, Javitch JA, Rudnick G (2003) Characterization of a functional bacterial homologue of sodium-dependent neurotransmitter transporters. J Biol Chem 278:12703-12709
Barker EL, Perlman MA, Adkins EM, Houlihan WJ, Pristupa ZB, Niznik HB, Blakely RD (1998) High affinity recognition of serotonin transporter antagonists defined by species-scanning mutagenesis—an aromatic residue in transmembrane domain I dictates species-selective recognition of citalopram and mazindol. J Biol Chem 273:19459-19468
Barker EL, Moore KR, Rakhshan F, Blakely RD (1999) Transmembrane domain I contributes to the permeation pathway for serotonin and ions in the serotonin transporter. J Neurosci 19:4705-4717
Cao Y, Li M, Mager S, Lester HA (1998) Amino acid residues that control pH modulation of transport-associated current in mammalian serotonin transporters. J Neurosci 18:7739-7749
Cao YW, Mager S, Lester HA (1997) H^+ permeation and pH regulation at a mammalian serotonin transporter. J Neurosci 17:2257-2266
Chen JG, Rudnick G (2000) Permeation and gating residues in serotonin transporter. Proc Natl Acad Sci U S A 97:1044-1049
Chen JG, Liu-Chen S, Rudnick G (1997a) External cysteine residues in the serotonin transporter. Biochemistry 36:1479-1486
Chen JG, Sachpatzidis A, Rudnick G (1997b) The third transmembrane domain of the serotonin transporter contains residues associated with substrate and cocaine binding. J Biol Chem 272:28321-28327
Chen JG, Liu-Chen S, Rudnick G (1998) Determination of external loop topology in the serotonin transporter by site-directed chemical labeling. J Biol Chem 273:12675-12681
Chen R, Han DD, Gu HH (2005) A triple mutation in the second transmembrane domain of mouse dopamine transporter markedly decreases sensitivity to cocaine and methylphenidate. J Neurochem 94:352-359
Farhan H, Korkhov VM, Paulitschke V, Dorostkar MM, Scholze P, Kudlacek O, Freissmuth M, Sitte HH (2004) Two discontinuous segments in the carboxyl terminus are required for membrane targeting of the rat γ-aminobutyric acid transporter-1 (GAT1). J Biol Chem 279:28553-28563
Gu H, Wall SC, Rudnick G (1994) Stable expression of biogenic amine transporters reveals differences in inhibitor sensitivity, kinetics, and ion dependence. J Biol Chem 269:7124-7130
Gu HH, Wu XH, Giros B, Caron MG, Caplan MJ, Rudnick G (2001) The NH2-terminus of norepinephrine transporter contains a basolateral localization signal for epithelial cells. Mol Biol Cell 12:3797-3807

Henry LK, Adkins EM, Han Q, Blakely RD (2003) Serotonin and cocaine-sensitive inactivation of human serotonin transporters by methanethiosulfonates targeted to transmembrane domain I. J Biol Chem 278:37052–37063

Keller PC 2nd, Stephan M, Glomska H, Rudnick G (2004) Cysteine-scanning mutagenesis of the fifth external loop of serotonin transporter. Biochemistry 43:8510–8516

Keynan S, Kanner BI (1988) γ-Aminobutyric acid transport in reconstituted preparations from rat brain: coupled sodium and chloride fluxes. Biochemistry 27:12–17

Khoshbouei H, Wang HW, Lechleiter JD, Javitch JA, Galli A (2003) Amphetamine-induced dopamine efflux—a voltage-sensitive and intracellular Na^+-dependent mechanism. J Biol Chem 278:12070–12077

Lin F, Lester HA, Mager S (1996) Single-channel currents produced by the serotonin transporter and analysis of a mutation affecting ion permeation. Biophys Chem 71:3126–3135

Mager S, Min C, Henry DJ, Chavkin C, Hoffman BJ, Davidson N, Lester HA (1994) Conducting states of a mammalian serotonin transporter. Neuron 12:845–859

Melamed N, Kanner BI (2004) Transmembrane domains I and II of the γ-aminobutyric acid transporter GAT-4 contain molecular determinants of substrate specificity. Mol Pharmacol 65:1452–1461

Mitchell SM, Lee E, Garcia ML, Stephan MM (2004) Structure and function of extracellular loop 4 of the serotonin transporter as revealed by cysteine-scanning mutagenesis. J Biol Chem 279:24089–24099

Muth TR, Ahn J, Caplan MJ (1998) Identification of sorting determinants in the C-terminal cytoplasmic tails of the γ-aminobutyric acid transporters Gat-2 and Gat-3. J Biol Chem 273:25616–25627

Nelson PJ, Rudnick G (1979) Coupling between platelet 5-hydroxytryptamine and potassium transport. J Biol Chem 254:10084–10089

Ni YG, Chen JG, Androutsellis-Theotokis A, Huang CJ, Moczydlowski E, Rudnick G (2001) A lithium-induced conformational change in serotonin transporter alters cocaine binding, ion conductance, and reactivity of cys-109. J Biol Chem 276:30942–30947

Perego C, Bulbarelli A, Longhi R, Caimi M, Villa A, Caplan MJ, Pietrini G (1997) Sorting of two polytopic proteins, the gamma-aminobutyric acid and betaine transporters, in polarized epithelial cells. J Biol Chem 272:6584–6592

Petersen CI, DeFelice LJ (1999) Ionic interactions in the Drosophila serotonin transporter identify it as a serotonin channel. Nat Neurosci 2:605–610

Quick MW (2003) Regulating the conducting states of a mammalian serotonin transporter. Neuron 40:537–549

Roux MJ, Supplisson S (2000) Neuronal and glial glycine transporters have different stoichiometries. Neuron 25:373–383

Rudnick G (1977) Active transport of 5-hydroxytryptamine by plasma membrane vesicles isolated from human blood platelets. J Biol Chem 252:2170–2174

Rudnick G (1998) Bioenergetics of neurotransmitter transport. J Bioenerg Biomembr 30:173–185

Rudnick G (2002) Mechanisms of biogenic amine neurotransmitter transporters. In: Reith MEA (ed) Neurotransmitter transporters, structure, function, and regulation. Humana Press, Totowa, pp 25–52

Rudnick G, Clark J (1993) From synapse to vesicle: the reuptake and storage of biogenic amine neurotransmitters. Biochim Biophys Acta 1144:249–263

Rudnick G, Nelson PJ (1978) Platelet 5-hydroxytryptamine transport an electroneutral mechanism coupled to potassium. Biochemistry 17:4739–4742

Saier MH (1999) A functional-phylogenetic system for the classification of transport proteins. J Cell Biochem Suppl 32–33:84–94

Sato Y, Zhang YW, Androutsellis-Theotokis A, Rudnick G (2004) Analysis of transmembrane domain 2 of rat serotonin transporter by cysteine scanning mutagenesis. J Biol Chem 279:22926–22933

Schuldiner S, Steiner-Mordoch S, Yelin R, Wall SC, Rudnick G (1993) Amphetamine derivatives interact with both plasma membrane and secretory vesicle biogenic amine transporters. Mol Pharmacol 44:1227–1231

Sen N, Shi L, Beuming T, Weinstein H, Javitch JA (2005) A pincer-like configuration of TM2 in the human dopamine transporter is responsible for indirect effects on cocaine binding. Neuropharmacology 49:780–790

Smicun Y, Campbell SD, Chen MA, Gu H, Rudnick G (1999) The role of external loop regions in serotonin transport. Loop scanning mutagenesis of the serotonin transporter external domain. J Biol Chem 274:36058–36064

Spencer RH, Rees DC (2002) The alpha-helix and the organization and gating of channels. Annu Rev Biophys Biomol Struct 31:207–233

Stephan MM, Chen MA, Penado KM, Rudnick G (1997) An extracellular loop region of the serotonin transporter may be involved in the translocation mechanism. Biochemistry 36:1322–1328

Talvenheimo J, Fishkes H, Nelson PJ, Rudnick G (1983) The serotonin transporter-imipramine 'receptor': different sodium requirements for imipramine binding and serotonin translocation. J Biol Chem 258:6115–6119

Tate C, Blakely R (1994) The effect of N-linked glycosylation on activity of the Na^+- and Cl^--dependent serotonin transporter expressed using recombinant baculovirus in insect cells. J Biol Chem 269:26303–26310

Wall SC, Gu H, Rudnick G (1995) Biogenic amine flux mediated by cloned transporters stably expressed in cultured cell lines: amphetamine specificity for inhibition and efflux. Mol Pharmacol 47:544–550

Yamashita A, Singh SK, Kawate T, Jin Y, Gouaux E (2005) Crystal structure of a bacterial homologue of Na^+/Cl^--dependent neurotransmitter transporters. Nature 437:215–223

Zhang YW, Rudnick G (2005) Cysteine scanning mutagenesis of serotonin transporter intracellular loop 2 suggests an alpha-helical conformation. J Biol Chem 280:30807–30813

Zhou Y, Bennett ER, Kanner BI (2004) The aqueous accessibility in the external half of transmembrane domain I of the GABA transporter GAT-1 is modulated by its ligands. J Biol Chem 279:13800–13808

The Importance of Company: Na$^+$ and Cl$^-$ Influence Substrate Interaction with SLC6 Transporters and Other Proteins

M. E. A. Reith[1,2] (✉) · J. Zhen[1,2] · N. Chen[2]

[1]Department of Biological Sciences, Illinois State University, Normal IL, 61656, USA
maarten.reith@med.nyu.edu

[2]Department of Psychiatry, New York University School of Medicine, New York NY, 10016, USA

1	Introduction	76
2	Cotransport of Na$^+$ and Cl$^-$ with Substrate in SLC6 Family	76
3	Residues Controlling Na$^+$ Modulation	77
3.1	SLC6 Family	77
3.2	SLC1A2 in SLC1 Family	81
3.3	PutP and SGLT1 in SGLT Family	83
3.4	Other Transporters or Pumps	84
4	Cl$^-$ Binding to Proteins and Cl$^-$ Modulation of Transport Proteins	86
5	Comments on Transporter Residues Governing Na$^+$ and Cl$^-$ Modulation of Transport	88
	References	89

Abstract SLC6 transporters, which include transporters for γ-aminobutyric acid (GABA), norepinephrine, dopamine, serotonin, glycine, taurine, L-proline, creatine, betaine, and neutral cationic amino acids, require Na$^+$ and Cl$^-$ for their function, and this review covers the interaction between transporters of this family with Na$^+$ and Cl$^-$ from a structure-function standpoint. Because detailed structure-function information regarding ion interactions with SLC6 transporters is limited, we cover other proteins cotransporting Na$^+$ or Cl$^-$ with substrate (SLC1A2, PutP, SLC5A1, melB), or ion binding to proteins in general (rhodanese, ATPase, LacY, thermolysine, angiotensin-converting enzyme, halorhodopsin, CFTR). Residues can be involved in directly binding Na$^+$ or Cl$^-$, in coupling ion binding to conformational changes in transporter, in coupling Na$^+$ or Cl$^-$ movement to transport, or in conferring ion selectivity. Coordination of ions can involve a number of residues, and portions of the substrate and coupling ion binding sites can be distal in space in the tertiary structure of the transporter, with other portions that are close in space thought to be crucial for the coupling process. The reactivity with methanethiosulfonate reagents of cysteines placed in strategic positions in the transporter provides a readout for conformational changes upon ion or substrate binding. More work is needed to establish the relationships between ion interactions and oligomerization of SLC6 transporters.

Keywords Neurotransmitter transporters · Na$^+$ · Cl$^-$ · Ion coordination · Substrate permeation

1
Introduction

The focus of this chapter is on ion interactions for members of the Na$^+$,Cl$^-$-dependent neurotransmitter transporter family [SLC6; for gene nomenclature used here see the special issue on solute carriers in the *European Journal of Physiology* (Hediger et al. 2004)] from the standpoint of structure-function: Which residues bind Na$^+$ or Cl$^-$ to influence substrate interaction? Genes of the SLC6 family encode proteins including Na$^+$,Cl$^-$-dependent transporters for γ-aminobutyric acid (GABA) (GAT1, SLC6A1; GAT2, SLC6A13; GAT3, SLC6A11), norepinephrine (NET, SLC6A2), dopamine (DAT, SLC6A3), serotonin (SERT, SLC6A4), glycine (GLYT2, SLC6A5), taurine (TAUT, SLC6A6), L-proline (PROT, SLC6A7), creatine (CT1, SLC6A8; CT2, SLC6A10), glycine (GLYT1, SLC6A9, betaine (BGT1, SLC6A12), and neutral cationic amino acids [ATB(0+), SLC6A14] (For a recent review, see Chen et al. 2003). Because detailed structure-function information regarding ion interactions with SLC6 transporters is limited (please see "Note Added in Proof" on p. 89 for recently crystallized LeuT), we will cover other proteins cotransporting Na$^+$ or Cl$^-$ with substrate, or ion binding to proteins in general. Over the years, our laboratory has been interested in the DAT, and we will include information on DAT residues playing a role in the impact of ions on substrate and blocker affinity.

2
Cotransport of Na$^+$ and Cl$^-$ with Substrate in SLC6 Family

Extensive information has been collected on the cotransport of Na$^+$ and Cl$^-$ in the SLC6 family. Recent reviews have dealt with the various approaches to measure stoichiometry (Rudnick 2002) and a generally agreed upon stoichiometry—depending on the SLC6 member—of 1–3 Na$^+$:1 Cl$^-$:1 substrate (with Cl$^-$ in one case being required but perhaps not translocated; Chen et al. 2003). For DAT there is evidence for different multisubstrate mechanisms in rat tissue compared with human DAT (hDAT) expressed in HEK-293 cells (McElvain and Schenk 1992; Povlock and Schenk 1997; Earles and Schenk 1999). Thus, in the same laboratory where DAT in rat striatum and nucleus accumbens appears to bind DA or Na$^+$ first (randomly) followed by Cl$^-$ last (McElvain and Schenk 1992; Povlock and Schenk 1997; however see Amejdki-Chab et al. 1992b), hDAT in HEK-293 cells binds two Na$^+$ ions first, DA second, and Cl$^-$ last (Earles and Schenk 1999). In addition to the possibility of a species difference in such DAT properties, one should consider that neuronal DAT expressed in a nonneuronal environment might display different properties, potentially involving posttranslational modifications as a function of expression environment.

3
Residues Controlling Na$^+$ Modulation

3.1
SLC6 Family

Within the SLC6 transporter family of which DAT is a member, residues in SERT (SLC6A4) and GAT1 (SLC6A1) have been implicated in Na$^+$ modulation. Penado et al. (1998) found that mutation of 7Asn368, 7Gly376, or 7Phe380 (first numbers denote transmembrane domain (TM)) in rat SERT (rSERT) impaired 5-HT (serotonin) transport at both low (15 mM) and high (150 mM) outside [Na$^+$], whereas mutation of many surrounding residues selectively impaired uptake at low [Na$^+$]. This is consonant with a role of these residues in Na$^+$ binding or coupling of Na$^+$ binding to other steps in 5-HT translocation. Alternatively, these residues interact with other residues that govern the proper conformation of the Na$^+$ binding site. Mager et al. (1996) reported that Trp68 at the top of TM1 in GAT1 affects Na$^+$ sensitivity of transport of the substrate GABA, with W68L mutation shifting the EC$_{50}$ for Na$^+$ in supporting transport-associated current from 44 to 6.8 µM. In addition, the interaction between Na$^+$ and GAT1 was measured in oocytes expressing GAT1 by either [Na$^+$]- or voltage-jumping to elicit charge movement resulting from the synchronized removal of Na$^+$ from the transporters (Mager et al. 1996). The symmetry between the current observed with the jump and that when the original [Na$^+$] or voltage is restored reflects the reversibility of the process and is in agreement with the current representing ion binding/unbinding and not ion permeation (Mager et al. 1998); for an additional different model see Grossman and Nelson (2003). The EC$_{50}$ for Na$^+$ in supporting charge movement shifted from 43 to 2.7 mM by W68L mutation; because the maximal charge movement was similar between WT and W68L, it was concluded that the mutant bound Na$^+$ more tightly. One possibility is that intracellular [Na$^+$] could exceed the K_d for Na$^+$ binding to the intracellular face of the mutant GAT1, preventing efficient release of Na$^+$ into the cytoplasm and thereby efficient transport. In our studies on the corresponding mutant W84L of hDAT (SLC6A3), we also found that Na$^+$ was more potent in stimulating binding of the DAT blocker 2β-carbomethoxy-3β-(4-fluorophenyl)tropane (CFT) (see Chen et al. 2002; and compare panels A and B in Fig. 1).

However, the plateau of binding was elevated in W84L DAT (Fig. 1B vs A), and we concluded the *coupling* between Na$^+$ and CFT binding, not Na$^+$ binding itself, was enhanced by the mutation. It is possible that DAT and GAT1, and

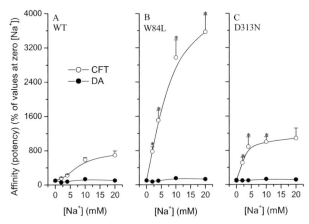

Fig. 1 A–C CFT affinity and DA inhibitory potency as a function of Na^+ concentration. **A** Wildtype (WT); **B** W84L; **C** D313N. Cell membranes were incubated with 4 nM [^{3}H]CFT for 15 min at 21 °C in HEPES-Tris buffer containing various concentrations of Na^+-ISE. Results are shown as the percentage of the binding affinity ($1/K_D$) or inhibitory potency ($1/K_i$) at zero [Na^+]. The K_d values for CFT at 0 and 20 mM [Na^+] were 95.3 and 11.6 nM at WT, 74.5 and 2.22 nM at W84L, and 38.8 and 2.45 nM at D313. The K_i values for DA at 0 and 20 mM [Na^+] were 7.3 and 6.27 µM at WT, 7.0 and 5.04 µM at W84L, and 6.3 and 4.25 µM at D313. Values represent means ± SEM (standard error of mean) of 6 or more experiments performed in triplicate. *$p < 0.05$ versus the corresponding value for WT group (Dunnett's test). (Reprinted with permission from Chen et al. 2002, Fig. 4 therein)

the role of the Trp at the top of TM1, differ in the interaction with Na^+, but it should also be kept in mind that the kinds of measurement of Na^+ interaction in the current experiments on DAT are not the same as those in the study of Mager et al. (1996) on GAT1.

Our study (Chen et al. 2002) and that of Mager et al. (1996) agree on one important point: For DAT, the Na^+ stimulation of CFT binding—along with data on the effect of uptake blockers and Zn^{2+} (see Chen et al. 2002, 2004b; and below)—suggests that W84L mutation hinders the ability of DAT to adopt the inward-facing state, whereas for GAT1, GAT1 W68L mutation appears to block GAT1 at a point in the transport cycle after the binding of Na^+ but before that of GABA, hindering the completion of the transport cycle with release of Na^+ into the cytosol (Mager et al. 1996). Consonant with both sets of data is the idea that mutation of Trp at the top of TM1 locks the transporter in the outward-facing state. Thus, the set point for the conformational equilibrium of the transporter can be perturbed by removal of a residue such as Trp at the top of TM1, generating more of the outward-facing state that is poised for binding Na^+. For DAT, this state also binds CFT with higher affinity. The gradual increase in CFT potency upon raising [Na^+] (Fig. 1) can be interpreted according to the two- or multiple-state model proposed for G protein-coupled receptors (Leff 1995; Wade et al. 2001; Parmentier et al.

2002): The apparent affinity of a ligand is a function of the distribution between active and inactive states, and we measure a time-averaged affinity for the continuously interconverting states changing back and forth numerous times within the time window of the binding assay. The gradual increase in affinity upon raising [Na$^+$] occurs with not only CFT, but also other blockers of DAT, and the Na$^+$ dependency curves are more steep for the W84L mutant (Fig. 2).

As mentioned above, it appears that W84L DAT favors the outward-facing conformation. In consonance, 10 µM Zn^{2+}—a concentration constraining DAT in an outward-facing-like state (Loland et al. 1999, 2002)—is less effective in enhancing [^{3}H]CFT binding in W84L than in WT (Fig. 3A), which is to be expected if W84L were already prone to be in the outward-facing state prior to treatment with Zn^{2+}. The Zn^{2+}-induced conformational changes impede [^{3}H]DA uptake to the same extent in WT and W84L (Fig. 3B).

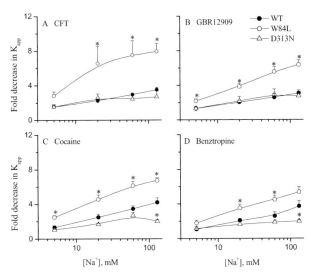

Fig. 2 Relationship between Na$^+$ and K_{app} (apparent equilibrium dissociation constant) for uptake inhibitors as measured by inhibition of [^{3}H]CFT binding. Cells were incubated with 2 or 4 nM [^{3}H]CFT and various concentrations of the indicated compounds for 15 min. The Na$^+$ concentration in the assay was varied by isotonically replacing NaCl with N-methyl-D-glucosamine (NMDG)-Cl. Data are expressed as fold decrease in the K_{app} upon addition of Na$^+$. The fold decrease is calculated as the ratio of the K_{app} at 0 mM Na$^+$ to the K_{app} at 5, 20, 60, or 130 mM Na$^+$. Shown are means ± SE of 4–8 experiments performed in triplicate. The K_{app} values in the absence of Na$^+$ were 56.6, 238, 514, and 304 nM for CFT, GBR 12909, cocaine, and benztropine, respectively, in WT. The corresponding values were 17.3, 542, 98.7, and 966 nM at W84L, and 15.6, 123, 94.1, and 394 nM at D313N. *$p < 0.05$ versus the corresponding value for WT (Dunnett's test). (Reprinted with permission from Chen et al. 2004b; Fig. 4 therein)

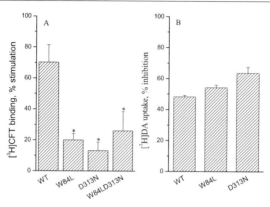

Fig. 3 A,B Relative response of [^{3}H]CFT binding and [^{3}H]DA uptake to Zn^{2+} upon mutation of W84 and D313. **A** [^{3}H]CFT binding. Cells were incubated with 4 nM [^{3}H]CFT and 10 µM Zn^{2+} in a Na^+-free and NMDG-Cl substituted buffer for 15 min. Data are expressed as percentage stimulation of the binding in the absence of Zn^{2+}. **B** [^{3}H]DA uptake. Cells were incubated with 10 nM [^{3}H]DA and 10 µM Zn^{2+} in regular Na^+-containing buffer for 5 min. Data are expressed as percentage inhibition of the uptake in the absence of Zn^{2+}. Shown are means±SE of three experiments performed in triplicate. *$p < 0.05$ versus the value at WT (Dunnett's test). (Reprinted with permission from Chen et al. 2004b, Fig. 8 therein)

Another DAT construct with a mutation in the top of TM6, D313N, also appears to favor the outward-facing conformation as suggested by the diminished effect of Zn^{2+} on [^{3}H]CFT binding (Fig. 3A). Both W84L and D313N display higher affinity for CFT and cocaine in the absence of Na^+ (Chen et al. 2004b), consonant with both mutants being enriched with transporters occurring in the outward-facing state preferred by cocaine and its analogs. Na^+ can still further increase the proportion of outward-facing mutant transporters, as shown by the gradual increase in affinity of DAT blockers for D313N and W84L (Fig. 2).

At a high Na^+ concentration of 130 mM, we conceptualize most DATs to be in the outward-facing conformation (Fig. 4, compare panels A and B). In order to accommodate the observation that D313N hDAT cells display a lower affinity for DA than WT, we depict an obstruction in the external entry path for DA (gray area in Fig. 4C). Thus, fewer of the outward-facing-like states can accept extracellular DA, causing a lower affinity for DA measured for the average of continuously interconverting states.

Our results on DA recognition in hDAT cells point to the importance of considering *access* of DA in addition to DA's affinity for its binding site. As illustrated in Fig. 4, in intact cells (panels A, B, and C), limitation of access alters apparent DA binding, in contrast to broken membrane preparations (panels D and E) where DA can access either outward- or inward-facing DATs in medium bathing both sides of the membrane.

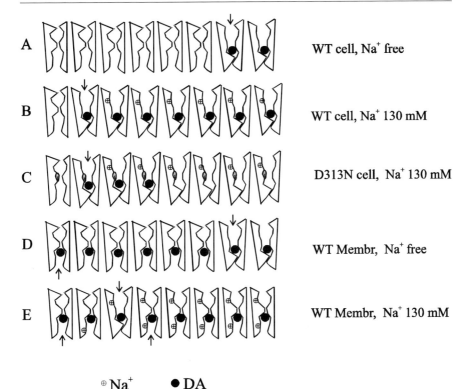

Fig. 4 A–E Cartoon of access model for DA at the DAT, illustrating the initial conformational state for ligand binding. **A** WT cells in Na^+-free media. Most transporters reside in inward-facing states where DA binding sites are inaccessible from the extracellular side. **B** WT cells at 130 mM Na^+. Binding of Na^+ to a more extracellularly located site increases the fraction of the outward-facing state, stimulating the binding of externally applied DA by enhancing external access. **C** D313N cell at 130 mM Na^+. The *gray region* making the external entry path narrower symbolizes the problem in external access to DA binding site. Thus, fewer of outward-facing-like states can accept extracellular DA. **D** WT membranes in Na^+-free media. Although most transporters reside in a state resembling the inward-facing state adopted in these membranes when actually part of the intact cell, this does not affect DA access, because DA can approach its binding site from both sides of the membrane. **E** WT membranes at 130 mM Na^+. Na^+ binding causes conformational redistribution: Intracellular Na^+ binds to a separate site, increasing the fraction of the inward-facing state. This does not affect DA binding because it has access from both sides. Features of DA binding to membrane preparations of D313N resemble those described for WT

3.2
SLC1A2 in SLC1 Family

Kanner's group has extensively studied Na^+ modulation of the high-affinity glutamate transporter GLT-1 (also known as EAAT2), member (SLC1A2) of the SLC1 glutamate/neutral amino acid transporter family (Kavanaugh et al. 1997;

Zhang et al. 1998; Zhang and Kanner 1999). All this work took place before the high-resolution structure came out in 2004 of the crystallized Glt$_{Ph}$, a glutamate transporter homolog from the archaeon *Pyrococcus horikoshii* by Yernool et al. (2004). In the following paragraph, we will compare the findings from the biochemical approaches summarized here with the anatomical features revealed by the crystal. GLT-1 cotransports three Na$^+$ ions with glutamate, followed by countertransport of K$^+$; it has been proposed that K$^+$ binds at a site previously occupied by one of the Na$^+$ ions (Kavanaugh et al. 1997). Whereas Glu404 in TM7 influences K$^+$ coupling, the neighboring residue Tyr403 conveys both K$^+$ and Na$^+$ affinity (Zhang et al. 1998). The data were interpreted in the context of nonidentical but partially overlapping binding sites for potassium and sodium. It is important to note that in this case conservative mutations were studied, E404D and Y403F, showing the need for the precise nature of the side chain of the acidic or aromatic amino acid for the cation interaction. Interestingly, Y403C can be protected against methanethiosulfonate (MT) by transportable substrates but is sensitized to it by the nontransportable dihydrokainate, suggesting alternating access for the 403 position (Eliasof and Jahr 1996). Another residue, Ser440, is implicated in Na$^+$ selectivity (Zhang and Kanner 1999). Ser440 is located at the extracellular face of the re-entrant loop between TM7 and 8, likely close to the binding site for glutamate (see Kanai and Hediger 2004, and see below: the crystal structure of the archaeal glutamate transporter homolog). S440G mutation enabled not only Na$^+$, but also Li$^+$ to drive transport, depending on the nature of the residue at position 443 (Zhang and Kanner 1999). S440C and S443C were sensitive to (2-aminoethyl)methanethiosulfonate (MTSEA) and [(2-trimethylammonium)ethyl]MT (MTSET), both positively charged, and (2-sulfonatoethyl)MT (MTSES, negatively charged); in S440C, glutamate, but not Na$^+$ alone, protected. Likely, glutamate obstructed access of MT to the 440 position, consonant with close vicinity of position 440 to the substrate binding site. Position 443 was not protected by glutamate, suggesting it was further away from the glutamate site. Because (1) S440C was protected against MTSET by D-aspartate and dihydrokainate in both sodium and lithium media, and (2) lithium impaired substrate translocation in WT, it appears that a step after substrate binding conveys sodium selectivity. Zhang and Kanner (1999) propose that one of the carboxyl groups of the transported acidic amino acid participates in liganding the cotransporting Na$^+$, thereby conferring Na$^+$ selectivity. The phenomenon of substrate providing one of the liganding groups is seen also in other cases of cation binding to proteins (see Sect. 3.4).

Many of the above findings with biochemical approaches are in excellent agreement with the anatomical indicators from the crystallized glutamate transporter homolog Glt$_{Ph}$ mentioned above (Yernool et al. 2004). The findings regarding Ser440 constitute one example. The eukaryotic Ser440 corresponds to archaeal Gly354 in Glt$_{Ph}$, and Gly354 is located in the tip of helical hairpin 2 (HP2) in the crystal (Yernool et al. 2004); the aligned eukaryotic Ser440 in GLT-1 is depicted by Kanai and Hediger (2004) in the B-portion of the re-entrant loop between TM7 and 8. Gly354 in crystallized Glt$_{Ph}$ is within 5 Å of the substrate binding site hinted upon in the crystal by the contours of non-protein electron density near the interface between HP1 and 2 conforming to the size of a glutamate molecule. According to this depiction of the substrate binding site, Arg397 is on the polar face of TM8, poised to interact with the γ-carboxy group of glutamate. Arg397 corresponds to Arg447 in eukaryotic glutamate transporters, and the guanidinium group of this residue in EAAT3 (also referred to as EAAC-1, or SLC1A1) has been proposed to interact with the γ-carboxy group of the substrate glutamate based on biochemical evidence (Bendahan et al. 2000). Again in agreement with biochemical findings, Arg397 in the crystal structure interacts by a π-cation mechanism with Tyr317, aligned with eukaryotic Tyr403. As detailed above, Tyr403 is crucial for both Na$^+$ and K$^+$ affinity (Zhang et al. 1998). Arg397 in Glt$_{Ph}$ is also close to Gln318, corresponding to eukaryotic Glu404 which is involved in K$^+$ coupling. Thus, the combined biochemical and crystal information points to a common pathway for transported substrate and ions, with the crystal results providing a tremendously enhanced view of the transporter parts lining the substrate pathway. It is also important to note here that the crystal information for Glt$_{Ph}$ indicates the existence of the transporter in a tetramer, with each protomer carrying its separate set of sites interacting with ions and substrate.

3.3
PutP and SGLT1 in SGLT Family

PutP is a sodium/proline transporter, a member of the SGLT gene family which is also referred to as the sodium/substrate symporter family, or SSSF gene family. SLC5 transporters, such as SGLT1 (gene *SLC5A1*) discussed in the next paragraph, are part of this larger 220-member SGLT gene family. Properties, ligand specificity, and cell or tissue expression vary widely within the family, with some members even showing substrate permeation that is not coupled to sodium transport (Wright and Turk 2004). PutP transports proline coupled to Na$^+$ in a 1:1 stoichiometry according to an ordered binding mechanism (Wright 2001). Asp55 at the top of TM2 appeared to be essential for transport, and only replacement with Glu rescued some uptake activity (Quick and Jung 1997). This D55E mutant displayed a 50-fold reduced affinity for sodium while transporting proline with relatively unaltered K_m, consonant with Asp55 being directly involved in Na$^+$ binding. Highly decreased apparent affinity for sodium was also observed upon removal of the hydroxyl groups at the positions of either Ser340 or Thr341, suggesting that TM9 contributes to an ion translocation pathway of PutP (Jung 2002). In consonance, the nearby Cys344 in TM9 is thought to be involved in the binding of Na$^+$ and proline (Hanada et al. 1992). Substitution of a Ser residue residing in TM2 close to Asp55, Ser57, reduces the apparent affinity of both proline and sodium, but increasing [pro] restores the sodium affinity to a near-normal value, suggesting close cooperativity between substrate and sodium sites, and, possibly close proximity for these sites in the tertiary structure of PutP (Quick et al. 1996; Pirch et al. 2002; Jung 2002). Asp55 has been proposed to interact directly with the coupling ion Na^+, and position 57 becomes more accessible to sulfhydryl reagents in the presence of Na^+, likely indicating a conformational change in this region of the protein induced by Na^+ (Quick et al. 1999). Arg40, at the bottom of TM2, is thought to be close to the ion binding site and involved in the coupling of downhill Na^+ transport with proline accumulation, but probably does not make direct contact with either Na^+ or proline (Quick et al. 1999). A common motif, which includes Gly328, Ala366, Leu371, Gly375, and Arg376 in PutP of *Escherichia coli*, is present in several Na^+ symport carriers, and it has been proposed to be essential for Na^+ recognition or binding (Deguchi et al. 1990). However, Arg376 does not appear to reside at the sodium binding site, according to a study by Yamato et al. (1994), casting doubt regarding a specific role of this entire motif in Na^+ interaction with Na^+ symport carriers.

As PutP, the Na^+–glucose cotransporter SGLT1 is a member of the SSSF gene family; SGLT1 belongs to the SLC5 family, which is part of SSSF. Appreciable effort has been devoted to uncovering residues controlling Na^+ modulation of sugar transport by this transporter. Thus, Lo and Silverman (1998a, b) and Vayro et al. (1998) point to the region 162–173 [EL (extracellular loop) connecting TM4 and 5] as forming the pathway that external Na^+ takes to reach its binding site. From experiments on sensitivity of Cys replacements to sulfhydryl reagents of the MT type, residues 163, 166, 170, and 173 are likely facing externally in the SGLT1 folded structure. Meinild et al. (2001) reexamined position 166 (Ala), measuring transport function electrophysiologically in oocytes expressing SGLT1. Compared with WT, A166C had a slightly increased apparent affinity for Na^+, but decreased affinity for sugars. Reacting A166C with positively charged MTSET or MTSEA prevented sugar transport but not binding; the results combined led to the conclusion that Ala166 is not directly involved in Na^+ binding and that the conformational change in the Na^+ pathway induced by Na^+ binding proceeds normally. Rather, position 166 was proposed to be involved in the interaction between the Na^+ and sugar pathways. Ala166 in EL2 is likely close to Asp454 in EL5 (connecting TM10 and 11) in the tertiary structure of SGLT1, and the negative charge on Asp454 appears to play an important role in coupling Na^+ and sugar uptake (Diez-Sampedro et al. 2004). For cation *selectivity* in SGLT1, Asp204 is crucial in the intracellular loop (IL)3 (connecting TM5 and 6; Quick et al. 2001), as is the aligned Asp187 in the related transporter PutP (Quick and Jung 1998).

3.4
Other Transporters or Pumps

The melibiose carrier of *E. coli* (melB) cotransports sodium and melibiose or other α-galactosides (Wilson and Ding 2001). Although the melibiose permease prefers Na^+, it can also use H^+ or Li^+ for cotransport. From studies in which Asp residues were replaced with Ser, Tyr, Leu, Cys, Asn, Gln, or Glu, and the resultant mutants were examined for Na^+ or H^+/melibiose cotransport, a hypothetical Na^+ binding site was proposed in which Na^+ was coordinated by 1Asp10, 2Asp55, 2Asp59, and 4Asp124 (again first numbers denote TM) (Wilson and Ding 2001).

In a number of cases where proteins were susceptible to crystallization, Na^+ binding has been studied by X-ray diffraction of crystals soaked in the absence and presence of Na^+; identification of cation sites is based on anomalous difference Fourier peaks. With this technique, Hol, Kalk, and colleagues (Lijk et al. 1984; Kooystra et al. 1988) characterized binding sites for Na^+, Cs^+, and NH_4^+ in rhodanese, a thiosulfate sulfurtransferase; this enzyme binds divalent metal ions, possibly to these same cation sites, although this is not required for catalytic activity. The Na^+ binding site is composed of the main-chain carbonyl oxygens of Glu71, Val72, and Lys249 (Lijk et al. 1984), whereas the Cs^+ site has the additional main-chain carbonyl oxygen of Asp272 as well as the side-chain oxygen of the same Asp, equidistantly oriented around the cation (Kooystra et al. 1988).

The same approach when applied to bovine Hsc70 ATPase fragment, points to coordination of Na^+ by six ligands: the carboxyl of Asp10, the carbonyl oxygen of Tyr15, two H_2O molecules of the Mg^{2+} cluster, one internal H_2O, and the oxygen of β-phosphate of ADP (the substrate) (Wilbanks and McKay 1995). The latter is an example of a substrate itself providing liganding to a cation. The enzymatic activity of Hsc70 ATPase is facilitated by monovalent cations, with K^+ more effectively activating than Na^+; in consonance, the difference Fourier X-ray analysis shows a repositioning of the β-phosphate with Na^+ compared to K^+, which affects the interaction between cation and substrate (Wilbanks and McKay 1995).

In a final example of cation binding uncovered by X-ray crystallography, the lactose permease of *E. coli* (LacY) has been studied by Abramson et al. (2003). This 12-TM transporter uses the downhill electrochemical proton gradient to achieve uphill substrate accumulation. Cotransport of cation (H^+) and substrate (lactose) is thought to occur in an ordered fashion, with protonation followed by substrate binding, outward- to inward-state conversion, and release of first substrate and then proton. Crystals were obtained with a LacY mutant arrested in the inward-facing conformation, and difference Fourier analysis was applied to delineate the binding pocket for the high-affinity lactose analog β-D-galactopyranosyl-1-thio-β-D-galactopyranoside (TDG) (Abramson et al. 2003). Because proton sites cannot be visualized directly in crystals, results from studies with LacY mutants were used to depict residues in the crystal involved in proton translocation and coupling. Sites interacting with substrate and sites interacting with coupling ions appear to involve appreciably distal domains, with TMs 1, 4, 5, and 11 providing residues interacting with substrate, and TMs 8, 9, and 10 providing residues for H^+ translocation and coupling. The areas comprising the substrate-interactive residues in TMs 1, 4 are removed in space from that occupied by ion-reactive residues in TMs 9–10, and are depicted on the separate arms of the portions of the transporter that form the outward- and inward-facing conformations of the transporter (Abramson et al. 2003). Other portions of the substrate- and ion-interacting areas are close by. Thus, residues Arg144 in TM5 and Glu269 in TM8 are thought to be coupled with a salt bridge when substrate binds (interacting with Arg144 through hydrogen bonding) and the transporter adopts the inward-facing conformation. Upon release of substrate and proton in the cytoplasm, the outward-facing conformation is re-formed with breakage of the 144–269 salt bridge.

4
Cl⁻ Binding to Proteins and Cl⁻ Modulation of Transport Proteins

Mechanisms underlying the interaction between Cl^- and transporters have been studied less intensely than those for Na^+ interaction. For DAT, it is known that Cl^- stimulates substrate binding (Amejdki-Chab et al. 1992a, b; Wall et al. 1993; Gu et al. 1994; Li and Reith 1999) and that the effect depends on the type of substrate (Li et al. 2002). Chimera studies have advanced a role in Cl^- interactions for TMs 1–2 and TMs 9–12 (Syringas et al. 2000, 2001; see end of this section), but it is not known which residues are involved. Residues of proteins interacting with Cl^- could themselves bind Cl^-, or be involved in other ways in modulating Cl^- sensitivity. As far as anion binding to proteins is concerned, such binding depends on the charge density of the anion in turn regulating the ability to bind water molecules and interact with basic groups on proteins (Di Stasio 2004). Small ions of high charge density (kosmotropes, among anions: F^-) bind water strongly, whereas large monovalent ions of low-charge density (chaotropes, among anions: Cl^-, Br^-, I^-) bind water molecules weakly. In general, oppositely charged ions of equal water affinity can form inner sphere ion pairs (Collins 2004). The positive charges on proteins [derivatives of ammonium ion (lysine), guanidinium ion (arginine), and imidazolium ion (histidine)] are all weakly hydrated and thus can form inner sphere ion pairs with the weakly hydrated anion chloride. In comparison, strongly hydrated anions such as fluoride interact with proteins indirectly, via intervening water molecules. Anions which have a lower charge density than Cl^- (in order of progressively lower charge density: Br^-, I^-) are more weakly hydrated and bind more tightly to proteins (reverse Hofmeister series). The partial positive charge on amide NH groups (Asn, Gln) is also weakly hydrated and can bind weakly hydrated anions (Collins 2004).

Although the binding of anions such as Cl^- to various proteins appears to be site-specific, little is known about their possible binding site(s) (Di Stasio 2004). The interaction between Cl^- and the metalloproteinase thermolysin causes a lowering of the K_m for hydrolysis of tripeptides such as Gly-Phe-Ala (Yang et al. 1994). This is thought to result from binding of Cl⁻ to Arg203, which competes for the salt bridge between Arg 203 and Asp170, thereby allowing substrate binding (Yang et al. 1994). In another metalloproteinase, carboxypeptidase A, Cl⁻ is thought to compete for substrate binding by binding to Arg145, with this arginine being the recognition site for the C-terminal carboxylate of the substrate peptide (Williams and Auld 1986). Again, within the category of metalloproteinases, Cl⁻ enhances substrate binding to angiotensin-converting enzyme, likely by binding to a critical Lys residue (Shapiro and Riordan 1983; Bunning and Riordan 1987). A Cl⁻–Arg108 interaction (in this case, hydrogen-bonded) is postulated for halorhodopsin, which uses light energy to pump halide ions (Braiman et al. 1994), and clinical evidence implicates Arg117 in Cl⁻ transport by cystic fibrosis transmembrane conductance regulator (CFTR) (Dean et al. 1990). A binding site for Cl⁻ has been described by Wilbanks and McKay (1995) in crystals of the same Hsc70 ATPase fragment discussed above for Na⁺ binding (Sect. 3.4). In this crystal, the Cl⁻ binding site is composed of Asp32, Gln33, and Lys126, involving the amide, the εN, and the ζN of the respective amino acids (Wilbanks and McKay 1995).

Transporter residues could be involved in the impact of Cl⁻ on transporter function without themselves binding Cl⁻. For example, residues could play a role in coupling the conformational change induced by ion and substrate binding to the conformational change associated with substrate translocation. According to the alternate access model, Cl⁻ binds with Na⁺ and substrate to the outward-facing state, a reorientation takes place exposing the solutes to the internal compartment, and ions along with substrate dissociate rapidly from the inward-facing state (Cao et al. 1998; Rudnick 2002). The reorientation step in the former sequence is often (also in this text) referred to as the translocation step, although the transporter may not actually be "moving" the substrate appreciably, as viewed parallel to the plasma membrane. An example of this can be seen in the comparison of the inward-facing state of the LacY transporter of *E. coli* with bound substrate (as determined by crystallography) and the outward-facing state with substrate (as deduced from crosslinking experiments and molecular modeling; see Fig. 5 in Abramson et al. 2003). In this alternate access model, residues could affect the distribution between inward-facing and outward-facing states.

For hDAT, Loland and colleagues (2002, 2004) and our group (Chen et al. 2004a) have advanced evidence for a role of residues Lys264, Tyr335, Asp345, and Asp436 in the conformational equilibrium between outward- and inward-facing conformations, with a mutation of them leading to a greater proportion of DAT residing in the inward-facing state. Both the N-terminal portion (TMs 1–2) of DAT and SERT have been implicated in Cl⁻ sensitivity of substrate uptake (Syringas et al. 2000, 2001; Barker et al. 1999), as well as the C-terminal portion (TMs 9–12) of DAT (Syringas et al. 2000, 2001). It is not known whether Cl⁻ permeation, or just recognition, is required for these interactions. For the

SERT, there is evidence for a common pathway for 5-HT and Cl^- (and Na^+) involving TM1 and Asp 98 in particular (Barker et al. 1999).

5
Comments on Transporter Residues Governing Na^+ and Cl^- Modulation of Transport

In considering the above studies on Na^+ modulation, one can recognize a number of salient issues. Thus, residues can be involved in directly binding Na^+, in coupling Na^+ binding to conformational changes in the transporter, in coupling the driving force of the transmembrane Na^+ gradient (i.e., Na^+ *movement*) to transport, or in conferring cation selectivity. Coordination of cations such as sodium can involve a number of residues, up to six. Binding sites for substrate and coupling ions can be distal in space in the tertiary structure of the transporter, with other segments of the transporter being responsible for the coupling process. The reactivity with MT reagents of cysteines placed in strategic positions in the transporter can provide a readout for conformational changes upon ion or substrate binding. In general, residues that have been shown to bind Na^+ include: acidic amino acids donating electronegative oxygen atoms for cation interaction; aromatics (Trp, Tyr, Phe) coordinating cation binding through π electrons; and main chain carbonyl oxygens of various amino acids. In addition, water molecules and substrate oxygen have been implicated in Na^+ binding in some cases.

From the above-discussed studies on Cl^- modulation, again it is clear that more than one residue can be involved in binding of this anion. Portions of transporter domains interacting with substrate and chloride can be close in space. For the SERT, the Blakely group has advanced evidence for a common pathway for the substrate 5-HT and the ions Na^+ and Cl^- involving TM1 and Asp98, in particular (Barker et al. 1999). No information is available distinguishing residues involved in directly binding Cl^-, in coupling Cl^- binding to conformational changes in transporter, in coupling Cl^- *movement* to transport, or in conferring anion selectivity. To our knowledge, no studies are available on protection by Cl^- against reactivity with MT reagents of cysteines placed in strategic positions in monoamine or other transporters. Also, no information is available as to whether Cl^- binds more tightly to the outward- than inward-facing state. In general, residues shown to bind Cl^- include: Lys, Arg, and His, which—as weakly hydrated positive charges on protein—bind the weakly hydrated anion Cl^-; weakly hydrated partial positive charges on amide NHs of Asn and Gln, which can also bind Cl^-; and the main-chain amide of other amino acids.

It is becoming increasingly clear that transporters in the SLC6 family occur in oligomeric form (Hastrup et al. 2001, 2003; Torres et al. 2003; Baucum et al. 2003; Sitte et al. 2004). It is not yet known whether multimeric organization in

SLC6 is required for transport activity. Experiments with coexpression of different proportions of a SERT construct sensitive to MT compounds and another insensitive construct point to a functional interaction between SERT protomers (Kilic and Rudnick 2000). Seidel et al. (2005) advance evidence for influx and efflux of substrate through coupled but separate protomers in monoamine transporters oligomers. The GLUT1 (Zottola et al. 1995) and LacS transporter (Veenhoff et al. 2001) occur as oligomers (tetramers and dimers, respectively) in the plasma membrane, and in both cases it is thought that each subunit in the complex sustains cotransport of substrate and Na^+. However, there is cooperativity between transporter units (dimers for GLUT1 and monomers for LacS) that is evident for ion gradient-driven uptake of substrate but not for substrate exchange. This can be explained if the reorientation of the empty carrier in one unit is coupled to translocation of a fully loaded carrier in another unit (Zottola et al. 1995; Veenhoff et al. 2001). More work is needed to establish the relationships between ion interactions and oligomerization of SLC6 transporters. Even if transport itself is not affected by oligomerization, the effect of transport blockers may be impacted (Hastrup et al. 2003), which could influence other transporter properties such as the ability to carry ion currents (Rudnick 2002). If defects in SLC6 transporter oligomerization turn out to be involved in specific diseases, it would add a new potential target for therapeutic intervention.

Note Added in Proof This article was written before the recently puplished crystal structure of a bacterial homologue of the Na^+, Cl^--dependent neurotransmitter transports from Aquifex aeolicus, the $LeuT_{AB}$ which cotransports leucine and Na^+ (Yamashita et al., Nature 2005 [Sept]:437:215–223). Located halfway across the membrane bilayer, at the unwound regions of TM1 and 6, are two Na^+ binding sites with interacting residues in TM1, 6, 7, and 8.LeuT does not co-transport $ClNa^-$. Rore more details on Na^+ binding sites on LeuT in relation to those on mamallan counterparts, please see the chapter by Rudnick in this volume (pp 59–73).

Acknowledgements We are grateful to the National Institute on Drug Abuse for support (DA 08379 and 13261) for experiments performed in our laboratory (discussed here in text and shown in figures).

References

Abramson J, Smirnova I, Kasho V, Verner G, Kaback HR, Iwata S (2003) Structure and mechanism of the lactose permease of Escherichia coli. Science 301:610–615

Amejdki-Chab N, Benmansour S, Costentin J, Bonnet JJ (1992a) Effects of several cations on the neuronal uptake of dopamine and the specific binding of [3H]GBR 12783: attempts to characterize the Na^+ dependence of the neuronal transport of dopamine. J Neurochem 59:1795–1804

Amejdki-Chab N, Costentin J, Bonnet JJ (1992b) Kinetic analysis of the chloride dependence of the neuronal uptake of dopamine and effect of anions on the ability of substrates to compete with the binding of the dopamine uptake inhibitor GBR 12783. J Neurochem 58:793–800

Barker EL, Moore KR, Rakhshan F, Blakely RD (1999) Transmembrane domain I contributes to the permeation pathway for serotonin and ions in the serotonin transporter. J Neurosci 19:4705–4717

Baucum II AJ, Rau KS, Hanson JE, Fleckenstein AE (2003) Neurotoxic regimens of methamphetamine increase dopamine transporter oligomer formation. Soc Neurosci Abstr 29:253.13

Bendahan A, Armon A, Madani N, Kavanaugh MP, Kanner BI (2000) Arginine-447 plays a pivotal role in substrate interactions in a neuronal glutamate transporter. J Biol Chem 275:37436–37442

Braiman MS, Walter TJ, Briercheck DM (1994) Infrared spectroscopic detection of light-induced change in chloride-arginine interaction in halorhodopsin. Biochemistry 33:1629–1635

Bunning P, Riordan JF (1987) Sulfate potentiation of the chloride activation of angiotensin converting enzyme. Biochemistry 26:3374–3377

Cao Y, Li M, Mager S, Lester HA (1998) Amino acid residues that control pH modulation of transport-associated current in mammalian serotonin transporters. J Neurosci 18:7739–7749

Chen N, Sun L, Reith MEA (2002) Cationic interactions at the human dopamine transporter reveal binding conformations for dopamine distinguishable from those for the cocaine analog 2 beta-carbomethoxy-3 beta (4-fluorophenyl)tropane. J Neurochem 81:1383–1393

Chen N, Rickey J, Berfield JL, Reith MEA (2004a) Aspartate 345 of the dopamine transporter is critical for conformational changes in substrate translocation and cocaine binding. J Biol Chem 279:5508–5519

Chen N, Zhen J, Reith MEA (2004b) Mutation of Trp84 and Asp313 of the dopamine transporter reveals similar mode of binding interaction for GBR 12909 and benztropine as opposed to cocaine. J Neurochem 89:853–864

Chen NH, Reith MEA, Quick MW (2003) Synaptic uptake and beyond: the sodium- and chloride-dependent neurotransmitter transporter family SLC6. Pflugers Arch 447:519–531

Collins KD (2004) Ions from the Hofmeister series and osmolytes: effects on proteins in solution and in the crystallization process. Methods 34:300–311

Dean M, White MB, Amos J, Gerrard B, Stewart C, Khaw KT, Leppert M (1990) Multiple mutations in highly conserved residues are found in mildly affected cystic fibrosis patients. Cell 61:863–870

Deguchi Y, Yamato I, Anraku Y (1990) Nucleotide sequence of gltS, the Na^+/glutamate symport carrier gene of Escherichia coli B. J Biol Chem 265:21704–21708

Di Stasio E (2004) Anionic regulation of biological systems: the special role of chloride in the coagulation cascade. Biophys Chem 112:245–252

Diez-Sampedro A, Loo DD, Wright EM, Zampighi GA, Hirayama BA (2004) Coupled sodium/glucose cotransport by SGLT1 requires a negative charge at position 454. Biochemistry 43:13175–13184

Earles C, Schenk JO (1999) Multisubstrate mechanism for the inward transport of dopamine by the human dopamine transporter expressed in HEK cells and its inhibition by cocaine. Synapse 33:230–238

Eliasof S, Jahr CE (1996) Retinal glial cell glutamate transporter is coupled to an anionic conductance. Proc Natl Acad Sci U S A 93:4153–4158

Grossman TR, Nelson N (2003) Effect of sodium lithium and proton concentrations on the electrophysiological properties of the four mouse GABA transporters expressed in Xenopus oocytes. Neurochem Int 43:431–443

Gu H, Wall SC, Rudnick G (1994) Stable expression of biogenic amine transporters reveals differences in inhibitor sensitivity, kinetics, and ion dependence. J Biol Chem 269:7124–7130

Hanada K, Yoshida T, Yamato I, Anraku Y (1992) Sodium ion and proline binding sites in the Na$^+$/proline symport carrier of Escherichia coli. Biochim Biophys Acta 1105:61–66

Hastrup H, Karlin A, Javitch JA (2001) Symmetrical dimer of the human dopamine transporter revealed by cross-linking Cys-306 at the extracellular end of the sixth transmembrane segment. Proc Natl Acad Sci U S A 98:10055–10060

Hastrup H, Sen N, Javitch JA (2003) The human dopamine transporter forms a tetramer in the plasma membrane: cross-linking of a cysteine in the fourth transmembrane segment is sensitive to cocaine analogs. J Biol Chem 278:45045–45048

Hediger MA, Romero MF, Peng JB, Rolfs A, Takanaga H, Bruford EA (2004) The ABCs of solute carriers: physiological, pathological and therapeutic implications of human membrane transport proteins. Pflugers Arch 447:465–468

Jung H (2002) The sodium/substrate symporter family: structural and functional features. FEBS Lett 529:73–77

Kanai Y, Hediger MA (2004) The glutamate/neutral amino acid transporter family SLC1: molecular, physiological and pharmacological aspects. Pflugers Arch 447:469–479

Kavanaugh MP, Bendahan A, Zerangue N, Zhang Y, Kanner BI (1997) Mutation of an amino acid residue influencing potassium coupling in the glutamate transporter GLT-1 induces obligate exchange. J Biol Chem 272:1703–1708

Kilic F, Rudnick G (2000) Oligomerization of serotonin transporter and its functional consequences. Proc Natl Acad Sci U S A 97:3106–3111

Kooystra PJ, Kalk KH, Hol WG (1988) Soaking in Cs2SO4 reveals a caesium-aromatic interaction in bovine-liver rhodanese. Eur J Biochem 177:345–349

Leff P (1995) The two-state model of receptor activation. Trends Pharmacol Sci 16:89–97

Li LB, Reith MEA (1999) Modeling of the interaction of Na$^+$ and K$^+$ with the binding of dopamine and [3H]WIN 35,428 to the human dopamine transporter. J Neurochem 72:1095–1109

Li LB, Cui XN, Reith MEA (2002) Is Na(+) required for the binding of dopamine, amphetamine, tyramine, and octopamine to the human dopamine transporter? Naunyn Schmiedebergs Arch Pharmacol 365:303–311

Lijk LJ, Torfs CA, Kalk KH, De Maeyer MC, Hol WG (1984) Differences in the binding of sulfate, selenate and thiosulfate ions to bovine liver rhodanese, and a description of a binding site for ammonium and sodium ions. An X-ray diffraction study. Eur J Biochem 142:399–408

Lo B, Silverman M (1998a) Cysteine scanning mutagenesis of the segment between putative transmembrane helices IV and V of the high affinity Na$^+$/Glucose cotransporter SGLT1. Evidence that this region participates in the Na$^+$ and voltage dependence of the transporter. J Biol Chem 273:29341–29351

Lo B, Silverman M (1998b) Replacement of Ala-166 with cysteine in the high affinity rabbit sodium/glucose transporter alters transport kinetics and allows methanethiosulfonate ethylamine to inhibit transporter function. J Biol Chem 273:903–909

Loland CJ, Norregaard L, Gether U (1999) Defining proximity relationships in the tertiary structure of the dopamine transporter. Identification of a conserved glutamic acid as a third coordinate in the endogenous Zn(2+)-binding site. J Biol Chem 274:36928–36934

Loland CJ, Norregaard L, Litman T, Gether U (2002) Generation of an activating Zn(2+) switch in the dopamine transporter: mutation of an intracellular tyrosine constitutively alters the conformational equilibrium of the transport cycle. Proc Natl Acad Sci U S A 99:1683–1688

Loland CJ, Granas C, Javitch JA, Gether U (2004) Identification of intracellular residues in the dopamine transporter critical for regulation of transporter conformation and cocaine binding. J Biol Chem 279:3228–3238

Mager S, Kleinberger-Doron N, Keshet GI, Davidson N, Kanner BI, Lester HA (1996) Ion binding and permeation at the GABA transporter GAT1. J Neurosci 16:5405–5414

Mager S, Cao Y, Lester HA (1998) Measurement of transient currents from neurotransmitter transporters expressed in Xenopus oocytes. Methods Enzymol 296:551–566

McElvain JS, Schenk JO (1992) A multisubstrate mechanism of striatal dopamine uptake and its inhibition by cocaine. Biochem Pharmacol 43:2189–2199

Meinild AK, Loo DD, Hirayama BA, Gallardo E, Wright EM (2001) Evidence for the involvement of Ala 166 in coupling Na(+) to sugar transport through the human Na(+)/glucose cotransporter. Biochemistry 40:11897–11904

Parmentier ML, Prezeau L, Bockaert J, Pin JP (2002) A model for the functioning of family 3 GPCRs. Trends Pharmacol Sci 23:268–274

Penado KM, Rudnick G, Stephan MM (1998) Critical amino acid residues in transmembrane span 7 of the serotonin transporter identified by random mutagenesis. J Biol Chem 273:28098–28106

Pirch T, Quick M, Nietschke M, Langkamp M, Jung H (2002) Sites important for Na^+ and substrate binding in the Na^+/proline transporter of Escherichia coli, a member of the Na^+/solute symporter family. J Biol Chem 277:8790–8796

Povlock SL, Schenk JO (1997) A multisubstrate kinetic mechanism of dopamine transport in the nucleus accumbens and its inhibition by cocaine. J Neurochem 69:1093–1105

Quick M, Jung H (1997) Aspartate 55 in the Na^+/proline permease of Escherichia coli is essential for Na^+-coupled proline uptake. Biochemistry 36:4631–4636

Quick M, Jung H (1998) A conserved aspartate residue, Asp187, is important for Na^+-dependent proline binding and transport by the Na^+/proline transporter of Escherichia coli. Biochemistry 37:13800–13806

Quick M, Tebbe S, Jung H (1996) Ser57 in the Na^+/proline permease of Escherichia coli is critical for high-affinity proline uptake. Eur J Biochem 239:732–736

Quick M, Stolting S, Jung H (1999) Role of conserved Arg40 and Arg117 in the Na^+/proline transporter of Escherichia coli. Biochemistry 38:13523–13529

Quick M, Loo DD, Wright EM (2001) Neutralization of a conserved amino acid residue in the human Na^+/glucose transporter (hSGLT1) generates a glucose-gated H^+ channel. J Biol Chem 276:1728–1734

Rudnick G (2002) Mechanisms of biogenic amine neurotransmitter transporters. In: Reith M (ed) Neurotransmitter transporters: structure, function, and regulation. Humana Press, Totowa, pp 25–52

Seidel S, Singer EA, Just H, Farhan H, Scholze P, Kudlacek O, Holy M, Koppatz K, Krivanek P, Freissmuth M, Sitte HH (2005) Amphetamines take two to tango: an oligomer-based counter-transport model of neurotransmitter transport explores the amphetamine action. Mol Pharmacol 67:140–151

Shapiro R, Riordan JF (1983) Critical lysine residue at the chloride binding site of angiotensin converting enzyme. Biochemistry 22:5315–5321

Sitte HH, Farhan H, Javitch JA (2004) Sodium-dependent neurotransmitter transporters: oligomerization as a determinant of transporter function and trafficking. Mol Interv 4:38–47

Syringas M, Janin F, Mezghanni S, Giros B, Costentin J, Bonnet JJ (2000) Structural domains of chimeric dopamine-Noradrenaline human transporters involved in the Na(+)- and Cl(−)-dependence of dopamine transport. Mol Pharmacol 58:1404–1411

Syringas M, Janin F, Giros B, Costentin J, Bonnet JJ (2001) Involvement of the NH2 terminal domain of catecholamine transporters in the Na(2+) and Cl(−)-dependence of a [3H]-dopamine uptake. Br J Pharmacol 133:387–394

Torres GE, Carneiro A, Seamans K, Fiorentini C, Sweeney A, Yao WD, Caron MG (2003) Oligomerization and trafficking of the human dopamine transporter. Mutational analysis identifies critical domains important for the functional expression of the transporter. J Biol Chem 278:2731–2739

Vayro S, Lo B, Silverman M (1998) Functional studies of the rabbit intestinal Na^+/glucose carrier (SGLT1) expressed in COS-7 cells: evaluation of the mutant A166C indicates this region is important for Na^+-activation of the carrier. Biochem J 332:119–125

Veenhoff LM, Heuberger EH, Poolman B (2001) The lactose transport protein is a cooperative dimer with two sugar translocation pathways. EMBO J 20:3056–3062

Wade SM, Lan K, Moore DJ, Neubig RR (2001) Inverse agonist activity at the alpha(2A)-adrenergic receptor. Mol Pharmacol 59:532–542

Wall SC, Innis RB, Rudnick G (1993) Binding of the cocaine analog 2 beta-carbomethoxy-3 beta-(4-[125I]iodophenyl)tropane to serotonin and dopamine transporters: different ionic requirements for substrate and 2 beta-carbomethoxy-3 beta-(4-[125I]iodophenyl) tropane binding. Mol Pharmacol 43:264–270

Wilbanks SM, McKay DB (1995) How potassium affects the activity of the molecular chaperone Hsc70. II. Potassium binds specifically in the ATPase active site. J Biol Chem 270:2251–2257

Williams AC, Auld DS (1986) Kinetic analysis by stopped-flow radiationless energy transfer studies: effect of anions on the activity of carboxypeptidase A. Biochemistry 25:94–100

Wilson TH, Ding PZ (2001) Sodium-substrate cotransport in bacteria. Biochim Biophys Acta 1505:121–130

Wright EM (2001) Renal Na(+)-glucose cotransporters. Am J Physiol Renal Physiol 280:F10–F18

Wright EM, Turk E (2004) The sodium/glucose cotransport family SLC5. Pflugers Arch 447:510–518

Yamato I, Kotani M, Oka Y, Anraku Y (1994) Site-specific alteration of arginine 376, the unique positively charged amino acid residue in the mid-membrane-spanning regions of the proline carrier of Escherichia coli. J Biol Chem 269:5720–5724

Yang JJ, Artis DR, Van Wart HE (1994) Differential effect of halide anions on the hydrolysis of different dansyl substrates by thermolysin. Biochemistry 33:6516–6523

Yernool D, Boudker O, Jin Y, Gouaux E (2004) Structure of a glutamate transporter homologue from Pyrococcus horikoshii. Nature 431:811–818

Zhang Y, Kanner BI (1999) Two serine residues of the glutamate transporter GLT-1 are crucial for coupling the fluxes of sodium and the neurotransmitter. Proc Natl Acad Sci U S A 96:1710–1715

Zhang Y, Bendahan A, Zarbiv R, Kavanaugh MP, Kanner BI (1998) Molecular determinant of ion selectivity of a (Na^+ + K^+)-coupled rat brain glutamate transporter. Proc Natl Acad Sci U S A 95:751–755

Zottola RJ, Cloherty EK, Coderre PE, Hansen A, Hebert DN, Carruthers A (1995) Glucose transporter function is controlled by transporter oligomeric structure. A single, intramolecular disulfide promotes GLUT1 tetramerization. Biochemistry 34:9734–9747

Currents in Neurotransmitter Transporters

K. Gerstbrein · H. H. Sitte (✉)

Institute of Pharmacology, Center for Biomolecular Medicine and Pharmacology, Medical University Vienna, Währingerstrasse 13a, 1090 Vienna, Austria
harald.sitte@meduniwien.ac.at

1	Introduction	95
2	Electrophysiological Background in a Nutshell	98
3	Coupled and Uncoupled Currents	99
4	The Single-File Model	100
5	Transient Currents	101
6	Leak Current	102
7	Is There a Physiological Role for Transporter-Associated Currents?	103
8	Currents and Amphetamines	105
	References	108

Abstract Traditionally, substrate translocation by neurotransmitter transporters has been described by the alternate access model. Recent structural data obtained with three distantly related transporters have also been interpreted as supportive of this model, because conformational correlates were visualized (inward-facing conformation, occluded state). However, the experimental evidence is overwhelmingly in favour of a more complex mode of operation: Transporters also exist in conformations that do not seal the permeation pathway. These conformations support a channel-like activity, including random permeation of substrate and co-substrate ions in a single-file mode. It is likely that the channel-like activity is modified by the interaction of the transporters with accessory proteins and regulatory kinases. Finally, channel-like activity is instrumental to understand the mechanism of action of amphetamines.

Keywords Channel mode · Current · Amphetamine · Cocaine · Oligomer-based countertransport model · Depolarization

1
Introduction

Synaptic transmission is achieved by Ca^{2+}-dependent release of vesicular neurotransmitters, such as serotonin (5-HT), dopamine (DA), norepinephrine

(NE), γ-aminobutyric acid (GABA) and the excitatory amino acid glutamate (EAA). Transporters for neurotransmitters play a key role in removing neurotransmitters from the synaptic cleft, hence terminating the actions of these substances on their corresponding receptors and limiting their effects both in magnitude and duration.

Early experiments yielded insights into the sodium- and chloride-dependent transport mechanism and established the concept that the sodium gradient provided the driving force. In 1966, Jardetzky proposed the "alternate access model" for the function of the transporters, in which a binding site for substrates and co-substrates is alternately accessible from either the extracellular space or the cytoplasm (Jardetzky 1966). In 2003, the crystal structure of lactose permease of *Escherichia coli* became available; this transporter is an intensely studied member of the major facilitator superfamily of transporters (Abramson et al. 2003).

The structural data were consistent with an alternative access model as the mechanism of transport, and a "rock-and-switch" model was put forth to account for the structural rearrangement required for the conformational transition between inward- and outward-facing conformation. In this latter model, binding of substrates is proposed to trigger the movement of two helices; this is transmitted to the two lobes of the protein, which follow suit by rigid body motion and thus either seal access to the cell interior (outward-facing conformation) or to the external milieu (inward-facing conformation). Subsequently, the structure of a bacterial homologue of a neuronal glutamate transporter was solved (Yernool et al. 2004).

While this structure was completely different from all known structures (including that of other transporters), it also displayed features that were consistent with an alternate access model; in this structure, however, substrate access is most readily controlled by two hairpin-regions, which may act as gates on either side of the membrane and thus block the aqueous pathway. Most recently, LeuT, the leucine transporter of *Aquifex aeolicus* was crystallized (Yamashita et al. 2005); LeuT is a homologue of neuronal sodium chloride-dependent neurotransmitter transporters. Accordingly, the protein consists of 12 membrane-spanning helices, with the first 10 transmembrane domains (TM) forming the protein core. TM1 and TM6 are not continuous helices; they both contain an unwound region in the helical structure.

Together with TM3 and TM8, these unwound regions were visualized as binding sites for substrate and sodium ions. Indeed, residues in these areas are highly conserved throughout the neurotransmitter transporter families, and mutations abolish both substrate binding and transport. By using cysteine mutagenesis experiments, it was possible to make predictions about the tertiary structure of transporters, and together with random mutagenesis experiments it was possible to identify residues important for substrate binding and translocation (Barker et al. 1994, 1999; Barker and Blakely 1998; Bismuth et al. 1997; Chen et al. 1997; Goldberg et al. 2003; Ponce et al. 2000). The crystallized structure captured the binding site in an occluded state, separated from both the extracellular space and the cytoplasm. The mechanistic implication for the transport process was proposed by the authors as an "alternate access model" with at least three conformational states: open to outside, occluded and open to inside.

Thus, these three snapshots of transporters highlight the fact that the problem of alternate access may be solved by very different and unrelated structural means. Interestingly, in the structure of LeuT, the substrate binding side was devoid of water molecules, and there wasn't any other evidence for an additional aqueous pathway through the transporter protein. This was surprising because, in spite of the analogy of the structures and the predicted (alternate access) model, several properties of NSS (neurotransmitter:Na^+ symporter) cannot be explained by a simple alternating access model (Adams and DeFelice 2003; DeFelice and Blakely 1996; Sitte et al. 2001; Sonders and Amara 1996). Patch-clamp experiments revealed that neurotransmitter transporters do not only facilitate the transport of neurotransmitter into the cell, they also support an excessive ion flux (Fairman et al. 1995; Galli et al. 1995; Mager et al. 1993, 1994; Wadiche et al. 1995). This high conductance of neurotransmitter transporters is reminiscent of a ligand-gated ion channel, although the currents of the transporters are much smaller than those of ion channels. This view of the transporter as a "pump with a channel" does not exclude the alternating access model for transport activity, but it renders the transport mechanism more complicated than previously expected (Sonders et al. 2005). The issue is succinctly summarized by Accardi and Miller who concluded: "transporters and channels may be separated by an exceedingly fine line" (2004). In fact, in the structures of both—LeuT and the glutamate transporter—segments of the proteins can also be conceptually treated as gates that may open simultaneously and thus convert the transporter into a channel.

Subtle differences in primary structure suffice to cross the line between transporter and channel: This is best exemplified by a comparison of the sodium-glucose transporter SGLT-3. In pigs, this transporter does translocate glucose, but in people it is a glucose-activated Na^+-channel and thus a glucose sensor (Diez-Sampedro et al. 2003). Conversely, both bacterial (Accardi and Miller 2004) and mammalian chloride channels are actually H^+/Cl^--antiporters (Scheel et al. 2005). The 5-HT transporter of the fruit fly *Drosophila melanogaster* differs substantially from its mammalian counterpart. It displays an electrogenic uptake and extraordinarily high currents, with approximately 50 charges carried per 5-HT molecule (at −80 mV) compared to 5–12 charges for their mammalian orthologues. This channel-like mode can also be observed for the substrate translocation: At high concentrations, substrate permeation is consistent with a single-file model (see Sect. 4), which is indicative of diffusion through a pore (Petersen and DeFelice 1999).

2
Electrophysiological Background in a Nutshell

One common property of the family of neurotransmitter transporters is the dependence on sodium ions for the transport process (Mager et al. 1994, 1996; Rudnick 1997; Storck et al. 1992). Reuptake of substrate is indirectly coupled to the hydrolysis of ATP, for it is contingent on the movement of sodium ions down its concentration gradient. Thus, the sodium gradient provides the net driving force for this secondary active transport process. Since ions are carriers of charges, the reuptake of neurotransmitter is always associated with a flow of electrical charges across the cell membrane.

Note that, in theory, a flow of charges will not necessarily result in a macroscopic steady-state current: If the transport cycle of serotonin transporter (SERT) translocated one positively charged sodium ion from the outside of the cell to the inside and subsequently counter-transported one positively charged potassium ion from the cytoplasm out of the cell, the prediction would be that no current can be measured. There would be a flow of charges, but the net flow of the overall process would be zero. Theoretically, this movement of ions may be revealed in clamp experiments by an increased noise upon transporter activity.

Generally the term "electrical current" (I) describes the movement of charges (dQ) in a certain time interval (dt):

$$I = \frac{dQ}{dt} \qquad (1)$$

The amount of moving particles depends on the potential difference, which provides the driving force for the movement, and by the resistance which opposes the movement. This connection between current (I), voltage (E) and resistance (R) is described by Ohm's law:

$$I = \frac{E}{R} \qquad (2)$$

More commonly the conductance (g) is used to describe Ohm's law, which is the reciprocal of the resistance:

$$I = g^*E \qquad (3)$$

The charge of an ion is a multiple of the elementary charge of an electron, which amounts to 1.6^{-19} C (coulomb). Assuming one monovalent ion crossing the membrane every second, a theoretical current of 1.6^{-19} A is measured.

3
Coupled and Uncoupled Currents

The stoichiometry of the reuptake process in neurotransmitter transporters has been thoroughly investigated (Rudnick 1998). Over the past few years, i.e. in particular after cloning of the transporter DNA made heterologous expression possible, our knowledge has considerably increased concerning the ions involved in various currents. Generally speaking, the charge that shifts across the cell membrane through one transport cycle varies from zero (electroneutrality) to a maximum of 2 charges, depending on the transporter. The expected transport-associated current (I_T) can be calculated by multiplying the maximal uptake velocity (V_{max}), the net flow of electrons per transport cycle (z) and the charge of one electron (e):

$$I_T = V_{max}{}^* z^* e \tag{4}$$

To estimate the transport-associated current elicited by one transporter, it is necessary to know the turnover rate of the transporter, which can be determined by dividing the maximal uptake velocity of the transporter by the number of transporters on the cell surface. However, this is contingent on the availability of high-affinity inhibitors which can be used as radioligands. In their absence, it is possible to estimate the number of transporters at the cell surface by analysis of the transient currents, which are elicited by voltage jumps. A prerequisite for this method is the presence of capacitive currents, which is the case in rat GABA transporter 1 (GAT) (Mager et al. 1993). Depending on the transporter and the analysed system, approximately 0.3–20 substrate molecules are carried back into the cell cytoplasm per second. The current (I_T) can be calculated by substitution of V_{max} with the number of transporters (N) multiplied by the transport rate (T):

$$I_T = N^* T^* z^* e \tag{5}$$

For neurotransmitter transporters, both the turnover rate and the numbers of transporters on the surface are comparatively small (that is, by comparison to other pumps), and the expected ion flux produced by the transporter would be very small and hardly detectable with available electrophysiological equipment. Nevertheless, all of the neurotransmitter transporters, even those with predicted electroneutrality of the overall process, show a measurable current in clamped experiments. Measurements of substrate-elicited current greatly exceed the predicted size, and therefore the transport of substrate alone cannot directly account for the recorded current levels. Accordingly, the former theoretical considerations needed further modifications (Galli et al. 1995; Lin et al. 1996).

In contrast to the transport-associated currents with fixed stoichiometry, evidence suggests that this "current in excess" (I_C) is caused by formation of a pore within the transporter molecule. This novel view of transporters was a motivation to apply a wide range of electrophysiological approaches that are used to investigate ion channels. Hence, the analogies between neurotransmitter transporter and ion channels were subsequently documented by determining single-channel events, channel conductance, reversal potentials and mole fraction, as well as by noise analysis (Cammack and Schwartz 1996; Fairman et al. 1995; Fairman and Amara 1999; Galli et al. 1996; Petersen and DeFelice 1999). Nevertheless, currents elicited by ion channels are vast compared to currents associated with transporters.

The movement of ions through a pore is not coupled to a fixed stoichiometry, but dependent on the conductance of the channel (g), the probability to find the channel in an open state (p), the number of channels—or in our case transporters (N)—and the driving force for the ion, which is the difference of the membrane potential (E_m) and the reversal potential for the ion (E_x):

$$I_C = g^* p^* N^* (E_m - E_x) \tag{6}$$

The macroscopic current, which is measured in clamp experiments as the sum of both the transport-associated current I_T and the uncoupled channel-like current I_C. Application of substrate and subsequent binding to the transporter results in an increase of the turnover rate of the transporter (I_T), and it increases the open probability (I_C) or the channel conductance.

The opening of a channel can be described as a small burst of charge-flux through the membrane, comparable to a rectangular impulse. The opening probability reflects both the duration of one burst—thus the mean open time of a single channel—and the frequency of occurrence of the events.

4
The Single-File Model

The observations summarized above can be reconciled with an extended alternate access model by assuming that the transporter functions as a sodium-driven pump that displays some additional channel-like features. In other words, the uncoupled current I_C through the channel reflects the noise of the transport process. In principle, however, the uptake of substrate is mediated by a sodium gradient, and one transport cycle has a well-defined and fixed stoichiometry.

For some transporters, it has nevertheless been proposed that the transporter-associated pore not only conducts ions, but it is also permeable to the substrate of the transporter. This means the same rules apply for neurotransmitters as for other channels: The substrate follows its electrochemical gradient. In this model, three factors contribute to the rapid clearing of extracellular neurotransmitter:

1. Monoamines are positively charged molecules, therefore the membrane potential drives them into the cell.
2. Storage of neurotransmitter in vesicles keeps the intracellular concentration of transmitter low, thus keeping the reversal potential for the neurotransmitter positive.
3. The pore is also permeable to sodium (and chloride), and no ion can slip past another ion.

This would causes an additional driving force for neurotransmitter uptake by the sodium gradient, because each neurotransmitter molecule which enters the pore is consecutively pushed through the channel by sodium and chloride ions (Adams and DeFelice 2002). The single-file model shows no well-defined transport cycle and consequently no fixed stoichiometry. In this case, speaking of "stoichiometry of the transport process" reflects a mean value of transported ions per molecule substrate.

This model is backed by observations in potassium channels, where the crystallographic structure of the highly conserved selectivity filter of the channel showed a single-file, permeating pore (Mackinnon 2003).

5
Transient Currents

The membrane of cells provides an electrically isolating layer between the cytoplasm and the external environment of the cell. This configuration is equivalent to a capacitor.

The insulating layer of a capacitor separates charges, which results in a potential difference across the insulating medium. The connection between the amount of separated charges (Q) and the resulting potential difference (E) is the capacitance (C).

$$Q = C^*E \tag{7}$$

The capacitance is determined by the geometry of the layer, and by a material constant of the isolating media, the dielectric constant.

Combining Eq. 1 and Eq. 7 gives the current/voltage relation of a capacitor:

$$I = C^* \frac{dE}{dt} \tag{8}$$

With Eq. 8 in mind, it is possible to estimate the current response of a membrane when applying a rectangular voltage jump. The term dE/dt describes the slope of the membrane potential. For a rectangular impulse, the slope is very high for a short time as the signal changes, and zero at all other times. Thus, the expected current response is an infinitesimal short spike of infinite height as the potential changes.

If resistance is introduced into the model, the current spike induced by the voltage pulse becomes smaller and broader, because the maximal current is limited to E/R. Substitution of I with E/R in Eq. 8 and solving the differential equation gives the resulting time course of the current. Like many other processes in nature, it decays exponentially with the time constant τ. The time constant (τ) of this decay is calculated by multiplication of the resistance (R) and the capacitance:

$$\tau = R^*C \qquad (9)$$

A voltage jump on the membrane as performed in patch-clamp experiments clamps the membrane to a new potential. Initially, the capacitor (the membrane) holds the former voltage. The newly applied voltage drives charges to the insulating layer, thus recharging the membrane. The membrane potential changes exponentially to the new applied voltage with the time constant as described in Eq. 9. A voltage jump from a negative potential to a positive potential shifts exactly the same amount of charge as the returning jump from positive to negative, but the algebraic sign changes.

Taken together, the transient currents are the result of charge movements, with the capacitance of the membrane as storage for charges. Upon application of voltage jumps, neurotransmitter transporters reveal additional charge movements in the electrical field of the membrane, which are detected as transient currents. These charges can be assigned to either rapid binding or unbinding of ions to the transporter. Alternatively, these transient currents may be the result of a movement of charged residues within the transporter molecule.

For SERT, a special behaviour of the transient currents has been described, which is distinct from the capacitive transient currents of other neurotransmitter transporters (Mager et al. 1994). Capacitive transient currents resemble charge movements, but charged particles do not cross the insulating membrane. In SERT, a voltage jump from positive to negative potential gives rise to larger current than the corresponding jump from the negative to positive potentials. The behaviour may be explained by a short, burst-like opening of a channel, resulting in an additional flow of charges across the membrane bilayer. These transients are referred to as "resistive" transient currents, because the additional charges can be explained by a resistor placed in parallel to the capacitor.

6
Leak Current

In some neurotransmitters, a detectable current can be found even under resting conditions that is in the absence of neurotransmitter substrate. Under physiological conditions, these tonically active conductances are carried by Na^+ or Cl^- (Lin et al. 1996; Mager et al. 1994; Sonders et al. 1997). Because the leak current is not associated with any substrate translocation, it can be interpreted only as the result of channel openings. This leak current highlights the similarity between transporter and ion channel, because it can be accounted for only by the presence of a conducting pore within the transporter. In addition, the permeation pathway can also conduct other ions, e.g. lithium or potassium.

Inhibitors of the uptake process do not block only the uptake of substrate, but they also reduce the macroscopic current, thereby revealing the size of the leak current. In the absence of Na^+, substrate can still bind to the transporter but it cannot be translocated. The leak conductance is also seen in Na^+-free media, an observation which also stresses the existence of a distinct operating mode of the transporter. Application of lithium to SERT does not only permit lithium to permeate through the transporter, it also induces an additional conformational change, which contributes to the higher single-channel conductance for lithium than for sodium (Ni et al. 2001). Usually the size of the leak amounts to 10%–20% of the maximal substrate-induced current. The physiological relevance of the leak current has been questioned, because it is not detected in neurons. Therefore the leak current has been ascribed to the artificial conditions arising from heterologous expression of transporters in non-neuronal cells: It is, for instance, conceivable that the absence of syntaxin 1A (or related neuron-specific syntaxins) unmasks a leak current which may be suppressed by the tight association of syntaxin to the aminoterminus of the transporter (Beckman et al. 1998; Deken et al. 2000; Horton and Quick 2001; Quick 2003). However, in neurons the leak current may escape detection because it is too small by comparison to the many additional interfering conductances (Ingram et al. 2002).

7
Is There a Physiological Role for Transporter-Associated Currents?

Up to now, the physiological role of the uncoupled currents remains enigmatic for most neurotransmitter transporters. Nevertheless, it is safe to conclude that the uncoupled currents can indeed have an influence on transmembrane potential, as shown for dopamine transporter (DAT) (Carvelli et al. 2004; Sonders and Amara 1996) and the glutamate transporter (see chapter by R.J. Vandenberg, this volume). The uncoupled currents of the transporters for 5-HT, GABA and NE are carried by Na^+ ions, hence the conducted charges depolarizes the cell. The channels of the glutamate transporters and the dopamine transporter are permeable for anions (usually chloride; Ingram et al. 2002; Meinild et al. 2004), therefore the membrane potential is shifted in the direction of the Cl^- reversal potential by a switch of these transporters to their channel mode. Because the range of the membrane potential overlaps with the range of the reversal potential for Cl$^-$, the net effect can either result in hyperpolarization or depolarization of the cell membrane, depending on the actual membrane potential and the reversal potential for Cl$^-$ (Ingram et al. 2002; Meinild et al. 2004; Sulzer and Galli 2003). Monoamine substrates are translocated as charged species (Berfield et al. 1999). Hence, a negative membrane potential provides an additional driving force, which is underscored by a mutation in DAT (DAT-Y335A; see chapter by K. Nørgaard-Nielsen and U. Gether, this volume). This mutated transporter displays large Na$^+$-leak currents, which clamp the membrane potential to depolarized levels. Accordingly, uptake of dopamine by DAT-Y335A is abolished, but it can be restored in the presence of Zn^{2+}. Zn^{2+} is likely to be a physiological ligand of DAT (Norregaard et al. 1998; Richfield 1993) and induces an additional anion conductance through the transporter pore (Meinild et al. 2004). Finally, substrate-associated uncoupled currents have been shown to affect the membrane potential and thus modulate the excitability in neurons (Falkenburger et al. 2001; Ingram et al. 2002).

Drugs of abuse (e.g. cocaine and amphetamine) and therapeutically used antidepressants target neurotransmitter transporters (Iversen 2000). Elevation of synaptic monoamine concentrations is thought to induce such clinical effects. However, it is clear that these compounds, in addition, affect the transporter-associated currents. As inhibitors of the transporter, antidepressants and cocaine block the substrate uptake and, thus, transport-associated currents. Interestingly, the leak current can also be blocked by administration of these drugs (Khoshbouei et al. 2004; Sitte et al. 1998; Sonders et al. 1997) and by substrates (including amphetamines, see following section; Sonders et al. 1997). It is not known if the altered conductance of the transporter has any impact on the reinforcing and addictive properties of cocaine or on the mood-stabilizing effects of antidepressants. The beneficial clinical effect of antidepressants (that is, the resolution of abnormal mood states) is only observed after a prolonged treatment (2–3 weeks), while occupancy of the transporter and, hence, inhibition of neurotransmitter reuptake is seen immediately after initiation of treatment. The mechanistic basis for this delay is not understood, but it is thought to arise from adaptive changes in gene expression caused by chronic blockage of SERT and NET (see chapters by H. Bönisch and M. Brüss, and by K.-P. Lesch and R. Mössner, this volume). It is conceivable that the altered electrical properties of the neurons (resulting from long-term suppression of leak currents through SERT and NET) contribute to the adaptation induced by antidepressants.

More recently, doubts have been cast on the physiological relevance of transporter-associated currents, because they have been considered an artefact arising from heterologous expression. In neurons, transporters interact with several proteins (Bjerggaard et al. 2004; Farhan et al. 2004; Torres et al. 2001); these regulated associations are thought to be important for targeting and delivery of transporters to their destination, i.e. the synaptic specialization where they are enriched in a ring peripheral to the active zone. They may also affect internalization and endosomal recycling. In addition, these interacting proteins are likely to affect the electrophysiological properties of the transporter, because their binding to intracellular domains are likely to impinge on the structure of the transporter. This is exemplified by syntaxin 1A, which associates with the amino terminus of several transporters (GAT1, SERT, NET; see chapter by M.W. Quick, this volume). Upon binding to SERT, syntaxin eliminates sodium-driven leak current and abolishes substrate-induced excess current, but it does not interfere with the velocity of substrate translocation. Thus, in the presence of syntaxin the current is coupled stoichiometrically to the transport cycle (Quick 2003; for a detailed account, the reader is also referred to the chapter by M.W. Quick in this volume).

8
Currents and Amphetamines

Amphetamine and amphetamine congeners [e.g., 3,4-methylene-dioxy-metamphetamine (MDMA), "ecstasy"], methamphetamine ("ice") or amphetamine ("speed") target the monoamine transporters DAT, NET and SERT (Sulzer et al. 2005). Amphetamines induce transport reversal and lead to a pronounced efflux of neurotransmitter via the transporter (Pifl et al. 2005). Thus, in vivo, amphetamines cause non-exocytotic release of monoamines—that is, DA, 5-HT and NE/epinephrine. The proportion of the effect depends on the relative efficacy of each compound in inducing reverse transport in DAT, SERT and NET, thus accounting for the difference in biological effects elicited by different compounds.

For the past 25 years, amphetamine-induced carrier-mediated outward transport has been explained by the "facilitated exchange diffusion hypothesis" (Fischer and Cho 1979). In this model, release of substrate is contingent on prior uptake of the amphetamines by the transporter. Therefore the potency of a drug to elicit efflux is predicted to be proportional to the affinity of the compound for the transporter. Contrary to this prediction, it has been shown that the releasing action of different amphetamine derivates does not correlate with their affinity for the individual transporter (Hilber et al. 2005).

There are many more observations that are not accounted for by the model of facilitated exchange diffusion, not the least of which is the fact that in neurons the substrate (i.e. the neurotransmitter) is actually not available for countertransport because it is stored in vesicles. Under physiological conditions, there are different processes that keep the cytosolic concentration of monoamine neurotransmitters low. (1) The reserpine-sensitive, proton-driven vesicular monoamine transporter (VMAT-1 and VMAT-2) retrieves neurotransmitters into the synaptic vesicles. Monoamines that escape the action of VMAT are rapidly degraded by the action of monoamine oxidases (MAO-A and MAO-B) and to a lesser extent by catechol O-methyltransferase (COMT; Seiden et al.

1993). Hence, amphetamines must also exert an action on synaptic vesicles and/or degradation. The "weak base hypothesis" provides a unifying concept that also takes vesicular uptake and degradation of neurotransmitter into account: In this model, amphetamine exerts three types of actions that contribute to transport reversal (Sulzer et al. 1993; Sulzer et al. 1995). (2) Amphetamines enter the cell via the transporter as charged substrates. In the unprotonated form, amphetamines can subsequently diffuse through the membrane back into the synaptic cleft and thus be available for another round of transporter-mediated uptake. Therefore one molecule of amphetamine can cycle repeatedly between the interior of the cell and the extracellular space; this cycle is thought to lead to accumulation of Na^+, Cl^- and H^+ in the cytoplasm (Rudnick 1997). (3) Amphetamines interfere with vesicular uptake of monoamines by competing for VMAT and, upon accumulation in the vesicles, by their buffering action, which dissipates the pH gradient. (4) Finally, amphetamines also inhibit the degradation of monoamines by MAO.

The weak base hypothesis effectively explains how substrate becomes available on the cytoplasmic side of the transporter. However, it does not readily account for the propensity of the transporter to mediate efflux of substrate. In fact, while transporters can function bidirectionally, substrate influx is the preferred mode, and this is evident from the fact that the K_M for inward transport is several orders of magnitude lower than that for outward transport (Sitte et al. 2001, 2002).

There are three effects that may contribute to amphetamine-induced substrate efflux:

1. The action of amphetamines is—in part—contingent on protein kinase C (PKC)-mediated (Kantor and Gnegy 1998) and Ca^{2+}/calmodulin-dependent phosphorylation (Kantor et al. 1999); this was first observed for DAT but subsequently confirmed for NET (Kantor et al. 2001) and SERT (Seidel et al. 2005). It is at not clear how amphetamines cause activation of PKC (Giambalvo 1992a, b) and of Ca^{2+}/calmodulin-dependent protein kinase, but the stimulation is dependent on intracellular Ca^{2+} (Giambalvo 2003). The pleiotropic actions of kinases may allow for the recruitment of many different mechanisms that afford transport reversal, e.g. phosphorylation-dependent inhibition of Na^+/K^+-ATPase (Kazanietz et al. 2001) which favours intracellular accumulation of Na^+. However, there is circumstantial evidence for a direct action of PKC isoforms on transporters. Truncation of the 22 N-terminal residues of DAT eliminates the ability of DAT to undergo PKC-dependent phosphorylation (Granas et al. 2003) and suppresses reverse transport induced by amphetamine (Khoshbouei et al. 2004); the amphetamine-induced reverse transport is also blunted after mutational replacement of the five serine residues in this segment of DAT by alanine (Khoshbouei et al. 2004). Most recently, the relevant isoform was identified as PKC β(II) (Johnson et al. 2005).

2. Amphetamines induce a pronounced Na^+-current; there is a good correlation between the size of the elicited currents and the potency of different amphetamine congeners to release DA or 5-HT (Gerstbrein et al. 2005; Hilber et al. 2005; Sitte et al. 1998). In the alternate access model (see also chapter by G. Rudnick, this volume), the local accumulation of intracellular Na^+ is predicted to favour the accumulation of transporters in the inward-facing conformation and thus render the transporter more prone to support substrate efflux. In addition, in the presence of amphetamine, the membrane potential is shifted to positive values such that the electrochemical gradient favours outward transport (Khoshbouei et al. 2003).

3. Finally, binding of amphetamines may favour the channel mode; in other words, in the presence of amphetamine, the pore may be stabilized and support efflux. Evidence for this mode has recently been obtained by amperometry (Kahlig et al. 2005): Brief spike-like elevations of extracellular neurotransmitter can be detected in the presence of amphetamine but not of substrate dopamine. These spikes are consistent with fast release of neurotransmitter through a channel. Because of the inherent limits of amperometry, it is difficult to quantitate the relative proportion of dopamine that is released by spikes and by continuous efflux. The ratio is at 9:1 in favour of continuous efflux (Kahlig et al. 2005); this estimate suggests that amphetamine does allow the transporter to cycle efficiently through outward- and inward-facing conformations and that the channel-like mode is a comparably rare event. However, the very fact that one, in principle, can detect a contribution of a channel mode to outward transport also supports the concept that oligomeric assembly of neurotransmitter transporters is relevant to amphetamine-induced reverse transport. Evidently, a single transporter moiety can hardly allow for permeation of amphetamine and simultaneously support the efflux of dopamine. It is much more plausible to assume that amphetamine triggers the channel mode in an adjacent transporter moiety within the oligomeric complex (Seidel et al. 2005).

Thus, more than a century after their first use in people, amphetamines still represent a pharmacological enigma. However, these compounds have been instrumental in uncovering the various modes in which transporters can operate. In addition, the large diversity of regulatory inputs that impinge on monoamine transporters have been—in part—unravelled in the quest to understand how they elicit their actions.

Acknowledgements We gratefully acknowledge support from the Austrian Science Foundation (grant P17076 to H.H.S.).

References

Abramson J, Smirnova I, Kasho V, Verner G, Iwata S, Kaback HR (2003) The lactose permease of Escherichia coli: overall structure, the sugar-binding site and the alternating access model for transport. FEBS Lett 555:96–101

Accardi A, Miller C (2004) Secondary active transport mediated by a prokaryotic homologue of ClC Cl⁻ channels. Nature 427:803–807

Adams SV, DeFelice LJ (2002) Flux coupling in the human serotonin transporter. Biophys J 83:3268–3282

Adams SV, DeFelice LJ (2003) Ionic currents in the human serotonin transporter reveal inconsistencies in the alternating access hypothesis. Biophys J 85:1548–1559

Barker EL, Blakely RD (1998) Structural determinants of neurotransmitter transport using cross-species chimeras: studies on serotonin transporter. Methods Enzymol 296:475–498

Barker EL, Kimmel HL, Blakely RD (1994) Chimeric human and rat serotonin transporters reveal domains involved in recognition of transporter ligands. Mol Pharmacol 46:799–807

Barker EL, Moore KR, Rakhshan F, Blakely RD (1999) Transmembrane domain I contributes to the permeation pathway for serotonin and ions in the serotonin transporter. J Neurosci 19:4705–4717

Beckman ML, Bernstein EM, Quick MW (1998) Protein kinase C regulates the interaction between a GABA transporter and syntaxin 1A. J Neurosci 18:6103–6112

Berfield JL, Wang LC, Reith ME (1999) Which form of dopamine is the substrate for the human dopamine transporter: the cationic or the uncharged species? J Biol Chem 274:4876–4882

Bismuth Y, Kavanaugh MP, Kanner BI (1997) Tyrosine 140 of the gamma-aminobutyric acid transporter GAT-1 plays a critical role in neurotransmitter recognition. J Biol Chem 272:16096–16102

Bjerggaard C, Fog JU, Hastrup H, Madsen K, Loland CJ, Javitch JA, Gether U (2004) Surface targeting of the dopamine transporter involves discrete epitopes in the distal C terminus but does not require canonical PDZ domain interactions. J Neurosci 24:7024–7036

Cammack JN, Schwartz EA (1996) Channel behavior in a gamma-aminobutyrate transporter. Proc Natl Acad Sci U S A 93:723–727

Carvelli L, McDonald PW, Blakely RD, DeFelice LJ (2004) Dopamine transporters depolarize neurons by a channel mechanism. Proc Natl Acad Sci U S A 101:16046–16051

Chen JG, Sachpatzidis A, Rudnick G (1997) The third transmembrane domain of the serotonin transporter contains residues associated with substrate and cocaine binding. J Biol Chem 272:28321–28327

DeFelice LJ, Blakely RD (1996) Pore models for transporters? Biophys J 70:579–580

Deken SL, Beckman ML, Boos L, Quick MW (2000) Transport rates of GABA transporters: regulation by the N-terminal domain and syntaxin 1A. Nat Neurosci 3:998–1003

Diez-Sampedro A, Hirayama BA, Osswald C, Gorboulev V, Baumgarten K, Volk C, Wright EM, Koepsell H (2003) A glucose sensor hiding in a family of transporters. Proc Natl Acad Sci U S A 100:11753–11758

Fairman WA, Amara SG (1999) Functional diversity of excitatory amino acid transporters: ion channel and transport modes. Am J Physiol 277:F481–F486

Fairman WA, Vandenberg RJ, Arriza JL, Kavanaugh MP, Amara SG (1995) An excitatory amino-acid transporter with properties of a ligand-gated chloride channel. Nature 375:599–603

Falkenburger BH, Barstow KL, Mintz IM (2001) Dendrodendritic inhibition through reversal of dopamine transport. Science 293:2465–2470

Farhan H, Korkhov VM, Paulitschke V, Dorostkar MM, Scholze P, Kudalcek O, Freissmuth M, Sitte HH (2004) Two discontinuous segments in the carboxy terminus are required for membrane targeting of the rat GABA transporter-1 (GAT1). J Biol Chem 279:28553–28563

Fischer JF, Cho AK (1979) Chemical release of dopamine from striatal homogenates: evidence for an exchange diffusion model. J Pharmacol Exp Ther 208:203–209

Galli A, Defelice LJ, Duke BJ, Moore KR, Blakely RD (1995) Sodium-dependent norepinephrine-induced currents in norepinephrine-transporter-transfected HEK-293 cells blocked by cocaine and antidepressants. J Exp Biol 198:2197–2212

Galli A, Blakely RD, Defelice LJ (1996) Norepinephrine transporters have channel modes of conduction. Proc Natl Acad Sci U S A 93:8671–8676

Gerstbrein K, Sandtner W, Scholze P, Holy M, Wiborg O, Singer EA, Freissmuth M, Sitte HH (2005) Mutants of the serotonin transporter reveal amino acids important for outward transport upon amphetamine-application. Abstracts of the 11th Symposium of the Austrian Pharmacological Society (APHAR), Vienna, 24-26 November 2005, p 358

Giambalvo CT (1992a) Protein kinase C and dopamine transport-1. Effects of amphetamine in vivo. Neuropharmacology 31:1201–1210

Giambalvo CT (1992b) Protein kinase C and dopamine transport-2. Effects of amphetamine in vitro. Neuropharmacology 31:1211–1222

Giambalvo CT (2003) Differential effects of amphetamine transport vs. dopamine reverse transport on particulate PKC activity in striatal synaptoneurosomes. Synapse 49:125–133

Goldberg NR, Beuming T, Weinstein H, Javitch JA (2003) Structure and function of neurotransmitter transporters. In: Bräuner-Osborne H, Schousboe A (eds) Molecular neuropharmacology: strategies and methods. Humana Press, Totowa, pp 213–234

Granas C, Ferrer J, Loland CJ, Javitch JA, Gether U (2003) N-terminal truncation of the dopamine transporter abolishes phorbol ester- and substance P receptor-stimulated phosphorylation without impairing transporter internalization. J Biol Chem 278:4990–5000

Hilber B, Scholze P, Dorostkar MM, Sandtner W, Holy M, Boehm S, Singer EA, Sitte HH (2005) Serotonin-transporter mediated efflux: a pharmacological analysis of amphetamines and non-amphetamines. Neuropharmacology 49:811–819

Horton N, Quick MW (2001) Syntaxin 1A up-regulates GABA transporter expression by subcellular redistribution. Mol Membr Biol 18:39–44

Ingram SL, Prasad BM, Amara SG (2002) Dopamine transporter-mediated conductances increase excitability of midbrain dopamine neurons. Nat Neurosci 5:971–978

Iversen L (2000) Neurotransmitter transporters: fruitful targets for CNS drug discovery. Mol Psychiatry 5:357–362

Jardetzky O (1966) Simple allosteric model for membrane pumps. Nature 211:969–970

Johnson LA, Guptaroy B, Lund D, Shamban S, Gnegy ME (2005) Regulation of amphetamine-stimulated dopamine efflux by protein kinase C beta. J Biol Chem 280:10914–10919

Kahlig KM, Binda F, Khoshbouei H, Blakely RD, McMahon DG, Javitch JA, Galli A (2005) Amphetamine induces dopamine efflux through a dopamine transporter channel. Proc Natl Acad Sci U S A 102:3495–3500

Kantor L, Gnegy ME (1998) Protein kinase C inhibitors block amphetamine-mediated dopamine release in rat striatal slices. J Pharmacol Exp Ther 284:592–598

Kantor L, Hewlett GH, Gnegy ME (1999) Enhanced amphetamine- and K^+-mediated dopamine release in rat striatum after repeated amphetamine: differential requirements for Ca^{2+}- and calmodulin-dependent phosphorylation and synaptic vesicles. J Neurosci 19:3801–3808

Kantor L, Hewlett GH, Park YH, Richardson-Burns SM, Mellon MJ, Gnegy ME (2001) Protein kinase C and intracellular calcium are required for amphetamine-mediated dopamine release via the norepinephrine transporter in undifferentiated PC12 cells. J Pharmacol Exp Ther 297:1016–1024

Kazanietz MG, Caloca MJ, Aizman O, Nowicki S (2001) Phosphorylation of the catalytic subunit of rat renal Na^+, K^+-ATPase by classical PKC isoforms. Arch Biochem Biophys 388:74–80

Khoshbouei H, Wang H, Lechleiter JD, Javitch JA, Galli A (2003) Amphetamine-induced dopamine efflux. A voltage-sensitive and intracellular Na^+-dependent mechanism. J Biol Chem 278:12070–12077

Khoshbouei H, Sen N, Guptaroy B, Johnson L, Lund D, Gnegy ME, Galli A, Javitch JA (2004) N-terminal phosphorylation of the dopamine transporter is required for amphetamine-induced efflux. PLoS Biol 2:E78

Lin F, Lester HA, Mager S (1996) Single-channel currents produced by the serotonin transporter and analysis of a mutation affecting ion permeation. Biophys J 71:3126–3135

Mackinnon R (2003) Potassium channels. FEBS Lett 555:62–65

Mager S, Naeve J, Quick M, Labarca C, Davidson N, Lester HA (1993) Steady states, charge movements, and rates for a cloned GABA transporter expressed in Xenopus oocytes. Neuron 10:177–188

Mager S, Min C, Henry DJ, Chavkin C, Hoffman BJ, Davidson N, Lester HA (1994) Conducting states of a mammalian serotonin transporter. Neuron 12:845–859

Mager S, Kleinberger DN, Keshet GI, Davidson N, Kanner BI, Lester HA (1996) Ion binding and permeation at the GABA transporter GAT1. J Neurosci 16:5405–5414

Meinild AK, Sitte HH, Gether U (2004) Zinc potentiates an uncoupled anion conductance associated with the dopamine transporter. J Biol Chem 279:49671–49679

Ni YG, Chen JG, Androutsellis-Theotokis A, Huang CJ, Moczydlowski E, Rudnick G (2001) A lithium-induced conformational change in serotonin transporter alters cocaine binding, ion conductance, and reactivity of Cys-109. J Biol Chem 276:30942–30947

Norregaard L, Frederiksen D, Nielsen EO, Gether U (1998) Delineation of an endogenous zinc-binding site in the human dopamine transporter. EMBO J 17:4266–4273

Petersen CI, DeFelice LJ (1999) Ionic interactions in the Drosophila serotonin transporter identify it as a serotonin channel. Nat Neurosci 2:605–610

Pifl C, Nagy G, Berenyi S, Kattinger A, Reither H, Antus S (2005) Pharmacological characterization of ecstasy synthesis byproducts with recombinant human monoamine transporters. J Pharmacol Exp Ther 314:346–354

Ponce J, Biton B, Benavides J, Avenet P, Aragon C (2000) Transmembrane domain III plays an important role in ion binding and permeation in the glycine transporter GLYT2. J Biol Chem 275:13856–13862

Quick MW (2003) Regulating the conducting states of a mammalian serotonin transporter. Neuron 40:537–549

Richfield EK (1993) Zinc modulation of drug binding, cocaine affinity states, and dopamine uptake on the dopamine uptake complex. Mol Pharmacol 43:100–108

Rudnick G (1997) Mechanism of biogenic amine neurotransmitter transporters. In: Reith M (ed) Neurotransmitter transporters: structure, function, and regulation. Humana Press Inc, Totowa, pp 73–100

Rudnick G (1998) Bioenergetics of neurotransmitter transport. J Bioenerg Biomembr 30:173–185

Scheel O, Zdebik AA, Lourdel S, Jentsch TJ (2005) Voltage-dependent electrogenic chloride/proton exchange by endosomal CLC proteins. Nature 436:424–427

Seidel S, Singer EA, Just H, Farhan H, Scholze P, Kudlacek O, Holy M, Koppatz K, Krivanek P, Freissmuth M, Sitte HH (2005) Amphetamines take two to tango: an oligomer-based counter-transport model of neurotransmitter transport explores the amphetamine action. Mol Pharmacol 67:140–151

Seiden LS, Sabol KE, Ricaurte GA (1993) Amphetamine: effects on catecholamine systems and behavior. Annu Rev Pharmacol Toxicol 33:639–677

Sitte HH, Huck S, Reither H, Boehm S, Singer EA, Pifl C (1998) Carrier-mediated release, transport rates, and charge transfer induced by amphetamine, tyramine, and dopamine in mammalian cells transfected with the human dopamine transporter. J Neurochem 71:1289–1297

Sitte HH, Hiptmair B, Zwach J, Pifl C, Singer EA, Scholze P (2001) Quantitative analysis of inward and outward transport rates in cells stably expressing the cloned human serotonin transporter: inconsistencies with the hypothesis of facilitated exchange diffusion. Mol Pharmacol 59:1129–1137

Sitte HH, Singer EA, Scholze P (2002) Bi-directional transport of GABA in human embryonic kidney (HEK-293) cells stably expressing the rat GABA transporter GAT-1. Br J Pharmacol 135:93–102

Sonders MS, Amara SG (1996) Channels in transporters. Curr Opin Neurobiol 6:294–302

Sonders MS, Zhu SJ, Zahniser NR, Kavanaugh MP, Amara SG (1997) Multiple ionic conductances of the human dopamine transporter: the actions of dopamine and psychostimulants. J Neurosci 17:960–974

Sonders MS, Quick M, Javitch JA (2005) How did the neurotransmitter cross the bilayer? A closer view. Curr Opin Neurobiol 15:296–304

Storck T, Schulte S, Hofmann K, Stoffel W (1992) Structure, expression, and functional analysis of a Na(+)-dependent glutamate/aspartate transporter from rat brain. Proc Natl Acad Sci U S A 89:10955–10959

Sulzer D, Galli A (2003) Dopamine transport currents are promoted from curiosity to physiology. Trends Neurosci 26:173–176

Sulzer D, Maidment NT, Rayport S (1993) Amphetamine and other weak bases act to promote reverse transport of dopamine in ventral midbrain neurons. J Neurochem 60:527–535

Sulzer D, Chen TK, Lau YY, Kristensen H, Rayport S, Ewing A (1995) Amphetamine redistributes dopamine from synaptic vesicles to the cytosol and promotes reverse transport. J Neurosci 15:4102–4108

Sulzer D, Sonders MS, Poulsen NW, Galli A (2005) Mechanisms of neurotransmitter release by amphetamines: a review. Prog Neurobiol 75:406–433

Torres GE, Yao WD, Mohn AR, Quan H, Kim KM, Levey AI, Staudinger J, Caron MG (2001) Functional interaction between monoamine plasma membrane transporters and the synaptic PDZ domain-containing protein PICK1. Neuron 30:121–134

Wadiche JI, Amara SG, Kavanaugh MP (1995) Ion fluxes associated with excitatory amino acid transport. Neuron 15:721–728

Yamashita A, Singh SK, Kawate T, Jin Y, Gouaux E (2005) Crystal structure of a bacterial homologue of Na^+/Cl^--dependent neurotransmitter transporters. Nature 437:215–223

Yernool D, Boudker O, Jin Y, Gouaux E (2004) Structure of a glutamate transporter homologue from Pyrococcus horikoshii. Nature 431:811–818

Mutational Analysis of Glutamate Transporters

R. J. Vandenberg

Department of Pharmacology, Institute for Biomedical Research, University of Sydney, 2006 New South Wales, Australia
robv@med.usyd.edu.au

1	Glutamate Transporters	114
2	Ion/Flux Coupling Determines the Concentrating Capacity of the Transporters	114
3	Uncoupled Ion Currents Associated with Glutamate Transporters	116
4	The Structure of a Bacterial Glutamate Transporter	117
5	Mutational Studies of Mammalian Glutamate Transporters	118
5.1	General Considerations and Approaches	118
5.2	Structural Studies Using Site-Directed Mutagenesis	119
5.2.1	Cysteine-Scanning Mutagenesis	119
5.2.2	Cross-Linking of Cysteine Mutants	121
5.3	Substrate Recognition and Translocation	122
5.3.1	The Use of Chimeras to Define Functional Domains	122
5.3.2	Glutamate and Ion Binding Sites	123
5.3.3	Glutamate Binding Site in the Crystal Structure of Gltph	126
5.4	Zn^{2+} Binding Sites on EAAT1 and EAAT4	127
5.5	The Chloride Channel Within the Transporter	127
5.5.1	Different Conformational States Regulate the Two Functions	128
5.5.2	Residues that Line the Chloride Channel	129
5.5.3	Opening the Chloride Channel Gate	130
6	A Structural Model for the Transport and Chloride Channel Functions of Glutamate Transporters	130
	References	132

Abstract Glutamate transporters are a family of transporters that regulate extracellular glutamate concentrations so as to maintain a dynamic and high-fidelity cell signalling process in the brain. Site-directed mutagenesis has been used to investigate various aspects of the structural and functional properties of these transporters to gain insights into how they work. This field of research has recently undergone a major development with the determination of the crystal structure of a bacterial glutamate transporter, and this chapter relates the results from mutagenesis experiments with what we now know about glutamate transporter structure.

Keywords Glutamate transport · Neutral amino acid transport · Site-directed mutagenesis · Chloride channel

1
Glutamate Transporters

Glutamate is the predominant excitatory neurotransmitter in the mammalian central nervous system and activates a wide range of ionotropic and metabotropic receptors to mediate a complex array of functions. The extracellular glutamate concentration is tightly controlled by a family of glutamate transporters (EAATs)—expressed in both neurons and glia—that serve to maintain a dynamic signalling system between neurons. The failure or down-regulation of EAAT function will lead to elevations in extracellular glutamate concentrations, causing excessive stimulation of glutamate receptors which, if prolonged, will result in excitotoxicity and cell death. These processes are thought to underlie the pathogenesis of ischaemic brain damage following a stroke and also various neurodegenerative disorders, such as amyotrophic lateral sclerosis and Alzheimer's disease.

This chapter shall review the current understanding of the functions of glutamate transporters and the closely related neutral amino acid transporters. It will then describe some of the site-directed mutagenesis studies that have been used to better understand the molecular basis for how these proteins work. This field of research has entered a very interesting phase because the crystal structure of a bacterial glutamate transporter has recently been determined. This development will greatly facilitate a better understanding of how these proteins work and, in discussing the mutagenesis work, this review will attempt to relate the conclusions to what is now known of the structure of these proteins.

Five human EAAT subtypes have been cloned and are termed EAAT1–5 (Arriza et al. 1994, 1997; Fairman et al. 1995), but in other species the homologues are termed GLAST1 (rat equivalent of EAAT1) (Storck et al. 1992), GLT1 (rat equivalent of EAAT2) (Pines et al. 1992) and EAAC1 (rabbit equivalent of EAAT3) (Kanai and Hediger 1992). In addition to L-glutamate, all five EAATs transport L-aspartate and D-aspartate; EAAT3 also transports L-cysteine. This family of transporters also includes the neutral amino acid carriers, ASCT1 and ASCT2, which are selective for alanine, serine, cysteine and threonine (Arriza et al. 1993; Shafqat et al. 1993; Utsunomiya-Tate et al. 1996). The five cloned glutamate transporter subtypes share 50%–60% amino acid sequence identity with each other and 30%–40% identity with the two neutral amino acid carriers (see Fig. 1). In addition, various splice variants of the transporters have been described (Sullivan et al. 2004; Utsunomiya-Tate et al. 1997).

2
Ion/Flux Coupling Determines the Concentrating Capacity of the Transporters

Glutamate transport is coupled to the co-transport of three Na^+ ions and one H^+ followed by the counter-transport of one K^+ ion (Fig. 2). Theoretically,

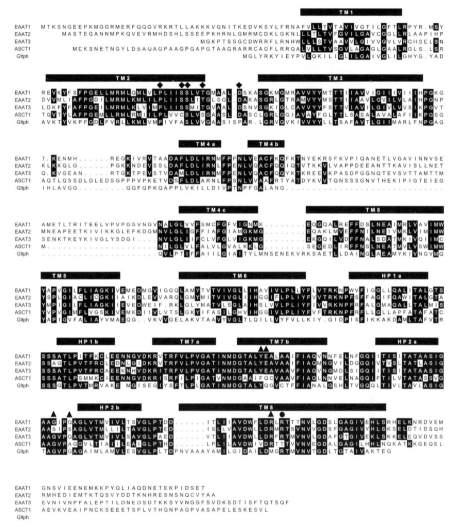

Fig. 1 Amino acid sequence alignment of human glutamate transporters (EAAT1, EAAT2, EAAT3), the human neutral amino acid transporter (ASCT1) and the bacterial glutamate transporter (Gltph). Transmembrane domains 1 through 8 and the hairpin loops 1 and 2 are indicated by *bars* above the sequence. Amino acid residues that are discussed in this chapter are highlighted with *circles* (glutamate binding), *triangles* (cation binding) and *diamonds* (chloride permeation)

this coupling ratio can support a transmembrane glutamate concentration gradient exceeding 10^6 under equilibrium conditions, and allows the transporters to continue removing glutamate over a wide range of ionic conditions (Zerangue and Kavanaugh 1996b). In addition to contributing to the driving force of transport, countertransport of K^+ ions by the EAATs appears to be

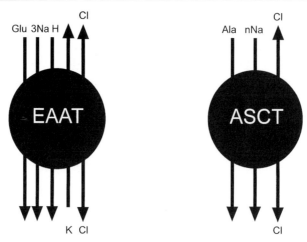

Fig. 2 Schematic representation of the stoichiometry of ion flux coupling of EAATs and ASCTs. The number of Na^+ co-transported with neutral amino acids has not been determined. The ion coupling properties of Gltph have not been determined

necessary to facilitate the re-orientation of the transporter from an inward- to outward-facing conformation, so as to allow the transporter to continue the cycle (Kavanaugh et al. 1997).

Although ASCT1 and ASCT2 are members of the same family of transporters, they differ in their ion-coupling mechanism. Uptake of alanine by ASCT1 is coupled to the movement of Na^+ ions, but in contrast to the EAATs, ASCT1 does not utilize K^+ ions or protons (Zerangue and Kavanaugh 1996a). The lack of K^+ and H^+ coupling may explain why the ASCT1 functions as an exchanger as apposed to a transporter. After the substrate is taken across the membrane via ASCT1 and into the cell with the obligatory Na^+ ion(s), the substrate is released, but instead of binding K^+ and completing the cycle, the ASCT1 re-binds Na^+ and another neutral amino acid and returns the amino acid to extracellular solution.

3
Uncoupled Ion Currents Associated with Glutamate Transporters

In addition to this stoichiometric conductance, sodium-dependent glutamate binding to the EAATs activates an anion conductance which has the following lyotropic selectivity sequence: thiocyanate $(SCN^-)>ClO_4^->NO_3^->I^->Br^->Cl^->F^- >>$gluconate where SCN^- is the most permeant anion (Fairman et al. 1995; Wadiche et al. 1995a). Although this uncoupled anion conductance requires glutamate and Na^+ binding to the transporter, the direction of anion flux is independent of the rate or direction of glutamate transport. EAATs also allow a glutamate-independent chloride leak conductance (Otis and Jahr

1998; Vandenberg et al. 1995) that differs in its anion selectivity sequence from the substrate-activated chloride conductance (Melzer et al. 2003; Ryan et al. 2004). The physiological relevance of these uncoupled anion conductances is not fully understood, but it has been suggested that the chloride conductance may act as a voltage clamp. The two positive charges that move into the cell with each transport cycle will depolarize the cell membrane and reduce the driving force of glutamate into the cell. If chloride ions enter the cell during glutamate transport, the degree of depolarization may be limited or they may even cause hyperpolarization, depending on the chloride ion gradient across the membrane. This may occur for a number of reasons: first, to maintain the membrane potential so as to maintain an optimal rate of transport; second, to stabilize the membrane potential and reduce the excitability of the cell; third, to reduce energy consumption by the cell that otherwise would be required to re-establish the resting membrane potential.

Although the neutral amino acid transporters ASCT1 and ASCT2 have different cation coupling mechanisms, they both allow uncoupled chloride conductances (Grewer and Grabsch 2004; Zerangue and Kavanaugh 1996a). Given the high degree of amino acid sequence identity between the EAATs and the AS-CTs, it is highly likely that that the transport mechanism and chloride channel properties of the two subfamilies of transporters will be related.

4
The Structure of a Bacterial Glutamate Transporter

The mammalian glutamate transporters show 20%–35% sequence identity with a number of Na^+- and H^+-dependent glutamate transporters from bacteria, with the degree of identity most striking in the transmembrane domain structures (Fig. 1). The crystal structure of a bacterial glutamate transporter (Gltph) has recently been determined; and based on the degree of sequence identity with the mammalian counterparts, it is very likely that they will form similar structures (Yernool et al. 2004). The transporter is a homomeric trimer in the shape of a bowl sitting within the membrane (Fig. 3A). It contains an aqueous basin facing the extracellular solution and a pointed wedge structure facing the intracellular solution. Each protomer is wedge shaped and comes together to form a structure with a three-fold axis of symmetry (Fig. 3B). The protomers contain a number of unusual secondary structure features. There are eight transmembrane (TM) domains that vary in length from 19 to 49 residues and two helix-turn-helix motifs or hairpin loops termed HP1 and HP2. The C-terminal half of each protomer is the most highly conserved region between bacterial and mammalian homologues and contains HP1 and HP2, which have been implicated in forming the glutamate recognition site (Fig. 3B, see discussion below). HP1 and HP2 of each protomer do not come into close contact with the HPs of the other protomers, which suggests that

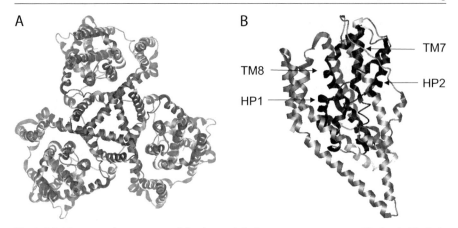

Fig. 3 A,B The crystal structure of the bacterial glutamate transporter Gltph. A Gltph is a homotrimeric complex with a three-fold axis of symmetry. This view of the trimer from the extracellular surface looks down through the transporter. **B** The structure of a single protomer. HP1, HP2, TM7 and TM8 are indicated with *arrows*

each protomer contains a separate glutamate recognition site and is capable of transport.

5
Mutational Studies of Mammalian Glutamate Transporters

5.1
General Considerations and Approaches

The following sections shall give an overview of the site-directed mutagenesis studies that have provided insight into the structural and functional properties of glutamate transporters. Most of these studies were carried out without knowledge of the three-dimensional structure of the transporters, and the following sections shall go through some of these studies and also highlight how they fit in with what we now know about the structure. Mutagenesis can be a very powerful tool to identify crucial amino acid residues responsible for structure and function, but there are a number of potential pitfalls that can generate misleading results. In particular, there are multiple reasons why a mutation may cause a loss of function. These include: loss of critical interactions between the substrate and the protein; disruption of interactions between amino acid residues that prevent crucial conformational changes required for function; disruption of interactions between residues that prevent formation of a stable protein; disruption of interactions that are required for protein–protein interactions that are necessary for formation of the oligomeric complex or the targeting of the protein to the correct location within the cell.

The most reliable interpretations of data from mutant proteins are where some of these possibilities can be ruled out. For example, for a non-functional mutant transporter it would be desirable to show that the protein is expressed at the cell surface and that at least some functional aspect of the transporter is maintained, such as the ability to function as a chloride channel. This would give greater confidence that the mutation causes a subtle structural change that is crucial for full function but does not cause large-scale structural changes in the protein.

In many instances, the mutagenesis results for glutamate transporters fit in well with what we now know about the structure, but there are others which do not. It should be borne in mind that the crystal structure represents one conformation of the protein, yet the transporter must undergo significant conformational changes to perform the function of transport. In some instances, mutagenesis experiments that do not fit with the static structure may yield information about possible conformational changes that are required for transport.

5.2
Structural Studies Using Site-Directed Mutagenesis

In the initial papers describing the amino acid sequences of three different but closely related glutamate transporters, hydropathy plots were used to predict the TM topology. Despite very similar plots, the authors of the three papers came up with three different models (Kanai and Hediger 1992; Pines et al. 1992; Storck et al. 1992). The hydropathy plots were complicated because of the presence of a number of charged residues within the mainly hydrophobic C-terminal domain. The C-terminal half of the protein is particularly well conserved between transporter subtypes, including the bacterial glutamate transporters; and based on this observation, it was speculated that the C-terminal half of the protein contained the glutamate and ion binding sites. Most structure–function studies have focused on this region.

5.2.1
Cysteine-Scanning Mutagenesis

Cysteine-scanning mutagenesis (Akabas et al. 1992, 1994) is a very useful technique for obtaining structural information of integral membrane proteins. Single cysteine residues are introduced at defined sites and their accessibility probed with a variety of sulfhydryl-reactive reagents. Residues in a protein that are accessible from the extracellular side of the membrane can be modified by both permeable and impermeable reagents in whole cells, whereas residues that are accessible from the cytoplasmic side of the membrane can only be modified with membrane-permeable reagents in whole cells. Residues that are buried within the protein structure are not modified at all, and those facing the lipid bilayer are usually not modified because the reagents are most reactive in an aqueous environment or are poorly soluble in the lipid phase.

To gain insights into the structure of glutamate transporters, the groups of Amara and Kanner used cysteine-scanning mutagenesis with a particular emphasis on the C-terminal half of the protein. Working with EAAT1, Seal and Amara (1998) mutated each residue between A395 to A414 to cysteine and, after checking the functional status of each mutant to confirm that the protein generated formed the correct structure, they probed the accessibility of the residues to sulfhydryl reagents. Both A395C and A414C are accessible to the bulky charged MTS reagents MTSET ([2-(trimethylammonium)ethyl]methanethiosulfonate) and MTSES (2-sulfonatoethyl-methanethiosulfonate) and were suggested to lie at the extracellular surface. Many of the residues in between were less reactive, but react with the membrane permeable reagent MTSEA [(2-aminoethyl)methanethiosulfonate], which suggests that the intervening residues are less exposed and partially buried within the membrane region. This region is not sufficiently long to form an α-helix that is capable of traversing the membrane, and from this it was suggested that this region may form a "re-entrant" loop structure (Seal and Amara 1998). Grunewald et al. (1998) conducted similar studies on GLT1 and also found evidence for a re-entrant loop, but the location of this loop is somewhat different. Further extensive cysteine-scanning mutagenesis has been carried out by both groups (Grunewald and Kanner 2000; Grunewald et al. 2002; Seal et al. 2000), and whilst similar models were developed, there are a number of interesting differences. The topology diagrams presented in Fig. 4A and B summarize a large body of evidence. The main difference between the two models is in the location and number of re-entrant loops and the degree with which the flexible loop structures extend into the membrane region. It is very interesting to compare these models with the topology as determined from the crystal structure of the bacterial glutamate transporter (Yernool et al. 2004). Although a number of features of the topology were predicted, the exact locations of the re-entrant loops and TM regions do not all fit with the crystal structure. The two re-entrant loops turn out to be two α-helical hairpin structures connected by flexible loop structures (see Fig. 4C). It is noteworthy that some of the TM domains are considerably longer than was predicted by the cysteine-scanning methods. In particular TM2, TM3 and TM5 are tilted and are up to 50 amino

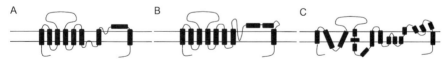

Fig. 4 A–C Membrane topology models of glutamate transporters. **A** The model predicted by Seal et al. (2002) and **B** the model predicted by Grunewald et al. (2002). These models were proposed based on cysteine-scanning mutagenesis studies (see text for details). **C** The membrane topology of Gltph as determined by X-ray crystallography (Yernool et al. 2004)

acids. The reasons for the discrepancies between the models can also be put down to the unusual topology observed in the crystal structure, with a large aqueous accessible basin structure that extends half way through the membrane. Given that there are no precedents for such an unusual structure, it is not surprising that the predictions based on cysteine-scanning were open to multiple interpretations.

5.2.2
Cross-Linking of Cysteine Mutants

Proximity relationships between different domains of transporters have been determined by cross-linking pairs of cysteine residues introduced at various sites. In Fig. 4A there are two re-entrant loops containing residues that are accessible from the extracellular surface, and the extent of modification of these residues with sulfhydryl reagents is reduced in the presence of glutamate. This suggests that residues from both loops may be in close proximity, and Brocke et al. (2002) tested this idea in GLT1 by introducing cysteine pairs within the two loop regions and investigating whether a variety of pairs can come in sufficiently close contact to allow disulfide bonds to form. A412C (TM7 in crystal structure) and V427C (first arm of HP2 in crystal structure) formed a disulfide bond, and A364C (loop region within HP1 in crystal structure) formed a disulfide bond with S440C (loop region with HP2 in crystal structure). Furthermore, the formation of disulfide bonds in both cases could be prevented by either dihydrokainate or glutamate, which suggests that the hairpin loops are in close proximity and are involved in glutamate/dihydrokainate binding to the transporter. In the crystal structure of the bacterial glutamate transporter, the two hairpin loops and TM7 are in close proximity, which is consistent with the cross-linking studies.

Our group has also used a similar strategy to identify regions that may be in close proximity to the chloride channel domain of EAAT1 (Ryan et al. 2004 also see Sect. 6 below). In this study, we attempted to identify residues that were in close proximity to TM2. Cysteine mutants of TM2 residues were paired with cysteine mutants in TM7, the second arm of HP2 and TM8 and the only combinations that led to disulfide bond formation were between Q93C (TM2) and V452C (HP2) and between R90C (TM2) and V452C (HP2). Thus, TM2 must come into close proximity to HP2, which suggests that the chloride channel may be in close proximity to the glutamate translocation domain (see Sect. 6). However, in the crystal structure of the bacterial glutamate transporter, the equivalent residues are not sufficiently close to form a disulfide bond. At first glance these results would appear to be contradictory, but it should be remembered that the crystal structure represents a single conformational state of the transporter, and we know that transporters must undergo significantly conformational changes during the transport process. Therefore, it is conceivable that both results are correct, but reflect different conformational states. The conformational changes that the protein undergoes will be discussed further toward the end of this chapter.

5.3
Substrate Recognition and Translocation

A number of site-directed mutagenesis approaches have been used to identify the glutamate and ion binding sites on glutamate transporters. Three different mutagenesis methods that have generated insights into the molecular basis for transporter function are: the study of chimeras between closely related, but pharmacological and functionally different transporters; site-directed mutagenesis of highly conserved charged or polar residues in the C-terminal domain; and introduction of cysteine residues and then using sulfhydryl-reactive agents to probe the glutamate recognition site. This review will provide examples of how these approaches have been used and relate these results to what we now know of the structure of the bacterial glutamate transporter.

5.3.1
The Use of Chimeras to Define Functional Domains

The first approach to define substrate and ion binding sites involved using chimeric transporters derived from EAAT1 and EAAT2 (Mitrovic et al. 1998; Vandenberg et al. 1995). The glutamate transporters EAAT1 and EAAT2 show some distinct pharmacological and electrophysiological differences. Some of the pharmacological differences include: Kainate, dihydrokainate and 3-methylglutamate are potent blockers of transport by EAAT2, but have no effect on the other EAATs (Arriza et al. 1994; Vandenberg et al. 1997); 4-Methylglutamate is a blocker of EAAT2, but is a substrate for EAAT1 (Vandenberg et al. 1997); and the $K_{0.5}$ for L-serine-O-sulfate transport by EAAT1 is tenfold lower than the $K_{0.5}$ for EAAT2 (Arriza et al. 1994; Mitrovic et al. 1998; Vandenberg et al. 1998b). The glutamate-activated chloride conductance mediated by EAAT1 is also significantly larger in EAAT1 compared to EAAT2 (Wadiche et al. 1995b). These differences were exploited in attempts to define ligand binding domains and the chloride channel of glutamate transporters (Mitrovic et al. 1998; Vandenberg et al. 1995). A number of chimeras were generated, but only a small proportion formed functional transporters. This is not surprising given what we now know about the homo-trimeric structure of the transporters. Whilst it is not possible at this stage to assign pharmacological differences to single amino acid differences between the transporters, the use of chimeras suggested that HP2 and TM8 play a role in conferring differences in substrate selectivity, and HP1, HP2, TM7 and TM8 combine to determine sensitivity to transport blockers such as kainate and 3-methylglutamate (Fig. 5). The observations that 3-methylglutamate and 4-methylglutamate are blockers of EAAT2 whilst 4-methylglutamate is a substrate for EAAT1 are intriguing

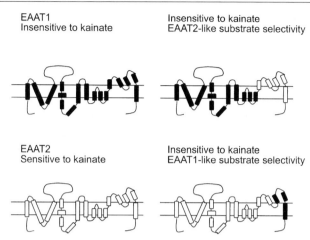

Fig. 5 Definition of functional domains of glutamate transporters through the use of chimeras generated from EAAT1 and EAAT2. The membrane topology model is adapted from the crystal structure of the bacterial glutamate transporter, Gltph. HP2 determines substrate selectivity, whereas the combination of HP1, HP2, TM7 and TM8 determines the sensitivity toward transport blockers, such as kainate

because they suggests that there must be quite subtle differences in the way in the ligands bind to the two transporters. At this stage, there is no clear understanding of the molecular basis for these differences. However, the determination of the crystal structure, with a possible glutamate bound to the transporter, should make it possible to begin to address this question in a more directed manner. The first sites to be targeted would include residues within the HP structures and TM7 and TM8.

5.3.2
Glutamate and Ion Binding Sites

For glutamate and the co-transported ions to pass through the membrane, they must interact with polar or charged residues within the TM domains. A number of groups have targeted for mutagenesis charged or polar residues that are highly conserved between glutamate transporter subtypes. Pines et al. (1995) identified five negatively charged residues in the C-terminal half of GLT1, which they postulated might be involved in the transport process. Glutamate 404 within TM7 has received particular attention. Glutamate binding to the E404D transporter was not affected, but transport was greatly impaired. In contrast, transport of D- or L-aspartate is similar to the rate of transport by wild-type transporters. These observations were initially interpreted as meaning that E404 plays a crucial role in discriminating how bound substrates are transported through the protein (Pines et al. 1995). However, further investigations into the electrophysiological properties of the E404D mutant have identified additional features that provide insight into the molecular events required for transport (Grewer et al. 2003; Kavanaugh et al. 1997). The E404D mutation renders the transporter insensitive to K^+, but still allows Na^+-dependent exchange and the chloride channel to function (Kavanaugh et al. 1997). Thus, in the wild-type transporter, K^+ binds to E404 to allow the transporter to complete the cycle. The mutant does not bind K^+, and so the transporter cannot complete the cycle but does allow exchange. It is interesting to note that the neutral amino acid transporter ASCT1 contains a glutamine residue at the position corresponding to E404 of the GLT1. ASCT1 behaves as an exchanger, as opposed to a transporter, and also does not require K^+ to function (Zerangue and Kavanaugh 1996a). The single amino acid switch at this crucial site may explain the insensitivity to K^+ and also why it behaves as an exchanger as opposed to a transporter.

An additional role for E404 has been proposed from the work of Grewer and colleagues (Grewer et al. 2003; Watzke et al. 2000). Working with EAAC1 (EAAT3) these investigators demonstrated that a H^+ binds to the transporter before glutamate and then the complex is translocated through the protein. The binding of glutamate is strongly pH-dependent, and it was concluded that there is a crucial residue with an ionizable sidechain with a pK_a of approximately 8. Following translocation of glutamate through the membrane, dissociation of glutamate from the transporter is controlled by a shift in the pK_a of the residue by at least 1.5 pK units. Presumably such a residue would lie in close proximity to the glutamate binding site. The residue corresponding to E404 in EAAC1 is E373 and Grewer et al. (2003) investigated in more detail the pH-dependent transport properties of EAAC1-E373Q. The mutant transporter does not catalyse net transport of glutamate, but does allow Na^+-dependent glutamate exchange, which is consistent with the observations of Pines et al. (1995) and Kavanaugh et al. (1997). In contrast to wild-type EAAC1, glutamate binding to the mutant is pH-independent. This suggests that E373 may provide the pH-sensitive site required for H^+ binding and transport. Furthermore, to complete the transport cycle, the H^+ needs to be released before K^+ can bind. The E373Q mutant does not release a proton, and the sidechain remains uncharged after release of glutamate. Thus, the reasons why the EAAC1-E373Q and GLT1-E404N are not capable of net transport are twofold. First, the E sidechain needs to be protonated prior to binding glutamate from the external surface and deprotonated after release of glutamate to the intracellular side. Second, the de-protonated sidechain may then bind K^+ and re-orientate the empty transporter such that the glutamate binding site is facing the extracellular surface.

The roles of amino acids in the immediate vicinity of E404 in GLT1 have also been investigated (Zhang et al. 1998). Y403 appears to contribute to K^+ coupling as well as influence Na^+ selectivity. The Y403F mutant is insensitive to K^+ and functions as an exchanger with many similarities to the E404N mutant. The Y403F also shows higher affinity for Na^+, and in contrast to wild-type GLT1, the alternate cations Li^+ and Cs^+ can substitute for Na^+ in catalysing exchange. This suggests that the Na^+ recognition site is likely to be in close proximity to this part of the transporter. Two residues within the loop structure of HP2 also play a role on ion discrimination. In GLT1, mutations of S440 or S443 generate transporters in which Li^+ can partially substitute for Na^+ in supporting transport. In EAAT1, 3, 4 and 5 these residues are glycine (at position equivalent to 440) and glutamine (at position equivalent to 443). Single mutations of these residues in GLT1 have the greatest level of Li^+-supported transport, whereas the double mutant S440G/S443Q has considerably less Li^+-supported transport. These observations suggest that the combination of residues is required to confer ion selectivity and that this may be achieved with two serines or a glycine with a glutamine (Zhang and Kanner 1999).

Glutamate contains two negatively charged carboxyl groups, and it is reasonable to postulate that positively charged residues may be required to provide specificity in catalysing the transfer of glutamate through the transporter. The roles of two arginine residues within TM8 have been investigated (Bendahan et al. 2000; Borre and Kanner 2004). Mutation of R447 in EAAC1 (EAAT3) to C, E, G or S renders the transporter unable to transport D-aspartate, but the mutants are able to transport L-cysteine and L-serine. L-Cysteine is transported by both the wild-type EAAC1 and the R447C mutant, but, in contrast to the wild-type, acidic amino acids are unable to compete with L-cysteine for binding to the mutant transporter. Thus, R447 appears to form part of the initial binding site for glutamate, and the positively charged sidechain at this position may be required to bind to the negatively charged carboxyl group of glutamate. In addition to the disruption of glutamate recognition by the R447C mutant transporter, the transporter shows impaired K^+ coupling. From these observations it has been proposed that R447 sequentially binds glutamate and K^+ ions and thereby enables the coupling of the two fluxes (Bendahan et al. 2000). R445 in EAAC1 also appears to influence cation coupling of transport. Application of ^{3}H-glutamate to the R445S mutant under voltage clamp at -60 mV generates a charge-to-flux ratio that is approximately 30-fold greater than the ratio observed for wild-type EAAC1. The extra current is due to uncoupled Na^+ and K^+ ion fluxes through the transporter. Thus, the R445S appears to relax the coupling of Na^+/K^+ and glutamate and allows additional cations to flow through the transporter in an uncoupled manner (Borre and Kanner 2004).

Cysteine-scanning mutagenesis has also been used to identify potential glutamate and/or Na^+ binding sites on the transporter. In the series of studies from the Amara and Kanner groups (see Sect. 5.2.1), a number of approaches were applied to probe the accessibility of mutant cysteine residues in the region encompassing TM7 and HP2. Various cysteine mutants within these regions showed differential sensitivity to sulfhydryl reagents in the presence and absence of Na^+ and glutamate. Two interpretations of this type of data are possible. First, these residues may form part of the glutamate binding site, so that when glutamate and Na^+ are bound the sulfhydryl reagent cannot get access to the site. Second, glutamate and Na bound to the transporter may induce conformational changes to another site in the transporter which makes the site less accessible and consequently less reactive. In EAAT1, A395C in TM7 is protected from MTS reagents by the presence of glutamate, and both Y405C and E406C in TM7 show reduced reactivity with MTS reagents in the presence of Na^+ (Seal et al. 1998, 2000). Zarbiv also demonstrated that the reactivity of Y403C in GLT1 (equivalent to Y405 in EAAT1) was sensitive to Na^+ ions and furthermore that glutamate was not able to protect the Y403C from reacting with N-ethylmaleimide (Zarbiv et al. 1998). N-Ethylmaleimide is highly permeant through lipid membranes and was suggested to gain access to Y403C from the internal surface. Thus, Y403 may be alternately accessible from both sides of the membrane, which is consistent with the prediction that it plays a crucial role in the translocation process.

Residues in both arms of HP2 have also been investigated in this manner. In GLT1, a number of cysteine mutants between G422C and S443—in the first arm of HP2—showed reduced reactivity to MTS reagents in the presence of the EAAT2/GLT1 selective transport blocker dihydrokainate (Grunewald et al. 2002). In EAAT1, similar reductions in reactivity were observed in the presence of glutamate for cysteine mutants between L448C and V458C within the second arm of HP2. These results collectively demonstrate that TM7 and HP2 contribute to the formation of the glutamate binding site of the transporters and that many residues within these domains undergo significant conformational changes during the transport process.

5.3.3
Glutamate Binding Site in the Crystal Structure of Gltph

Non-protein electron density, which may be due to glutamate, has been detected in close proximity to TM7, TM8, HP1 and HP2 (Yernool et al. 2004). However, the currently achievable resolution of the crystal structure is not sufficient to allow visualization of glutamate or any of its molecular contacts with the transporter. The putative glutamate molecule in the crystal structure of Gltph lies beneath the surface of the basin and within 5 Å of HP2. HP2 contains residues that correspond to S440 and S443 of GLT1 that have been implicated in Na^+ binding (see Sect. 5.2.2), which suggests that the Na^+ and glutamate binding sites are intimately linked. On the other side of the putative binding site are residues from TM7 which include the residues equivalent to Y403 and E404 of GLT1, which have been implicated in K^+ and H^+ binding. The residue equivalent to R447 of GLT1 in TM8 also sits in close proximity to the putative glutamate molecule, but at present it is not clear whether it is the γ-carboxyl group of glutamate that binds to the R sidechain. Thus, although much of the mutagenesis data are consistent with the crystal structure, there are a number of details concerning the precise contacts between glutamate and the transporter that remain to be established.

5.4
Zn^{2+} Binding Sites on EAAT1 and EAAT4

Zn^{2+} is co-released with glutamate and modulates the activity of a number of proteins expressed in excitatory synapses, including glutamate transporters. Application of Zn^{2+} to oocytes expressing EAAT1 causes a reduction in the rate of glutamate transport, but has no apparent effect on the chloride conductance (Vandenberg et al. 1998a). Zn^{2+} has no effect on the EAAT2 and EAAT3 transporters, but it inhibits the chloride conductance of EAAT4 with little effect on the rate of glutamate transport (Mitrovic et al. 2001). The molecular basis for these differences was put down to the presence or absence of a pair of histidine residues at the extracellular edge of TM3. Mutation of either of these histidine residues in EAAT1 (H146 or H156) to alanine abolished the effects of Zn^{2+} and similarly for the corresponding residues in EAAT4. The first of these histidine residues is conserved in EAAT2, but the residue corresponding to the second histidine is a glycine residue. Wild-type EAAT2 is insensitive to Zn^{2+}, but mutation of the glycine residue to histidine generates a transporter that is sensitive to Zn^{2+} (Mitrovic et al. 2001; Vandenberg et al. 1998a). The binding of Zn^{2+} to EAAT1 is non-competitive with respect to glutamate and Na^+ binding/transport, and therefore the Zn^{2+} binding site is unlikely to overlap with the glutamate recognition site, but is capable of modulating the conformational changes required for transport. In the crystallographic structure, the region corresponding to the Zn^{2+} binding site is located in the loop region immediately after TM3 and near the top of a crevice that Yernool et al. (2004) speculated may form a lipid modulatory site. Thus, the Zn^{2+} binding may serve to limit the flexibility of this loop region that in turn may restrict the movements of HP1, HP2, TM7 and TM8, which may be required for transport.

5.5
The Chloride Channel Within the Transporter

Glutamate and Na^+ binding to glutamate transporters activates a chloride conductance through the transporter (Fairman et al. 1995; Wadiche et al. 1995a). The observation that the rate and direction of chloride flux is uncoupled from the transport process raised the question as to whether the two functions are mediated by the same or different pores through the protein. A number of models have been proposed to explain the dual functions of glutamate transporters.

The first model proposed by Wadiche et al. (1995a) suggested that chloride ions permeate the same pore of the transporter as glutamate and the co-transported ions, and that the binding sites for the various ions and substrates are closely related (Fig. 6A). Larsson et al. (1996) proposed a similar model with a single pore, but suggested that the chloride ion binding sites are separate

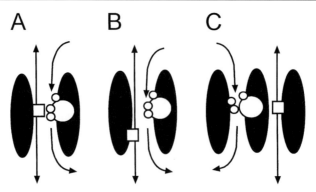

Fig. 6 A–C Three schematic models for the dual functions of glutamate transporters. **A** A single pore model where glutamate/Na^+/H^+/K^+ binding to the transporter forms the Cl^- ion binding site. **B** A single pore model with separate molecular determinants for glutamate/Na^+/H^+/K^+ and Cl^-. **C** A dual pore model with separate pores for glutamate/Na^+/H^+/K^+ transport and the Cl^- channel. Key: *large circle* is glutamate, *small circles* are Na^+ ions, *square* is Cl^-; K^+ and H^+ ions have been omitted

from the glutamate and co-transported ion binding sites (Fig. 6B). Sonders and Amara (1996) suggested that there may be a separate pore through the transporter for chloride ions and that this extra pore may be an intrinsic property of the transporter or may arise through the association of additional proteins (Fig. 6C).

5.5.1
Different Conformational States Regulate the Two Functions

There is evidence to suggest that different conformational states of the transporter mediate the transport and chloride channel functions, but whether this represents different conformational states of the same pore or different pores altogether has not been established. Whilst investigating the properties of the extracellular hydrophobic linker region (now identified as the second half of HP2 in the crystal structure), the groups of Amara (Seal et al. 2001), Kanner (Borre et al. 2002) and this author's group (Ryan and Vandenberg 2002) made the observations that modification of cysteine mutants of residues within this domain abolished glutamate transport, but retained glutamate-dependent activation of the chloride channel. These results suggest that glutamate binding and chloride channel activation are not affected, but translocation of glutamate through the membrane is impaired. Thus, it may be concluded that different conformational states of the protein are required for transport compared to channel activation. In support of this conclusion was an earlier observation by Wadiche and Kavanaugh (1998) that at 4°C, activation of the chloride channel could occur but transport does not proceed. This suggests that there are temperature-dependent conformational changes in the transporter required for transport that are distinct from the conformational changes that are sufficient for opening of the chloride channel.

5.5.2
Residues that Line the Chloride Channel

In order to distinguish between the single and dual pore models, it is necessary to identify the molecular determinants for transport and also the chloride channel. From the previous discussion, the glutamate and $Na^+/K^+/H^+$ binding sites are located in the two hairpin loops and TM7 and TM8, but it is interesting to note that many of the mutations that disrupt the transport process do not affect chloride ion permeation through the transporter, which suggests that there may be separate molecular determinants.

In solution, ions will be hydrated or surrounded by water molecules, and for an ion to permeate the lipid membrane of a cell it needs to pass through an aqueous pore. The selectivity of ion channels is determined by the selectivity filter, a narrow part of the pore where the ion interacts with the lining of the channel. The hydrated ion is often larger than the selectivity filter, and to allow the ion to pass through the filter some or all of the water molecules must be removed from the ion and be replaced by surrogate water molecules. The sidechains of serine, threonine and tyrosine can perform this role and form binding sites for the permeating ions.

TM2 is highly conserved between transporter subtypes and contains a number of positively charged residues at the intracellular and extracellular edges and also serine and threonine residues in the TM segment. Mutations of these residues affect various properties of the chloride channel, but do not affect the transport of glutamate. First, mutations that change the polarity of residues, but maintain the size of the residue (such as S102A, S103A, S103V or T106A) change the relative anion permeability. The S103V mutant has a particularly striking effect on anion permeation. In the wild-type EAAT1, nitrate is more permeant through the channel than iodide followed by bromide and then chloride ions. In the S103V mutant, iodide is the most permeant anion followed by bromide, nitrate and chloride ions. This change in order of relative permeability is indicative of a change in the way anions interact with the lining of the channel, and from this observation we made the suggestion that S103 forms part of the selectivity filter of the channel (Ryan et al. 2004).

Many of the residues within TM2 are accessible to the aqueous environment, as judged by the reactivity of cysteine mutants of these residues to sulfhydryl-reactive reagents. The nature of the reactivity has been used to gain further insight into the structure and also confirm the role of this region in formation of the anion channel. Cysteine mutants of residues 90 and 93 show a greater reactivity with the negatively charged MTSES reagent compared to the positively charged MTSET, which is consistent with this region forming a channel structure that selects anions over cations. The minimum pore diameter of the chloride channel has been estimated to be 5 Å (Wadiche et al. 1995a), whereas the size of the MTSET and MTSES reagents is 5.8 Å. Therefore the MTS reagents are unlikely to penetrate deep into the pore, which may explain why mutants beyond V96C are unreactive to MTS reagents. However, it was possible to probe further into the channel by using Hg^{2+}, which is significantly smaller than the MTS reagents. L99C, S102C and S103C were all modified with Hg^{2+}, which suggests that these residues also form part of an aqueous pore. Thus, a number of residues between R90 and T106 are likely to form an aqueous pore that influences anion permeation.

5.5.3
Opening the Chloride Channel Gate

Aspartate 112 is in the intracellular loop between TM2 and TM3 (see Fig. 3) and may also play an important role in chloride channel function. The D112A mutant has a greatly reduced glutamate-activated chloride current, but a greatly enhanced leak current. This observation suggests that the leak current is mediated by the same pore as the glutamate-activated chloride current and also that D112 may form part of a structure that causes opening and closing of the channel. It has been postulated that a glutamate residue in the ClC channel sits within the pore of the channel when in the closed state and prevents chloride permeation. Upon activation of the channel, the glutamate residue swings out of the pore to allow chloride ions to pass (Dutzler et al. 2003). D112 may play a similar role in EAAT1. In the absence of glutamate and Na^+ ions, D112 may sit within the pore, but upon binding of glutamate and Na^+, the D112 may move to allow chloride ions to pass. In the D112A mutant, the channel may be locked in the open state, so that when glutamate binds there is no additional chloride conductance. If the leak and glutamate-activated chloride conductances are mediated by the same pore structure, it is noteworthy that the relative anion permeabilities of the two states are different. For the anion leak current, I^- is more permeant than NO^{3-}, whereas for the glutamate-activated anion current, NO^{3-} is more permeant than I^- (Melzer et al. 2003; Ryan et al. 2004). These differences suggest that glutamate binding alters the conformational state of the protein and changes the way anions permeate the channel.

6
A Structural Model for the Transport and Chloride Channel Functions of Glutamate Transporters

The determination of the crystal structure of a bacterial glutamate transporter has been a major step in beginning to understand how glutamate transporters work, but it should be remembered that the structure represents only one conformation and that many structural changes are required for the protein to carry out the function of transport. The last section of this chapter shall use the current structural information and results from a range of site-directed mutagenesis approaches to speculate on the molecular basis for transport and chloride channel functions of glutamate transporters.

In 1966, Jardetzky (1966) proposed an alternating access mechanism to explain how transporters can effectively couple ion gradients to the movement of substrates across the membrane. The mechanism involves a structure with two gates that flank the substrate binding site. Opening of the external gate will allow the substrate to bind within the transporter; and after the external gate closes, the internal gate opens and allows the substrate to pass through to the internal solution. Yernool et al. (2004) have suggested that HP2 forms the external gate and HP1 forms the internal gate of glutamate transporters, with sections of TM7 and TM8 forming the glutamate and ion recognition sites (Fig. 7). They suggest that, in the crystal structure, the transporter has both gates in the closed state. The HP2 gate is sitting on top of the putative glutamate molecule and thereby locking it into place. They also suggest that the HP1 gate in the crystal structure is also closed. A number of the studies of mutant transporters described above can be used to speculate on the nature of the conformational transitions required for transport and chloride channel function. As a starting point, HP2 is likely to undergo considerable conformational changes upon glutamate and Na^+ binding (Grunewald et al. 2002; Leighton et al. 2002). Furthermore, in the crystal structure, the residue equivalent to Y403 of GLT1 (in TM7) is buried within the protein; yet we know from studies of the Y403C mutant that it changes its aqueous accessibility in the presence of Na^+ and also that it is accessible from both sides of the membrane (Grunewald et al. 2002; Seal et al. 2000). Therefore, it is reasonable to suggest that HP2 may flip between open and closed states (and change its accessibility in the process) to transiently expose the glutamate and Na^+ binding sites formed by TM7 and TM8. The movements of HP2 and the binding of glutamate and Na^+ ions to TM7 and TM8 is likely to impact on the way in which these regions interact with other parts of the protein. In the crystal structure, HP2 lies approximately 20 Å from TM2; yet—in a study of cross-linking cysteine mutants—these two regions can come sufficiently close to form a disulfide bond. Thus, HP2 must move substantially inward toward TM2 and in so doing is likely to distort the position of HP1. These types of protein movements may provide the impulse to open the internal gate and allow glutamate release into the internal environment. Once glutamate is released, the unoccupied and negatively charged E404 (GLT1 numbering) may bind K^+. Binding of K^+ must then cause the closing of the internal gate followed by opening of the external gate and release of the K^+ ion to the extracellular surface.

Another consequence of the movements of HP1 and HP2 is that they would generate an additional aqueous pore along the side of TM2, which may be the chloride channel. Although there are clearly separate molecular determinants for transport and channel functions, the two processes do seem to be physically

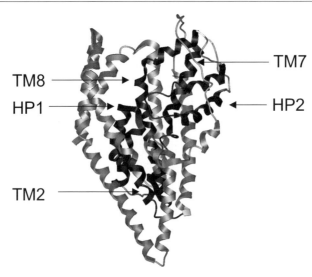

Fig. 7 Proposed transport and channel domains of glutamate transporters. HP1 and HP2 may form the internal and external gates of the transporter with TM7 and TM8 forming the glutamate/Na^+/H^+/K^+ binding sites. TM2 may form part of the Cl^- channel. See text for details on the suggested movements of the various regions responsible for transport and channel activation

linked. For example, S103 is likely to form part of a chloride binding site and is approximately 6 Å from the second arm of HP1. Furthermore, TM7 is likely to undergo significant conformational changes, and the intracellular edge of TM7 is in close proximity to D112 of EAAT1 that may form part of the chloride channel gate. Movements of TM7 relative to TM2 and HP1 may change the conformation of D112 to allow the separate gating process for the chloride channel.

The ideas on the conformational changes required for the transport and chloride channel functions are speculative, and there are many details that require clarification or investigation. In many respects, this is an extremely exciting time for transport researchers because with knowledge of the crystal structure we can now begin to put some of these ideas to the test. The use of site-directed mutagenesis approaches together with the backup of a structural framework will no doubt continue to provide very instructive tools to better understand how these fascinating proteins work.

References

Akabas MH, Stauffer DA, Xu M, Karlin A (1992) Acetylcholine receptor channel structure probed in cysteine-substitution mutants. Science 258:307–310

Akabas MH, Kaufmann C, Archdeacon P, Karlin A (1994) Identification of acetylcholine receptor channel-lining residues in the entire M2 segment of the alpha subunit. Neuron 13:919–927

Arriza JL, Kavanaugh MP, Fairman WA, Wu YN, Murdoch GH, North RA, Amara SG (1993) Cloning and expression of a human neutral amino acid transporter with structural similarity to the glutamate transporter gene family. J Biol Chem 268:15329–15332

Arriza JL, Fairman WA, Wadiche JI, Murdoch GH, Kavanaugh MP, Amara SG (1994) Functional comparisons of three glutamate transporter subtypes cloned from human motor cortex. J Neurosci 14:5559–5569

Arriza JL, Eliasof S, Kavanaugh MP, Amara SG (1997) Excitatory amino acid transporter 5, a retinal glutamate transporter coupled to a chloride conductance. Proc Natl Acad Sci U S A 94:4155–4160

Bendahan A, Armon A, Madani N, Kavanaugh MP, Kanner BI (2000) Arginine 447 plays a pivotal role in substrate interactions in a neuronal glutamate transporter. J Biol Chem 275:37436–37442

Borre L, Kanner BI (2001) Coupled, but not uncoupled, fluxes in a neuronal glutamate transporter can be activated by lithium ions. J Biol Chem 276:40396–40401

Borre L, Kanner BI (2004) Arginine 445 controls the coupling between glutamate and cations in the neuronal transporter EAAC-1. J Biol Chem 279:2513–2519

Borre L, Kavanaugh MP, Kanner BI (2002) Dynamic equilibrium between coupled and uncoupled modes of a neuronal glutamate transporter. J Biol Chem 277:13501–13507

Brocke L, Bendahan A, Grunewald M, Kanner BI (2002) Proximity of two oppositely oriented reentrant loops in the glutamate transporter GLT-1 identified by paired cysteine mutagenesis. J Biol Chem 277:3985–3992

Dutzler R, Campbell EB, MacKinnon R (2003) Gating the selectivity filter in ClC chloride channels. Science 300:108–112

Fairman WA, Vandenberg RJ, Arriza JL, Kavanaugh MP, Amara SG (1995) An excitatory amino-acid transporter with properties of a ligand-gated chloride channel. Nature 375:599–603

Grewer C, Grabsch E (2004) New inhibitors for the neutral amino acid transporter ASCT2 reveal its Na^+-dependent anion leak. J Physiol 557:747–759

Grewer C, Watzke N, Rauen T, Bicho A (2003) Is the glutamate residue Glu-373 the proton acceptor of the excitatory amino acid carrier 1? J Biol Chem 278:2585–2592

Grunewald M, Kanner BI (2000) The accessibility of a novel reentrant loop of the glutamate transporter GLT-1 is restricted by its substrate. J Biol Chem 275:9684–9689

Grunewald M, Bendahan A, Kanner BI (1998) Biotinylation of single cysteine mutants of the glutamate transporter GLT-1 from rat brain reveals its unusual topology. Neuron 21:623–632

Grunewald M, Menaker D, Kanner BI (2002) Cysteine-scanning mutagenesis reveals a conformationally sensitive reentrant pore-loop in the glutamate transporter GLT-1. J Biol Chem 277:26074–26080

Jardetzky O (1966) Simple allosteric model for membrane pumps. Nature 211:969–970

Kanai Y, Hediger MA (1992) Primary structure and functional characterization of a high-affinity glutamate transporter [see comments]. Nature 360:467–471

Kavanaugh MP, Bendahan A, Zerangue N, Zhang Y, Kanner BI (1997) Mutation of an amino acid residue influencing potassium coupling in the glutamate transporter GLT-1 induces obligate exchange. J Biol Chem 272:1703–1708

Larsson HP, Picaud SA, Werblin FS, Lecar H (1996) Noise analysis of the glutamate-activated current in photoreceptors. Biophys J 70:733–742

Leighton BH, Seal RP, Shimamoto K, Amara SG (2002) A hydrophobic domain in glutamate transporters forms an extracellular helix associated with the permeation pathway for substrates. J Biol Chem 277:29847–29855

Melzer N, Biela A, Fahlke C (2003) Glutamate modifies ion conduction and voltage-dependent gating of excitatory amino acid transporter-associated anion channels. J Biol Chem 278:50112–50119

Mitrovic AD, Amara SG, Johnston GA, Vandenberg RJ (1998) Identification of functional domains of the human glutamate transporters EAAT1 and EAAT2. J Biol Chem 273:14698–14706

Mitrovic AD, Plesko F, Vandenberg RJ (2001) Zn(2+) inhibits the anion conductance of the glutamate transporter EAAT4. J Biol Chem 276:26071–26076

Otis TS, Jahr CE (1998) Anion currents and predicted glutamate flux through a neuronal glutamate transporter. J Neurosci 18:7099–7110

Pines G, Danbolt NC, Bjoras M, Zhang Y, Bendahan A, Eide L, Koepsell H, Storm-Mathisen J, Seeberg E, Kanner BI (1992) Cloning and expression of a rat brain L-glutamate transporter [published erratum appears in Nature 1992 Dec 24–31;360(6406):768] [see comments]. Nature 360:464–467

Pines G, Zhang Y, Kanner BI (1995) Glutamate 404 is involved in the substrate discrimination of GLT-1, a ($Na^+ + K^+$)-coupled glutamate transporter from rat brain. J Biol Chem 270:17093–17097

Ryan RM, Vandenberg RJ (2002) Distinct conformational states mediate the transport and anion channel properties of the glutamate transporter EAAT-1. J Biol Chem 277:13494–13500

Ryan RM, Mitrovic AD, Vandenberg RJ (2004) The chloride permeation pathway of a glutamate transporter and its proximity to the glutamate translocation pathway. J Biol Chem 279:20742–20751

Seal RP, Amara SG (1998) A reentrant loop domain in the glutamate carrier EAAT1 participates in substrate binding and translocation. Neuron 21:1487–1498

Seal RP, Daniels GM, Wolfgang WJ, Forte MA, Amara SG (1998) Identification and characterization of a cDNA encoding a neuronal glutamate transporter from Drosophila melanogaster. Receptors Channels 6:51–64

Seal RP, Leighton BH, Amara SG (2000) A model for the topology of excitatory amino acid transporters determined by the extracellular accessibility of substituted cysteines. Neuron 25:695–706

Seal RP, Shigeri Y, Eliasof S, Leighton BH, Amara SG (2001) Sulfhydryl modification of V449C in the glutamate transporter EAAT1 abolishes substrate transport but not the substrate-gated anion conductance. Proc Natl Acad Sci U S A 98:15324–15329

Shafqat S, Tamarappoo BK, Kilberg MS, Puranam RS, McNamara JO, Guadano-Ferraz A, Fremeau RT Jr (1993) Cloning and expression of a novel Na(+)-dependent neutral amino acid transporter structurally related to mammalian Na^+/glutamate cotransporters. J Biol Chem 268:15351–15355

Sonders MS, Amara SG (1996) Channels in transporters. Curr Opin Neurobiol 6:294–302

Storck T, Schulte S, Hofmann K, Stoffel W (1992) Structure, expression, and functional analysis of a Na(+)-dependent glutamate/aspartate transporter from rat brain. Proc Natl Acad Sci U S A 89:10955–10959

Sullivan R, Rauen T, Fischer F, Wiessner M, Grewer C, Bicho A, Pow DV (2004) Cloning, transport properties, and differential localization of two splice variants of GLT-1 in the rat CNS: implications for CNS glutamate homeostasis. Glia 45:155–169

Utsunomiya-Tate N, Endou H, Kanai Y (1996) Cloning and functional characterization of a system ASC-like Na^+-dependent neutral amino acid transporter. J Biol Chem 271:14883–14890

Utsunomiya-Tate N, Endou H, Kanai Y (1997) Tissue specific variants of glutamate transporter GLT-1. FEBS Lett 416:312–316

Vandenberg RJ, Arriza JL, Amara SG, Kavanaugh MP (1995) Constitutive ion fluxes and substrate binding domains of human glutamate transporters. J Biol Chem 270:17668–17671

Vandenberg RJ, Mitrovic AD, Chebib M, Balcar VJ, Johnston GA (1997) Contrasting modes of action of methylglutamate derivatives on the excitatory amino acid transporters, EAAT1 and EAAT2. Mol Pharmacol 51:809–815

Vandenberg RJ, Mitrovic AD, Johnston GA (1998a) Molecular basis for differential inhibition of glutamate transporter subtypes by zinc ions. Mol Pharmacol 54:189–196

Vandenberg RJ, Mitrovic AD, Johnston GA (1998b) Serine-O-sulphate transport by the human glutamate transporter, EAAT2. Br J Pharmacol 123:1593–1600

Wadiche JI, Kavanaugh MP (1998) Macroscopic and microscopic properties of a cloned glutamate transporter/chloride channel. J Neurosci 18:7650–7661

Wadiche JI, Amara SG, Kavanaugh MP (1995a) Ion fluxes associated with excitatory amino acid transport. Neuron 15:721–728

Wadiche JI, Arriza JL, Amara SG, Kavanaugh MP (1995b) Kinetics of a human glutamate transporter. Neuron 14:1019–1027

Watzke N, Rauen T, Bamberg E, Grewer C (2000) On the mechanism of proton transport by the neuronal excitatory amino acid carrier 1. J Gen Physiol 116:609–622

Yernool D, Boudker O, Jin Y, Gouaux E (2004) Structure of a glutamate transporter homologue from Pyrococcus horikoshii. Nature 431:811–818

Zarbiv R, Grunewald M, Kavanaugh MP, Kanner BI (1998) Cysteine scanning of the surroundings of an alkali-ion binding site of the glutamate transporter GLT-1 reveals a conformationally sensitive residue. J Biol Chem 273:14231–14237

Zerangue N, Kavanaugh MP (1996a) ASCT-1 is a neutral amino acid exchanger with chloride channel activity. J Biol Chem 271:27991–27994

Zerangue N, Kavanaugh MP (1996b) Flux coupling in a neuronal glutamate transporter. Nature 383:634–637

Zhang Y, Kanner BI (1999) Two serine residues of the glutamate transporter GLT-1 are crucial for coupling the fluxes of sodium and the neurotransmitter. Proc Natl Acad Sci U S A 96:1710–1715

Zhang Y, Bendahan A, Zarbiv R, Kavanaugh MP, Kanner BI (1998) Molecular determinant of ion selectivity of a ($Na^+ + K^+$)-coupled rat brain glutamate transporter. Proc Natl Acad Sci U S A 95:751–755

The Diverse Roles of Vesicular Glutamate Transporter 3

R. P. Seal · R. H. Edwards (✉)

Departments of Neurology and Physiology, UCSF School of Medicine, 600 16th St., GH-N272B, San Francisco CA, 94143-2140, USA
edwards@itsa.ucsf.edu

1	Introduction	137
2	Vesicular Glutamate Transport	138
2.1	Identification of the Vesicular Glutamate Transporters	139
2.2	Catecholamine Neurons Express VGLUT2	140
2.3	Glutamate Release by Dopamine Neurons In Vivo	141
3	Identification of VGLUT3 in Neurons and Glia	142
3.1	VGLUT3 in Monoamine Neurons	143
3.2	Cholinergic Interneurons of the Striatum	144
3.3	Inhibitory Interneurons of the Hippocampus, Cortex, and Retina	144
3.4	Dendritic Release of Glutamate	145
3.5	Expression of VGLUT3 Transiently During Development	146
4	Conclusion	146
References		147

Abstract The expression of vesicular glutamate transporters (VGLUTs) 1 and 2 accounts for the ability of most traditionally accepted excitatory neurons to release glutamate by exocytosis. However, several cell populations (serotonin and dopamine neurons) have been demonstrated to release glutamate in vitro and do not obviously express these transporters. Rather, these neurons express a novel, third isoform that in fact appears confined to neurons generally associated with a transmitter other than glutamate. They include serotonin and possibly dopamine neurons, cholinergic interneurons in the striatum, and GABAergic interneurons of the hippocampus and cortex. Although the physiological role of VGLUT3 remains largely conjectural, several observations in vivo suggest that the glutamate release mediated by VGLUT3 has an important role in synaptic transmission, plasticity, and development.

Keywords Glutamate release · Vesicular glutamate transport · VGLUT3 · Neurotransmitter co-release · Dendritic glutamate release

1
Introduction

A number of observations have suggested the release of glutamate by neurons not traditionally considered to use glutamate as a neurotransmitter. When grown individually on islands of glia, serotonin neurons form glutamatergic synapses onto themselves known as autapses (Johnson 1994). Since the cells are grown in isolation, the glutamatergic synapses must be formed by serotonin neurons. Dopamine neurons also form glutamatergic autapses in culture (Sulzer et al. 1998). Interestingly, the processes staining for tyrosine hydroxylase have appeared distinct from those staining for glutaminase, the enzyme generally associated with glutamate production (Curthoys and Watford 1995), suggesting segregation of release sites for the two transmitters into distinct processes. In addition, many reports have described the release of glutamate from astrocytes in vitro (Parpura et al. 1994; Araque et al. 1998; Bezzi et al. 1998; Araque et al. 2000). Outside the nervous system, glutamate has been suggested to act as a signal produced by mitochondria to stimulate the release of insulin from pancreatic β-cells (Maechler et al. 1997; Maechler and Wollheim 1999). Glutamate would presumably act as an intracellular signal in this case, but the effect apparently requires uptake of the glutamate into vesicles. Although other work disputes the role of glutamate in insulin release (MacDonald and Fahien 2000), glutamate has been demonstrated to act as an independent extracellular signal from multiple endocrine cells including pancreatic islet cells and pinealocytes (Moriyama and Yamamoto 1995; Hayashi et al. 2001; Hayashi et al. 2003; Morimoto et al. 2003). However, it has until recently remained unclear whether the observations of glutamate release in vitro reflect a normal function or artifact of culture, with dedifferentiation resulting in a glutamatergic phenotype.

2
Vesicular Glutamate Transport

The ability to release glutamate as an extracellular signal depends on its storage inside neurosecretory vesicles that undergo regulated exocytosis. All cells contain glutamate for its role in protein synthesis and intermediary metabolism, so transport into secretory vesicles defines the cells capable of glutamate release. Vesicular glutamate transport activity thus identifies the cells that release glutamate as an extracellular signal.

Like the transport of other classical transmitters into neurosecretory vesicles, vesicular glutamate transport depends on a H^+ electrochemical gradient produced by the vacuolar H^+ pump (Forgac 2000). Transport involves the exchange of lumenal H^+ for cytoplasmic transmitter; however, the vesicular transport of different transmitters differs in its dependence on the chemical (ΔpH) and electrical ($\Delta \Psi$) components of this gradient (Schuldiner et al. 1995; Liu and Edwards 1997a). Vesicular monoamine and acetylcholine transport relies primarily on ΔpH, while vesicular γ-aminobutyric acid (GABA) transport relies more evenly on ΔpH and $\Delta \Psi$. In contrast, vesicular glutamate transport depends almost entirely on $\Delta \Psi$. Vesicular glutamate transport also has several other distinctive features: specificity for glutamate rather than aspartate;

and a biphasic dependence on chloride with 2–10 mM chloride optimal for transport, apparently due to an allosteric action of chloride at the transporter (Hartinger and Jahn 1993; Wolosker et al. 1996). These differences in function presumably account for the lack of sequence similarity among the three major families of vesicular neurotransmitter transporters.

2.1
Identification of the Vesicular Glutamate Transporters

The vesicular glutamate transporters (VGLUTs) belong to a larger family of polytopic membrane proteins that were previously implicated in the transport of inorganic phosphate (Werner et al. 1991). More recently, it has become apparent that these proteins normally function to transport organic anions, with one member of the family mediating the transport of sialic acid out of lysosomes (Morin et al. 2004; Wreden et al. 2005). Mutations in the protein sialin indeed cause a lysosomal sialic acid storage disorder (Mancini et al. 1991; Verheijen et al. 1999). Other family members have been demonstrated to confer a large chloride conductance modulated by substrates (Busch et al. 1996; Bröer et al. 1998), and VGLUT1 has also been shown to confer a substantial chloride conductance in the membranes from transfected cells (Bellocchio et al. 2000). The identification of VGLUT1 and 2 has also provided the first marker for glutamatergic neurons (Bellocchio et al. 2000; Takamori et al. 2000; Bai et al. 2001; Fremeau et al. 2001; Herzog et al. 2001; Takamori et al. 2001; Varoqui et al. 2002). Indeed, the expression of VGLUT1 and 2 in inhibitory neurons confers the potential for quantal glutamate release (Takamori et al. 2000; Takamori et al. 2001).

VGLUT1 and 2 show a striking, complementary pattern of expression in the adult vertebrate brain. Glutamate neurons in the cerebral cortex, hippocampus, and cerebellar cortex generally express VGLUT1, whereas neurons in the thalamus, deep cerebellar nuclei, and brainstem express VGLUT2 (Fremeau et al. 2004b). VGLUT1 and 2 exhibit very similar functional characteristics, including specificity for glutamate, not aspartate, and a biphasic dependence on chloride. It has therefore remained unclear how the two isoforms might differ. During early postnatal development of the cortex, hippocampus, and cerebellar granule cells, VGLUT2 actually precedes the expression of VGLUT1, then disappears (Miyazaki et al. 2003; Fremeau et al. 2004a). Consistent with this, VGLUT1 knockout mice show substantial residual synaptic transmission in the hippocampus and cerebellar parallel fiber synapses of 1-week-old animals that disappears as the animals age (Fremeau et al. 2004a). The knockouts also become distinguishable in behavior from wildtype littermates at the same time (2–3 weeks after birth) that VGLUT2 levels decline in regions normally destined to express only VGLUT1 in adulthood. After this time, the knockouts require considerable care, but can survive for several months. The analysis of VGLUT1 knockout mice in hippocampus thus enables us to compare the properties of VGLUT2 with those of VGLUT1 (the predominant isoform in wildtype animals). Importantly, the analysis in vivo indicates that VGLUT1 and 2 release sites behave distinctly, with no reduction in quantal size or evoked response in the knockout (Fremeau et al. 2004a). Rather, the VGLUT1 synapses are completely silent in the knockout, and the remaining VGLUT2 synapses appear entirely normal. However, the excitatory transmission depresses more rapidly after high-frequency stimulation in VGLUT1 knockouts than in wildtype, suggesting that the two isoforms differ in their recycling at the nerve terminal.

2.2
Catecholamine Neurons Express VGLUT2

Although VGLUT1 and 2 account for the release of glutamate by essentially all known glutamatergic neurons, they were not obviously expressed by the cell populations (e.g., serotonin and dopamine neurons) suggested to release glutamate in vitro. However, it has become clear that at least some catecholamine neurons in the caudal brainstem express VGLUT2. Double labeling for VGLUT2 messenger RNA (mRNA) and tyrosine hydroxylase protein (the rate-limiting step in catecholamine biosynthesis) has shown that most of the catecholamine neurons in the C1–3, A2 cell groups, and the area postrema express VGLUT2 (Stornetta et al. 2002a). Although the potential role of glutamate released by many of these cells has remained unexplored, bulbospinal C1 neurons of the rostral ventral lateral medulla (RVLM), which control blood pressure by regulating sympathetic tone, had been suggested to signal through glutamate (Morrison et al. 1991; Huangfu et al. 1994; Deuchars et al. 1995). Consistent with this role, hypotension induces c-fos expression within the VGLUT2$^+$ RVLM cells (Stornetta et al. 2002b). Further, bulbospinal RVLM cells differ from other catecholamine populations by failing to express an identified plasma membrane catecholamine transporter (Lorang et al. 1994), suggesting that glutamate may in fact be their primary transmitter. A number of catecholamine populations thus express VGLUT2, but it has not so far been possible to detect VGLUT2 in vivo in noradrenergic neurons of the locus coeruleus or in serotonergic neurons (Stornetta et al. 2002a). The data with regard to dopamine neurons have been conflicting. In vitro, midbrain dopamine neurons express VGLUT2 and this may be independent of the time grown in culture (Dal Bo et al. 2004), suggesting expression in vivo. Recent work has also shown expression of VGLUT2 mRNA by midbrain neurons projecting to the cortex (Hur and Zaborszky 2005), but whether these are dopamine neurons has remained unclear.

Supporting a more generalized role for signaling by glutamate, spinal motor neurons have recently been shown to form glutamatergic as well as cholinergic synapses (Nishimaru et al. 2005). The peripheral neuromuscular junction is purely cholinergic, as indicated from classical studies. However, paired recordings of motor neurons and inhibitory Renshaw cells or other motoneurons show postsynaptic currents mediated by activation of glutamate as well as acetylcholine receptors (Nishimaru et al. 2005). Motor neurons indeed express VGLUT2, which localizes specifically to the terminals formed onto Renshaw cells and motoneurons, not at the neuromuscular junction. Further, VGLUT2 does not colocalize at these synapses with vesicular acetylcholine transporter (VAChT), indicating segregation of the release sites for glutamate and acetylcholine.

2.3
Glutamate Release by Dopamine Neurons In Vivo

Despite the difficulty identifying a VGLUT isoform expressed by dopamine neurons, recent work has begun to indicate a physiological role for glutamate release by dopamine neurons in vivo. Taking advantage of a slice preparation that includes both the midbrain and the ventral striatum, and the labeling of live dopamine neurons with green fluorescent protein (GFP), it has been possible to stimulate ventral tegmental area (VTA) neurons and observe fast, ionotropic, glutamate-mediated excitatory responses in nucleus accumbens neurons (Chuhma et al. 2004). The parallel expression of VGLUT2 by caudal brainstem catecholamine neurons in the RVLM and the expression of VGLUT2 by VTA dopamine neurons in vitro suggests that VGLUT2 may be responsible, but this remains to be demonstrated.

If dopamine neurons release glutamate in vivo, what roles do the two transmitters play? A recent study of neurons in the prefrontal cortex (PFC) indicates that the stimulation of dopaminergic afferents from the VTA first produces a series of rapid excitatory and inhibitory responses at 10–40 ms that is blocked by glutamate but not dopamine receptor antagonists, suggesting direct activation of both excitatory and inhibitory neurons in the PFC (Lavin et al. 2005). Although not inhibited by dopamine receptor antagonists, lesion of the dopamine system with 6OH-dopamine eliminates the response, indicating that it is carried by dopamine projections. The acute response to midbrain stimulation is then followed by an inhibition of spontaneous and potentiation of evoked PFC firing rates that last minutes—much longer than the increases in dopamine measured by voltammetry—but are still blocked by dopamine receptor antagonists (Lavin et al. 2005). Thus, glutamate may convey a rapid signal about the timing of phasic dopamine neuron activation; whereas dopamine conveys longer-lasting modulatory effects that persist long after the actual increase in dopamine. A full understanding of these two roles will require careful behavioral analysis and identification of the VGLUT isoform expressed by dopamine neurons. Despite the circumstantial evidence supporting a role for VGLUT2, additional evidence of VGLUT expression by midbrain dopamine neurons in vivo implicates another isoform: By in situ hybridization, neurons in the substantial nigra pars compacta and VTA label weakly but distinctly for the third isoform, VGLUT3 (Fremeau et al. 2002).

3
Identification of VGLUT3 in Neurons and Glia

The inability to detect VGLUT1 or 2 expression by multiple neuronal populations suggested to release glutamate in vitro led to the search for additional isoforms. In contrast to VGLUT1 and 2, VGLUT3 is expressed at extremely low levels in the brain and was therefore detected both as a rare cross-hybridizing sequence and as a closely related sequence in the available genomic databases (Fremeau et al. 2002; Gras et al. 2002; Schafer et al. 2002; Takamori et al. 2002). Indeed, the absence of VGLUT3 from complementary DNA (cDNA) expressed sequence tag (EST) databases attests to its low abundance. VGLUT3 is also expressed outside the nervous system in liver and possibly kidney, unlike the other isoforms. In terms of transport activity, however, it has a specificity for glutamate and biphasic dependence on chloride similar to VGLUT1 and 2. It has been suggested to exhibit a slightly different dependence on ΔpH and $\Delta\Psi$ (Gras et al. 2002; Takamori et al. 2002), but this may simply reflect a difference in subcellular location, with different components of the H^+ electrochemical gradient driving glutamate transport in different vesicle populations and hence an apparent rather than real difference in ionic coupling.

Inside the nervous system, VGLUT3 is expressed by a very specific subset of cells in multiple brain regions. In situ hybridization for VGLUT3 mRNA shows expression by scattered GABAergic interneurons in the hippocampus and cortex, particularly layers 2 and 6; cholinergic interneurons of the striatum; serotonergic neurons in the dorsal raphe; a novel subset of amacrine cells; and a small subset of astrocytes (Fremeau et al. 2002; Gras et al. 2002; Schafer et al. 2002). However, recent work has now revealed the low-level expression in astrocytes of VGLUT1 and 2 as well as VGLUT3 (Bezzi et al. 2004; Montana et al. 2004; Zhang et al. 2004). The much stronger expression of VGLUT1 and 2 by neurons had apparently made detection of the low-level glial expression difficult. Different astrocytes also appear to express different isoforms (Bezzi et al. 2004; Montana et al. 2004; Zhang et al. 2004). Immunoelectron microscopy showing expression of VGLUT3 protein on astrocytic endfeet (Fremeau et al. 2002) suggests a particular role for this isoform in the blood–brain barrier, whereas the others presumably contribute to signaling between neurons and astrocytes (Araque et al. 1999; Bezzi et al. 2004).

Within neurons, VGLUT3 has a different subcellular location than VGLUT1 and 2. VGLUT1 and 2 localize essentially exclusively to axon terminals, whereas VGLUT3 also resides on cell bodies and dendrites of particular neuronal populations (GABAergic interneurons of the cortex and hippocampus and striatal cholinergic interneurons, but not dopamine or serotonin neurons) (Fremeau et al. 2002). Because different vesicles may exhibit different ΔpH and $\Delta\Psi$, the differences in trafficking of VGLUT3 from VGLUT1 and 2 may account for the apparent differences in ionic coupling. As noted in Sect. 2, the dependence of VGLUT isoforms on these two components of the H^+ electrochemical gradient may reflect their subcellular location rather than their intrinsic ionic coupling (Gras et al. 2002; Takamori et al. 2002).

3.1
VGLUT3 in Monoamine Neurons

In situ hybridization for VGLUT3 shows perhaps the strongest labeling in midline and dorsal raphe, suggesting expression by serotonergic neurons (Fremeau et al. 2002; Gras et al. 2002). However, VGLUT3 is not detectable in raphe cell bodies by immunocytochemistry, and thus it remained possible that non-serotonergic neurons in this region may express VGLUT3. To address this, we immunostained single-cell, autaptic cultures of raphe neurons, and found that at least 70% of the 5-HT$^+$ cells also expressed VGLUT3 (Fremeau et al. 2002). Also, Schafer et al. demonstrated by double in situ hybridization in mouse brain slices that a majority of serotonin neurons [vesicular monoamine transporter (VMAT) mRNA$^+$ and TH mRNA$^-$] in the rostral brainstem raphe nuclei express VGLUT3 mRNA (Schafer et al. 2002). Consistent with these findings, VGLUT3 colocalizes with serotonin, the serotonin transporter (SERT), and the neuronal VMAT2 in large boutons bordering stratum radiatum and stratum lacunosum-moleculare of the hippocampus (Schafer et al. 2002). In other locations, however, 5-HT and SERT, which often exhibit a linear appearance, do not colocalize with the punctate VGLUT3 (Somogyi et al. 2004). VGLUT3 thus localizes to a subset of serotonergic processes, at least in part because the protein is expressed by only a subset of serotonergic neurons; but it remains possible that VGLUT3 confers glutamate release from a distinct population of secretory vesicles within serotonergic neurons that do not also contain serotonin.

VGLUT3 is also expressed in caudal brainstem raphe nuclei. In contrast to neurons of the RVLM that regulate cardiovascular function, neurons in the rostral medullary raphe pallidus (RPa) and raphe magnus (RMg) control body temperature through regulation of thermogenesis in brown adipose tissue (BAT) (Morrison 1999; Morrison 2003). Neurons in the RPa and RMg also express VGLUT3, and these cells show increased c-fos expression after treatment with prostaglandin E$_2$ to induce fever (Nakamura et al. 2004). By anterograde and retrograde tract tracing, these neurons project to preganglionic sympathetic neurons in the intermediolateral cell column (IML) of the thoracic spinal cord. Glutamate application in the IML activates thermogenesis in BAT, and glutamate receptor antagonists block BAT thermogenesis produced by activation of RPa neurons (Nakamura et al. 2004), suggesting that glutamate release conferred by the expression of VGLUT3 in RPa neurons contributes to an important autonomic response. It is also tempting to consider this descending pathway from medullary raphe nuclei analogous to the ascending serotonergic, VGLUT3$^+$ projections from dorsal and medial raphe nuclei in the upper brainstem. Despite the presence of serotonin neurons in the rostral medulla, however, only a minority of the VGLUT3$^+$ cells in RPa are serotonergic (Nakamura et al. 2004). In this case as well, glutamate may thus be the principal transmitter for these VGLUT3$^+$ neurons.

VGLUT3 may also confer glutamate release from dopamine neurons. As noted in Sect. 2.2, the expression of VGLUT2 by brainstem catecholamine neurons and the analysis of midbrain dopamine neurons in culture suggest a role for VGLUT2 (Dal Bo et al. 2004; Hur and Zaborszky 2005). However, in situ hybridization also provides evidence for expression of VGLUT3, at least in the adult (Fremeau et al. 2002).

3.2
Cholinergic Interneurons of the Striatum

In the striatum, tonically active cholinergic interneurons express VGLUT3 (Fremeau et al. 2002; Gras et al. 2002). In both dorsal and ventral striatum, VGLUT3 colocalizes essentially completely with the biosynthetic enzyme choline acetyltransferase (ChAT) and VAChT in cell bodies as well as processes. The abundant expression of VGLUT3 by these cells predicts an important role for glutamate in their signaling. However, many basic questions remain about the mode of glutamate release: In particular, do VGLUT3 and VAChT colocalize on the same synaptic vesicles? By electron microscopy, immunoperoxidase staining for VAChT with immunogold staining for VGLUT3 supports colocalization to the same processes suggested by the light microscopy (Gras et al. 2002), but the diffuse nature of the peroxidase stain limits conclusions about localization to intracellular membranes. The segregation of VAChT and VGLUT2 in motor neurons indicates the potential for segregation of VAChT from VGLUT3, but the ultrastructural and biochemical evidence does not yet provide definitive information about the trafficking in striatal interneurons.

3.3
Inhibitory Interneurons of the Hippocampus, Cortex, and Retina

The initial in situ hybridization for VGLUT3 indicated expression in scattered cell bodies of the hippocampus and superficial layers (2 and 3) of the cortex, which is suggestive of interneurons (Fremeau et al. 2002). Although certain interneurons are excitatory rather than inhibitory, antibody staining indeed confirmed colocalization of VGLUT3 with the biosynthetic enzyme for GABA, glutamic acid decarboxylase, in both cell bodies and processes. Unlike the striatal interneurons, however, VGLUT3 colocalizes with only a fraction of the GABAergic interneurons. These constitute the cholecystokinin$^+$ (CCK$^+$) subset that also expresses the CB1 cannabinoid receptor (Katona et al. 1999) and preprotachykinin B (Hioki et al. 2004), but does not stain for parvalbumin, calretinin, vasoactive intestinal peptide, or somatostatin (Somogyi et al. 2004). The labeled processes form perisomatic synapses in the pyramidal cell layer of the hippocampus and within layers 2 and 3 of the cortex (Fremeau et al.

2002; Hioki et al. 2004; Somogyi et al. 2004). Thus, inhibitory interneurons may release glutamate as well as GABA, and consistent with this, the hippocampal terminals labeling for VGLUT3 form symmetric rather than the asymmetric synapses typical of excitatory connections (Fremeau et al. 2002). Interestingly, previous work had suggested the co-release of glutamate and GABA from synaptosomes (Docherty et al. 1987), but the physiological role of glutamate release by interneurons remains a mystery.

A subset of amacrine cells in the retina, generally considered inhibitory interneurons, also expresses VGLUT3 (Fremeau et al. 2002; Fyk-Kolodziej et al. 2004; Johnson et al. 2004). Many amacrine cells use glycine as their transmitter, and VGLUT3 indeed colocalizes with glycine immunoreactivity in the cell bodies of selected amacrine cells. However, the processes labeling for VGLUT3 contain little if any detectable glycine, and the VGLUT3 at these sites does not generally colocalize with vesicular GABA transporter (VGAT) (Johnson et al. 2004). VGLUT3 may thus define a subset of glutamatergic amacrine cells, and the protein does appear in proximity to the metabotropic glutamate receptor mGluR4. Indeed, the activation of a metabotropic receptor may confer the inhibitory signal anticipated from interneuronal populations, even though it is mediated by glutamate.

3.4
Dendritic Release of Glutamate

The dendritic localization of VGLUT3 also has the potential to account for previous observations of dendritic glutamate release from pyramidal neurons in the cortex (Zilberter 2000). Although endocannabinoids mediate depolarization-induced suppression of inhibition at certain synapses in the hippocampus, cerebellum, and cortex (Wilson and Nicoll 2002), a similar fast suppression of inhibition has been reported to occur in an endocannabinoid-independent manner. In paired recordings of the synapses formed by fast-spiking cortical interneurons onto pyramidal cells, a cannabinoid agonist does not affect inhibitory potentials, nor does a cannabinoid antagonist block the suppression of these potentials induced by postsynaptic stimulation (Harkany et al. 2004). To test a role for retrograde release of glutamate in this suppression, the authors took advantage of the unusual features of VGLUT activity. In particular, they introduced the cell-impermeant, VGLUT inhibitor Evans blue directly into the postsynaptic cell, and blocked the suppression produced by postsynaptic stimulation. Because Evans blue is not particularly potent or specific, they also took advantage of the unusual chloride sensitivity of the VGLUTs, and blocked the suppression by increasing the chloride concentration in the postsynaptic neuron (Harkany et al. 2004). Further, they demonstrated the presence of VGLUT3 immunoreactivity in the dendrites of pyramidal cells in the cortex. Weaker than the staining in interneurons, this labeling had not previously been appreciated, very similar to the expression of VGLUT1 and 2 in glia, which is overshadowed by the neuronal expression. Immunoelectron microscopy confirms the expression of VGLUT3 in cortical dendrites (Fremeau et al. 2002; Harkany et al. 2004). This work indicates yet another locus for physiological signaling mediated by VGLUT3.

3.5
Expression of VGLUT3 Transiently During Development

Similar to the transient expression of VGLUT2 during early postnatal development of the cortex and hippocampus, VGLUT3 appears transiently at selected locations. Scattered cells migrating from the surface of the ventricles into the neuropil express VGLUT3, suggesting a role in neural progenitors (Boulland et al. 2004). Consistent with this possibility, cultures enriched in progenitor cells also show co-expression of VGLUT3 with nestin, a marker for progenitor cells. In the cerebellum, Purkinje cells, which are generally considered GABAergic, express VGLUT3 transiently at their terminals in the deep cerebellar nuclei (Boulland et al. 2004). Although the role of glutamate release early in development remains unknown in most of these cases, recent work in the auditory system has begun to suggest one possibility.

To determine sound localization, the lateral superior olive (LSO) computes interaural differences in sound intensity by integrating excitatory input from the cochlear nucleus with inhibitory input from the medial nucleus of the trapezoid body (MNTB). During development, the MNTB–LSO synapse undergoes a functional refinement that increases the precision of the tonotopic map (Kim and Kandler 2003). In addition, the synapse undergoes a switch in transmitter from GABA to glycine (Kotak et al. 1998). Although GABA is often excitatory early in development, the MNTB terminals also release glutamate, both in response to electrical stimulation and, to avoid activation of passing fibers, in response to uncaging of glutamate in the MNTB (Gillespie et al. 2005). Further, this co-release of glutamate with GABA is transient, and coincides with the expression of VGLUT3 at the MNTB–LSO synapse. Although the developmental role of transient glutamate release at this synapse remains unclear, the results suggest yet another synapse where the glutamate release mediated by VGLUT3 has a physiological role.

4
Conclusion

In summary, work in a variety of systems has suggested the co-release of glutamate with a variety of other transmitters including serotonin, dopamine, GABA, and acetylcholine. In many cases, VGLUT3 appears responsible for the ability of these cell types to release glutamate. However, the physiological and developmental roles of this release remain unknown.

References

Araque A, Sanzgiri RP, Parpura V, Haydon PG (1998) Calcium elevation in astrocytes causes an NMDA receptor-dependent increase in the frequency of miniature synaptic currents in cultured hippocampal neurons. J Neurosci 18:6822–6829

Araque A, Parpura V, Sanzgiri RP, Haydon PG (1999) Tripartite synapses: glia, the unacknowledged partner. Trends Neurosci 22:208–215

Araque A, Li N, Doyle RT, Haydon PG (2000) SNARE protein-dependent glutamate release from astrocytes. J Neurosci 20:666–673

Bai L, Xu H, Collins JF, Ghishan FK (2001) Molecular and functional analysis of a novel neuronal vesicular glutamate transporter. J Biol Chem 276:36764–36769

Bellocchio EE, Reimer RJ, Fremeau RTJ, Edwards RH (2000) Uptake of glutamate into synaptic vesicles by an inorganic phosphate transporter. Science 289:957–960

Bezzi P, Carmignoto G, Pasti L, Vesce S, Rossi D, Rizzini BL, Pozzan T, Volterra A (1998) Prostaglandins stimulate calcium-dependent glutamate release in astrocytes. Nature 391:281–285

Bezzi P, Gundersen V, Galbete JL, Seifert G, Steinhauser C, Pilati E, Volterra A (2004) Astrocytes contain a vesicular compartment that is competent for regulated exocytosis of glutamate. Nat Neurosci 7:613–620

Boulland JL, Qureshi T, Seal RP, Rafiki A, Gundersen V, Bergersen LH, Fremeau RT Jr, Edwards RH, Storm-Mathisen J, Chaudhry FA (2004) Expression of the vesicular glutamate transporters during development indicates the widespread corelease of multiple neurotransmitters. J Comp Neurol 480:264–280

Bröer S, Schuster A, Wagner CA, Bröer A, Forster I, Biber J, Murer H, Werner A, Lang F, Busch AE (1998) Chloride conductance and Pi transport are separate functions induced by the expression of NaPi-1 in Xenopus oocytes. J Membr Biol 164:71–77

Busch AE, Schuster A, Waldegger S, Wagner CA, Zempel G, Broer S, Biber J, Murer H, Lang F (1996) Expression of a renal type I sodium/phosphate transporter (NaPi-1) induces a conductance in Xenopus oocytes permeable for organic and inorganic anions. Proc Natl Acad Sci USA 93:5347–5351

Chuhma N, Zhang H, Masson J, Zhuang X, Sulzer D, Hen R, Rayport S (2004) Dopamine neurons mediate a fast excitatory signal via their glutamatergic synapses. J Neurosci 24:972–981

Curthoys NP, Watford M (1995) Regulation of glutaminase activity and glutamine metabolism. Annu Rev Nutr 15:133–159

Dal Bo G, St-Gelais F, Danik M, Williams S, Cotton M, Trudeau LE (2004) Dopamine neurons in culture express VGLUT2 explaining their capacity to release glutamate at synapses in addition to dopamine. J Neurochem 88:1398–1405

Deuchars SA, Morrison SF, Gilbey MP (1995) Medullary-evoked EPSPs in neonatal rat sympathetic preganglionic neurones in vitro. J Physiol 487:453–463

Docherty M, Bradford HF, Wu JY (1987) Co-release of glutamate and aspartate from cholinergic and GABAergic synaptosomes. Nature 330:64–66

Forgac M (2000) Structure, mechanism and regulation of the clathrin-coated vesicle and yeast vacuolar H(+)-ATPases. J Exp Biol 203:71–80

Fremeau RT Jr, Troyer MD, Pahner I, Nygaard GO, Tran CH, Reimer RJ, Bellocchio EE, Fortin D, Storm-Mathisen J, Edwards RH (2001) The expression of vesicular glutamate transporters defines two classes of excitatory synapse. Neuron 31:247–260

Fremeau RT Jr, Burman J, Qureshi T, Tran CH, Proctor J, Johnson J, Zhang H, Sulzer D, Copenhagen DR, Storm-Mathisen J, Reimer RJ, Chaudhry FA, Edwards RH (2002) The identification of vesicular glutamate transporter 3 suggests novel modes of signaling by glutamate. Proc Natl Acad Sci U S A 99:14488–14493

Fremeau RT Jr, Kam K, Qureshi T, Johnson J, Copenhagen DR, Storm-Mathisen J, Chaudhry FA, Nicoll RA, Edwards RH (2004a) Vesicular glutamate transporters 1 and 2 target to functionally distinct synaptic release sites. Science 304:1815–1819

Fremeau RT Jr, Voglmaier S, Seal RP, Edwards RH (2004b) VGLUTs define subsets of excitatory neurons and suggest novel roles for glutamate. Trends Neurosci 27:98–103

Fyk-Kolodziej B, Dzhagaryan A, Qin P, Pourcho RG (2004) Immunocytochemical localization of three vesicular glutamate transporters in the cat retina. J Comp Neurol 475:518–530

Gillespie DC, Kim G, Kandler K (2005) Inhibitory synapses in the developing auditory system are glutamatergic. Nat Neurosci 8:332–338

Gras C, Herzog E, Bellenchi GC, Bernard V, Ravassard P, Pohl M, Gasnier B, Giros B, El Mestikawy S (2002) A third vesicular glutamate transporter expressed by cholinergic and serotoninergic neurons. J Neurosci 22:5442–5451

Harkany T, Holmgren C, Hartig W, Qureshi T, Chaudhry FA, Storm-Mathisen J, Dobszay MB, Berghuis P, Schulte G, Sousa KM, Fremeau RT Jr, Edwards RH, Mackie K, Ernfors P, Zilberter Y (2004) Endocannabinoid-independent retrograde signaling at inhibitory synapses in layer 2/3 of neocortex: involvement of vesicular glutamate transporter 3. J Neurosci 24:4978–4988

Hartinger J, Jahn R (1993) An anion binding site that regulates the glutamate transporter of synaptic vesicles. J Biol Chem 268:23122–23127

Hayashi M, Otsuka M, Morimoto R, Hirota S, Yatsushiro S, Takeda J, Yamamoto A, Moriyama Y (2001) Differentiation-associated Na+-dependent inorganic phosphate cotransporter (DNPI) is a vesicular glutamate transporter in endocrine glutamatergic systems. J Biol Chem 276:43400–43406

Hayashi M, Yamada H, Uehara S, Morimoto R, Muroyama A, Yatsushiro S, Takeda J, Yamamoto A, Moriyama Y (2003) Secretory granule-mediated co-secretion of L-glutamate and glucagon triggers glutamatergic signal transmission in islets of Langerhans. J Biol Chem 278:1966–1974

Herzog E, Bellenchi GC, Gras C, Bernard V, Ravassard P, Bedet C, Gasnier B, Giros B, El Mestikaway S (2001) The existence of a second vesicular glutamate transporter specifies subpopulations of glutamatergic neurons. J Neurosci 21:RC181

Hioki H, Fujiyama F, Nakamura K, Wu SX, Matsuda W, Kaneko T (2004) Chemically specific circuit composed of vesicular glutamate transporter 3- and preprotachykinin B-producing interneurons in the rat neocortex. Cereb Cortex 14:1266–1275

Huangfu D, Hwang LJ, Riley TA, Guyenet PG (1994) Role of serotonin and catecholamines in sympathetic responses evoked by stimulation of rostral medulla. Am J Physiol 266:R338–352

Hur EE, Zaborszky L (2005) Vglut2 afferents to the medial prefrontal and primary somatosensory cortices: a combined retrograde tracing in situ hybridization. J Comp Neurol 483:351–373

Johnson J, Sherry DM, Liu X, Fremeau RT Jr, Seal RP, Edwards RH, Copenhagen DR (2004) Vesicular glutamate transporter 3 expression identifies glutamatergic amacrine cells in the rodent retina. J Comp Neurol 477:386–398

Johnson MD (1994) Synaptic glutamate release by postnatal rat serotonergic neurons in microculture. Neuron 12:433–442

Katona I, Sperlagh B, Sik A, Kafalvi A, Vizi ES, Mackie K, Freund TF (1999) Presynaptically located CB1 cannabinoid receptors regulate GABA release from axon terminals of specific hippocampal interneurons. J Neurosci 19:4544–4558

Kim G, Kandler K (2003) Elimination and strengthening of glycinergic/GABAergic connections during tonotopic map formation. Nat Neurosci 6:282–290

Kotak VC, Korada S, Schwartz IR, Sanes DH (1998) A developmental shift from GABAergic to glycinergic transmission in the central auditory system. J Neurosci 18:4646–4655

Lavin A, Nogueira L, Lapish CC, Wightman RM, Phillips PE, Seamans JK (2005) Mesocortical dopamine neurons operate in distinct temporal domains using multimodal signaling. J Neurosci 25:5013–5023

Liu Y, Edwards RH (1997a) The role of vesicular transport proteins in synaptic transmission and neural degeneration. Annu Rev Neurosci 20:125–156

Lorang D, Amara SG, Simerly RB (1994) Cell-type-specific expression of catecholamine transporters in the rat brain. J Neurosci 14:4903–4914

MacDonald MJ, Fahien LA (2000) Glutamate is not a messenger in insulin secretion. J Biol Chem 275:34025–34027

Maechler P, Wollheim CB (1999) Mitochondrial glutamate acts as a messenger in glucose-induced insulin exocytosis [see comments]. Nature 402:685–689

Maechler P, Kennedy ED, Pozzan T, Wollheim CB (1997) Mitochondrial activation directly triggers the exocytosis of insulin in permeabilized pancreatic beta-cells. EMBO J 16:3833–3841

Mancini GM, Beerens CE, Aula PP, Verheijen FW (1991) Sialic acid storage diseases. A multiple lysosomal transport defect for acidic monosaccharides. J Clin Invest 87:1329–1335

Miyazaki T, Fukaya M, Shimizu H, Watanabe M (2003) Subtype switching of vesicular glutamate transporters at parallel fibre-Purkinje cell synapses in developing mouse cerebellum. Eur J Neurosci 17:2563–2572

Montana V, Ni Y, Sunjara V, Hua X, Parpura V (2004) Vesicular glutamate transporter-dependent glutamate release from astrocytes. J Neurosci 24:2633–2642

Morimoto R, Hayashi M, Yatsushiro S, Otsuka M, Yamamoto A, Moriyama Y (2003) Co-expression of vesicular glutamate transporters (VGLUT1 and VGLUT2) and their association with synaptic-like microvesicles in rat pinealocytes. J Neurochem 84:382–391

Morin P, Sagne C, Gasnier B (2004) Functional characterization of wild-type and mutant human sialin. EMBO J 23:4560–4570

Moriyama Y, Yamamoto A (1995) Vesicular L-glutamate transporter in microvesicles from bovine pineal glands. Driving force, mechanism of chloride anion activation, and substrate specificity. J Biol Chem 270:22314–22320

Morrison SF (1999) RVLM and raphe differentially regulate sympathetic outflows to splanchnic and brown adipose tissue. Am J Physiol 276:R962–R973

Morrison SF (2003) Raphe pallidus neurons mediate prostaglandin E2-evoked increases in brown adipose tissue thermogenesis. Neuroscience 121:17–24

Morrison SF, Callaway J, Milner TA, Reis DJ (1991) Rostral ventrolateral medulla: a source of the glutamatergic innervation of the sympathetic intermediolateral nucleus. Brain Res 562:126–135

Nakamura K, Matsumura K, Hubschle T, Nakamura Y, Hioki H, Fujiyama F, Boldogkoi Z, Konig M, Thiel HJ, Gerstberger R, Kobayashi S, Kaneko T (2004) Identification of sympathetic premotor neurons in medullary raphe regions mediating fever and other thermoregulatory functions. J Neurosci 24:5370–5380

Nishimaru H, Restrepo CE, Ryge J, Yanagawa Y, Kiehn O (2005) Mammalian motor neurons corelease glutamate and acetylcholine at central synapses. Proc Natl Acad Sci U S A 102:5245–5249

Parpura V, Basarsky TA, Liu F, Jeftinija K, Jeftinija S, Haydon PG (1994) Glutamate-mediated astrocyte-neuron signalling [see comments]. Nature 369:744–747

Schafer MK, Varoqui H, Defamie N, Weihe E, Erickson JD (2002) Molecular cloning and functional identification of mouse vesicular glutamate transporter 3 and its expression in subsets of novel excitatory neurons. J Biol Chem 277:50734–50748

Schuldiner S, Shirvan A, Linial M (1995) Vesicular neurotransmitter transporters: from bacteria to humans. Physiol Rev 75:369–392

Somogyi J, Baude A, Omori Y, Shimizu H, Mestikawy SE, Fukaya M, Shigemoto R, Watanabe M, Somogyi P (2004) GABAergic basket cells expressing cholecystokinin contain vesicular glutamate transporter type 3 (VGLUT3) in their synaptic terminals in hippocampus and isocortex of the rat. Eur J Neurosci 19:552–569

Stornetta RL, Sevigny CP, Guyenet PG (2002a) Vesicular glutamate transporter DNPI/VGLUT2 mRNA is present in C1 and several other groups of brainstem catecholaminergic neurons. J Comp Neurol 444:191–206

Stornetta RL, Sevigny CP, Schreihofer AM, Rosin DL, Guyenet PG (2002b) Vesicular glutamate transporter DNPI/VGLUT2 is expressed by both C1 adrenergic and nonaminergic presympathetic vasomotor neurons of the rat medulla. J Comp Neurol 444:207–220

Sulzer D, Joyce MP, Lin L, Geldwert D, Haber SN, Hattori T, Rayport S (1998) Dopamine neurons make glutamatergic synapses in vitro. J Neurosci 18:4588–4602

Takamori S, Rhee JS, Rosenmund C, Jahn R (2000) Identification of a vesicular glutamate transporter that defines a glutamatergic phenotype in neurons. Nature 407:189–194

Takamori S, Rhee JS, Rosenmund C, Jahn R (2001) Identification of differentiation-associated brain-specific phosphate transporter as a second vesicular glutamate transporter. J Neurosci 21:RC182

Takamori S, Malherbe P, Broger C, Jahn R (2002) Molecular cloning and functional characterization of human vesicular glutamate transporter 3. EMBO Rep 3:798–803

Varoqui H, Schafer MK, Zhu H, Weihe E, Erickson JD (2002) Identification of the differentiation-associated Na+/Pi transporter as a novel vesicular glutamate transporter expressed in a distinct set of glutamatergic synapses. J Neurosci 22:142–155

Verheijen FW, Verbeek E, Aula N, Beerens CE, Havelaar AC, Joosse M, Peltonen L, Aula P, Galjaard H, van der Spek PJ, Mancini GM (1999) A new gene, encoding an anion transporter, is mutated in sialic acid storage diseases. Nat Genet 23:462–465

Werner A, Moore ML, Mantei N, Biber J, Semenza G, Murer H (1991) Cloning and expression of cDNA for a Na/Pi cotransport system of kidney cortex. Proc Natl Acad Sci USA 88:9608–9612

Wilson RI, Nicoll RA (2002) Endocannabinoid signaling in the brain. Science 296:678–682

Wolosker H, de Souza DO, de Meis L (1996) Regulation of glutamate transport into synaptic vesicles by chloride and proton gradient. J Biol Chem 271:11726–11731

Wreden CC, Wlizla M, Reimer RJ (2005) Varied mechanisms underlie the free sialic acid storage disorders. J Biol Chem 280:1408–1416

Zhang Q, Pangrsic T, Kreft M, Krzan M, Li N, Sul JY, Halassa M, Van Bockstaele E, Zorec R, Haydon PG (2004) Fusion-related release of glutamate from astrocytes. J Biol Chem 279:12724–12733

Zilberter Y (2000) Dendritic release of glutamate suppresses synaptic inhibition of pyramidal neurons in rat neocortex. J Physiol 528:489–496

Extraneuronal Monoamine Transporter and Organic Cation Transporters 1 and 2: A Review of Transport Efficiency

E. Schömig · A. Lazar · D. Gründemann (✉)

Department of Pharmacology, University of Cologne,
Glueler Straße 24, 50931 Cologne, Germany
dirk.gruendemann@uni-koeln.de

1	Introduction	152
1.1	Structure and Function	152
1.2	Localization	153
1.3	Genes, Polymorphisms and Knock-Out Mice	153
2	Substrate Specificity	155
2.1	Analysis of Transport Efficiency	155
2.2	OCT1	157
2.3	OCT2	162
2.4	EMT	169
3	Roundup	172
	References	175

Abstract The extraneuronal monoamine transporter (EMT) corresponds to the classical steroid-sensitive monoamine transport mechanism that was first described as "uptake$_2$" in rat heart with noradrenaline as substrate. The organic cation transporters OCT1 and OCT2 are related to EMT. The three carriers share basic structural and functional characteristics. Hence, EMT, OCT1 and OCT2 constitute a group referred to as non-neuronal monoamine transporters or organic cation transporters. After a brief general introduction, this review focuses on the critical analysis of substrate specificity. We calculate from the available literature and compare consensus transport efficiency (clearance) data for human and rat EMT, OCT1 and OCT2, expressed in transfected cell lines. From the plethora of inhibitors that have been tested, the casual observer likely gets the impression that these carriers indiscriminately transport very many compounds. However, our knowledge about actual substrates is rather limited. 1-Methyl-4-phenylpyridinium (MPP$^+$) is an excellent substrate for all three carriers, with clearances typically in the range of 20–50 µl min^{-1} mg protein^{-1}. The second-best general substrate is tyramine with a transport efficiency (TE) range relative to MPP$^+$ of 20%–70%. The TEs of OCT1 and OCT2 for dopamine, noradrenaline, adrenaline and 5-HT in general are rather low, in the range relative to MPP$^+$ of 5%–15%. This suggests that OCT1 and OCT2 are not primarily dedicated to transport these monoamine transmitters; only EMT may play a significant role in catecholamine inactivation. For many substrates, such as tetraethylammonium, histamine, agmatine, guanidine, cimetidine, creatinine, choline and acetylcholine, the transport efficiencies are markedly different among the carriers.

Keywords Extraneuronal monoamine transporter · Organic cation transporter · Transport efficiency · Substrate specificity · Heterologous expression

1
Introduction

1.1
Structure and Function

The extraneuronal monoamine transporter (EMT) was cloned in 1998 (Gründemann et al. 1998c; Kekuda et al. 1998). It corresponds to the classical steroid-sensitive monoamine transport mechanism that was first described as "uptake$_2$" in rat heart with noradrenaline as substrate (Iversen 1965; Trendelenburg 1988). It is considered, in particular, responsible for the non-neuronal inactivation of circulating catecholamines (Eisenhofer 2001; Eisenhofer et al. 1996). The organic cation transporters OCT1 (Gründemann et al. 1994) and OCT2 (Okuda et al. 1996) are related in amino acid sequence to EMT. The three carriers share basic structural and functional characteristics. Hence, EMT, OCT1 and OCT2 constitute a group referred to as non-neuronal monoamine transporters or organic cation transporters (Gründemann et al. 1994; Gründemann et al. 1999; Gründemann and Schömig 2003). EMT from rat has also been labelled "OCT3" (Wu et al. 1998). However, EMT does not accept typical organic cations such as tetraethylammonium (TEA), guanidine, creatinine and choline as substrates (see Sect. 2.4). Further, evolutionary distances of EMT to OCT1 and to OCT2 are considerably greater than between OCT1 and OCT2 (see Sect. 3). Thus, the preferred and well-established designation is extraneuronal monoamine transporter.

Hydropathy analysis predicts that OCT1, OCT2 and EMT are integral membrane proteins with 12 α-helical transmembrane segments. All carriers are around 555 amino acids long and reside in the plasma membrane. Common structural features are a large extracellular loop with multiple potential N-glycosylation sites between transmembrane segments 1 and 2 and multiple potential intracellular phosphorylation sites. Both the N- and C-terminus probably face the cytosol.

Heterologous expression in oocytes from *Xenopus laevis* and mammalian cell lines has revealed common functional qualities: (1) A single positive charge is required on substrates; uncharged, doubly charged or negatively charged solutes are not transported. (2) Decrease of extracellular pH and depolarization of the plasma membrane reduce velocity of uptake. (3) Transport is independent from gradients of Na$^+$ and Cl$^-$. (4) It has been proved by *trans*-stimulation that all three proteins are transporters, not just channels. Transport works in both directions (uptake and efflux). Thus, depending on substrate gradients across the cell membrane, OCT1, OCT2 and EMT may mediate electrogenic uniport of a substrate or electroneutral exchange of two substrates (antiport). (5) By contrast to the neuronal carriers of catecholamines, the affinity of EMT and relatives for e.g. noradrenaline is rather low ($K_m > 0.5$ mmol/l). This is compensated by high capacity (high turnover number). (6) EMT and relatives are relatively resistant to "neuronal" uptake inhibitors such as desipramine or reserpine.

1.2
Localization

Although our knowledge of transporter tissue distribution is incomplete and sometimes inconsistent, marked differences are certain. OCT1 is expressed primarily in liver, and in rodents additionally, and strongly in kidney and intestine. There is good evidence that OCT1 resides in the sinusoidal membrane of hepatocytes and that OCT1 from rat is located at the basolateral membrane of S1 and S2 segments of renal proximal tubules. The strong sinusoidal expression in liver suggests that OCT1 may determine systemic exposure to drug and xenobiotic substrates.

OCT2 is strongly and almost exclusively expressed in kidney, in the straight part of the proximal tubule (outer medulla, predominantly S3 segment, perhaps some late S2) (Gründemann et al. 1998b). Much less mRNA has been spotted in parts of the brain. OCT2 has been detected both in apical (luminal) (Dudley et al. 2000; Gründemann et al. 1997; Sweet et al. 2001) and in basolateral membrane compartments (Karbach et al. 2000; Sugawara-Yokoo et al. 2000). In male rat kidneys, OCT2 is expressed stronger by a factor of about 1.5–3 (depending on the detection method) vs females (Urakami et al. 1999). No such differences have been found for OCT1 and EMT. The significance of this difference is unclear.

EMT is expressed, with substantial species differences, in many but not all tissues. Expression levels vary widely between organs and during organ development (e.g. in placenta). Consistently high expression has been reported for placenta and heart. Recent reports indicate strong expression of EMT in the area postrema (Haag et al. 2004; Vialou et al. 2004). Expression in the kidney is universally low. Expression of EMT in glia cells is well documented (Inazu et al. 2003; Russ et al. 1996). However, there are two recent reports which suggest expression also in neurons (Kristufek et al. 2002; Shang et al. 2003). EMT co-localizes with MAO-A in mouse placenta (Verhaagh et al. 2001) and with MAO-B in rat area postrema (Haag et al. 2004).

1.3
Genes, Polymorphisms and Knock-Out Mice

OCT1, OCT2 and EMT are members of the amphiphilic solute facilitator (ASF) family of transporters (gene symbol group *SLC22*) (Eraly and Nigam 2002; Schömig et al. 1998). The genes of OCT1 (*SLC22A1*), OCT2 (*SLC22A2*) and EMT (*SLC22A3*) are most likely descendants from a single original gene, since they code for related amino acid sequences and they cluster closely together in a conserved region, on chromosome 6 in human (6q26–27) and on chromosome 17 in mouse. The intron–exon structures of all human genes have been elucidated (Gründemann and Schömig 2000b; Hayer et al. 1999).

Gene and transcript variability in humans has been investigated in some detail. Mutations could considerably affect absorption, distribution and elimination of drug substrates. However, no data on the clinical consequences of polymorphisms of the *SLC22A1–3* genes are available so far.

Alternative splicing of OCT1h (h, human; r, rat; m, mouse) has been reported; the implications are unclear (Hayer et al. 1999). Several single amino acid substitutions have been found for OCT1h that reduce or eliminate uptake of 1-methyl-4-phenylpyridinium (MPP$^+$) into *Xenopus* oocytes, i.e. Arg61Cys, Cys88Arg, Gly220Val, Pro341Leu, Gly401Ser and Gly465Arg (Kerb et al. 2002; Shu et al. 2003). The frequent Phe120Leu mutation and the deletion of Met420 did not affect transport activity. The Ser14Phe variant exhibited increased transport activity.

For OCT2h, a splice variant has been presented, again with unclear consequences (Urakami et al. 2002). Eight non-synonymous mutations (Pro54Ser, Phe161Leu, Met165Ile, Met165Val, Ala270Ser, Ala297Gly, Arg400Cys and Lys432Gln) have been detected (Leabman et al. 2002). Some of the mutant carriers have been analysed for K_m of uptake of MPP$^+$ into oocytes, but a comparison of transport efficiencies is not available.

In contrast to OCT1 and OCT2, the human gene of EMT is highly conserved. Direct sequencing revealed three synonymous mutations and four intronic polymorphisms but no amino acid exchanges in 100 Caucasians (Lazar et al. 2003). These findings suggest strong evolutionary pressure to preserve the amino acid sequence of EMTh.

The functions of OCT1, OCT2 and EMT have also been investigated with transgenic mice. In OCT1 knock-out mice, hepatic uptake and intestinal excretion of organic cations was reduced (Jonker et al. 2001). No OCT2 knock-out mice are available (Jonker and Schinkel 2004), but in OCT1/OCT2 double knock-out mice, renal secretion of organic cations was abolished (Jonker et al. 2003). EMT knock-out mice are viable and fertile (Zwart et al. 2001). It is not entirely clear whether the available EMT knock-out represents a full knock-out, since an aberrant EMT transcript was detected. In adult hearts, MPP$^+$ levels were reduced by 72% for knock-out mice. After injection of MPP$^+$ into pregnant females, accumulation of MPP$^+$ was reduced threefold for homozygous mutant embryos but not for placenta or amniotic fluid. EMT has also been implicated in the regulation of salt-intake, since EMT-deficient mice exhibited increased ingestion of saline under thirst and salt appetite conditions (Vialou et al. 2004).

2
Substrate Specificity

2.1
Analysis of Transport Efficiency

To understand the physiological purpose and the pathophysiological significance of a carrier protein, two major characteristics must be scrutinized: substrate specificity and localization. In this review we focus on substrate specificity. Many papers and reviews deal exhaustively with inhibitors (Koepsell 2004; Koepsell et al. 2003; Suhre et al. 2005). Unfortunately, the reported K_i-values are not as consistent as one would expect, e.g. K_is of OCT2r for decynium22 have been reported that differ more than 1,000-fold (14 µmol/l (Okuda et al. 1999) vs 12 nmol/l (Gründemann et al. 1998b). Moreover, it must be stressed that it is not possible to tell whether an inhibitor is actually a substrate. Since the chief function of a carrier is to transport substrates rather than to be the target of inhibitors, the differential analysis of the substrate spectrum is much more important.

In order to discriminate between good and poor substrates, one must determine the transport efficiency (TE), which is defined, analogous to catalytic efficiency for enzymes, by k_{cat}/K_m. This ratio takes into account the affinity (K_m) and turnover number (k_{cat}) of a carrier for a particular substrate. A good signal-to-noise ratio is important to accurately measure or even detect transport activity, but it is not related to TE. Noise corresponds to non-specific uptake, which covers diffusion, endocytosis, the activity of endogenous carrier(s) and extracellular binding. Non-specific uptake is measured by using (1) sufficient specific inhibitor to completely block the carrier, (2) a control cell line without expression of the particular transporter or (3) ice to freeze the carrier. Signal corresponds to specific uptake (also termed "expressed uptake" in heterologous expression systems), which is calculated from total uptake minus non-specific uptake to gauge uptake of solute mediated exclusively by the expressed carrier. Thus, a good signal-to-noise ratio must not indicate at all that the carrier runs at full capacity, it may simply reflect low uptake into control cells (as for TEA).

It was our aim here to systematically analyse available literature for transport efficiency of EMT, OCT1 and OCT2, both from human and rat. It is not possible to extract TEs true to the definition, since not a single turnover number has been determined for these carriers so far. For this, one would have to determine the number of active carriers (E_{total}) in the plasma membrane. We can, nevertheless, determine TE from V_{max}/K_m, which is a measure directly proportional to k_{cat}/K_m (since $V_{max} = k_{cat}*E_{total}$). Any day-to-day or lab-to-lab variation can be compensated if the TE of a reference substrate is measured in parallel (Gründemann et al. 1999). However, this has hardly been done in published work from most groups, so this was not an option for our analysis.

It then follows that TEs in principle are not strictly comparable between different groups because of different transporter densities. However, many results are quite similar across groups, probably because of high-level transporter expression in different expression systems.

A quantitative comparison of substrate quality on the basis of transport efficiency can provide important insights. Our basic and simple assumption is that relevant substrates—because of evolutionary pressure—are transported with high efficiency. For the physiological substrates, our experience suggests that transport efficiency should be in the order of 50 µl min^{-1} mg protein^{-1} or higher in mammalian cell lines with regular transporter overexpression based on strong viral promoters. If the TE is below 1 µl min^{-1} mg protein^{-1}, then the substrate is probably irrelevant. The power of this approach has recently become apparent from the discovery of the ergothioneine transporter (ETT; gene symbol *SLC22A4*). For years, this carrier, formerly called "novel organic cation transporter", OCTN1, was considered a polyspecific transporter of organic cations with the principal substrate TEA (Tamai et al. 1997; Yabuuchi et al. 1999) or, another carnitine transporter because of the close sequence homology with the carnitine transporter OCTN2 (*SLC22A5*) (Tamai et al. 1998; Tamai et al. 2000). In our experiments, uptake of TEA was detectable at a reasonable signal-to noise ratio (5.5); however, the TEs were consistently very low for both TEA (0.8 µl min^{-1} mg protein^{-1}) and carnitine (0.6 µl min^{-1} mg protein^{-1}), which suggested that the real substrate had yet to be defined. A substrate lead, stachydrine (alias proline betaine), was discovered by liquid chromatography (LC)-mass spectrometry (MS) difference shading (Gründemann et al. 2005), with a TE of about 20 µl min^{-1} mg protein^{-1}. Since this TE still was not as high as expected for a true substrate, we searched until we perceived ergothioneine, which displays TEs of 70–195 µl min^{-1} mg protein^{-1}.

For the following tables in this review we have only considered experiments where the uptake of substrate (which is usually labelled with a radioisotope) into cells [either transfected cell lines or complementary (c)RNA-injected *Xenopus laevis* oocytes] was measured directly, as opposed to detection based on e.g. electrophysiology or fluorescence, which we deem less specific. TE was calculated preferably as V_{max}/K_m. Alternatively, TE was calculated as v/S (where v represents initial rates of specific or expressed uptake, see above; it follows from the Michaelis–Menten equation $v = V_{max}*S /(K_m + S)$ that $v/S = V_{max}/K_m$ provided that S is much smaller than K_m; if this condition was clearly violated, TE was calculated as $v*(1 + x)/S$, where x represents S/K_m). The ratio v/S is also termed clearance, since it denotes the imaginary volume of incubation buffer cleared completely from substrate by transport activity per unit time. As the last option, TE was extracted from time courses of uptake (as the k_{in} kinetic constant, corrected for uptake into control cells). All TE values were converted to units of µl min^{-1} mg protein^{-1} where necessary. Estimated conversion factors were as follows: (1) 10^6 cells correspond to 0.25 mg protein and (2) 1 cm² cell culture dish corresponds to 0.035 mg protein. We have compiled separate tables for rat and human orthologues and for *Xenopus* oocyte and cell-line expression systems.

The oocyte system certainly has its virtues, but by now we do not consider it first-choice anymore. In addition to the well-known seasonality and batch-to-batch variations of transporter expression levels, this is due to the propensity of the oocyte to generate artefacts. For example, a cRNA was convincingly shown to induce transport activity for choline (Mayser et al. 1992). However, in a later paper (Schloss et al. 1994) from the same group it was shown that the carrier—now called CT1 (*SLC6A8*)—in fact transports creatine. We have seen solid uptake of glutamate—which clearly is no substrate of this carrier (Gründemann et al. 1998a)—induced by the injection of EMTh cRNA. EMTh has never expressed for us in oocytes, whereas the rat orthologue easily does (under the same conditions). We have not observed similar oddities in transfected cell lines. In several instances, the oocyte data do not match the cell-line data at all (cf. e.g. 5-HT and histamine for OCT2h). Hence, in the following sections, oocyte data tables are not discussed further and consensus TEs have been computed for cell-line data only.

2.2
OCT1

The data available on the transport efficiency of human OCT1 expressed in transfected cell lines is shown in Table 1. Table 2 lists TEs of OCT1h expressed in oocytes (note the different units). The data from Table 1 were condensed, as explained in Box 1, into a consensus series of relative TEs (Fig. 1). It is apparent

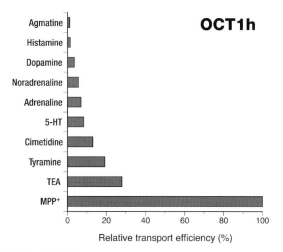

Fig. 1 Consensus TEs for OCT1h

Table 1 OCT1h expressed in cell lines[a]

Substrate	OCT1h – transfected cell line Clearance (µl min^{-1} mg protein^{-1})	Exp.
MPP$^+$	48	v/S
Tyramine	9.2	v/S
5-HT	4.0	v/S
Adrenaline	3.3	v/S
Noradrenaline	2.7	v/S
Dopamine	1.7	v/S
Histamine	0.7	v/S
	293 cells, stably transfected (Gründemann and Schömig 2000a)	
MPP$^+$	1.5	v/S
TEA	0.4	V_{max}/K_m
Cimetidine	0.2	v/S
	HeLa cells, transiently transfected; 30 min uptake at 22 °C (Zhang et al. 1998)	
MPP$^+$	53	V_{max}/K_m
Agmatine	0.6	v/S
	293 cells, stably transfected (Gründemann et al. 2003)	
MPP$^+$	0.02	v/S
	293 cells, stably transfected (Hayer et al. 1999)	
MPP$^+$	0.1	v/S
	293 cells, stably transfected (Hayer-Zillgen et al. 2002)	

[a]We have omitted prostaglandins (Kimura et al. 2002) from the table, since our own experiments suggest that OCTs do not transport PGE$_2$ at all (Harlfinger et al. 2005). We have also excluded v/S data of OCT1h on transport of MPP$^+$ (TE = 540 µl min^{-1} mg protein^{-1}) and acetylcholine (TE = 13.2 µl min^{-1} mg protein^{-1}), determined with resuspended CHO cells (Lips et al. 2005), since, in these experiments, uptake times of 1 s were used. This ultra short uptake phase would not seem to correspond to the linear phase of continuous uptake (= initial rate of uptake), which usually lasts for up to 5 min (Suhre et al. 2005). It could instead represent an initial "burst" of uptake activity caused if empty carriers are at first predominantly oriented with the substrate binding site to the outside. With the membrane charged negatively at the inside and the carrier charged negatively to attract positively charged substrates, the empty carrier is expected to favour the outside orientation. Exp., kind of experiment; MPP$^+$, 1-methyl-4-phenylpyridinium; TEA, tetraethylammonium

that MPP$^+$ is a very good substrate (about 50 µl min^{-1} mg protein^{-1}), 3.6-fold better than TEA (28% relative to MPP$^+$). Note, however, that much lower TEs for MPP$^+$ have also been reported (Table 1), perhaps due to inefficient expression. Cimetidine (13%) is a notable substrate. The best physiological substrate is tyramine (19%). The best transmitter substrate for OCT1h is 5-HT (8%); the catecholamines adrenaline (7%), noradrenaline (6%) and dopamine (4%) are

Table 2 OCT1h expressed in *Xenopus* oocytes

Substrate	OCT1h – *Xenopus* oocyte Clearance (µl h^{-1} oocyte^{-1})	Exp.
Tributylmethylammonium	3.2	v/S
N-Methyl-quinidine	2.3	v/S
N-Methyl-quinine	1.8	v/S
N-(4,4-Azo-n-pentyl)-21-deoxyajmalinium	1.6	v/S
N-(4,4-Azo-n-pentyl)-quinuclidine	1.1	v/S
Azidoprocainamide methiodide	0.5	v/S
	(Van Montfoort et al. 2001)	
Bamet-UD2	1.0	V_{max}/K_m
Bamet-R2	0.3	V_{max}/K_m
Cisplatin	0	v/S
	(Briz et al. 2002)	
MPP$^+$	0.2	v/S
TEA	0.1	v/S
	(Sakata et al. 2004)	
MPP$^+$	11	v/S
Acetylcholine	0.2	v/S
	(Lips et al. 2005)	
MPP$^+$	0.3	V_{max}/K_m
	(Zhang et al. 1997)	

Exp., kind of experiment; MPP$^+$, 1-methyl-4-phenylpyridinium; TEA, tetraethylammonium

transported inefficiently compared with MPP$^+$. Histamine and agmatine (both at 1%) are virtually not substrates of OCT1h.

> **Box 1: Calculation of Consensus Transport Efficiencies**
> The consensus series of relative TEs (Figs. 1–6) were estimated as follows: (1) Only reports with more than one TE were considered. (2) A substrate pool was selected with substrates interrelated in more than one report; e.g. for OCT2h, MPP$^+$ has been measured against TEA and cimetidine (Urakami et al. 2002), TEA against cimetidine (Barendt and Wright 2002) and MPP$^+$ against cimetidine (Dudley et al. 2000). (3) The following function (numbered from 1 to n) was defined for each substrate couple separately for each report: $y = \log[\text{ratio}/(\text{TE1}/\text{TE2})]$, where the ratio is calculated from experimental data, and TE1 and TE2 represent the consensus TEs we want to determine. This function assumes a log normal distribution of clearances. If all experimental data were perfectly consistent, then y would be 0 after

fitting of TEs. For fitting of TEs by non-linear regression (pro Fit 6.0.3 software, Quantum Soft, Switzerland; Levenberg-Marquardt fitting, y-errors unknown), mock data were tabulated in the x column as 1...n, and 0 in the y column. During fitting, the TE of MPP$^+$ was set fixed at 100%. The software attempts to minimize the sum of square errors (the errors are given by the y-values) by changing all TEs simultaneously. This procedure generates a single best set of TEs to describe all reported ratios. (4) With the consensus TEs, remaining values were adjusted within individual reports as above (with all consensus TEs set to fixed) if multiple ratios had to be considered or simply proportionally for just a single ratio.

More data are available for OCT1 from rat expressed in transfected cell lines than for the human orthologue. The transport efficiencies are shown in Table 3. Table 4 lists oocyte data for OCT1r. Figure 2 shows the derived consensus series of relative TEs for OCT1r expressed in cell lines. MPP$^+$ (100%) again is transported 3.7-fold more efficiently than TEA (27%). Surprisingly, the antidiabetic metformin, a biguanide, is transported as efficiently (26%) as TEA. The derivates buformin (41%) and phenformin (67%) are transported even better. Transport of metformin may be clinically relevant, since there is

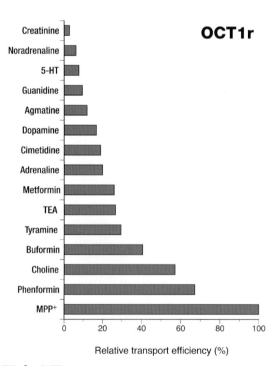

Fig. 2 Consensus TEs for OCT1r

Table 3 OCT1r expressed in cell lines[a]

Substrate	OCT1r – transfected cell line Clearance (µl min^{-1} mg protein^{-1})	Exp.
MPP$^+$	13	v/S
Tyramine	7.5	v/S
Adrenaline	7.2	v/S
TEA	7.0	v/S
Dopamine	5.0	v/S
5-HT	3.7	v/S
Noradrenaline	1.0	v/S
	293 cells, stably transfected (Breidert et al. 1998)	
MPP$^+$	8.2	v/S
Tyramine	1.5	v/S
Dopamine	0.8	v/S
Adrenaline	0.7	v/S
Noradrenaline	0.6	v/S
5-HT	0.3	v/S
Histamine	0	v/S
	293 cells, stably transfected (Gründemann and Schömig 2000a)	
MPP$^+$	8.9	v/S
Choline	6.0	v/S
TEA	3.3	v/S
Cimetidine	2.0	v/S
Guanidine	1.0	v/S
Creatinine	0.3	v/S
Histamine	0	v/S
	293 cells, stably transfected (Gründemann et al. 1999)	
Phenformin	9.6	V_{max}/K_m
Buformin	5.8	V_{max}/K_m
TEA	3.8	V_{max}/K_m
Metformin	3.7	V_{max}/K_m
	CHO-K1 cells (Wang et al. 2002)	
MPP$^+$	22	v/S
Agmatine	2.8	v/S
	293 cells, stably transfected (Gründemann et al. 2003)	
TEA	11	V_{max}/K_m
	HRPE cells, vaccinia virus expression system (Wu et al. 1998)	

Table 3 (continued)

Substrate	OCT1r – transfected cell line Clearance (µl min^{-1} mg protein^{-1})	Exp.
MPP$^+$	2.5	V_{max}/K_m
	293 cells, stably transfected (Martel et al. 1996)	
TEA	6.3	V_{max}/K_m
	MDCK cells, stably transfected (Urakami et al. 1998)	

[a]There are two somewhat incongruent data sets from our group, the older (Breidert et al. 1998) suggesting considerably higher TEs for transmitters than the more recent (Gründemann and Schömig 2000a). Since the reason for the discrepancies (cf. e.g. adrenaline vs noradrenaline) is unclear, we have admitted both data sets into consensus estimation. We have excluded v/S data of OCT1r on transport of dopamine (TE = 100 µl min^{-1} mg protein^{-1}), determined with resuspended 293 cells (Busch et al. 1996a), since, again, ultra short uptake times of 2 s were used (see legend to Table 1). Alanine, glutamate, 3-O-methylglucose and vincristine are not substrates of OCT1r expressed in 293 cells (Gründemann et al. 1998a). Exp., kind of experiment; HRPE, human retinal pigment epithelial; MDCK, Madin–Darby canine kidney; MPP$^+$, 1-methyl-4-phenylpyridinium; TEA, tetraethylammonium

good evidence that OCT1 is involved in lactic acidosis caused by metformin (Wang et al. 2003). Unfortunately, no data on transport of biguanides by OCT1 from human or by OCT2 or EMT are available.

A report from our group suggests efficient transport by OCT1r of choline (57%). Similar to OCT1h, tyramine (30%) and cimetidine (19%) are notable substrates. As explained in the legend of Table 3, the TEs for adrenaline (20%), dopamine (17%) and noradrenaline (6%) may require revision to lower values, since a large difference for adrenaline and noradrenaline seems implausible. Anyway, the catecholamines are not leading substrates. Compared to OCT2 and EMT (see Sects. 2.3 and 2.4, below), the TE for 5-HT (8%) is again rather high. Finally, agmatine (12%) and guanidine (10%) are mediocre substrates and creatinine and histamine are virtually not substrates.

2.3
OCT2

The data available on the transport efficiency of OCT2 from human expressed in transfected cell lines are shown in Table 5. Table 6 shows TEs of OCT2h expressed in oocytes. The data from Table 5 were again condensed into a consensus series of relative TEs (Fig. 3). It is apparent that MPP$^+$ is a very good substrate (100%), about twofold better than TEA (47%). Surprisingly, data from a single report (Dudley et al. 2000) suggest that the non-selective β-antagonist propranolol (370%) is transported about 4 times more efficiently than MPP$^+$.

Table 4 OCT1r expressed in *Xenopus* oocytes[a]

Substrate	OCT1r – *Xenopus* oocyte Clearance (µl h^{-1} oocyte^{-1})	Exp.
Tributylmethylammonium	3.4	v/S
N-Methyl-quinidine	2.7	v/S
N-Methyl-quinine	2.6	v/S
N-(4,4-Azo-n-pentyl)-quinuclidine	1.6	v/S
Azidoprocainamide methiodide	1.4	v/S
N-(4,4-Azo-n-pentyl)-21-deoxyajmalinium	0.2	v/S
	(Van Montfoort et al. 2001)	
MPP$^+$	19	V_{max}/K_m
NMN	1.3	V_{max}/K_m
Choline	0.5	V_{max}/K_m
	Preincubation with choline (Busch et al. 1996b)	
2'-Deoxytubercidin	3.2	V_{max}/K_m
TEA	1.6	V_{max}/K_m
	(Chen and Nelson 2000)	
MPP$^+$	16	v/S
Acetylcholine	0.4	v/S
	(Lips et al. 2005)	
Dopamine	12	V_{max}/K_m
	Preincubation with choline (Busch et al. 1996a)	
MPP$^+$	5.3	v/S
	(Nagel et al. 1997)	
Choline	1.3	V_{max}/K_m
	(Sweet et al. 2001)	
Pramipexole	4.7	V_{max}/K_m
	(Ishiguro et al. 2005)	
Amantadine	3.5	v/S
	(Goralski et al. 2002)	
TEA	0.8	V_{max}/K_m
	(Gründemann et al. 1994)	

[a] A recent report (Ishiguro et al. 2005) suggests that the dopamine D2 receptor agonist pramipexole may be a substrate, but no known substrate was assayed in parallel. Some large hydrophobic inhibitors such as cyanine 863, quinine and tubocurarine were initially claimed to be substrates based on electrophysiological measurements (Busch et al. 1996b), but that view was abandoned after experiments with radiolabelled quinine and quinidine revealed no transport at all (Nagel et al. 1997). To explain the electrophysiological artefacts, it was suggested that the inhibitors had blocked the efflux of choline which had been used to preincubate oocytes. Exp., kind of experiment; NMN, N^1-methylnicotinamide; MPP$^+$, 1-methyl-4-phenylpyridinium; TEA, tetraethylammonium

Table 5 OCT2h expressed in cell lines[a]

Substrate	OCT2h – transfected cell line Clearance (µl min^{-1} mg protein^{-1})	Exp.
MPP$^+$	30	v/S
Tyramine	19	v/S
Histamine	5.8	v/S
Noradrenaline	3.2	v/S
Adrenaline	3.0	v/S
Dopamine	2.3	v/S
5-HT	0.9	v/S
	293 cells, stably transfected (Gründemann and Schömig 2000a)	
MPP$^+$	22	v/S
TEA	8.7	V_{max}/K_m
Guanidine	7.6	v/S
Cimetidine	5.7	v/S
	293 cells, transiently transfected (Urakami et al. 2002)	
Propranolol	25	v/S
MPP$^+$	5.5	V_{max}/K_m
Cimetidine	2.2	v/S
	OCT2h+tag; mouse IMCD3 cells, transiently transfected (Dudley et al. 2000)	
MPP$^+$	47	V_{max}/K_m
Agmatine	8.2	V_{max}/K_m
	293 cells, stably transfected (Gründemann et al. 2003)	
TEA	7.2	V_{max}/K_m
Cimetidine	2.7	V_{max}/K_m
	CHO-K1 cells, stably transfected (Barendt and Wright 2002)	
TEA	5.2	v/S
Creatinine	1.8	v/S
	293 cells, transiently transfected (Urakami et al. 2004)	
TEA	4.9	V_{max}/K_m
	CHO cells, stably transfected (Suhre et al. 2005)	

[a] Again, we have omitted prostaglandins (Kimura et al. 2002) from the table (see Table 1). We have also excluded data on transport of MPP$^+$ (TE = 169 µl min^{-1} mg protein^{-1}) and dopamine (TE = 59 µl min^{-1} mg protein^{-1}), determined from V_{max}/K_m with resuspended 293 cells, since, again, in those experiments uptake times of 1 s were used (Busch et al. 1998). AZT and levofloxacin are no substrates of OCT2h (Urakami et al. 2002). Exp., kind of experiment; MPP$^+$, 1-methyl-4-phenylpyridinium; TEA, tetraethylammonium

Table 6 OCT2h expressed in *Xenopus* oocytes

Substrate	OCT2h – *Xenopus* oocyte Clearance (µl h^{-1} oocyte^{-1})	Exp.
5-HT	4.5	V_{max}/K_m
Memantine	3.0	v/S
Dopamine	1.5	V_{max}/K_m
Histamine	0.2	V_{max}/K_m
Noradrenaline	0.2	V_{max}/K_m
	Preincubation with choline (Busch et al. 1998)	
MPP$^+$	11	V_{max}/K_m
TEA	8.3	V_{max}/K_m
Choline	6.7	V_{max}/K_m
NMN	4.0	V_{max}/K_m
	Preincubation with choline (Gorboulev et al. 1997)	
MPP$^+$	14	v/S
Acetylcholine	0.6	v/S
	(Lips et al. 2005)	
Choline	2.1	v/S
	(Sweet et al. 2001)	
MPP$^+$	3.0	V_{max}/K_m
	(Leabman et al. 2002)	

Exp., kind of experiment; NMN, N^1-methylnicotinamide; MPP$^+$, 1-methyl-4-phenylpyridinium; TEA, tetraethylammonium

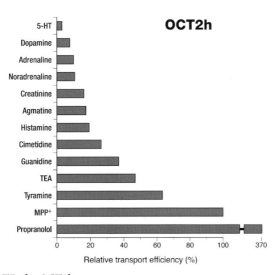

Fig. 3 Consensus TEs for OCT2h

If this ratio can be confirmed, then β-antagonists represent an interesting direction of further substrate search. Guanidine (37%) and cimetidine (27%) are prominent substrates. The best physiological substrate is tyramine (63%); this compound, however, has a relatively high passive permeability with cellular membranes (Gründemann et al. 2003), it may not need a carrier's help. Further physiological substrates are histamine (19%), agmatine (17%) and creatinine (16%). The catecholamines dopamine (8%), noradrenaline (11%) and adrenaline (10%), and especially 5-HT (3%) are, in fact, not leading substrates of this carrier.

The data available on the transport efficiency of OCT2 from rat expressed in transfected cell lines are shown in Table 7. We have reported authentic OCT2r, i.e. with a correct C-terminus (Gründemann et al. 1999). It is unclear if the artificial C-terminus used by other groups affects transport efficiency of OCT2r. Figure 4 shows the derived consensus series of relative TEs. MPP$^+$ (100%) again is transported about twofold more efficiently than TEA (56%). With the rat carrier, histamine (57%) and particularly guanidine (95%) are even better substrates than TEA. Again, tyramine (41%), cimetidine (40%) and agmatine (39%) are good substrates. A report (Goralski et al. 2002) with resuspended 293 cells suggests that amantadine (47%) is a good substrate too. Creatinine (11%) and especially the monoamines dopamine (4%), noradrenaline (2%), adrenaline (3%) and 5-HT (2%) fall off against the other substrates. According to oocyte studies, choline is a notable substrate of both OCT2h (Table 6) and OCT2r (Table 8). However, we have not been able to detect OCT2r-mediated

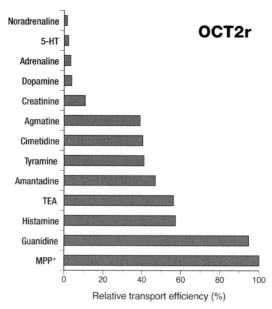

Fig. 4 Consensus TEs for OCT2r

Table 7 OCT2r expressed in cell lines

Substrate	OCT2r – transfected cell line Clearance (µl min^{-1} mg protein^{-1})	Exp.
MPP$^+$	50	V_{max}/K_m
Guanidine	47	v/S
Histamine	28	v/S
TEA	28	v/S
Cimetidine	20	v/S
Creatinine	5.3	v/S
Dopamine	1.9	V_{max}/K_m
293 cells, stably transfected (Gründemann et al. 1999)		
MPP$^+$	41	v/S
Histamine	16	v/S
Tyramine	12	v/S
Dopamine	1.4	v/S
Adrenaline	0.7	v/S
5-HT	0.6	v/S
Noradrenaline	0.4	v/S
293 cells, stably transfected (Gründemann and Schömig 2000a)		
MPP$^+$	18	v/S
Adrenaline	3.4	v/S
5-HT	1.4	v/S
Noradrenaline	1.0	v/S
Dopamine	1.0	v/S
293 cells, stably transfected (Gründemann et al. 1998b)		
MPP$^+$	41	V_{max}/K_m
Agmatine	16	V_{max}/K_m
293 cells, stably transfected (Gründemann et al. 2003)		
TEA	18	v/S
Amantadine	15	v/S
293 cells, stably transfected, resuspended; 3 s uptake; standard buffer (Goralski et al. 2002)		
TEA	5.1	v/S
293 cells, stably transfected (Gründemann et al. 1997)		
TEA	2.7	V_{max}/K_m
MDCK cells, stably transfected (Urakami et al. 1998)		
TEA	4.7	V_{max}/K_m
NIH3T3 cells, stably transfected (Pan et al. 1999)		
TEA	0.7	V_{max}/K_m
MDCK cells, stably transfected (Sweet and Pritchard 1999)		

Table 7 (continued)

Substrate	OCT2r – transfected cell line Clearance (µl min^{-1} mg protein^{-1})	Exp.
Choline	2.8	v/S
	293 cells, transiently transfected (Arndt et al. 2001)	

Exp., kind of experiment; MPP$^+$, 1-methyl-4-phenylpyridinium; TEA, tetraethylammonium

Table 8 OCT2r expressed in *Xenopus* oocytes

Substrate	OCT2r – *Xenopus* oocyte Clearance (µl h^{-1} Oocyte^{-1})	Exp.
Guanidine	4.2	V_{max}/K_m
MPP$^+$	4.0	V_{max}/K_m
TEA	3.6	V_{max}/K_m
Histamine	2.0	V_{max}/K_m
NMN	1.4	V_{max}/K_m
Choline	1.2	V_{max}/K_m
	Relative V_{max} converted to absolute V_{max} with choline data (Arndt et al. 2001)	
TEA	0.8	V_{max}/K_m
MPP$^+$	3.1	v/S
	(Okuda et al. 1999)	
MPP$^+$	15	v/S
Acetylcholine	0.6	v/S
	(Lips et al. 2005)	
TEA	0.09	v/S
	(Okuda et al. 1996)	
TEA	1.7	V_{max}/K_m
	(Sweet and Pritchard 1999)	
Choline	0.9	V_{max}/K_m
	(Sweet et al. 2001)	
Amantadine	5.0	v/S
	(Goralski et al. 2002)	
Pramipexole	0.8	V_{max}/K_m
	(Ishiguro et al. 2005)	

Exp., kind of experiment; NMN, N^1-methylnicotinamide; MPP$^+$, 1-methyl-4-phenylpyridinium; TEA, tetraethylammonium uptake of choline into 293 cells (Gründemann et al. 1999). There is one report (Arndt et al. 2001) where low uptake of choline (2.8 µl min^{-1} mg protein^{-1}) into transiently transfected 293 cells has been observed, but no comparison to a good substrate was made.

2.4
EMT

The data available on the transport efficiency of EMT from human expressed in transfected cell lines are shown in Table 9. There is only one report of expression of EMTh in oocytes (Table 10). Figure 5 shows the consensus series of relative TEs for EMTh. MPP$^+$ again is the reference substrate (100%). In contrast to OCT1 and OCT2, TEA is not a substrate at all (Gründemann et al. 1998c). Guanidine (0.8%), 5-HT (0.7%), creatinine (0.6%) and dopamine (0.4%) are also close to zero. Histamine (41%), tyramine (25%), adrenaline (18%), agmatine (13%) and noradrenaline (10%) are prominent physiological substrates. Cimetidine (7%) is not as good a substrate for EMTh as for OCT2.

For EMT from rat—as for OCT1h—information on transport efficiency determined with transfected cell lines is surprisingly scarce. Table 11 lists available data. Table 12 shows TEs of EMTr expressed in oocytes. Figure 6 shows the consensus series of relative TEs for EMTr. Compared with EMTh, we have dramatically higher TEs for noradrenaline (84% relative to MPP$^+$) and adrenaline (66%). Tyramine (72%), histamine (66%) and agmatine (32%) are high again. TEs for 5-HT (6%) and dopamine (28%) remain relatively low, but dopamine in particular—by contrast to EMTh—is markedly transported.

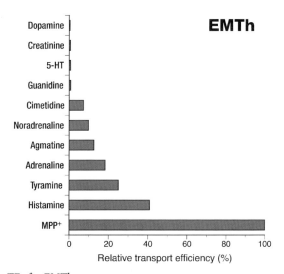

Fig. 5 Consensus TEs for EMTh

Table 9 EMTh expressed in cell lines[a]

Substrate	EMTh – transfected cell line Clearance (µl min^{-1} mg protein^{-1})	Exp.
MPP$^+$	60	v/S
Histamine	31	v/S
Tyramine	20	v/S
Adrenaline	16	v/S
Noradrenaline	7.6	V_{max}/K_m
5-HT	1.3	v/S
TEA	0	v/S
Dopamine	0	v/S
	293 cells, stably transfected (Gründemann et al. 1998c)	
MPP$^+$	31	v/S
Histamine	11	v/S
Tyramine	6.6	v/S
Adrenaline	4.6	v/S
Noradrenaline	2.7	v/S
5-HT	0.1	v/S
Dopamine	0.1	v/S
	293 cells, stably transfected (Gründemann and Schömig 2000a)	
MPP$^+$	44	v/S
Histamine	21	v/S
Cimetidine	3.5	v/S
Guanidine	0.4	v/S
Creatinine	0.3	v/S
Choline	0	v/S
	293 cells, stably transfected (Gründemann et al. 1999)	
MPP$^+$	33	v/S
Agmatine	4.2	v/S
	293 cells, stably transfected (Gründemann et al. 2003)	

[a]We have omitted a data set (in units of µl min^{-1} mg protein^{-1}: MPP: 8.3, TEA: 1.4, guanidine: 0.9) with EMTh expressed via the vaccinia virus system in human retinal pigment epithelial (HRPE) cells (Wu et al. 2000), because it is unclear if initial rates of uptake were determined (uptake time 15 min), and the conditions of measurement were non-physiological (Na$^+$-free uptake buffer, pH 8.5, room temperature instead of 37 °C). We have also excluded data (Hasannejad et al. 2004) on transport (all with units of µl min^{-1} mg protein^{-1}) of cardiac antiarrhythmics quinidine (16), lidocaine (180), disopyramide (640) and procainamide (1029), since e.g. the last clearance is exceedingly high (if 0.25 ml uptake buffer was applied per well of cells, then after 5 min and with 0.05 mg protein per well the uptake solution would be completely free of substrate), so that we must assume an error in that report. The chemotherapeutic agent 2-chloroethyl-3-sarcosinamide-1-nitrosourea (SarCNU) has also been suggested as substrate of EMTh (Chen et al. 1999; Panasci et al. 1996). However, EMT-mediated uptake of SarCNU in a model of heterologous expression has not been measured yet. Exp., kind of experimen; MPP$^+$, 1-methyl-4-phenylpyridiniumt; TEA, tetraethylammonium

Table 10 EMTh expressed in *Xenopus* oocytes

Substrate	EMTh – *Xenopus* oocyte Clearance ($\mu l\, h^{-1}\, oocyte^{-1}$)	Exp.
MPP$^+$	11	v/S
Acetylcholine	0.02	v/S
	(Lips et al. 2005)	

Exp., kind of experiment; MPP$^+$, 1-methyl-4-phenylpyridinium

Table 11 EMTr expressed in cell lines[a]

Substrate	EMTr – transfected cell line Clearance ($\mu l\, min^{-1}\, mg\, protein^{-1}$)	Exp.
MPP$^+$	32	v/S
Noradrenaline	27	v/S
Tyramine	23	v/S
Adrenaline	21	v/S
Histamine	21	v/S
Dopamine	9.0	v/S
5-HT	1.9	v/S
	293 cells, stably transfected (Gründemann and Schömig 2000a)	
MPP$^+$	29	v/S
Agmatine	9.2	V_{max}/K_m
	293 cells, stably transfected (Gründemann et al. 2003)	
TEA	0.08	V_{max}/K_m
Guanidine	1.3	v/S
	HeLa cells, pH 8.5, vaccinia virus system (Kekuda et al. 1998)	
MPP$^+$	50	V_{max}/K_m
	293 cells, stably transfected (Gründemann et al. 2002)	

[a]We have excluded data on transport of MPP$^+$ (7.8 $\mu l\, min^{-1}\, mg\, protein^{-1}$; V_{max}/K_m), dopamine (1,333 $\mu l\, min^{-1}\, mg\, protein^{-1}$; v/S) and TEA (0.0085 $\mu l\, min^{-1}\, mg\, protein^{-1}$; V_{max}/K_m) obtained with EMTr expressed in the HRPE cell line via the vaccinia virus expression system (Wu et al. 1998). The TE for dopamine indicates an error, since it corresponds with an uptake time of 30 min to a cleared volume of 40 ml mg protein^{-1}; however, we gather (Wu et al. 2000) that 0.25 ml was applied per well, which translates into about 5 ml mg protein^{-1} only. Moreover, there is no time course to judge whether uptake after 30 min still represents an initial rate or is already close to equilibrium. Finally, non-physiological uptake conditions were used (Na$^+$-free uptake buffer, pH 8.5, room temperature instead of 37°C). These shortcomings also apply to another report, where EMTr was expressed in HeLa cells (Kekuda et al. 1998). It was included in the table, however, to illustrate the low TE for TEA. Exp., kind of experiment; MPP$^+$, 1-methyl-4-phenylpyridinium; TEA, tetraethylammonium

Table 12 EMTr expressed in *Xenopus* oocytes

Substrate	EMTr – *Xenopus* oocyte Clearance (μl h^{-1} oocyte^{-1})	Exp.
TEA	0.9	v/S
Guanidine	0.9	v/S
		pH 8.5 (Kekuda et al. 1998)
MPP$^+$	11	v/S
Acetylcholine	0	v/S
		(Lips et al. 2005)

Exp., kind of experiment; MPP$^+$, 1-methyl-4-phenylpyridinium; TEA, tetraethylammonium

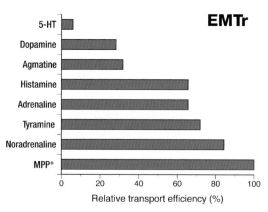

Fig. 6 Consensus TEs for EMTr

3
Roundup

We are convinced that it is important to quantitatively compare the power of a transporter for various substrates to approach its true nature [rather than to sort solutes into "yes", "perhaps", or "no" substrate categories (Jonker and Schinkel 2004; van Montfoort et al. 2003)]. The data from Figs. 1–6 suggest some common features of EMT, OCT1 and OCT2 from human and rat, but also some prominent differences. It is obvious that the neurotoxin MPP$^+$ without exception is an excellent substrate, with clearances typically in the range of 20–50 μl min^{-1} mg protein^{-1}. With the still unconfirmed exceptions of propranolol (tested only with OCT2h; Dudley et al. 2000) and of agmatine$^+$—a notional substrate with a single positive charge (Gründemann et al. 2003)—there is no better substrate known. MPP$^+$ is also a good substrate of the neuronal catecholamine carriers NET and DAT and the vesicular monoamine transporters

(VMATs). However, it is not a physiological substrate. The second-best general substrate is tyramine—a physiological substrate acquired e.g. from food and a "trace amine" in the central nervous system (Borowsky et al. 2001)—with a TE range relative to MPP$^+$ of 20%–70%. However, the passive permeability of tyramine with cell membranes is rather high, i.e. in time, even cells without carrier will accumulate substantial amounts.

Although there are discordant data sets for OCT1r, it appears that TEs of OCT1 and OCT2 for dopamine, noradrenaline, adrenaline and 5-HT in general are rather low, in the range relative to MPP$^+$ of 5%–15%, or 0.5–4 µl min^{-1} mg protein^{-1} in absolute terms of clearance. This suggests that OCT1 and OCT2 are not primarily dedicated to transport these monoamine transmitters. By contrast, EMT from rat has high TEs for the catecholamines (30%–80% relative to MPP$^+$, 9–27 µl min^{-1} mg protein^{-1}). Note, however, that we have observed a phenomenon of transporter activation for EMTr and EMTm— but not for EMTh—upon stable heterologous expression of these carriers in 293 cells (Gründemann et al. 2002). Thus, transporter activation could be responsible for the high catecholamine TEs of EMTr. The human orthologue shows TEs for adrenaline and noradrenaline intermediate between EMTr and OCT1/OCT2, i.e. 10%–20% relative to MPP$^+$, or 5–10 µl min^{-1} mg protein^{-1}. There is virtually no EMTh-mediated transport of dopamine and 5-HT. Thus, the available TE data suggest that only EMT may play a significant role in catecholamine inactivation. Interestingly, although expression of EMT in the lung seems to be relatively low (Kekuda et al. 1998), inhibition of uptake of noradrenaline by EMT into airway smooth muscle cells may contribute to acute bronchial vasoconstriction caused by inhaled glucocorticoids (Horvath et al. 2001; Horvath et al. 2003; Lips et al. 2005).

TEA at about 30%–50% relative to MPP$^+$, or 5–25 µl min^{-1} mg protein^{-1}, is a fine substrate for both OCTs. It is not, however, a substrate for EMTh, and, at 0.08 µl min^{-1} mg protein^{-1}, only a poor substrate for EMTr (Kekuda et al. 1998).

Histamine (at 20%–60% relative to MPP$^+$, or 10–30 µl min^{-1} mg protein^{-1}) and agmatine (similar to histamine for OCT2, half as good as histamine for EMT) consistently are good substrates for OCT2 and EMT, but not for OCT1. This opens the possibility that OCT2 and EMT play a vital role in inactivating or releasing these transmitters. Histamine is involved in inflammation and allergy, and in the regulation of vigilance and sleep-wake rhythm (Huang et al. 2001). Agmatine has been implicated in control of cellular proliferation and tumour suppression and may function as a neurotransmitter (Satriano et al. 1999). Guanidine (40%–90% relative to MPP$^+$) and its derivatives cimetidine (25%–40% relative to MPP$^+$) and creatinine (10%–20% relative to MPP$^+$) are accepted particularly well as substrates by OCT2, but not so much by EMTh and OCT1r.

A number of reports suggest transport of choline in the oocyte expression system, with large differences in TEs relative to MPP$^+$: Table 6, OCT2h,

75% (Gorboulev et al. 1997); Table 8, OCT2r, 30% (Arndt et al. 2001); Table 4, OCT1r, 3% (Busch et al. 1996b). However, we bear in mind the oocyte choline artefact outlined in Sect 2.1 above (Schloss et al. 1994). There is only a single report of (low) choline transport in a transfected cell line, mediated by OCT2r (Arndt et al. 2001). We have not observed choline transport for OCT2r and EMTh in 293 cells (Gründemann et al. 1999), but we have detected transport via OCT1r (57% relative to MPP$^+$). Interestingly, in a recent report (Lips et al. 2005), transport of acetylcholine in the oocyte system was consistently low for OCT1h (2.2% relative to MPP$^+$), OCT1r (2.6%), OCT2h (4.3%) and OCT2r (4.2%), and negligible for EMTh (0.2%) and EMTr (no transport). Clearly, transport of choline and acetylcholine should be re-examined systematically for OCT1, OCT2 and EMT in transfected cell lines.

From the plethora of inhibitors that have been tested on EMT, OCT1 and OCT2, a casual observer might get the impression that these carriers indiscriminately transport very many compounds. However, our analysis of transport efficiencies shows that our knowledge about actual substrates is rather limited. Except for cimetidine, no confirmed data are available on transport efficiencies of clinically relevant drugs by the human carriers. This is surprising, since many drugs are cationic and thus potential carrier substrates.

It is possible to regard the carriers as fairly tolerant towards structural details of their substrates, as long as some basic requirements are met, in particular a single positive charge. This is similar with the vesicular monoamine transporters which are not specific for a single transmitter. Thus, the carriers may transport anything that somehow fits; they may function as a flexible means to transport many substrates with low to moderate transport efficiency. This corresponds to a notion of redundant isoenzymes for the elimination of metabolites and xenobiotics with marginal differences.

Alternatively, we hold it entirely possible that the endogenous key substrates of the carriers have yet to be discovered. The best physiological substrates known so far, i.e. tyramine and histamine, use only a fraction of the transport capacity we see for MPP$^+$. And the TEs for MPP$^+$ (20–50 µl min^{-1} mg protein^{-1}) must not be the end of the line in terms of transport power. Moreover, amino acid sequence statistics indicate that OCT1 and OCT2 have an evolutionary distance (70% identity for the human carriers) that is larger than the distance (77% identity) between the highly specific transporters for ergothioneine (ETT) and carnitine (OCTN2). The distance of EMT to both OCT1 and OCT2 is still larger (each pair at 50% identity). It is thus conceivable that each carrier has its specific key substrate; other compounds may be transported of course, albeit at reduced efficiencies (cf. ETT). The differences outlined above for histamine, agmatine, TEA and choline support this hypothesis. Moreover, pharmacological characterization has revealed significant differences in sensitivity to inhibitors, e.g. OCT1 is relatively resistant to some steroids (such as 17β-estradiol and testosterone) and cyanine inhibitors, which are the most potent inhibitors of OCT2 and EMT (Russ et al. 1993) and O-methylisoprenaline is more effective on EMT than on OCT1 and OCT2.

In conclusion, a focus on analysis of transport efficiency, integrated with the respective expression profile, may help substantially to direct further research and to establish the precise purpose of each carrier. Eventually, the substrate specificity is key to understanding the physiological and pathophysiological significance of a transporter.

References

Arndt P, Volk C, Gorboulev V, Budiman T, Popp C, Ulzheimer-Teuber I, Akhoundova A, Koppatz S, Bamberg E, Nagel G, Koepsell H (2001) Interaction of cations, anions, and weak base quinine with rat renal cation transporter rOCT2 compared with rOCT1. Am J Physiol Renal Physiol 281:F454–468

Barendt WM, Wright SH (2002) The human organic cation transporter (hOCT2) recognizes the degree of substrate ionization. J Biol Chem 277:22491–22496

Borowsky B, Adham N, Jones KA, Raddatz R, Artymyshyn R, Ogozalek KL, Durkin MM, Lakhlani PP, Bonini JA, Pathirana S, Boyle N, Pu X, Kouranova E, Lichtblau H, Ochoa FY, Branchek TA, Gerald C (2001) Trace amines: identification of a family of mammalian G protein-coupled receptors. Proc Natl Acad Sci USA 98:8966–8971

Breidert T, Spitzenberger F, Gründemann D, Schömig E (1998) Catecholamine transport by the organic cation transporter type 1 (OCT1). Br J Pharmacol 125:218–224

Briz O, Serrano MA, Rebollo N, Hagenbuch B, Meier PJ, Koepsell H, Marin JJ (2002) Carriers involved in targeting the cytostatic bile acid-cisplatin derivatives cis-diammine-chloro-cholylglycinate-platinum(II) and cis-diammine-bisursodeoxycholate-platinum(II) toward liver cells. Mol Pharmacol 61:853–860

Busch AE, Quester S, Ulzheimer JC, Gorboulev V, Akhoundova A, Waldegger S, Lang F, Koepsell H (1996a) Monoamine neurotransmitter transport mediated by the polyspecific cation transporter rOCT1. FEBS Lett 395:153–156

Busch AE, Quester S, Ulzheimer JC, Waldegger S, Gorboulev V, Arndt P, Lang F, Koepsell H (1996b) Electrogenetic properties and substrate specificity of the polyspecific rat cation transporter rOCT1. J Biol Chem 271:32599–32604

Busch AE, Karbach U, Miska D, Gorboulev V, Akhoundova A, Volk C, Arndt P, Ulzheimer JC, Sonders MS, Baumann C, Waldegger S, Lang F, Koepsell H (1998) Human neurons express the polyspecific cation transporter hOCT2, which translocates monoamine neurotransmitters, amantadine, and memantine. Mol Pharmacol 54:342–352

Chen R, Nelson JA (2000) Role of organic cation transporters in the renal secretion of nucleosides. Biochem Pharmacol 60:215–219

Chen Z-P, Wang G, Huang Q, Sun Z-F, Zhou L-Y, Wang A-D, Panasci LC (1999) Enhanced antitumor activity of SarCNU in comparison to BCNU in an extraneuronal monoamine transporter positive human glioma xenograft model. J Neurooncol 44:7–14

Dudley AJ, Bleasby K, Brown CDA (2000) The organic cation transporter OCT2 mediates the uptake of beta-adrenoceptor antagonists across the apical membrane of renal LLC-PK1 cell monolayers. Br J Pharmacol 131:71–79

Eisenhofer G (2001) The role of neuronal and extraneuronal plasma membrane transporters in the inactivation of peripheral catecholamines. Pharmacol Ther 91:35–62

Eisenhofer G, McCarty R, Pacak K, Russ H, Schömig E (1996) Disprocynium24, a novel inhibitor of the extraneuronal monoamine transporter, has potent effects on the inactivation of circulating noradrenaline and adrenaline in conscious rat. Naunyn Schmiedebergs Arch Pharmacol 354:287–294

Eraly SA, Nigam SK (2002) Novel human cDNAs homologous to Drosophila Orct and mammalian carnitine transporters. Biochem Biophys Res Commun 297:1159–1166

Goralski KB, Lou G, Prowse MT, Gorboulev V, Volk C, Koepsell H, Sitar DS (2002) The cation transporters rOCT1 and rOCT2 interact with bicarbonate but play only a minor role for amantadine uptake into rat renal proximal tubules. J Pharmacol Exp Ther 303:959–968

Gorboulev V, Ulzheimer JC, Akhoundova A, Ulzheimer-Teuber I, Karbach U, Quester S, Baumann C, Lang F, Busch AE, Koepsell H (1997) Cloning and characterization of two human polyspecific organic cation transporters. DNA Cell Biol 16:871–881

Gründemann D, Schömig E (2000a) Efficiency of transport of monoamine transmitters by non-neuronal transporters. Pharmacol Toxicol 87:34

Gründemann D, Schömig E (2000b) Gene structures of the human non-neuronal monoamine transporters EMT and OCT2. Hum Genet 106:627–635

Gründemann D, Schömig E (2003) Organic cation transporters. In: Rosenthal W, Offermanns S (eds) Encyclopedic reference of molecular pharmacology. Springer Verlag, Heidelberg Berlin New York, pp 696–701

Gründemann D, Gorboulev V, Gambaryan S, Veyhl M, Koepsell H (1994) Drug excretion mediated by a new prototype of polyspecific transporter. Nature 372:549–552

Gründemann D, Babin-Ebell J, Martel F, Örding N, Schmidt A, Schömig E (1997) Primary structure and functional expression of the apical organic cation transporter from kidney epithelial LLC-PK1 cells. J Biol Chem 272:10408–10413

Gründemann D, Breidert T, Spitzenberger F, Schömig E (1998a) Molecular structure of the carrier responsible for hepatic uptake of catecholamines. In: Goldstein DS, Eisenhofer G, McCarty R (eds) Advances in pharmacology vol. 42: catecholamines: bridging basic science with clinical medicine. Academic Press, San Diego, pp 346–349

Gründemann D, Köster S, Kiefer N, Breidert T, Engelhardt M, Spitzenberger F, Obermüller N, Schömig E (1998b) Transport of monoamine transmitters by the organic cation transporter type 2 (OCT2). J Biol Chem 273:30915–30920

Gründemann D, Schechinger B, Rappold GA, Schömig E (1998c) Molecular identification of the corticosterone-sensitive extraneuronal monoamine transporter. Nat Neurosci 1:349–351

Gründemann D, Liebich G, Kiefer N, Köster S, Schömig E (1999) Selective substrates for non-neuronal monoamine transporters. Mol Pharmacol 56:1–10

Gründemann D, Koschker A-C, Haag C, Honold C, Zimmermann T, Schömig E (2002) Activation of the extraneuronal monoamine transporter (EMT) from rat expressed in 293 cells. Br J Pharmacol 137:910–918

Gründemann D, Hahne C, Berkels R, Schömig E (2003) Agmatine is efficiently transported by non-neuronal monoamine transporters EMT and OCT2. J Pharmacol Exp Ther 304:810–817

Gründemann D, Harlfinger S, Golz S, Geerts A, Lazar A, Berkels R, Jung N, Rubbert A, Schömig E (2005) Discovery of the ergothioneine transporter. Proc Natl Acad Sci USA 102:5256–5261

Haag C, Berkels R, Gründemann D, Lazar A, Taubert D, Schömig E (2004) The localisation of the extraneuronal monoamine transporter (EMT) in rat brain. J Neurochem 88:291–297

Harlfinger S, Fork C, Lazar A, Schömig E, Gründemann D (2005) Are Organic Cation Transporters capable of transporting prostaglandins? Naunyn Schmiedebergs Arch Pharmacol 372:125–130

Hasannejad H, Takeda M, Narikawa S, Huang XL, Enomoto A, Taki K, Niwa T, Jung SH, Onozato ML, Tojo A, Endou H (2004) Human organic cation transporter 3 mediates the transport of antiarrhythmic drugs. Eur J Pharmacol 499:45–51

Hayer M, Bönisch H, Brüss M (1999) Molecular cloning, functional characterization and genomic organization of four alternatively spliced isoforms of the human organic cation transporter 1 (hOCT1/ SLC22A1). Ann Hum Genet 63:473–482

Hayer-Zillgen M, Brüss M, Bönisch H (2002) Expression and pharmacological profile of the human organic cation transporters hOCT1, hOCT2 and hOCT3. Br J Pharmacol 136:829–836

Horvath G, Lieb T, Conner GE, Salathe M, Wanner A (2001) Steroid sensitivity of norepinephrine uptake by human bronchial arterial and rabbit aortic smooth muscle cells. Am J Respir Cell Mol Biol 25:500–506

Horvath G, Sutto Z, Torbati A, Conner GE, Salathe M, Wanner A (2003) Norepinephrine transport by the extraneuronal monoamine transporter in human bronchial arterial smooth muscle cells. Am J Physiol Lung Cell Mol Physiol 285:L829–837

Huang ZL, Qu WM, Li WD, Mochizuki T, Eguchi N, Watanabe T, Urade Y, Hayaishi O (2001) Arousal effect of orexin A depends on activation of the histaminergic system. Proc Natl Acad Sci U S A 98:9965–9970

Inazu M, Takeda H, Matsumiya T (2003) Expression and functional characterization of the extraneuronal monoamine transporter in normal human astrocytes. J Neurochem 84:43–52

Ishiguro N, Saito A, Yokoyama K, Morikawa M, Igarashi T, Tamai I (2005) Transport of the dopamine D2 agonist pramipexole by rat organic cation transporters OCT1 and OCT2 in kidney. Drug Metab Dispos 33:495–499

Iversen LL (1965) The uptake of catechol amines at high perfusion concentrations in the rat isolated heart: a novel catechol amine uptake process. Br J Pharmacol 25:18–33

Jonker JW, Schinkel AH (2004) Pharmacological and physiological functions of the polyspecific organic cation transporters: OCT1, 2, and 3 (SLC22A1–3). J Pharmacol Exp Ther 308:2–9

Jonker JW, Wagenaar E, Mol CA, Buitelaar M, Koepsell H, Smit JW, Schinkel AH (2001) Reduced hepatic uptake and intestinal excretion of organic cations in mice with a targeted disruption of the organic cation transporter 1 (oct1). Mol Cell Biol 21:5471–5477

Jonker JW, Wagenaar E, Van Eijl S, Schinkel AH (2003) Deficiency in the organic cation transporters 1 and 2 (Oct1/Oct2 [Slc22a1/Slc22a2]) in mice abolishes renal secretion of organic cations. Mol Cell Biol 23:7902–7908

Karbach U, Kricke J, Meyer-Wentrup F, Gorboulev V, Volk C, Loffing-Cueni D, Kaissling B, BAchmann S, Koepsell H (2000) Localization of organic cation transporters OCT1 and OCT2 in rat kidney. Am J Physiol 279:F679–F687

Kekuda R, Prasad PD, Wu X, Wang H, Fei Y-J, Leibach FH, Ganapathy V (1998) Cloning and functional characterization of a potential-sensitive, polyspecific organic cation transporter (OCT3) most abundantly expressed in placenta. J Biol Chem 273:15971–15979

Kerb R, Brinkmann U, Chatskaia N, Gorbunov D, Gorboulev V, Mornhinweg E, Keil A, Eichelbaum M, Koepsell H (2002) Identification of genetic variations of the human organic cation transporter hOCT1 and their functional consequences. Pharmacogenetics 12:591–595

Kimura H, Takeda M, Narikawa S, Enomoto A, Ichida K, Endou H (2002) Human organic anion transporters and human organic cation transporters mediate renal transport of prostaglandins. J Pharmacol Exp Ther 301:293–298

Koepsell H (2004) Polyspecific organic cation transporters: their functions and interactions with drugs. Trends Pharmacol Sci 25:375–381

Koepsell H, Schmitt BM, Gorboulev V (2003) Organic cation transporters. Rev Physiol Biochem Pharmacol 150:36–90

Kristufek D, Rudorfer W, Pifl C, Huck S (2002) Organic cation transporter mRNA and function in the rat superior cervical ganglion. J Physiol 543:117–134

Lazar A, Gründemann D, Berkels R, Taubert D, Zimmermann T, Schömig E (2003) Genetic variability of the extraneuronal monoamine transporter EMT (SLC22A3). J Hum Genet 48:226–230

Leabman MK, Huang CC, Kawamoto M, Johns SJ, Stryke D, Ferrin TE, DeYoung J, Taylor T, Clark AG, Herskowitz I, Giacomini KM (2002) Polymorphisms in a human kidney xenobiotic transporter, OCT2, exhibit altered function. Pharmacogenetics 12:395–405

Lips KS, Volk C, Schmitt BM, Pfeil U, Arndt P, Miska D, Ermert L, Kummer W, Koepsell H (2005) Polyspecific cation transporters mediate luminal release of acetylcholine from bronchial epithelium. Am J Respir Cell Mol Biol 33:79–88

Martel F, Vetter T, Russ H, Gründemann D, Azevedo I, Koepsell H, Schömig E (1996) Transport of small organic cations in the rat liver—the role of the organic cation transporter OCT1. Naunyn Schmiedebergs Arch Pharmacol 354:320–326

Mayser W, Schloss P, Betz H (1992) Primary structure and functional expression of a choline transporter expressed in the rat nervous system. FEBS Lett 305:31–36

Nagel G, Volk C, Friedrich T, Ulzheimer JC, Bamberg E, Koepsell H (1997) A reevaluation of substrate specificity of the rat cation transporter rOCT1. J Biol Chem 272:31953–31956

Okuda M, Saito H, Urakami Y, Takano M, Inui K (1996) cDNA cloning and functional expression of a novel rat kidney organic cation transporter, OCT2. Biochem Biophys Res Commun 224:500–507

Okuda M, Urakami Y, Saito H, Inui K-I (1999) Molecular mechanisms of organic cation transport in OCT2-expressing Xenopus oocytes. Biochim Biophys Acta 1417:224–231

Pan BF, Sweet DH, Pritchard JB, Chen R, Nelson JA (1999) A transfected cell model for the renal toxin transporter, rOCT2. Toxicol Sci 47:181–186

Panasci LC, Marcantonio D, Noe AJ (1996) SarCNU (2-chloroethyl-3-sarcosinamide-1-nitrosourea): a novel analogue of chloroethylnitrosourea that is transported by the catecholamine uptake2 carrier, which mediates increased cytotoxicity. Cancer Chemother Pharmacol 37:505–508

Russ H, Engel W, Schömig E (1993) Isocyanines and pseudoisocyanines as a novel class of potent noradrenalin transport inhibitors: synthesis, detection, and biological activity. J Med Chem 36:4208–4213

Russ H, Staudt K, Martel F, Gliese M, Schömig E (1996) The extraneuronal transporter for monoamine transmitters exists in cells derived from human central nervous system glia. Eur J Neurosci 8:1256–1264

Sakata T, Anzai N, Shin HJ, Noshiro R, Hirata T, Yokoyama H, Kanai Y, Endou H (2004) Novel single nucleotide polymorphisms of organic cation transporter 1 (SLC22A1) affecting transport functions. Biochem Biophys Res Commun 313:789–793

Satriano J, Kelly CJ, Blantz RC (1999) An emerging role for agmatine. Kidney Int 56:1252–1253

Schloss P, Mayser W, Betz H (1994) The putative rat choline transporter CHOT1 transports creatine and is highly expressed in neural and muscle-rich tissue. Biochem Biophys Res Commun 198:637–645

Schömig E, Spitzenberger F, Engelhardt M, Martel F, Örding N, Gründemann D (1998) Molecular cloning and characterization of two novel transport proteins from rat kidney. FEBS Lett 425:79–86

Shang T, Uihlein AV, Van Asten J, Kalyanaraman B, Hillard CJ (2003) 1-Methyl-4-phenylpyridinium accumulates in cerebellar granule neurons via organic cation transporter 3. J Neurochem 85:358–367

Shu Y, Leabman MK, Feng B, Mangravite LM, Huang CC, Stryke D, Kawamoto M, Johns SJ, DeYoung J, Carlson E, Ferrin TE, Herskowitz I, Giacomini KM (2003) Evolutionary conservation predicts function of variants of the human organic cation transporter, OCT1. Proc Natl Acad Sci U S A 100:5902–5907

Sugawara-Yokoo M, Urakami Y, Koyama H, Fujikura K, Masuda S, Saito H, Naruse T, Inui K, Takata K (2000) Differential localization of organic cation transporters rOCT1 and rOCT2 in the basolateral membrane of rat kidney proximal tubules. Histochem Cell Biol 114:175–180

Suhre WM, Ekins S, Chang C, Swaan PW, Wright SH (2005) Molecular determinants of substrate/inhibitor binding to the human and rabbit renal organic cation transporters hOCT2 and rbOCT2. Mol Pharmacol 67:1067–1077

Sweet DH, Pritchard JB (1999) rOCT2 is a basolateral potential-driven carrier, not an organic cation/proton exchanger. Am J Physiol 277:F890–F898

Sweet DH, Miller DS, Pritchard JB (2001) Ventricular choline transport: a role for organic cation transporter 2 expressed in choroid plexus. J Biol Chem 276:41611–41619

Tamai I, Yabuuchi H, Nezu J-I, Sai Y, Oku A, Shimane M, Tsuji A (1997) Cloning and characterization of a novel human pH-dependent organic cation transporter, OCTN1. FEBS Lett 419:107–111

Tamai I, Ohashi R, Nezu J, Yabuuchi H, Oku A, Shimane M, Sai Y, Tsuji A (1998) Molecular and functional identification of sodium ion-dependent, high affinity human carnitine transporter OCTN2. J Biol Chem 273:20378–20382

Tamai I, Ohashi R, Nezu JI, Sai Y, Kobayashi D, Oku A, Shimane M, Tsuji A (2000) Molecular and functional characterization of organic cation/carnitine transporter family in mice. J Biol Chem 275:40064–40072

Trendelenburg U (1988) The extraneuronal uptake and metabolism of catecholamines. In: Trendelenburg U, Weiner N (eds) Catecholamines I. Springer-Verlag, Heidelberg Berlin New York, pp 279–319

Urakami Y, Okuda M, Masuda S, Saito H, Inui K-I (1998) Functional characteristics and membrane localization of rat multispecific organic cation transporters, OCT1 and OCT2, mediating tubular secretion of cationic drugs. J Pharmacol Exp Ther 287:800–805

Urakami Y, Nakamura N, Takahashi K, Okuda M, Saito H, Hashimoto Y, Inui K (1999) Gender differences in expression of organic cation transporter OCT2 in rat kidney. FEBS Lett 461:339–342

Urakami Y, Akazawa M, Saito H, Okuda M, Inui K (2002) cDNA cloning, functional characterization, and tissue distribution of an alternatively spliced variant of organic cation transporter hOCT2 predominantly expressed in the human kidney. J Am Soc Nephrol 13:1703–1710

Urakami Y, Kimura N, Okuda M, Inui K (2004) Creatinine transport by basolateral organic cation transporter hOCT2 in the human kidney. Pharm Res 21:976–981

Van Montfoort JE, Müller M, Groothuis GMM, Meijer DKF, Koepsell H, Meier PJ (2001) Comparison of "type I" and "type II" organic cation transport by organic cation transporters and organic anion-transporting polypeptides. J Pharmacol Exp Ther 298:110–115

van Montfoort JE, Hagenbuch B, Groothuis GM, Koepsell H, Meier PJ, Meijer DK (2003) Drug uptake systems in liver and kidney. Curr Drug Metab 4:185–211

Verhaagh S, Barlow DP, Zwart R (2001) The extra-neuronal monoamine transporter SLC22A3/ORCT3 co-localizes with the Maoa metabolizing enzyme in mouse placenta. Mech Dev 100:127–130

Vialou V, Amphoux A, Zwart R, Giros B, Gautron S (2004) Organic cation transporter 3 (Slc22a3) is implicated in salt-intake regulation. J Neurosci 24:2846–2851

Wang DS, Jonker JW, Kato Y, Kusuhara H, Schinkel AH, Sugiyama Y (2002) Involvement of organic cation transporter 1 in hepatic and intestinal distribution of metformin. J Pharmacol Exp Ther 302:510–515

Wang DS, Kusuhara H, Kato Y, Jonker JW, Schinkel AH, Sugiyama Y (2003) Involvement of organic cation transporter 1 in the lactic acidosis caused by metformin. Mol Pharmacol 63:844–848

Wu X, Kekuda R, Huang W, Fei Y-J, Leibach FH, Chen J, Conway SJ, Ganapathy V (1998) Identity of the organic cation transporter OCT3 as the extraneuronal monoamine transporter (uptake2) and evidence for the expression of the transporter in the brain. J Biol Chem 273:32776–32786

Wu X, Huang W, Ganapathy ME, Wang H, Kekuda R, Conway SJ, Leibach FH, Ganapathy V (2000) Structure, function, and regional distribution of the organic cation transporter OCT3 in the kidney. Am J Physiol 279:F449–F458

Yabuuchi H, Tamai I, Nezu J-I, Sakamoto K, Oku A, Shimane M, Sai Y, Tsuji A (1999) Novel membrane transporter OCTN1 mediates multispecific, bidirectional, and pH-dependent transport of organic cations. J Pharmacol Exp Ther 289:768–773

Zhang L, Dresser MJ, Gray AT, Yost SC, Terashita S, Giacomini KM (1997) Cloning and functional expression of a human liver organic cation transporter. Mol Pharmacol 51:913–921

Zhang L, Schaner ME, Giacomini KM (1998) Functional characterization of an organic cation transporter (hOCT1) in a transiently transfected human cell line (HeLa). J Pharmacol Exp Ther 286:354–361

Zwart R, Verhaagh S, Buitelaar M, Popp-Snijders C, Barlow DP (2001) Impaired activity of the extraneuronal monoamine transporter system known as uptake-2 in Orct3/Slc22a3-deficient mice. Mol Cell Biol 21:4188–4196

The Role of SNARE Proteins in Trafficking and Function of Neurotransmitter Transporters

M. W. Quick

Department of Biological Sciences, University of Southern California,
HNB 228, 3641 Watt Way, Los Angeles CA, 90089-2520, USA
mquick@usc.edu

1	Introduction	182
1.1	SNARE Proteins	182
1.2	Syntaxin 1A Regulation of Ion Channels	184
1.3	Initial Evidence for SNARE Regulation of Neurotransmitter Transporters	185
2	Overview of Syntaxin 1A Regulation of Neurotransmitter Transporters	186
2.1	Glycine Transporters	186
2.2	Amine Transporters	187
2.3	Glutamate Transporters	187
3	Details of Syntaxin 1A Regulation of GAT1	188
3.1	Regulation of GAT1 Trafficking	188
3.2	Regulation of Intrinsic Properties of GAT1	189
4	Syntaxin 1A Regulation of SERT Conducting States	191
5	Conclusions	193
	References	193

Abstract The SNARE hypothesis of vesicle fusion proposes that a series of protein–protein interactions governs the delivery of vesicles to various membrane targets such as the Golgi network and the plasma membrane. Key players in this process include members of the syntaxin family of membrane proteins. The first member identified in this family, syntaxin 1A, plays an essential role in the docking and fusion of neurotransmitter-containing vesicles to the presynaptic membrane of neurons. Syntaxin 1A and other syntaxin family members have also been shown to interact with, and directly regulate, a variety of ion channels. More recently, the family of plasma membrane neurotransmitter transporters, proteins that function in part to control transmitter levels in brain, have been shown to be direct targets of syntaxin 1A regulation. This regulation involves both the trafficking of transporters as well as the control of ion and transmitter flux through transporters. In this chapter, the functional effects of syntaxin-transporter interactions are reviewed, and how such interactions may regulate neuronal signaling are considered.

Keywords Neurotransmitter uptake · Protein–protein interactions · Regulation · Syntaxin 1A · Transmitter transport

1
Introduction

Reliable chemical neurotransmission requires the precise control of spatial and temporal activity of transmitter in the extracellular space, and inappropriate transmitter levels are associated with a variety of neurological disorders. For example, low levels of biogenic amine transmitters, such as serotonin, are associated with depression (Nutt 2002); high levels of glutamate can result in spillover of transmitter to neighboring synapses, desensitization of glutamate receptors, excitotoxicity, and neural degeneration (Arundine and Tymianski 2003). Plasma membrane neurotransmitter transporters play multiple roles in the central nervous system, many of which are just now being discovered, but a central role for transporters is in the control of extracellular transmitter levels. Transporters manage this role in several ways. First, transporters can be expressed at very high levels near transmitter release sites on neurons and glia (Chaudhry et al. 1995; Chiu et al. 2002), where they sequester transmitter away from receptors (Diamond and Jahr 1997) and regulate spillover from the synaptic cleft (Ichinose and Lukasiewicz 2002; Overstreet and Westbrook 2003). Second, although ionic conditions typically favor the uptake of transmitter from the extracellular space, transporters can operate in reverse, and nonvesicular efflux of transmitter contributes to ambient extracellular transmitter levels (Rudnick 1998). Indeed, efflux is a principal mode of calcium-independent neurotransmitter release in some systems (Schwartz 1987). Third, the uptake of transmitter is tied to the electrochemical gradient and the stoichiometry of the uptake process. The half-maximal effective concentrations for transport occur in the nanomolar to low micromolar range.

Because transmitter levels need to be accurately regulated, it is not surprising that cells have developed means to regulate transporters. Much research in the field of transporters over the past decade has been aimed at understanding the initial triggers and subcellular pathways that control transporter function. Such regulation impacts not only the trafficking of transporters to and from the plasma membrane, but the flux of transmitter through individual transporters. Many of these regulators of transport are described in other chapters in this book. In this chapter, the role that the plasma membrane SNARE syntaxin 1A plays in regulating plasma membrane neurotransmitter transporters is examined.

1.1
SNARE Proteins

Cell growth, development, and differentiation require the delivery of cell constituents to appropriate subcellular locations. For membrane-bound complexes, this requires specific targeting of the complex to the correct membrane (e.g. Golgi, endosome, plasma membrane) and insertion of the complex into the membrane without destroying cell integrity. This is accomplished by incorporating the complex into a vesicular membrane, delivering the vesicle to the target membrane, and fusing the vesicular phospholipid bilayer to the target membrane bilayer. As discussed by Jahn and Südhof (1999), this fusion reaction is highly bioenergetically unfavorable, and so specific proteins exist that enhance the fusion process.

Insights into the proteins involved in the fusion process began with in vitro, cell-free experiments identifying two cytosolic proteins that were positive regulators of vesicle docking and fusion (Rothman 1994). These are the ATPase N-ethylmaleimide-sensitive factor (NSF) and its associated partner, soluble NSF attachment protein (SNAP), which permits NSF to attach to target membranes. Soon thereafter, proteins in complex with NSF and SNAP were identified (Sollner et al. 1993). These proteins reside on vesicle and target membranes and serve as soluble NSF attachment protein receptors (SNAREs). Interestingly, there are specific SNAREs for specific membranes, and this finding gave rise to the SNARE hypothesis of vesicle docking and fusion (Rothman and Warren 1994).

One of the most well-studied vesicle fusion events in cells is that of the fusion of neurotransmitter-containing vesicles to the plasma membrane of neurons, resulting in transmitter release into the extracellular space. The first SNAREs to be characterized were plasma membrane-resident, 35-kDa proteins highly enriched at synapses (and also found in other nonneuronal secretory cells). They were called syntaxin 1A and 1B (Bennett et al. 1992). There are now 15 known syntaxin isoforms (in humans). Consistent with the SNARE hypothesis, specific syntaxin isoforms are found on specific subcellular target membranes in all cell types, and they are thought to be positive regulators of specific membrane fusion events. Syntaxin isoforms have also been well characterized in yeast budding and fusion events, supporting a conserved role for SNAREs in the basic cellular events of all organisms (Bennett and Scheller 1993).

At the nerve terminal, syntaxin 1A forms a highly stable complex with two other SNAREs: the plasma membrane SNARE SNAP-25, and the vesicle SNARE VAMP-2 (vesicle-associated membrane protein 2; synaptobrevin). High-resolution structural analysis reveals that four parallel helices make up the core complex: VAMP-2 and syntaxin 1A contribute one helix each, and SNAP-25 contributes two (Sutton et al. 1998). This arrangement thus brings the vesicle in direct opposition to the plasma membrane and readies it for the fusion event.

Structurally, syntaxin 1A is a 288-amino-acid protein with a single C-terminal transmembrane domain (TMD); the remainder of the protein is oriented to the cytoplasm. Proximal to the TMD is a helical region that participates in the core complex. It is known as the SNARE motif or the H3 domain, and it extends approximately from residues 189 to 258. The residues in the H3 domain are arranged in seven sequential heptad repeats. Residues 1–189 of syntaxin 1A contain three additional helical domains. These additional domains (HA, HB, HC) likely participate in syntaxin 1A binding to other binding partners (Weimbs et al. 1997). For example, Munc-18, a negative regulator of membrane fusion, exerts its inhibition of SNARE complex formation through protein–protein interactions with the HA, HB, HC, and H3 domains of syntaxin 1A (Dulubova et al. 1999).

1.2
Syntaxin 1A Regulation of Ion Channels

If the SNARE hypothesis is correct, then the delivery of vesicles containing membrane proteins to neuronal plasma membranes is dependent upon syntaxin 1A and other components of the fusion core complex. And thus, factors that regulate membrane fusion might be expected to be regulators of the number of membrane proteins present on the cell surface. However, the story became much more interesting with evidence showing that syntaxin 1A directly interacts and functionally regulates calcium channels (for review, see Zamponi 2003).

Voltage-gated calcium channels are composed of four separate TMD regions that are linked together. One of the linker regions contains a "synprint," a stretch of amino acid residues in which SNARE proteins such as syntaxin 1A and SNAP-25 bind (Sheng et al. 1994). This interaction can be regulated by a number of different signaling cascades including protein kinase C (PKC), G proteins, and calcium. The association between syntaxin 1A and voltage-gated calcium channels raises the possibility that the association may play a role in synaptic transmission by localizing calcium channels (and calcium) near to the site of membrane fusion, which is a calcium-dependent process. However, some data suggest that the interaction does not play a role in localizing this signaling complex (Zamponi 2003).

Interestingly, syntaxin 1A inhibits calcium channel function directly. Coexpression of these two proteins in *Xenopus* oocytes causes a shift in the current–voltage relationship for calcium channel inactivation. The end result of this shift is that a larger number of channels are likely to be in the inactive state, and will be unavailable for opening in response to a nerve terminal voltage change (Bezprozvanny et al. 1995). Using botulinum toxins, which act to cleave syntaxin 1A near its TMD, Stanley (2003) has recently shown that similar inactivation occurs in neurons. Because calcium channels (and calcium) are intimately related to the fusion of neurotransmitter-containing vesicles to the plasma membrane, such a direct interaction and regulation of voltage-gated calcium channels by syntaxin 1A, although unexpected, was perhaps not too surprising. However, this was only the first evidence for channel interactions with syntaxin 1A. The cystic fibrosis transmembrane regulator (CFTR) chloride channel was soon shown to be another.

Syntaxin 1A physically interacts with CFTR and reduces CFTR-mediated currents both in *Xenopus* oocytes and in epithelial cells that normally express these proteins. The physical and functional interactions are inhibited by the high affinity syntaxin-binding protein Munc-18 (Naren et al. 1997). Syntaxin 1A stoichiometrically binds to the N-terminal cytoplasmic tail of CFTR. The modulation of CFTR currents by syntaxin 1A is eliminated by deletion of the tail or by using the tail as a blocking peptide. The CFTR binding site on syntaxin 1A maps to the H3 domain. The chloride current activity of recombinant ΔF508 CFTR (i.e., the most common disease-causing cystic fibrosis mutant) is potentiated by disrupting its interaction with syntaxin 1A in cultured epithelial cells (Naren et al. 1998). Interestingly, the N-terminal cytoplasmic tail of CFTR controls protein kinase A-dependent channel gating through a physical interaction with the regulatory (R) domain of CFTR. Since syntaxin 1A binds the N-terminal tail of CFTR, it regulates CFTR channel gating through inter-domain interactions (Naren et al. 1999).

Accumulating evidence shows that ion channel regulation by SNAREs is a recurring theme. In addition to voltage-gated calcium channels and CFTR chloride channels, SNARES have been shown to regulate multiple K^+ channels and epithelial Na^+ channels. Additionally, syntaxin 1A regulates plasma membrane neurotransmitter transporters.

1.3
Initial Evidence for SNARE Regulation of Neurotransmitter Transporters

Two lines of evidence converged to suggest to us that SNARE proteins may be involved in the regulation of neurotransmitter transporters. First, we had recently shown that the rat brain γ-aminobutyric acid (GABA) transporter GAT1 was subject to functional regulation by PKC, and that the change in function was correlated with a redistribution of the transporter to and from intracellular locations to the plasma membrane (Corey et al. 1994). Second, it had been shown that GLUT4, the insulin-sensitive glucose transporter, requires syntaxin 4 and cellubrevin (a nonneuronal isoform of VAMP) for delivery to the plasma membrane in adipocytes (Cheatham et al. 1996). Thus, we reasoned that in order for GAT1 to be regulated by trafficking to and from the plasma membrane, it might require SNARE-mediated vesicle fusion.

Initial tests of this hypothesis were done in *Xenopus* oocytes expressing GAT1 alone or in the presence of syntaxin 1A (Quick et al. 1997). Changes in the surface expression of GAT1 induced by syntaxin 1A were prevented by acute injection of botulinum toxin C1 (BoTx C1) or chronic injection of syntaxin 1A antisense. Stronger evidence for a functional interaction came from experiments in dissociated hippocampal neuron cultures that endogenously express both GAT1 and syntaxin 1A (Beckman et al. 1998). Disruption of syntaxin 1A with BoTx or by using syntaxin 1A antisense treatments altered GABA uptake. More interestingly, coimmunoprecipitation experiments suggested that GAT1 and syntaxin 1A were part of a macromolecular complex in neurons.

2
Overview of Syntaxin 1A Regulation of Neurotransmitter Transporters

Since the initial studies on GAT1 and syntaxin 1A interactions, interactions with other neurotransmitter transporters have been investigated. A number of outstanding laboratories have made important advances in our understanding of these protein complexes with regard to the nature of their interaction, their functional effects, and their physiological relevance. These findings are discussed in this section.

2.1
Glycine Transporters

GLYT1 is the predominant glial glycine transporter. In addition to expression at glycinergic synapses, GLYT1 is also expressed at glutamatergic synapses. Thus, it is likely that GLYT1 plays a role in regulating the glycine levels necessary for the activation of the N-methyl-D-aspartate (NMDA)-subtype of glutamate receptor. GLYT2 is the "neuronal" glycine transporter isoform, in which expression is restricted predominantly to neurons at glycinergic nerve terminals. In COS-7 cells, Geerlings et al. (2000) showed that cotransfection of either glycine transporter isoform with syntaxin 1A resulted in an approximate 40% reduction in glycine transport; this reduction could be reversed by cotransfection with Munc-18. The loss in function correlated with a decrease in surface transporter expression.

In further analysis of syntaxin 1A regulation of GLYT2, Geerlings et al. (2001) went on to show that, in purified synaptosomes from brain, GLYT2 immunoreactivity can be found on small synaptic-like vesicles. Under conditions that cause exocytosis of neurotransmitter-containing vesicles, GLYT2 was found to rapidly traffic to the cell surface and then internalize. In the presence of BoTx C1, which inactivates syntaxin 1A, GLYT2 was no longer able to reach the plasma membrane, but it did internalize normally. These data are consistent with the idea that GLYT2 might reside on neurotransmitter-containing vesicles in the nerve terminal, and that their delivery to the cell surface occurs in parallel with transmitter release, in a syntaxin 1A-dependent manner. Alternatively, GLYT2 may reside on a separate vesicle that is subject to the same forms of regulation as that of neurotransmitter-containing vesicles, as suggested for the GABA transporter GAT1 (Deken et al. 2003).

More recently, it has been shown that GLYT2 interacts with the PDZ domain protein syntenin-1 (Ohno et al. 2004). Syntenin-1 is a syntaxin 1A binding partner. Thus, these three proteins may represent a complex required for the trafficking and localization of GLYT2 to glycinergic nerve terminals.

2.2
Amine Transporters

Similar to GAT1 and glycine transporters, serotonin (5-hydroxytryptamine; 5-HT) uptake is reduced when the 5-HT transporter SERT is coexpressed with syntaxin 1A. Pull-down assays from brain homogenates have revealed a physical interaction involving the N-terminal tail of SERT (Haase et al. 2001). In synaptosomes, inhibition of p38 mitogen-activated protein (MAP) kinase signaling or activation of PKC decreases the interaction of SERT with syntaxin 1A (Samuvel et al. 2005). Thus, the action of these second messengers on SERT trafficking may be in part through regulation of SERT-syntaxin 1A interactions. Our work on SERT and syntaxin 1A will be presented in detail in Sect. 4.

To date, little has been done on dopamine transporter (DAT) and SNARE interactions. Lee et al. (2004), using yeast two-hybrid approaches, identified interactions between DAT, syntaxin 1A, and the receptor for activated C kinases (RACK1). This association involves the intracellular N-terminal domain of DAT. They suggest that such a protein complex would be important in DAT trafficking and regulation by PKC.

More is known about the consequences of norepinephrine transporter (NET)-syntaxin 1A interactions (Sung et al. 2003). NET and syntaxin 1A colocalize in axon terminals; in cells stably expressing NET, BoTx or syntaxin 1A antisense treatment reduces NET transport activity. As with SERT, the N-terminal cytoplasmic domain of NET is necessary for the physical interaction with syntaxin 1A. PKC activators decrease the levels of NET-syntaxin 1A complexes. Since PKC activation is correlated with an increase in intracellularly localized transporters, and since syntaxin 1A is a positive regulator of transporter surface expression, this inverse relationship between PKC activity and transporter-syntaxin 1A complexes makes mechanistic sense.

As will also be discussed herein for GAT1, and as was discussed in Sect. 1.2 for calcium channels and CFTR chloride channels, syntaxin 1A regulates not only surface availability of excitability proteins but their intrinsic properties as well. For NET, single channel current activity measured in inside-out patches is reduced 80% in the presence of a syntaxin 1A cytoplasmic domain fusion protein (Sung et al. 2003). Thus, syntaxin 1A is a positive regulator of surface NET expression but a negative regulator of intrinsic NET activity.

2.3
Glutamate Transporters

All of the accumulating evidence suggests that members of the family of Na^+- and Cl^--dependent plasma membrane neurotransmitter transporters are regulated by syntaxin 1A. How about members of the family of Na^+-dependent plasma membrane glutamate transporters? In oocytes, Zhu et al. (2005) have recently shown that, in the presence of syntaxin 1A, expression of the glutamate transporter EAAC1 is increased on the plasma membrane. However, glutamate transport is reduced and the anion conductance associated with the transporter is significantly decreased. These results mirror those found for NET and GAT1.

3
Details of Syntaxin 1A Regulation of GAT1

As mentioned in Sect. 1.3, our initial investigations using activators and inhibitors of PKC showed that GABA uptake could be altered; the PKC effect did not alter the apparent affinity of the transporter for GABA but did change the maximal transport capacity (Corey et al. 1994). This result suggested a change in the number of functional transporters on the plasma membrane, and so we began to test the hypothesis that GAT1 trafficking occurred in a manner similar, if not identical, to the trafficking of neurotransmitter-containing small synaptic vesicles.

3.1
Regulation of GAT1 Trafficking

Redistribution of GAT1 by acute application of PKC agonists and antagonists implied the trafficking of GAT1 via transporter-containing vesicles. Immunoelectron microscopy had revealed GABA transporters (Barbaresi et al. 2001) on vesicle-like structures in presynaptic GABAergic terminals. We isolated vesicle fractions from synaptosomes and found GAT1 immunoreactivity in fractions of similar buoyant density to that of neurotransmitter-containing small synaptic vesicles (Deken et al. 2003). Purified GAT1-containing vesicles are clear and 50 nm in diameter. Based upon immunoblot analysis of known vesicle proteins, GAT1-containing vesicles contain no synaptophysin, a marker for neurotransmitter-containing vesicles. Conversely, synaptophysin-containing vesicles appear to contain little or no GAT1. In contrast to synaptophysin-positive vesicles, GAT1-positive vesicles lack SV2, synaptotagmin isoforms 1 and 2, and the vesicular GABA transporter. Both vesicle populations share the small G protein rab3a, and the SNARE proteins VAMP and syntaxin 1A. The lack of the vesicular GABA transporter on GAT1-positive vesicles makes them unlikely to contain GABA. We hypothesize that GAT1 resides on a vesicle separate from neurotransmitter-containing vesicles either because GAT1 traffics at regions distinct from release, or because neurons need to regulate transporter expression independent of release. The presence of SNAREs on the GAT1 vesicle are consistent with a role of these proteins in regulating GAT1 expression.

Using BoTx to cleave and inactivate specific SNARE proteins showed that these SNAREs regulate GAT1 endogenously (Horton and Quick 2001). Cleavage of syntaxin 1A causes a decrease in the amount of GAT1 present on the plasma membrane of neurons in hippocampal cultures, and a corresponding increase in cytoplasmic transporter. Thus, syntaxin 1A acts as a positive regulator of GAT1 surface expression. It is unclear if syntaxin 1A regulation of GAT1 trafficking is due to its conventional role as a SNARE. BoTx B, which cleaves VAMP, has little effect on GAT1 trafficking (Deken et al. 2000, Horton and Quick 2001).

It could be that syntaxin 1A regulation of GAT1 trafficking is due to direct interactions between these proteins, as the H3 domain of syntaxin 1A directly interacts with GAT1 (to be discussed in detail the following section). However, the contribution of this interaction to GAT1 trafficking is unclear. Expression of various syntaxin 1A constructs along with GAT1 in oocytes reveals domains in addition to the H3 domain that are required to positively regulate GAT1 expression (Horton and Quick 2001).

Syntaxin 1A forms multiple multimeric complexes in brain, and so networks of protein–protein interactions could modulate GAT1 trafficking. In expression systems, we can reverse syntaxin 1A's regulation of GAT1 trafficking by overexpression of Munc-18 (Beckman et al. 1998). Phosphorylation of Munc-18, which inhibits its binding to syntaxin 1A, promotes syntaxin 1A's effects on GAT1. Not only can syntaxin 1A regulate GAT1, but indeed the reverse situation is also possible. GAT1 could regulate the availability of other proteins, such as calcium channels and other components of the docking and fusion apparatus, to interact with syntaxin 1A. If syntaxin 1A availability is limiting, then syntaxin 1A's participation in vesicle release might preclude it from downregulating transporter function. Thus, transporter function would be positively linked to neurotransmitter release.

3.2
Regulation of Intrinsic Properties of GAT1

Regulating transporter trafficking is one method for modulating extracellular transmitter levels; another way would be to regulate the transport process directly. Syntaxin 1A does this through its direct interactions with GAT1. The association of GAT1 with syntaxin 1A is mediated predominantly by amino acids 30–54 in the N-terminal cytoplasmic tail of GAT1 and the H3 domain of syntaxin 1A. Site-directed mutagenesis that eliminates three aspartic acid residues (D40, D43, D45) in the N-terminal tail of GAT1 eliminates syntaxin 1A binding. Measurements of GAT1 function show that syntaxin 1A reduces rates of GABA flux from 7 per second to 2 per second (Deken et al. 2000).

How might syntaxin 1A inhibit GAT1 turnover rates? Injection of a peptide corresponding to the fourth intracellular loop of GAT1 (IL4) into oocytes expressing GAT1 greatly reduces GABA transport rates, similar to that seen with syntaxin 1A. Mixing the IL4 peptide with a peptide corresponding to the N-terminal cytoplasmic tail of GAT1 reverses the inhibition. The inhibition, and direct binding between the two domains, is prevented by mutagenesis of charged residues in either the IL4 or N-terminal domains. Syntaxin 1A is unable to regulate the GAT1 IL4 mutant, suggesting that the GAT1 IL4 domain serves as a barrier for transport, and this barrier can be regulated by syntaxin 1A via interactions with the N-terminal tail (Hansra et al. 2004). These interactions are diagrammed in Fig. 1.

What step in the transport cycle is being affected by syntaxin 1A? In an alternating access model of transport (Hilgemann and Lu 1999), GAT1 exists in a conformation in which it binds two Na^+, one Cl^-, and one GABA molecule. Next, a conformational change occurs to promote translocation across the membrane. After releasing its substrates, the unbound transporter then reorients. Syntaxin 1A causes reductions in forward and reverse transport; there is

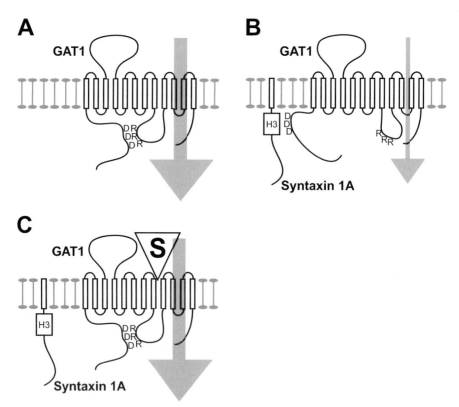

Fig. 1 A–C Model of syntaxin 1A regulation of GABA flux rates through GAT1. The IL4 domain of GAT1 inhibits transport. **A** Interactions between Asp residues (*D*) in the N-terminal tail and Arg residues (*R*) in the IL4 domain positively regulate GABA flux. **B** In the presence of syntaxin 1A, the H3 domain binds Asp residues (*D*) in the N-terminal tail of GAT1, allowing the IL4 domain to inhibit transport. **C** The chronic presence of GAT1 substrates (*S*) dissociates GAT1 from syntaxin 1A, reversing the inhibition of GABA flux. The *size of the gray arrow* illustrates the amount of GABA transport

also reduction in GABA exchange, a step in transport that bypasses reorientation (Bennett et al. 2000). Thus, syntaxin 1A likely affects a step in translocation after substrate binding but which involves both unidirectional transport and transmitter exchange (Wang et al. 2003).

The reason why syntaxin 1A is a positive modulator of GAT1 surface expression but a negative regulator of GAT1 transport is unclear. One possibility is that neurons use syntaxin 1A not only to increase surface expression of GAT1 but also to keep the transporter functionally suppressed until a time when removal of GABA from the synaptic cleft is required. If true, then substrates of GAT1 may regulate flux in a syntaxin 1A-dependent manner. Application of transporter substrates increases GAT1 flux rates, essentially reversing syntaxin 1A inhibition (Fig. 1). The substrate-induced rate change requires the presence of syntaxin 1A, and substrate application results in a decrease in the fraction of syntaxin 1A that is bound to GAT1 on a time scale comparable to the substrate-induced change in flux rates (Quick 2002).

4
Syntaxin 1A Regulation of SERT Conducting States

Our recent work has focused on an analysis of SERT-syntaxin 1A interactions (Quick 2003). Using a number of different biochemical assays, we showed, as with GAT1, that the H3 domain of syntaxin 1A is required for SERT binding, and that charged residues in the N-terminal tail of SERT are necessary for direct syntaxin 1A interactions. Our attention then turned to how this direct interaction regulates SERT function.

For mammalian SERT, early biochemical approaches using radiolabeled flux and antagonist binding assays suggested that each transport cycle was electroneutral, with one Na^+, one Cl^-, and one positively charged 5-HT cotransported, and one K^+ counter-transported (Rudnick and Clark 1993). After cloning of the SERT gene, expression studies and high-resolution electrophysiological approaches revealed more complex permeation properties. Rat brain SERT demonstrates at least four separate conductances when heterologously expressed (Cao et al. 1997): (1) a 5-HT- and Na^+-independent, H^+-dependent leak conductance that is activated below pH 7.0 and which is blocked by SERT antagonists; (2) a 5-HT-independent, Na^+-dependent leak conductance that is blocked by SERT antagonists; (3) a large 5-HT-independent transient conductance that is revealed during hyperpolarization and is blocked by 5-HT and SERT antagonists; and (4) a 5-HT- and voltage-dependent conductance that is associated with 5-HT transport. This latter finding was surprising given the supposed electroneutral transport process, and it resulted from variable transport of approximately 7–12 additional charges with each transport cycle.

Under voltage clamp, we examined radiolabeled substrate fluxes in individual oocytes expressing SERT and syntaxin 1A. The results indicated a substrate stoichiometry of 1 5-HT:1 Na^+:1 Cl^-, and thus suggested an electroneutral transport process. When syntaxin 1A was cleaved by BoTx, the stoichiometry varied with voltage (1 5-HT:7 Na^+:1 Cl^- at −40 mV; 1 5-HT:12 Na^+:1 Cl^- at −80 mV). In addition, with BoTx treatment, a 5-HT-independent, Na^+-dependent leak conductance became apparent. Thus, in the presence of syntaxin 1A, 5-HT uptake is tightly coupled to ion flux, and 5-HT-induced ionic currents and 5-HT-independent Na^+ leak currents are absent. We refer to this state as SERT's coupled flux (CF) mode. When syntaxin 1A is not interacting with SERT, 5-HT uptake is loosely coupled to ion flux, and 5-HT-induced ionic currents and 5-HT-independent Na^+ leak currents are present. We refer to this state as SERT's uncoupled flux (UF) mode. These modes are diagrammed in Fig. 2.

We then showed that the SERT CF and UF modes are present in thalamocortical neurons that endogenously express SERT and syntaxin 1A. With a SERT N-terminal tail peptide in the pipette to disrupt SERT-syntaxin 1A interactions, immediately after gaining cell access, application of 5-HT failed to elicit an inward current different from control. However, 15 min after gaining access, 5-HT-induced currents were seen. This revelation of 5-HT-sensitive currents

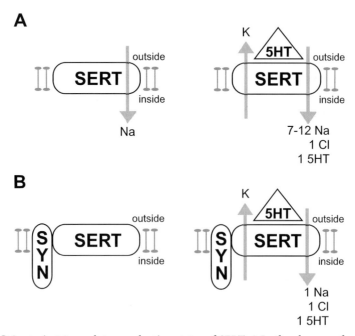

Fig. 2 A,B Syntaxin 1A regulates conducting states of SERT. **A** In the absence of serotonin, SERT shows Na^+ leak currents. In the presence of serotonin (*5HT*), substrate translocation is electrogenic. The current is carried by a variable, voltage-dependent flux of Na^+ ions. **B** In the presence of syntaxin 1A (*SYN*), the Na^+ leak conductance is eliminated, and transport is electroneutral

(and leak currents) did not occur when a scrambled N-tail peptide was included in the pipette. These data suggest that SERT exhibits CF and UF modes endogenously, and that these states are regulated by syntaxin 1A interactions.

5-HT-independent leak currents and 5-HT-induced transport-associated currents mediated by SERT act to depolarize the cell's membrane potential. Interaction of SERT with syntaxin 1A acts to inhibit this depolarization. The significance of this depolarization on (1) cell excitability, (2) other physiological effects of SERT ion fluxes, and (3) the extent to which syntaxin 1A and its binding partners modulate these effects has yet to be determined.

5
Conclusions

The transporter field has barely scratched the surface with regard to SNARE regulation of neurotransmitter transport. We know that many members of the family of sodium- and chloride-dependent plasma membrane neurotransmitter transporters are regulated by syntaxin 1A, and we have begun to map the sites of interaction. We know that the interaction can regulate both the trafficking of transporters and their unitary functional properties. We know that the transporter can exist in large multimeric protein complexes, not only with syntaxin 1A but with factors that can regulate syntaxin 1A-transporter interactions. However, many fundamental questions remain. For example, we do not know which forms of regulation by syntaxin 1A are related to syntaxin acting in its role as a SNARE, and which are attributable to its direct interactions with transporters. We do not know if transporters can contribute protein domains to a core complex for fusion. We do not know whether syntaxin 1A-transporter complexes are restricted to the plasma membrane or whether these complexes find each other in other subcellular compartments. We have only begun to examine other SNAREs for their capacity to regulate transporters, and very importantly, we are only beginning to understand the physiological consequences of interactions between SNAREs and transporters and whether reagents can be developed that act on SNARE-transporter interactions for the treatment of disorders related to abnormal transmitter levels in brain.

Acknowledgements Supported by Public Health Service Grants DA-10509, MH-61468, and MH-74034

References

Arundine M, Tymianski M (2003) Molecular mechanisms of calcium-dependent neurodegeneration in excitotoxicity. Cell Calcium 34:325–337

Barbaresi P, Gazzanelli G, Malatesta M (2001) γ-Aminobutyric acid transporters in the cat periaqueductal gray: a light and electron microscopic immunocytochemical study. J Comp Neurol 429:337–354

Beckman ML, Bernstein EM, Quick MW (1998) Protein kinase C regulates the interaction between a GABA transporter and syntaxin 1A. J Neurosci 18:6103–6112

Bennett ER, Su H, Kanner BI (2000) Mutation of arginine 44 of GAT-1, a [Na(+)+Cl(−)]-coupled gamma-aminobutyric acid transporter from rat brain, impairs net flux but not exchange. J Biol Chem 275:34106–34113

Bennett MK, Scheller RH (1993) The molecular machinery for secretion is conserved from yeast to neurons. Proc Natl Acad Sci U S A 90:2559–2563

Bennett MK, Calakos N, Scheller RH (1992) Syntaxin: a synaptic protein implicated in docking of synaptic vesicles at presynaptic active zones. Science 257:255–259

Bezprozvanny I, Scheller RH, Tsien RW (1995) Functional impact of syntaxin on gating of N-type and Q-type calcium channels. Nature 378:363–366

Cao Y, Mager S, Lester HA (1997) H^+ permeation and pH regulation at a mammalian serotonin transporter. J Neurosci 17:2257–2266

Chaudhry FA, Lehre KP, Campagne M, Ottersen OP, Danbolt NC, Storm-Mathisen J (1995) Glutamate transporters in glial plasma membranes: highly differentiated localizations revealed by quantitative ultrastructural immunocytochemistry. Neuron 15:711–720

Cheatham B, Volchuk A, Kahn CR, Wang L, Rhodes CJ, Klip A (1996) Insulin-stimulated translocation of GLUT4 glucose transporters requires SNARE-complex proteins. Proc Natl Acad Sci U S A 93:15169–15173

Chiu CS, Jensen K, Sokolova I, Wang D, Li M, Deshpande P, Davidson N, Mody I, Quick MW, Quake SR, Lester HA (2002) Number, density, and surface/cytoplasmic distribution of GABA transporters at presynaptic structures of knock-in mice carrying GABA transporter subtype1-green fluorescent protein fusions. J Neurosci 22:10251–10266

Corey JL, Davidson N, Lester HA, Brecha N, Quick MW (1994) Protein kinase C modulates the activity of a cloned gamma-aminobutyric acid transporter expressed in Xenopus oocytes via regulated subcellular redistribution of the transporter. J Biol Chem 269:14759–14767

Deken SL, Beckman ML, Boos L, Quick MW (2000) Transport rates of GABA transporters: regulation by the N-terminal domain and syntaxin 1A. Nat Neurosci 3:998–1003

Deken SL, Wang D, Quick MW (2003) Plasma membrane GABA transporters reside on distinct vesicles and undergo rapid regulated recycling. J Neurosci 23:1563–1568

Diamond JS, Jahr CE (1997) Transporters buffer synaptically released glutamate on a sub-millisecond time scale. J Neurosci 17:4672–4687

Dulubova I, Sugita S, Hill S, Hosaka M, Fernandez I, Sudhof TC, Rizo J (1999) A conformational switch in syntaxin during exocytosis: role of munc18. EMBO J 18:4372–4382

Geerlings A, Lopez-Corcuera B, Aragon C (2000) Characterization of the interactions between the glycine transporters GLYT1 and GLYT2 and the SNARE protein syntaxin 1A. FEBS Lett 470:51–54

Geerlings A, Nunez E, Lopez-Corcuera B, Aragon C (2001) Calcium- and syntaxin 1-mediated trafficking of the neuronal glycine transporter GLYT2. J Biol Chem 276:17584–17590

Haase J, Killian AM, Magnani F, Williams C (2001) Regulation of the serotonin transporter by interacting proteins. Biochem Soc Trans 29:722–728

Hansra N, Arya S, Quick MW (2004) Intracellular domains of a rat brain GABA transporter that govern transport. J Neurosci 24:4082–4087

Hilgemann DW, Lu CC (1999) GAT1 (GABA:Na^+:Cl^-) cotransport function. Database reconstruction with an alternating access model. J Gen Physiol 114:459–475

Horton N, Quick MW (2001) Syntaxin 1A up-regulates GABA transporter expression by subcellular redistribution. Mol Membr Biol 18:39–44

Ichinose T, Lukasiewicz PD (2002) GABA transporters regulate inhibition in the retina by limiting GABA(C) receptor activation. J Neurosci 22:3285–3292

Jahn R, Südhof TC (1999) Membrane fusion and exocytosis. Annu Rev Biochem 68:863–911

Lee KH, Kim MY, Kim DH, Lee YS (2004) Syntaxin 1A and receptor for activated C kinase interact with the N-terminal region of human dopamine transporter. Neurochem Res 29:1405–1409

Naren AP, Nelson DJ, Xie W, Jovov B, Pevsner J, Bennett MK, Benos DJ, Quick MW, Kirk KL (1997) Regulation of CFTR chloride channels by syntaxin and Munc18 isoforms. Nature 390:302–305

Naren AP, Quick MW, Collawn JF, Nelson DJ, Kirk KL (1998) Syntaxin 1A inhibits CFTR chloride channels by means of domain-specific protein–protein interactions. Proc Natl Acad Sci U S A 95:10972–10977

Naren AP, Cormet-Boyaka E, Fu J, Villain M, Blalock JE, Quick MW, Kirk KL (1999) CFTR chloride channel regulation by an interdomain interaction. Science 286:544–548

Nutt DJ (2002) The neuropharmacology of serotonin and noradrenaline in depression. Int Clin Psychopharmacol 17:S1–S12

Ohno K, Koroll M, El Far O, Scholze P, Gomeza J, Betz H (2004) The neuronal glycine transporter 2 interacts with the PDZ domain protein syntenin-1. Mol Cell Neurosci 26:518–529

Overstreet LS, Westbrook GL (2003) Synapse density regulates independence at unitary inhibitory synapses. J Neurosci 23:2618–2626

Quick MW (2002) Substrates regulate gamma-aminobutyric acid transporters in a syntaxin 1A-dependent manner. Proc Natl Acad Sci USA 99:5686–5691

Quick MW (2003) Regulating the conducting states of a mammalian serotonin transporter. Neuron 40:537–549

Quick MW, Corey JL, Davidson N, Lester HA (1997) Second messengers, trafficking-related proteins, and amino acid residues that contribute to the functional regulation of the rat brain GABA transporter GAT1. J Neurosci 17:2967–2979

Rothman JE (1994) Mechanisms of intracellular protein transport. Nature 372:55–63

Rothman JE, Warren G (1994) Implications of the SNARE hypothesis for intracellular membrane topology and dynamics. Curr Biol 4:220–233

Rudnick G (1998) Bioenergetics of neurotransmitter transport. J Bioenerg Biomembr 30:173–185

Rudnick G, Clark J (1993) From synapse to vesicle: the reuptake and storage of biogenic amine neurotransmitters. Biochim Biophys Acta 1144:249–263

Samuvel DJ, Jayanthi LD, Bhat NR, Ramamoorthy S (2005) A role for p38 mitogen-activated protein kinase in the regulation of the serotonin transporter: evidence for distinct cellular mechanisms involved in transporter surface expression. J Neurosci 25:29–41

Schwartz EA (1987) Depolarization without calcium can release γ-aminobutyric acid from a retinal neuron. Science 238:350–355

Sheng Z, Rettig T, Takahashi M, Catterall WA (1994) Identification of a syntaxin-binding site on N-type calcium channels. Neuron 13:1303–1313

Sollner T, Whiteheart SW, Brunner M, Erdjument-Bromage H, Geromanos S, Tempst P, Rothman JE (1993) SNAP receptors implicated in vesicle targeting and fusion. Nature 362:318–324

Stanley EF (2003) Syntaxin I modulation of presynaptic calcium channel inactivation revealed by botulinum toxin C1. Eur J Neurosci 17:1303–1305

Sung U, Apparsundaram S, Galli A, Kahlig KM, Savchenko V, Schroeter S, Quick MW, Blakely RD (2003) A regulated interaction of syntaxin 1A with the antidepressant-sensitive norepinephrine transporter establishes catecholamine clearance capacity. J Neurosci 23:1697–1709

Sutton RB, Fasshauer D, Jahn R, Brunger AT (1998) Crystal structure of a SNARE complex involved in synaptic exocytosis at 2.4 A resolution. Nature 395:347–353

Wang D, Deken SL, Whitworth TL, Quick MW (2003) Syntaxin 1A inhibits GABA flux, efflux, and exchange mediated by the rat brain GABA transporter GAT1. Mol Pharmacol 64:905–913

Weimbs T, Low SH, Chapin SJ, Mostov KE, Bucher P, Hofmann K (1997) A conserved domain is present in different families of vesicular fusion proteins: a new superfamily. Proc Natl Acad Sci U S A 94:3046–3051

Zamponi GW (2003) Regulation of presynaptic calcium channels by synaptic proteins. J Pharmacol Sci 92:79–83

Zhu Y, Fei J, Schwarz W (2005) Expression and transport function of the glutamate transporter EAAC1 in Xenopus oocytes is regulated by syntaxin 1A. J Neurosci Res 79:503–508

Regulation of the Dopamine Transporter by Phosphorylation

J. D. Foster · M. A. Cervinski · B. K. Gorentla · R. A. Vaughan (✉)

Department of Biochemistry and Molecular Biology, University of North Dakota School of Medicine and Health Sciences, 501 North Columbia Road, Grand Forks ND, 58203, USA
rvaughan@medicine.nodak.edu

1	Introduction to the Dopamine Transporter	198
1.1	The Na^+/Cl^--Dependent Neurotransmitter Transporter Family	198
1.2	Interactions with Psychostimulants, Therapeutic Drugs, and Neurotoxins	199
2	Dopamine Transporter Phosphorylation	200
2.1	Effects of Kinases and Phosphatases	201
2.2	Phosphorylation Sites	204
2.3	Transporter Downregulation	205
2.4	Effects of Substrates	206
2.5	Effects of Transport Blockers	208
3	Future Perspectives	209
	References	210

Abstract The dopamine transporter (DAT) is a neuronal phosphoprotein and target for psychoactive drugs that plays a critical role in terminating dopaminergic transmission by reuptake of dopamine from the synaptic space. Control of DAT activity and plasma membrane expression are therefore central to drug actions and the spatial and temporal regulation of synaptic dopamine levels. DATs rapidly traffic between the plasma membrane and endosomal compartments in both constitutive and protein kinase C-dependent manners. Kinase activators, phosphatase inhibitors, and transported substrates modulate DAT phosphorylation and activity, but the underlying mechanisms and role of phosphorylation in these processes are poorly understood. Complex adaptive changes in DAT function potentially related to these processes are also induced by psychostimulant and therapeutic transport blockers such as cocaine and methylphenidate. This chapter provides an overview of the current state of knowledge regarding DAT phosphorylation and its relationship to transporter activity and trafficking. A better understanding of how dopaminergic neurons regulate DAT function and the role of phosphorylation may lead to the identification of novel therapeutic targets for the treatment and prevention of dopaminergic disorders.

Keywords Dopamine transporter · Phosphorylation · Protein kinase C · Methamphetamine · Cocaine

1
Introduction to the Dopamine Transporter

Dopaminergic synaptic transmission controls a number of neurobiological functions including motor activity, mood, and reward, and abnormalities of dopamine (DA) signaling are associated with a variety of psychiatric disorders such as depression, schizophrenia, attention deficit hyperactivity disorder, drug abuse, and Parkinson's disease (Carlsson 1987; Roth and Elsworth 1995; Greengard 2001). Dopaminergic signaling initiated by release of neurotransmitter into the synaptic space is followed by termination of transmission via reuptake of DA by the action of DA transporters (DATs) (Giros et al. 1996). Modulation of DA clearance by regulation of DAT activity and/or surface expression would therefore greatly affect DA signaling parameters. Many studies have now documented the acute regulation of DA uptake, and in some cases this has been demonstrated to occur concomitantly with altered DAT surface expression. The steady-state level of DATs in the plasma membrane is the sum of the transporter reaching the cell surface via the biosynthetic pathway, leaving the surface via endocytosis, and returning to the surface via recycling from the endosomal pools, but the molecular mechanisms controlling these processes are unknown. Dysregulation of any of these steps may therefore be involved in the etiology of dopaminergic disorders by leading to inappropriate DAT surface expression and consequent aberrant clearance of synaptic DA. Another potential mechanism for regulating clearance is alteration of the intrinsic activity of surface transporters through changes in substrate affinity or turnover rate. Since the cloning of DATs over a decade ago, there has been growing interest in the possibility that DAT activity might be regulated by phosphorylation-mediated processes that could provide a rapid mechanism for modulating DA clearance in response to specific physiological stimuli. Here we present an overview of the progress that has been made regarding the role of protein kinases and phosphatases, and transport blockers and substrates in the phosphorylation and regulation of DAT.

1.1
The Na^+/Cl^--Dependent Neurotransmitter Transporter Family

DAT belongs to the SLC6 family of Na^+/Cl^--dependent neuronal membrane transporters that includes carriers for the neurotransmitter and amino acid substrates norepinephrine (NET), serotonin (SERT), γ-aminobutyric acid (GAT), glycine (GLYT), taurine, proline, betaine, and creatine (Gainetdinov and Caron 2003; Chen et al. 2004). The topological structure of these transporters is believed to consist of 12 transmembrane-spanning domains (TMs) with the N- and C-terminal domains located in the cytoplasm. Considerable sequence homology is present within the TM domains, which may reflect a common mechanism for substrate translocation, while the connecting loops and cytoplasmic tails are more divergent and may represent sites for unique characteristics and functions. Substrate translocation by transporters from this family is driven by the electrochemical gradient generated by the plasma membrane Na^+/K^+ ATPase. The transport mechanism involves the sequential binding of Na^+, Cl^-, and DA, and movement of the ions down their concentration gradient provides the energy to translocate DA against its concentration gradient (Gu et al. 1994, 1996; Rudnick 1998). Transport has been proposed to occur by an alternating access mechanism in which external and internal gates regulate substrate movements (Jardetzky 1966; Rudnick 1998), but the molecular mechanism is not understood and channel-like and oligomeric transport mechanisms have also been suggested (Rudnick 1998; Pifl and Singer 1999; Sitte et al. 2001; Carvelli et al. 2004; Seidel et al. 2005). Transport-driven accumulation of intracellular substrate generates the appropriate thermodynamic conditions to induce reverse transport, resulting in the efflux of substrates from the intracellular to the extracellular compartments (Fischer and Cho 1979; Liang and Rutledge 1982).

1.2
Interactions with Psychostimulants, Therapeutic Drugs, and Neurotoxins

The monoamine transporters DAT, NET, and SERT are targets for many pharmacological agents that alter brain function, including abused psychostimulant drugs, antidepressants, and neurotoxins (Barker and Blakely 1995; Amara and Sonders 1998; Miller et al. 1999). Because these transporters are found only in neurons that contain the cognate transmitter, the drugs that affect them specifically impact those neuronal pathways. Cocaine and related drugs are competitive transport inhibitors that bind to DAT and block transmitter reuptake but are not transported. In contrast, amphetamine (AMPH) and methamphetamine (METH) are substrates for DAT that inhibit transmitter reuptake by competing with DA for transport and induce efflux of intracellular DA through transport reversal. The net effect of these drugs is increased synaptic DA concentrations and prolonged stimulation of downstream neurons. Although cocaine and amphetamines similarly affect all of the monoamine transporters, their reinforcing properties are largely dependent on their interaction with DAT (Kuhar 1992; Wise 1996).

Monoamine transporters are also the sites of action of several therapeutic compounds that display varying degrees of specificity. Serotonin selective reuptake inhibitors such as fluoxetine (Prozac, Eli Lilly and Company, Indianapolis), citalopram (Celexa, Forest Pharmaceuticals, St. Louis), paroxetine (Paxil, GlaxoSmithKline, Research Triangle Park), and sertraline (Zoloft, Pfizer, New York) and nonselective tricyclics such as bupropion (Wellbutrin, GlaxoSmithKline) and nomifensine (Merital, Hoechst-Roussel, Somerville) are used to treat depression. The nonselective reuptake inhibitors methylphenidate (Ritalin, Novartis, Basel) and mazindol (Mazanor, Wyeth-Ayerst, Collegeville) are used to treat attention deficit disorder and eating disorders, respectively. In these cases, the increased levels of neurotransmitters induced by the drugs ameliorate the effects of the disorders.

In addition, DATs can function as gateways for cell-specific entry of synthetic and endogenous neurotoxins. At high concentrations, DA and its oxidized metabolites such as 6-hydroxydopamine, as well as AMPH and METH, generate free radicals that disrupt cellular respiration and lead to neuronal death (Miller et al. 1999; Storch et al. 2004). Similar effects are induced by the synthetic compound 1-methyl-4-phenylpyridinium (MPP^+) and structurally related environmental chemicals such as the pesticide paraquat that are transported into dopaminergic neurons by DAT (Javitch and Snyder 1984; Gainetdinov et al. 1997). It is thought that cumulative toxic insults from such compounds may lead to the selective neuronal death found in Parkinson's disease and other dopaminergic neuropathologies (Miller et al. 1999; Storch et al. 2004).

2
Dopamine Transporter Phosphorylation

The elucidation of the primary amino acid sequence of DAT revealed the presence of several potential phosphorylation sites for protein kinases (Giros et al. 1991; Shimada et al. 1991; Usdin et al. 1991). The predicted intracellular domains of the human DAT contain 15 serine, 9 threonine, and 5 tyrosine residues, many of which are found in consensus sequences for protein kinase C (PKC), cAMP-dependent protein kinase (PKA), cGMP-dependent protein kinase (PKG), and calcium calmodulin-dependent protein kinase (CaMK). Some of these sites are highly conserved throughout the neurotransmitter transporter family, including a PKC/PKG consensus sequence in intracellular loop 2 and several sites in the N- and C-terminal tails close to TMs 1 and 12 (Vaughan 2004).

Metabolic phosphorylation of DAT with $^{32}PO_4$ has demonstrated that the human, rat, and mouse isoforms of DATs are phosphoproteins. In both striatal tissue and cultured cells, DATs exhibit basal phosphorylation that is stimulated by treatment with PKC activators such as phorbol 12-myristate 13-acetate (PMA), protein phosphatase inhibitors such as okadaic acid (OA) or the substrates AMPH and METH (Huff et al. 1997; Vaughan et al. 1997; Cowell et al. 2000; Granas et al. 2003; Lin et al. 2003; Cervinski et al. 2005). These treatments also alter DAT cell surface expression and activity, and the relationship between phosphorylation and DAT regulation is an active area of investigation. The most extensively studied kinase with respect to regulation of DAT is PKC, but several studies indicate contributions from other kinases. DAT activity has been shown to be modulated by activators and inhibitors of PKA, CaMK, phosphatidylinositol 3-kinase (PI3K), protein kinase B (PKB/Akt), mitogen-activated protein kinase (MAPK), and protein tyrosine kinases (Uchikawa et al. 1995; Batchelor and Schenk 1998; Doolen and Zahniser 2001; Carvelli et al. 2002; Lin et al. 2003; Moron et al. 2003; Page et al. 2004; Garcia et al. 2005), but only scant evidence has been obtained to date pertaining to their potential involvement in DAT phosphorylation. In addition, the physiological stimuli and receptors that regulate DAT phosphorylation and activity in vivo are unknown. Activation of the PKC-linked glutamate receptor (mGluR) in striatal tissue leads to DAT downregulation (Page et al. 2001), but attempts in our laboratory to detect altered DAT phosphorylation with mGluR-agonist treatments have been negative (M. Cervinski and R. Vaughan, unpublished result). In human (h)DAT-transfected EM4 cells, substance P receptors (hNK-1) are coupled via PKC to the downregulation of DA transport activity and increased DAT phosphorylation (Granas et al. 2003), but this association has not yet been demonstrated in neuronal tissue.

2.1
Effects of Kinases and Phosphatases

DAT phosphorylation has been characterized by metabolic labeling in rat and mouse striatal tissue and hDAT and rat (r)DAT heterologously expressing cells (Fig. 1). In the absence of exogenous treatments, expressed and native DATs display a basal level of metabolic phosphorylation that demonstrates the presence of constitutive phosphorylation and dephosphorylation. Treatment

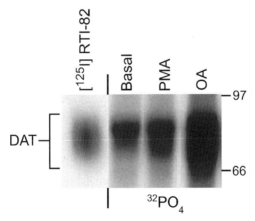

Fig. 1 DAT phosphorylation in rat striatal tissue. Rat striatal slices were metabolically labeled with $^{32}PO_4$ and homogenized by passage through a 26-gauge needle. Homogenates were treated with vehicle (basal), 10 µM PMA or 10 µM OA for 15 min followed by DAT solubilization, immunoprecipitation, sodium dodecyl sulfate-polyacrylamide gel electrophoresis (SDS-PAGE) and autoradiography. [^{125}I]-RTI-82 photoaffinity-labeled DATs were immunoprecipitated and electrophoresed in parallel. Molecular mass markers are shown in kilodaltons, and the position of DAT migration is indicated by the *bracket*

of cells or tissue with PMA leads to a concentration-dependent increase in DAT phosphorylation that begins within minutes and reaches a plateau by 10–20 min (Huff et al. 1997; Vaughan et al. 1997). This increase is blocked by PKC inhibitors such as staurosporine and bisindolylmaleimide and does not occur with the inactive phorbol isomer 4α-PDD, substantiating the involvement of PKC. Conventional PKC isoforms (α, βI, βII, γ) have been implicated in PMA-dependent regulation of hDATs expressed in *Xenopus* oocytes (Doolen and Zahniser 2002), and DATs from rat striatal tissue form co-immunoprecipitation complexes with $PKC\beta_I$ and $PKC\beta_{II}$, but not PKCα or PKCγ (Johnson et al. 2005). Consistent with these findings, attempts in our laboratory to demonstrate in vitro phosphorylation of purified DAT with PKCα as well as PKA, PKG, and CaMK have been negative. Collectively, these results suggest that PKCβ and/or downstream kinases are responsible for phosphorylation and/or regulation of DATs. However, the signaling pathways and identity of kinases involved in the direct phosphorylation of DAT remain to be directly demonstrated, leaving important questions unanswered in this field.

The significant role of protein phosphatases in regulation of DAT phosphorylation levels was established by treatment of rat striatal tissue or heterologously expressing cells with OA, a broad-spectrum protein phosphatase inhibitor. This compound produces robust increases in the DAT phosphorylation level within 5–10 min (Vaughan et al. 1997). In striatal tissue, side-by-side comparisons show that OA alone produces significantly greater DAT phosphorylation increases than PMA alone (Fig. 1). The phosphorylation increase produced by OA in the absence of exogenous kinase activators reflects the rate of basal phosphorylation and indicates that, in resting neurons, significant tonic phosphatase activity not only maintains DAT in a state of relative dephosphorylation but can also partially counteract the effects of one of the strongest known exogenous PKC activators. This robust dephosphorylation pressure may hinder the identification of endogenous signals promoting in vivo DAT phosphorylation.

The identity of the phosphatase(s) involved in DAT dephosphorylation has been examined using phosphatase selective inhibitors. In striatal synaptosomes, DAT phosphorylation is strongly increased by OA and calyculin at doses that inhibit both protein phosphatase 1 (PP1) and protein phosphatase 2A (PP2A), but is only modestly increased by the PP2A-specific inhibitor microcystin (Vaughan et al. 1997), compatible with catalysis of dephosphorylation by both phosphatases. Similarly, in striatal broken cell homogenates, DAT phosphorylation is stimulated strongly by a peptide inhibitor highly specific for PP1 and less strongly by a PP2A-specific inhibitor (Fig. 2; Foster et al. 2003b). Purified PP1 can also dephosphorylate purified $^{32}PO_4$-labeled DAT in vitro at physiological substrate and enzyme concentrations, demonstrating the potential for this enzyme to act directly on DAT in vivo. $^{32}PO_4$-labeled DAT is not dephosphorylated in vitro by PP2A (Foster et al. 2003b) although the catalytic subunit of PP2A co-immunoprecipitates with rat striatal DAT (Bau-

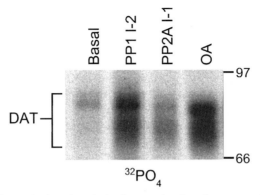

Fig. 2 Inhibition of DAT dephosphorylation by protein phosphatase-specific inhibitors. Rat striatal slices were metabolically labeled with $^{32}PO_4$ and homogenized by passage through a 26-gauge needle. Homogenates were treated with vehicle (basal), 20 nM PP1 inhibitor 2 (*I-2*), 65 nM PP2A inhibitor 1 (*I-1*), or 10 μM OA for 30 min followed by DAT solubilization, immunoprecipitation, SDS-PAGE, and autoradiography. Molecular mass markers are shown in kilodaltons, and the position of DAT migration is indicated by the *bracket*

man et al. 2000). Other phosphatases do not appear to be involved, as in vivo DAT phosphorylation was not affected by the protein phosphatase 2B (PP2B) inhibitor cyclosporine, and neither PP2B nor protein tyrosine phosphatases are able to dephosphorylate $^{32}PO_4$-labeled DAT in vitro (Foster et al. 2003b). Together these results provide evidence for the involvement of both PP1 and PP2A but not PP2B or protein tyrosine phosphatases in dephosphorylation of DAT. Although further work is needed to verify the action of PP1 and PP2A on DAT, it is intriguing to speculate that the involvement of two phosphatases may indicate enzyme-specific roles in various conditions or at distinct DAT phosphorylation sites. However, the specific residues dephosphorylated by these phosphatases and in vivo regulation of their activity with respect to DAT remain unknown.

Far less is known about basal phosphorylation of DAT and its potential functions. An important unresolved issue is the stoichiometry of basal phosphorylation and its relationship to stimulated conditions. It is currently not known if PMA-stimulated DAT phosphorylation represents additional numbers of DATs phosphorylated on the same residues as the basal state or if it occurs via phosphorylation on distinct serines. The identity of the kinase that catalyzes basal phosphorylation is also not known, but does not appear to be PKC or PKA, as basal phosphorylation is not suppressed by PKC inhibitors or increased by PKA activators (Vaughan et al. 1997; Cervinski et al. 2005). Some preliminary experiments suggest that constitutive DAT phosphorylation is partially suppressed by inhibitors of PI3K and MAPK; but this remains to be verified (Lin et al. 2003).

2.2
Phosphorylation Sites

A vital step in elucidating the functional significance of DAT phosphorylation is the identification of phosphorylation sites. Phosphoamino acid analysis of $^{32}PO_4$-labeled DAT prepared from either heterologously expressing cells or rat striatal tissue has demonstrated that basal, PMA-stimulated, and OA-stimulated phosphorylation occurs primarily on serine (Foster et al. 2002a, b). A low level of phosphothreonine has been observed, but phosphotyrosine has not been detected by $^{32}PO_4$ labeling or by anti-phosphotyrosine immunoblotting (R. Vaughan, unpublished result). The phosphorylated region on DAT was first identified by peptide mapping of $^{32}PO_4$-labeled rat striatal transporters. These studies demonstrated that N-terminal tail proteolytic fragments of DAT contain basal, PKC-stimulated, and OA-stimulated phosphorylation, while no evidence was obtained for $^{32}PO_4$ labeling of domains containing the C-terminal tail or interhelical loops (Foster et al. 2002b). The N-terminal tail of rDAT contains eight widely spaced serines, six of which are conserved in hDAT (see Fig. 3), and additional mapping indicated that the majority of the $^{32}PO_4$ labeling in rDAT occurs in the serine cluster located near the distal end of the protein. This finding was later confirmed by truncation of the first 22 or 21 residues in human or rat DATs, respectively (Fig. 3), which abolishes basal and PMA-stimulated transporter phosphorylation (Granas et al. 2003; Cervinski et al. 2005). These results indicate that phosphorylation of expressed and native protein occurs in the same domain, and the positive detection of N-terminal phosphorylation by peptide mapping strongly precludes a potential interpretation of mutagenesis results that the N-terminal tail is not the site of phosphorylation but its truncation prevents phosphorylation of another domain. Simultaneous mutation of three PKC consensus sites at serines 262 and 586 and threonine 613 in the hDAT internal loop 2 and C-terminal tail was reported to prevent PMA-stimulated DAT phosphorylation (Chang et al. 2001). This result is inconsistent with phosphorylation occurring at N-terminal sites,

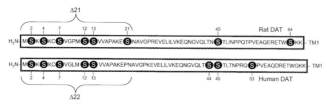

Fig. 3 Schematic depiction of DAT N-terminal tail phosphorylation sites. The N-terminal amino acid sequences of rat (*upper*) and human (*lower*) DATs through transmembrane domain 1 (*TM1*) are depicted with potential serine phosphorylation sites highlighted with *black circles* and *white letters*. Mutants with the deletion of 21 N-terminal amino acids in rat DAT (Δ21) and 22 N-terminal amino acids in human DAT (Δ22) are indicated by the brackets

but may be explained by the extended PMA treatment time used in this study or by the induction of a DAT conformation resulting from mutagenesis that does not permit phosphorylation at N-terminal sites.

The N-terminal serine cluster contains five serines at positions 2, 4, 7, 12, and 13 that are identical in rDAT and hDAT, as well as serine 21, which is unique to rDAT; but the number and position of sites phosphorylated are unknown. Serines 7 and 21 are the only PKC consensus sites in this cluster, and some experiments have indicated a loss of PMA-stimulated phosphorylation after mutation of rDAT serine 7 (Lin et al. 2003). However, rDAT mutants in which all five distal serines are individually changed to alanine undergo easily detectable basal and PMA-stimulated phosphorylation (Foster et al. 2003a), suggesting that PKC-mediated phosphorylation occurs at more than one site and/or at noncanonical sites. These results suggest the potential for DAT phosphorylation to be hierarchical or for mutation of a normally phosphorylated site to induce compensatory phosphorylation at an alternate site.

Another unresolved issue is the identity of the phosphothreonine detected in phosphoamino acid analysis. No threonines are present in the deleted region of the 21/22 residue N-terminal truncation mutants, but additional serine and threonine residues have been examined with respect to downregulation of DAT activity by mutagenesis in the background of an N-terminal truncation (Granas et al. 2003). The potential role of phosphothreonine in the function and regulation of the transporter is not known.

2.3
Transporter Downregulation

PMA and OA treatments that stimulate DAT phosphorylation also lead to reductions in DA transport activity (reviewed in Zahniser and Doolen 2001; Vaughan 2004). Two mechanisms postulated to explain transport downregulation are internalization of DAT, which would remove its access to extracellular substrate, and regulation of intrinsic activity of surface transporters. DATs rapidly traffic between the plasma membrane and endosomal compartments in both constitutive and PKC-stimulated manners, and it has been proposed that increased internalization of DAT by PKC is due to accelerated endocytosis and attenuated return to the plasma membrane (Daniels and Amara 1999; Melikian and Buckley 1999; Loder and Melikian 2003). The correlation of DAT phosphorylation with decreased cell-surface expression and activity led to the early hypothesis that phosphorylation was a signal for transporter internalization in a manner analogous to G protein-coupled receptor (GPCR) desensitization (Gainetdinov et al. 2004). However, the deletion of the hDAT and rDAT N-terminal serine clusters that abolishes transporter phosphorylation does not affect PKC-induced downregulation or internalization, demonstrating that transporter downregulation can occur in the absence of stimulated phosphorylation (Granas et al. 2003; Cervinski et al. 2005). More recently, motifs that mediate constitutive and PKC-regulated endocytosis but do not contain serines or threonines have been identified in the DAT C-terminal tail (Holton et al. 2005). These findings demonstrate that DAT phosphorylation is not an internalization signal as initially hypothesized and that induction of PKC-dependent downregulation and endocytosis must therefore occur via phosphorylation of endocytic components or other DAT-associated proteins. DAT phosphorylation may therefore serve a different role, such as control of cell surface transporter function or regulation of later trafficking events.

It is currently not clear if downregulation of transport activity occurs solely via removal of DATs from the plasma membrane or if other mechanisms that regulate activity of surface DATs are also involved. Some experiments indicate that PKC-induced downregulation of DA transport occurs more rapidly than transporter internalization, suggesting the existence of regulatory events prior to transporter endocytosis (Loder and Melikian 2003). However, electrophysiological studies suggest that endocytosed transporters are active during internalization, arguing against catalytic inactivation prior to sequestration (Kahlig et al. 2004). The resolution of these issues may have important implications for understanding the function of DAT phosphorylation.

DAT activity is also rapidly modulated both positively and negatively (Zahniser and Doolen 2001) by several other signaling molecules/pathways in addition to PKC that may provide the potential for integrative regulation of DAT functions by multiple signals. DA transport is increased in a CaMK-dependent manner by PKA activators (Batchelor and Schenk 1998; Page et al. 2004) and Ca^{2+} (Uchikawa et al. 1995) without evidence for altered DAT phosphorylation (Vaughan et al. 1997; Vaughan 2004). MAPK and PI3K have been implicated in upregulation of DAT function (Carvelli et al. 2002; Moron et al. 2003) as well as basal DAT phosphorylation (Lin et al. 2003). Decreased DA transport velocity has been found after treatment with protein tyrosine kinase inhibitors (Doolen and Zahniser 2001) or nitric oxide (NO) (Lonart and Johnson 1994; Pogun et al. 1994), although NO-induced increases in DAT activity have also been reported (reviewed in West et al. 2002; Volz and Schenk 2004). These results suggest a potential role for these kinases, NO signaling, and PKC in reciprocal regulation of DAT cell surface levels, possibly by both transporter phosphorylation-dependent and -independent mechanisms.

2.4
Effects of Substrates

In addition to exogenous phosphorylation modulators, another class of treatments that regulates DAT activity includes the transported substrates DA, AMPH, and METH. Pretreatments of cells or tissue with these compounds induce rapid reductions in transport activity and dynamin-dependent transporter internalization, both of which are blocked by cocaine and PKC inhibitors (Saunders et al. 2000; Sandoval et al. 2001; Chi and Reith 2003; Cervinski et al. 2005). The regulation of DAT by substrates differs from that of some of the other SLC6 members such as GAT1 and SERT, which undergo substrate-induced homeostatic upregulation that increases transmitter clearance under conditions of elevated transport demand (Ramamoorthy and Blakely 1999). The downregulation of DAT in the presence of substrates may represent a neuroprotective mechanism that limits the intracellular accumulation of toxic substrates such as METH, AMPH, and 6-hydroxydopamine.

The similarities between substrate- and PMA-induced effects suggests that the processes may share common mechanisms. Our laboratory has found that in the absence of exogenous kinase or phosphatase treatments, AMPH and METH stimulate significant increases in DAT metabolic phosphorylation in both striatal tissue and heterologously expressing cells (Cervinski et al. 2005). METH-induced DAT phosphorylation is prevented by cocaine—suggesting a requirement for METH binding or transport—and is blocked by PKC inhibitors (Cervinski et al. 2005). Significantly, injection of rats with METH leads to increased ex vivo DAT phosphorylation, demonstrating the physiological relevance of this finding. METH-induced phosphorylation is not detected in the 21-amino-acid N-terminal truncation mutant of DAT (Fig. 3; Cervinski et al. 2005), indicating that it occurs on the same serine cluster as PMA-induced phosphorylation. However, the N-terminal truncation mutant retains METH-induced downregulation, further supporting a mechanistic link between substrate- and PKC-induced regulatory effects.

While PKC-dependent phosphorylation of DAT is not required for PMA- or METH-induced downregulation of DA transport, it is necessary for substrate-induced transport reversal. The involvement of PKC in efflux has been extensively characterized by Gnegy and colleagues, who have shown that AMPH-induced DA efflux is stimulated by PKC activators and blocked by PKCβ inhibitors (Kantor and Gnegy 1998; Johnson et al. 2005). Recently, it has been shown that AMPH-induced DA efflux is lost by N-terminal truncation of DAT or replacement of all five N-terminal serines with alanines, and is restored by changing the serines to aspartic acids, which mimics the negative charge produced by phosphorylation (Khoshbouei et al. 2004). Thus, the stimulation of DAT $^{32}PO_4$-labeling by AMPH and METH (Cervinski et al. 2005) may reflect phosphorylation required for DA efflux. Within the N-terminal serine to alanine background, individual replacement of residues 7 or 12 with aspartic acid substantially restored efflux, suggesting that these residues may be phosphorylation sites, but this has not been confirmed by direct phosphorylation methods.

Amphetamines induce reverse transport of DA via a two-component mechanism comprising a slow exchange element and a rapid channel-like mode (Kahlig et al. 2005), which leads to efflux of cytosolic DA into the synapse (Fischer and Cho 1979; Liang and Rutledge 1982). The slow exchange process is consistent with a facilitated exchange-diffusion mechanism (Fischer and Cho 1979) in which amphetamines induce the accumulation of an inward-facing transporter conformation that can bind intracellular substrate and translocate it to the extracellular space. A counter-exchange mechanism has also been proposed in which influx and efflux pathways are postulated to reside in separate components of oligomeric transporters (Seidel et al. 2005). In either mechanism, phosphorylation of the transporter may result in a conformation that favors substrate efflux. Because DA efflux may regulate excitability of dopaminergic neurons (Falkenburger et al. 2001), DAT phosphorylation could have a significant impact on intrinsic dopaminergic processes.

The blockade of METH-induced phosphorylation and downregulation by PKC inhibitors indicates a requirement for activation of PKC, but the mechanism by which this occurs is currently not clear. Stoichiometric or channel-like movement of ions through DAT during transport that induce alterations in intracellular levels of Na^+, Ca^{2+}, and H^+ have been proposed to activate phospholipases C and A_2, leading to production of lysophospholipids and diacylglycerol that could activate PKC (Giambalvo 2004). Transported amphetamines may also increase the fraction of PKC bound to membranes, providing the enzyme with increased access to the transporter (Kramer et al. 1998). An alternate hypothesis is that conformational changes occurring in DAT during the transport cycle alter the accessibility of the N-terminal tail to kinases or phosphatases that results in increased phosphorylation. However, $^{32}PO_4$-labeling of DAT is not significantly stimulated by transport of DA (Gorentla and Vaughan 2005), indicating that transport per se cannot explain the METH results unless DA and amphetamine transport differ in a way related to these differential effects.

2.5
Effects of Transport Blockers

It is well established that chronic cocaine induces adaptive regulation of many dopaminergic neuronal components including DATs, and elucidating these changes is a primary task in drug abuse research. While many studies have identified changes in DAT expression, binding, and transport as a result of chronic treatment with cocaine and other transport blockers, far less has been done to examine acute cocaine effects or their relationship to DAT phosphorylation/regulation processes. For the few studies that have been done, the current evidence is contradictory. Short-term pretreatments with cocaine have been reported to either increase (Daws et al. 2002) or have no effect (Sandoval et al. 2001; Chi and Reith 2003; Cervinski et al. 2005) on DAT cell surface expression and activity; and in striatal synaptosomes, cocaine suppresses PMA-induced DAT phosphorylation (Cowell et al. 2000). These results suggest the potential for drug-induced regulation of DATs to be related to DAT phosphorylation. However, our laboratory has found no evidence that acute pretreatments of rDAT Lewis lung carcinoma porcine kidney ($LLC-PK_1$) cells with cocaine or the therapeutically used drugs mazindol and methylphenidate affect levels of basal or PKC-stimulated phosphorylation of DATs (Gorentla and Vaughan 2005). The results from isolated cells indicate that DAT phosphorylation is not altered simply by conformational changes induced by drug binding and present the possibility that effects observed in tissue may be due to indirect processes induced by cocaine. We have also found that acute cocaine pretreatment of LLC-PK$_1$ cells does not affect DAT transport activity or PMA-induced transport downregulation (Gorentla and Vaughan 2005). Thus, the preponderance of evidence currently available indicates little potential for acute cocaine effects to directly influence DAT processes regulated by PKC, but the conflicting results from the various studies indicate the need for further study of these issues.

3
Future Perspectives

The only process identified to date that requires DAT phosphorylation is substrate-induced efflux (Granas et al. 2003; Khoshbouei et al. 2004). The mechanism by which this occurs remains to be determined, but presumably phosphorylation induces a DAT conformation and/or interaction with binding partners that is required for reverse transport to occur. Currently, there are no findings that relate to DAT conformational changes or intramolecular domain interactions regulated by phosphorylation, and although several DAT binding partners have been identified, none of them has been demonstrated to be involved with efflux or DAT phosphorylation. DAT binding partners identified to date include: the multiple LIM domain-containing adaptor protein Hic-5, syntaxin 1A, and the receptor for activated C kinase (RACK-1), which bind to the N-terminal tail (Carneiro et al. 2002; Lee et al. 2004); α-synuclein and the postsynaptic density-95/discs large/zona occludens-1 (PDZ) domain-containing PKC-interacting protein-1 (PICK-1), which complex with the C-terminal tail (Lee et al. 2001; Torres et al. 2001); and PP2A, calmodulin, and synaptosome-associated protein-25 (SNAP-25), whose interaction sites are currently unknown (Bauman et al. 2000; Torres and Caron 2005). Interestingly, syntaxin 1A and SNAP-25 not only interact with DAT but can form stable complexes with each other (Misura et al. 2001), and both PICK-1 and RACK-1 interact with PKC (Ron et al. 1994; Mochly-Rosen et al. 1995; Staudinger et al. 1995, 1997). Scaffolds formed by these or other proteins may facilitate the formation of complexes involved in DAT regulation, phosphorylation, or DA efflux, although the involvement of DAT phosphorylation in such interactions remains to be demonstrated.

Many other important aspects of DAT phosphorylation and regulation remain to be elucidated, including the identification of the kinase(s) involved in the direct phosphorylation of the transporter, the signaling pathways controlling DAT kinases and phosphatases in vivo, the stoichiometry of basal and stimulated phosphorylation, and the subcellular compartments in which phosphorylation and dephosphorylation occur. Elucidation of these properties and other molecular mechanisms regulating DAT phosphorylation and function may lead to the identification of novel therapeutic targets for the treatment and prevention of dopaminergic disorders including schizophrenia, depression, attention deficit hyperactivity disorder, Parkinson's disease, and drug abuse.

Acknowledgements We gratefully acknowledge support from the National Institute on Drug Abuse to RAV (R01 DA13147) and the National Science Foundation ND EPSoR to RAV and JDF (IIP-SG).

References

Amara SG, Sonders MS (1998) Neurotransmitter transporters as molecular targets for addictive drugs. Drug Alcohol Depend 51:87–96

Barker EL, Blakely RD (1995) Norepinephrine and serotonin transporters: molecular targets of antidepressant drugs. In: Bloom FE, Kupfer DJ (eds) Psychopharmacology: the fourth generation of progress. Raven Press, New York, pp 321–333

Batchelor M, Schenk JO (1998) Protein kinase A activity may kinetically upregulate the striatal transporter for dopamine. J Neurosci 18:10304–10309

Bauman AL, Apparsundaram S, Ramamoorthy S, Wadzinski BE, Vaughan RA, Blakely RD (2000) Cocaine and antidepressant-sensitive biogenic amine transporters exist in regulated complexes with protein phosphatase 2A. J Neurosci 20:7571–7578

Carlsson A (1987) Perspectives on the discovery of central monoaminergic neurotransmission. Annu Rev Neurosci 10:19–40

Carneiro AM, Ingram SL, Beaulieu JM, Sweeney A, Amara SG, Thomas SM, Caron MG, Torres GE (2002) The multiple LIM domain-containing adaptor protein Hic-5 synaptically colocalizes and interacts with the dopamine transporter. J Neurosci 22:7045–7054

Carvelli L, Moron JA, Kahlig KM, Ferrer JV, Sen N, Lechleiter JD, Leeb-Lundberg LM, Merrill G, Lafer EM, Ballou LM, Shippenberg TS, Javitch JA, Lin RZ, Galli A (2002) PI 3-kinase regulation of dopamine uptake. J Neurochem 81:859–869

Carvelli L, McDonald PW, Blakely RD, Defelice LJ (2004) Dopamine transporters depolarize neurons by a channel mechanism. Proc Natl Acad Sci U S A 101:16046–16051

Cervinski MA, Foster JD, Vaughan RA (2005) Endogenous and psychoactive substrates stimulate dopamine transporter phosphorylation by a transport and protein kinase C dependent mechanism. J Biol Chem 280:40442–40449

Chang MY, Lee SH, Kim JH, Lee KH, Kim YS, Son H, Lee YS (2001) Protein kinase C-mediated functional regulation of dopamine transporter is not achieved by direct phosphorylation of the dopamine transporter protein. J Neurochem 77:754–761

Chen NH, Reith ME, Quick MW (2004) Synaptic uptake and beyond: the sodium- and chloride-dependent neurotransmitter transporter family SLC6. Pflugers Arch 447:519–531

Chi L, Reith ME (2003) Substrate-induced trafficking of the dopamine transporter in heterologously expressing cells and in rat striatal synaptosomal preparations. J Pharmacol Exp Ther 307:729–736

Cowell RM, Kantor L, Hewlett GH, Frey KA, Gnegy ME (2000) Dopamine transporter antagonists block phorbol ester-induced dopamine release and dopamine transporter phosphorylation in striatal synaptosomes. Eur J Pharmacol 389:59–65

Daniels GM, Amara SG (1999) Regulated trafficking of the human dopamine transporter. Clathrin-mediated internalization and lysosomal degradation in response to phorbol esters. J Biol Chem 274:35794–35801

Daws LC, Callaghan PD, Moron JA, Kahlig KM, Shippenberg TS, Javitch JA, Galli A (2002) Cocaine increases dopamine uptake and cell surface expression of dopamine transporters. Biochem Biophys Res Commun 290:1545–1550

Doolen S, Zahniser NR (2001) Protein tyrosine kinase inhibitors alter human dopamine transporter activity in Xenopus oocytes. J Pharmacol Exp Ther 296:931–938

Doolen S, Zahniser NR (2002) Conventional protein kinase C isoforms regulate human dopamine transporter activity in Xenopus oocytes. FEBS Lett 516:187–190

Falkenburger BH, Barstow KL, Mintz IM (2001) Dendrodendritic inhibition through reversal of dopamine transport. Science 293:2465–2470

Fischer JF, Cho AK (1979) Chemical release of dopamine from striatal homogenates: evidence for an exchange diffusion model. J Pharmacol Exp Ther 208:203–209

Foster JD, Blakely RD, Vaughan RA (2002a) Basal and stimulated phosphorylation sites on striatal and recombinant dopamine transporters. Abstract Viewer/Itinerary Planner. Society for Neuroscience Program, Washington, DC, No. 442.18

Foster JD, Pananusorn B, Vaughan RA (2002b) Dopamine transporters are phosphorylated on N-terminal serines in rat striatum. J Biol Chem 277:25178–25186

Foster JD, Blakely RD, Vaughan RA (2003a) Mutational analysis of potential phosphorylation sites in the N-terminal tail of the rat dopamine transporter. Abstract Viewer/Itinerary Planner. Society for Neuroscience Program, Washington, DC, No. 167.12

Foster JD, Pananusorn B, Cervinski MA, Holden HE, Vaughan RA (2003b) Dopamine transporters are dephosphorylated in striatal homogenates and in vitro by protein phosphatase 1. Brain Res Mol Brain Res 110:100–108

Gainetdinov RR, Caron MG (2003) Monoamine transporters: from genes to behavior. Annu Rev Pharmacol Toxicol 43:261–284

Gainetdinov RR, Fumagalli F, Jones SR, Caron MG (1997) Dopamine transporter is required for in vivo MPTP neurotoxicity: evidence from mice lacking the transporter. J Neurochem 69:1322–1325

Gainetdinov RR, Premont RT, Bohn LM, Lefkowitz RJ, Caron MG (2004) Desensitization of G protein-coupled receptors and neuronal functions. Annu Rev Neurosci 27:107–144

Garcia BG, Wei Y, Moron JA, Lin RZ, Javitch JA, Galli A (2005) Akt is essential for insulin modulation of amphetamine-induced human dopamine transporter cell-surface redistribution. Mol Pharmacol 68:102–109

Giambalvo CT (2004) Mechanisms underlying the effects of amphetamine on particulate PKC activity. Synapse 51:128–139

Giros B, el Mestikawy S, Bertrand L, Caron MG (1991) Cloning and functional characterization of a cocaine-sensitive dopamine transporter. FEBS Lett 295:149–154

Giros B, Jaber M, Jones SR, Wightman RM, Caron MG (1996) Hyperlocomotion and indifference to cocaine and amphetamine in mice lacking the dopamine transporter. Nature 379:606–612

Gorentla BK, Vaughan RA (2005) Differential effects of dopamine and psychoactive drugs on dopamine transporter phosphorylation and regulation. Neuropharmacology 49:759–768

Granas C, Ferrer J, Loland CJ, Javitch JA, Gether U (2003) N-terminal truncation of the dopamine transporter abolishes phorbol ester- and substance P receptor-stimulated phosphorylation without impairing transporter internalization. J Biol Chem 278:4990–5000

Greengard P (2001) The neurobiology of slow synaptic transmission. Science 294:1024–1030

Gu H, Wall SC, Rudnick G (1994) Stable expression of biogenic amine transporters reveals differences in inhibitor sensitivity, kinetics, and ion dependence. J Biol Chem 269:7124–7130

Gu HH, Wall S, Rudnick G (1996) Ion coupling stoichiometry for the norepinephrine transporter in membrane vesicles from stably transfected cells. J Biol Chem 271:6911–6916

Holton KL, Loder MK, Melikian HE (2005) Nonclassical, distinct endocytic signals dictate constitutive and PKC-regulated neurotransmitter transporter internalization. Nat Neurosci 8:881–888

Huff RA, Vaughan RA, Kuhar MJ, Uhl GR (1997) Phorbol esters increase dopamine transporter phosphorylation and decrease transport V_{max}. J Neurochem 68:225–232

Jardetzky O (1966) Simple allosteric model for membrane pumps. Nature 211:969–970

Javitch JA, Snyder SH (1984) Uptake of MPP+ by dopamine neurons explains selectivity of parkinsonism-inducing neurotoxin, MPTP. Eur J Pharmacol 106:455–456

Johnson LA, Guptaroy B, Lund D, Shamban S, Gnegy ME (2005) Regulation of amphetamine-stimulated dopamine efflux by protein kinase C beta. J Biol Chem 280:10914–10919

Kahlig KM, Javitch JA, Galli A (2004) Amphetamine regulation of dopamine transport. Combined measurements of transporter currents and transporter imaging support the endocytosis of an active carrier. J Biol Chem 279:8966–8975

Kahlig KM, Binda F, Khoshbouei H, Blakely RD, McMahon DG, Javitch JA, Galli A (2005) Amphetamine induces dopamine efflux through a dopamine transporter channel. Proc Natl Acad Sci U S A 102:3495–3500

Kantor L, Gnegy ME (1998) Protein kinase C inhibitors block amphetamine-mediated dopamine release in rat striatal slices. J Pharmacol Exp Ther 284:592–598

Khoshbouei H, Sen N, Guptaroy B, Johnson L, Lund D, Gnegy ME, Galli A, Javitch JA (2004) N-terminal phosphorylation of the dopamine transporter is required for amphetamine-induced efflux. PLoS Biol 2:E78

Kramer HK, Poblete JC, Azmitia EC (1998) Characterization of the translocation of protein kinase C (PKC) by 3,4-methylenedioxymethamphetamine (MDMA/ecstasy) in synaptosomes: evidence for a presynaptic localization involving the serotonin transporter (SERT). Neuropsychopharmacology 19:265–277

Kuhar MJ (1992) Molecular pharmacology of cocaine: a dopamine hypothesis and its implications. Ciba Found Symp 166:81–89; discussion 89–95

Lee FJ, Liu F, Pristupa ZB, Niznik HB (2001) Direct binding and functional coupling of alpha-synuclein to the dopamine transporters accelerate dopamine-induced apoptosis. FASEB J 15:916–926

Lee KH, Kim MY, Kim DH, Lee YS (2004) Syntaxin 1A and receptor for activated C kinase interact with the N-terminal region of human dopamine transporter. Neurochem Res 29:1405–1409

Liang NY, Rutledge CO (1982) Evidence for carrier-mediated efflux of dopamine from corpus striatum. Biochem Pharmacol 31:2479–2484

Lin Z, Zhang PW, Zhu X, Melgari JM, Huff R, Spieldoch RL, Uhl GR (2003) Phosphatidylinositol 3-kinase, protein kinase C, and MEK1/2 kinase regulation of dopamine transporters (DAT) require N-terminal DAT phosphoacceptor sites. J Biol Chem 278:20162–20170

Loder MK, Melikian HE (2003) The dopamine transporter constitutively internalizes and recycles in a protein kinase C-regulated manner in stably transfected PC12 cell lines. J Biol Chem 278:22168–22174

Lonart G, Johnson KM (1994) Inhibitory effects of nitric oxide on the uptake of [3H]dopamine and [3H]glutamate by striatal synaptosomes. J Neurochem 63:2108–2117

Melikian HE, Buckley KM (1999) Membrane trafficking regulates the activity of the human dopamine transporter. J Neurosci 19:7699–7710

Miller GW, Gainetdinov RR, Levey AI, Caron MG (1999) Dopamine transporters and neuronal injury. Trends Pharmacol Sci 20:424–429

Misura KM, Gonzalez LC Jr, May AP, Scheller RH, Weis WI (2001) Crystal structure and biophysical properties of a complex between the N-terminal SNARE region of SNAP25 and syntaxin 1a. J Biol Chem 276:41301–41309

Mochly-Rosen D, Smith BL, Chen CH, Disatnik MH, Ron D (1995) Interaction of protein kinase C with RACK1, a receptor for activated C-kinase: a role in beta protein kinase C mediated signal transduction. Biochem Soc Trans 23:596–600

Moron JA, Zakharova I, Ferrer JV, Merrill GA, Hope B, Lafer EM, Lin ZC, Wang JB, Javitch JA, Galli A, Shippenberg TS (2003) Mitogen-activated protein kinase regulates dopamine transporter surface expression and dopamine transport capacity. J Neurosci 23:8480–8488

Page G, Peeters M, Najimi M, Maloteaux JM, Hermans E (2001) Modulation of the neuronal dopamine transporter activity by the metabotropic glutamate receptor mGluR5 in rat striatal synaptosomes through phosphorylation mediated processes. J Neurochem 76:1282–1290

Page G, Barc-Pain S, Pontcharraud R, Cante A, Piriou A, Barrier L (2004) The up-regulation of the striatal dopamine transporter's activity by cAMP is PKA-, CaMK II- and phosphatase-dependent. Neurochem Int 45:627–632

Pifl C, Singer EA (1999) Ion dependence of carrier-mediated release in dopamine or norepinephrine transporter-transfected cells questions the hypothesis of facilitated exchange diffusion. Mol Pharmacol 56:1047–1054

Pogun S, Baumann MH, Kuhar MJ (1994) Nitric oxide inhibits [3H]dopamine uptake. Brain Res 641:83–91

Ramamoorthy S, Blakely RD (1999) Phosphorylation and sequestration of serotonin transporters differentially modulated by psychostimulants. Science 285:763–766

Ron D, Chen CH, Caldwell J, Jamieson L, Orr E, Mochly-Rosen D (1994) Cloning of an intracellular receptor for protein kinase C: a homolog of the beta subunit of G proteins. Proc Natl Acad Sci U S A 91:839–843

Roth RH, Elsworth JD (1995) Biochemical pharmacology of midbrain dopamine neurons. In: Bloom FE, Kupfer DJ (eds) Psychopharmacology: the fourth generation of progress. Raven Press, New York, pp 227–243

Rudnick G (1998) Bioenergetics of neurotransmitter transport. J Bioenerg Biomembr 30:173–185

Sandoval V, Riddle EL, Ugarte YV, Hanson GR, Fleckenstein AE (2001) Methamphetamine-induced rapid and reversible changes in dopamine transporter function: an in vitro model. J Neurosci 21:1413–1419

Saunders C, Ferrer JV, Shi L, Chen J, Merrill G, Lamb ME, Leeb-Lundberg LM, Carvelli L, Javitch JA, Galli A (2000) Amphetamine-induced loss of human dopamine transporter activity: an internalization-dependent and cocaine-sensitive mechanism. Proc Natl Acad Sci U S A 97:6850–6855

Seidel S, Singer EA, Just H, Farhan H, Scholze P, Kudlacek O, Holy M, Koppatz K, Krivanek P, Freissmuth M, Sitte HH (2005) Amphetamines take two to tango: an oligomer-based counter-transport model of neurotransmitter transport explores the amphetamine action. Mol Pharmacol 67:140–151

Shimada S, Kitayama S, Lin CL, Patel A, Nanthakumar E, Gregor P, Kuhar M, Uhl G (1991) Cloning and expression of a cocaine-sensitive dopamine transporter complementary DNA. Science 254:576–578

Sitte HH, Hiptmair B, Zwach J, Pifl C, Singer EA, Scholze P (2001) Quantitative analysis of inward and outward transport rates in cells stably expressing the cloned human serotonin transporter: inconsistencies with the hypothesis of facilitated exchange diffusion. Mol Pharmacol 59:1129–1137

Staudinger J, Zhou J, Burgess R, Elledge SJ, Olson EN (1995) PICK1: a perinuclear binding protein and substrate for protein kinase C isolated by the yeast two-hybrid system. J Cell Biol 128:263–271

Staudinger J, Lu J, Olson EN (1997) Specific interaction of the PDZ domain protein PICK1 with the COOH terminus of protein kinase C-alpha. J Biol Chem 272:32019–32024

Storch A, Ludolph AC, Schwarz J (2004) Dopamine transporter: involvement in selective dopaminergic neurotoxicity and degeneration. J Neural Transm 111:1267–1286

Torres GE, Caron MG (2005) Approaches to identify monoamine transporter interacting proteins. J Neurosci Methods 143:63–68

Torres GE, Yao WD, Mohn AR, Quan H, Kim KM, Levey AI, Staudinger J, Caron MG (2001) Functional interaction between monoamine plasma membrane transporters and the synaptic PDZ domain-containing protein PICK1. Neuron 30:121–134

Uchikawa T, Kiuchi Y, Yura A, Nakachi N, Yamazaki Y, Yokomizo C, Oguchi K (1995) Ca^{2+}-dependent enhancement of [3H]dopamine uptake in rat striatum: possible involvement of calmodulin-dependent kinases. J Neurochem 65:2065–2071

Usdin TB, Mezey E, Chen C, Brownstein MJ, Hoffman BJ (1991) Cloning of the cocaine-sensitive bovine dopamine transporter. Proc Natl Acad Sci U S A 88:11168–11171

Vaughan RA (2004) Phosphorylation and regulation of psychostimulant-sensitive neurotransmitter transporters. J Pharmacol Exp Ther 310:1–7

Vaughan RA, Huff RA, Uhl GR, Kuhar MJ (1997) Protein kinase C-mediated phosphorylation and functional regulation of dopamine transporters in striatal synaptosomes. J Biol Chem 272:15541–15546

Volz TJ, Schenk JO (2004) L-arginine increases dopamine transporter activity in rat striatum via a nitric oxide synthase-dependent mechanism. Synapse 54:173–182

West AR, Galloway MP, Grace AA (2002) Regulation of striatal dopamine neurotransmission by nitric oxide: effector pathways and signaling mechanisms. Synapse 44:227–245

Wise RA (1996) Neurobiology of addiction. Curr Opin Neurobiol 6:243–251

Zahniser NR, Doolen S (2001) Chronic and acute regulation of Na^+/Cl^--dependent neurotransmitter transporters: drugs, substrates, presynaptic receptors, and signaling systems. Pharmacol Ther 92:21–55

The Dopamine Transporter: A Vigilant Border Control for Psychostimulant Action

J. M. Williams (✉) · A. Galli

Department of Molecular Physiology and Biophysics,
Center for Molecular Neuroscience, Vanderbilt University Medical Center,
465 21st Ave. S., 7124 MRB III, Nashville TN, 37232, USA
jason.m.williams@vanderbilt.edu

1	Introduction: Dopamine Transmission and Psychostimulants	216
2	Ion Channel-Like Behavior of the DAT	218
3	Regulation of DAT Function by Its Substrates	221
4	Regulation of DAT Function by Its Inhibitors	224
5	Dopamine D2 Receptor Modulation of DAT Function	225
6	Summary	226
	References	227

Abstract Neurotransmission within the mesocorticolimbic dopamine system has remained the central focus of investigation into the molecular, cellular and behavioral properties of psychostimulants for nearly three decades. The primary means by which dopamine transmission in the synapse is terminated is via the dopamine transporter (DAT), the presynaptic plasmalemmal protein that is responsible for the reuptake of released dopamine. Numerous abused as well as clinically important drugs have important pharmacological interactions with DAT. In general, these compounds fall into two categories: those that block dopamine transport (e.g., cocaine, methylphenidate) and those that serve as substrates for transport [e.g., dopamine, amphetamine and 3,4-methylenedioxymethamphetamine (MDMA or "ecstasy")]. Recent data from in vitro and in vivo studies have suggested that DAT, like other biogenic amine transporters, share several characteristics with classical ligand-gated ion channels. In addition, substrates for transport promote redistribution of DAT away from the plasma membrane, while transport inhibitors such as cocaine disrupt this process. In addition, presynaptic autoreceptors for dopamine have been implicated in the modulation of DAT surface expression and function. The present chapter summarizes some of the recent discoveries pertaining to the electrogenic properties of DAT and their potential relevance to the effects of amphetamine-like stimulants on DAT function. Although there are a number of intracellular and extracellular modulatory influences on dopamine clearance that may play particular roles in psychostimulant action, we specifically focus on the differential direct modulation of DAT function by transport substrates and inhibitors, and we also discusses the role of presynaptic D2 receptors in transport regulation.

Keywords Dopamine transporter · Amphetamine · Cocaine · Ion channel · Trafficking

1
Introduction: Dopamine Transmission and Psychostimulants

The dysfunction of dopaminergic pathways in the CNS has been strongly implicated in the etiology and progression of several psychiatric and neurodegenerative disorders, including schizophrenia, attention-deficit hyperactivity disorder, Parkinson's disease and drug addiction. Furthermore, over three decades of intense research have established a predominant role for dopamine in the reinforcing and locomotor effects of psychostimulants such as cocaine, amphetamine and methamphetamine. The primary neuroanatomical targets for these compounds are the long-length dopaminergic circuits that project from the ventral tegmental area and substantia nigra to several limbic and cortical structures involved in cognition, emotion, goal-directed movement and reward processes (Deutch et al. 1988; Le Moal and Simon 1991; Oades and Halliday 1987). Most notable among the regions innervated by dopamine-releasing neurons are the caudate putamen (dorsal striatum), nucleus accumbens (ventral striatum) and prefrontal cortex, which have all been shown to serve critical functions in the neurocellular and behavioral underpinnings of psychostimulant addiction (Koob 1992; Wise 2002).

Evidence for the involvement of central dopaminergic transmission in reinforcement began to amass in the early 1970s, when the now-classic studies using the neurotoxin 6-hydroxydopamine (6-OHDA) demonstrated that selective depletion of dopamine interfered with appetitive behavior (Breese and Traylor 1970; Fibiger et al. 1973). Other studies revealed specific roles for mesolimbic and mesostriatal dopamine systems in amphetamine-stimulated locomotor activity (Asher and Aghajanian 1974; Creese and Iversen 1974; Kelly et al. 1975). In addition, lesioning of the nucleus accumbens with 6-OHDA, as well as intra-accumbal administration of dopamine-depleting agents such as α-methyltyrosine or dopamine receptor antagonists, were each shown to attenuate cocaine (de la Garza and Johanson 1982; Pettit et al. 1984; Roberts et al. 1977) or amphetamine (Davis and Smith 1973; Lorrain et al. 1999; Lyness et al. 1979) self-administration. Consistent with these data were the findings that animals will self-administer the selective dopamine reuptake inhibitor nomifensine directly into the nucleus accumbens (Carlezon et al. 1995) and that psychostimulant reinforcement is significantly and positively correlated with the ability of these drugs to inhibit dopamine clearance (Ritz et al. 1987). Collectively, these and many other findings established the framework for the emerging hypothesis that the expression of psychostimulant reinforcement was intricately coupled to dopamine homeostasis within mesocorticolimbic "reward circuits" (Dackis and O'Brien 2001; Koob and Bloom 1988; Kuhar et al. 1991).

The strength of dopaminergic signaling depends on the amplitude and frequency of neuronal activity leading to the release of dopamine and dopamine functional interaction with pre- and postsynaptic receptors, as well as transmitter longevity in the synapse. The latter process is regulated by enzymatic degradation, by diffusion away from the synapse and by high-affinity reuptake of dopamine into the presynaptic neuron (Glowinski et al. 1966). Of these, reuptake appears to be the principal means of controlling the duration over which dopamine is present in the extracellular space and is thus the primary mechanism for terminating dopamine transmission (Giros et al. 1996; Hertting and Axelrod 1961). This clearance process is mediated by the dopamine transporter (DAT), a Na^+/Cl^--dependent symport protein containing 12 putative hydrophobic transmembrane domains, intracellular amino and carboxy termini, a glycosylated extracellular loop and several intracellular sites for phosphorylation by a variety of protein kinases (Giros et al. 1991; Kilty et al. 1991; Shimada et al. 1991). The importance of dopamine uptake in dopamine clearance has been further evinced by studies in DAT knockout mice (Gainetdinov et al. 2002; Torres et al. 2003; Uhl et al. 2002). These mice are spontaneously hyperactive and have basal extracellular levels of dopamine that persist within the synapse at least 100 times longer when compared to their wild-type littermates (Giros et al. 1996). Furthermore, acute administration of amphetamine or cocaine has no effect on locomotor activity or dopamine release and uptake in these DAT knockout mice (Giros et al. 1996). In light of these and many other findings, DAT appears to be one of the key determinants of the fidelity of dopaminergic synaptic transmission and a major molecular substrate for the actions of psychostimulants.

Rapid and efficient clearance of dopamine prevents overstimulation of dopamine receptors and reduces the metabolic demands on the presynaptic neuron for new synthesis and vesicular storage of dopamine. In previous years, DAT was regarded as a static, unregulated component of the presynaptic membrane that operated at a relatively constant activity to remove excess dopamine. However, a number of compelling discoveries in the transporter field have emerged over the last decade, forcing researchers to re-evaluate the importance of regulation of DAT function, both at the acute and chronic levels, to psychostimulant pharmacology. While DAT inhibitors such as cocaine block the reuptake of released dopamine from the synaptic cleft, amphetamine-like substrates for DAT are transported inside the neuron, where they then deplete vesicular dopamine and induce transmitter release (Sulzer et al. 2005). Adding another layer of complexity to the regulation of dopamine clearance is the relatively recent discovery that transporters for dopamine, like those for other biogenic amines, undergo constitutive trafficking between intracellular compartments and the plasma membrane through processes that are differentially modulated by substrates for, and inhibitors of, transmitter uptake (Beckman and Quick 1998; Blakely and Bauman 2000; Kahlig and Galli 2003; Melikian 2004; Mortensen and Amara 2003; Saunders et al. 2000; Schmitz et al. 2003;

Zahniser and Doolen 2001). These processes are thought to be phosphorylation and activity dependent, and they appear to involve several different intracellular signaling systems. Thus, in light of the importance of DAT in maintaining appropriate extracellular dopamine levels for optimal synaptic communication, the number and complexity of the regulatory influences on dopamine reuptake may not be surprising.

While the overall relevance of DAT trafficking and functional regulation to psychostimulant abuse remains to be fully clarified, it seems obvious that elucidating the dynamic processes that govern dopamine clearance, as well as how these mechanisms are altered by exposure to psychostimulants, will supply additional insights that may prove crucial for our understanding of the neural underpinnings of drug dependence and the potential development of pharmacotherapeutic strategies for its treatment and prevention. The current chapter surveys and discusses the ion channel-like properties of DAT as well as the regulatory signals affecting transporter function and/or surface expression, with an emphasis on the direct effects of DAT ligands and the indirect effects of presynaptic dopamine receptors.

2
Ion Channel-Like Behavior of the DAT

Cloning and characterization of DAT from a variety of species over the last 15 years have yielded a wealth of information regarding its structural, biophysical, pharmacological and functional properties (Giros et al. 1991, 1992; Jayanthi et al. 1998; Kilty et al. 1991; Porzgen et al. 2001; Shimada et al. 1991; Usdin et al. 1991; Wu and Gu 1999). Within these model systems, DAT activity is non-competitively blocked by psychostimulants such as cocaine and methylphenidate. In addition, DAT-mediated uptake of dopamine is inhibited in a competitive fashion by pseudosubstrates such as amphetamine, methamphetamine and 3,4-methylenedioxymethamphetamine (MDMA or "ecstasy"). Substrates for the DAT are transported inside the presynaptic dopaminergic neuron; but unlike dopamine, amphetamine-like pseudosubstrates can induce dopamine release from synaptic vesicles, causing its accumulation within the cytosol (Sulzer et al. 1992, 1995). They can in turn evoke the efflux of dopamine through the DAT into the perisynaptic space (Gnegy et al. 2004; Kahlig et al. 2005; Khoshbouei et al. 2004; Khoshbouei et al. 2003; Seidel et al. 2005).

DATs, like most Na^+/Cl^--dependent transporters, mediate the translocation of substrates across the plasma membrane and inside the cell. Since dopamine must be moved against its concentration gradient, DAT thermodynamically couples substrate translocation to the inward flux of ions, from which DAT derives its electrogenicity (Amara et al. 1998; Kahlig and Galli 2003; Sonders and Amara 1996). A so-called *permeation pathway* of co-transported ions has been suggested for many years now to derive in part from specific substrate-induced conformational changes in internal and external "gates" within or affiliated with transporter substructure (Jardetzky 1966). The binding of each molecule of dopamine to DAT and its subsequent translocation into the cytosol of the presynaptic neuron require the coordinated binding and cotransport of two sodium ions and one chloride ion (Giros and Caron 1993; Gu et al. 1994; Krueger 1990; Sonders et al. 1997). Proper DAT function is therefore mediated by coupling the movement of sodium ions down their electrochemical gradient to the transmembrane translocation of dopamine (Amara and Kuhar 1993; Amara et al. 1998; DeFelice and Galli 1998; Lester et al. 1994; Sulzer and Galli 2003). At physiological pH, dopamine is positively charged; thus, the translocation of a single dopamine molecule along with two sodium ions and one chloride ion is predicted to be electrogenic, with the translocation of two positive charges for every transport cycle.

Until recently, it had been presumed that the ionic conductances affiliated with DAT serve only as passive components necessary to thermodynamically promote dopamine translocation. However, recent intriguing findings from electrophysiological experiments have supported the notion that transporters for biogenic amines exhibit biophysical characteristics akin to those of classical voltage-gated and ligand-gated ion channels (Ingram et al. 2002; Sonders and Amara 1996; Sulzer and Galli 2003). Within *Xenopus laevis* oocytes and heterologous expression systems, DAT function appears to be coupled to inward current in response to application of dopamine or pseudosubstrates such as amphetamine (Kahlig et al. 2004, 2005; Khoshbouei et al. 2003, 2004; Sitte et al. 1998; Sonders et al. 1997). However, there is an apparent disparity between the theoretical and the empirically determined charge-to-flux ratio for DAT, in that each dopamine transport cycle is coupled with a differential movement of charge across the plasma membrane (Lester et al. 1996; Sonders and Amara 1996). The charge-to-flux ratios observed for DAT are actually greater than those predicted from the proposed stoichiometric models for DAT-coupled ion permeation, suggesting that some component of the transport-associated currents is uncoupled from the process of dopamine translocation. Thus, it has been suggested that DAT mediates electrical currents in two ways: one that is coupled to dopamine translocation (transporter-like) and one that is uncoupled (channel-like) but is partially sustained by dopamine movement. DAT inhibitors such as cocaine block both components, which are stimulated by substrates for transport such as dopamine or amphetamine (Kahlig et al. 2004, 2005; Sonders et al. 1997). The channel-like state has been suggested to affect cellular excitability (Carvelli et al. 2004; Ingram et al. 2002) and to represent a significant portion of the amphetamine-induced dopamine efflux (Kahlig et al. 2005; Kahlig et al. 2004).

Patch clamp recordings combined with amperometry, a technique that allows the identification and measurement of the number of dopamine molecules by their oxidation/reduction potentials, has previously elucidated a channel-like activity in detached patches containing the norepinephrine transporter

(NET) (Galli et al. 1998). This activity correlated with amperometric spikes reflecting a channel mode of norepinephrine conduction that may be gated by the substrate. Thus, it was proposed that norepinephrine was able to move across the plasma membrane down its electrochemical gradient by way of a NET aqueous pore (Galli et al. 1998). Similarly, recent amperometric recordings in outside-out patches obtained from cells stably expressing the human DAT (hDAT) and isolated midbrain dopamine neurons in culture have been used to examine the electrogenic nature of dopamine transport (Kahlig et al. 2005). In these studies, ion concentrations in the patch pipette were adjusted to those that are observed during amphetamine stimulation. Under these conditions, it was evident that DAT-mediated efflux of dopamine occurred through distinct transporter-like and channel-like activities, demonstrating for the first time that these two states of transport coexist (Kahlig et al. 2005). Transporter-like activity occurred over a relatively slow process that was consistent with an exchange model, whereas the channel-like mode of efflux occurred rapidly and accounted for the efflux of substantial quantities of dopamine that approximated those of a typical exocytotic release from dopamine vesicles (Kahlig et al. 2005). This channel-like mode may contribute to the bursting pattern of synaptic dopamine that is seen in midbrain neurons that are known to become hyperactive in response to amphetamine-like psychostimulants.

While the precise ionic composition and physiological significance of macroscopic DAT-mediated currents remains a matter of intense investigation, recent work from Ingram and colleagues (2002) suggests that these currents may be involved in the regulation of neuronal excitability. When examined in rat mesencephalic cultures, dopamine was able to elicit sodium-dependent inward currents that produced a depolarization leading to an enhanced neuronal firing rate. This activation occurred at concentrations that are below the threshold for activation of dopamine D2 autoreceptors and was likewise mimicked by amphetamine. Surprisingly, these studies identified the primary ionic conductance responsible for this excitation as chloride flux. Recent studies from Carvelli and collaborators (2004) have extended these findings by examining single-channel activity associated with transporters expressed in cultured dopaminergic neurons from *Caenorhabditis elegans* (DAT-1). This study was consistent with the findings of Ingram et al. (2002) in that they found that chloride ions permeate DAT-mediated single-channel currents. Taken together, these studies suggest that, in addition to their primary role in the uptake of released dopamine, DAT-mediated ion conductances stimulated by amphetamine may actually participate in the maintenance of neuronal excitation and thus be able to modulate the vesicular release of dopamine itself. In this context it is important to point out that ionic conductances coupled to monoamine transport may also influence the intracellular ionic milieu. For example, Gnegy and colleagues (2004) demonstrated that amphetamine increases intracellular calcium concentration in cells expressing hDAT. They also utilized the patch-clamp technique in the whole-cell configuration to examine the role of internal calcium in the ability of amphetamine to stimulate dopamine efflux (Gnegy et al. 2004). In this study, it was demonstrated that chelation of intracellular calcium produced a voltage-dependent decrease in the amphetamine-induced current and dopamine efflux, which is consistent with previous findings obtained in PC12 cells expressing the NET (Kantor et al. 2001).

Recent studies suggested that the cytosolic sodium concentrations might also be an important ionic regulator of amphetamine action at the DAT. For instance, recent work from our laboratory (Khoshbouei et al. 2003) has suggested that a rise in intracellular sodium levels may be necessary to drive the DAT in reverse in response to amphetamine-like substrates. Using whole-cell patch-clamping combined with amperometry, these studies showed that DAT-mediated efflux of dopamine was cocaine-sensitive, voltage-dependent, electrogenic and increased in proportion to the intracellular sodium level within the recording electrode (Khoshbouei et al. 2003). Confocal microscopic measurements of a sodium-sensitive fluorescent dye also indicated that intracellular sodium accumulation increased with exposure to amphetamine (Khoshbouei et al. 2003). This study proposed a novel two-step mechanism for amphetamine action whereby: (1) Amphetamine first binds to and is transported by DAT, which leads to an influx of sodium ions; (2) The amphetamine-induced inward current increases the availability of sodium ions inside the cell having access to DAT, which—together with the transmembrane potential—can promote the reversal of the transport cycle and induce dopamine release. Previous findings from Sitte and colleagues (1998) first suggested this possibility by demonstrating that the releasing properties of DAT substrates were significantly and positively correlated with their ability to stimulate inward currents through the transporter. Thus, based on this model, amphetamine-evoked dopamine efflux is dependent on the ability of amphetamine to elevate intracellular sodium concentration (Khoshbouei et al. 2003).

3
Regulation of DAT Function by Its Substrates

Amphetamine, like dopamine, serves as a substrate for DAT (as well as NET) and also inhibits dopamine reuptake in a competitive manner (Heikkila et al. 1975; Jones et al. 1999; Khoshbouei et al. 2003). However, probably the most well-known pharmacological property of amphetamine is its ability to promote the reversal of dopamine transport, inducing the release of transmitter into the synapse (Pierce and Kalivas 1997; Sulzer et al. 2005). In 1979, Fischer and Cho proposed a model to explain this characteristic of amphetamine, which has come to be known as the *facilitated exchange diffusion* model (Fischer and Cho 1979). According to this model, amphetamine first binds to DAT, which is primarily in an outward (extracellular)-facing confirmation; but as it gets transported into the neuron, there is a conformational shift such that more transporters are oriented inward toward the cytosol. Transporters in this inward confirmation are then able to bind intracellular dopamine, upon which DAT confirmation reverts back to the outward-facing orientation to release dopamine into the extracellular space. A parallel process described by the *weak base* or *vesicle depletion* model to explain amphetamine's dopamine-releasing actions has also emerged in more recent years (Sulzer et al. 1992; Sulzer and Rayport 1990). Sulzer and colleagues suggest that amphetamine, acting as a weak base, causes the release of dopamine from synaptic vesicles, which raises intracellular dopamine to levels favoring its passive diffusion down its concentration gradient through DAT and outside the cell. At present there is ample evidence that at least some aspects of both models are contributing to the myriad pharmacological actions of amphetamine (Sulzer et al. 2005).

The regulation of DAT function by amphetamine-like psychostimulant substrates has been an area of intense and fruitful investigation in recent years (Gulley and Zahniser 2003; Kahlig and Galli 2003; Saunders et al. 2000; Sulzer and Galli 2003). It was demonstrated in work from Fleckenstein and colleagues that a single systemic injection of amphetamine (Fleckenstein et al. 1999), MDMA (Metzger et al. 1998) or methamphetamine (Fleckenstein et al. 1997) could induce a significant attenuation in dopamine uptake into striatal synaptosomes when prepared within 1 h after administration. These effects are reversible and appear to result from a decrease in V_{max} for dopamine uptake (transporter affinity for dopamine was generally unaffected), suggesting that the number of DATs on the plasma membrane could have been downregulated by acute drug exposure. Notably, the total DAT protein level in the cell does not appear to be affected by this acute treatment (Kokoshka et al. 1998). In vitro incubation of striatal synaptosomes with methamphetamine also caused a dose- and time-dependent reduction in dopamine uptake that required the activity of protein kinase C (Kim et al. 2000; Sandoval et al. 2001). These data are consistent with those collected recently by Gulley and colleagues (2002), who showed that a repeated, intermittent, 1-min exposure of amphetamine to oocytes expressing hDAT resulted in a downregulation in transporter-associated currents. This report also demonstrated using high-speed chronoamperometry that clearance of exogenously applied dopamine was reduced in the dorsal striatum (but not nucleus accumbens) of anesthetized rats (Gulley et al. 2002).

In addition to direct competition for substrate binding sites, the ability of substrates for DAT to directly modulate transporter expression on the plasma membrane surface represents a short-term, rapid and dynamic regulation of DAT function that affords them the capacity to acutely respond to momentary fluctuations in synaptic dopamine. The fact that this regulation often results from a loss of DAT activity, observed as a reduced V_{max} with typically little or no change in the apparent K_m, has led many investigators to speculate that these acute effects on dopamine uptake derive from alterations in transporter trafficking. Saunders and colleagues (2000) showed that substrates for transport induce internalization of cell surface transporters in cells expressing hDAT. Within the first 20 min of an hour-long exposure to 2 µM amphetamine, a redistribution of DAT immunofluorescence from the plasma membrane surface to intracellular compartments was clearly visible and was blocked by inclusion of the DAT inhibitors cocaine, nomifensine or mazindol in the incubation medium (Saunders et al. 2000). Trafficking of the DAT, which appeared to be mediated by endocytic mechanisms, coincided with a significant functional decrease in [^{3}H]dopamine uptake (Chi and Reith 2003) and transporter-associated ionic conductances (Chi and Reith 2003; Saunders et al. 2000). Thus, these data contributed to the emerging view that, similar to other neurotransmitter transporters (Ramamoorthy et al. 1998), DAT functions as a dynamic molecule within the membrane bilayer whose acute regulation via trafficking reduces the number of transporters available to clear extracellular dopamine and thereby exacerbates the cytotoxic effects of hyperdopaminergia.

In addition to the amphetamine-like psychostimulants, it appears that the endogenous substrate dopamine is able to acutely attenuate DAT levels on the cell surface. In the studies by Saunders and colleagues above, cell-surface redistribution of DAT was demonstrated by redistribution of hDAT-associated immunofluorescence after a 1-h incubation with 10 µM dopamine and was associated with a significant attenuation in the uptake of [^{3}H]dopamine (Saunders et al. 2000). In other studies from Chi and Reith (2003), 10–100 µM dopamine exhibited a concentration-related diminution in the subsequent V_{max} for dopamine transport. These effects were apparent after only 10 min of incubation with dopamine and persisted for up to 1 h (Chi and Reith 2003). Similar to the findings of Saunders et al. (2000), these studies found that the reduced V_{max} that occurred after incubation with 100 µM dopamine corresponded with decreases in plasma membrane DAT. Using cleavable biotinylation methods, the downregulation of DAT from the cell surface appeared to result from an increased rate of endocytosis (Chi and Reith 2003). Recent studies from our laboratory have extended these findings using electrophysiological measures of DAT function combined with imaging techniques that assess this DAT redistribution process (Kahlig et al. 2004). These studies suggested that the amphetamine-evoked loss in hDAT function was mediated by a removal of active transporters from the cell surface (Kahlig et al. 2004).

In vivo studies have also suggested that DAT may undergo a relatively rapid rate of constitutive recycling to and from the plasma membrane. Brief, repeated exogenous application of dopamine resulted in a substantial reduction in dopamine clearance as measured by in vivo voltammetry within the dorsolateral striatum of anesthetized rats (Gulley et al. 2002). This effect was apparent after dopamine was applied every 2 min, when the amount of extracellular dopamine oxidized at the carbon fiber electrode was elevated by approximately two- to threefold above baseline. When longer intervals between repeated dopamine applications were tested, this effect was not observed; thus, these data indicate a relatively short time course for dopamine-related electrochemical measures suggestive of a rapid reduction in DAT clearance capacity in vivo. These data are also consistent with the ability of 10 µM dopamine to reduce hDAT labeling with the cocaine analog [^{3}H]WIN35428 in oocytes (Gulley et al. 2002), suggesting that the dopamine-induced attenuation in transporter activity resulted from a reduction in DAT levels on the cell surface. However, since these reports utilized relatively high concentrations of exogenously applied dopamine, the relevance of these effects to the regulation of synaptic dopaminergic tone remains to be fully clarified.

4
Regulation of DAT Function by Its Inhibitors

While exposure to DAT substrates appears to induce internalization and recycling of DAT, there are also reports that inhibitors of dopamine reuptake have the ability to promote an acute upregulation of transporter accumulation at the plasma membrane surface. For instance, exposure of cells stably expressing hDAT to 10 µM cocaine for 10 min results in an increased expression of cell-surface DAT biotinylation and a 30% upregulation of dopamine transport, suggesting that dopaminergic synapses may be capable of rapid alterations in uptake capacity (Daws et al. 2002). In further support for the possibility that this rapid upregulation in DAT function is taking place in vivo, a single injection of cocaine (30 mg/kg, i.p.) in rats produces a substantial (56%) increase the V_{max} for dopamine uptake into nucleus accumbens synaptosomes, with no significant change in substrate affinity for the transporter (Daws et al. 2002). These studies also showed, using in vivo amperometric measurements in anesthetized rats, that intrastriatal application of low cocaine concentrations enhanced the clearance rate of dopamine (Daws et al. 2002). Collectively, these findings are consistent with the novel hypothesis that inhibitors of dopamine uptake are able to acutely mobilize DAT to the cell surface in the initial stages of transport blockade. This acute upregulation was also observed in studies by Little and colleagues (2002), who showed that in murine neuroblastoma neurons stably transfected with hDAT, increases in [^{3}H]WIN35428 binding, and [^{3}H]dopamine uptake were apparent after the cells had incubated (12 and 3 h, respectively) with 1 µM cocaine. These changes in DAT function appeared to result from upregulation of cell surface DAT, while total protein levels and messenger RNA (mRNA) abundance were not affected by cocaine treatment (Little et al. 2002). These data suggest that inhibitors of dopamine uptake may play opposing roles compared with substrates for transport.

It is currently unclear whether the effects of cocaine on DAT activity described above, which occur on an acute timescale, play a significant role in the behavioral or addictive properties of psychostimulants. However, as has been well-studied in animal models, long-term regulation of DAT function may manifest as a consequence of repeated (i.e., chronic) cocaine treatment. For instance, exposure of male Long–Evans rats to repeated daily cocaine (15 mg/kg per day, i.p.) for 2 to 3 weeks led to robust increases in [^{3}H]cocaine binding throughout the striatum and cortex, which the authors speculated underlies a supersensitivity to the pharmacological and behavioral effects of cocaine (Alburges et al. 1993). Similarly, on the last day of chronic unlimited-access cocaine self-administration, male Wistar rats displayed nearly a 70% increase in [^{3}H]WIN35428 binding in the nucleus accumbens (Wilson et al. 1994). Cocaine-overdose victims have also displayed a potentiation in DAT binding throughout the basal ganglia (Little et al. 1993; Staley et al. 1994) and enhanced DAT function, as determined in cryopreserved striatal synaptosomes obtained within 24 h after autopsy (Mash et al. 2002). In contrast to the acute treatment, it is not currently known if the above chronic changes reflect functional alterations in DAT trafficking and redistribution from the cell surface.

5
Dopamine D2 Receptor Modulation of DAT Function

Of all the receptors systems shown to influence DAT function, the dopamine D2 receptor has received the most attention, in keeping with the well-established role of these receptors in mediating many of the behavioral effects of amphetamine-like psychostimulants (Amit and Smith 1992; Clark et al. 1991; Furmidge et al. 1991; Lynch and Wise 1985; Yokel and Wise 1975). While presynaptic D2 autoreceptors have been known to downregulate transmitter synthesis and release (Langer 1997), a number of in vivo, ex vivo, and in vitro studies have suggested that these receptors may also serve to augment DAT function. This novel property of D2 receptors is consistent with the general theory of the function that is inherent to receptor-mediated autoregulation: namely, that its net effect is to reduce dopamine signaling in the synapse. Over 12 years ago, Parsons and colleagues demonstrated that pretreatment with the selective D2 receptor antagonist pimozide resulted in attenuation in the ability of repeated daily exposure to cocaine DAT function (Parsons et al. 1993). It was also shown that when dopamine is exogenously applied within the striatum, nucleus accumbens, or prefrontal cortex, blockade of D2 (but not D1) receptors attenuates dopamine clearance rates as determined by in vivo voltammetry (Cass and Gerhardt 1994). Furthermore, the neuroleptic drug haloperidol, also a D2-like receptor antagonist, has also been shown to inhibit the uptake of exogenously applied (Cass and Gerhardt 1994) or electrically evoked (Benoit-Marand et al. 2001) dopamine in the dorsolateral striatum. Other studies using rotating disk voltammetry revealed that the velocity of dopamine transport within minced striatal tissue is enhanced by the D2-like receptor agonist quinpirole (Batchelor and Schenk 1998; Meiergerd et al. 1993). In synaptosomes prepared from the ventral striatum (i.e., nucleus accumbens), quinpirole dose-dependently enhanced dopamine uptake, an effect that was blocked by the D2-selective, irreversible alkylator *N*-isothiocyanatophenethyl spiperone (Thompson et al. 2001).

Recent data collected in D2 receptor knockout mice add further support for the involvement of D2 receptors in promoting dopamine uptake. Compared to their wild-type littermates, D2 receptor knockout mice have a reduced clearance of exogenously applied dopamine (Dickinson et al. 1999). While basal and evoked levels of dopamine release were similar between controls and D2 receptor knockout mice, their response to the D2 antagonist raclopride were markedly different: wild-type mice displayed a drug-induced attenuation in dopamine uptake, whereas mutant mice were unchanged (Dickinson et al. 1999). These data provide compelling evidence that D2 receptors could positively regulate DAT function to enhance dopamine clearance capacity (however, see Prasad and Amara 2001). One potential explanation for these data is that D2 receptors may be able to mobilize DATs to the plasma membrane surface through alterations in transporter trafficking (discussed in Gulley and Zahniser 2003). For instance, in oocytes co-expressing the human D2 receptor and hDAT, activation of D2 receptors led to an increase in [^{3}H]WIN35428 binding and a corresponding increase in [^{3}H]dopamine uptake that was voltage-independent and pertussis-toxin sensitive and which resulted from a potentiated V_{max} (Mayfield and Zahniser 2001). These data suggest that D2 receptors can redistribute DATs from inside the cell to the membrane surface through processes that require the rapid involvement of inhibitory G protein ($G_{i/o}$) signaling. In addition to the ability of D2 receptors to directly influence DAT function, it was also recently reported by Kimmel and colleagues (2001) that D2 receptors can regulate the turnover and degradation rates for DAT. In that study, quinpirole treatment resulted in a decrease in transporter half-life within the striatum, while having the opposite effect in the nucleus accumbens (Kimmel et al. 2001). Thus far, these findings as well as those from other studies suggest that direct modulation of DAT function may be mediated by D2 dopamine receptor activation. It is possible that some component of this modulatory signal may somehow become dysregulated as a consequence on long-term exposure to psychostimulant drugs acting on the DAT.

6
Summary

The reuptake of dopamine, mediated by DAT, is a critical determinant of the efficiency of dopamine transmission and is the primary target for the actions of psychostimulants. Several lines of evidence support the notion that DAT activity, like that of other biogenic amine transporters, is dynamically influenced by a number of regulatory mechanisms. Many of these processes occur acutely and transiently and involve rapid modulation of the ionic conductances associated with dopamine transport. DAT exhibits a channel-like activity that may afford it the ability to rapidly influence the intracellular ionic environment and potentially modify the signaling cascades coupled to transport function. Furthermore, psychostimulants appear to have differential effects on the acute process of redistribution of DAT cell-surface expression, depending on how these compounds interact with the transporter. While amphetamine-like substrates have been shown to reduce cell-surface DAT, cocaine and other DAT inhibitors tend to transiently upregulate transporter levels on the plasma membrane. Although there are many receptor-effector systems that have been shown to modulate the clearance of dopamine, one of the most thoroughly studied processes involves the activation of presynaptic, $G_{i/o}$-coupled D2 dopamine receptors. There is recent evidence to suggest that, in addition to their well-established autoregulatory effects on dopamine release and synthesis, D2 receptors may also stimulate an increase in DAT function by promoting the expression of more transporters on the cell membrane and possibly regulating the turnover of DAT molecules.

References

Alburges ME, Narang N, Wamsley JK (1993) Alterations in the dopaminergic receptor system after chronic administration of cocaine. Synapse 14:314–323

Amara SG, Kuhar MJ (1993) Neurotransmitter transporters: recent progress. Annu Rev Neurosci 16:73–93

Amara SG, Sonders MS, Zahniser NR, Povlock SL, Daniels GM (1998) Molecular physiology and regulation of catecholamine transporters. Adv Pharmacol 42:164–168

Amit Z, Smith BR (1992) Remoxipride, a specific D2 dopamine antagonist: an examination of its self-administration liability and its effects on d-amphetamine self-administration. Pharmacol Biochem Behav 41:259–261

Asher IM, Aghajanian GK (1974) 6-Hydroxydopamine lesions of olfactory tubercles and caudate nuclei: effect on amphetamine-induced stereotyped behavior in rats. Brain Res 82:1–12

Batchelor M, Schenk JO (1998) Protein kinase A activity may kinetically upregulate the striatal transporter for dopamine. J Neurosci 18:10304–10309

Beckman ML, Quick MW (1998) Neurotransmitter transporters: regulators of function and functional regulation. J Membr Biol 164:1–10

Benoit-Marand M, Borrelli E, Gonon F (2001) Inhibition of dopamine release via presynaptic D2 receptors: time course and functional characteristics in vivo. J Neurosci 21:9134–9141

Blakely RD, Bauman AL (2000) Biogenic amine transporters: regulation in flux. Curr Opin Neurobiol 10:328–336

Breese GR, Traylor TD (1970) Effect of 6-hydroxydopamine on brain norepinephrine and dopamine: Evidence for selective degeneration of catecholamine neurons. J Pharmacol Exp Ther 174:413–420

Carlezon WA Jr, Devine DP, Wise RA (1995) Habit-forming actions of nomifensine in nucleus accumbens. Psychopharmacology (Berl) 122:194–197

Carvelli L, McDonald PW, Blakely RD, Defelice LJ (2004) Dopamine transporters depolarize neurons by a channel mechanism. Proc Natl Acad Sci U S A 101:16046–16051

Cass WA, Gerhardt GA (1994) Direct in vivo evidence that D2 dopamine receptors can modulate dopamine uptake. Neurosci Lett 176:259–263

Chi L, Reith ME (2003) Substrate-induced trafficking of the dopamine transporter in heterologously expressing cells and in rat striatal synaptosomal preparations. J Pharmacol Exp Ther 307:729–736

Clark D, Furmidge LJ, Petry N, Tong ZY, Ericsson M, Johnson D (1991) Behavioural profile of partial D2 dopamine receptor agonists. 1. Atypical inhibition of d-amphetamine-induced locomotor hyperactivity and stereotypy. Psychopharmacology (Berl) 105:381–392

Creese I, Iversen SD (1974) The role of forebrain dopamine systems in amphetamine induced stereotyped behavior in the rat. Psychopharmacologia 39:345–357

Dackis CA, O'Brien CP (2001) Cocaine dependence: a disease of the brain's reward centers. J Subst Abuse Treat 21:111–117

Davis WM, Smith SG (1973) Blocking effect of alpha-methyltyrosine on amphetamine based reinforcement. J Pharm Pharmacol 25:174–177

Daws LC, Callaghan PD, Moron JA, Kahlig KM, Shippenberg TS, Javitch JA, Galli A (2002) Cocaine increases dopamine uptake and cell surface expression of dopamine transporters. Biochem Biophys Res Commun 290:1545–1550

de la Garza R, Johanson CE (1982) Effects of haloperidol and physostigmine on self-administration of local anesthetics. Pharmacol Biochem Behav 17:1295–1299

DeFelice LJ, Galli A (1998) Electrophysiological analysis of transporter function. Adv Pharmacol 42:186–190

Deutch AY, Goldstein M, Baldino F Jr, Roth RH (1988) Telencephalic projections of the A8 dopamine cell group. Ann N Y Acad Sci 537:27–50

Dickinson SD, Sabeti J, Larson GA, Giardina K, Rubinstein M, Kelly MA, Grandy DK, Low MJ, Gerhardt GA, Zahniser NR (1999) Dopamine D2 receptor-deficient mice exhibit decreased dopamine transporter function but no changes in dopamine release in dorsal striatum. J Neurochem 72:148–156

Fibiger HC, Zis AP, McGeer EG (1973) Feeding and drinking deficits after 6-hydroxydopamine administration in the rat: similarities to the lateral hypothalamic syndrome. Brain Res 55:135–148

Fischer J, Cho A (1979) Chemical release of dopamine from striatal homogenates: evidence for an exchange diffusion model. J Pharmacol Exp Ther 208:203–209

Fleckenstein AE, Metzger RR, Wilkins DG, Gibb JW, Hanson GR (1997) Rapid and reversible effects of methamphetamine on dopamine transporters. J Pharmacol Exp Ther 282:834–838

Fleckenstein AE, Haughey HM, Metzger RR, Kokoshka JM, Riddle EL, Hanson JE, Gibb JW, Hanson GR (1999) Differential effects of psychostimulants and related agents on dopaminergic and serotonergic transporter function. Eur J Pharmacol 382:45–49

Furmidge LJ, Exner M, Clark D (1991) Role of dopamine D1 and D2 receptors in mediating the d-amphetamine discriminative cue. Eur J Pharmacol 202:191–199

Gainetdinov RR, Sotnikova TD, Caron MG (2002) Monoamine transporter pharmacology and mutant mice. Trends Pharmacol Sci 23:367–373

Galli A, Blakely RD, DeFelice LJ (1998) Patch-clamp and amperometric recordings from norepinephrine transporters: channel activity and voltage-dependent uptake. Proc Natl Acad Sci U S A 95:13260–13265

Giros B, Caron MG (1993) Molecular characterization of the dopamine transporter. Trends Pharmacol Sci 14:43–49

Giros B, el Mestikawy S, Bertrand L, Caron MG (1991) Cloning and functional characterization of a cocaine-sensitive dopamine transporter. FEBS Lett 295:149–154

Giros B, el Mestikawy S, Godinot N, Zheng K, Han H, Yang-Feng T, Caron MG (1992) Cloning, pharmacological characterization, and chromosome assignment of the human dopamine transporter. Mol Pharmacol 42:383–390

Giros B, Jaber M, Jones SR, Wightman RM, Caron MG (1996) Hyperlocomotion and indifference to cocaine and amphetamine in mice lacking the dopamine transporter. Nature 379:606–612

Glowinski J, Axelrod J, Iversen LL (1966) Regional studies of catecholamines in the rat brain. IV. Effects of drugs on the disposition and metabolism of H3-norepinephrine and H3-dopamine. J Pharmacol Exp Ther 153:30–41

Gnegy ME, Khoshbouei H, Berg KA, Javitch JA, Clarke WP, Zhang M, Galli A (2004) Intracellular Ca^{2+} regulates amphetamine-induced dopamine efflux and currents mediated by the human dopamine transporter. Mol Pharmacol 66:137–143

Gu H, Wall SC, Rudnick G (1994) Stable expression of biogenic amine transporters reveals differences in inhibitor sensitivity, kinetics, and ion dependence. J Biol Chem 269:7124–7130

Gulley JM, Zahniser NR (2003) Rapid regulation of dopamine transporter function by substrates, blockers and presynaptic receptor ligands. Eur J Pharmacol 479:139–152

Gulley JM, Doolen S, Zahniser NR (2002) Brief, repeated exposure to substrates downregulates dopamine transporter function in Xenopus oocytes in vitro and rat dorsal striatum in vivo. J Neurochem 83:400–411

Heikkila R, Orlansky H, Cohen G (1975) Studies on the distinction between uptake inhibition and release of [3H]dopamine in rat brain tissue slices. Biochem Pharmacol 24:847–852

Hertting G, Axelrod J (1961) Fate of tritiated noradrenaline at the sympathetic nerve endings. Nature 192:172–173

Ingram SL, Prasad BM, Amara SG (2002) Dopamine transporter-mediated conductances increase excitability of midbrain dopamine neurons. Nat Neurosci 5:971–978

Jardetzky O (1966) Simple allosteric model for membrane pumps. Nature 211:969–970

Jayanthi LD, Apparsundaram S, Malone MD, Ward E, Miller DM, Eppler M, Blakely RD (1998) The Caenorhabditis elegans gene T23G5.5 encodes an antidepressant- and cocaine-sensitive dopamine transporter. Mol Pharmacol 54:601–609

Jones SR, Joseph JD, Barak LS, Caron MG, Wightman RM (1999) Dopamine neuronal transport kinetics and effects of amphetamine. J Neurochem 73:2406–2414

Kahlig KM, Galli A (2003) Regulation of dopamine transporter function and plasma membrane expression by dopamine, amphetamine, and cocaine. Eur J Pharmacol 479:153–158

Kahlig KM, Javitch JA, Galli A (2004) Amphetamine regulation of dopamine transport. Combined measurements of transporter currents and transporter imaging support the endocytosis of an active carrier. J Biol Chem 279:8966–8975

Kahlig KM, Binda F, Khoshbouei H, Blakely RD, McMahon DG, Javitch JA, Galli A (2005) Amphetamine induces dopamine efflux through a dopamine transporter channel. Proc Natl Acad Sci U S A 102:3495–3500

Kantor L, Hewlett GH, Park YH, Richardson-Burns SM, Mellon MJ, Gnegy ME (2001) Protein kinase C and intracellular calcium are required for amphetamine-mediated dopamine release via the norepinephrine transporter in undifferentiated PC12 cells. J Pharmacol Exp Ther 297:1016–1024

Kelly P, Seviour P, Iversen S (1975) Amphetamine and apomorphine responses in the rat following 6-OHDA lesions of the nucleus accumbens septi and corpus striatum. Brain Res 94:507–522

Khoshbouei H, Wang H, Lechleiter JD, Javitch JA, Galli A (2003) Amphetamine-induced dopamine efflux. A voltage-sensitive and intracellular Na+-dependent mechanism. J Biol Chem 278:12070–12077

Khoshbouei H, Sen N, Guptaroy B, Johnson L, Lund D, Gnegy ME, Galli A, Javitch JA (2004) N-terminal phosphorylation of the dopamine transporter is required for amphetamine-induced efflux. PLoS Biol 2:E78

Kilty JE, Lorang D, Amara SG (1991) Cloning and expression of a cocaine-sensitive rat dopamine transporter. Science 254:578–579

Kim S, Westphalen R, Callahan B, Hatzidimitriou G, Yuan J, Ricaurte GA (2000) Toward development of an in vitro model of methamphetamine-induced dopamine nerve terminal toxicity. J Pharmacol Exp Ther 293:625–633

Kimmel HL, Joyce AR, Carroll FI, Kuhar MJ (2001) Dopamine D1 and D2 receptors influence dopamine transporter synthesis and degradation in the rat. J Pharmacol Exp Ther 298:129–140

Kokoshka JM, Vaughan RA, Hanson GR, Fleckenstein AE (1998) Nature of methamphetamine-induced rapid and reversible changes in dopamine transporters. Eur J Pharmacol 361:269–275

Koob GF (1992) Drugs of abuse: anatomy, pharmacology and function of reward pathways. Trends Pharmacol Sci 13:177–184

Koob GF, Bloom FE (1988) Cellular and molecular mechanisms of drug dependence. Science 242:715–723

Krueger BK (1990) Kinetics and block of dopamine uptake in synaptosomes from rat caudate nucleus. J Neurochem 55:260–267

Kuhar MJ, Ritz MC, Boja JW (1991) The dopamine hypothesis of the reinforcing properties of cocaine. Trends Neurosci 14:299–302

Langer SZ (1997) 25 years since the discovery of presynaptic receptors: present knowledge and future perspectives. Trends Pharmacol Sci 18:95–99

Le Moal M, Simon H (1991) Mesocorticolimbic dopaminergic network: functional and regulatory roles. Physiol Rev 71:155–234

Lester HA, Mager S, Quick MW, Corey JL (1994) Permeation properties of neurotransmitter transporters. Annu Rev Pharmacol Toxicol 34:219–249

Lester HA, Cao Y, Mager S (1996) Listening to neurotransmitter transporters. Neuron 17:807–810

Little KY, Kirkman JA, Carroll FI, Clark TB, Duncan GE (1993) Cocaine use increases [3H]WIN35428 binding sites in human striatum. Brain Res 628:17–25

Little KY, Elmer LW, Zhong H, Scheys JO, Zhang L (2002) Cocaine induction of dopamine transporter trafficking to the plasma membrane. Mol Pharmacol 61:436–445

Lorrain DS, Arnold GM, Vezina P (1999) Mesoaccumbens dopamine and the self-administration of amphetamine. Ann N Y Acad Sci 877:820–822

Lynch MR, Wise RA (1985) Relative effectiveness of pimozide, haloperidol and trifluoperazine on self-stimulation rate-intensity functions. Pharmacol Biochem Behav 23:777–780

Lyness WH, Friedle NM, Moore KE (1979) Destruction of dopaminergic nerve terminals in nucleus accumbens: effect on d-amphetamine self-administration. Pharmacol Biochem Behav 11:553–556

Mash DC, Pablo J, Ouyang Q, Hearn WL, Izenwasser S (2002) Dopamine transport function is elevated in cocaine users. J Neurochem 81:292–300

Mayfield RD, Zahniser NR (2001) Dopamine D2 receptor regulation of the dopamine transporter expressed in Xenopus laevis oocytes is voltage-independent. Mol Pharmacol 59:113–121

Meiergerd SM, Patterson TA, Schenk JO (1993) D2 receptors may modulate the function of the striatal transporter for dopamine: kinetic evidence from studies in vitro and in vivo. J Neurochem 61:764–767

Melikian HE (2004) Neurotransmitter transporter trafficking: endocytosis, recycling, and regulation. Pharmacol Ther 104:17–27

Metzger RR, Hanson GR, Gibb JW, Fleckenstein AE (1998) 3-4-Methylenedioxymethamphetamine-induced acute changes in dopamine transporter function. Eur J Pharmacol 349:205–210

Mortensen OV, Amara SG (2003) Dynamic regulation of the dopamine transporter. Eur J Pharmacol 479:159–170

Oades RD, Halliday GM (1987) Ventral tegmental (A10) system: neurobiology. 1. Anatomy and connectivity. Brain Res 434:117–165

Parsons LH, Schad CA, Justice JB Jr (1993) Co-administration of the D2 antagonist pimozide inhibits up-regulation of dopamine release and uptake induced by repeated cocaine. J Neurochem 60:376–379

Pettit HO, Ettenberg A, Bloom FE, Koob GF (1984) Destruction of dopamine in the nucleus accumbens selectively attenuates cocaine but not heroin self-administration in rats. Psychopharmacology (Berl) 84:167–173

Pierce RC, Kalivas PW (1997) Repeated cocaine modifies the mechanism by which amphetamine releases dopamine. J Neurosci 17:3254–3261

Porzgen P, Park SK, Hirsh J, Sonders MS, Amara SG (2001) The antidepressant-sensitive dopamine transporter in Drosophila melanogaster: a primordial carrier for catecholamines. Mol Pharmacol 59:83–95

Prasad BM, Amara SG (2001) The dopamine transporter in mesencephalic cultures is refractory to physiological changes in membrane voltage. J Neurosci 21:7561–7567

Ramamoorthy S, Giovanetti E, Qian Y, Blakely RD (1998) Phosphorylation and regulation of antidepressant-sensitive serotonin transporters. J Biol Chem 273:2458–2466

Ritz MC, Lamb RJ, Goldberg SR, Kuhar MJ (1987) Cocaine receptors on dopamine transporters are related to self-administration of cocaine. Science 237:1219–1223

Roberts DC, Corcoran ME, Fibiger HC (1977) On the role of ascending catecholaminergic systems in intravenous self-administration of cocaine. Pharmacol Biochem Behav 6:615–620

Sandoval V, Riddle EL, Ugarte YV, Hanson GR, Fleckenstein AE (2001) Methamphetamine-induced rapid and reversible changes in dopamine transporter function: an in vitro model. J Neurosci 21:1413–1419

Saunders C, Ferrer JV, Shi L, Chen J, Merrill G, Lamb ME, Leeb-Lundberg LM, Carvelli L, Javitch JA, Galli A (2000) Amphetamine-induced loss of human dopamine transporter activity: an internalization-dependent and cocaine-sensitive mechanism. Proc Natl Acad Sci U S A 97:6850–6855

Schmitz Y, Benoit-Marand M, Gonon F, Sulzer D (2003) Presynaptic regulation of dopaminergic neurotransmission. J Neurochem 87:273–289

Seidel S, Singer EA, Just H, Farhan H, Scholze P, Kudlacek O, Holy M, Koppatz K, Krivanek P, Freissmuth M, Sitte HH (2005) Amphetamines take two to tango: an oligomer-based counter-transport model of neurotransmitter transport explores the amphetamine action. Mol Pharmacol 67:140–151

Shimada S, Kitayama S, Lin CL, Patel A, Nanthakumar E, Gregor P, Kuhar M, Uhl G (1991) Cloning and expression of a cocaine-sensitive dopamine transporter complementary DNA. Science 254:576–578

Sitte HH, Huck S, Reither H, Boehm S, Singer EA, Pifl C (1998) Carrier-mediated release, transport rates, and charge transfer induced by amphetamine, tyramine, and dopamine in mammalian cells transfected with the human dopamine transporter. J Neurochem 71:1289–1297

Sonders MS, Amara SG (1996) Channels in transporters. Curr Opin Neurobiol 6:294–302

Sonders MS, Zhu SJ, Zahniser NR, Kavanaugh MP, Amara SG (1997) Multiple ionic conductances of the human dopamine transporter: the actions of dopamine and psychostimulants. J Neurosci 17:960–974

Staley JK, Hearn WL, Ruttenber AJ, Wetli CV, Mash DC (1994) High affinity cocaine recognition sites on the dopamine transporter are elevated in fatal cocaine overdose victims. J Pharmacol Exp Ther 271:1678–1685

Sulzer D, Galli A (2003) Dopamine transport currents are promoted from curiosity to physiology. Trends Neurosci 26:173–176

Sulzer D, Rayport S (1990) Amphetamine and other psychostimulants reduce pH gradients in midbrain dopaminergic neurons and chromaffin granules: a mechanism of action. Neuron 5:797–808

Sulzer D, Pothos E, Sung HM, Maidment NT, Hoebel BG, Rayport S (1992) Weak base model of amphetamine action. Ann N Y Acad Sci 654:525–528

Sulzer D, Chen TK, Lau YY, Kristensen H, Rayport S, Ewing A (1995) Amphetamine redistributes dopamine from synaptic vesicles to the cytosol and promotes reverse transport. J Neurosci 15:4102–4108

Sulzer D, Sonders MS, Poulsen NW, Galli A (2005) Mechanisms of neurotransmitter release by amphetamines: a review. Prog Neurobiol 75:406–433

Thompson TL, Bridges SR, Weirs WJ (2001) Alteration of dopamine transport in the striatum and nucleus accumbens of ovariectomized and estrogen-primed rats following N-(p-isothiocyanatophenethyl) spiperone (NIPS) treatment. Brain Res Bull 54:631–638

Torres GE, Gainetdinov RR, Caron MG (2003) Plasma membrane monoamine transporters: structure, regulation and function. Nat Rev Neurosci 4:13–25

Uhl GR, Hall FS, Sora I (2002) Cocaine, reward, movement and monoamine transporters. Mol Psychiatry 7:21–26

Usdin TB, Mezey E, Chen C, Brownstein MJ, Hoffman BJ (1991) Cloning of the cocaine-sensitive bovine dopamine transporter. Proc Natl Acad Sci U S A 88:11168–11171

Wilson JM, Nobrega JN, Carroll ME, Niznik HB, Shannak K, Lac ST, Pristupa ZB, Dixon LM, Kish SJ (1994) Heterogeneous subregional binding patterns of 3H-WIN35428 and 3H-GBR12935 are differentially regulated by chronic cocaine self-administration. J Neurosci 14:2966–2979

Wise RA (2002) Brain reward circuitry: insights from unsensed incentives. Neuron 36:229–240

Wu X, Gu HH (1999) Molecular cloning of the mouse dopamine transporter and pharmacological comparison with the human homologue. Gene 233:163–170

Yokel RA, Wise RA (1975) Increased lever pressing for amphetamine after pimozide in rats: implications for a dopamine theory of reward. Science 187:547–549

Zahniser NR, Doolen S (2001) Chronic and acute regulation of Na^+/Cl^--dependent neurotransmitter transporters: drugs, substrates, presynaptic receptors, and signaling systems. Pharmacol Ther 92:21–55

Oligomerization of Neurotransmitter Transporters: A Ticket from the Endoplasmic Reticulum to the Plasma Membrane

H. Farhan · M. Freissmuth · H. H. Sitte (✉)

Institute of Pharmacology, Centre for Biomolecular Medicine and Pharmacology, Medical University Vienna, Währingerstrasse 13a, 1090 Vienna, Austria
harald.sitte@meduniwien.ac.at

1	Oligomerization of Neurotransmitter Transporters	234
1.1	The Structural Basis of Oligomer Formation	234
2	The ER Export	237
2.1	Selective Export Vs Selective Retention	238
3	Oligomerization and ER Export	240
3.1	Sensors of Oligomeric Assembly	242
3.2	Oligomerization and Exit from the Golgi	242
3.3	Life Visualization of Oligomer Formation in the Secretory Pathway	243
3.4	The Importance of Oligomer Formation for the Action of Amphetamine	243
	References	245

Abstract Cellular localization of neurotransmitter transporters is important for the precise control of synaptic transmission. By removing the neurotransmitters from the synaptic cleft, these transporters terminate signalling and affect duration and intensity of neurotransmission. Thus, a lot of work has been invested in the determination of the cellular compartment to which neurotransmitter transporters localize. In particular, the polarized distribution has received substantial attention. However, trafficking of transporters in the early secretory pathway has been largely ignored. Oligomer formation is a prerequisite for newly formed transporters to pass the stringent quality control mechanisms of the endoplasmic reticulum (ER), and this quaternary structure is also the preferred state which transporters reside in at the plasma membrane. Only properly assembled transporters are able to recruit the coatomer coat proteins that are needed for ER-to-Golgi trafficking. In this review, we will start with a brief description on transporter oligomerization that underlies ER-to-Golgi trafficking, followed by an introduction to ER-to-Golgi trafficking of neurotransmitter transporters. Finally, we will discuss the importance of oligomer formation for the pharmacological action of the illicitly used amphetamines and its derivatives.

Keywords Na^+/Cl^--dependent neurotransmitter transporters · Oligomer formation · Quaternary structure · Endoplasmic reticulum · Golgi apparatus · Export mechanisms

1
Oligomerization of Neurotransmitter Transporters

Neurotransmitter:sodium symporters (NSS) (Busch and Saier 2002) exist as constitutive oligomers at the plasma membrane of living cells (Sitte et al. 2004; Sitte and Freissmuth 2003; Torres et al. 2003b). This notion is in line with observations in related neurotransmitter transporter families (e.g. for amino acids like glutamate: Seal and Amara 1999) or transmembrane (TM) transporters of different species, for instance bacteria (Veenhoff et al. 2002). Most recently, the crystal structure has been solved for a bacterial homologue of neuronal monoamine transporters in mammals, i.e. a leucine transporter (LeuT) derived from *Aquifex aeolicus* (Yamashita et al. 2005). The level of homology between TM spanning helices is remarkably high in some segments, although a direct sequence alignment shows little overall similarity to NSS [(similarity analysis performed at www.expasy.org; result given as a percentage) for example: 26.9 to human dopamine transporter (hDAT; gb-S44626), 26.8 to the rat γ-aminobutyric acid transporter (rGAT1; gb-M59742)]. It is of obvious importance for the present chapter that the LeuT has been crystallized in a dimeric—i.e. the smallest possible—oligomeric complex (Yamashita et al. 2005). Hence, the first section will focus on earlier experiments designed to map possible contact sites in different NSS members; the resulting—largely circumstantial—evidence for specific contact sites will be reviewed for plausibility by comparing the predictions with the new structural information obtained from the LeuT.

1.1
The Structural Basis of Oligomer Formation

The search for contact sites in NSS was driven by the existence of different sequence motifs that resembled motifs described in either phospholamban or glycophorin A (GXXXG; MacKenzie and Engelman 1998; Popot and Engelman 2000; White and Wimley 1999) or a leucine repeat region found to underlie oligomerization of transcription factors (Heldin 1995). The GXXXG motif has been described in the sixth TM domain of the dopamine transporter (DAT) (Hastrup et al. 2001) and has been shown to be present in different other NSS members (Sitte et al. 2004). Similarly, a leucine heptad repeat has been shown to be more or less well conserved among different NSS members (Amara and Kuhar 1993), as well as a previously reported bacterial NSS homologue, a transporter for tryptophan that has been found in *Symbiobacterium thermophilum* (Androutsellis-Theotokis et al. 2003). The γ-aminobutyric acid (GABA) transporter GAT1 contains a canonical leucine heptad repeat in the second TM domain; four leucines (in positions 83, 90, 97 and 107; Scholze et al. 2002) are separated by six intervening amino acid residues each, and this is also true for the DAT (Torres et al. 2003a). Mutational exchange of these leucines leads to intracellular retention of the mutant GAT1 and DAT constructs. In the case of GAT1, this retention was accompanied by a loss in the efficiency of fluorescence resonance energy transfer (FRET) between fluorescently tagged transporters (Scholze et al. 2002). These results were consistent with the conclusion that the leucine heptad repeat in TM domain 2 of GAT1 supports oligomer formation. Importantly, the intracellularly retained transporters are not misfolded, because they still afford substrate transport with a reasonably well-preserved affinity for substrate (Scholze et al. 2002). A second leucine-rich repeat has been described in TM9 of the hDAT (Torres et al. 2003a); however, obliteration of that motif affected neither the activity nor localization of the transporter.

As succinctly argued by Engelman's group, a leucine-rich region cannot establish a self-sustained protein–protein contact in the hydrophobic milieu of the membrane (Zhou et al. 2001). Polar residues are needed to stabilize these interactions by establishing hydrogen bonds. Hence, we searched for possible partners sandwiched in between the leucine-rich stretch to support the oligomeric contacts between single TM2s in GAT1 (Korkhov et al. 2004). Based on results obtained with FRET and β-lactamase complementation assay (Galarneau et al. 2002), we identified a tyrosine residue in position 86 and proposed a glutamate residue in position 101 which were apparently required for oligomer formation (Korkhov et al. 2004). Glu^{101} apparently fulfilled the criteria required for its serving as the hydrophilic glue which stabilized the helix–helix interaction by acting as a hydrogen bond donor (Korkhov et al. 2004). In contrast, the tyrosine residue was interpreted to participate in hydrophobic stacking (e.g. π–π stacking interactions) because it could be replaced by phenylalanine.

Before examining this interpretation against the published LeuT-crystal structure, one has to take into account that the obvious heterogeneity of the NSS family (Nelson 1998). Alignment of TM2 shows that a canonical leucine heptad repeat is not present in the serotonin transporter (SERT) (Sitte and Freissmuth 2003); with alanine in position 125, TM2 of SERT cannot possibly function as a leucine zipper because there would be a large cavity. While TM2 of DAT and GAT1 apparently play important roles in the assembly of proper transporter oligomers (Scholze et al. 2002; Torres et al. 2003a), this is obviously not the case for SERT; mutational exchange of two additional leucines to alanines in TM2 of SERT do not impede the correct localization of SERT to the plasma membrane (H. Just, H.H. Sitte and J. Freissmuth, unpublished observation). Hence, oligomerization is achieved by different means in SERT. This notion is also supported by a survey of contact sites in SERT that clearly pointed to an additional, third contact site within the TMs 11 and 12 (Just et al. 2004).

How do the interaction domains and the discrepancies described above fit with the recently resolved crystal structure? As shown in Fig. 1A, the highlighted TM2 of LeuT is only partially accessible from the lateral side (that is from the lipid phase of the membrane, see white curvature to indicate accessibility). There are two bulky masses (Fig. 1, white arrows) near the edges of the plasma membrane that somewhat sterically hinder a vertical, non-angled interaction of two adjacent, oligomerizing transporters. In particular, there is the large extracellular loop 3 that connects TMs 5 and 6 at the outward-facing plane of the membrane and intracellular loop 5 that forms the connection between TMs 10 and 11. Given that TM2 is thus only partially accessible, it appears far-fetched to assume that a completely buried glutamic acid side chain is involved in extensive hydrogen bonding between subunits. Nevertheless, if inspected more closely, the sequences in TM2 diverge at several positions; this also true for the bulky extra- and intracellular segments that limit the movement of TM2 in SERT (Fig. 1). Thus, it is conceivable that TM2 supports the

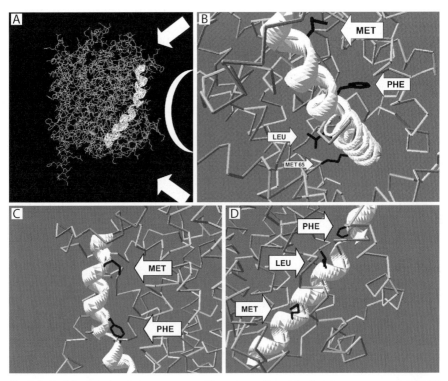

Fig. 1 A Model of the leucine transporter from *A. aeolicus*. The leucine transporter (LeuT) was discovered as a bacterial homologue to the human monoamine transporters. Shown is a side view of the LeuT to illustrate that the α-helical transmembrane domain 2 (represented as *solid, white ribbon*) is—at the very least—partially accessible to the lipid phase of the membrane. The *arrows* point to bulky protein masses of LeuT that may pose obstacles for any protein–protein interaction (but that are absent in eukaryotic NSS). **b–D** The leucine heptad repeat in TM2 of LeuT. Close-up view of transmembrane domain 2 (as seen from top in *B*, rear in *C* and front in *d*): The amino acids corresponding to the "leucine heptad repeat" in GAT1 and DAT is depicted by showing the amino acid side chains in *black* (indicated by *arrows*). TM2 is shown as *white ribbon*; the accessible space (indicated in Fig. 1A as curvature) is located at the *right side*

assembly of the quaternary structure of several NSS members, while it clearly does not do so in LeuT; the sequence of LeuT's TM2 shows that it contains two methionines, a phenylalanine and one leucine in the positions that in GAT1 are occupied only by leucines. In fact, LeuT resembles SERT in this respect much more than any other member of the NSS family. Furthermore, as shown in Fig. 1B, the amino acid side chains are only partially accessible in TM2. Hence, at the present state, it is safe to conclude that (1) TM2 does not mediate direct subunit contacts that drive oligomer formation in SERT (and LeuT) and (2) the current evidence does not allow us to decide whether the effect of TM2 mutations on oligomer formation result from a direct impact on the oligomeric interface or a structural rearrangement that acts at a distance.

As mentioned, SERT oligomerization is stabilized by a hypothetical contact site in a region comprising TM11/TM12 (Just et al. 2004). In the LeuT crystal, the dimer interface is formed by TM9 and TM12. While it is tempting to incorporate the structural information of LeuT into a model of the interface in the SERT quaternary structure (Hirai and Subramaniam 2004), a glance at the primary structure is sobering, for it is readily evident that the sequences of these TM domains are completely divergent (Yamashita et al. 2005).

2
The ER Export

Oligomer formation plays an important role in trafficking of NSS, in particular for their export from the endoplasmic reticulum. Hence, we will briefly summarize of the current model of membrane protein export from the endoplasmic reticulum.

After synthesis of membrane proteins by ribosomes of the rough endoplasmic reticulum, proteins are integrated into the membrane of the endoplasmic reticulum (ER). A stringent quality control ensures that only correctly folded proteins are allowed to exit the ER. The ER provides an environment that is optimized for protein folding and maturation. Like other organelles of the secretory pathway, the lumen of the ER is topologically equivalent to the extracellular milieu. The ER lumen differs from the cytosol with respect to ionic concentration and the complement of molecular chaperons. The first posttranslational modifications also take place in the ER such as N-glycosylation, disulphide bond formation, signal peptide cleavage, glycophosphatidylinositol (GPI)-anchor addition and proline hydroxylation (Ellgaard and Helenius 2003). Typically, proteins are concentrated into sub-domains of the rough ER that are free of ribosomes and in which protrusions can be seen that resemble budding vesicles (Orci et al. 1991). These are referred to as ER exit sites (ERES). Proteins in ERES are still subject to ER-based quality control (Mezzacasa and Helenius 2002). ERES are coated by components of the coat protein (COP)II coat complex. COPII-coated vesicles mediate the first step of ER export. The core components of the COPII coat are: Sar1, the Sec23–Sec24 complex and the Sec13–Sec31 complex. Sar1 is a small (21 kDa) Ras-like GTPase. Guanine nucleotide exchange is achieved by Sec12, which is a type II TM protein.

Activated Sar1 is recruited to the membrane of the ER by means of a hydrophobic sequence in its N-terminus. Subsequently, the Sec23–Sec24 dimer is recruited; this results in the formation of an oligomeric assembly which is referred to as the pre-budding complex. While Sec23 is the GAP (GTPase activating protein) for Sar1 (Yoshihara et al. 1999), the main function attributed to Sec24 is to recognize cargo (Miller et al. 2003; Mossessova et al. 2003). The outer layer of the COPII coat is formed by the Sec13–Sec31 heterotetramer, which accelerates GTP hydrolysis by Sar1. The coat lattice is thought to induce a membrane curvature that forces the budding of vesicles. In yeast, COPII vesicles reach the Golgi. In mammalian cells, COPII vesicles fuse with the intermediate compartment (ERGIC). Export from the latter compartment to the Golgi is regulated by another type of coat called COPI.

We have recently investigated the role of the carboxyl terminus of GAT1 in intracellular trafficking (Farhan et al. 2004). Previous work has suggested a role of the last 39 amino acids in determining surface expression of the transporter (Bendahan and Kanner 1993). On the other hand, a later report suggested that the last 36 amino acids are dispensable for GAT1 trafficking (Perego et al. 1997). However, we were not able to reproduce the results of Perego et al. (1997). In fact, we found a Sec24D interaction motif in the carboxyl terminus of GAT1; this motif is located between amino acids L^{553}–Q^{572}. This assignment is based on observations obtained by using fluorescence resonance energy transfer. In the meantime, we verified the finding by employing a GST pull-down assay (H. Farhan, V. Reiterer, M. Freissmuth, unpublished result). The exact role of this interaction in trafficking of GAT1 is currently under investigation.

2.1
Selective Export Vs Selective Retention

In principle there are two mechanisms by which ER export can contribute to the control of surface expression: selective export and selective retention. In the case of selective export, ER export motifs mediate the interaction with components of the COPII machinery that drive their sorting into COPII coated vesicles. There are two types of ER export motifs, di-acidic and di-hydrophobic motifs. In addition, several atypical motifs have been reported (e.g. NPF in Sed5p and R/K-X-R/K in the galactosyltransferase). ER exit via selective export implies that proteins are retained in the ER if these motifs become inaccessible. Recently, the molecular mechanism of ER retention of cystic fibrosis transmembrane regulator (CFTR)-ΔF508 was uncovered (Wang et al. 2004). The authors found that a 563YKDAD567 motif in the nucleotide binding domain 1 mediates the interaction with Sec24. In CFTR-ΔF508 this motif becomes inaccessible to Sec24. Therefore the protein is retained. In line with this finding is the observation that CFTR-ΔF508 escapes to the plasma membrane when cells are incubated at reduced temperature (Denning et al. 1992), a condition that facilitates folding.

Selective retention is based on the exposure of retention motifs by newly synthesized proteins. The best-characterized retention motif is RXR (where X is any amino acid). The protein can only escape from the ER when this motif is marked. The $GABA_B$ receptor is a prominent example. This receptor is composed of two subunits: GB1 and GB2. GB1 exposes an RXR motif, which causes ER retention. GB1 is produced always in excess. The rate-limiting factor for the cell-surface expression of $GABA_B$ receptors is the production of GB2, which hetero-oligomerizes with GB1 and hides its retention motif. In addition, GB2 provides an ER export motif. The same principle is true for the NR1 and NR2 subunits of the N-methyl-D-aspartate (NMDA) receptor (Standley et al. 2000). The nature of the selective retention mode implies that these proteins have to escape the ER as oligomers such that one partner covers the retention motif of the other. So far there is no direct evidence that neurotransmitter transporters use this mode of export and that this underlies their oligomeric assembly. However, Kalandadze et al. (2004) recently reported that a leucine-based motif in the glutamate transporter GLT-1 mediates ER export. This motif is composed of six leucines and it does not match any trafficking motif reported in the literature. Mutation of these residues to alanine (L6/A6) led to retention of the transporter in the ER. In addition, an arginine-based motif adjacent to the export motif mediates ER retention. When the RXR motif was mutated, surface

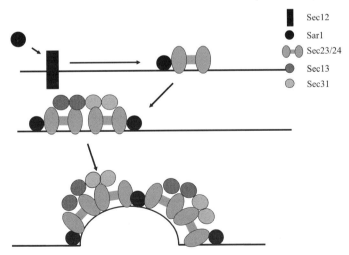

Fig. 2 Model of COPII coat assembly and budding from the ER. The Sar1p GTPase is activated at export sites by the guanine nucleotide exchange factor Sec12p. Activated Sar1 together with Sec23/24 forms the pre-budding complex. Subsequently the Sec13/31 heterotetramer is recruited. This leads to coat polymerization and membrane deformation

expression of the L6/A6 mutant was partially restored. The authors concluded that the mode of GLT-1 ER export was selective retention. According to the available architecture of GLT-1 at the time of publication, both motifs were supposed to lie in an extracellular loop. In other words, this loop was predicted to lie in the ER lumen; it was therefore conceptually difficult to grasp the significance of the observations, because this extracellular loop would not have any access to cytosolic components that mediate incorporation into COPII vesicles. The crystal structure of a bacterial glutamate transporter was resolved recently and revealed a number of unusual features, which were not appreciated by algorithms used for structural predictions based on sequence information (Yernool et al. 2004). When the sequences of GLT-1 and the bacterial homologue are compared (Fig. 2), the leucine-based motif is partially located in the extracellular loop and partially in TM8. The RXR motif is confined to TM8. This TM domain is not part of the oligomeric interface. Therefore, it is not covered upon trimerization of the transporter. It is therefore likely that the mutations of the leucine-based and of the RXR motif summarized above impinged on ER-export by indirect effects on the structure of GLT-1.

3
Oligomerization and ER Export

We (Schmid et al. 2001b) and others (Sorkina et al. 2003) reported that neurotransmitter transporters exist as constitutive oligomers in living cells. Oligomers form already in the ER (Scholze et al. 2002; Sorkina et al. 2003). Based on these observations, we put forth the "oligomerization hypothesis" which postulates that oligomerization is a prerequisite for ER-export. The hypothesis makes several predictions that have been verified:

1. Mutations that disrupt oligomer formation result in ER retention of transporters (Scholze et al. 2002; Torres et al. 2003a).

2. Proteins are retained in the ER because they lack an export signal or because they carry a retention signal (see Sect. 2.1). Mechanistically, oligomerization must result either in complementation of the export signal or in the masking of a retention signal. As mentioned above (Sect. 2), the amino acid residues L^{553}–Q^{572} in the carboxyl terminus of GAT1 support biding of Sec24D and are therefore a bona fide export signal (Farhan et al. 2004). However, oligomeric assembly of GAT1 is not required for the interaction with Sec24D; when co-expressed with an appropriately tagged Sec24D, an oligomerization-deficient mutant of GAT1 (GAT1-L2A) still supported a robust FRET that was visualized over the endoplasmic reticulum. Nevertheless, GAT1-L2A failed to exit the ER. This finding can be rationalized in a model where oligomers are required for efficient recruitment of Sec24D and subsequent coat assembly because the export motifs in the carboxyl-

termini are presented in optimal orientation. This arrangement affords the rapid assembly of the COPII coat and thus budding of COPII vesicles.

3. The acid test for a concept is its ability to explain the biological consequence of a mutation, in particular the phenotypic consequences of a clinically relevant disease. Familial orthostatic intolerance was recently found to be caused by a mutation in one allele of the norepinephrine transporter (NET-A457P) (Shannon et al. 2000). This mutation is associated with a loss in more than 90% of NET activity. Thus, only one allele needs to be affected to essentially abrogate norepinephrine transport and heterozygous individuals are affected. This can only be accounted for by a dominant-negative effect of the mutant protein on the NET encoded by the normal gene. The "oligomerization hypothesis" predicts this dominant-negative effect. A mutated protein which is retained in the endoplasmic reticulum is phenotypically dominant if it associates with the protein encoded by the normal allele. This was the case; when co-expressed with wild-type NET, NET-A457P prevents the insertion of the wild-type transporter into the plasma membrane. Both wild type and mutant NET accumulate in the ER (Hahn et al. 2003). Cellular uptake of norepinephrine is obviously contingent on the presence of the transporter on the plasma membrane. If transporters were dimers, stochastic association of NET and NET-A457P would result in residual transport activity of about 25%. The fact that the expression of NET-A457P depresses residual transport activity of NET to a substantially larger extent indicates that transporters are higher order oligomers; additional arguments for this conjecture are listed elsewhere (Sitte and Freissmuth 2003).

The oligomerization hypothesis outlined above relies on the assumption that the COPII component is capable of sensing the orientation (and the concentration) of Sec24D binding sites (see Sect. 3.1, below). A crucial role for the orientation of export motifs has been suggested by Otte and Barlowe (Otte and Barlowe 2002). They found that the cargo receptors Erv41p and Erv46p must heterodimerize in order to exit the ER. This is dependent on export motifs in their cytoplasmic carboxyl termini. If the C-terminus of Erv41p is exchanged with that of Erv46p (and vice versa), the heterodimer fails to exit the ER. Nevertheless, if the oligomerization hypothesis is to be more than hand waving, it requires an explanation that is consistent with the biochemical activity of the components. We believe that the GTPase activity of Sar1 provides the crucial link; previously, Sato and Nakano (2003) showed that oligomerization of the TM cargo receptors Emp46p and Emp47p directed their incorporation into COPII vesicles. The failure of Emp46p/47p monomers to exit the ER can be overcome by using non-hydrolyzable GTP analogues. This led to a kinetic proof-reading model (Sato and Nakano 2004). According to this model, oligomeric assembly of cargo in the pre-budding complex keeps Sar1 in the GTP-bound state. The full hydrolytic activity of the Sar1-GTPase is triggered, if the coat is assembled—that is, if Sec13/Sec31 enhances the GAP

(GTPase activating protein) activity of Sec23 (Antonny and Schekman 2001). Finally, it is important to note that there is precedent for homo-oligomerization in ER-export, a prominent example being ERGIC53, a mammalian lectin that cycles between ER and ERGIC. Hexamerization of ERGIC53 is necessary for its export from the ER (Nufer et al. 2003).

3.1
Sensors of Oligomeric Assembly

The importance of export motif orientation implies that there is a sensor for oligomeric assembly. The carboxyl terminus of the K(ATP) channel α-subunit Kir6.2 contains an arginine-based ER retention motif. Yuan et al. (2003) found that the 14-3-3 isoforms ε and ζ specifically recognize this signal; 14-3-3 bound better to dimers than to monomers and better to tetramers than to dimers. This interaction leads to release of Kir6.2 from the ER. Therefore 14-3-3 proteins qualify as oligomer sensors. Recently the amino terminus of NET was found to interact with 14-3-3, and this interaction was suggested to regulate trafficking of the transporter (Sung et al. 2005).

Yeast Shr3p is another example for such a sensor; it is an integral membrane protein with four TM segments. Deletion of Shr3p leads to retention of amino acid permeases (AAPs) in the ER (Ljungdahl et al. 1992). AAPs are transporters that are distantly related to neurotransmitter transporters and share the same topology (that is 12 TM segments). COPII components interact with Shr3p; hence, it has been suggested that Shr3p initiates budding of vesicles en route to the Golgi in the vicinity of AAPs Gap1p and Hip1p (Gilstring et al. 1999; Kuehn et al. 1996). More recently, Shr3p was shown to prevent aggregation of AAPs, thereby enabling them to fold correctly (Kota and Ljungdahl 2005). Assembly of AAPs into oligomers would lead to the release from Shr3p and exit from the ER.

3.2
Oligomerization and Exit from the Golgi

Evidence presented so far indicates that oligomerization is a pre-requisite for ER export. Sorkina et al. (2003) showed that the dopamine transporter forms constitutive oligomers. There is no indication so far that the oligomeric state of a transporter is necessary for its trafficking beyond the ER. However, there is a precedent for a role of protein oligomerization in export from the Golgi. GPI-anchored proteins (GPI-APs) have to become associated with detergent-resistant microdomains (DRMs) or rafts in order to be sorted to the apical compartment in the epithelial cell line MDCK (Madin–Darby canine kidney cells). However, this step, although necessary, is not sufficient (Sarnataro et al. 2002). The major determinant that specifies apical sorting of GPI-AP is their oligomerization which is initiated in the medial Golgi (Paladino et al. 2004).

3.3
Life Visualization of Oligomer Formation in the Secretory Pathway

FRET microscopy has proved to be a powerful method to visualize protein oligomers in living cells (Schmid and Sitte 2003). However, FRET microscopy has several limitations, not the least of which is a poor time resolution. Recently, the yellow fluorescent protein (YFP) protein complementation assay was shown to be useful in detecting low-affinity interactions in the secretory pathway (Nyfeler et al. 2005). In this assay, two fragments of YFP are attached to two proteins. In case of interaction, the YFP molecule is complemented and YFP fluorescence indicates the cellular compartment in which the interaction takes place. In addition, the intensity of YFP fluorescence may indicate the strength of the interaction. The mammalian lectin ERGIC53 was fused to one partner, and cathepsin C or cathepsin Z (both are known to interact with ERGIC53) were tagged with the second fragment. In addition, the two YFP fragment were fused to ERGIC53. YFP fluorescence could be seen in the ER and in the ERGIC but not in the Golgi. The intensity of YFP fluorescence was much stronger in the case of ERGIC53 homo-oligomers than in the case of ERGIC53 cathepsin C/Z interaction. This is explained by the fact that ERGIC53 forms constitutive hexamers, while only one molecule of cathepsin C/Z binds to ERGIC53. Taken together, the YFP protein complementation assay and FRET microscopy provide powerful tools for investigations on the dynamics of neurotransmitter transporter oligomers and the compartment in which oligomerization takes place. YFP protein complementation offers the advantage that allows researchers to follow a protein over time (because it obviates the change in filters, splitting of beams, etc.). FRET microscopy, however, is more likely to provide reliable quantitative information, because the efficiency of energy transfer can be calculated and hence distance parameters can be extracted, provided that the caveats and pitfalls are kept in mind (Schmid et al. 2001a).

3.4
The Importance of Oligomer Formation for the Action of Amphetamine

Monoaminergic neurotransmitter transporters are the prime target of the amphetamines; they comprise a large variety of widely abused, psychotropic substances including D-amphetamine ("speed"), methamphetamine ("ice") and methylene-dioxymethamphetamine (MDMA; "ecstasy"). Amphetamines were originally marketed as appetite suppressants (before they were withdrawn on a worldwide basis because of their addictive potential and because some compounds caused pulmonary hypertension). At present, the amphetamine derivative methylphenidate is used in hyperactive children (see related chapters in this volume by Z. Lin and B.K. Madras, as well as M.S. Mazei-Robison and R.D. Blakely). All amphetamine-like compounds induce carrier-mediated efflux (Levi and Raiteri 1993) and thereby elevate synaptic monoamine concentrations. Hence, a prolonged stimulation of postsynaptic receptors ensues. Amphetamines differ in their affinities for monoamine transporters: for instance, D-amphetamine predominantly targets DAT while MDMA acts mainly on SERT (Seiden et al. 1993; Green et al. 2003). Many insights have been generated that shed some light on the mechanism of action of amphetamines. However, the mechanistic basis for their effect remains enigmatic at the molecular level.

Monoamine transporters exist in quaternary structures at the plasma membrane as outlined in Sect. 1.1 (Kilic and Rudnick 2000; Kocabas et al. 2003; Schmid et al. 2001b; Sorkina et al. 2003); we therefore hypothesized that amphetamine-induced outward movement of substrate might be contingent on the arrangement in these oligomeric complexes. To test this hypothesis, we generated head-to-tail fusion proteins that consisted of an amphetamine-resistant GAT1 and an amphetamine-susceptible SERT (Seidel et al. 2005). Amphetamines target and affect other cellular components as well [e.g. protein kinase C (Khoshbouei et al. 2004) and vesicular transporters (Sulzer et al. 1995)]. We therefore subjected the SERT-GAT1 concatamer to a pharmacological characterization (Seidel et al. 2005).

The distinct transporter moieties in SERT-GAT1 were fully functional. Consistent with previously published observations on SERT-NET concatamers (Horschitz et al. 2003), they did apparently work independently from one another, provided that only one transporter was challenged with substrate. However, indirect effects were clearly visible; by means of electrophysiological measurements and the patch-clamp technique in the whole-cell configuration, it was possible to record transporter-mediated currents that clearly indicated current in excess (see also related chapter in this volume by K. Gerstbrein and H.H. Sitte). These currents are driven by sodium (Adams and DeFelice 2002; Galli et al. 1996; Petersen and DeFelice 1999; Sitte et al. 1998; Sonders et al. 1997), and therefore they elevate intracellular sodium (Saunders et al. 2000). This causes a shift in the overall transporter conformation. The probability that the transporter resides more in an inward-facing conformation—and is thus inaccessible for external substrate—rises as a function of the increase in intracellular Na^+. Likewise, this reasoning was fully borne out by the results with SERT-GAT1-expressing cells: GABA uptake was diminished in the presence of para-chloroamphetamine and vice versa—serotonin uptake was diminished in the presence of GABA (Seidel et al. 2005).

In an oligomer-based counter-transport model, the substrate is taken up by one moiety and transporter-mediated release is achieved by the neighbouring moiety (see Fig. 3). However, from a pharmacological perspective, the occupancy of the single transporter moieties in an oligomeric complex by substrate must reduce the probable efflux; in other words, the more substrate available at the cell exterior, the fewer transporters can be available for efflux—hence, amphetamine-induced 5-HT efflux ought to subside when the amphetamine concentrations rise above a certain level. A biphasic concentration response curve again reflected these predictions (Seidel et al. 2005).

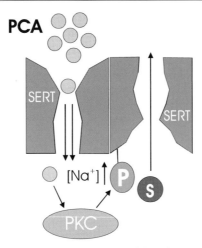

Fig. 3 The oligomer-based counter-transport model. Schematic representation of the oligomer-based counter-transport model. The model shows the effect of the amphetamine congener parachloroamphetamine (PCA) at low concentrations. PCA is transported via the serotonin transporter (SERT, shown in a dimeric state) into the cell and activates protein kinase C with concomitant rise of the sodium concentration. The other SERT moiety in the oligomeric complex is subject to phosphorylation (*P*) and thereby primed for outward transport of substrate (*S*)

Most importantly, we hypothesized that carrier-mediated efflux relies on the spatial proximity within a transporter oligomer. To verify this conjecture, we examined the amphetamine effects on GABA-preloaded SERT-GAT1-expressing cells. A tiagabine-sensitive, concentration-dependent GABA efflux was triggered by amphetamine application. This finding confirmed the oligomer-based counter-transport model.

Acknowledgements We gratefully acknowledge support from the Austrian Science Foundation (P19072-B09 to H.F. and P17076 to H.H.S.) and from the Austrian National Bank to H.F. (10507).

References

Adams SV, DeFelice LJ (2002) Flux coupling in the human serotonin transporter. Biophys J 83:3268–3282

Amara SG, Kuhar MJ (1993) Neurotransmitter transporters: recent progress. Annu Rev Neurosci 16:73–93

Androutsellis-Theotokis A, Goldberg NR, Ueda K, Beppu T, Beckman ML, Das S, Javitch JA, Rudnick G (2003) Characterization of a functional bacterial homologue of sodium-dependent neurotransmitter transporters. J Biol Chem 278:12703–12709

Antonny B, Schekman R (2001) ER export: public transportation by the COPII coach. Curr Opin Cell Biol 13:438–443

Bendahan A, Kanner BI (1993) Identification of domains of a cloned rat brain GABA transporter which are not required for its functional expression. FEBS Lett 318:41–44

Busch W, Saier MH Jr (2002) The transporter classification (TC) system, 2002. Crit Rev Biochem Mol Biol 37:287–337

Denning GM, Anderson MP, Amara JF, Marshall J, Smith AE, Welsh MJ (1992) Processing of mutant cystic fibrosis transmembrane conductance regulator is temperature-sensitive. Nature 358:761–764

Ellgaard L, Helenius A (2003) Quality control in the endoplasmic reticulum. Nat Rev Mol Cell Biol 4:181–191

Farhan H, Korkhov VM, Paulitschke V, Dorostkar MM, Scholze P, Kudalcek O, Freissmuth M, Sitte HH (2004) Two discontinuous segments in the carboxy terminus are required for membrane targeting of the rat GABA transporter-1 (GAT1). J Biol Chem 279:28553–28563

Galarneau A, Primeau M, Trudeau LE, Michnick SW (2002) Beta-lactamase protein fragment complementation assays as in vivo and in vitro sensors of protein protein interactions. Nat Biotechnol 20:619–622

Galli A, Blakely RD, DeFelice LJ (1996) Norepinephrine transporters have channel modes of conduction. Proc Natl Acad Sci U S A 93:8671–8676

Gilstring CF, Melin-Larsson M, Ljungdahl PO (1999) Shr3p mediates specific COPII coatomer-cargo interactions required for the packaging of amino acid permeases into ER-derived transport vesicles. Mol Biol Cell 10:3549–3565

Green AR, Mechan AO, Elliott JM, O'Shea E, Colado MI (2003) The pharmacology and clinical pharmacology of 3,4-methylenedioxymethamphetamine (MDMA, "ecstasy"). Pharmacol Rev 55:463–508

Hahn MK, Robertson D, Blakely RD (2003) A mutation in the human norepinephrine transporter gene (SLC6A2) associated with orthostatic intolerance disrupts surface expression of mutant and wild-type transporters. J Neurosci 23:4470–4478

Hastrup H, Karlin A, Javitch JA (2001) Symmetrical dimer of the human dopamine transporter revealed by cross-linking Cys-306 at the extracellular end of the sixth transmembrane segment. Proc Natl Acad Sci U S A 98:10055–10060

Heldin CH (1995) Dimerization of cell surface receptors in signal transduction. Cell 80:213–223

Hirai T, Subramaniam S (2004) Structure and transport mechanism of the bacterial oxalate transporter OxlT. Biophys J 87:3600–3607

Horschitz S, Hummerich R, Schloss P (2003) Functional coupling of serotonin and noradrenaline transporters. J Neurochem 86:958–965

Just H, Sitte HH, Schmid JA, Freissmuth M, Kudlacek O (2004) Identification of an additional interaction domain in transmembrane domains 11 and 12 that supports oligomer formation in the human serotonin transporter. J Biol Chem 279:6650–6657

Kalandadze A, Wu Y, Fournier K, Robinson MB (2004) Identification of motifs involved in endoplasmic reticulum retention-forward trafficking of the GLT-1 subtype of glutamate transporter. J Neurosci 24:5183–5192

Khoshbouei H, Sen N, Guptaroy B, Johnson L, Lund D, Gnegy ME, Galli A, Javitch JA (2004) N-terminal phosphorylation of the dopamine transporter is required for amphetamine-induced efflux. PLoS Biol 2:E78

Kilic F, Rudnick G (2000) Oligomerization of serotonin transporter and its functional consequences. Proc Natl Acad Sci U S A 97:3106–3111

Kocabas AM, Rudnick G, Kilic F (2003) Functional consequences of homo- but not hetero-oligomerization between transporters for the biogenic amine neurotransmitters. J Neurochem 85:1513–1520

Korkhov VM, Farhan H, Freissmuth M, Sitte HH (2004) Oligomerization of the γ-aminobutyric acid transporter-1 is driven by an interplay of polar and hydrophobic interactions in transmembrane helix II. J Biol Chem 279:55728–55736

Kota J, Ljungdahl PO (2005) Specialized membrane-localized chaperones prevent aggregation of polytopic proteins in the ER. J Cell Biol 168:79–88

Kuehn MJ, Schekman R, Ljungdahl PO (1996) Amino acid permeases require COPII components and the ER resident membrane protein Shr3p for packaging into transport vesicles in vitro. J Cell Biol 135:585–595

Levi G, Raiteri M (1993) Carrier-mediated release of neurotransmitters. Trends Neurosci 16:415–419

Ljungdahl PO, Gimeno CJ, Styles CA, Fink GR (1992) SHR3: a novel component of the secretory pathway specifically required for localization of amino acid permeases in yeast. Cell 71:463–478

MacKenzie KR, Engelman DM (1998) Structure-based prediction of the stability of transmembrane helix-helix interactions: the sequence dependence of glycophorin A dimerization. Proc Natl Acad Sci U S A 95:3583–3590

Mezzacasa A, Helenius A (2002) The transitional ER defines a boundary for quality control in the secretion of tsO45 VSV glycoprotein. Traffic 3:833–849

Miller EA, Beilharz TH, Malkus PN, Lee MC, Hamamoto S, Orci L, Schekman R (2003) Multiple cargo binding sites on the COPII subunit Sec24p ensure capture of diverse membrane proteins into transport vesicles. Cell 114:497–509

Mossessova E, Bickford LC, Goldberg J (2003) SNARE selectivity of the COPII coat. Cell 114:483–495

Nelson N (1998) The family of Na^+/Cl^- neurotransmitter transporters. J Neurochem 71:1785–1803

Nufer O, Kappeler F, Guldbrandsen S, Hauri HP (2003) ER export of ERGIC-53 is controlled by cooperation of targeting determinants in all three of its domains. J Cell Sci 116:4429–4440

Nyfeler B, Michnick SW, Hauri HP (2005) Capturing protein interactions in the secretory pathway of living cells. Proc Natl Acad Sci U S A 102:6350–6355

Orci L, Ravazzola M, Meda P, Holcomb C, Moore HP, Hicke L, Schekman R (1991) Mammalian Sec23p homologue is restricted to the endoplasmic reticulum transitional cytoplasm. Proc Natl Acad Sci U S A 88:8611–8615

Otte S, Barlowe C (2002) The Erv41p-Erv46p complex: multiple export signals are required in trans for COPII-dependent transport from the ER. EMBO J 21:6095–6104

Paladino S, Sarnataro D, Pillich R, Tivodar S, Nitsch L, Zurzolo C (2004) Protein oligomerization modulates raft partitioning and apical sorting of GPI-anchored proteins. J Cell Biol 167:699–709

Perego C, Bulbarelli A, Longhi R, Caimi M, Villa A, Caplan MJ, Pietrini G (1997) Sorting of two polytopic proteins, the gamma-aminobutyric acid and betaine transporters, in polarized epithelial cells. J Biol Chem 272:6584–6592

Petersen CI, DeFelice LJ (1999) Ionic interactions in the Drosophila serotonin transporter identify it as a serotonin channel. Nat Neurosci 2:605–610

Popot JL, Engelman DM (2000) Helical membrane protein folding, stability, and evolution. Annu Rev Biochem 69:881–922

Sarnataro D, Paladino S, Campana V, Grassi J, Nitsch L, Zurzolo C (2002) PrPC is sorted to the basolateral membrane of epithelial cells independently of its association with rafts. Traffic 3:810–821

Sato K, Nakano A (2003) Oligomerization of a cargo receptor directs protein sorting into COPII-coated transport vesicles. Mol Biol Cell 14:3055–3063

Sato K, Nakano A (2004) Reconstitution of coat protein complex II (COPII) vesicle formation from cargo-reconstituted proteoliposomes reveals the potential role of GTP hydrolysis by Sar1p in protein sorting. J Biol Chem 279:1330–1335

Saunders C, Ferrer JV, Shi L, Chen J, Merrill G, Lamb ME, Leeb-Lundberg LM, Carvelli L, Javitch JA, Galli A (2000) Amphetamine-induced loss of human dopamine transporter activity: an internalization-dependent and cocaine-sensitive mechanism. Proc Natl Acad Sci U S A 97:6850–6855

Schmid JA, Sitte HH (2003) Fluorescence resonance energy transfer in the study of cancer pathways. Curr Opin Oncol 15:55–64

Schmid JA, Just H, Sitte HH (2001a) Impact of oligomerization on the function of the human serotonin transporter. Biochem Soc Trans 29:732–736

Schmid JA, Scholze P, Kudlacek O, Freissmuth M, Singer EA, Sitte HH (2001b) Oligomerization of the human serotonin transporter and of the rat GABA transporter 1 visualized by fluorescence resonance energy transfer microscopy in living cells. J Biol Chem 276:3805–3810

Scholze P, Freissmuth M, Sitte HH (2002) Mutations within an intramembrane leucine heptad repeat disrupt oligomer formation of the rat GABA transporter 1. J Biol Chem 277:43682–43690

Seal RP, Amara SG (1999) Excitatory amino acid transporters: a family in flux. Annu Rev Pharmacol Toxicol 39:431–456

Seidel S, Singer EA, Just H, Farhan H, Scholze P, Kudlacek O, Holy M, Koppatz K, Krivanek P, Freissmuth M, Sitte HH (2005) Amphetamines take two to tango: an oligomer-based counter-transport model of neurotransmitter transport explores the amphetamine action. Mol Pharmacol 67:140–151

Seiden LS, Sabol KE, Ricaurte GA (1993) Amphetamine: effects on catecholamine systems and behavior. Annu Rev Pharmacol Toxicol 33:639–677

Shannon JR, Flattem NL, Jordan J, Jacob G, Black BK, Biaggioni I, Blakely RD, Robertson D (2000) Orthostatic intolerance and tachycardia associated with norepinephrine-transporter deficiency. N Engl J Med 342:541–549

Sitte HH, Freissmuth M (2003) Oligomer formation by Na^+-Cl^--coupled neurotransmitter transporters. Eur J Pharmacol 479:229–236

Sitte HH, Huck S, Reither H, Boehm S, Singer EA, Pifl C (1998) Carrier-mediated release, transport rates, and charge transfer induced by amphetamine, tyramine, and dopamine in mammalian cells transfected with the human dopamine transporter. J Neurochem 71:1289–1297

Sitte HH, Farhan H, Javitch JA (2004) Sodium-dependent neurotransmitter transporters: oligomerization as a determinant of transporter function and trafficking. Mol Interv 4:38–47

Sonders MS, Zhu SJ, Zahniser NR, Kavanaugh MP, Amara SG (1997) Multiple ionic conductances of the human dopamine transporter: the actions of dopamine and psychostimulants. J Neurosci 17:960–974

Sorkina T, Doolen S, Galperin E, Zahniser NR, Sorkin A (2003) Oligomerization of dopamine transporters visualized in living cells by FRET microscopy. J Biol Chem 278:28274–28283

Standley S, Roche KW, McCallum J, Sans N, Wenthold RJ (2000) PDZ domain suppression of an ER retention signal in NMDA receptor NR1 splice variants. Neuron 28:887–898

Sulzer D, Chen TK, Lau YY, Kristensen H, Rayport S, Ewing A (1995) Amphetamine redistributes dopamine from synaptic vesicles to the cytosol and promotes reverse transport. J Neurosci 15:4102–4108

Sung U, Jennings JL, Link AJ, Blakely RD (2005) Proteomic analysis of human norepinephrine transporter complexes reveals associations with protein phosphatase 2A anchoring subunit and 14-3-3 proteins. Biochem Biophys Res Commun 333:671–678

Torres GE, Carneiro A, Seamans K, Fiorentini C, Sweeney A, Yao WD, Caron MG (2003a) Oligomerization and trafficking of the human dopamine transporter. Mutational analysis identifies critical domains important for the functional expression of the transporter. J Biol Chem 278:2731–2739

Torres GE, Gainetdinov RR, Caron MG (2003b) Plasma membrane monoamine transporters: structure, regulation and function. Nat Rev Neurosci 4:13–25

Veenhoff LM, Heuberger EH, Poolman B (2002) Quaternary structure and function of transport proteins. Trends Biochem Sci 27:242–249

Wang X, Matteson J, An Y, Moyer B, Yoo JS, Bannykh S, Wilson IA, Riordan JR, Balch WE (2004) COPII-dependent export of cystic fibrosis transmembrane conductance regulator from the ER uses a di-acidic exit code. J Cell Biol 167:65–74

White SH, Wimley WC (1999) Membrane protein folding and stability: physical principles. Annu Rev Biophys Biomol Struct 28:319–365

Yamashita A, Singh SK, Kawate T, Jin Y, Gouaux E (2005) Crystal structure of a bacterial homologue of Na^+/Cl^--dependent neurotransmitter transporters. Nature 437:215–223

Yernool D, Boudker O, Jin Y, Gouaux E (2004) Structure of a glutamate transporter homologue from Pyrococcus horikoshii. Nature 431:811–818

Yoshihara M, Ueda A, Zhang D, Deitcher DL, Schwarz TL, Kidokoro Y (1999) Selective effects of neuronal-synaptobrevin mutations on transmitter release evoked by sustained versus transient Ca^{2+} increases and by cAMP. J Neurosci 19:2432–2441

Yuan H, Michelsen K, Schwappach B (2003) 14-3-3 dimers probe the assembly status of multimeric membrane proteins. Curr Biol 13:638–646

Zhou FX, Merianos HJ, Brunger AT, Engelman DM (2001) Polar residues drive association of polyleucine transmembrane helices. Proc Natl Acad Sci U S A 98:2250–2255

Acute Regulation of Sodium-Dependent Glutamate Transporters: A Focus on Constitutive and Regulated Trafficking

M. B. Robinson

Departments of Pediatrics and Pharmacology, Children's Hospital of Philadelphia, University of Pennsylvania, 502 AbramsonResearch Building, 3615 Civic Center Blvd., Philadelphia PA, 19104-4318, USA
Robinson@pharm.med.upenn.edu

1	Introduction	252
2	Characteristics of Na^+-Dependent Glutamate Transporters	254
2.1	Localization	254
2.2	Contributions to Synaptic Transmission	254
2.3	Role in Toxicity	256
3	Rapid Regulation of Glutamate Transporters	257
3.1	General Comments Regarding Plasma Membrane Expression of Glutamate Transporters	260
3.2	Constitutive Trafficking of Glutamate Transporters	262
3.3	EAAC1/EAAT3	264
3.4	GLT-1/EAAT2	267
3.5	GLAST/EAAT1	269
4	Conclusions	270
	References	270

Abstract The acidic amino acid glutamate activates a family of ligand-gated ion channels to mediate depolarization that can be as short-lived as a few milliseconds and activates a family of G protein-coupled receptors that couple to both ion channels and other second messenger pathways. Glutamate is the predominant excitatory neurotransmitter in the mammalian central nervous system and is required for essentially all motor, sensory, and cognitive functions. In addition, glutamate-mediated signaling is required for development and the synaptic plasticity thought to underlie memory formation and retrieval. The levels of glutamate in brain approach 10 mmol/kg and most cells in the CNS express at least one of the receptor subtypes. Unlike acetylcholine that mediates "rapid" excitatory neurotransmission at the neuromuscular junction, there is no evidence for extracellular inactivation of glutamate. Instead, glutamate is cleared by a family of Na^+-dependent transport systems that are found on glial processes that sheath the synapse and found on the pre- and postsynaptic elements of neurons. These transporters ensure crisp excitatory transmission by maintaining synaptic concentrations below those required for tonic activation of glutamate receptors under baseline conditions ($\sim$1 μM) and serve to limit activation of glutamate receptors after release. During the past few years, it has become clear that like many of the other neurotransmitter transporters discussed in this volume of *Handbook of Experimental Pharmacology*, the activity of these transporters can be rapidly regulated by a variety of effectors. In this chapter, a broad overview of excitatory signaling will be followed by a brief introduction to the family of Na^+-dependent glutamate transporters and a detailed discussion of our current understanding of the mechanisms that control transporter activity. The focus will be on our current understanding of the mechanisms that could regulate transporter activity within minutes, implying that this regulation is independent of transcriptional or translational control mechanisms. The glutamate transporters found in forebrain are regulated by redistributing the proteins to or from the plasma membrane; the signals involved and the net effects on transporter activity are being defined. In addition, there is evidence to suggest that the intrinsic activity of these transporters is also regulated by mechanisms that are independent of transporter redistribution; less is known about these events. As this field progresses, it should be possible to determine how this regulation affects physiologic and pathologic events in the CNS.

Keywords Glutamate · Transporters · Trafficking · GLT-1 · GLAST · EAAC1 · EAAT

1
Introduction

Glutamate is the predominant excitatory neurotransmitter in the brain with levels that approach 10 mmol/kg wet weight, or levels that are 1,000- to 10,000-fold higher than those observed for many other important neurotransmitters (e.g., serotonin, norepinephrine, and dopamine). After release from neurons (for a review, see Vizi and Kiss 1998) or glia (for a recent discussion, see Zhang et al. 2004), it interacts with ligand-gated ion channels or G protein-coupled receptors. The ligand-gated ion channels are named for exogenous, subtype-selective agonists, including N-methyl-D-aspartate (NMDA), α-amino-3-hydroxy-5-methyl-4-isoxazolepropionate (AMPA), and kainate. Each of these subtypes of receptors is composed of families of different subunits that assemble into multimers (usually heteromultimers), and with varying degrees of selectivity they are permeable to Na^+, Ca^{2+}, and K^+. In addition, at least some of the ligand-gated ion channels also couple directly or indirectly to intracellular signaling pathways (for a recent review, see Sheng and Pak 2000). The family of G protein-coupled receptors, also called metabotropic glutamate receptors (mGluRs), includes eight different gene products that couple to second messenger systems, such as inhibition of adenylate cyclase, activation of phospholipase C, and regulation of ion channels (for reviews, see Schoepp and Conn 1993; Blasi et al. 2001; Spooren et al. 2003). These receptors regulate critical aspects of both excitatory and inhibitory synaptic transmission by both pre- and postsynaptic mechanisms. Signaling through these families of receptors is required for essentially all physiologic processes, such as receipt of sensory information, motor control, autonomic function, and cognition, and is also critical for synapse development and synaptic plasticity that underlies memory formation/retrieval.

While it is clear that glutamate is critical for normal brain development and function, there is strong evidence that excessive activation of glutamate receptors can kill neurons and other cells that express glutamate receptors. In fact, it has been known for quite some time that low micromolar concentrations of glutamate effectively kill neurons in culture (Choi 1987) and also kill oligodendroglial precursors (McDonald et al. 1998). This process of excitotoxicity is thought to contribute to the brain damage that accompanies acute insults to the CNS, such as head trauma, stroke, and hypoglycemia. There is also evidence that this process of "excitotoxicity" may contribute to chronic neurodegenerative diseases, including Alzheimer's disease and amyotrophic lateral sclerosis (ALS) (for reviews, see Choi 1992; Doble 1999). The fact that glutamate receptor antagonists are protective in animal models of the acute insults provides the most compelling evidence that excitotoxicity contributes to the damage associated with these insults. However, it should be noted that in spite of this strong evidence in animal models, clinical trials of excitatory amino acid antagonists in humans have not observed significant beneficial effects (for reviews, see Obrenovitch et al. 2000; Ikonomidou and Turski 2002). Therefore, with the exception of riluzole and amantadine (which also has dopaminergic activity) for which there is some evidence that they have anti-glutamatergic activity, there are essentially no clinically used drugs that are known to directly influence the glutamate system. Given that glutamate has such a fundamental role in mammalian nervous system function and it can clearly kill cells that express certain subtypes of receptors, it seems likely that sophisticated systems evolved to protect the normal nervous system from toxicity.

Unlike many other neurotransmitters, there is no evidence for extracellular metabolism of glutamate, instead signaling is terminated by a combination of diffusion out of the synapse, clearance by Na^+-dependent transport activity, and by receptor desensitization (for review, see Huang and Bergles 2004). Na^+-dependent transport activity is mediated by a family of five proteins that share approximately 50% sequence similarity (for reviews, see Sims and Robinson 1999; Danbolt 2001; Shigeri et al. 2004). Originally the non-human homologs of three of these transporters were termed GLAST (glutamate aspartate transporter), GLT-1 (glutamate transporter subtype 1), and EAAC1 (excitatory amino acid carrier 1); the human homologs were named EAAT1 (for excitatory amino acid transporter 1), EAAT2, and EAAT3, respectively. In addition, there are two other members of the family, EAAT4 and EAAT5. There are also variants of some of these transporters that arise by alternate messenger RNA (mRNA) splicing. These transporters are heterogeneously distributed on glial cells, presynaptic termini, and—somewhat surprisingly—on dendritic spines, positioning each of these transporters to uniquely affect different aspects of signaling.

In this review, the properties of these transporters will be briefly summarized, including their role in synaptic transmission and toxicity. Over the last few years it has become increasingly clear that the activity of these transporters can be acutely (within min) regulated by mechanisms that are at least in part associated with a redistribution of transporters to or from the plasma membrane. Our current understanding of this regulation will be discussed.

2
Characteristics of Na^+-Dependent Glutamate Transporters

2.1
Localization

The five transporters are differentially distributed throughout the nervous system. At least a few of the transporters are found in the periphery (for reviews, see Sims and Robinson 1999; Danbolt 2001; Shigeri et al. 2004). Expression of GLAST/EAAT1 is found on glial cells throughout the nervous system, but is particularly enriched in apparently specialized glia, such as the Bergmann glia in the cerebellum, supporting glia in the vestibular end organ and Muller cells of the retina. GLT-1/EAAT2 is also found on astrocytes with some variation in expression throughout the brain and somewhat lower levels in the spinal cord (Rothstein et al. 1994). Neither GLT-1 nor GLAST appears to be uniformly distributed on the membrane, suggesting there are mechanisms that restrict distribution to specific domains on the plasma membrane (Chaudhry et al. 1995). Remarkably, the levels of both of these transporters approach several thousand molecules per cubic micrometer (Lehre and Danbolt 1998), and GLT-1 has been estimated to represent up to 1% of brain protein (for discussion, see Danbolt 2001). There are variants of GLT-1 that arise from alternate mRNA splicing, and at least one of these variants is found on presynaptic nerve terminals (Schmitt et al. 2002; Chen et al. 2004). EAAC1/EAAT3 is found on neurons and some white matter cells, particularly developing oligodendroglia. EAAC1 expression is enriched in neurons that are primarily associated with glutamatergic signaling but is found on perisynaptic regions of dendrites and on cell bodies (Coco et al. 1997; Conti et al. 1998; He et al. 2000, 2001). EAAC1 is also found on γ-aminobutyric acidergic (GABAergic) neurons (for review, see Danbolt 2001). In contrast to the other glutamate transporters, EAAC1 immunoreactivity is found in the cytoplasm of cells throughout the nervous system, implying that there is a pool of transporters available for trafficking to the plasma membrane (see Sect. 3). EAAT4 is enriched in Purkinje cells of the cerebellum where expression is highest on the cell body and spines, but there are also low levels of EAAT4 in some forebrain neurons (Furuta et al. 1997; Tanaka et al. 1997a). EAAT5 is expression is restricted to retina (Arriza et al. 1997).

2.2
Contributions to Synaptic Transmission

These transporters are generally assumed to function in alternating access mode where glutamate, 2 or 3 molecules of Na^+, and a proton bind to the extracellular face of the transporter. After a conformational change and release of these substrates, K^+ binds and the transporter reorients. The coupling of multiple molecules of Na^+ provides additional electrochemical energy, such that a very large concentration gradient of glutamate (up to 10^6) can be maintained across the membrane (Zerangue and Kavanaugh 1996). It is not possible to directly measure synaptic concentrations of glutamate, though it seems likely that they should be below those that would tonically activate receptors (≤ 1 μM). Yet in microdialysis experiments, the levels of extracellular glutamate can be as high as 10–50 μM. While it is possible that the prediction regarding synaptic concentrations of glutamate is not correct, it seems more likely that in microdialysis experiments the measures of glutamate do not reflect the local environment of the synapse, implying that they sample the extrasynaptic environment (for review/recent discussion, Timmerman and Westerink 1997; Baker et al. 2002). Regardless, the levels of glutamate in brain are as much as 10,000-fold higher than those that would be expected in the synapse, and the Na^+-dependent transporters are the only transporters that are on the plasma membrane that can concentratively drive the accumulation of glutamate. Therefore, these transporters would seemingly play an important role in limiting "noise" so that a "signal" can be detected by receptors upon release of glutamate.

The stoichiometric mode of transporter operation is slow compared to the duration of rapid excitatory signaling, with a complete cycle requiring approximately 10–75 ms (Wadiche et al. 1995; Bergles and Jahr 1997). Compared to the enzyme that degrades acetylcholine, acetylcholine esterase, which is nearly as fast as diffusion (10^4 cycles per second; for discussion, see Johnson and Moore 2000), this rate of 10–100 cycles per second is extremely slow, essentially glacial. Since ligand-gated ion channel activation can result in a maximal response within milliseconds of glutamate release, and this depolarization can be as short-lived as a few milliseconds, transporter cycling is not fast enough to impact on the extremely rapid excitatory responses. In contrast, metabotropic receptor signaling can be much slower, and cycling of the transporters can control the duration of these responses (for review, see Huang and Bergles 2004).

Over the past few decades, several groups have examined the impact of decreased transporter function on synaptic signaling. These studies have largely relied on pharmacologic inhibition of transporter function or genetic deletion of the transporters. A range of effects has been observed, and it appears that the impact of decreased transporter function is influenced by a wide variety of mechanisms (for reviews, see Conti and Weinberg 1999; Huang and Bergles 2004). At some synapses, the high density of transporters on glial membranes effectively buffers the amount of glutamate that is available for activation of postsynaptic receptors, such that if binding of glutamate to transporters is prevented with competitive inhibitors, the amplitude of the rapid postsynaptic response increases (Tong and Jahr 1994). This is a somewhat surprising observation, but it suggests that the enrichment of glial transporters in specific processes is critical to support excitatory signaling. In fact, one might expect that mechanisms analogous to those used to enrich receptors in the postsynaptic density (for review, see Sheng and Sala 2001) might tether transporters to the appropriate membrane domains. There are examples where transport inhibition prolongs the duration of the excitatory response (Otis et al. 1996). In this case, the duration of the postsynaptic response is still too fast for a transporter to undergo a cycle, suggesting that inhibition of transport activity allows spillover between neighboring synapses and prolonged receptor activation. In fact, it has been suggested that the postsynaptic neuronal transporters might limit synaptic spillover in hippocampus (Diamond 2001). The impact of transporter function on synaptic activity is also likely to depend on the frequency at which glutamate is being released, because one assumes that tonic activation might result in saturation of the transporters. Finally, several groups have found that transporter activity is important for controlling the intensity of mGluR activation, resulting in decreased transmitter release through activation of presynaptic mGluRs and/or changes in postsynaptic signaling (for example, see Otis et al. 2004). Together, these studies show that decreased transporter function can clearly influence glutamate signaling and that the effects are likely to be synapse dependent. To date, the effect of enhanced transporter function on excitatory signaling has not been specifically examined.

2.3
Role in Toxicity

As is true for our understanding of the effect of transport function on synaptic responses, our understanding of the role of transporters in toxicity has historically depended on pharmacologic agents that are used to block transport activity and/or genetic approaches to decrease/eliminate transporter protein. From these studies, there is very strong evidence that decreased transporter function can be directly toxic to neurons and can exacerbate the toxicity observed in insults caused by exogenous glutamate (Rosenberg et al. 1992). There is also evidence that impaired transporter function can increase the sensitivity of neurons to acute insults, such as hypoxia/ischemia or acute trauma (Dugan et al. 1995). From studies with mice genetically deleted of the transporters (Peghini et al. 1997; Tanaka et al. 1997b; Watase et al. 1998), one would conclude that loss of GLT-1/EAAT2 is the most important mechanism for limiting excitotoxicity, although selective antisense knockdown of different transporters suggests that the other transporters also have a role in limiting excitotoxicity (Rothstein et al. 1996).

At a minimum, acute insults that result in failure of transporter function, because of a loss of electrochemical driving forces, are likely to result in an extracellular accumulation of glutamate and excitotoxicity. In addition, because transport is a reversible process, glutamate will likely move down its concentration gradient when the electrochemical driving forces no longer favor influx (for discussions, see Attwell et al. 1993; Levi and Raiteri 1993; Rossi et al. 2000). Since astroglia are selectively endowed with glutamine synthetase, it has been argued that cytoplasmic pools of glutamate are lower in astrocytes than in neurons. In fact, hippocampal slices prepared from mice deleted of the neuronal transporter EAAC1 show a delayed anoxic depolarization shift, a measure of glutamate release (Gebhardt et al. 2002). In contrast, the anoxic depolarization shift is not altered in mice deleted of GLT-1 (Hamann et al. 2002). Together these studies suggest that reversed operation of these transporters (1) contributes to the rise in extracellular glutamate observed when the electrochemical gradients collapse and (2) may be a significant source of the rise in extracellular glutamate that occurs under these conditions.

In a couple of recent studies, the effects of overexpression of transporters on excitotoxicity have been examined. For example, transgenic mice overexpressing EAAT2 have been developed (Guo et al. 2003). Primary neuronal cultures from these mice are less sensitive to exogenous glutamate toxicity. When these mice were crossed with a mouse model of ALS, a delay in onset of one of the motor impairments observed in this model of ALS was observed. Interestingly, β-lactam antibiotics were recently shown to increase expression of GLT-1 in vitro and in vivo (Rothstein et al. 2005). This was accompanied by a decrease in the sensitivity of neuronal cultures to oxygen-glucose deprivation, and treatment of a mouse model of ALS with one of these antibiotics diminished or delayed many of the symptoms observed in these mice. These studies suggest that increasing transporter function may be neuroprotective in the absence of impaired energy metabolism.

3
Rapid Regulation of Glutamate Transporters

Given that glutamate transporters can impact both normal physiologic responses to glutamate as well as glutamate toxicity, it is not surprising that their function is regulated both acutely and chronically. Although our understanding of chronic regulation (or regulation that involves a net change in the total number of transporters) continues to evolve (for reviews, see Sims and Robinson 1999; Danbolt 2001; Gegelashgvili et al. 2001), this chapter will focus on the regulation of glutamate transporters that can occur within minutes.

A wide number of treatments rapidly change glutamate transport activity in a variety of preparations. Many of these effects are independent of changes in the total number of transporters, or there is the implication that the effects are independent of de novo transporter synthesis because they occur very rapidly, and it is assumed that the process of transcription, translation, assembly, and glycosylation will take longer than a few minutes.

In many cases, it can be quite difficult to define the mechanism by which rapid changes in activity occur, and without a mechanism it is nearly impossible to determine if the effect is likely to be physiologically or pathologically relevant. For example, simple depolarization could have profound effects on the clearance of glutamate, but effects observed in model systems may not occur physiologically, raising questions regarding relevance. Certainly, as indicated in the previous section, during a hypoxic/ischemic insult to the nervous system, the loss of ATP and the subsequent collapse of the electrochemical gradients not only results in failure of inward transport but will likely result in reversed operation of the transporters contributing to the rise in extracellular glutamate that accompanies these insults (see previous section).

A number of ions interact with the transporter during a single cycle (H^+, Na^+, and K^+), and changes in any one of these ions could affect transporter function. A common theme in biology involves the assembly of different proteins into macromolecular complexes to increase specificity of effects through the creation of microenvironments. For example, the postsynaptic density includes receptors, a number of scaffolding proteins, and signaling molecules (for review, see Sheng and Sala 2001). This not only ensures that receptors are appropriately positioned to respond to synaptic glutamate, but also facilitates the coupling of receptors to specific signals. The two transporters that are currently thought to mediate the bulk of glutamate clearance in the CNS, GLT-1 and GLAST, are not uniformly distributed on the glial membrane, and interestingly, GLT-1 colocalizes with the Na^+/K^+ ATPase that maintains the Na^+ and K^+ electrochemical gradients (Cholet et al. 2002). Based on this observation, it is easy to imagine that there could be global regulation in electrochemical gradients throughout a cell that might occur as a function of development or during a pathologic insult; but in addition, local changes in ion gradients might serve to regulate glutamate transport activity.

Although changes in the K_m for glutamate could have profound effects on clearance of glutamate, particularly under conditions when the transporters are not saturated, the measurement of a K_m value is complicated by several factors. First, below saturating concentrations of Na^+, the K_m value for glutamate is reduced by increasing Na^+ (Robinson and Dowd 1997; Danbolt 2001). Second, there are very large differences (up to 10- to 50-fold) in the K_m values for the same glutamate transporter in different systems (for review, see Robinson and Dowd 1997). This may imply that cellular constituents (interacting proteins or the absence thereof) influence the K_m value. Finally, the K_m value might be affected by an artifact that was first discussed in context of glutamate receptor activation for compounds that are cleared by these transporters in brain slice preparations (Garthwaite 1985). In some systems (like astrocyte cultures), we (and others) have observed remarkable capacities; with V_{max} values that are as high as 30 nmol/mg protein per minute (Garlin et al. 1995). Since these cells are growing as an adherent monolayer and substrate is placed over the cells, it is at least possible that there is a local environment close to the cells where the rate at which glutamate is being cleared by the transporters is faster than the rate at which glutamate diffuses from the bulk medium into the local microenvironment. This would cause an apparent increase in the K_m

value because the concentration of glutamate in the bulk environment required to half-maximally saturate the transporter would be higher than the concentration near the cells (for further discussion, see Schlag et al. 1998). Although there are examples of treatments that affect the K_m value for glutamate, these have not been pursued beyond the initial reports.

The other kinetic constant that is readily measured is maximum velocity (V_{max}), which equals the number of transporters multiplied by their turnover number (or catalytic efficiency). As discussed above, altering electrochemical gradients can also influence the V_{max} values, which may or may not be interesting. One way to attempt to control for this possibility is to determine if the changes are specific to Na^+-dependent glutamate transport by testing the impact of a treatment on other Na^+-dependent transport systems found in the same preparation. Of course, if the effects are specific to glutamate, it only provides indirect evidence that the changes are not due to nonspecific effects on the electrochemical gradients. In the discussion of the signals that regulate the various glutamate transporters, some examples of presumed specific effects on catalytic efficiency will be described. However, the mechanisms for these changes have not been defined with the kind of precision with which Quick and his colleagues have demonstrated that syntaxin 1A interactions with one of the subtypes of GABA transporter can change the catalytic efficiency of this transporter by up to sixfold (see M.W. Quick in this volume).

Two distinct mechanisms could contribute to changes in the numbers of transporters available on the plasma membrane. First, regulation of transcriptional or translational processes could easily change the number of transporters, and there are many examples of altered glutamate transporter activity that have been associated with changes in the total number of transporters expressed in a particular preparation (for reviews, see Sims and Robinson 1999; Danbolt 2001; Gegelashgvili et al. 2001). Frequently these changes are associated with altered messenger RNA (mRNA) levels, implicating regulation at the level of changes in transcription or mRNA stability. It is possible that transporters will be rapidly transcribed and translated to produce rapid changes in the number of transporters on the cell surface. However, it is generally assumed that these processes would take longer than a few minutes, because transporters are folded, assembled into multimers, and partially glycosylated in the endoplasmic reticulum, followed by transport to the Golgi for terminal glycosylation and export to the plasma membrane.

The second mechanism to regulate the number of transporters on the plasma membrane involves redistributing transporters to or from the plasma membrane without changing the total number of transporters expressed per cell. Membrane proteins are generally thought to recycle on and off the plasma membrane, providing opportunities to regulate the processes involved in either endocytosis or delivery of transporters (for a recent review, see Royle and Murrell-Lagnado 2002). There are several classical examples of membrane proteins that are regulated in this manner. For example, desensitization of many G protein-coupled receptors occurs as a result of agonist-induced activation of specific kinases that phosphorylate the receptor. β-Arrestin is then recruited into a complex with the receptor that is internalized by an endocytic pathway (for reviews, see Ferguson 2001; Claing et al. 2002; Marchese et al. 2003). Another classic example involves the insulin-dependent regulation of a subtype of glucose transporter (GLUT4) that is important for plasma glucose homeostasis (Bryant et al. 2002). In this case, there is a very large intracellular pool of transporter that is thought to reside in a specific subcellular compartment; activation of the insulin receptor is thought result in delivery of this "regulated" pool of transporter to the plasma membrane. In addition, there is evidence that there is a separate pool of GLUT4 that constitutively recycles between the plasma membrane and intracellular compartments (for reviews, see Czech and Corvera 1999; Watson et al. 2000). It was originally thought that the recycling pool of transporter was sorted to different compartments and was, at least in part, required for loading of the regulated pool of transporter. However, more recent studies provide evidence that the regulated pool is directly loaded from newly synthesized transporter that is exported from the Golgi (Watson et al. 2004). In the next few sections, our current understanding of glutamate transporter trafficking, which is not nearly as sophisticated as it is for G protein-coupled receptors and GLUT4, will be summarized.

3.1
General Comments Regarding Plasma Membrane Expression of Glutamate Transporters

Many integral membrane proteins continuously recycle on and off the plasma membrane. Of course, the number of proteins on the plasma membrane at steady state is dependent upon the rate of internalization (endocytosis of transporter) relative to the rate of delivery of transporter to the plasma membrane (for review, see Royle and Murrell-Lagnado 2002). If net delivery to the plasma membrane is faster than internalization, one would expect most or all of the transporter to be located at the plasma membrane and vice versa. While it is theoretically possible that regulation results in redistribution of transporters to or away from the plasma membrane, regulated redistribution to the plasma membrane is obviously dependent upon the presence of an intracellular pool of transporter.

In a number of studies, several different groups have examined the subcellular distribution of three of the glutamate transporters (GLT-1, GLAST, and EAAC1). For most of these studies, cells are incubated with a membrane-impermeable, biotin-containing reagent that reacts with amino groups on proteins. This incubation is performed at 4 °C to halt all membrane trafficking events, essentially providing a measure of the steady-state levels of transporter on the plasma membrane (Daniels and Amara 1998). After cell lysis, biotinylated proteins are batch extracted using avidin-coated beads. It is possible to isolate three separate fractions (lysate, non-biotinylated, and biotinylated) using this strategy. We generally dilute all of the fractions to the same extent so that the percentage of transporter in the biotinylated (cell surface) fraction can be semiquantitatively analyzed by Western blot. In these studies, it is important to monitor for cell lysis by probing for a cytoplasm-specific marker, such as actin, in the same experiments.

Generally, in a cell line that endogenously expresses only EAAC1 (C6 glioma), we find that between 20% and 30% of the transporter is in the biotinylated fraction, suggesting that only a small percentage of the transporter is on the cell surface (Sims et al. 2000). We have performed similar experiments in primary neuronal cultures from cortex and observed a similar percentage of transporter in the biotinylated fraction (González et al. 2002; Fournier et al. 2004). However, these percentages should be interpreted with some caution because they vary depending on several factors, including the presence or absence of serum (Sims et al. 2000), the number of days primary neurons are maintained in culture, and the brain region used to prepare primary neurons (M.I. González, E.A. Waxman, D.R. Lynch, and M.B. Robinson, unpublished observations). As neurons in culture form functional contacts, it seems possible that neuronal signaling influences surface expression, but this has not been systematically examined. While primary neuronal cultures provide a cellular milieu that is more likely to reflect that observed in vivo, it is important to determine if there is an intracellular pool of EAAC1 in vivo. Based on both light and electron microscopic analysis (Rothstein et al. 1994; Coco et al. 1997; Conti et al. 1998; Kugler and Schmitt 1999; He et al. 2000; He et al. 2001), there appears to be a fairly substantial pool of EAAC1 immunoreactivity in the cytoplasm in vivo.

In primary cultures of astrocytes that express GLT-1 or mixed cultures of neurons and astrocytes, between 60% and 80% of the total GLT-1 or GLAST immunoreactivity is found in the biotinylated/cell surface fraction (Schlag et al. 1998; Kalandadze et al. 2002; Susarla et al. 2004). In electron microscopic analyses of GLAST and GLT-1 immunoreactivity in vivo, there is no evidence of a significant intracellular pool of either of these transporters (Chaudhry et al. 1995; Lehre et al. 1995). However, these transporters are present on very fine astrocytic processes; therefore, one cannot rule out the possibility that some fraction of the transporters resides just below the surface of the plasma membrane. Although the differences in cell surface expression of EAAC1 and these two glial transporters might initially be attributed to the fact that these transporters are found in different cell populations, we have also transiently transfected C6 glioma with either GLT-1 or epitope-tagged EAAC1. We found similar differences in the cell surface expression with GLT-1, mostly on the cell surface, and EAAC1, mostly in intracellular pools (Kalandadze et al. 2002). This differential distribution of these transporters in the same cellular milieu argues that primary sequence differences between these two transporters influences either delivery of the transporter to the plasma membrane and/or constitutive endocytosis.

It should be noted that others have observed evidence of intracellular GLT-1 or GLAST (Duan et al. 1999; Guillet et al. 2005). As one will see below, it is becoming clear that many signaling pathways regulate trafficking of the transporters either to or away from the plasma membrane. Therefore, it seems likely that differences in the percentage of transporter at the cell surface are simply a reflection of endogenous activation/inhibition of the signaling pathways under different cell culture conditions.

3.2
Constitutive Trafficking of Glutamate Transporters

One approach to determining how activation or inhibition of signaling pathways regulates cell surface expression of the transporters is to develop an understanding of the kinetics of delivery and endocytosis of the transporters. Since many of the steps in recycling, endocytosis, shuttling through intracellular compartments, and fusion back to the plasma membrane require energy, it has been argued that the rate of recycling of various proteins is likely to be related to cellular need to regulate the amount of protein at the plasma membrane (Royle and Murrell-Lagnado 2002). To measure delivery, cells are incubated with the same membrane-impermeable biotinylating reagent mentioned in the previous section. The incubations with biotinylating reagent occur at physiologic temperatures (37 °C) for varying lengths of time. Under these conditions, the amount of biotinylated transporter equals the amount of transporter on the cell surface at time point 0 (steady state) plus the amount of transporter delivered to the cell surface during incubation. The amount of transporter on the cell surface at steady state can be measured under non-trafficking permissive conditions (4 °C) in a parallel experiment. The change in biotinylated immunoreactivity is proportional to the rate of delivery of the transporters to the plasma membrane. This assumes that the rate of the biotinylation reaction is much faster than transporter recycling and that biotinylating transporters or other surface proteins does not affect recycling of the transporters. Using this approach, we found that the amount of biotinylated transporter, compared to that observed at steady state, approximately doubles within 15 min in either C6 glioma or primary neuronal cultures (Fournier et al. 2004). Since others have found that the half-life for degradation of newly synthesized transporters is in the order of several hours (Yang and Kilberg 2002), this increase in biotinylated transporter cannot be attributed to synthesis of new transporter. Instead, these data strongly suggest that EAAC1 recycles on and off the plasma membrane and are consistent with transporters having a half-life at the plasma membrane of no more than 5 to 7 min. This might be an overestimate of the half-life, because this measure assumes that the reaction with the biotinylating reagent is essentially instantaneous. Of course, one implication of having most of the transporter on the cell surface (e.g., GLT-1 and GLAST) is that this measure of delivery cannot be used. With longer incubations (up to 1 h), most of the EAAC1 immunoreactivity is biotinylated, suggesting that most of the intracellular pool of EAAC1 is in compartments that ultimately cycle to the plasma membrane.

To measure endocytosis, plasma membrane proteins are labeled at a temperature that prevents trafficking (4 °C) with a membrane-impermeable biotinylating reagent that contains a disulfide bond. After washing out and quenching the biotinylating reagent, cells are warmed to 37 °C for varying periods of time. After cooling the cells to 4 °C again, the biotinylating reagent is stripped using a membrane-impermeable, disulfide-reducing reagent (Loder and Melikian 2003; Fournier et al. 2004; Wang and Quick 2005). Under these conditions, the amount of biotinylated material that remains is a measure of the amount of transporter that has been internalized. Based on our experience, these experiments are more difficult than the other biotinylation approaches. The number of washes required can result in washing away of cells, and one of the more commonly used disulfide reducing reagents (2-mercaptoethanesulfonic acid, also called MesNa) is so unstable that freshly made stocks are required with each wash. Since there is a relatively small pool of EAAC1 on the cell surface under baseline conditions, we found it difficult to measure endocytosis (Fournier et al. 2004). Therefore, we redistributed a pool of transporters to the plasma membrane by activating protein kinase C (PKC) and washed the cells extensively for a long period. Under these conditions, approximately 50% of the transporters become inaccessible to the disulfide-reducing reagent within 5 min, consistent with a half-life of 5 min and with the measures of delivery. One could argue that activating PKC, which causes redistribution of EAAC1 from a subcellular compartment to the plasma membrane, could be affecting the kinetics of endocytosis; but if anything, it seems to slow the kinetics of endocytosis (Fournier et al. 2004).

The measure of delivery is dependent upon there being an intracellular pool that can become biotinylated at 37 °C. With the measure of endocytosis, no biotinylated transporter will accumulate inside the cell if there is essentially no intracellular pool of transporter under baseline conditions. Although instantaneously blocking recycling with a pharmacologic agent such as monensin can help circumvent this problem, it presupposes the transporters recycle through a specific compartment and that the agent is selective (for an example, see Sorkina et al. 2005). Under the experimental conditions used in our laboratory, most GLT-1 and GLAST is on the cell surface. Therefore, it is unclear if GLT-1 or GLAST rapidly cycle on and off the plasma membrane or if they are relatively stably maintained on the plasma membrane.

In subsequent sections, the effects of different signaling pathways on surface expression of the three main forebrain glutamate transporters will be summarized.

3.3
EAAC1/EAAT3

Several years ago, we found that direct activation of PKC causes a relatively robust increase (~twofold) in Na^+-dependent glutamate transport in C6 glioma, a cell line that appears to only express the EAAC1 subtype of transporter (Dowd et al. 1996; Dowd and Robinson 1996; Najimi et al. 2002). This increase occurs within minutes and is associated with redistribution of transporter from a subcellular compartment to the cell surface as defined using biotinylation (Davis et al. 1998). The effect of PKC activation on EAAC1 surface expression has also been observed in primary neuronal cultures (González et al. 2002; Fournier et al. 2004; Guillet et al. 2005). In hippocampal slices, activation of PKC also increases EAAC1 surface expression and Na^+-dependent transport activity (Levenson et al. 2001). However, this regulated redistribution of EAAC1 may not be recapitulated in other model systems. We have preliminarily tested a cell line that is commonly used for transient transfections (HEK293) and found no evidence for a PKC-dependent redistribution of EAAC1 to the plasma membrane. In contrast, others have found that in *Xenopus* oocytes injected with EAAC1 complementary RNA (cRNA) and in Madin-Darby canine kidney cells stably transfected with EAAC1, PKC activation decreases EAAC1-mediated activity by a mechanism that is consistent with a redistribution of transporter to an intracellular compartment (Trotti et al. 2001). In this same study, the authors were able to detect an increase in transport activity in C6 glioma. This suggests that at least some of the effects on trafficking of EAAC1 are dependent upon the cellular milieu. While this might be troubling, it should be remembered that it has long been recognized that the insulin-dependent regulation of the GLUT4 glucose transporter is only observed in a few cellular systems, such as adipocytes and muscle cell lines (Martin et al. 1999). Since this regulated redistribution of EAAC1 to the plasma membrane occurs in neuronal cultures and brain slice preparations, it seems possible that PKC activation might increase EAAC1 surface expression under physiologic or pathologic conditions in vivo.

A variety of observations, including the fact that activation of PKC increases EAAC1-mediated activity more than it increases cell surface expression, prompted us to test the idea that PKC might have more than one effect on transporter activity. Using subtype-selective inhibitors and downregulation of specific PKC subtypes, we were able to develop evidence that two different PKCs might regulate EAAC1 (see Fig. 1A for a schematic). PKCα appears to be required for redistribution of EAAC1 to the plasma membrane, and PKCε appears to increase transport activity by a mechanism that is independent of an increase in the surface expression of EAAC1 (González et al. 2002). The effect of PKCε is not associated with a change in Na^+-dependent glycine transport, providing indirect evidence that this effect cannot simply be attributed to a change in the Na^+-electrochemical gradient, but this has not been mechanis-

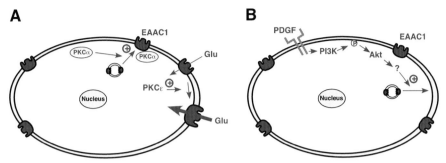

Fig. 1 A,B Schematic illustration of some of the signals that regulate activity and cell surface expression of the neuronal glutamate transporter, EAAC1. **A** There is evidence that PKC has two distinct effects on EAAC1. The first effect is dependent upon activation of PKCα and involves a redistribution of EAAC1 from a subcellular compartment to the cell surface. This effect is associated with formation of immunoprecipitable complexes of EAAC1 and PKCα. Activation of PKC increases activity to a greater extent than it increases the cell surface expression of EAAC1. This secondary effect seems to be related to an increase in the catalytic efficiency of the transporters and appears to depend on activation of PKCε (González et al. 2002, 2003). **B** Platelet-derived growth factor (PDGF) also increases EAAC1 activity and cell surface expression. The effects of PDGF are blocked by inhibitors of phosphatidylinositol 3-kinase (PI3-K) or by a dominant-negative variant of Akt. The effects of PDGF are also mimicked by a constitutively active variant of Akt (Sims et al. 2000; Krizman-Genda et al. 2005). The effects of either PKC or PDGF on cell surface expression appear to be due to increases in the rate of transporter delivery to the plasma membrane (Fournier et al. 2004); but at present, it is not clear if these effects are due to direct transporter modification or modification of accessory proteins

tically pursued further. Although PKC activation is associated with formation of immunoprecipitable complexes between EAAC1 and PKCα (González et al. 2003), at present it is not clear why PKCα forms a complex with EAAC1. Perhaps the simplest explanation is that formation of the complex increases the likelihood that activated PKCα will selectively phosphorylate either the transporter or other associated proteins that are required for redistribution. Alternatively, formation of the complex may provide a tag to sort transporters to specific vesicles or membrane domains to change the kinetics of transporter movement to or from the plasma membrane. In a series of interesting studies, another group has found that the general anesthetic, isoflurane, increases EAAC1 activity and surface expression by a mechanism that depends on PKC activation. They have used complementary techniques to show that PKCα may be required for isoflurane-induced redistribution of EAAC1 and that isoflurane also stimulates formation of EAAC1-PKCα complexes (Huang and Zuo 2005).

Platelet-derived growth factor (PDGF) receptor activation also increases Na^+-dependent EAAC1-mediated transport activity in C6 glioma (Sims et al. 2000 see Fig. 1B). As is observed with the effects of PKC activation, the increase occurs within a few minutes and is associated with a redistribution of EAAC1 from a subcellular compartment to the plasma membrane. Although somewhat more modest than the effects of PKC activation, PDGF also causes an increase in surface expression of EAAC1 in primary neuronal cultures (Fournier et al. 2004). In contrast to the effects of PKC activation, the increase in surface expression essentially correlates with the increase in activity, suggesting that redistribution of transporter is sufficient to explain the increase in activity. The effects of PDGF receptor activation are not blocked by inhibitors of PKC, but the effects of PKC activation and PDGF receptor activation are not additive (Sims et al. 2000). This suggests either that these two signaling pathways converge on a common downstream signaling molecule/pathway or that they regulate the same finite pool of transporter. The effects of PDGF are blocked by inhibitors of the phosphatidylinositol 3-kinase (PI3-K) or by a dominant-negative variant of the serine threonine kinase called protein kinase B (or Akt). Furthermore, a constitutively active variant of Akt increases both the activity and surface expression of EAAC1 (Krizman-Genda et al. 2005) and increases activity of EAAC1 upon expression in *Xenopus* oocytes (Schniepp et al. 2004). These studies suggest that both PI3-K and Akt are required for PDGF-induced redistribution of EAAC1. In fact, a similar pathway is required for insulin-dependent regulation of GLUT4 (Bryant et al. 2002).

The effects of both PKC activation and PDGF receptor activation on the kinetics of endocytosis and delivery of EAAC1 to the plasma membrane were examined (Fournier et al. 2004). Based on an analysis of the rate of delivery of EAAC1 to the plasma membrane, it appears that activation of PKC or PDGF receptor increases the rate of delivery of EAAC1 to the plasma membrane (Fig. 1). In contrast to PDGF, activation of PKC also completely blocks the appearance of internalized transporters after reversible biotinylation, suggesting that it also inhibits endocytosis. At present, we have no evidence that redistribution of EAAC1 by these signals requires a direct modification of the transporters (A.L. Sheldon, M.I. González, and M.B. Robinson, unpublished). In fact, even though GLUT4 trafficking has been the subject of intense scrutiny for almost two decades, there is no evidence that direct transporter modification is required for the effects of insulin. This implies that these signals cause redistribution of the transporters by modifying an accessory protein that is either cytoplasmic or is associated with transporter-containing vesicles. By combining this information with a definition of the organelles and molecular machinery involved in recycling of EAAC1, it may be possible to determine how these signals effect changes in the cell surface expression of the transporters.

Other signals also change EAAC1 activity and/cell surface expression, including activation of a neurotensin receptor that belongs to the family of class A G protein-coupled receptors that is generally thought to couple to activation of phospholipase C, mobilization of intracellular Ca^{2+}, and activation of PKC (Najimi et al. 2002). As is observed after activation of PKC or PDGF receptor, the increase in activity is associated with an increase in biotinylated transporter. Surprisingly, the neurotensin-induced increase in activity is not blocked by inhibitors of PKC nor is it blocked by an inhibitor of PI3-K in C6 glioma, providing evidence that the effects are independent of these two signaling pathways. In a subsequent study, this same group showed that activation of the endothelin-1 receptor increases EAAC1 surface expression and activity by a mechanism that appears to be independent of PKC or PI3-K, but is blocked by pertussis toxin, suggesting that a $G_{i/o}$ subtype of G protein may be involved (Najimi et al. 2005). The effects of neurotensin or endothelin-1 are blocked by cytochalasin, suggesting that cytoskeletal elements are required for these effects. In neuronal cultures, an inhibitor of protein kinase A (PKA), N-[2-(p-bromocinnamylamino)-ethyl]-5-(isoquinolinesulfonamide), also called H89, reduces biotinylated EAAC1 (Guillet et al. 2005), suggesting that endogenous activation of PKA increases cell surface expression of EAAC1. Constitutively active serum- and glucocorticoid-inducible kinase (SGK1) increases EAAC1 activity in *Xenopus* oocytes, but it is not known if this effect is dependent on a change in surface expression of EAAC1 (Schniepp et al. 2004). Finally, genetic deletion of presenilin—one of the proteins implicated in familial Alzheimer's disease that is required for amyloid precursor protein metabolism or inhibition of γ-secretase—decreases surface expression of EAAC1 in neuronal cultures (Yang et al. 2004). This effect was mimicked by depletion of intracellular calcium stores. Since it appears that presenilin deficiency also reduces calcium levels in the endoplasmic reticulum, Yang and colleagues suggested that calcium might also have a critical role in regulating EAAC1 surface expression. Together, these studies provide compelling evidence that the surface expression of EAAC1 is regulated by a variety of different signals in neurons.

As mentioned earlier, EAAC1 is enriched in brain regions and cell types that are both remarkably plastic and exquisitely sensitive to excitotoxic insults. It seems reasonable to consider the possibility that the regulated trafficking of EAAC1 is important for these phenomena. In fact, Levenson and colleagues have shown that EAAC1 redistributes from one membrane fraction (intracellular) to another (plasma membrane) in area CA1 of hippocampal slices that are subjected to high-frequency stimulation that causes long-term potentiation or in slices prepared from animals that have undergone contextual fear conditioning (Levenson et al. 2001). This associated change in transporter expression suggests that, under certain conditions, EAAC1 trafficking may be important for memory formation.

3.4
GLT-1/EAAT2

In contrast to EAAC1, activation of PKC decreases the activity and surface expression of GLT-1/EAAT2 by about 30%–50%. This effect has been observed by a number of different groups both in transfected cells and in primary cultures derived from brain tissue (Kalandadze et al. 2002; Zhou and Sutherland 2004; Guillet et al. 2005). In fact, in the same cellular system in which EAAC1 is redistributed to the plasma membrane (C6 glioma), GLT-1 is internalized in response to activation of PKC (Kalandadze et al. 2002). This provides strong evidence that it should be possible to identify differences in primary structure that govern this differential regulation of the two transporters. Using a family of chimeras, we were able to identify a 43-amino-acid domain that was required for PKC-dependent internalization of GLT-1. This domain contains five serine residues, and although mutation of serine-486 to alanine partially blocked internalization, we were unable to detect a decrease in transporter phosphorylation in this mutant transporter (Kalandadze et al. 2002). This suggests that direct transporter phosphorylation is not required for internalization. However, it is also possible that the transporter is phosphorylated at several sites in response to PKC activation and that mutation of serine 486 does not result in a sufficient decrease to be detected in these studies. The effect of PKC activation is also associated with an increase in the number of GLT-1 clusters at or near the plasma membrane, and this clustering is blocked by cytochalasin D, suggesting that actin is required (Zhou and Sutherland 2004). Interestingly, it appears that neuronal release of glutamate results in a similar redistribution of GLT-1 to clusters that seem to localize near neurons (Poitry-Yamate et al. 2002), suggesting that this clustering may be important for shaping excitatory signaling.

Vermeiren and colleagues observed an effect of metabotropic receptor activation on GLT-1-mediated activity (Vermeiren et al. 2005). In astrocytes that express GLT-1 and mGluR5, GLT-1 seems to be nonfunctional under baseline conditions, and activation of mGluR5 changes the pharmacologic properties of the transport activity to resemble that observed for GLT-1 within seconds. This effect is mimicked by activation of PKC and blocked by inhibitors of PKC. Furthermore, this effect was independent of a redistribution of GLT-1 from a subcellular compartment to the plasma membrane. Together, these studies suggest that nonfunctional GLT-1 can be activated by mGluR5 through a PKC-dependent mechanism.

Although we have not observed a significant amount of non-biotinylated GLT-1, others have observed a robust non-biotinylated amount of GLT-1 immunoreactivity, suggesting that intracellular pools of transporter can accumulate under specific conditions (Kalandadze et al. 2002; Guillet et al. 2005). Under conditions in which there is a relatively robust intracellular pool, H89, the PKA inhibitor that decreases EAAC1 surface expression, increases cell surface expression of GLT-1, suggesting that endogenous activation of PKA decreases GLT-1 surface expression (Guillet et al. 2005). We find that PDGF has no effect on GLT-1 cell surface expression upon transient transfection of GLT-1 into C6 glioma (A.L. Sheldon, M.I. González, and M.B. Robinson, unpublished observations). However, inhibition of one of the downstream effectors of PDGF, PI3-K, decreases surface expression of GLT-1 in primary neuronal cultures (Guillet et al. 2005).

3.5
GLAST/EAAT1

Over a decade ago, activation of PKC was shown to increase Na^+-dependent glutamate transport activity in astrocytes (Casado et al. 1991). It now seems likely that in these cultures essentially all of the transport activity is mediated by GLAST (Swanson et al. 1997; Schlag et al. 1998). In these same types of cultures, we found a similar PKC-dependent increase in transport activity. This effect was specific for glutamate uptake, with no effect or a modest decrease in Na^+-dependent glycine transport (Susarla et al. 2004). However, we did not detect any intracellular (non-biotinylated) pool of GLAST in these experiments. In fact, activation of PKC paradoxically decreased the amount of GLAST immunoreactivity in both the biotinylated fraction and in total cell lysates under the same conditions that resulted in an increase in transporter activity. Using a variety of different strategies, we concluded that activation of PKC might result in masking of epitopes that are required for antibody recognition. The effects of PKC on GLAST-mediated expression seem to vary depending on the type of primary astrocyte used for these studies, with decreases observed on retinal Muller cells (Wang et al. 2003). In forebrain cultures that are enriched in neurons, activation of PKC or inhibition of PKA decreases GLAST immunoreactivity in the biotinylated fraction, and neither has an effect on total GLAST immunoreactivity, providing evidence for a PKC-dependent internalization and PKA-dependent distribution to the plasma membrane (Guillet et al. 2005).

Under conditions in which there is a relatively robust intracellular pool, glutamate increases the amount of biotinylated GLAST, and it appears that this increase in surface expression is somehow mediated by an interaction of glutamate (or other glutamate transporter substrates) with GLAST (Duan et al. 1999). As was observed with GLT-1, neuronal release of glutamate also caused a clustering of GLAST in astrocytes, suggesting that common mechanisms may regulate both of these astrocytic transporters (Poitry-Yamate et al. 2002). GLAST surface expression is also increased by insulin-like growth factor-1 (Gamboa and Ortega 2002). Finally, GLAST activity is downregulated by co-expression with the ubiquitin ligase, Nedd4-2, upon expression in *Xenopus* oocytes (Boehmer et al. 2003). This inhibition is blunted by co-expression of a constitutively active variant of serum and glucocorticoid responsive kinase (SGK1) and by constitutively active Akt. These studies imply that ubiquitination of GLAST is involved in GLAST regulation and may alter surface expression of GLAST, but this was not explored.

Endothelin also acutely decreases GLAST-mediated activity in primary astrocyte cultures by a mechanism that depends on an accumulation of intracellular Ca^{2+} and is blocked by a PKC antagonist, but the effect on transport activity was thought to be related to astrocyte depolarization (Leonova et al. 2001).

4
Conclusions

The sodium-dependent glutamate transporters serve a critical function in the mammalian CNS, both clearing the most abundant neurotransmitter and limiting the toxic potential of this transmitter. Several studies have now documented that the trafficking of the three predominant forebrain glutamate transporters to and from the plasma membrane can be regulated. Furthermore, there are hints that the intrinsic activity of these transporters can also be regulated. These events would seemingly provide a mechanism to regulate transporter function on a timescale that is much faster than would be possible by transcriptional and translational events. It seems likely that at least a few of the effects are going to depend on cellular context that can result in both qualitative and quantitative differences in transporter regulation. Although the field is beginning to define the diverse signaling molecules that are involved, still very little is known about the mechanisms involved, and even less is known about how these regulatory events are triggered under physiologic or pathologic conditions.

Acknowledgements Work in the author's laboratory is supported by the NIH.

References

Arriza JL, Eliasof S, Kavanaugh MP, Amara SG (1997) Excitatory amino acid transporter 5, a retinal glutamate transporter coupled to a chloride conductance. Proc Natl Acad Sci USA 94:4155–4160

Attwell D, Barbour B, Szatkowski M (1993) Nonvesicular release of neurotransmitter. Neuron 11:401–407

Baker DA, Xi ZX, Shen H, Swanson CJ, Kalivas PW (2002) The origin and neuronal function of in vivo glutamate. J Neurosci 22:9134–9141

Bergles DE, Jahr CE (1997) Synaptic activation of glutamate transporters in hippocampal astrocytes. Neuron 19:1297–1308

Blasi AD, Conn PJ, Pin J, Nicoletti F (2001) Molecular determinants of metabotropic glutamate receptor signaling. Trends Pharmacol Sci 22:114–120

Boehmer C, Henke G, Schniepp R, Palmada M, Rothsetin JD, Broer S, Lang F (2003) Regulation of the glutamate transporter EAAT1 by the ubiquitin ligase Nedd4-2 and the serum and glucocorticoid-inducible kinase isoforms SGK1/3 and protein kinase B. J Neurochem 86:1181–1188

Bryant NJ, Govers R, James DE (2002) Regulated transport of the glucose transporter GLUT4. Nat Rev Mol Cell Biol 3:267–277

Casado M, Zafra F, Aragón C, Giménez C (1991) Activation of high-affinity uptake of glutamate by phorbol esters in primary glial cell cultures. J Neurochem 57:1185–1190

Chaudhry FA, Lehre KP, Campagne MVL, Ottersen OP, Danbolt NC, Storm-Mathisen J (1995) Glutamate transporters in glial plasma membranes: highly differentiated localizations revealed by quantitative ultrastructural immunocytochemistry. Neuron 15:711–720

Chen W, Mahadomrongkul V, Berger UV, Bassan M, DeSilva T, Tanaka K, Irwin N, Aoki C, Rosenberg PA (2004) The glutamate transporter GLT1a is expressed in excitatory terminals of mature hippocampal neurons. J Neurosci 24:1136–1148

Choi DW (1987) Ionic dependence of glutamate neurotoxicity. J Neurosci 7:369–379

Choi DW (1992) Excitotoxic cell death. J Neurobiol 23:1261–1276

Cholet N, Pellerin L, Magistretti PJ, Hamel E (2002) Similar perisynaptoic glial localization for the Na^+,K^+-ATPase a2 subunit and the glutamate transporters GLAST and GLT-1 in the rat somatosensory cortex. Cereb. Cortex 12:515–525

Claing A, Laporte SA, Caron MG, Lefkowitz RJ (2002) Endocytosis of G protein-coupled receptors; roles of G protein-coupled receptor kinases and β-arrestin proteins. Prog Neurobiol 66:61–79

Coco S, Verderio C, Trotti D, Rothstein JD (1997) Non-synaptic localization of the glutamate transporter EAAC1 in cultured hippocampal neurons. Eur J Neurosci 9:1902–1910

Conti F, Weinberg RJ (1999) Shaping excitation at glutamatergic synapses. Trends Neurosci 22:451–458

Conti F, DiBiasi S, Minelli A, Rothstein JD, Melone M (1998) EAAC1, a high-affinity glutamate transporter, is localized to astrocytes and GABAergic neurons besides pyramidal cells in the rat cerebral cortex. Cereb. Cortex 8:108–116

Czech MP, Corvera S (1999) Signaling mechanisms that regulate glucose transport. J Biol Chem 274:1865–1868

Danbolt NC (2001) Glutamate uptake. Prog Neurobiol 65:1–105

Daniels GM, Amara SG (1998) Selective labeling of neurotransmitter transporters at the cell surface. Methods Enzymol 296:307–318

Davis KE, Straff DJ, Weinstein EA, Bannerman PG, Correale DM, Rothstein JD, Robinson MB (1998) Multiple signaling pathways regulate cell surface expression and activity of the excitatory amino acid carrier 1 subtype of Glu transporter in C6 glioma. J Neurosci 18:2475–2485

Diamond JS (2001) Neuronal glutamate transporters limit activation of NMDA receptors by neurotransmitter spillover on CA1 pyramidal cells. J Neurosci 21:8328–8338

Doble A (1999) The role of excitotoxicity in neurodegenerative disease: Implications for therapy. Pharmacol Ther 81:163–221

Dowd LA, Robinson MB (1996) Rapid stimulation of EAAC1-mediated Na^+-dependent L-glutamate transport activity in C6 glioma by phorbol ester. J Neurochem 67:508–516

Dowd LA, Coyle AJ, Rothstein JD, Pritchett DB, Robinson MB (1996) Comparison of Na^+-dependent glutamate transport activity in synaptosomes, C6 glioma, and Xenopus Oocytes expressing excitatory amino acid carrier 1 (EAAC1). Mol Pharmacol 49:465–473

Duan S, Anderson CM, Stein BA, Swanson RA (1999) Glutamate induces rapid upregulation of glutamate transport and cell-surface expression of GLAST. J Neurosci 19:10193–10200

Dugan LL, Bruno VMG, Amagasu SM, Giffard RG (1995) Glia modulate the response of murine cortical neurons to excitotoxicity: Glia exacerbate AMPA neurotoxicity. J Neurosci 15:4545–4555

Ferguson SSG (2001) Evolving concepts in G protein-coupled receptor endocytosis: the role in receptor desensitization and signaling. Pharmacol Rev 53:1–24

Fournier KM, González MI, Robinson MB (2004) Rapid trafficking of the neuronal glutamate transporter, EAAC1: Evidence for distinct trafficking pathways differentially regulated by protein kinase C and platelet-derived growth factor. J Biol Chem 279:34505–34513

Furuta A, Martin LJ, Lin CL, Dykes-Hoberg M, Rothstein JD (1997) Cellular and synaptic localization of the neuronal glutamate transporters EAAT3 and EAAT4. Neuroscience 81:1031–1042

Gamboa C, Ortega A (2002) Insulin-like growth factor-1 increases activity and surface levels of the GLAST subtype of glutamate transporter. Neurochem Int 40:397–403

Garlin AB, Sinor AD, Sinor JD, Jee SH, Grinspan JB, Robinson MB (1995) Pharmacology of sodium-dependent high-affinity L-[3H]glutamate transport in glial cultures. J Neurochem 64:2572–2580

Garthwaite J (1985) Cellular uptake disguises action of L-glutamate on N-methyl-D-aspartate receptors. Br J Pharmacol 85:297–307

Gebhardt C, Körner R, Heinemann U (2002) Delayed anoxic depolarizations in hippocampal neurons of mice lacking the excitatory amino acid carrier 1. J Cereb Blood Flow Metab 22:569–575

Gegelashgvili G, Robinson MB, Trotti D, Rauen T (2001) Regulation of glutamate transporters in health and disease. Prog Brain Res 132:267–286

González MI, Kazanietz MG, Robinson MB (2002) Regulation of the neuronal glutamate transporter excitatory amino acid carrier-1 (EAAC1) by different protein kinase C subtypes. Mol Pharmacol 62:901–910

González MI, Bannerman PG, Robinson MB (2003) Phorbol myristate acetate-dependent interaction of protein kinase Cα and the neuronal glutamate transporter EAAC1. J Neurosci 23:5589–5593

Guillet BA, Velly LJ, Canolle B, F MM, Nieoullon AL, Pisano P (2005) Differential regulation by protein kinases of activity and cell surface expression of glutamate transporters in neuron-enriched cultures. Neurochem Int 46:337–346

Guo H, Lai L, Butchbach ME, Stockinger MP, Shan X, Bishop GA, Lin CL (2003) Increased expression of the glial glutamate transporter EAAT2 modulates excitotoxicity and delays the onset but not the outcome of ALS in mice. Hum Mol Genet 12:2519–2532

Hamann M, Rossi DJ, Marie H, Attwell D (2002) Knocking out the glial glutamate transporter GLT-1 reduces glutamate uptake but does not affect hippocampal glutamate dynamics in early simulated ischemia. Eur J Neurosci 15:308–314

He Y, Janssen WGM, Rothstein JD, Morrison JH (2000) Differential synaptic localization of the glutamate transporter EAAC1 and glutamate receptor subunit GluR2 in the rat hippocampus. J Comp Neurol 418:255–269

He Y, Hof PH, Janssen WGM, Rothstein JD, Morrison JH (2001) Differential synaptic localization of GluR2 and EAAC1 in the macaque monkey entorhinal cortex: a postembedding immunogold study. Neurosci Lett 311:161–164

Huang Y, Zuo Z (2005) Isoflurane induces a protein kinase Cα-dependent increase in cell surface protein level and activity of glutamate transporter type 3. Mol Pharmacol 67:1522–1533

Huang YH, Bergles DE (2004) Glutamate transporters bring competition to the synapse. Curr Opin Neurobiol 14:346–352

Ikonomidou C, Turski L (2002) Why did NMDA receptor antagonists fail clinical trials for stroke and traumatic injury? Lancet Neurol 1:383–386

Johnson G, Moore SW (2000) Cholinesterase-like catalytic antibodies: reaction with substrates and inhibitors. Mol Immunol 37:707–719

Kalandadze A, Wu Y, Robinson MB (2002) Protein kinase C activation decreases cell surface expression of the GLT-1 subtype of glutamate transporter. Requirement of a carboxyl-terminal domain and partial dependence on serine 486. J Biol Chem 277:45741–45750

Krizman-Genda E, Gonzalez MI, Zelenaia O, Robinson MB (2005) Evidence that Akt mediates platelet-derived growth factor-dependent increases in activity and surface expression of the neuronal glutamate transporter, EAAC1. Neuropharmacology 49:872–882

Kugler P, Schmitt A (1999) Glutamate transporter EAAC1 is expressed in neurons and glial cells in the rat nervous system. Glia 27:129–142

Lehre KP, Danbolt NC (1998) The number of glutamate transporter subtype molecules at glutamatergic synapses: chemical and stereological quantification in young adult rat brain. J Neurosci 18:8751–8757

Lehre KP, Levy LM, Ottersen OP, Storm-Mathisen J, Danbolt NC (1995) Differential expression of two glial glutamate transporters in the rat brain: Quantitative and immunocytochemical observations. J Neurosci 15:1835–1853

Leonova J, Thorlin T, ND A, Eriksson PS, Ronnback L, Hansson E (2001) Endothelin-1 decreases glutamate uptake in primary cultured rat astrocytes. Am J Physiol Cell Physiol 281:C1495–C1503

Levenson J, Weeber E, Selcher JC, Kategaya LS, Sweatt JD, Eskin A (2001) Long-term potentiation and contextual fear conditioning increase neuronal glutamate uptake. Nat Neurosci 5:155–161

Levi G, Raiteri M (1993) Carrier-mediated release of neurotransmitters. Trends Neurosci 16:415–419

Loder MK, Melikian HE (2003) The dopamine transporter constitutively internalizes and recycles in a protein kinase C-regulated manner in stably transfected PC12 cell lines. J Biol Chem 278:22168–22174

Marchese A, Chen C, Kim Y-M, Benovic JL (2003) The ins and outs of G-protein-coupled receptor trafficking. Trends Biochem Sci 28:369–376

Martin S, Slot JW, James DE (1999) GLUT4 trafficking in insulin-sensitive cells. Cell Biochem Biophys 30:89–113

McDonald JW, Althomsons SP, Hyrc KL, Choi DW, Goldberg MP (1998) Oligodendrocytes from forebrain are highly vulnerable to AMPA/kainate receptor-mediated excitotoxicity. Nat Med 4:291–297

Najimi M, Maloteaux JM, Hermans E (2002) Cytoskeleton-related trafficking of the EAAC1 glutamate transporter after activation of the G(q/11)-coupled neurotensin receptor NTS1. FEBS Lett 523:224–228

Najimi M, Maloteaux J-M, Hermans E (2005) Pertussis toxin-sensitive modulation of glutamate transport by endothelin-1 type A receptors in glioma cells. Biochim Biophys Acta 1668:195–202

Obrenovitch TP, Urenjak J, Zilkha E, Jay TM (2000) Excitotoxicity in neurological disorders—the glutamate paradox. Int J Dev Neurosci 18:281–287

Otis TS, Wu Y-C, Trussell LO (1996) Delayed clearance of transmitter and the role of glutamate transporters at synapses with multiple release sites. J Neurosci 16:1634–1644

Otis TS, Brasnjo G, Dzubay JA, Pratap M (2004) Interactions between glutamate transporters and metabotropic glutamate receptors at excitatory synapses in the cerebellar cortex. Neurochem Int 45:537–544

Peghini P, Janzen J, Stoffel W (1997) Glutamate transporter EAAC-1-deficient mice develop dicarboxylic aminoaciduria and behavioral abnormalities but no neurodegeneration. EMBO J 16:3822–3832

Poitry-Yamate CL, Vutskits L, Rauen T (2002) Neuronal-induced and glutamate-dependent activation of glial glutamate transporter function. J Neurochem 82:987–997

Robinson MB, Dowd LA (1997) Heterogeneity and functional properties of subtypes of sodium-dependent glutamate transporters in the mammalian central nervous system. Adv Pharmacol 37:69–115

Rosenberg PA, Amin S, Leitner M (1992) Glutamate uptake disguises neurotoxic potency of glutamate agonists in cerebral cortex in dissociated cell culture. J Neurosci 12:56–61

Rossi DJ, Oshima T, Attwell D (2000) Glutamate release in severe brain ischaemia is mainly by reversed uptake. Nature 403:316–321

Rothstein JD, Martin L, Levey AI, Dykes-Hoberg M, Jin L, Wu D, Nash N, Kuncl RW (1994) Localization of neuronal and glial glutamate transporters. Neuron 13:713–725

Rothstein JD, Dykes-Hoberg M, Pardo CA, Bristol LA, Jin L, Kuncl RW, Kanai Y, Hediger M, Wang Y, Schielke JP, Welty DF (1996) Knockout of glutamate transporters reveals a major role for astroglial transport in excitotoxicity and clearance of glutamate. Neuron 16:675–686

Rothstein JD, Patel S, Regan MR, Haenggeli C, Huang YH, Bergles DE, Jin L, Hoberg MD, Vidensky S, Chung DS, Toan SV, Bruijn LI, Su Z-Z, Gupta P, Fisher PB (2005) B-Lactam antibiotics offer neuroprotection by increasing glutamate transporter expression. Nature 433:73–77

Royle SJ, Murrell-Lagnado RD (2002) Constitutive cycling: a general mechanism to regulate cell surface proteins. Bioessays 25:39–46

Schlag BD, Vondrasek JR, Munir M, Kalandadze A, Zelenaia OA, Rothstein JD, Robinson MB (1998) Regulation of the glial Na^+-dependent glutamate transporters by cyclic AMP analogs and neurons. Mol Pharmacol 53:355–369

Schmitt A, Asan E, Lesch K-P, Kugler P (2002) A splice variant of glutamate transporter GLT1/EAAT2 expressed in neurons: cloning and localization in rat nervous system. Neuroscience 109:45–61

Schniepp R, Kohler K, Ladewig T, Guenther E, Henke G, Palmada M, Boehmer C, Rothsetin JD, Broer S, Lang F (2004) Retinal colocalization and in vitro interaction of the glutamate receptor EAAT3 and the serum- and glucocorticoid-inducible kinase SGK1. Invest Ophthalmol 45:1442–1449

Schoepp DD, Conn PJ (1993) Metabotropic glutamate receptors in brain function and pathology. Trends Pharmacol Sci 14:13–20

Sheng M, Pak DT (2000) Ligand-gated ion channel interactions with cytoskeletal and signaling proteins. Annu Rev Physiol 62:755–778

Sheng M, Sala C (2001) PDZ domains and the organization of supramolecular complexes. Annu Rev Neurosci 24:1–29

Shigeri Y, Seal RP, Shimamoto K (2004) Molecular pharmacology of glutamate transporters, EAATs and VGLUTs. Brain Res Brain Res Rev 45:250–265

Sims KD, Robinson MB (1999) Expression patterns and regulation of glutamate transporters in the developing and adult nervous system. Crit Rev Neurobiol 13:169–197

Sims KD, Straff DJ, Robinson MB (2000) Platelet-derived growth factor rapidly increases activity and cell surface expression of the EAAC1 subtype of glutamate transporters through activation of phosphatidylinositol 3-kinase. J Biol Chem 274:5228–5327

Sorkina T, Hoover BR, Zahniser NR, Sorkin A (2005) Constitutive and protein kinase C-induced internalization of the dopamine transporter is mediated by a clathrin-dependent mechanism. Traffic 6:157–170

Spooren W, Ballard T, Gasparini F, Amalric M, Mutel V, Schreiber R (2003) Insight into the function of group I and group II metabotropic glutamate (mGluR) receptors: behavioural characterization and implications for treatment of CNS disorders. Behav Pharmacol 14:257–277

Susarla BS, Seal RP, Zelenaia O, Watson DJ, Wolfe JH, Amara SG, Robinson MB (2004) Differential regulation of GLAST immunoreactivity and activity by protein kinase C: evidence for modification of amino and carboxy termini. J Neurochem 91:1151–1163

Swanson RA, Liu J, Miller JW, Rothstein JD, Farrell K, Stein BA, Longuemare MC (1997) Neuronal regulation of glutamate transporter subtype expression in astrocytes. J Neurosci 17:932–940

Tanaka J, Ichikawa R, Watanabe M, Tanaka K, Inoue Y (1997a) Extra-junctional localization of glutamate transporter EAAT4 at excitatory Purkinje cell synapses. Neuroreport 8:2461–2464

Tanaka K, Watase K, Manabe T, Yamada K, Watanabe M, Takahashi K, Iwama H, Nishikawa T, Ichihara N, Kikuchi T, Okuyama S, Kawashima N, Hori S, Takimoto M, Wada K (1997b) Epilepsy and exacerbation of brain injury in mice lacking the glutamate transporter GLT-1. Science 276:1699–1702

Timmerman W, Westerink BHC (1997) Brain microdialysis of GABA amd glutamate: What does it signify? Synapse 27:242–261

Tong G, Jahr CE (1994) Block of glutamate transporters potentiates postsynaptic excitation. Neuron 13:1195–1203

Trotti D, Peng J-B, Dunlop J, Hediger MA (2001) Inhibition of the glutamate transporter EAAC1 expressed in Xenopus oocytes by phorbol esters. Brain Res 914:196–203

Vermeiren C, Najimi M, Vanhoutte N, Tilleux S, Hemptinne Id, Maloteaux J-M, Hermans E (2005) Acute up-regulation of glutamate uptake mediated by mGluR5a in reactive astrocytes. J Neurochem 94:405–416

Vizi ES, Kiss JP (1998) Neurochemistry and pharmacology of the major hippocampal transmitter systems: synaptic and nonsynaptic interactions. Hippocampus 8:566–607

Wadiche JI, Arriza JL, Amara SG, Kavanaugh MP (1995) Kinetics of a human glutamate transporter. Neuron 14:1019–1027

Wang D, Quick MW (2005) Trafficking of the plasma membrane gamma-aminobutyric acid transporter, GAT1. Size and rates of an acutely recycling pool. J Biol Chem 280:18703–18709

Wang Z, Li W, Mitchell CK, Carter-Dawson L (2003) Activation of protein kinase C reduces GLAST in the plasma membrane of rat Muller cells in primary culture. Vis Neurosci 20:611–619

Watase K, Hashimoto K, Kano M, Yamada K, Watanabe M, Inoue Y, Okuyama S, Sakagawa T, Ogawa S-i, Kawachima N, Hori S, Takimoto M, Wada K, Tanaka K (1998) Motor discoordination and increased susceptibility to cerebellar injury in GLAST mutant mice. Eur J Neurosci 10:976–988

Watson RT, Kanzaki M, Pessin JE (2000) Regulated membrane trafficking of the insulin-responsive glucose transporter 4 in adipocytes. Endocr Rev 25:177–204

Watson RT, Khan AH, Furukawa M, Hou JC, Li L, Kanzaki M, Okada S, Kandror KV, Pessin JE (2004) Entry of newly synthesized GLUT4 into the insulin-responsive storage compartment is GGA dependent. EMBO J 23:2059–2070

Yang W, Kilberg MS (2002) Biosynthesis, intracellular targeting, and degradation of the EAAC1 glutamate/aspartate transporter in C6 glioma cells. J Biol Chem 277:38350–38357

Yang Y, Kinney GA, Spain WJ, Breitner JCS, Cook DG (2004) Presenilin-1 and intracellular calcium stores regulate neuronal glutamate uptake. J Neurochem 88:1361–1372

Zerangue N, Kavanaugh MP (1996) Flux coupling in a neuronal glutamate transporter. Nature 383:634–637

Zhang Q, Fukuda M, Bockstaele EV, Pascual O, Haydon PG (2004) Synaptotagmin IV regulates glial glutamate release. Proc Natl Acad Sci U S A 101:9441–9446

Zhou J, Sutherland ML (2004) Glutamate transporter cluster formation in astrocytic processes regulates glutamate uptake activity. J Neurosci 24:6301–6306

Regulation and Dysregulation of Glutamate Transporters

R. Sattler · J. D. Rothstein (✉)

Department of Neurology, Johns Hopkins University, 600 N Wolfe Street, Meyer 6-109, Baltimore MD, 21287, USA
jrothste@jhmi.edu

1	Introduction	278
1.1	Glutamate	278
1.2	Glutamate Transporter	280
2	Regulation of Glutamate Transporters	281
2.1	Transcriptional and Translational Regulation	281
2.1.1	EAAT2/GLT-1	281
2.1.2	EAAT1/GLAST	284
2.2	Posttranslational Modification	285
2.2.1	Transporter Protein Maturation	286
2.2.2	Membrane Targeting and Stabilization	287
2.2.3	Transporter Protein Trafficking	289
2.2.4	Transporter Modification	293
3	Dysregulation of Glutamate Transporters	294
3.1	Dysregulation Through Altered Transcriptional Regulation	295
3.2	Dysregulation Through Altered Transporter Targeting	296
3.3	Dysregulation Through Altered Transporter Modification	296
4	Conclusions	296
	References	297

Abstract Glutamate is the primary excitatory neurotransmitter in the central nervous system. During synaptic activity, glutamate is released into the synaptic cleft and binds to glutamate receptors on the pre- and postsynaptic membrane as well as on neighboring astrocytes in order to start a number of intracellular signaling cascades. To allow for an efficient signaling to occur, glutamate levels in the synaptic cleft have to be maintained at very low levels. This process is regulated by glutamate transporters, which remove excess extracellular glutamate via a sodium-potassium coupled uptake mechanism. When extracellular glutamate levels rise to about normal, glutamate overactivates glutamate receptors, triggering a multitude of intracellular events in the postsynaptic neuron, which ultimately results in neuronal cell death. This phenomenon is known as excitotoxicity and is the underlying mechanisms of a number of neurodegenerative diseases. A dysfunction of the glutamate transporter is thought to contribute to cell death during excitotoxicity. Therefore, efforts have been made to understand the regulation of glutamate transporter function. Transporter activity can be regulated in different ways, including through gene expression, transporter protein targeting and trafficking and through posttranslational modifications of the transporter protein. The identification of these mechanisms has helped to understand the role of glutamate transporters during pathology and will aid in the development of therapeutic strategies with the transporter as a desirable target.

Keywords Glutamate transporter · Transporter regulation · Trafficking · Gene expression · Disease

1
Introduction

To ensure efficient neurotransmission in the brain, it is critical that extracellular neurotransmitter levels are rapidly returned to baseline following regulated secretion. Most extracellular neurotransmitter concentrations are tightly controlled by high-affinity plasma membrane transporters that serve to maintain a dynamic signaling system between neurons. The majority of these uptake processes occur either presynaptically or on surrounding astrocytes. Given the essential role in controlling the extracellular levels of neurotransmitter, it becomes obvious that dysfunction of these transporters will easily lead to a disturbance of synaptic transmission and thereby contribute to various neurological disorders. With L-glutamate as the major excitatory neurotransmitter in the brain, glutamate transporter function and regulation is crucial in maintaining a low extracellular concentration of glutamate. Due to the central role of glutamate in the central nervous system (CNS), we will focus in the present chapter on the regulation of glutamate transporters and on the mechanisms of dysregulation of these transporters during disease.

1.1
Glutamate

L-Glutamate is the major excitatory neurotransmitter in the CNS, and activation of the corresponding receptors mediates rapid synaptic transmission. While glutamate receptor stimulation is involved in processes of learning and memory as well as in other plastic changes in the CNS such as synapse induction and elimination during development, excessive accumulation of extracellular glutamate and overactivation of glutamate receptors contribute to neuronal cell death. This phenomenon is known as glutamate excitotoxicity and is thought to underlie a number of acute CNS diseases including CNS ischemia and trauma as well as chronic neurodegenerative disorders such as Huntington's disease (HD), Alzheimer's disease (AD), and amyotrophic lateral sclerosis (ALS).

Under physiological conditions, most glutamate is present intracellularly, while the glutamate concentration in the surrounding extracellular fluid is up to a million-fold less. This steep concentration gradient is necessary for efficient glutamate receptor stimulation upon presynaptic glutamate release during rapid synaptic transmission. To keep the levels of glutamate in the synaptic cleft sufficiently low, glutamate has to be removed from the extracellular space.

There are no known enzymes extracellularly that are able to metabolize and thereby convert glutamate to an inactive form. Consequently, the only way to remove glutamate rapidly from the synaptic cleft is by cellular uptake. This uptake is carried out by sodium-dependent glutamate transporters that are present on the plasma membrane of perisynaptic astrocytes and, to a lesser degree, on neurons themselves (see Fig. 1). Glutamate uptake becomes particularly important under pathological conditions, when glutamate in the synaptic cleft rises to a great extent above normal levels, high enough to trigger excitotoxicity in the postsynaptic neuron (Maragakis and Rothstein 2004; Rothstein et al. 1996). This is when brain tissue requires very high glutamate uptake activity to protect itself against glutamate toxicity. Once glutamate is taken up, it is either used for metabolic purposes or reused as a transmitter by entering the glutamate–glutamine cycle (Danbolt 2001).

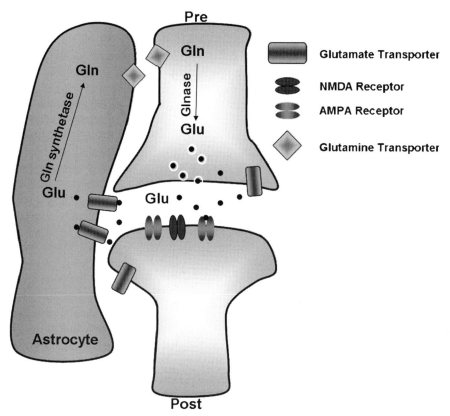

Fig. 1 Glutamate (*Glu*) transporters at the central nervous system synapse: Glutamate is released from the presynaptic (*pre*) neuron into the synaptic cleft. From there, it either activates postsynaptic (*post*) glutamate receptors or is removed from the synaptic cleft by glutamate transporters to feed into the glutamine (*gltn*)–glutamate cycle

1.2
Glutamate Transporter

Glutamate uptake via plasma membrane glutamate transporters involves an active transport of neurotransmitter against its concentration gradient and is driven by an electrochemical gradient of both Na^+ and K^+. Five plasma membrane glutamate transporter subtypes have been identified thus far: excitatory amino acid transporters EAAT1–EAAT5 (human nomenclature) or GLAST/GLT-1/EAAC1/EAAT4/EAAT5 (rodent nomenclature for the five respective human transporters) (see Table 1). The subtypes differ in their distribution pattern with regard to cell type and brain region. EAAT2/GLT-1 (Shashidharan et al. 1994) is primarily found on astrocytes and is the major glutamate transporter in the forebrain (Furuta et al. 1997b). EAAT1/GLAST (Shashidharan and Plaitakis 1993) is the major glutamate transporter present on astrocytes in the cerebellum (Furuta et al. 1997b), the inner ear (Furness and Lehre 1997), circumventricular organs (Berger and Hediger 2000), and retina (Derouiche et al. 1995; Rauen 2000). EAAT3/EAAC1 is a neuronal glutamate transporter with highest concentrations found in the hippocampus, cerebellum, and basal ganglia (Furuta et al. 1997a, b). EAAT4 is for the most part expressed in the Purkinje cells of the cerebellar molecular layer (Barpeled et al. 1997; Furuta et al. 1997a) while EAAT5 is mainly expressed in the retina (Arriza et al. 1997).

Table 1 Nomenclature and expression pattern of glutamate transporter subtypes

Glutamate transporter subtype	Human homolog	Cell type	Anatomic localization
GLAST	EAAT1	Astrocytes, oligodendrocytes	Cerebellum, cortex, spinal cord
GLT1	EAAT2	Astrocytes	Throughout brain and spinal cord
GLT1b	EAAT2b	Astrocytes and neurons	Throughout brain and spinal cord
EAAC1	EAAT3	Neurons	Hippocampus, cerebellum, striatum
EAAT4	EAAT4	Purkinje cells	Cerebellum
EAAT5	EAAT5	Photoreceptors and bipolar Cells	Retina

2
Regulation of Glutamate Transporters

As discussed above, glutamatergic transmission is involved in many important brain functions, while elevated concentrations of extracellular glutamate can cause severe excitotoxic damage to the receiving neurons. It is therefore crucial to maintain efficient glutamate uptake. Recent experiments using antisense oligonucleotides and targeted gene disruption confirmed the critical involvement of glutamate transporter protein in maintaining a healthy glutamate homeostasis (Rothstein et al. 1996; Tanaka et al. 1997). Mice lacking GLT-1 developed increased extracellular glutamate levels, excitotoxic neurodegeneration, and progressive paralysis. In addition, knocking down EAAC1 in vivo led to epileptic seizures (Rothstein et al. 1996). These studies emphasize the need for an understanding of glutamate transporter regulation and illustrate how this knowledge may facilitate the use of the glutamate transporter system as a pharmaceutical target for the treatment of neurodegenerative diseases. In this chapter, we will highlight mechanisms of glutamate transporter regulation and the consequences of malfunction during disease.

The mechanisms of glutamate transporter regulation are not well defined, and to date little is known about the factors that are responsible for regulating protein expression and transporter activity. Regulation can occur on multiple levels, including DNA transcription and protein translation, as well as posttranslational modification, which, in turn, may affect glutamate transporter protein targeting, localization, and transport activity. While DNA transcription and protein expression events require considerable time (several hours) to show effects on transporter activity, posttranslational modification of transporter protein can occur very acutely within minutes. As with most biological events, it is likely that a combination of all of these mechanisms is important for efficient glutamate uptake to occur.

2.1
Transcriptional and Translational Regulation

Early studies showed glutamate uptake in pure astrocyte cultures increases if the astrocytes are treated with media derived from neuronal cultures (Drejer et al. 1983; Gegelashvili et al. 1997; Schlag et al. 1998; Swanson et al. 1997). In the absence of neurons, astrocytes express only GLAST. When co-cultured with neurons, expression of GLT-1 in astrocytes is induced while basal expression of GLAST is slightly enhanced (Gegelashvili et al. 1997; Swanson et al. 1997). These data suggested that there are neuronal soluble factors that increase glutamate transporter protein and messenger RNA (mRNA). This direct upregulation is more prominent for GLT-1 as compared to GLAST, suggesting that transcriptional activation of astrocytic glutamate transporters may be subtype-specific.

2.1.1
EAAT2/GLT-1

While the soluble factors present in neuron conditioned media (NCM) have not yet been identified, progress has been made toward elucidating their signal transduction pathways. It has been proposed that upregulation of GLT-1 by NCM depends on the activation of the p42/44 MAP kinases via the tyrphostin-sensitive receptor tyrosine kinase (RTK) signaling pathway (Gegelashvili et al. 2000, 2001; Swanson et al. 1997). NCM also phosphorylates transcription factors CREM-1 and ATF-1 (Gegelashvili et al. 2000; Swanson et al. 1997), and inhibitors of phosphatidylinositol 3-kinase (PI3K), tyrosine kinase, or nuclear transcription factor κB (NF-κB) almost completely blocked NCM-induced upregulation of GLT-1 protein expression (Swanson et al. 1997; Zelenaia et al. 2000).

In the search for the soluble factors that are responsible for transporter upregulation, growth factors are the most prominent group of physiological molecules studied. Epidermal growth factor (EGF) as well as transforming growth factor-α (TGF-α) induced strong upregulation of GLT-1 in cultured astrocytes, while platelet-derived growth factor (PDGF) showed no effect (Zelenaia et al. 2000). Nor was there an effect seen using insulin, basic fibroblast growth factor (bFGF), and nerve growth factor (NGF), while there was a significant upregulation of GLT-1 caused by dibutyryl-cAMP (db-cAMP) (Eng et al. 1997; Schlag et al. 1998; Swanson et al. 1997; Zelenaia et al. 2000). The effects of EGF and dbcAMP were blocked by inhibitors of PI3K and NF-κB, similar to what was seen with NCM (Zelenaia et al. 2000). However, inhibition of the EGF receptor during treatment with NCM did not block the increased expression of GLT-1 (Zelenaia et al. 2000). This suggests that there are independent signaling pathways leading to transcriptional upregulation of GLT-1 that seem to converge at some level along the cascades.

Early studies indicated the presence of a consensus sequence for NF-κB binding within the 5'-untranslated region of the GLT-1/EAAT2 complementary DNA (cDNA) clones (Meyer et al. 1996). These findings were confirmed when the EAAT2 promoter was cloned. The EAAT2 sequence revealed a number of potential regulatory transcription factor-binding elements, including NF-κB, N-myc, and NFAT (nuclear factor of activated T cells) (Su et al. 2003). Furthermore, the authors demonstrated that EGF, TGF-α, and dbcAMP, as well as bromo-cAMP, increased EAAT2 mRNA expression in primary human fetal astrocyte (PHFA) cultures, whereas TNF-α decreased expression. Using nuclear run-on assays, the authors verified that these alterations of EAAT2 in PHFA occur on a transcriptional level. Overexpression of promoter deletion constructs indicated the involvement of NF-κB in EAAT2 transcriptional regulation. Pharmacological characterization of defined biochemical pathways leading to transcriptional regulation of EAAT2 in PHFA confirmed the convergence of independent signaling pathways, similar to what was found earlier in cultured astrocytes (Zelenaia et al. 2000).

For example, EGF-dependent EAAT2 upregulation was blocked by inhibitors of tyrosine kinase, NF-κB, PI3K, and mitogen-activated kinase (MEK)1/2, but unaffected by inhibitors of PKA. On the other hand, dbcAMP-dependent activation was blocked by inhibitors of PKA, NF-κB, PI3K, and MEK1/2, but not by tyrosine kinase inhibitors (Su et al. 2003). At the same time, blocking NF-κB restored EAAT2 promoter activity during treatment with TNF-α. These data confirm a transcriptional regulation of EAAT2 expression and the involvement of specific regions of the EAAT2 promoter, such as the NF-κB transcription factor binding site. How NF-κB can act both positively and negatively depending on the agent administered is still unclear. Sitcheran et al. (2005) examined this phenomenon in more detail. Using the EAAT2 promoter fragment constructs, they showed that EGF-induced NF-κB activation is independent of IκB degradation, which is a common step in the signaling cascade upstream of NF-κB activation (Yamamoto and Gaynor 2004). Also, TNFα-mediated repression is dependent on the recruitment of N-myc, another transcription factor whose binding site is present on the EAAT2 promoter sequence (Sitcheran et al. 2005; Su et al. 2003).

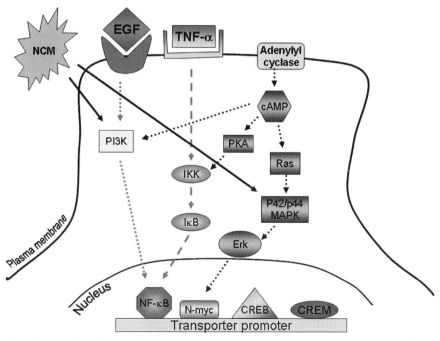

Fig. 2 Suggested pathways effecting glutamate transporter promoter activation (for abbreviations of individual factors, please refer to text)

Another neuron-derived factor that has been demonstrated to regulate the expression of GLT-1 (and GLAST) is pituitary adenylate cyclase-activating polypeptide (PACAP). Exposure of cortical glial cultures to PACAP increased glutamate uptake and protein expression of GLT-1 and GLAST as well as glutamine synthetase (Figiel and Engele 2000). The effects were inhibited by PACAP-inactivating antibodies or by PACAP receptor (PAC1) antagonists. The increase in GLT-1 expression occurred via activation of protein kinase A (PKA) and PKC pathways (Fig. 2), while GLAST upregulation was only dependent on activation of PKA. Knocking down PAC1 in vivo showed a significant reduction of GLAST mRNA in the dentate gyrus, but not in the cortex, while mRNA levels of GLT-1 seemed unaffected by the loss of PAC1 (Zink et al. 2004).

2.1.2
EAAT1/GLAST

While GLT-1/EAAT2 is activated directly by soluble factors released by neurons, activation of GLAST seems to be mediated through different signaling pathways. Early studies identified L-glutamate as one soluble neuron-derived factor acting as a GLAST regulator (Gegelashvili et al. 1996). Glutamate induced an increase in glutamate uptake in cultured astroglia accompanied by increased expression of GLAST protein. Inhibitors of α-amino-3-hydroxy-5-methyl-4-isoxazolepropionic acid (AMPA)/kainate receptors were able to block this upregulation, which was specific to GLAST and did not affect GLT-1. The glutamate-mediated increase in GLAST protein levels was not accompanied by an increase in mRNA levels, suggesting translational regulation instead of transcriptional regulation. On the other hand, when cells were treated with dbcAMP, both mRNA and protein levels were significantly elevated compared to non-treated cultures (Gegelashvili et al. 1996; Swanson et al. 1997). This illustrates again that the regulation of glutamate transporters occurs on different levels and through different signaling pathways depending on the stimulating factor.

More detailed examination of glutamate-mediated GLAST upregulation showed that activation of group II metabotropic glutamate receptors (mGluRs) caused a significant upregulation of GLAST protein levels in astroglial cultures supplemented with NCM, while activation of group I mGluRs led to a decrease in GLAST protein (Gegelashvili et al. 2001). This study did not examine the mRNA level of GLAST, and therefore is not clear if the mGluR-induced changes occur on a transcriptional and/or translational level.

When Zelenaia and colleagues studied the pathways of GLT-1 transcriptional upregulation, they also looked at the regulation of GLAST under several different treatment paradigms (Zelenaia and Robinson 2000). Similarly to GLT-1, dbcAMP, EGF, and TGF-α significantly increased GLAST protein levels and mRNA levels in cultured astrocytes. The authors did not further examine the detailed signaling pathways for GLAST in the same way they did for GLT-1.

Exposure of primary astroglial cells to estrogen caused a significant increase in GLAST (and GLT-1) mRNA levels as well as protein levels, which was accompanied by increased glutamate uptake (Pawlak et al. 2005). Similar observations were made with human cultured astrocytes derived from the cortex of AD patients (Liang et al. 2002). The authors suggested that upregulation of glutamate transporters may be one mechanism by which estrogens provide neuroprotection against excitotoxic glutamate overflow (Liang et al. 2002; Pawlak et al. 2005).

Cloning of the EAAT1 promoter revealed multiple putative transcription factor binding sites including NF-κB, cAMP responsive element binding protein (CREB), activating protein 1 (AP1), GC-box elements, and others (Kim et al. 2003). The region of the promoter required for basal EAAT1 promoter activity was found to include the following transcription factor binding sites: GC-box for stimulating protein (Sp)1 and Sp3 transcription factors, X-box for protein RFX1 as well as gut-enriched Kruppel-like factors, serum response factor (SRF), Atp1a1 regulatory element binding factor (AREB)6, and upstream stimulating factor (USF). Gel shift and supershift analyses on this promoter indicated that the GC-box-binding Sp1 and Sp3 transcription factors are responsible for most of the basal activity of the human EAAT1 gene (Kim et al. 2003).

The authors also looked at factors inducing transcriptional activation. EGF, TGF-α, and 8-bromo cAMP induced a significant increase in EAAT1 promoter activity when overexpressed in PHFA, while glutamate, 12-myristate 13-acetate (PMA), and dbcAMP had no effect. Furthermore, EGF and TGF-α increased endogenous EAAT1 mRNA levels in human fetal astrocyte cultures, while TNF-α decreased mRNA levels (Kim et al. 2003). Finally, the increase in mRNA levels led to increased EAAT1-specific transport activity in primary astrocytes. It follows, therefore, that biological modulators initiate activation of glutamate transporters at the transcriptional level, eventually leading to an increase in functional protein levels.

2.2
Posttranslational Modification

Studies on the molecular structure of the glutamate transporter protein and its membrane topology have become important, as they provide the essential information on how the transporters can be regulated via posttranslational modifications (for review see Sonders et al. 2005). Briefly, the transporters are composed of eight helical transmembrane domains (TMD) and two helical hairpins right before and after TMD7. Both, the N- and C-terminus are cytoplasmic, while there is a large extracellular hydrophilic region between the third and fourth TMD (Slotboom et al. 1999). The recent high-resolution crystal structure of a bacterial glutamate transporter homolog Glt_{Ph} revealed that the transporter is assembled as a trimer (Yernool et al. 2004). Modes of regulation based on this structure are discussed below.

2.2.1
Transporter Protein Maturation

During maturation of nascent transporter polypeptides in the endoplasmic reticulum (ER) and Golgi complex, posttranslational modifications occur that can influence the functional properties of the mature protein. One such modification is the glycosylation of extracellular domains of the transporter protein. Glycosylation sites were identified on GLAST in the extracellular loop of transmembrane helices 3 and 4 at Asn206 and Asn216 (Conradt et al. 1995). Using site-directed mutagenesis followed by overexpression of wildtype and glycosylation-deficient GLAST in *Xenopus* oocytes, the authors showed that N-glycosylation had no effect on the transport activity of GLAST. Similar observations were made using overexpressed GLT-1 in BHK cells (Raunser et al. 2005). Both the glycosylated and non-glycosylated form of GLT-1 were transported to the plasma membrane with equal efficiency and showed similar uptake activities upon reconstitution into liposomes.

Some proteins are retained in the ER and therefore develop into an immature, non-glycosylated protein. Kalandadze et al. (2004) identified an extracellular leucine-based motif in GLT-1 that suppresses a downstream arginine-based motif (RXR) that functions as an ER retention signal. Mutation of the leucine motif leads to retention of the transporter in the ER, where it co-localizes with ER chaperone protein GRP78. As a result, the transporter protein is not terminally glycosylated and represents an immature form of the transporter. The region containing these motifs is located in the extracellular carboxyl-terminal domain of the transporter between transmembrane domain 7 and 8 and is required for phosphorylation-dependent events of GLT-1 redistribution, as discussed in Sect. 2.2.3 (Kalandadze et al. 2002). There is a high degree of homology in this region between the different transporter subtypes, which suggests that similar processes of posttranslational modification may occur among all subtypes (Kalandadze et al. 2004).

A very recent study reported the identification of a splice variant of the human glutamate transporter EAAT1, EAAT1ex9skip (Vallejo-Illarramendi et al. 2005). This splice variant lacks the entire exon 9 of EAAT1, and its mRNA is translated into a truncated protein localized to the ER. Furthermore, when co-expressed with full-length EAAT1, EAAT1ex9skip acts as a negative regulator of the former. This is quite similar to previous reports of alternate splice products of EAAT2 described by Lin et al. (1998). Interestingly, exon 9 includes the leucine-based motif described by Kalandadze et al. and excludes the RXR motif. This confirms the repressor properties of the leucine-rich motif on RXR-induced ER retention and may explain the ER-restricted localization and functional inactivity of EAAT1ex9skip.

Intracellular proteins have been identified as interacting with this carboxyl-terminal region (A.M. Ruggerio, unpublished observations; A. Watanabe, unpublished observations). At least one of these proteins, glutamate transporter interacting protein 3-18 (GTRAP3-18), plays a role in transporter maturation and could possibly represent a chaperone for glutamate transporters. This interacting protein appears to regulate the ER-to-Golgi maturation of oligomeric transporter proteins, and, as such, ultimately alters the cell surface presence of glutamate transporters. These actions, in part, appear to be mediated thru the ER exit and glycosylation of the nascent transporters.

2.2.2
Membrane Targeting and Stabilization

As discussed earlier, in order to achieve efficient synaptic transmission, the level of extracellular glutamate needs to be kept low at all times. This requires a highly ordered arrangement of specialized membrane domains containing the players involved. These include presynaptically localized vesicles releasing glutamate and postsynaptic glutamate receptors binding glutamate as well as astrocytic or peri-synaptically localized glutamate transporters taking up glutamate. The close vicinity of all of these membrane domains is critical in shaping the amplitude of postsynaptic responses, and the processes involved in targeting the glutamate transporter to appropriate membrane structures are thought to be highly dynamic events that are activity dependent (Jackson et al. 2001; Zhou and Sutherland 2004).

These specialized membrane domains include lipid rafts, which are lipid-protein microdomains of the plasma membrane that are enriched with cholesterol and glycosphingolipids (Simons and Toomre 2000). They participate in a wide variety of cellular processes including regulation of trafficking and clustering of membrane-associated proteins and their intracellular signaling molecules (Becher et al. 2001; Fallon et al. 2002; Hering et al. 2003; Suzuki et al. 2001). Butchbach et al. (2004) showed that depletion of membrane cholesterol by methyl-β-cyclodextrin reduced Na^+-dependent glutamate uptake in primary cortical cultures for both astrocytic transporter EAAT2 and neuronal transporter EAAT3, although to a lesser extent for the latter. Biochemical analysis further confirmed the association of glutamate transporters with cholesterol-rich lipid raft microdomains of the plasma membrane (Butchbach et al. 2004). The reduced glutamate uptake in the absence of cholesterol was accompanied by a decrease of transporter protein at the cell surface, which was blocked by a non-specific inhibitor of receptor internalization. This suggests that endocytosis of the transporters may be increased when there is no membrane cholesterol present to stabilize the transporters at the cell surface, resulting in decreased glutamate uptake.

Another study proposed the localization of glutamate transporters to specialized membrane domains, although based on transcriptional regulation (see Sect. 2.1) rather than on posttranslational modification. Zschocke et al. (2005) studied the effects of cAMP and TGF-α on caveolin gene expression in correlation with GLT-1 gene expression. Caveolin is the main structural constituent of caveolae, a subset of lipid rafts. The authors hypothesized that caveolin and GLT-1 may be regulated through a common PI3K-dependent pathway. Biochemical analyses using pharmacological blockers suggested a correlation of a reciprocal regulation of caveolins and GLT-1. Activation of a PI3K-dependent pathway using either TGF-α or cAMP decreased caveolin expression, while GLT-1 protein expression was upregulated. Immunocytochemistry and analysis of membrane fractions confirmed a localization of GLT-1 to non-caveolar lipid raft microdomains of cortical astroglial cells.

Both of these studies suggest that glutamate transporters are localized to specific membrane domains where they can take up glutamate most efficiently. Future work will be needed to examine how this localization is regulated under basal as well as pathophysiological conditions.

The targeting of transporter protein to specific sites was further suggested to be dependent on splice variations of the protein. A splice variant of GLT-1, GLT-1v/GLT-1b, with N-terminal splicing of the original GLT-1 but an alternative splicing at the carboxyl-terminal region (Chen et al. 2002; Schmitt et al. 2002), was studied in regards to its transport properties and localization compared to GLT-1 (Sullivan et al. 2004). While the functional properties of the splice variant were similar to GLT-1, GLT-1v was localized to glial processes in extrasynaptic regions. The authors conclude that while GLT-1v does not seem to be involved in clearing extracellular glutamate during fast synaptic transmission, it may be important to prevent glutamate spillover to adjacent synapses, and it may become a crucial player during pathological conditions, when extracellular glutamate rises highly above normal levels. It will be interesting to see whether these conditions will upregulate protein expression of this particular splice variant in order to provide better neuroprotection.

Localization and stabilization of membrane proteins is often dependent on their interaction with intracellular anchoring proteins. This phenomenon has been intensively studied in the field of neurotransmitter receptors, and a large number of receptor interacting proteins have been identified over the last decade (for review see Kim and Huganir 1999; Scannevin and Huganir 2000). Jackson et al. (2001) identified two interacting proteins, GTRAP41 (βIII spectrin) and GTRAP48 (PDZ RhoGef), that specifically bind to the intracellular carboxyl-terminal domain of EAAT4. When co-expressed in heterologous cells, the interaction of these proteins with EAAT4 increases glutamate transport in these cells. In addition, viral injection of GTRAP41 and GTRAP48 in vivo also resulted in increased glutamate uptake. The increased transport is associated with increased surface expression of the transporter at the plasma membrane (Jackson et al. 2001). GTRAP41 and GTRAP48 are suggested to stabilize EAAT4 at the plasma membrane, thereby reducing internalization and subsequent degradation. Interestingly, GTRAP3-18, an EAAC1 interacting protein, does not affect surface expression of EAAC1 at all (Lin et al. 2001). Similar to GTRAP41 and GTRAP48, GTRAP3-18 specifically binds to the C-terminal intracellular domain of EAAC1. Co-expression of GTRAP3-18 with EAAC1 in HEK293 cells decreased EAAC1-mediated glutamate uptake. As mentioned before, surface expression of EAAC1 was unaltered when co-expressed with GTRAP3-18, suggesting that loss of EAAC1-mediated glutamate uptake was not due to altered protein targeting. Instead, kinetic analyses of glutamate transport indicated decreased glutamate affinity of EAAC1 in the presence of overexpressed GTRAP3-18. To confirm the effects of GTRAP-18 in vivo, the authors knocked down GTRAP-18 gene expression by administering antisense oligomers intraventricularly, which led to increased cortical glutamate uptake (Lin et al. 2001).

In order for proteins to be targeted properly to their final destinations on the plasma membrane cell surface, a sorting process occurs as the proteins exit the trans-Golgi network. In general, neuronal proteins that are localized dendritically are concentrated on the basolateral surface when expressed in epithelial cells. The sorting of basolateral proteins depends on short motifs present in the cytoplasmic tails of the proteins. Apical sorting in epithelial cells correlates with axonal distribution in neurons. Mechanisms of apical sorting are not well characterized but are thought to involve glycosylation, raft association, and glycosylphosphatidylinositol linkage (Winckler and Mellman 1999). Cheng et al. (Cheng et al. 2002) identified a novel sorting motif in EAAT3 that directs the transporter to the apical membrane of epithelial cells and to the somatodendritic location in hippocampal neurons, a rather counterintuitive process. Deletion of the motif eliminated apical localization and impaired dendritic targeting in hippocampal neurons. The motif was sufficient to redirect the basolaterally localized EAAT1 and the nonpolarized EAAT2 to the apical surface. Clustering of EAAT3 was not affected by mutations of the sorting motif. The motif is very likely to bind to adaptor proteins that direct the transporter to its peri-synaptic location. Whether or not this sorting can be regulated by soluble factors or activity is still unknown and requires further investigation.

2.2.3
Transporter Protein Trafficking

One mechanism for regulating neurotransmitter transport involves changes in membrane trafficking of the transporter. Trafficking of membrane proteins between the plasma membrane and intracellular stores that is mediated by an interplay of internalization and membrane insertion can occur in a very short period and therefore allows for very fast regulation of functional protein activity. While trafficking of other membrane proteins such as neurotransmitter receptors has been studied extensively, the regulation of glutamate transporter trafficking is only starting to be understood. Two of the most prominent mechanisms for regulating membrane protein trafficking to and from the plasma membrane are direct phosphorylation of the membrane protein and phosphorylation of interacting scaffolding proteins.

2.2.3.1
EAAT2/GLT-1

Casado et al. (1991) were the first to observe increases in glutamate transport in glial cells when incubated with 12-O-tetradecanoylphorbol 13-acetate (TPA), a potent activator of PKC. A follow-up study from the same lab confirmed the increase in glutamate uptake and phosphorylation by PKC on serine residues of GLT-1 when virally overexpressed in HeLa cells (Casado et al. 1993). Interestingly, when GLT-1 was stably transfected into HeLa cells or two other peripheral cell lines, PKC had no regulatory effects on GLT-1 at all (Tan et al. 1999). Finally, activating PKC in Y-79 human retinoblastoma cells, which express endogenous EAAT2/GLT-1, inhibited glutamate uptake by decreasing the transporter's affinity for glutamate. The PKC-dependent decrease in GLT-1-mediated glutamate uptake was replicated using primary co-cultures of neurons and astrocytes that endogenously express GLT-1 as well as with overexpressed GLT-1 in C6 glioma cells (Kalandadze et al. 2002). Using these cultures models, the authors demonstrated a PKC-dependent decrease in GLT-1 surface expression that requires the presence of a 43-amino-acid long carboxyl-terminal domain of GLT-1 protein. Mutation analysis of serines and threonines in this domain indicated that serine 486 is partially responsible for PKC-dependent internalization of GLT-1. Further biochemical studies in C6 glioma cells as well as in crude rat brain synaptosomes identified PKCα as the required PKC subtype for PMA-induced GLT-1 internalization (Gonzalez et al. 2005). Immunoprecipitation of GLT-1 showed a complex formation with PKCα that was increased after PMA treatment and blocked by PKC inhibitors. Whether or not there is a direct interaction between GLT-1 and PKCα is unclear. However, these data strongly suggest that a formation of GLT-1-PKCα complexes is part of the internalization process.

Zhou and Sutherland (2004) confirmed the PKC-dependent internalization of GLT-1 in primary astrocyte cultures as well as C6 glioma cells. Using real-time imaging of green fluorescent protein (GFP)-tagged GLT-1 overexpressed in both cell systems, they visualized transporter internalization and noticed an increased cluster formation of GLT-1 upon PKC activation. Both events were blocked by PKC inhibitors as well as by overexpression of a dominant negative form of dynamin, suggesting that GLT-1 endocytosis is a clathrin-mediated process.

A recent study performed in neuron-enriched cultures confirmed a decrease in surface expression of GLT-1 in the presence of PMA (Guillet et al. 2005). The authors also identified a PI3K-dependent internalization of GLT-1. When cells were incubated with wortmannin, an inhibitor of PI3K, cell surface expression of GLT-1 was dramatically reduced. On the other hand, an inhibitor of PKA significantly increased surface expression of GLT-1 in these cultures. These data suggest that trafficking of glutamate transporters, and subsequently functional activity, can be regulated via independent signaling pathways.

The diverse effects of PKC activation on GLT-1 uptake activity as reported in varying cell model systems suggest that cell type-specific intracellular proteins may be required to allow for a PKC-dependent regulation of GLT-1 trafficking and thereby glutamate uptake activity upon stimulation.

2.2.3.2
EAAT3/EAAC1

Using C6 glioma cells, which endogenously express EAAC1, Dowd and Robinson (1996) reported a PKC-dependent increase in EAAC1-mediated glutamate transport activity by application of PMA. The increase in glutamate uptake occurred within minutes of PMA application and was independent of new protein synthesis. Follow-up studies from the same laboratory showed a correlation of increased uptake activity with increased surface expression of EAAC1 (Davis et al. 1998). PKC inhibitors blocked both PMA-induced increased activity and membrane insertion of EAAC1, while PI3K inhibitors only blocked PMA-induced membrane insertion. Using confocal microscopy, Davis et al. (1998) examined the increase in EAAC1 cell surface expression and confirmed increased membrane insertion as well as increased numbers of EAAC1 clusters at the cell surface. Although in the opposite direction, these results are reminiscent of what Zhou and Sutherland (2004) found with GLT-1 in astrocytes and C6 glioma cells. Similar increases in cell surface expression of EAAC1 were observed in C6 glioma cells when stimulated with PDGF (Sims et al. 2000). The increase in membrane insertion was paralleled by a decrease in intracellular EAAC1 protein and was blocked by PI3K inhibitors, but not by PKC inhibitors. No redistribution of EAAC1 was seen when cells were treated with other growth factors.

Guillet et al. (2005) confirmed a PKC-dependent increase in EAAC1 surface expression in primary neuron-enriched cultures. At the same time, inhibition of PKA or PI3K caused an internalization of EAAC1 in these cells. Similar to what Davis et al. (1998) found, the authors showed an increased formation of EAAC1 clusters in the intracellular compartment following wortmannin-induced internalization of the transporters.

Gonzalez et al. (2002, 2003) identified PKCε as the PKC subtype to mediate the PMA-induced increase in EAAC1 uptake activity, while activation of PKCα is responsible for the phorbol ester-induced membrane insertion of EAAC1. In addition, biochemical analysis of the activated EAAC1 protein showed the formation of EAAC1-PKCα complexes upon PMA treatment, which was blocked by PKC inhibitors. This observation was made in both C6 glioma cells and rat brain synaptosomes.

A more detailed kinetic analysis of EAAC1 trafficking in C6 glioma cells and primary neurons revealed PKC-dependent membrane insertion of EAAC1 in PMA-treated cultures within minutes (Fournier et al. 2004). Furthermore, PMA treatment reduced constitutive internalization of EAAC1, indicating that activation of PKC decreased EAAC1 endocytosis. When cells were incubated at lower temperatures to reduce intracellular membrane trafficking, PMA was not able to increase EAAC1 surface expression, while constitutive recycling of EAAC1 was unaffected. This suggests the presence of two distinct intracellular pools of EAAC1, one that is regulated by PMA-mediated PKC activation and one that is responsible for basal, constitutive protein recycling (Fournier et al. 2004).

2.2.3.3
EAAT1/GLAST

Overexpressed GLAST in *Xenopus* oocytes and HEK293 cells showed decreased transport activity when cells were treated with the PKC activator PMA (Conradt and Stoffel 1997). Site-directed mutagenesis of all putative PKC phosphorylation sites did not block the decrease in transport activity, suggesting that GLAST transporter activity is inhibited by phosphorylation at non-PKC consensus sites. Interestingly, immunostaining of surface protein did not reveal a change in surface expression of GLAST upon PMA treatment. Acute treatment of cerebellar chick Bergmann glial cells (BGC) with PKC activator phorbol 12-tetradecanoyl-13-acetate (TPA) confirmed a significant decrease in glutamate uptake activity without changes in protein levels of GLAST (Gonzalez et al. 1999; Gonzalez and Ortega 1997). However, when the authors prolonged the exposure to TPA to several hours, GLAST protein levels were decreased. They proposed that PKC activation might have an effect on GLAST gene expression through activation of transcription factor AP-1, which is found on the GLAST promoter sequence (Hagiwara et al. 1996; see also Sect. 2.1.2). While this study did not examine the effects of TPA on surface expression of GLAST in this culture system, Wang et al. (2003) found that treatment of BGC with PMA significantly decreased glutamate transport and GLAST cell surface expression as shown by biotinylation assays.

Using intact retinal tissue, inhibition of PKCδ activity decreased GLAST-mediated glutamate uptake (Bull and Barnett 2002). The authors did not examine whether inhibition of PKCδ caused internalization of the transporter. Meanwhile, working in neuron-enriched cultures, Guillet et al. (2005) reported a PMA-induced decrease in cell surface expression of GLAST. Similar results were obtained with inhibitors of PKA, while inhibitors of PI3K resulted in increased cell surface expression, suggesting again that independent signaling pathways are able to rapidly regulate transporter trafficking in response to acute intracellular signaling processes. PI3K inhibitors also blocked insulin-like growth factor-1-mediated increase in GLAST-dependent glutamate uptake in BGC (Gamboa and Ortega 2002).

In yet another study, acute treatment of primary astrocyte cultures with PMA caused an increase in glutamate transport activity, while long-term treatment had no effect (Susarla et al. 2004). The authors discovered modifications of several intracellular epitopes while attempting an immunohistochemical analysis of the treated cultures, which led to a loss of GLAST immunoreactivity. Using a different approach to study surface expression, they did not find a change upon PMA exposure. The opposing effects of PMA on uptake activity in this study compared to Guillet et al. (2005) remains unclear.

Another major mechanism responsible for subcellular membrane trafficking is changes in the cellular cytoskeleton. Amyloid β-protein (Aβ), a major constituent of amyloid plaques in AD, induced increased glutamate uptake and GLAST surface expression in cultured astrocytes, which was inhibited by actin-disrupting agents (Ikegaya et al. 2002). These data suggest the existence of actin-dependent mechanisms of GLAST redistribution. Similar changes in the cytoskeleton were achieved when primary astrocytes were treated with glutamate. Glutamate exposure produced an increase in GLAST surface expression that was blocked by cytochalasin B or cytochalasin D, both actin depolymerizing agents (Duan et al. 1999). Interestingly, inhibitors of PKC, PKA, or PI3K had no effect on glutamate-induced redistribution of GLAST (Duan et al. 1999).

2.2.4
Transporter Modification

Arachidonic acid (AA) is released from neurons during synaptic activity and can modulate synaptic transmission and therefore may be a good candidate as a transporter regulator (Linden 1998; Williams et al. 1989). Indeed, AA has been reported to inhibit the rate of glutamate uptake in neuronal synaptic terminals and astrocytes (Dorandeu et al. 1998; Lundy and McBean 1995; Manzoni and Mennini 1997; Volterra et al. 1992). Zerangue et al. (1995) identified a subtype-specific regulation of EAAT1 and EAAT2 by AA. While exposure of overexpressed EAAT1 in oocytes to AA led to decreased glutamate uptake, EAAT2-mediated glutamate transport in the same cell system was increased upon AA treatment. A second study from the same laboratory showed AA-induced increases of EAAT4-mediated transporter currents in cerebellar Purkinje neurons (Tzingounis et al. 1998). When they overexpressed EAAT4 in oocytes to study the mechanisms of increased uptake, they found that AA did not mediate an increase in the rate of glutamate transport but instead activated a proton-selective conductance. The authors suggest a mechanism by which synaptic activity may decrease intracellular pH in neurons where this transporter is localized. Whether or not these mechanisms are similar for GLT-1 or GLAST upregulation by AA is unknown.

Proteins that interact with the intracellular carboxyl-terminal domains of membrane proteins can control the subcellular localization of the protein (see Sect. 2.2.2 for GTRAPs), but they can also modulate activity. Marie and Attwell (1999) were able to demonstrate that the interaction of the last eight amino acids of the carboxyl terminal tail of GLAST with an as-yet-unknown intracellular protein modifies the affinity of GLAST for glutamate in retinal glia cells. Disruption of the interaction increased the glutamate affinity for GLAST and consequently increased the transporter currents by 40% at low glutamate concentrations. The same group recently identified the Lim protein Ajuba as an interacting protein for GLT-1 (Marie et al. 2002). When co-expressed in a heterologous system, transporter affinity and uptake velocity were unchanged. Thus, the functional significance of this interaction and how it may regulate glutamate transporter activity is not known.

Glutamate transporters are further modified and regulated by sulfhydryl-based redox mechanisms (extensively reviewed by Trotti et al. 1998). Exposure of cortical astrocytes to H_2O_2 or xanthine/xanthine oxidase decreased glutamate uptake dramatically, and the effect was blocked by free radical scavenger enzymes (Volterra et al. 1994). The same group later identified sulfhydryl (SH)-based redox modulatory sites (Trotti et al. 1997b). Mutation of three cysteine residues in GLAST (canine origin) showed oligomer formation, plasma-membrane localization, and transport kinetics similar to wildtype GLAST (Tamahara et al. 2002). Inhibition of glutamate transport by mercury was identical for mutant and wildtype GLAST when overexpressed in Cos-7 cells. The authors suggest that cysteine residues are not critical for the functional expression of GLAST. While oxidative processes play an important role during pathological conditions and could explain decreased functional activity of glutamate transporters in disease, the physiological role of these oxidative processes in regulation of glutamate transporter function is still unclear. In fact, as yet, there is little quantitative evidence that transporter disruptions in animal models can be accounted for by oxidative modifications.

3
Dysregulation of Glutamate Transporters

As mentioned in the introduction, glutamate transporters keep synaptic and extrasynaptic concentrations of glutamate low enough to prevent excitotoxicity. This was confirmed in early studies in which glutamate transporter expression was suppressed by either antisense oligonucleotide treatment or by genetically modified transporter knockout (KO) mice (Rothstein et al. 1996; Tanaka et al. 1997). In both studies, loss of glutamate transporter protein caused a significant increase in extracellular glutamate concentrations concomitant with increased neuronal cell death. The question of whether glutamate transporter dysfunction, leading to a rise in extracellular glutamate, plays a role in development and propagation of neurodegenerative diseases has only more recently been addressed. We will briefly summarize the data on dysregulation of glutamate transporters in regards to the topics we have discussed above. (For a more detailed discussion on glutamate transporters in disease, see reviews by reviews by Gegelashvili et al. 2001; Maragakis and Rothstein 2004.)

3.1
Dysregulation Through Altered Transcriptional Regulation

The most common dysfunction of glutamate transport in neurodegenerative disease seems to be caused by a decrease in transporter protein level. The mechanisms for loss of protein (transcriptional/translational) are generally not known. Protein downregulation is found in acute neurodegenerative diseases, such as ischemia/hypoxia (Chen et al. 2005; Fukamachi et al. 2001; Inage et al. 1998; Martin et al. 1997; Raghavendra Rao et al. 2000; Rao et al. 2001; Rothstein et al. 1996; Yeh et al. 2005) and a number of chronic neurodegenerative disorders, including HD (Behrens et al. 2002; Lievens et al. 2001), PD (Ginsberg et al. 1995; Levy et al. 1995), AD (Li et al. 1997; Masliah et al. 2000), and ALS (Bruijn et al. 1997; Howland et al. 2002; Rothstein et al. 1995). The downregulation of transporter protein was not always paralleled by decreases in mRNA levels. For example, transgenic mice expressing a mutant form of amyloid precursor protein, which plays a central role in AD, showed decreased glutamate uptake activity and decreased transporter protein levels, but had normal transporter mRNA levels (Masliah et al. 2000). The same results were obtained from a postmortem analysis of frontal cortex of AD patients (Li et al. 1997). A possible explanation for the loss in transporter protein without a loss in mRNA could be a dysregulation of transporter protein degradation processes.

Regardless of the mechanism for loss of transporter activity, overcoming this loss could be therapeutically relevant. Few attempts have been made so far to pharmacologically intervene with the loss of transporter protein in any of these diseases. The idea would be that bringing back the levels of transporter protein to physiological levels would prevent accumulation of extracellular glutamate and consequently protect against excitotoxic neuronal cell death. Two studies support this hypothesis. In both studies, transgenic mice have been generated that overexpress GLT-1 to varying degrees (1.5- to 5-fold over wildtype mice) (Guo et al. 2003; Sutherland et al. 2001). When these mice were crossed with transgenic mice overexpressing a mutant form of SOD1, an animal model for ALS, the animals showed a delayed onset of motor neuron degeneration and increased survival. This suggests that increasing the total number of GLT-1 molecules protects against neurodegeneration. Rothstein et al. 2005) have taken this idea one step further. The authors discovered that β-lactam antibiotics increase protein expression of GLT-1, both in vitro and in vivo. When they then treated SOD1 mutant mice with ceftriaxone, they reported a significant delay in disease onset as well as a significant increase in survival in the treated mice, confirming a neuroprotective effect of β-lactam antibiotics through upregulation of glutamate transporter protein (Rothstein et al. 2005).

Gene regulation can also be altered through abnormal splicing processes. For example, abnormal splicing of EAAT2 mRNA was found in motor cortex of

ALS patients (Lin et al. 1996). The altered splice product led to the production of truncated EAAT2 protein that they suggested might be responsible for reduced capacity for glutamate transport.

3.2
Dysregulation Through Altered Transporter Targeting

Very little is known about mistargeted glutamate transporter proteins in disease. It has been suggested that glutamate plays a role in glioma growth and invasion. One study, which was done in cell lines derived from human gliomas, indicates a mislocalization of EAAT1 to the nucleus, which consequently reduces glutamate uptake into these cells significantly (Ye and Sontheimer 1999). EAAT1 was also mislocalized in brain tissue from glioblastoma patients (Ye et al. 1999). The reduced glutamate uptake may lead to increased extracellular glutamate and may contribute to seizures that are common in glioma patients.

3.3
Dysregulation Through Altered Transporter Modification

There are very few studies reported that present direct proof on altered transporter modifications during disease. One can speculate that during an ischemic insult, where it is known that reactive oxygen species (and arachidonic acid) are produced and liberated in response to excessive glutamate receptor activation (Siesjo et al. 1989), glutamate uptake can be inhibited through direct oxidation of cysteine sulfhydryl groups (Trotti et al. 1997a, b). The same laboratory further showed that mutant SOD1 proteins catalyze oxidative reactions that target the carboxyl-terminal domain of GLT-1 and render it inactive (Trotti et al. 1999). The authors propose that toxic properties of SOD1 mutant protein lead to neuronal cell death via excitotoxic mechanisms in SOD1-linked familial ALS.

4
Conclusions

The awareness of the importance of glutamate transporters in health and disease has led to an increasing interest in the regulation of glutamate transporter function. We now know that transporters can be regulated very specifically via multiple mechanisms, starting from regulation of gene transcription by a number of different molecules to modulations of the transporter protein through posttranslational modifications. The understanding of these processes has helped us to identify mechanisms of dysfunction of the transporter under pathological conditions, when a loss of glutamate uptake leads to neurodegeneration in many neurological disorders.

Most importantly, the identification of these mechanisms has further advanced the development of new strategies for pharmacological intervention and therapy during diseases associated with a dysfunction of glutamate transporters.

References

Arriza JL, Eliasof S, Kavanaugh MP, Amara SG (1997) Excitatory amino acid transporter 5, a retinal glutamate transporter coupled to a chloride conductance. Proc Natl Acad Sci USA 94:4155–4160

Barpeled O, BenHur H, Biegon A, Groner Y, Dewhurst S, Furuta A, Rothstein JD (1997) Distribution of glutamate transporter subtypes during human brain development. J Neurochem 69:2571–2580

Becher A, White JH, McIlhinney RA (2001) The gamma-aminobutyric acid receptor B, but not the metabotropic glutamate receptor type-1, associates with lipid rafts in the rat cerebellum. J Neurochem 79:787–795

Behrens PF, Franz P, Woodman B, Lindenberg KS, Landwehrmeyer GB (2002) Impaired glutamate transport and glutamate-glutamine cycling: downstream effects of the Huntington mutation. Brain 125:1908–1922

Berger UV, Hediger MA (2000) Distribution of the glutamate transporters GLAST and GLT-1 in rat circumventricular organs, meninges, and dorsal root ganglia. J Comp Neurol 421:385–399

Bruijn L, Becher M, Lee M, Anderson K, Jenkins N, Copeland N, Sisodia S, Rothstein J, Borchelt D, Price D, Cleveland D (1997) ALS-linked SOD1 mutant G85R mediates damage to astrocytes and promotes rapidly progressive disease with SOD1-containing inclusions. Neuron 18:327–338

Bull ND, Barnett NL (2002) Antagonists of protein kinase C inhibit rat retinal glutamate transport activity in situ. J Neurochem 81:472–480

Butchbach ME, Tian G, Guo H, Lin CL (2004) Association of excitatory amino acid transporters, especially EAAT2, with cholesterol-rich lipid raft microdomains: importance for excitatory amino acid transporter localization and function. J Biol Chem 279:34388–34396

Casado M, Zafra F, Aragon C, Gimenez C (1991) Activation of high-affinity uptake of glutamate by phorbol esters in primary glial cell cultures. J Neurochem 57:1185–1190

Casado M, Bendahan A, Zafra F, Danbolt NC, Aragon C, Gimenez C, Kanner BI (1993) Phosphorylation and modulation of brain glutamate transporters by protein kinase C. J Biol Chem 268:27313–27317

Chen JC, Hsu-Chou H, Lu JL, Chiang YC, Huang HM, Wang HL, Wu T, Liao JJ, Yeh TS (2005) Down-regulation of the glial glutamate transporter GLT-1 in rat hippocampus and striatum and its modulation by a group III metabotropic glutamate receptor antagonist following transient global forebrain ischemia. Neuropharmacology 49:703–714

Chen W, Aoki C, Mahadomrongkul V, Gruber CE, Wang GJ, Blitzblau R, Irwin N, Rosenberg PA (2002) Expression of a variant form of the glutamate transporter GLT1 in neuronal cultures and in neurons and astrocytes in the rat brain. J Neurosci 22:2142–2152

Cheng C, Glover G, Banker G, Amara SG (2002) A novel sorting motif in the glutamate transporter excitatory amino acid transporter 3 directs its targeting in Madin-Darby canine kidney cells and hippocampal neurons. J Neurosci 22:10643–10652

Conradt M, Stoffel W (1997) Inhibition of the high-affinity brain glutamate transporter GLAST-1 via direct phosphorylation. J Neurochem 68:1244–1251

Conradt M, Storck T, Stoffel W (1995) Localization of N-glycosylation sites and functional role of the carbohydrate units of GLAST-1, a cloned rat brain L-glutamate/L-aspartate transporter. Eur J Biochem 229:682–687

Danbolt NC (2001) Glutamate uptake. Prog Neurobiol 65:1–105

Davis KE, Straff DJ, Weinstein EA, Bannerman PG, Correale DM, Rothstein JD, Robinson MB (1998) Multiple signaling pathways regulate cell surface expression and activity of the excitatory amino acid carrier 1 subtype of Glu transporter in C6 glioma. J Neurosci 18:2475–2485

Derouiche A, Rauen T (1995) Coincidence of L-glutamate/L-aspartate transporter (GLAST) and glutamine synthetase (GS) immunoreactions in retinal glia: evidence for coupling of GLAST and GS in transmitter clearance. J Neurosci Res 42:131–143

Dorandeu F, Antier D, Pernot-Marino I, Lapeyre P, Lallement G (1998) Venom phospholipase A2-induced impairment of glutamate uptake: an indirect and nonselective effect related to phospholipid hydrolysis. J Neurosci Res 51:349–359

Dowd LA, Robinson MB (1996) Rapid stimulation of EAAC1-mediated Na^+-dependent L-glutamate transport activity in C6 glioma cells by phorbol ester. J Neurochem 67:508–516

Drejer J, Meier E, Schousboe A (1983) Novel neuron-related regulatory mechanisms for astrocytic glutamate and GABA high affinity uptake. Neurosci Lett 37:301–306

Duan S, Anderson CM, Stein BA, Swanson RA (1999) Glutamate induces rapid upregulation of astrocyte glutamate transport and cell-surface expression of GLAST. J Neurosci 19:10193–10200

Eng DL, Lee YL, Lal PG (1997) Expression of glutamate uptake transporters after dibutyryl cyclic AMP differentiation and traumatic injury in cultured astrocytes. Brain Res 778:215–221

Fallon L, Moreau F, Croft BG, Labib N, Gu WJ, Fon EA (2002) Parkin and CASK/LIN-2 associate via a PDZ-mediated interaction and are co-localized in lipid rafts and postsynaptic densities in brain. J Biol Chem 277:486–491

Figiel M, Engele J (2000) Pituitary adenylate cyclase-activating polypeptide (PACAP), a neuron-derived peptide regulating glial glutamate transport and metabolism. J Neurosci 20:3596–3605

Fournier KM, Gonzalez MI, Robinson MB (2004) Rapid trafficking of the neuronal glutamate transporter, EAAC1: evidence for distinct trafficking pathways differentially regulated by protein kinase C and platelet-derived growth factor. J Biol Chem 279:34505–34513

Fukamachi S, Furuta A, Ikeda T, Ikenoue T, Kaneoka T, Rothstein JD, Iwaki T (2001) Altered expressions of glutamate transporter subtypes in rat model of neonatal cerebral hypoxia-ischemia. Brain Res Dev Brain Res 132:131–139

Furness DN, Lehre KP (1997) Immunocytochemical localization of a high-affinity glutamate-aspartate transporter, GLAST, in the rat and guinea-pig cochlea. Eur J Neurosci 9:1961–1969

Furuta A, Martin LJ, Lin CL, Dykes-Hoberg M, Rothstein JD (1997a) Cellular and synaptic localization of the neuronal glutamate transporters excitatory amino acid transporter 3 and 4. Neuroscience 81:1031–1042

Furuta A, Rothstein JD, Martin LJ (1997b) Glutamate transporter protein subtypes are expressed differentially during rat CNS development. J Neurosci 17:8363–8375

Gamboa C, Ortega A (2002) Insulin-like growth factor-1 increases activity and surface levels of the GLAST subtype of glutamate transporter. Neurochem Int 40:397–403

Gegelashvili G, Civenni G, Racagni G, Danbolt NC, Schousboe I, Schousboe A (1996) Glutamate receptor agonists up-regulate glutamate transporter GLAST in astrocytes. Neuroreport 8:261–265

Gegelashvili G, Danbolt NC, Schousboe A (1997) Neuronal soluble factors differentially regulate the expression of the GLT1 and GLAST glutamate transporters in cultured astroglia. J Neurochem 69:2612–2615

Gegelashvili G, Dehnes Y, Danbolt NC, Schousboe A (2000) The high-affinity glutamate transporters GLT1, GLAST, and EAAT4 are regulated via different signalling mechanisms. Neurochem Int 37:163–170

Gegelashvili G, Robinson MB, Trotti D, Rauen T (2001) Regulation of glutamate transporters in health and disease. Prog Brain Res 132:267–286

Ginsberg SD, Martin LJ, Rothstein JD (1995) Regional deafferentation down-regulates subtypes of glutamate transporter proteins. J Neurochem 65:2800–2803

Gonzalez I, Susarla BT, Robinson MB (2005) Evidence that protein kinase Calpha interacts with and regulates the glial glutamate transporter GLT-1. J Neurochem 94:1180–1188

Gonzalez MI, Ortega A (1997) Regulation of the Na^+-dependent high affinity glutamate/aspartate transporter in cultured Bergmann glia by phorbol esters. J Neurosci Res 50:585–590

Gonzalez MI, LopezColome AM, Ortega A (1999) Sodium-dependent glutamate transport in Muller glial cells: regulation by phorbol esters. Brain Res 831:140–145

Gonzalez MI, Kazanietz MG, Robinson MB (2002) Regulation of the neuronal glutamate transporter excitatory amino acid carrier-1 (EAAC1) by different protein kinase C subtypes. Mol Pharmacol 62:901–910

Gonzalez MI, Bannerman PG, Robinson MB (2003) Phorbol myristate acetate-dependent interaction of protein kinase Calpha and the neuronal glutamate transporter EAAC1. J Neurosci 23:5589–5593

Guillet BA, Velly LJ, Canolle B, Masmejean FM, Nieoullon AL, Pisano P (2005) Differential regulation by protein kinases of activity and cell surface expression of glutamate transporters in neuron-enriched cultures. Neurochem Int 46:337–346

Guo H, Lai L, Butchbach ME, Stockinger MP, Shan X, Bishop GA, Lin CL (2003) Increased expression of the glial glutamate transporter EAAT2 modulates excitotoxicity and delays the onset but not the outcome of ALS in mice. Hum Mol Genet 12:2519–2532

Hagiwara T, Tanaka K, Takai S, Maeno-Hikichi Y, Mukainaka Y, Wada K (1996) Genomic organization, promoter analysis, and chromosomal localization of the gene for the mouse glial high-affinity glutamate transporter Slc1a3. Genomics 33:508–515

Hering H, Lin CC, Sheng M (2003) Lipid rafts in the maintenance of synapses, dendritic spines, and surface AMPA receptor stability. J Neurosci 23:3262–3271

Howland D, Liu J, She Y, Goad B, Maragakis N, Kim B, Erickson J, Kulik J, DeVito L, Psaltis G, DeGennaro L, Cleveland D, Rothstein J (2002) Focal loss of the glutamate transporter EAAT2 in a transgenic rat model of SOD1 mutant-mediated amyotrophic lateral sclerosis (ALS). Proc Natl Acad Sci U S A 99:1604–1609

Ikegaya Y, Matsuura S, Ueno S, Baba A, Yamada MK, Nishiyama N, Matsuki N (2002) Beta-amyloid enhances glial glutamate uptake activity and attenuates synaptic efficacy. J Biol Chem 277:32180–32186

Inage YW, Itoh M, Wada K, Takashima S (1998) Expression of two glutamate transporters, GLAST and EAAT4, in the human cerebellum: their correlation in development and neonatal hypoxic-ischemic damage. J Neuropathol Exp Neurol 57:554–562

Jackson M, Song W, Liu MY, Jin L, Dykes-Hoberg M, Lin CI, Bowers WJ, Federoff HJ, Sternweis PC, Rothstein JD (2001) Modulation of the neuronal glutamate transporter EAAT4 by two interacting proteins. Nature 410:89–93

Kalandadze A, Wu Y, Robinson MB (2002) Protein kinase C activation decreases cell surface expression of the GLT-1 subtype of glutamate transporter. Requirement of a carboxyl-terminal domain and partial dependence on serine 486. J Biol Chem 277:45741–45750

Kalandadze A, Wu Y, Fournier K, Robinson MB (2004) Identification of motifs involved in endoplasmic reticulum retention-forward trafficking of the GLT-1 subtype of glutamate transporter. J Neurosci 24:5183–5192

Kim JH, Huganir RL (1999) Organization and regulation of proteins at synapses. Curr Opin Cell Biol 11:248–254

Kim SY, Choi SY, Chao W, Volsky DJ (2003) Transcriptional regulation of human excitatory amino acid transporter 1 (EAAT1): cloning of the EAAT1 promoter and characterization of its basal and inducible activity in human astrocytes. J Neurochem 87:1485–1498

Levy LM, Lehre KP, Walaas SI, Storm-Mathisen J, Danbolt NC (1995) Down-regulation of glial glutamate transporters after glutamatergic denervation in the rat brain. Eur J Neurosci 7:2036–2041

Li S, Mallory M, Alford M, Tanaka S, Masliah E (1997) Glutamate transporter alterations in Alzheimer disease are possibly associated with abnormal APP expression. J Neuropathol Exp Neurol 56:901–911

Liang Z, Valla J, Sefidvash-Hockley S, Rogers J, Li R (2002) Effects of estrogen treatment on glutamate uptake in cultured human astrocytes derived from cortex of Alzheimer's disease patients. J Neurochem 80:807–814

Lievens JC, Woodman B, Mahal A, Spasic-Boscovic O, Samuel D, Kerkerian-Le Goff L, Bates GP (2001) Impaired glutamate uptake in the R6 Huntington's disease transgenic mice. Neurobiol Dis 8:807–821

Lin CL, Bristol LA, Jin L, Dykes-Hoberg M, Crawford T, Clawson L, Rothstein JD (1998) Aberrant RNA processing in a neurodegenerative disease: the cause for absent EAAT2, a glutamate transporter, in amyotrophic lateral sclerosis. Neuron 20:589–602

Lin CLG, Orlov I, Ruggiero AM, Dykes-Hoberg M, Lee A, Jackson M, Rothstein JD (2001) Modulation of the neuronal glutamate transporter EAAC1 by the interacting protein GTRAP3-18. Nature 410:84–88

Lin G, Bristol LA, Rothstein JD (1996) An abnormal mRNA leads to downregulation of glutamate transporter EAAT2 (GLT-1) expression in amyotrophic lateral sclerosis. Ann Neurol 40:540–541

Linden DJ (1998) Synaptically evoked glutamate transport currents may be used to detect the expression of long-term potentiation in cerebellar culture. J Neurophysiol 79:3151–3156

Lundy DF, McBean GJ (1995) Pre-incubation of synaptosomes with arachidonic acid potentiates inhibition of [3H]D-aspartate transport. Eur J Pharmacol 291:273–279

Manzoni C, Mennini T (1997) Arachidonic acid inhibits 3H-glutamate uptake with different potencies in rodent central nervous system regions expressing different transporter subtypes. Pharmacol Res 35:149–151

Maragakis NJ, Rothstein JD (2004) Glutamate transporters: animal models to neurologic disease. Neurobiol Dis 15:461–473

Marie H, Attwell D (1999) C-terminal interactions modulate the affinity of GLAST glutamate transporters in salamander retinal glial cells. J Physiol 520:393–397

Marie H, Billups D, Bedford FK, Dumoulin A, Goyal RK, Longmore GD, Moss SJ, Attwell D (2002) The amino terminus of the glial glutamate transporter GLT-1 interacts with the LIM protein Ajuba. Mol Cell Neurosci 19:152–164

Martin LJ, Brambrink AM, Lehmann C, Portera-Cailliau C, Koehler R, Rothstein J, Traystman RJ (1997) Hypoxia-ischemia causes abnormalities in glutamate transporters and death of astroglia and neurons in newborn striatum. Ann Neurol 42:335–348

Masliah E, Alford M, Mallory M, Rockenstein E, Moechars D, Van Leuven F (2000) Abnormal glutamate transport function in mutant amyloid precursor protein transgenic mice. Exp Neurol 163:381–387

Meyer T, Speer A, Meyer B, Sitte W, Kuther G, Ludolph AC (1996) The glial glutamate transporter complementary DNA in patients with amyotrophic lateral sclerosis. Ann Neurol 40:456–459

Pawlak J, Brito V, Kuppers E, Beyer C (2005) Regulation of glutamate transporter GLAST and GLT-1 expression in astrocytes by estrogen. Brain Res Mol Brain Res 138:1–7

Raghavendra Rao VL, Rao AM, Dogan A, Bowen KK, Hatcher J, Rothstein JD, Dempsey RJ (2000) Glial glutamate transporter GLT-1 down-regulation precedes delayed neuronal death in gerbil hippocampus following transient global cerebral ischemia. Neurochem Int 36:531–537

Rao VL, Bowen KK, Dempsey RJ (2001) Transient focal cerebral ischemia down-regulates glutamate transporters GLT-1 and EAAC1 expression in rat brain. Neurochem Res 26:497–502

Rauen T (2000) Diversity of glutamate transporter expression and function in the mammalian retina. Amino Acids 19:53–62

Raunser S, Haase W, Bostina M, Parcej DN, Kuhlbrandt W (2005) High-yield expression, reconstitution and structure of the recombinant, fully functional glutamate transporter GLT-1 from Rattus norvegicus. J Mol Biol 351:598–613

Rothstein JD, Van Kammen M, Levey AI, Martin LJ, Kuncl RW (1995) Selective loss of glial glutamate transporter GLT-1 in amyotrophic lateral sclerosis. Ann Neurol 38:73–84

Rothstein JD, Dykes-Hoberg M, Pardo CA, Bristol LA, Jin L, Kuncl RW, Kanai Y, Hediger MA, Wang Y, Schielke JP, Welty DF (1996) Knockout of glutamate transporters reveals a major role for astroglial transport in excitotoxicity and clearance of glutamate. Neuron 16:675–686

Rothstein JD, Patel S, Regan MR, Haenggeli C, Huang YH, Bergles DE, Jin L, Dykes Hoberg M, Vidensky S, Chung DS, Toan SV, Bruijn LI, Su ZZ, Gupta P, Fisher PB (2005) Beta-Lactam antibiotics offer neuroprotection by increasing glutamate transporter expression. Nature 433:73–77

Scannevin RH, Huganir RL (2000) Postsynaptic organization and regulation of excitatory synapses. Nat Rev Neurosci 1:133–141

Schlag BD, Vondrasek JR, Munir M, Kalandadze A, Zelenaia OA, Rothstein JD, Robinson MB (1998) Regulation of the glial Na^+-dependent glutamate transporters by cyclic AMP analogs and neurons. Mol Pharmacol 53:355–369

Schmitt A, Asan E, Lesch KP, Kugler P (2002) A splice variant of glutamate transporter GLT1/EAAT2 expressed in neurons: cloning and localization in rat nervous system. Neuroscience 109:45–61

Shashidharan P, Plaitakis A (1993) Cloning and characterization of a glutamate transporter cDNA from human cerebellum. Biochim Biophys Acta 1216:161–164

Shashidharan P, Wittenberg I, Plaitakis A (1994) Molecular cloning of human brain glutamate/aspartate transporter II. Biochim Biophys Acta 1191:393–396

Siesjo BK, Agardh CD, Bengtsson F, Smith ML (1989) Arachidonic acid metabolism in seizures. Ann N Y Acad Sci 559:323–339

Simons K, Toomre D (2000) Lipid rafts and signal transduction. Nat Rev Mol Cell Biol 1:31–39

Sims KD, Straff DJ, Robinson MB (2000) Platelet-derived growth factor rapidly increases activity and cell surface expression of the EAAC1 subtype of glutamate transporter through activation of phosphatidylinositol 3-kinase. J Biol Chem 275:5228–5237

Sitcheran R, Gupta P, Fisher PB, Baldwin AS (2005) Positive and negative regulation of EAAT2 by NF-kappaB: a role for N-myc in TNFalpha-controlled repression. EMBO J 24:510–520

Slotboom DJ, Konings WN, Lolkema JS (1999) Structural features of the glutamate transporter family. Microbiol Mol Biol Rev 63:293–307

Sonders MS, Quick M, Javitch JA (2005) How did the neurotransmitter cross the bilayer? A closer view. Curr Opin Neurobiol 15:296–304

Su Z, Leszczyniecka M, Kang D, Sarkar D, Chao W, Volsky D (2003) Insights into glutamate transport regulation in human astrocytes: cloning of the promoter for excitatory amino acid transporter 2 (EAAT2). Proc Natl Acad Sci U S A 100:1955–1960

Sullivan R, Rauen T, Fischer F, Wiessner M, Grewer C, Bicho A, Pow DV (2004) Cloning, transport properties, and differential localization of two splice variants of GLT-1 in the rat CNS: Implications for CNS glutamate homeostasis. Glia 45:155–169

Susarla BT, Seal RP, Zelenaia O, Watson DJ, Wolfe JH, Amara SG, Robinson MB (2004) Differential regulation of GLAST immunoreactivity and activity by protein kinase C: evidence for modification of amino and carboxyl termini. J Neurochem 91:1151–1163

Sutherland ML, Martinowich K, Rothstein JD (2001) EAAT2 overexpression plays a neuroprotective role in the SOD1 G93A model of amyotrophic lateral sclerosis. Soc Neurosci Abstr 27:607.6

Suzuki T, Ito J, Takagi H, Saitoh F, Nawa H, Shimizu H (2001) Biochemical evidence for localization of AMPA-type glutamate receptor subunits in the dendritic raft. Brain Res Mol Brain Res 89:20–28

Swanson RA, Liu J, Miller JW, Rothstein JD, Farrell K, Stein BA, Longuemare MC (1997) Neuronal regulation of glutamate transporter subtype expression in astrocytes. J Neurosci 17:932–940

Tamahara S, Inaba M, Sato K, Matsuki N, Hikasa Y, Ono K (2002) Non-essential roles of cysteine residues in functional expression and redox regulatory pathways for canine glutamate/aspartate transporter based on mutagenic analysis. Biochem J 367:107–111

Tan J, Zelenaia O, Correale D, Rothstein JD, Robinson MB (1999) Expression of the GLT-1 subtype of Na^+-dependent glutamate transporter: pharmacological characterization and lack of regulation by protein kinase C. J Pharmacol Exp Ther 289:1600–1610

Tanaka K, Watase K, Manabe T, Yamada K, Watanabe M, Takahashi K, Iwama H, Nishikawa T, Ichihara N, Hori S, Takimoto M, Wada K (1997) Epilepsy and exacerbation of brain injury in mice lacking the glutamate transporter GLT-1. Science 276:1699–1702

Trotti D, Nussberger S, Volterra A, Hediger MA (1997a) Differential modulation of the uptake currents by redox interconversion of cysteine residues in the human neuronal glutamate transporter EAAC1. Eur J Neurosci 9:2207–2212

Trotti D, Rizzini BL, Rossi D, Haugeto O, Racagni G, Danbolt NC, Volterra A (1997b) Neuronal and glial glutamate transporters possess an SH-based redox regulatory mechanism. Eur J Neurosci 9:1236–1243

Trotti D, Danbolt NC, Volterra A (1998) Glutamate transporters are oxidant-vulnerable: a molecular link between oxidative and excitotoxic neurodegeneration? Trends Pharmacol Sci 19:328–334

Trotti D, Rolfs A, Danbolt NC, Brown RH Jr, Hediger MA (1999) SOD1 mutants linked to amyotrophic lateral sclerosis selectively inactivate a glial glutamate transporter. Nat Neurosci 2:848

Tzingounis AV, Lin CL, Rothstein JD, Kavanaugh MP (1998) Arachidonic acid activates a proton current in the rat glutamate transporter EAAT4. J Biol Chem 273:17315–17317

Vallejo-Illarramendi A, Domercq M, Matute C (2005) A novel alternative splicing form of excitatory amino acid transporter is a negative regulator of glutamate uptake. J Neurochem 95:341–348
Volterra A, Trotti D, Cassutti P, Tromba C, Salvaggio A, Melcangi RC, Racagni G (1992) High sensitivity of glutamate uptake to extracellular free arachidonic acid levels in rat cortical synaptosomes and astrocytes. J Neurochem 59:600–606
Volterra A, Trotti D, Floridi S, Racagni G (1994) Reactive oxygen species inhibit high-affinity glutamate uptake: molecular mechanism and neuropathological implications. Ann N Y Acad Sci 738:153–162
Wang Z, Li W, Mitchell CK, Carter-Dawson L (2003) Activation of protein kinase C reduces GLAST in the plasma membrane of rat Muller cells in primary culture. Vis Neurosci 20:611–619
Williams JH, Errington ML, Lynch MA, Bliss TV (1989) Arachidonic acid induces a long-term activity-dependent enhancement of synaptic transmission in the hippocampus. Nature 341:739–742
Winckler B, Mellman I (1999) Neuronal polarity: controlling the sorting and diffusion of membrane components. Neuron 23:637–640
Yamamoto Y, Gaynor RB (2004) IkappaB kinases: key regulators of the NF-kappaB pathway. Trends Biochem Sci 29:72–79
Ye ZC, Sontheimer H (1999) Glioma cells release excitotoxic concentrations of glutamate. Cancer Res 59:4383–4391
Ye ZC, Rothstein JD, Sontheimer H (1999) Compromised glutamate transport in human glioma cells: reduction-mislocalization of sodium-dependent glutamate transporters and enhanced activity of cystine-glutamate exchange. J Neurosci 19:10767–10777
Yeh TH, Hwang HM, Chen JJ, Wu T, Li AH, Wang HL (2005) Glutamate transporter function of rat hippocampal astrocytes is impaired following the global ischemia. Neurobiol Dis 18:476–483
Yernool D, Boudker O, Jin Y, Gouaux E (2004) Structure of a glutamate transporter homologue from Pyrococcus horikoshii. Nature 431:811–818
Zelenaia O, Schlag BD, Gochenauer GE, Ganel R, Song W, Beesley JS, Grinspan JB, Rothstein JD, Robinson MB (2000) Epidermal growth factor receptor agonists increase expression of glutamate transporter GLT-1 in astrocytes through pathways dependent on phosphatidylinositol 3-kinase and transcription factor NF-kappaB. Mol Pharmacol 57:667–678
Zelenaia OA, Robinson MB (2000) Degradation of glial glutamate transporter mRNAs is selectively blocked by inhibition of cellular transcription. J Neurochem 75:2252–2258
Zerangue N, Arriza JL, Amara SG, Kavanaugh MP (1995) Differential modulation of human glutamate transporter subtypes by arachidonic acid. J Biol Chem 270:6433–6435
Zhou J, Sutherland ML (2004) Glutamate transporter cluster formation in astrocytic processes regulates glutamate uptake activity. J Neurosci 24:6301–6306
Zink M, Schmitt A, Henn FA, Gass P (2004) Differential expression of glutamate transporters EAAT1 and EAAT2 in mice deficient for PACAP-type I receptor. J Neural Transm 111:1537–1542
Zschocke J, Bayatti N, Behl C (2005) Caveolin and GLT-1 gene expression is reciprocally regulated in primary astrocytes: association of GLT-1 with non-caveolar lipid rafts. Glia 49:275–287

Regulation of Vesicular Monoamine and Glutamate Transporters by Vesicle-Associated Trimeric G Proteins: New Jobs for Long-Known Signal Transduction Molecules

I. Brunk · M. Höltje · B. von Jagow · S. Winter · J. Sternberg · C. Blex · I. Pahner · G. Ahnert-Hilger (✉)

AG Funktionelle Zellbiologie, Institut für Integrative Neuroanatomie,
Centrum für Anatomie, Charité, Universitätsmedizin Berlin, Berlin, Germany
gudrun.ahnert@charite.de

1	Introduction	306
2	Secretory Vesicles and Their Neurotransmitter Transporters	306
2.1	Vesicular Monoamine Transporters	307
2.2	Vesicular Glutamate Transporters	307
3	Factors Influencing Transmitter Content of Individual Vesicles	309
4	Heterotrimeric G Proteins on Secretory Vesicles	310
5	Differences in the Regulation of VMAT and VGLUT Activities	312
5.1	The Role of the Electrochemical Gradient	312
5.2	Gαo2 and Gαq as Modulators of Vesicular Filling	314
5.2.1	G Protein-Mediated Regulation of VMAT Activity	314
5.2.2	Gαo2 Regulates VGLUT Activity by Changing the Chloride Dependence of the Transporter	317
6	Conclusions	320
	References	321

Abstract Neurotransmitters of neurons and neuroendocrine cells are concentrated first in the cytosol and then in either small synaptic vesicles of presynaptic terminals or in secretory vesicles by the activity of specific transporters of the plasma and the vesicular membrane, respectively. In the central nervous system the postsynaptic response depends—amongst other parameters—on the amount of neurotransmitter stored in a given vesicle. Neurotransmitter packets (quanta) vary over a wide range which may be also due to a regulation of vesicular neurotransmitter filling. Vesicular filling is regulated by the availability of transmitter molecules in the cytoplasm, the amount of transporter molecules and an electrochemical proton-mediated gradient over the vesicular membrane. In addition, it is modulated by vesicle-associated heterotrimeric G proteins, Gαo2 and Gαq. Gαo2 and Gαq regulate vesicular monoamine transporter (VMAT) activities in brain and platelets, respectively. Gαo2 also regulates vesicular glutamate transporter (VGLUT) activity by changing its chloride dependence. It appears that the vesicular content activates the G protein, suggesting a signal transduction from the luminal site which might be mediated by a vesicular

G protein-coupled receptor or as an alternative possibility by the transporter itself. Thus, G proteins control transmitter storage and thereby probably link the regulation of the vesicular content to intracellular signal cascades.

Keywords Vesicular transmitter transporter · G protein · Vesicular filling · Neurotransmitter · Storage

1
Introduction

Communication between neurons in the central nervous system mainly occurs at specialized areas, the synapses. Variations in the input and output at these areas are described as synaptic plasticity which is mediated by changes in the post- and presynaptic structures. Postsynaptically, the amount of receptors and ion channels regulate synaptic strength. At the presynapse, the availability and fusion competence of synaptic vesicles as well as the amount of neurotransmitter stored in individual vesicles determine the strength of the postsynaptic answer. With the advent of the molecular identification of the various vesicular neurotransmitter transporters, more attention is being paid to regulation of the neurotransmitter content. The present review will report on a new type of regulation of the vesicular filling mediated by vesicle-associated heterotrimeric G proteins.

2
Secretory Vesicles and Their Neurotransmitter Transporters

Synaptic vesicles are key organelles of neuronal and cellular communication. While a variety of proteins common to all of them regulate their transport to the specific release sites and the interaction with the plasma membrane, the various types of secretory vesicles concentrate different kinds of transmitters in their lumen. Small synaptic vesicles (SSV) in neuronal terminals and SSV analogues—also referred to as small synaptic-like microvesicles—in neuroendocrine cells store low-molecular-weight neurotransmitters. Dense-core and large dense-core vesicles occurring besides SSV in neurons and neuroendocrine cells contain various peptides as co-transmitters. Irrespective of the type of vesicle, vesicular neurotransmitter transporters concentrate low-molecular-weight neurotransmitters such as monoamines (by vesicular monoamine transporters, VMATs), acetylcholine (by vesicular acetylcholine transporters, VAChT), glutamate (by vesicular glutamate transporters, VGLUTs) or γ-aminobutyric acid (GABA) and glycine (by VGAT). The present review will focus on the regulation of the VMATs and the VGLUTs.

2.1
Vesicular Monoamine Transporters

Two structurally related but pharmacologically distinct VMATs have been cloned from PC12 cells (VMAT1, Liu et al. 1992) and from rat brain (VMAT2, Liu et al. 1992, 1994; Erickson et al. 1992). VMAT2 is the dominant transporter in brain but also occurs in a variety of peripheral cells like sympathetic neurons, enterochromaffin-like cells (Peter et al. 1995; Erickson et al. 1996) and blood platelets (Lesch et al. 1993; Höltje et al. 2003). VMAT1 appears to occur only in the periphery, at least in adult individuals, with the tissue-specific coding for one of the transporters being developmentally predetermined (Schütz et al. 1998). Both transporters accept monoamines such as serotonin, dopamine, noradrenaline and adrenaline at concentrations with micromolar K_m values for VMAT1 and submicromolar K_m values for VMAT2 in rat (Peter et al. 1994) and humans (Erickson et al. 1996). VMAT2 also transports histamine, barely recognized by VMAT1, and is 10 times (rat) (Peter et al. 1994) to 100 times (human) (Erickson et al. 1996) more sensitive to tetrabenazine. VMAT2 has a generally higher affinity for monoamines than VMAT1, a property that may be required for rapidly recycling SSV in brain compared to more slowly filling secretory granules in the adrenal medulla (Peter et al. 1994).

The vital importance of VMAT2 is highlighted by the development of VMAT2 deletion mutants by three independent groups. While homozygous mutants die postnatally, heterozygous individuals develop hypersensitivity towards amphetamine, cocaine and MPTP (1-methyl-4-phenyl-1,2,3,6-tetrahydropyridine) a substance causing parkinsonism in rodents (Takahashi et al. 1997; Wang et al. 1997; Fon et al. 1997). Especially the decreased resistance to MPTP toxicity emphasizes the protective role VMAT activity plays against oxidative stress by just removing these substances from the cytosol (Takahashi et al. 1997).

2.2
Vesicular Glutamate Transporters

Three VGLUTs, VGLUT1 (Bellocchio et al. 2000; Takamori et al. 2000), VGLUT2 (Fremeau et al. 2001; Bai et al. 2001; Takamori et al. 2001; Hayashi et al. 2001) and VGLUT3 (Gras et al. 2002; Takamori et al. 2002; Fremeau et al. 2002; Schäfer et al. 2002) have been cloned. VGLUT1 and 2 were originally described as brain-specific or differentiation-associated, Na^+-dependent, inorganic phosphate transporters, namely BNPI (Ni et al. 1994) or DNPI (Hisano et al. 2000), respectively, which appear to transport phosphate into the cytoplasm of nerve terminals (Ni et al. 1994) when integrated in the plasma membrane. They were later shown to transport glutamate into SSV with high avidity and specificity. So far, it is unknown how the decision between either function is regulated.

Generally, the three VGLUTs exhibit a higher K_m to their substrate than VMATs, being around 1–2 mM for both VGLUT1 and VGLUT2 (Bellocchio et al. 2000; Gras et al. 2002) and about 0.6 mM for VGLUT3 (Gras et al. 2002). All three VGLUTs are very specific for glutamate and do not accept other amino acids like aspartate (Takamori et al. 2000; 2001; Gras et al. 2002). The apparent affinity of glutamate to VGLUT is one to two orders of magnitude lower than to its various plasma membrane transporters. Nevertheless, the high cytoplasmic concentration of glutamate of around 100 mM guarantees a sufficient loading of vesicles with the transmitter (for review see Danbolt 2001). VGLUT1 and VGLUT2 have a distinct and mutually exclusive distribution in brain with VGLUT1 being the dominant transporter in cortex, hippocampus and cerebellum (Fremeau et al. 2001; Fujiyama et al. 2001; Kaneko and Fujiyama 2002) and VGLUT2 in thalamic and hypothalamic regions (Hisano et al. 2000; Sakata-Haga et al. 2001; Fujiyama et al. 2001; Fremeau et al. 2002; Kaneko and Fujiyama 2002). VGLUT2, in addition, is expressed in the pineal gland and in α-cells of Langerhans islets, suggesting a role in endocrine function (Hayashi et al. 2001). A strict separation between VGLUT1- and VGLUT2-containing terminals has been shown for cerebellar cortex, where parallel fibre terminals contain VGLUT1 whereas climbing fibre terminals express VGLUT2 (Fremeau et al. 2002). In contrast, VGLUT3 is found in serotonergic and cholinergic (Fremeau et al. 2002; Gras et al. 2002; Schäfer et al. 2002) as well as in GABAergic terminals (Fremeau et al. 2002), suggesting a role for glutamate as co-transmitter in these nerve terminals.

Despite this strict separation, VGLUT2 is the first transporter expressed in the nervous system and is exchanged in the respective neurons by VGLUT1 during postnatal development (Miyazaki et al. 2003). Deletion of VGLUT1, however, did not lead to a complete substitution by VGLUT2. The physiological importance of VGLUT1 is underscored by two recently published VGLUT1 knock-out models. In contrast to the VMAT2 deletion mutants, the VGLUT1 knock-outs survived for several weeks (Wojcik et al. 2004) or even months (Fremeau et al. 2004). Although the same molecule has been deleted, the two studies present different models for subcellular localization of VGLUT1 and for vesicular filling. Fremeau and colleagues concluded that VGLUT1 and VGLUT2 are segregated to separate synapses or release sites and that the loss of VGLUT1 did reduce the frequency of postsynaptic events but not the current amplitude. In their model, VGLUT1 and VGLUT2 do not appear to occur on the same vesicles, and the deletion of VGLUT1 leaves only VGLUT2 bearing vesicles to sustain glutamatergic transmission. These conclusions are contradicted by the data reported by Wojcik et al. (2004). These authors provide evidence that the amount of transporters per vesicle directly influences the amount of transmitter loaded and therefore the postsynaptic current. Using overexpression of VGLUT1 and FM1-43 labelling, they showed that vesicular filling is a reliable process and that there appears to be no checkpoint for filled or empty vesicles. However, since different experimental approaches have been used, it might well be that both interpretations are correct, depending on the developmental stage and the type of glutamatergic synapse investigated.

3
Factors Influencing Transmitter Content of Individual Vesicles

Data on the intravesicular concentrations of neurotransmitters mostly rely on the respective miniature inhibitory or excitatory postsynaptic currents, and thus are obscured by receptor affinity, activity of plasma membrane transporters, degrading enzymes and the morphology of the synaptic cleft. There are only reliable data for monoamine-storing vesicles where amperometry can be used to directly measure the release and thus the content of a single vesicle. By correlating amperometric charge and vesicular size, Bruns and colleagues showed that the variations in released quanta are due to variations in the volume of the respective secretory vesicles. On average, the transmitter concentrations in SSV and large dense-core vesicles (LDCV) are similar, being 270 mM with LDCV, while exhibiting a greater variability than SSV (Bruns et al. 2000). These data, however, were obtained with unstimulated neurons that do not have experienced changes in their environment and may have constitutively set their vesicular transmitter content (Bruns et al. 2000). Variations in quantal size have been documented at the frog neuromuscular junction, leading to the idea that vesicular filling may vary (van der Kloot 1991). These variations could be described either by a set point or by a steady-state model, suggesting that vesicles can only accept a certain amount or get more filled, respectively, if more neurotransmitter is available (Williams 1997). There is now a great body of evidence that vesicles with varying transmitter content have the same chance to fuse with the plasma membrane. In detail, this has been shown for cholinergic vesicles (van der Kloot et al. 2000) and for monoamine-storing secretory granules in VMAT2-deficient mice (Travis et al. 2000), as well as for synaptic vesicles in monoaminergic terminals of VMAT2-deletion mutants (Croft et al. 2005) and for glutamatergic vesicles (Wojcik et al. 2004). Even the distribution of synaptic vesicles and their morphology do not appear to be changed (or affected) irrespective of whether these vesicles are able to store transmitter or are designed to remain empty (van der Kloot et al. 2002; Croft et al. 2005). Overexpression of VAChT (Song et al. 1997), VMAT2 (Pothos et al. 2000) and VGLUT1 (Wojcik et al. 2004; Daniels et al. 2004) increases the stimulated release of neurotransmitters. In addition, changing the activity of synthesizing or degrading enzymes directly influenced the inhibitory postsynaptic current in GABAergic synapses without changing the overall morphology of the GABAergic vesicles (Engel et al. 2001). Tissue monoamine concentrations are increased in mice deficient for monoamine oxidase A (Cases et al. 1995). Offering the dopamine precursor L-dopa to PC12 cells (Pothos et al. 1998a) or midbrain dopaminergic neurons (Pothos et al. 1998b) increases the amount of transmitter released.

Thus, fusion competence and transmitter filling are regulated separately and vesicles can accept more transmitter if conditions permit.

Whether the transmitter concentration in the vesicles or the vesicular volume changes is not clear thus far, and controversial data have been obtained. In PC12 cells, the amount of stored transmitter appeared to be directly correlated to the volume of secretory granules (Colliver et al. 2000; Gong et al. 2003). In addition, overexpression of VGLUT in *Drosophila* is accompanied by an increase in quantal size and synaptic vesicle volume (Daniels et al. 2004). On the other hand, a 48-h incubation of leech neurons with reserpine did not change the morphology of serotonin-storing vesicles (Bruns et al. 2000), and deletion of VMAT2 did not affect the size of SSV (Croft et al. 2005). Accordingly, changes in the acetylcholine content did not change the size of secretory vesicles (van der Kloot et al. 2002). Probably, smaller vesicles such as SSV are not completely filled and may change their luminal transmitter concentration, depending on the environment. In this respect, differences between species (leech, *Drosophila*, frog, mouse) may be also relevant.

4
Heterotrimeric G Proteins on Secretory Vesicles

Heterotrimeric G proteins consisting of an α-, β- and γ-subunit are molecular switches coupling heptahelical receptors of the plasma membrane to intracellular effector systems. The 23 α-subunits, 5 β-subunits and 11 γ-subunits known so far may yield more than 1,000 combinations conferring specificity to G protein-mediated up- and downstream signalling. Generally, upon receptor activation the Gα-subunit exchanges a guanosine diphosphate (GDP) for a guanosine triphosphate (GTP) and dissociates from its βγ-subunit. The Gβγ dimers and the GTP-α complexes are presumed to regulate specific downstream effectors. So far only Gαo (Jiang et al. 1998) and Gαq (Offermanns 1999) have been associated with central nervous system defects in genetic ablation studies. Gαo-subunits are highly expressed in neuronal cells where they may reach 1%–2% of membrane protein (Sternweis and Robishaw 1984). However, effects mediated by the αo-subunits are amongst the least understood of the G protein α-subunits. This is complicated still further by the fact that there are two molecularly distinct Gαo proteins. Gαo1 and Gαo2 are the result of an alternative transcript splicing process that, after translation, produces α-subunits that differ in 25 of their 116 carboxyterminal amino acids. The carboxyterminal region of G protein α-subunits is engaged in receptor interaction and recognition in the context of the trimeric state (c.f. Grishna and Berlot 2000). Differences in receptor recognition, leading to the selective activation of either Go1 or Go2 regulating the same downstream effector system, were documented (Kleuss et al. 1991; Chen and Clarke 1996). The selective loss of retinal bipolar ON cell activation by glutamic acid from retinal photoreceptor cells in mice lacking Gαo1 (Dhingra et al. 2001), but not in mice lacking Gαo2 (Dhingra et al. 2002)—as well as the exclusive regulation of VMAT (Höltje et al. 2000) and VGLUT (Winter et al. 2005) activity by Gαo2 in brain—has provided further proof for the non-equivalence of the two Go proteins under in vivo conditions.

Apart from their localization at the plasma membrane, G proteins reside as functional heterotrimers on endomembranes such as secretory vesicles, where they also might well be activated from the lumenal side of organelles (Nürnberg and Ahnert-Hilger 1996; Ahnert-Hilger et al. 2003). Subclasses of SSV differ in their neurotransmitter transporter, which may also be reflected by different subsets of G proteins. Quantitative postembedding immunogold electron microscopy identified Gαo2-, Gαq- and Gβ-subunits, mostly Gβ2, on chromaffin granules of the rat adrenal medulla, where they colocalized with either VMAT1 or VMAT2 (Pahner et al. 2002). Since the Gβ-subunits 1–4 have to be

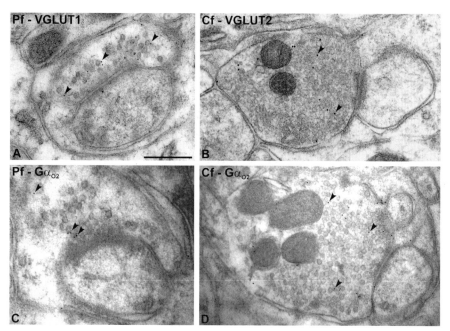

Fig. 1 A–D Gαo2-subunits on types of glutamatergic vesicles. Gαo2 immunogold signals were identified in two different types of glutamatergic terminals of the molecular layer of the rat cerebellum defined by either VGLUT1 or VGLUT2. A parallel fibre (*Pf*, **A**) and a climbing fibre profile (*Cf*, **B**) each contacting a spine of Purkinje cell dendrites but exhibiting immunoreactivity for either VGLUT1 or VGLUT2 (several immunogold particles are indicated by *arrowheads*), respectively, are shown. In both terminal types (*Pf*, **C**; *Cf*, **D**), which can be well identified according to morphological criteria (Palay and Chan-Palay 1974), vesicle-associated Gαo2 immunogold particles (several indicated by *arrowheads*) can be found. *Scale bar* represents 250 nm for **A**, 300 nm for **B** and 200 nm for **C, D**

tightly bound to a Gγ-subunit, these data indicate that chromaffin granules are equipped with functional G protein heterotrimers which regulate VMAT activity (Pahner et al. 2002). In addition, Gαo2 regulates VMAT2 activity in brain and is also seen on VMAT2-containing SSV in serotonergic terminals (Höltje et al. 2000). Synaptic terminals defined by the expression of either VGLUTs or VGAT contain different sets of G protein heterotrimers, and even glutamatergic vesicles differ in their G protein subunit profile, depending on the VGLUT subtype. While cerebellar parallel fibre terminals and Schaffer collateral terminals identified by the presence of VGLUT1 contain Gβ2, cerebellar climbing fibre terminals identified by the presence of VGLUT2 associate with another Gβ-subunits, Gβ1, 3 or 4 (Pahner et al. 2003). In both types of glutamatergic—as well as in GABAergic—terminals, equal amounts of Gγ7 were found to be associated with SSV (Pahner et al. 2003). SSV of GABAergic basket cell terminals exhibit less Gαo2 compared to the two glutamatergic terminals and also less Gβ2 compared to parallel fibre terminals (Pahner et al. 2003). In Fig. 1 an example of Gαo2 association to either VGLUT1 (parallel fibre, Fig. 1A, C) or VGLUT2 (climbing fiber, Fig. 1B, D) expressing glutamatergic terminals is given. The presence of G protein heterotrimers, α- and β-subunits, especially αo-subunits, is also confirmed by recent proteomic studies using either highly purified SSV (R. Jahn, personal communication) or clathrin-coated vesicles (Blondeau et al. 2004).

Together these data strongly indicate the presence of functional G protein heterotrimers on SSV, where they appear to be permanent constituents and not only transient residents. As judged from electron microscope (EM) studies, all transmitter transporter-expressing vesicle populations are associated with G proteins. Most notably, SSV of different types of glutamatergic and of GABAergic terminals differ in their respective G protein subunit combinations.

5
Differences in the Regulation of VMAT and VGLUT Activities

All vesicular neurotransmitter transporters have in common that their activity depends on the electrochemical gradient $\Delta\mu H^+$. In addition, at least for VMATs and VGLUTs, transporter activity is regulated by either Gαo2 with Gαq—depending on tissue—or by Gαo2 alone, respectively. Besides these general similarities, the specific parameters of regulation differ considerably between VMATs and VGLUTs.

5.1
The Role of the Electrochemical Gradient

The role of the electrochemical gradient ($\Delta\mu H^+$) over the vesicular membrane for monoamine uptake has been extensively described (Johnson 1988). The gradient consists of a proton gradient (ΔH^+) due to the activity of the vacuolar H^+-ATPase, resulting in an intravesicular pH of about 5.6 and an electrical gradient ($\Delta\psi$) that—due to the increase of positive charges in the vesicle lumen—results in a positive membrane potential towards the cytosol. Physiologically, ΔH^+ and $\Delta\psi$ drive as $\Delta\mu H^+$ the vesicular transmitter uptake. In SSV the differences in the requirements for either component of the electrochemical gradient optimal for GABA, glutamate or dopamine uptake have been realized before the molecular identification of the respective transporters (Hell et al. 1988; 1990; Maycox et al. 1990). VMAT and VAChT activity strictly depends on ΔH^+—with less influence of $\Delta\psi$—while VGLUTs depend more on $\Delta\psi$, and VGAT depends on both (Johnson 1988; Hell et al. 1990; Maycox et al. 1990; Schuldiner et al. 1995; Reimer et al. 1998). Additionally, synaptic vesicles are equipped with chloride channels which contribute to their acidification. Opening of vesicular chloride channels increases vesicular chloride concentration, thereby increasing ΔH^+ while decreasing $\Delta\psi$. A decrease in chloride intake increases $\Delta\psi$ and decreases ΔH^+ due to reduced proton accumulation, since under this condition the positive charges in the vesicular lumen are not neutralized by chloride ions (Maycox et al. 1990; Reimer et al. 2001). Genetic deletion of ClC3, a chloride channel expressed on synaptic vesicles, reduces acidification and ΔH^+ and consequently increases vesicular glutamate content (Stobrawa et al. 2001).

While VMATs transport one transmitter molecule in exchange for two protons (Johnson 1988; Schuldiner et al. 1995; Reimer et al. 2001), the situation is different for VGLUTs. Unlike positively charged monoamines, glutamate is negatively charged, which requires charge compensation by other ions. At low chloride, glutamate uptake is driven by $\Delta\psi$ (Hell et al. 1990; Maycox et al. 1990) and charge neutrality is maintained by proton co-transport. This results in a glutamate-dependent acidification of the vesicle interior. However, glutamate may not be taken up and stored as free glutamic acid under physiological conditions. Rather, charge balance may be maintained by chloride efflux through vesicular chloride channels (Maycox et al. 1990). Since recycling vesicles are likely to contain a high intravesicular chloride concentration, glutamate uptake would thus involve a net glutamate-chloride exchange. At high chloride, however, i.e. when ΔpH is high, glutamate uptake is coupled directly to proton exchange (Tabb et al. 1992; Wolosker et al. 1996). Under these conditions, transport is driven by ΔH^+. Low amounts of chloride between 4-6 mM increase transport activity in a manner that is independent of the chloride effects on the balance between $\Delta\psi$ and ΔH^+ and probably involves a direct activation of the transport system (Naito and Ueda 1985; Hartinger and Jahn 1993; Wolosker et al. 1996; Winter et al. 2005, see below). A large $\Delta\psi$ apparently increases the affinity of VGLUTs towards their substrate, but a large ΔH^+ positively influences the trapping of glutamate inside the vesicle (Wolosker et al. 1996). At a given number of transporter molecules, the transmitter concentration inside the vesicle thus depends on the availability of the transmitter and—depending on the respective transporter—on ΔH^+ and $\Delta\psi$.

Besides the difference of VMAT and VGLUT in relying more on ΔH^+ and $\Delta\psi$, respectively, monoamines and glutamate differ in their general properties, requiring a different handling by the respective neurons. Cytosolic concentrations of monoamines are toxic and kept low, giving a biological reason for the high affinity of VMATs for their substrates. Therefore, the highest gradient for transmitters is sustained by VMATs, which concentrate monoamines by a factor of 10^4 or 10^5. In resting LDCV and SSV of the leech synapse, an average serotonin concentration of 270 mM was measured (Bruns et al. 2000). In chromaffin granules, additional intravesicular components, i.e. the chromogranins, add to an effective trapping, yielding a catecholamine concentration over 1 M (Johnson 1988; Schuldiner et al. 1995). In contrast, extracellular glutamate is toxic, explaining the relatively high affinity of plasma membrane transporters for glutamate compared to the low affinity of the VGLUTs, while its cytosolic concentrations may reach up to 40 mM. VGLUTs concentrate the transmitters only tenfold or less than the cytoplasmic concentration, yielding an intravesicular concentration of not more than 100 mM (Danbolt 2001).

5.2
Gαo2 and Gαq as Modulators of Vesicular Filling

In addition to the above-mentioned regulators of transmitter storage in glutamatergic and monoaminergic vesicles, there is increasing evidence that vesicle-associated G proteins directly influence and fine tune the transmitter content by regulating the respective transporter activities.

5.2.1
G Protein-Mediated Regulation of VMAT Activity

The first indication for a regulation of VMAT activity was obtained in PC12 cells showing that Gαo2 downregulates VMAT1 activity (Ahnert-Hilger et al. 1998). This was confirmed in the human pancreatic carcinoid cell line BON, which exhibits neuroendocrine properties (Höltje et al. 2000), and in isolated chromaffin granules of the rat adrenal medulla (Pahner et al. 2002); both also express VMAT2 to a minor extent. VMAT2 is the predominant transporter in brain (Erickson et al. 1996). VMAT2 activity is also inhibited by the non-hydrolysable GTP-analogue guanylylimidodiphosphate GMP-P(NH)P, a G protein activator, or by activated Gαo2 applied to permeabilized raphe neurons in primary culture (Fig. 2A; Höltje et al. 2000) and when using crude synaptic vesicles (Höltje et al. 2000). Finally, the use of vesicular preparations from Gαo2 deletion mutants clearly indicates that Gαo2 is the regulator of VMAT activity in brain (Fig. 2B). Thus, both VMAT activities are regulated by Gαo2 irrespective of whether the transporters reside on large dense-core or small synaptic vesicles.

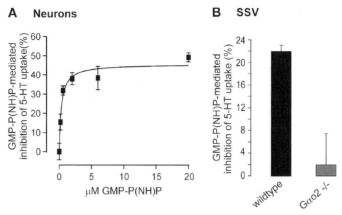

Fig. 2 A,B G protein-mediated inhibition of serotonin uptake into streptolysin O (SLO)-permeabilized raphe neurons or small synaptic vesicles. **A** Raphe neurons were permeabilized by SLO and subjected to serotonin uptake in the absence or presence of the indicated concentrations of GMP-P(NH)P (*abscissa*) as given (Höltje et al. 2000). Values (mean of three individual culture dishes) were corrected for unspecific binding obtained in the presence of reserpine and are expressed as GMP-P(NH)P-mediated inhibition (%). **B** Synaptic vesicle obtained from wildtype or $G\alpha o2^{-/-}$ mice were subjected to serotonin uptake in the absence or presence of 50 μM GMP-P(NH)P (Höltje et al. 2003). Values (mean of three samples) were corrected for unspecific binding obtained in the presence of reserpine and are expressed as GMP-P(NH)P-mediated inhibition (%)

5.2.1.1
Gαo2 or Gαq Regulate VMAT2 Activity Depending on Tissue

In addition to its occurrence in brain, VMAT2 is also expressed in platelets. Platelets are the major storage sites of serotonin apart from the CNS. They do not synthesize serotonin but take it up during their passage through the capillary network of the gut villi, where it is provided by enterochromaffin cells. Serotonin is transported by the plasma membrane transporter (SERT) and packaged into secretory vesicles by VMAT2 activity (Höltje et al. 2003). Platelets do not express Gαo2 but instead use mainly Gαq for their signal transduction. VMAT2 activity in platelets is also downregulated by the G protein activator GMP-P(NH)P irrespective of the species (Höltje et al. 2003; Fig. 3A). Using a Gαq deletion mutant we found that in platelets VMAT2 is regulated by Gαq. Gαq-mediated regulation of VMAT2 appears to be restricted to platelets since VMAT2 activity in brain-derived SSV was inhibited by Gαo2 and not affected in SSV from $G\alpha q^{-/-}$ mice (Höltje et al. 2003; Fig. 3B, C; see also Fig. 2B). Taken together, VMAT2 activity is regulated by different G proteins, depending on the respective tissue. Whether such promiscuity also applies to VMAT1 or other vesicular transmitter transporters is not known so far.

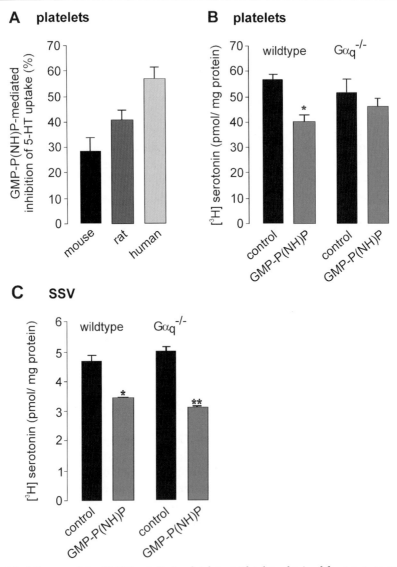

Fig. 3 A–C Gαq regulates VMAT activity in platelets. **A** Platelets obtained from mouse, rat or human were permeabilized with SLO, and serotonin uptake was performed in the absence or presence of 50 μM GMP-P(NH)P as given (Höltje et al. 2003). Values (mean of three samples) were corrected for unspecific binding obtained in the presence of reserpine and are expressed as GMP-P(NH)P-mediated inhibition (%). **B** Platelets obtained from wildtype or Gαq$^{-/-}$ mice were subjected to serotonin uptake in the absence or presence of GMP-P(NH)P. From Höltje et al. (2003), with permission. **C** Synaptic vesicle obtained from wildtype or Gαq$^{-/-}$ mice (Höltje et al. 2003) were subjected to serotonin uptake in the absence or presence of 50 μM GMP-P(NH)P. Values (mean of three samples) were corrected for unspecific binding obtained in the presence of reserpine. Note that Gαq regulates VMAT2 activity in platelets but not in brain

5.2.1.2
VMAT Activity Is Regulated by the Vesicular Content

A variety of experimental approaches indicate that secretory vesicles can accept more neurotransmitter if available (see Sect. 3). While these manipulations increase the transmitter content over a given amount, the opposite model characterized by empty vesicles is more difficult to find.

Using a genetic model characterized by the depletion of Tph1, a peripheral form of tryptophan hydroxylase (Walther et al. 2003)—resulting in platelets that are fully equipped for serotonin storage and release but contain only minute amounts of serotonin—we provide evidence that vesicular content is crucial for regulation of transmitter uptake. Naive platelets of Tph1$^{-/-}$ mice exhibited no G protein-mediated inhibition of serotonin uptake. Preloading of Tph1$^{-/-}$ platelets with serotonin or noradrenaline fully reconstituted the G protein-mediated inhibition (Höltje et al. 2003; Fig. 4A). This is as one would expect for a general regulation of VMAT, which transports all monoamines with an almost similar K_m. In addition, depleting intravesicular monoamine stores by treating BON cells or raphe neurons with the SERT inhibitor fluoxetine reduced the subsequent G protein-mediated regulation of vesicular serotonin uptake (Fig. 4B). These data indicate that signal transduction may start from the luminal site of the vesicle. Although we do not know the "receptor" working from the luminal site that senses the transmitter content, VMAT itself by one of its intravesicular loops may be a putative candidate (Ahnert-Hilger et al. 2003).

5.2.2
Gαo2 Regulates VGLUT Activity by Changing the Chloride Dependence of the Transporter

Besides VMAT, VGLUT is also regulated by Gαo2. GMP-P(NH)P downregulates glutamate uptake into SSV from rat brain (Pahner et al. 2003; Winter et al. 2005). Using mutant mice lacking various Gα-subunits including Gαo1, Gαo2, Gαq and Gα11, we found that VGLUTs are exclusively regulated by Gαo2 (Winter et al. 2005; Fig. 5). In addition, application of a Gαo2-specific monoclonal antibody prevented GMP-P(NH)P-mediated inhibition of VGLUT regulation in SSV obtained from either rat or mice brain, providing further evidence that VGLUTs are regulated by Gαo2 (Winter et al. 2005). Gαo2 exerts its action by specifically affecting the chloride dependence of VGLUTs. As mentioned earlier, VGLUTs show maximal activity at around 5 mM chloride. While at this chloride concentration, G protein activation results in a downregulation of VGLUT activity, VGLUT activity is increased following G protein activation when lower chloride concentrations are applied (Fig. 6A). Thus, activated Gαo2 shifts the maximum for chloride activation of VGLUT to lower chloride concentrations. In contrast, glutamate uptake by vesicles isolated from Gαo2$^{-/-}$

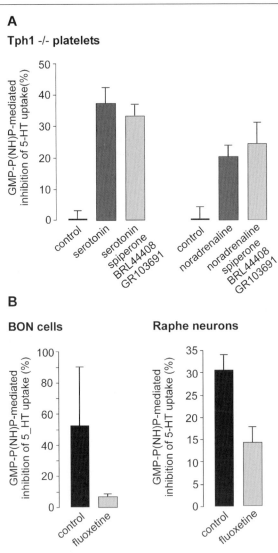

Fig. 4 A,B Vesicular filling is required for G protein-mediated regulation of VMAT activity. **A** Serotonin-diminished platelets obtained from Tph1$^{-/-}$ mice (Walther et al. 2003) were subjected to serotonin uptake in the absence or presence of GMP-P(NH)P following no treatment (*control*) or preloading with either serotonin or noradrenaline in the absence or presence of monoamine receptor inhibitors. Values were corrected for unspecific binding obtained in the presence of reserpine and are expressed as GMP-P(NH)P-mediated inhibition (%). From Höltje et al. (2003), with permission. **B** Bon cells or Raphe neurons were pretreated with 20 µM fluoxetine a selective blocker of the serotonin plasma membrane transporter for 24 h to diminish intracellular serotonin content before they were subjected to serotonin uptake in the absence or presence of GMP-P(NH)P. As in panel A, values were corrected for unspecific binding obtained in the presence of reserpine and are expressed as GMP-P(NH)P-mediated inhibition (%)

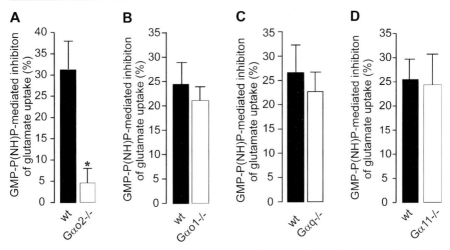

Fig. 5 Gαo2 regulates VGLUT activity. Glutamate uptake was performed into synaptic vesicles in the absence or presence of GMP-P(NH)P obtained from either wildtype (*wt*) or from the indicated G protein deletion mutants: Gαo2 A, Gαo1 B, Gαq C or Gα11 D. Values are the mean of two or three individual experiments and expressed as GMP-P(NH)P-mediated inhibition. From Winter et al. (2005), with permission

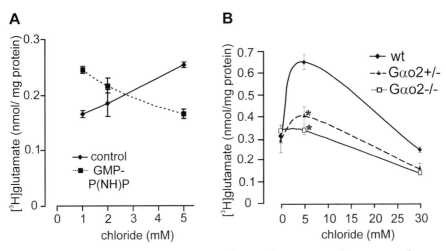

Fig. 6 A,B Gαo2 regulates the chloride dependence, of VGLUT. **A** Glutamate uptake was performed in synaptic vesicles in the absence or presence of GMP-P(NH)P using the chloride concentrations given at the *abscissa*. **B** Chloride dependence of VGLUT activity is reduced or diminished into SSV from heterozygous or knock-out animals, respectively. From Winter et al. (2005), with permission

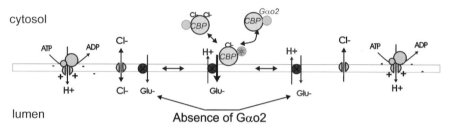

Fig. 7 Model describing the regulation of VGLUT activity by Gao2. For details see text

mice is no longer activated by chloride and is dramatically decreased in SSV obtained from heterozygous littermates (Winter et al. 2005; Fig. 6B). Thus, Gαo2 acts upon a putative regulatory chloride-binding domain that appears to modulate transport activity of VGLUT.

The data are best explained by the hypothetical presence of a separate chloride binding factor (CBF) that binds chloride and is controlled by Gαo2. In the absence of chloride, CBF is inactive and VGLUT is in a symport mode, with transport being exclusively driven by $\Delta\psi$. At low chloride, CBF is activated and in turn binds and thus activates VGLUT shifting it in a putative, so-called effective mode. Furthermore, there may already be a partial shift from the symport to the antiport mode under these conditions, with ΔH^+ contributing to the driving force (Tabb et al. 1992; Wolosker et al. 1996). Higher concentrations of chloride lead to a dissociation of CBF, with the transporter gradually shifting from the symport into the antiport mode. In this scenario, chloride regulation does not influence the transition between symport and antiport. However, it cannot be excluded that, in addition to activating VGLUTs, it also influences the symport–antiport equilibrium (Schuldiner et al. 1995). Activation of Gαo2 increases the chloride affinity of CBF, shifting the peak of binding and VGLUT activation to lower chloride concentrations. In Gαo2$^{-/-}$ mice, CBF is lost probably because Gαo2 is required for its stable expression or its protection from degradation. As a consequence, VGLUT cannot reach the putative effective mode (Fig. 7).

6
Conclusions

G protein activation may be a mechanism by which transmitter transporters are regulated, allowing the presynaptic terminal to modulate vesicular transmitter content. In brain, Gαo2 is the only Gα-subunit involved in the regulation of vesicular transmitter transporter activity that we know of. This activity is not shared by the closely related splice variant of Gαo1, underscoring the functional divergence of these two Go-proteins in neurons.

While G protein-mediated regulation of VMAT activity appears to be linked to the vesicular transmitter content, the situation is more complex for glutamatergic vesicles, as the efficacy of G protein regulation is linked to the cytoplasmic chloride concentration. If there is signalling from the vesicle lumen, the trigger could be either the vesicular glutamate or the vesicular proton concentration. In this respect, it is remarkable that proton-sensing G protein-coupled receptors have been recently identified which work best at pH 6.8 and are inhibited by pH over 7.8 or beneath 5.8 (Ludwig et al. 2003). Furthermore, it cannot be excluded that Gαo2 is regulated by cytosolic factors, thus linking it to intracellular signalling pathways as a means to regulate synaptic efficacy.

References

Ahnert-Hilger G, Nürnberg B, Exner T, Schäfer T, Jahn R (1998) The heterotrimeric G protein Go2 regulates catecholamine uptake by secretory vesicles. EMBO J 17:406–413

Ahnert-Hilger G, Höltje M, Pahner I, Winter S, Brunk I (2003) Regulation of vesicular transmitter transporter. Rev Physiol Biochem Pharmacol 150:140–160

Bai L, Xu H, Collins JF, Ghishan FK (2001) Molecular and functional analysis of a novel neuronal vesicular glutamate transporter. J Biol Chem 276:36764–36769

Bellocchio EE, Reimer RJ, Fremeau jr RT, Edwards RH (2000) Uptake of glutamate into synaptic vesicle by an inorganic phosphate transporter. Science 289:957–960

Blondeau F, Ritter B, Allaire PD, Wasiak S, Girard M, Hussain NK, Angers A, Legendre-Guillemin V, Roy L, Boismenu D, Kearney RE, Bell AW, Bergeron JJM, McPherson PS (2004) Tandem MS analysis of brain clathrin-coated vesicles reveals their critical involvement in synaptic vesicle recycling. Proc Natl Acad Sci USA 101:3833–3838

Bruns D, Riedel D, Klingauf J, Jahn R (2000) Quantal release of serotonin. Neuron 28:205–220

Cases O, Seif I, Grimsby J, Gaspar P, Chen K, Pournin S, Muller U, Aguet M, Babinet C, Chen Shih J, De Maeyer E (1995) Aggressive bahaviour and altered amounts of brain serotonin and norepinephrine in mice lacking MAOA. Science 208:1763–1766

Chen C, Clarke IJ (1996) Go-2 protein mediates the reduction in Ca^{2+} current by somatostatin in cultured ovine somatotrophs. J Physiol 491:21–29

Colliver TL, Pyott SJ, Achalabun M, Ewing AG (2000) VMAT-mediated changes in quantal size and vesicular volume. J Neurosci 20:5276–5282

Croft BG, Fortin GD, Corera AT, Edwards RH, Beaudet A, Trudeau L-E, Fon EA (2005) Normal biogenesis and cycling of empty synaptic vesicles in dopamine neurons of vesicular monoamine transporter 2 knockout mice. Mol Biol Cell 16:306–315

Danbolt NC (2001) Glutamate uptake. Prog Neurobiol 65:1–105

Daniels RW, Collins CA, Gelfand MV, Dant J, Brooks ES, Krantz DE, DiAntonio A (2004) Increased expression of the Drosophila vesicular glutamate transporter leads to excess glutamate release and a compensatory decrease in quantal content. J Neurosci 24:10466–10474

Dhingra A, Lyubarsky A, Jiang M, Pugh EN, Birnbaumer L, Sterling P, Vardi N (2001) The light responses of ON bipolar neurons require Gαo. J Neurosci 20:9053–9058

Dhingra A, Jiang M, Wang T-L, Lyubarsky A, Savchenko A, Bar-Yehuda T, Sterling P, Birnbaumer L, Vardi N (2002) Light response of retinal ON bipolar cells requires a specific splice variant of Gαo. J Neurosci 22:4878–4884

Engel D, Pahner I, Schulze K, Frahm C, Jarry H, Ahnert-Hilger G, Draguhn A (2001) Plasticity of central inhibitory synapses through GABA metabolism. J Physiol (Lond) 535:473–485

Erickson JD, Eiden LE, Hoffman BJ (1992) Expression cloning of a reserpine sensitive vesicular monoamine transporter. Proc Natl Acad Sci USA 89:10993–10997

Erickson JD, Schäfer MK, Bonner TI, Eiden LE, Weihe E (1996) Distinct pharmacological properties and distribution in neurons and endocrine cells of two isoforms of the human vesicular monoamine transporter. Proc Natl Acad Sci USA 93:5166–5171

Fon EA, Pothos EN, Sun B-C, Killeen N, Sulzer D, Edwards RH (1997) Vesicular transport regulates monoamine storage and release but is not essential for amphetamine action. Neuron 19:1271–1283

Fremeau jr RT, Matthew DT, Pahner I, Nygaard GO, Tran CH, Reimer RJ, Bellocchio EE, Fortin D, Storm-Mathisen J, Edwards RH (2001) The expression of vesicular glutamate transporters defines two classes of excitatory synapse. Neuron 31:247–260

Fremeau RT, Burman J, Qureshi T, Tran CH, Proctor J, Johnson J, Zhang H, Sulzer D, Copenhagen DR, Storm-Mathisen J, Reimer RJ, Chaudhry FH, Edwards RH (2002) The identification of vesicular glutamate transporter 3 suggests novel modes of signaling by glutamate. Proc Natl Acad Sci USA 99:14488–14493

Fremeau RT, Kam K, Qureshi T, Johnson J, Copenhagen DR, Storm-Mathisen J, Chaudry FA, Nicoll RA, Edwards RH (2004) Vesicular glutamate transporters 1 and 2 target to functionally distinct synaptic release sites. Science 304:1815–1819

Fujiyama F, Furuta T, Kaneko T (2001) Immunocytochemical localization of candidates for vesicular glutamate transporters the rat cerebral cortex. J Comp Neurol 435:379–387

Gong LW, Hafez I, Alvarez de Toledo G, Lindau M (2003) Secretory vesicles membrane area is regulated in tandem with quantal size in chromaffin cells. J Neurosci 23:7017–7921

Gras C, Herzog E, Bellenchi GC, Bernard V, Ravassard P, Pohl M, Gasnier B, Giros B, El Mestikawy S (2002) A third vesicular glutamate transporter expressed by cholinergic and serotoninergic neurons. J Neurosci 22:5442–5451

Grishina G, Berlot CH (2000) A surface exposed region of Gsα in which substitutions decrease receptor mediated activation and increase receptor affinity. Mol Pharmacol 57:1081–1092

Hartinger J, Jahn R (1993) An anion binding site that regulates the glutamate transporter of synaptic vesicles. J Biol Chem 268:23122–23127

Hayashi M, Otsuka M, Morimoto R, Hitota S, Yatsushiro S, Takeda J, Yamamoto A, Moriyama Y (2001) Differentiation-associated Na^+-dependent inorganic phosphate cotransporter (DNPI) is a vesicular glutamate transporter in endocrine glutamatergic systems. J Biol Chem 276:43400–43406

Hell JW, Maycox PR, Stadler H, Jahn R (1988) Uptake of GABA by rat brain synaptic vesicles isolated by a new procedure. EMBO J 7:3023–3029

Hell JW, Maycox PR, Jahn R (1990) Energy dependence and functional reconstitution of the γ-aminobutyric acid carrier from synaptic vesicles. J Biol Chem 265:2111–2117

Hisano S, Hoshi K, Ikeda Y, Maruyama D, Kanemoto M, Ichijo J, Kojima I, Takeda J, Nogami H (2000) Regional expression of a gene encoding a neuron-specific Na^+-dependent inorganic phosphate cotransporter (DNPI) in the rat forebrain. Mol Brain Res 83:34–43

Höltje M, von Jagow B, Pahner I, Lautenschlager M, Hörtnagl H, Nürnberg B, Jahn R, Ahnert-Hilger G (2000) The neuronal monoamine transporter VMAT2 is regulated by the trimeric GTPase Go2. J Neurosci 20:2131–2141

Höltje M, Winter S, Walther D, Pahner I, Hörtnagl H, Ottersen OP, Bader M, Ahnert-Hilger G (2003) The vesicular monoamine content regulates VMAT2 activity through Gαq in mouse platelets. Evidence for autoregulation of vesicular transmitter uptake. J Biol Chem 278:15850–15858

Jiang M, Gold MS, Boulay G, Spicher K, Pexton M, Brabet P, Srinivasan Y, Rudolph U, Ellison G, Birnbaumer L (1998) Multiple neurological abnormalities in mice deficient in the G protein Go. Proc Natl Acad Sci USA 95:3269–3274

Johnson jr RG (1988) Accumulation of biological amines into chromaffin granules: a model for hormone and neurotransmitter transport. Physiol Rev 68:232–307

Kaneko T, Fujiyama F (2002) Complementary distribution of vesicular glutamate transporter in the central nervous system. Neurosci Res 42:243–250

Kleuss C, Hescheler J, Ewel C, Rosenthal W, Schultz G, Wittig B (1991) Assignment of G-protein subtypes to specific receptors inducing inhibition of calcium currents. Nature 353:43–48

Lesch KP, Gross J, Wolozin BL, Murphy DL, Riederer P (1993) Extensive sequence divergence between the human and rat brain vesicular monoamine transporter: possible molecular basis for species differences in the susceptibility to MPP^+. J Neural Transm 93:75–82

Liu Y, Peter A, Roghani A, Schuldiner S, Prive GG, Eisenberg D, Brecha N, Edwards RH (1992) A cDNA that suppresses MPP^+ toxicity encodes a vesicular amine transporter. Cell 70:539–551

Liu Y, Schweitzer E, Nirenberg MJ, Pickel VM, Evans CJ, Edwards RH (1994) Preferential localization of a vesicular monoamine transporter to dense core vesicles in PC 12 cells. J Cell Biol 127:1419–1433

Ludwig M-G, Vanek M, Guerini D, Gasser JA, Jones CE, Junker U, Hofstetter H, Wolf RM, Seuwen K (2003) Proton-sensing G-protein-coupled receptors. Nature 425:93–98

Maycox PR, Hell JW, Jahn R (1990) Amino acid neurotransmission: spotlight on synaptic vesicles. Trends Neurosci 13:83–87

Miyazaki T, Fukaya M, Shimizu H, Watanabe M (2003) Subtype switching of vesicular glutamate transporters at parallel fibre-Purkinje cell synapses in developing mouse cerebellum. Eur J Neurosci 17:2563–2572

Naito S, Ueda T (1985) Characterization of glutamate uptake into synaptic vesicles. J Neurochem 44:99–109

Ni B, Rostock jr PR, Nadi NS, Paul SM (1994) Cloning and expression of a cDNA encoding a brain-specific Na^+-dependent inorganic phosphate cotransporter. Proc Natl Acad Sci USA 91:5607–5611

Nürnberg B, Ahnert-Hilger G (1996) Potential roles of heterotrimeric G proteins of the endomembrane system. FEBS Lett 389:61–65

Offermanns S (1999) New insights into the in vivo function of the heterotrimeric G-protein through gene deletion studies. Naunyn Schmiedebergs Arch Pharmacol 360:5–13

Pahner I, Höltje M, Winter S, Nürnberg B, Ottersen OP, Ahnert-Hilger G (2002) Subunit composition and functional properties of G-protein heterotrimers on rat chromaffin granules. Eur J Cell Biol 81:449–456

Pahner I, Höltje M, Winter S, Takamori S, Bellocchio EE, Spicher K, Laake P, Nürnberg B, Ottersen OP, Ahnert-Hilger G (2003) Functional G-protein heterotrimers are associated with vesicles of putative glutamatergic terminals: implications for regulation of transmitter uptake. Mol Cell Neurosci 23:398–413

Palay LS, Chan-Palay VC (1974) Cerebellar cortex. Cytology and organization. Springer-Verlag, Berlin Heidelberg New York

Peter D, Liu Y, Sternini C, de Giorgio R, Brecha N, Edwards RH (1995) Differential expression of two vesicular monoamine transporters. J Neurosci 15:6179–6188

Peter D Jimenez J, Liu Y, Kim J, Edwards RH (1994) The chromaffin granule and synaptic vesicle amine transporters differ in substrate recognition and sensitivity to inhibitors. J Biol Chem 269:7231–7237

Pothos EN, Przedborski S, Davila V, Schmitz Y, Sulzer D (1998a) D2-like dopamine receptor reduces quantal size in PC12 cells. J Neurosci 18:5575–5585

Pothos EN, Davila V, Sulzer D (1998b) Presynaptic recording of quanta from midbrain dopamine neurons and modulation of quantal size. J Neurosci 18:4106–4118

Pothos EN, Larsen KE, Krantz DE, Liu Y-j, Haycock JW, Setlik W, Gershon MD, Edwards RH, Sulzer D (2000) Synaptic vesicle transporter expression regulates vesicle phenotype and quantal size. J Neurosci 20:7297–7306

Reimer RJ, Fon EA, Edwards RH (1998) Vesicular neurotransmitter transport and the presynaptic regulation of quantal size. Curr Opin Neurobiol 8:405–412

Reimer RJ, Fremeau jr RT, Bellocchio EE, Edwards RH (2001) The essence of excitation. Curr Opin Cell Biol 13:417–421

Sakata-Haga H, Kanemoto M, Maruyama D, Hoshi K, Mogi K, Narita M, Okada N, Ikeda Y, Nogami H, Fukui Y, Kojima I, Takeda J, Hisano S (2001) Differential localization and colocalization of two neuron-types of sodium-dependent inorganic phosphate cotransporters in rat forebrain. Brain Res 902:142–155

Schäfer MKH, Varoqui H, Defamie N, Weihe E, Erickson JD (2002) Molecular cloning and functional identification of mouse vesicular glutamate transporter 3 and its expression in subsets of novel excitatory neurons. J Biol Chem 277:50734–50748

Schuldiner S, Shirvan A, Linial M (1995) Vesicular neurotransmitter transporters: from bacteria to humans. Physiol Rev 75:369–392

Schütz B, Schäfer MK, Eiden LE, Weihe E (1998) Vesicular amine transporter expression and isoforms selection in developing brain, peripheral nervous system and gut. Brain Res Dev Brain Res 106:181–204

Song H-j, Ming G-l, Fon E, Bellocchio E, Edwards E, Poo M-m (1997) Expression of a putative vesicular acetylcholine transporter facilitates quantal transmitter release. Neuron 18:815–826

Sternweis PC, Robishaw JD (1984) Isolation of two proteins with high affinity for guanine nucleotides from membranes of bovine brain. J Biol Chem 259:13806–13813

Stobrawa SM, Breiderhoff T, Takamori S, Engel D, Schweizer M, Zdebik AA, Bösl MR, Ruether K, Jahn H, Draguhn A, Jahn R, Jentsch TJ (2001) Disruption of ClC-3, a chloride channel expressed on synaptic vesicles, leads to a loss of the hippocampus. Neuron 29:185–196

Tabb JS, Kish PE, van Dyke R, Ueda T (1992) Glutamate transport into synaptic vesicles. Roles of membrane potential, pH gradient, and intravesicular pH. J Biol Chem 22:15412–15418

Takahashi N, Miner LI, Sora I, Ujike H, revay RS, Kostic V, Jackson-lewis V, Przedborski S, Uhl GR (1997) VMAT2 knockout mice: heterozygotes display reduced amphetamine-conditioned reward, enhanced amphetamine locomotion, and enhanced MPTP toxicity. Proc Natl Acad Sci USA 94:9938–9943

Takamori S, Rhee JS, Rosenmund C, Jahn R (2000) Identification of a vesicular glutamate transporter that defines a glutamatergic phenotype in neurons. Nature 407:189–194

Takamori S, Rhee JS, Rosenmund C, Jahn R (2001) Identification of differentiation-associated brain-specific phosphate transporter as a second vesicular glutamate transporter (VGLUT2). J Neurosci 21:RC182

Takamori S, Malherbe P, Broger C, Jahn R (2002) Molecular cloning and functional characterization of human vesicular glutamate transporter 3. EMBO Rep 3:798–803

Travis ER, Wang Y-M, Michael DJ, Caron MG, Wightman RM (2000) Differential quantal release of histamine and 5-hydroxytryptamine from mast cells of vesicular monoamine transporter 2 knock out mice. Proc Natl Acad Sci USA 97:162–167

Van der Kloot W (1991) The regulation of quantal size. Prog Neurobiol 36:93–103

Van der Kloot W, Colasante C, Cameron R, Malgó J (2000) Recycling and refilling of transmitter quanta at the frog neuromuscular junction. J Physiol 523:247–258

Van der Kloot W, Malgo J, Cameron R, Colasante C (2002) Vesicle size and transmitter release at the frog neuromuscular junction when quantal acetylcholine content is increased or decreased. J Physiol 541:385–389

Walther DJ, Peter JU, Bashammakh S, Hörtnagl H, Voits M, Fink H, Bader M (2003) Synthesis of serotonin by a second tryptophan hydroxylase isoform. Science 299:76

Wang Y-M, Gainetdinov PR, Fumagalli F, Xu F, Jones SR, Bock CB, Miller GW, Wightman RM, Caron MG (1997) Knockout of the vesicular monoamine transporter 2 gene results in neonatal death and supersensitivity to cocaine and amphetamine. Neuron 19:1285–1296

Williams J (1997) How does a vesicle know it is full? Neuron 18:683–686

Winter S, Brunk I, Walther DJ, Höltje M, Jiang M, Peter J-U, Takamori S, Jahn R, Birnbaumer L, Ahnert-Hilger G (2005) Gαo2 regulates VGLUT activity by changing its chloride dependence. J Neurosci 25:4672–4680

Wojcik SM, Rhee SE, Herzog E, Sigler A, Jahn R, Takamori S, Brose N, Rosenmund C (2004) An essential role for vesicular glutamate transporter 1 (VGLUT1) in postnatal development and control of quantal size. Proc Natl Acad Sci USA 101:7158–7163

Wolosker H, de Souza DO, de Meis L (1996) Regulation of glutamate transport into vesicles by chloride and proton gradient. J Biol Chem 271:11726–11731

Human Genetics and Pharmacology of Neurotransmitter Transporters

Z. Lin · B. K. Madras (✉)

Department of Psychiatry, Harvard Medical School, Division of Neurochemistry, New England Primate Research Center, 1 Pine Hill Drive, Southborough MA, 01772-9102, USA
bertha_madras@hms.harvard.edu

1	The Human Dopamine Transporter (*SLC6A3*)	328
1.1	Introduction	328
1.1.1	DAT Regulation and Adaptation	329
1.1.2	Transient Effects	330
1.1.3	DAT Regulation and the DAT Gene	330
1.2	DAT Genomic Variants	331
1.2.1	Introduction	331
1.2.2	Coding Region Variants	333
1.2.3	Non-coding Region Variants	333
1.3	DAT Protein Variants	334
1.4	Variations in DAT Expression Levels	339
1.5	DAT Association with Human Diseases	340
2	The Human Serotonin Transporter (*SLC6A4*)	343
2.1	Introduction	343
2.2	SERT Genomic Variants	344
2.2.1	Introduction	344
2.2.2	Coding Region Variants	344
2.2.3	Non-coding Variants	344
2.3	SERT Protein Variants	347
2.4	Variations in SERT Expression Levels	348
2.5	SERT Association with Human Diseases	350
3	The Human Norepinephrine Transporter (*SLC6A2*)	351
3.1	Introduction	351
3.2	NET Genomic VARIANTS	352
3.2.1	Non-coding Region Variants	352
3.2.2	Coding Region Variants	354
3.3	NET Protein Variants	354
3.4	Variations in hNET Expression Levels	355
3.5	NET Association with Human Diseases	356
4	Transporter Gene Knockout Mice: Implications	356
5	Summary	357
References		357

Abstract Biogenic amine neurotransmitters are released from nerve terminals and activate pre- and postsynaptic receptors. Released neurotransmitters are sequestered by transporters into presynaptic neurons, a major mode of their inactivation in the brain. Genetic studies of human biogenic amine transporter genes, including the dopamine transporter (hDAT; *SLC6A3*), the serotonin transporter (hSERT; *SLC6A4*), and the norepinephrine transporter (hNET; *SLC6A2*) have provided insight into how genomic variations in these transporter genes influence pharmacology and brain physiology. Genetic variants can influence transporter function by various mechanisms, including substrate affinities, transport velocity, transporter expression levels (density), extracellular membrane expression, trafficking and turnover, and neurotransmitter release. It is increasingly apparent that genetic variants of monoamine transporters also contribute to individual differences in behavior and neuropsychiatric disorders. This chapter summarizes current knowledge of transporters with a focus on genomic variations, expression variations, pharmacology of protein variants, and known association with human diseases.

Keywords Dopamine transporter · Serotonin transporter · Norepinephrine transporter · Single nucleotide polymorphisms · Transporter pharmacology

1
The Human Dopamine Transporter (*SLC6A3*)

1.1
Introduction

The neurotransmitter dopamine (DA) is implicated in a wide range of physiological processes (e.g., movement, cognition, memory, and reward) and pathophysiogical states. Deficits of brain DA are directly associated with motor impairment in Parkinson's disease and DA replacement therapy is the most effective treatment strategy for this disease. Conversely, D2 DA receptor antagonists are effective therapies for alleviating specific symptoms of schizophrenia, suggestive of excess DA neurotransmission in this psychiatric disorder. The human DA transporter (hDAT) is also a major target of therapeutic drugs and illicit drugs of abuse/addiction. DAT inhibitors (e.g., methylphenidate) or DAT substrates (e.g., amphetamine) effectively elevate extracellular DA levels and are widely used to treat attention deficit hyperactivity disorder (ADHD) and narcolepsy. A relatively weak DAT inhibitor bupropion is used therapeutically to treat depression in a subpopulation and to treat nicotine addiction in a wider cohort. Psychostimulant drugs of abuse, including cocaine, amphetamine, methamphetamine, and 3,4-methylene-dioxymethamphetamine (MDMA), target the DAT, albeit not exclusively (Madras et al. 1989; Verrico et al., 2005).

Cell bodies that produce DA are highly circumscribed in the substantia nigra, the ventral tegmental area and discrete regions of the hypothalamus, with terminals fanning out to the caudate nucleus, putamen nucleus accumbens, and frontal cortical regions. The DAT limits synaptic activity and diffusion mediated by DA, sequestering extracellular DA into neurons. The DAT is present on cell bodies, dendrites and axons, but apparently is not localized in the immediate active zone of the synapse (Hersch et al. 1997; Nirenberg et al. 1996). Accordingly, the DAT diminishes DA overflow into perisynaptic regions, but may not affect DA levels within the synapse. The DAT is expressed in all DA neurons, with expression levels high in neurons originating in the substantia nigra and ventral tegmental area (Ciliax et al. 1995) and projecting to the striatum, nucleus accumbens, prefrontal cortex, and hypothalamus. Importantly, relative concentrations of DA, DAT, and DA receptors are consistent in the caudate-putamen, nucleus accumbens, and substantia nigra, with the DAT likely to regulate DA signaling strength and duration in these brain regions. The ratio of DAT to DA receptor expression levels is lower in other brain regions (De La Garza and Madras 2000), a mismatch that could result in DA clearance by metabolism, diffusion, or by another transporter. The frontal cortex expresses much lower levels of the DAT and DA autoreceptors, stores less DA, and relies more on DA synthesis than on vesicular recycling for DA release. Consequently, DAT-mediated DA regulation in the striatum cannot be liberally extrapolated to frontal cortex.

The critical role of the DAT in regulating DA neurotransmission and presynaptic homeostasis in the basal ganglia is clearly apparent in mice with null mutations of the DAT. DAT-mutant mice display robust phenotypic changes in behavior (hyperactivity, impaired care by females for their offspring), appearance (small size, skeletal abnormalities), physiological function (pituitary hypoplasia, sleep dysregulation), and brain function (cognitive and sensorimotor gating deficits), in comparison with non-mutant mice. Adaptations in the striatum of the mutant mice include a reduction in the DA-synthesizing enzyme tyrosine hydroxylase, vesicular DA stores, stimulated DA release, and D_1 and D_2 (but not D_3) DA receptor densities and function (review: Gainetdinov and Caron 2003).

At the molecular level, the 12-membrane-spanning DAT protein contains a large extracellular loop with consensus sites for glycosylation that function to regulate DAT trafficking and stability (Li et al. 2004). Potential phosphorylation sites (serine, threonine, and tyrosine) may also acutely modulate DAT trafficking and activity (review: Mortensen and Amara 2003). DA is released from both dendrites and axons and may activate receptors locally or remotely through volume transmission. DAT generates three types of ion channel-like conductances (Ingram et al. 2002; Sonders et al. 1997). In the substantia nigra, substrate transport by the DAT initiates an excitatory DAT-mediated current, cell depolarization, and consequent augmentation of somatodendritic DA release (Falkenburger et al. 2001; Ingram et al. 2002). This control of DA release is region-specific, as in the striatum; DA clearance is a primary function of the DAT.

1.1.1
DAT Regulation and Adaptation

Regulation of DAT density and function is relevant to physiological and pathophysiological processes in brain. Physiological turnover of the DAT protein in rodent brain is approximately 2–3 days (Kimmel et al. 2000), but the DAT is also dynamically regulated. Ion gradients, phosphorylation activity, and other agents can modify DAT activity transiently or for extended periods, even after the agents are cleared from DAT (Gulley and Zahniser 2003; Mortensen and Amara 2003). Adaptive changes are reflected in density, activity, and cellular localization.

1.1.2
Transient Effects

The DAT can be regulated acutely by physiological, pharmacological, and activity-dependent mechanisms. Under normal conditions, the DAT is trafficked constitutively from the cell membrane to the intracellular milieu and then degraded or recycled to the cell surface (Daniels and Amara 1999; Loder and Melikian 2003). Activation of protein kinase C (PKC) pathways by phorbol esters reduces DA transport capacity and the number of binding sites on the cell surface (for reviews: Gulley and Zahniser 2003; Mortensen and Amara 2003). Acute exposure to substrates or inhibitors can also rapidly and reversibly modify DAT function. By promoting DAT trafficking (to the cell interior) and internalization, amphetamine acutely reduces DA transport capacity in rodent brain or cultured cells within 1 h of administration (Fleckenstein et al. 1999; Saunders et al. 2000). The DAT inhibitor cocaine, on the other hand, promotes upregulation of DAT density, function, DAT mobilization to the cell surface, and increased cell surface expression (Daws et al. 2002; Little et al. 2002). DA itself regulates the DAT, directly by downregulating surface expression, or indirectly by upregulating the DAT via D_2 DA autoreceptors (Gulley and Zahniser 2003). The DAT can act as a single unit, self-associate, or bind to other proteins. Oligomers of the DAT exist but again, their functional significance is not known (Sorkina et al. 2003; Torres et al. 2003).

1.1.3
DAT Regulation and the DAT Gene

DAT density may be altered by genetic abnormalities, but currently there are no candidate polymorphisms that correlate with DAT density. The 3'-untranslated region of the DAT gene has a fixed length repeat sequence that varies by number of repeats from 3–11 (Vandenbergh et al. 1992). Although the 10-repeat sequence is the most common form of the gene, inheritance of this repeat length in both alleles is associated with ADHD in some studies, but accounts for less than 4% of the variance (Cook et al. 1995; Madras et al. 2002; Waldman et al. 1998). The repeat number in this region of the gene and association with DAT density is controversial. One study demonstrated a higher DAT density with the 10/10 repeat genotype (Heinz et al. 2000); another study found a lower density in the 10/10 than the 9/10 repeat (Jacobsen et al. 2000);

a third study found no difference in DAT density (Martinez et al. 2001); and a fourth study found a higher DAT in the 9/9 repeat (van Dyck et al. 2005).

1.2
DAT Genomic Variants

1.2.1
Introduction

The large hDAT gene (51.7 kb) displays a high density of polymorphisms (>7/kb). It is estimated that more than 400 single nucleotide polymorphisms (SNPs) are located in this gene (51.7 kb from exons 1 to 15). More than 100 SNPs located in the 5′ region (16 kb from the upstream gene *FLJ12443* to exon 1) have been deposited into the National Center for Biotechnology Information (NCBI) dbSNP (Build 124). Less than 50 of the polymorphisms, including approximately 19 SNPs and one variable number of tandem repeat (VNTR) in the exons, have been documented in the literature (Greenwood et al. 2001; Greenwood and Kelsoe 2003; Grunhage et al. 2000; Vandenbergh et al. 1992, 2000) and most of them have been included in the dbSNP. This information suggests that *SLC6A3* is subject to extensive variations among individuals, and some of these variations perhaps confer risk factors for *SLC6A3*-related diseases. Listed in Fig. 1A are polymorphisms that have been used either in association studies and analyzed for effects on gene expression or cause missense mutations (indicated in the parentheses) in the hDAT protein. rs2975226, located immediately upstream of exon 1, is a T/A SNP and has been used in two association studies (see Sect. 1.5).

1.2.2
Coding Region Variants

The coding region of DAT is highly conserved. Re-sequencing of 551 Caucasian individuals for the coding region revealed 15 exon SNPs including 4 missense mutations, none of which have frequency exceeding 1% (Table 1; Grunhage et al. 2000; Vandenbergh et al. 2002). This information suggests that wildtype function of hDAT is required for human survival.

1.2.3
Non-coding Region Variants

The 3′ VNTR (40 bp repeats with 92% identity located from 2741 to 3144 in exon 15) contains at least 7–11 repeats of 40 bp, but the 9- and 10-repeats are the most frequent alleles (Doucette-Stamm et al. 1995; Nakatome et al. 1995). This 3′ VNTR has been the most popular genetic marker of this gene for more than 100 association studies during the last decade. Besides the 3′ VNTR, there are two additional islands of imperfect tandem repeats downstream. One is

2.6 repeats of 82 bp (90% identity located 475 bp downstream of the 3′ VNTR or from 3619 to 3830) and designated as 3′ TR2 here. Another is 5 repeats of 38 bp (75% identity located 486 bp downstream of 3′ VNTR or from 3630 to 3830) and designated 3′ TR3 here. 3′ TR2 and 3′ TR3 overlap each other for most of the regions. There is no information on whether 3′ TR2 and 3′ TR3 are variable.

Linkage disequilibrium (LD) between 5′ and 3′-ends of *SLC6A3* is low. The 5′ SNP rs11564749 (P+215 in Greenwood et al. 2001) displays D′ of less than 0.3 when pairing with eight different SNPs from exons 9 and 15, including the seven synonymous SNPs listed in Fig. 1A, based on analysis of hundreds of Caucasian chromosomes of northern European descent (Greenwood et al. 2002). There are two blocks of high LD, from exon 1 to intron 6 (5′ block) and from exons 9 to 15 (3′ block, see Fig. 1A; Greenwood et al. 2002). The HapMap database (http://www.hapmap.org/index.html.en, Release 16c.1, 15 June 2005) shows that LD between rs2617605 in intron 2 (not shown in Fig. 1A)

Fig. 1 A–C Variations of the human DAT (**A**, *SLC6A3*), serotonin transporter (SERT) (**B**, *SLC6A4*), and norepinephrine transporter (NET) (**C**, *SLC6A2*) genes, which are located on chromosomes 5p15.3, 17q11.1, and 16q12.2, respectively. *Horizontal gray lines* are chromosomal regions harboring the genes; the *crossing vertical bars* are exons (*black* for translated and *gray* for untranslated regions), and between exons are introns as *numbered underneath*. *Vertical lines* indicate the locations of SNPs, and wherever possible, dbSNPs (*rs*) are used; *vertical arrows* indicated genetic markers (*horizontal arrows*) of simple sequence repeats such as VNTR (*SLC6A3*, STin2 VNTR of *SLC6A4*) and indel (5-HTTLPR of *SLC6A4* and NETpPR of *SLC6A2*). For those that are not rs, coordinates of mRNAs (NM_001044 an entry on 2 May 2005 for *SLC6A3*, NM_001045 entered on 10 June 2005 for *SLC6A4* and AF061198 as reported in Kim et al. 1999 for *SLC6A2*) are used. *Parentheses* indicate amino acid variations caused by non-synonymous SNPs. *Bold* labels indicate markers (or haplotypes in *gray* of panel *a*) that displayed positive association with human diseases and *underlined* are variants that have been analyzed for transporter activity. *Italics* indicate these listed in the dbSNP database (Build 124) but not in the literature. Note for *SLC6A3*: two blocks of high LD are indicated as 5′ and 3′ blocks, based on LD analysis (Greenwood et al. 2002). Notes for *SLC6A2*: [a] originally reported as T-182C in (Zill et al. 2002); [b] originally reported as 1287G/A in (Stober et al. 1996); [c] originally reported as an intron 8 SNP by (Ono et al. 2003), but re-designated here because of the discovery of new exon 1 (Kim et al. 1999); three indicated blocks are haplotypes conserved in different populations, and block 1 may extends to 10.5 kb downstream of exon 15 (Belfer et al. 2004). Variation information came from the following references: Greenwood et al. (2001), Greenwood and Kelsoe (2003), Grunhage et al. (2000), and Vandenbergh et al. (2000) for *SLC6A3*; Battersby et al. (1996 and 1999), Delbruck et al. (1997 and 2001), Di Bella et al. (1996), Heils et al. (1996), Ogilvie et al. (1996), and Ozaki et al. (2003) for *SLC6A4*; and Hahn et al. (2005), Halushka et al. (1999), Iwasa et al. (2001), Ono et al. (2003), Shannon et al. (2000), Stober et al. (1996 and 1999), and Zill et al. (2002) for *SLC6A2*; and the NCBI databases (dbSNP Build 124 and MapView Build 35.1). At the *bottom* of each panel is the D′ value for LD in American Caucasians, based on HapMap (http://www.hapmap.org/index.html.en). See text for details on the SNPs used to calculate the D′ values. *Scale bar*: 2 kb

and rs27072 in exon 15 display D' = 0.375 for 59 Caucasians (CEPH, Utah residents with ancestry from North and Western Europe), 0.436 for 45 Chinese (Han Chinese Beijing), 0.18 for 43 Japanese (Tokyo), and 0.092 for 60 Africans (Yoruba Ibaden, Nigeria). The 3' block is diminished to some extent in the non-Caucasian populations. These data are consistent with the idea that there is very low LD between the 5' promoter region and 3' VNTR and that DAT gene variations differ significantly among different populations.

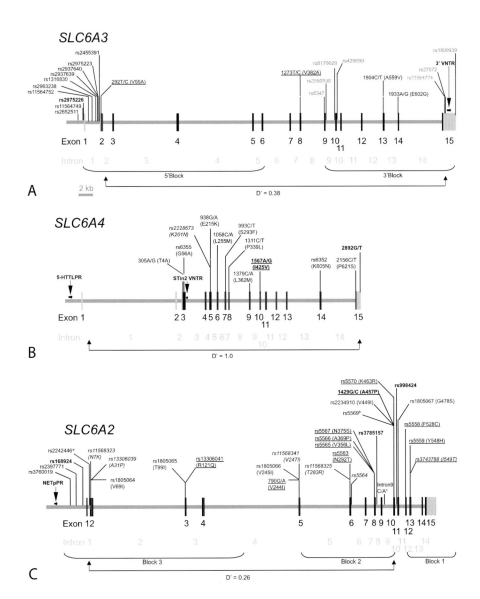

Table 1 Frequency (%) of human neurotransmitter transporter variants

hDAT	hSERT		hNET	
V55A 1%[a]	T4A 0.1%[c]	P621S 0.1%[c]	N7 K 0.2%[d]	V356L 4.3%[f]
V382A* 1%[a]	G56A 1.7%[d]		A31P ?	A369P** 2.3%[f]
A559V 0.5%[b]	K201N 15.7%[d]		V69I 0.3%[i]	N375S 4.3%[f]
E602G 0.5%[b]	E215K 0.1%[c]		T99I 1.8%[i]	V449I 0.07%[i]
	L255M 0.1%[h]		R121Q* 0.5%[e]	A457P** ~1%[g]
	S293F 0.1%[c]		V244I ?	K463R 10.0%[d]
	P339L 0.1%[c]		V245I 0.43%[i]	G478S 0.07%[i]
	L362M 0.1%[c]		V247I 0.2%[d]	F528C 5.5%[d]
	I425V#0.1%[c]		T283R 0.4%[d]	Y548H 4.7%[d]
	K605N 0.1%[c]		N292T* 4.3%[d]	I549T 1.4%[d]

*Uptake activity reduced by 37%–50%, **Uptake activity reduced by approximately 99% (Hahn et al. 2003, 2005; Lin and Uhl 2003), #Gain-of-function (Kilic et al. 2003; Runkel et al. 2000); see text for details, [a]Vandenbergh et al. 2000, [b]Grunhage et al. 2000, [c]Glatt et al. 2001, [d]dbSNP, [e]Iwasa et al. 2001, [f]Halushka et al. 1999 and dbSNP (Build 124), [g]Shannon et al. 2000, [h]Di Bella et al. 1996, [i]Runkel et al. 2000

1.3
DAT Protein Variants

To date, only 4 hDAT protein variants have been reported, including V55A, V382A, A559V, and E602G. The NCBI dbSNP database (Build 124) has not shown any additional protein variants yet. V55A and V382A have been stud-

Fig. 2 Conservation of variable amino acid residues of hDAT, hSERT, and hNET as demonstrated by alignment of 19 transporter amino acid sequences. *Black* and *gray* highlight the conservation of human transporter residues: *black* and *bold* with *clear background* for hDAT, *black with gray background* for hSERT, and *white with black background* for hNET; the corresponding variations are indicated on *top* of alignment. Transmembrane domains (*TMs*) are covered by *horizontal lines*. Prefix for transporter names: *h*, human; *r*, rat; *m*, mouse; *b*, bovine; *bf*, bullfrog (*Rana catesbeiana*); *dm*, *Drosophila melanogaster*; *dmcs*, *Drosophila melanogaster* Canton S; *ce*, Caenorhabditis elegans; *ms*, Manduca sexta. DAT, dopamine transporter; SERT, serotonin transporter; NET, norepinephrine transporter; ET, L-epinephrine transporter. Sequence sources as GenBank accession No.: hDAT, L24178; rDAT, M80570; mDAT, AF109072; bDAT, M80234; hNET, M65105; rNET, AB021971; mNET, U70306; bNET, U09198; fET, U72877; hSERT, L05568; rSERT, M79450; msSERT, AAN59781; bSERT, AF119122; dmSERT, U02296; dmcsSERT, U04809; ceDAT, AF079899; dmDAT, AAF76882; ceSERT, AAK84832; and msSERT, AAN59781. The last few C-terminal residues are irrelevant and not shown here

Human Genetics and Pharmacology of Neurotransmitter Transporters

Sequence alignment of neurotransmitter transporters showing conserved regions including TM1, TM2, and TM3, with highlighted residues T4A, G56A, V55A, N7K, A31P, T99I, V69I, and R121Q across hDAT, rDAT, mDAT, bDAT, ceDAT, dmDAT, hSERT, rSERT, mSERT, bSERT, dmSERT, dmcsSERT, ceSERT, nsSERT, hNET, rNET, mNET, bNET, and bfET sequences.

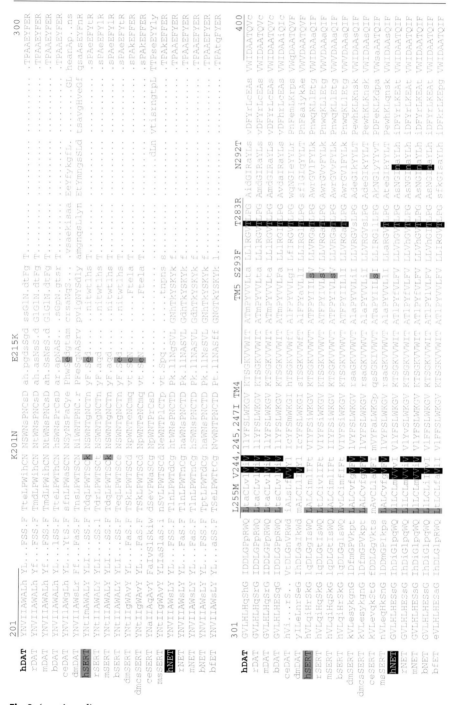

Fig. 2 (continued)

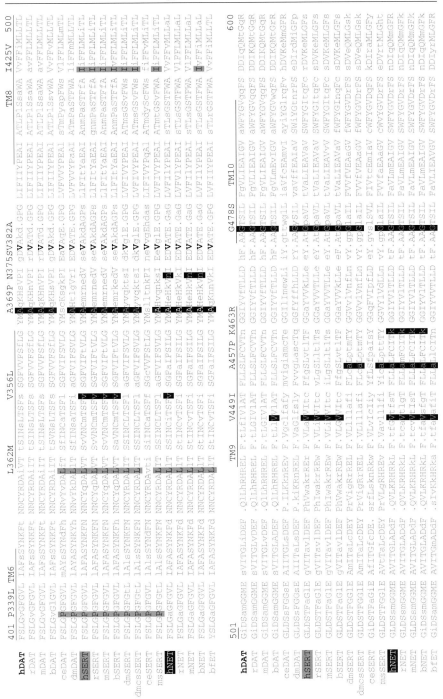

Fig. 2 (continued)

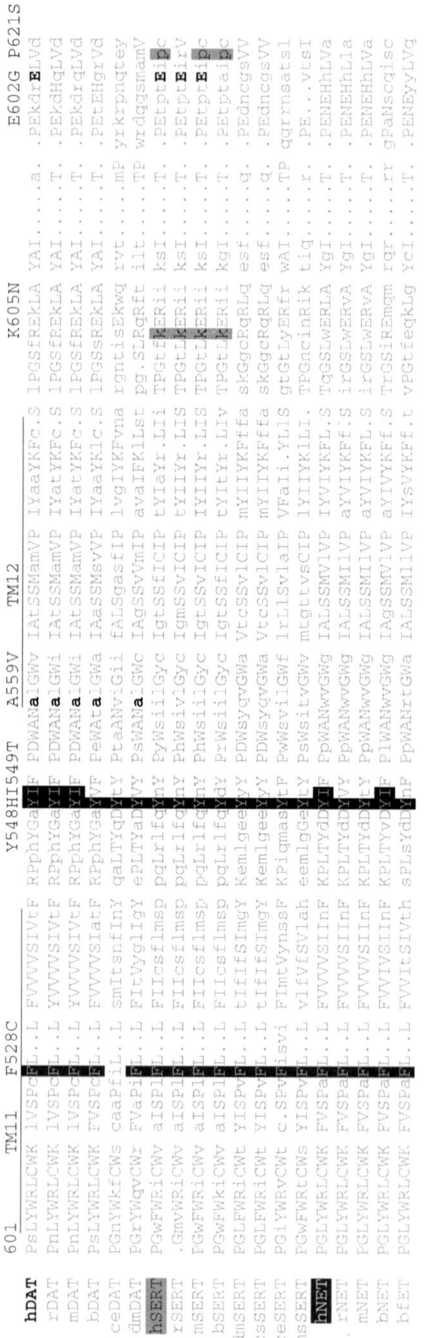

Fig. 2 (continued)

ied for pharmacological activities. The N-terminally located V55 is conserved in DATs from human, rat, and mouse but not from bovine, *Caenorhabditis elegans*, or *Drosophila*. This V55 is not present in any serotonin transporters (SERTs) or norepinephrine transporters (NETs) either (Fig. 2). Such modest conservation indicates lack of essentiality of V55 in DAT function. Consistently, functional analysis showed that V55A displays normal in vitro expression and pharmacological profiles in regards to DA uptake V_{max}, affinity for DA, cocaine, mazindol, and *d*-amphetamine in inhibition of [^{3}H]DA uptake experiments; the exception was that the DA uptake affinity decreased by 1.7-fold (Lin and Uhl 2003). By contrast, the extracellular loop 4 (ECL4)-located V382 is conserved in all of DATs, SERTs, and NETs cloned from seven different species so far, suggesting that V382 plays an important role in transport (Fig. 2). In vitro expression assays showed that Ala substitution for V382 disrupts plasma membrane targeting of DAT in COS-7 cells so that the DA uptake V_{max} and cocaine analog [^{3}H]2-β-carbomethoxy-3-β-(4-fluorophenyl) tropane (CFT) binding B_{max} values are reduced by 50% (uptake and binding affinities remain normal). Importantly, this disrupted expression is dominant to the wildtype protein expression. Besides the influence on plasma membrane expression levels, mutation V382A decreases affinity by 6.3-fold for DA in inhibiting [^{3}H]CFT binding but increases affinity by 1.3- to 1.6-fold for cocaine, benztropine, and GBR12909 in inhibition of [^{3}H]DA uptake experiments (Lin and Uhl, 2003; Lin et al. 1999).

A559V and E602G have not been examined yet. A559 is located at the boundary between extracellular space and transmembrane segment (TM)12 and is present in hDAT, rat (r)DAT, mouse (m)DAT and *Drosophila melanogaster* (dm)DAT but not in *Caenorhabditis elegans* (ce)DAT or any of the SERTs and NETs. At this position, a Val residue is present in ceDAT and rSERT. The C-terminally located E602 is present in only the hDAT out of the six DATs listed in Fig. 2. At this position, Gly is present in bovine (b)DAT and the dmSERTs. It is unlikely that E602G displays disrupted expression. Overall, the residues at these two positions are conserved for none of the SERTs and NETs either. It is reasonable to expect normal activity of both A559V and E602G.

1.4
Variations in DAT Expression Levels

hDAT expression levels differ significantly from individual to individual. Postmortem studies have shown that the differences in [^{3}H]WIN 35428 binding to hDAT could be as large as fourfold among different individuals, suggesting that hDAT expression differs significantly from individual to individual (Little et al. 1999). The question becomes whether *SLC6A3* markers are associated with such differences. It has been an interesting subject for a long time as to whether the 3′ VNTR, the most popular *SLC6A3* genetic marker, is correlated with hDAT expression levels in human brain. In vitro studies have shown that SV40 promoter-driven luciferase gene (luc^+) tagged with the 9-repeat allele at 3'-end expressed Luc activity 33% higher than that tagged with the 10-repeat allele in vitro (Miller and Madras 2002). The correlation between 3' VNTR and hDAT availability in human striatum using single-photon emission computed tomography (SPECT) imaging technology has been investigated. In 30 healthy subjects, individuals carrying the 9/10 repeat alleles displayed higher DAT availability than those carrying the 10/10 repeat alleles, which reached statistical significance (F = 9.43, df = 1, 24, p = 0.005) (Jacobsen et al. 2000). Five years later, they reported data from 96 healthy subjects suggesting that 9 repeat carriers (n = 41, 49.8 +/− 19.5 years) had a mean striatal DAT availability 8.9% higher than the 10 repeat homozygotes (n = 53, 49.9 +/− 19.2 years), which again reached statistical significance [F(1.93) = 6.25, p = 0.014] (van Dyck et al. 2005). These well-designed in vitro and in vivo studies have clearly shown the 10-repeat allele is associated with downregulated hDAT expression. Studies of smaller sample sizes in different populations have shown different results, reflecting the fact that hDAT availability in the human striatum varies significantly from individual to individual (Contin et al. 2004; Heinz et al. 2000; Martinez et al. 2001; Mill et al. 2002). These studies show clearly that hDAT availability declines significantly during aging. However, there is no evidence for individual differences in the declining rate.

Haplotypes conferred by the nine 5' SNPs (upstream of exon 2, see Fig. 1A) have been analyzed for promoter activity in vitro. A luciferase reporter construct containing haplotype TTTAAAGAC displayed promoter activity 1.4-fold higher than one containing CAGCGGAGT. Including the haplotypes conferred by the selected seven 3' SNPs (from intron 8 to exon 15, Fig. 1A) into the 5' reporter constructs increased promoter activity by approximately twofold (Greenwood and Kelsoe 2003).

1.5
DAT Association with Human Diseases

The DAT gene is associated with more than ten human diseases. The majority of clinical genotyping of hDAT is focused on ADHD, primarily because ADHD has a high genetic load and the DAT is a principal target of antihyperactivity medications. Indeed, among the 16 association studies using 3' VNTR as a DAT marker, 10 studies have demonstrated a positive association with ADHD, in Caucasians of American, Irish, British, and Canadian nationality and in Chinese populations (Table 2). A recent meta-analysis summarized genetic studies spanning 14 years (1991–2004) and found a positive association between the DAT gene and ADHD (Bobb et al. 2005b). A more recent SPECT study showed that reduced DAT availability in midbrain is found in ADHD patients, continuing to support this positive association (Jucaite et al. 2005). In vitro, in vivo imaging, and genetic studies, including DAT knockdown an-

Table 2 *SLC6A3* association with human diseases

Marker	Diseases	Association (No. of studies)		Reference(s)
3′ VNTR	ADHD	Yes	(10)	A
		No	(6)	B
		Yes by meta-analysis		C
	Poor response to Ritalin treatment	Yes	(3)	D
		No	(0)	
	Alcoholism	Yes	(10)	E
		No	(2)	F
	Illegal drug abuse	Yes	(3)	G
		no	(1)	H
	Smoking	Yes	(6)	I
		No	(1)	J
3′ Haplotypes[a]	Bipolar disorder	Yes	(1)	K
		No	(0)	
5′ SNP rs2975226	Schizophrenia	Yes	(1)	L
		No	(0)	
	Bipolar disorder	Yes	(1)	M
		No	(0)	

[a]Inferred from the seven SNPs (in gray) located from intron 8 to exon 15 in Fig. 1a, [b]Using rs27072 which is close to 3′ VNTR (Fig. 1a), A: Barr et al. 2001; Chen et al. 2003; Cook et al. 1995; Gill et al. 1997; Maher et al. 2002; Oh et al. 2003; Payton et al. 2001b; Qian et al. 2003; Rowe et al. 2001; Waldman et al. 1998, B: Curran et al. 2001; Holmes et al. 2000; Muglia et al. 2002; Payton et al. 2001a; Roman et al. 2001; Todd et al. 2001, C: Bobb et al. 2005b, D: Kirley et al. 2003; Ling et al. 2004; Roman et al. 2002; Winsberg and Comings 1999, E: Dobashi et al. 1997; Galeeva et al. 2001; Gorwood et al. 2003; Limosin et al. 2004; Muramatsu and Higuchi 1995; Sander et al. 1997; Schmidt et al. 1998; Ueno et al. 1999; Wernicke et al. 2002, F: Chen et al. 2001; Franke et al. 1999, G: Blum et al. 1997; Gelernter et al. 1994; Ujike et al. 2003, H: Hong et al. 2003, I: Audrain-McGovern et al. 2004; Erblich et al. 2004; Lerman et al. 1999; Ling et al. 2004; Vandenbergh et al. 2002, J: Jorm et al. 2000, K: Greenwood et al. 2001, L: Khodayari et al. 2004, M: Keikhaee et al. 2005

imal studies, have almost reached a consensus that hDAT expression levels in the midbrain/striatum may confer a risk factor for ADHD (Fischman and Madras 2005; Madras et al. 2005; Miller and Madras 2002; Zhuang et al. 2001). Furthermore, the 3′ VNTR 10-repeat carriers, who have lower DAT expression levels and high risk for ADHD, are prone to be resistant to methylphenidate treatments as shown by three studies in U.S., Brazil, and Ireland (Kirley et al. 2003; Roman et al. 2002; Winsberg and Comings 1999). In addition, the DAT gene appears to be associated with Tourette's syndrome in U.S. and German Caucasians (Comings et al. 1996; Muller-Vahl et al. 2000; Rowe et al. 1998).

The DAT has also been implicated in behavioral components of alcohol, cigarette, cocaine, and methamphetamine abuse (Table 2). The frequencies of positively associated 3′ VNTR 10-, 9-, and 7-repeat alleles are 5:8:2 (Blum et al. 1997; Dobashi et al. 1997; Erblich et al. 2004; Galeeva et al. 2001; Gelernter et al. 1994; Gorwood et al. 2003; Lerman et al. 2003; Muramatsu and Higuchi 1995; Sander et al. 1997; Schmidt et al. 1998; Ueno et al. 1999; Ujike et al. 2003; Vandenbergh et al. 2002; Wernicke et al. 2002). These data suggest that the 9-repeat allele is more involved than the 10-repeat allele in these behavioral paradigms, which is different from the findings from the ADHD studies in which the 10-repeat allele is implicated. Substance abusers are frequently co-morbid for neuropsychiatric disorders. Although complex and polygenetic in nature, it is of considerable interest to determine overlap between candidate genes associated with substance abuse and neuropsychiatric disorders.

Many studies using single markers, including the 3′ VNTR near the 3′-end of the DAT gene, have shown negative associations with several diseases. Of approximately ten association studies on bipolar disorder and the 3′-VNTR or SNPs located at the 3′-end of the DAT gene, none was indicative of an association. However, in a study using haplotypes conferred by SNPs located from exon 9 to 15 (Fig. 1A), significant association with bipolar disorder was discovered (Greenwood et al. 2001).

Sequence analysis of the coding region of the DAT gene from 551 Caucasian individuals revealed 15 exon SNPs including 4 missense mutations (Grunhage et al. 2000; Vandenbergh et al. 2000). None of these SNPs displayed any association with alcoholism, Tourette's syndrome, or bipolar disorder in these samples. The four missense mutations are rare, all with a frequency of less than 1% and none of them has been implicated in any disease yet. Nevertheless, it is still possible that other polymorphisms of low LD with the 3′-VNTR or these SNPs regulate DAT activity. Based on long-standing views on dopaminergic involvements in pathogenesis of schizophrenia and Parkinson's disease, more than 20 association studies have been done using the 3′ VNTR and the majority of them showed no association with these diseases.

Interestingly, two recent studies that used rs2975226 (T/A) located immediately upstream of exon 1 showed a strong association of the T allele with both bipolar disorder and schizophrenia (Keikhaee et al. 2005; Khodayari et al. 2004). These results suggest two things. First, since this T allele is located on a haplotype that displayed higher promoter activity in vitro, high promoter activity in certain brain regions might confer a risk factor for bipolar disorder and schizophrenia if these promoter marker-based findings can be validated in different studies. Second, markers in the 5′ promoter region should be used in future studies not only to clarify the inconsistencies of the current 3′ VNTR-based association findings, but also to investigate DAT as a risk factor for other DA-related diseases.

2
The Human Serotonin Transporter (*SLC6A4*)

2.1
Introduction

Serotonin contributes to and modulates a wide range of functions in the periphery, including vasoconstriction, gastrointestinal motility, and secretion. In normal brain states, serotonin is implicated in mood, sleep, appetite, and in the production of anxiety, fear, reward, and aggression. Dysfunction of brain serotonin signaling is implicated in several neuropsychiatric disorders, including depression, suicide, obsessive-compulsive disorder (OCD), migraine headaches, eating disorders, and autism. Neurons that produce and release serotonin are few in number, highly circumscribed in discrete raphe nuclei in the brainstem, and project to multiple brain regions (cortex, thalamus, basal ganglia, hippocampus, amygdala, and others), as well as the medulla and spinal cord (review: Jacobs and Azmitia 1992).

Serotonin activates more than 15 serotonin receptor subtypes that display unique distribution, structure, pharmacology, and signaling pathways. Implicated in mediating the wide array of physiological effects of serotonin, the activity of these receptors is regulated by SERT, which determines serotonin availability in brain. Encoded by a single gene, variations in SERT activity and density and SERT polymorphisms are implicated in depression, anxiety, OCD, suicide, autism, and substance abuse (Bastani et al. 1991; Malison et al. 1998; Newport et al. 2004; Owens and Nemeroff 1998; Paul et al. 1981). Not surprisingly, the SERT is a principal target of a range of therapeutic and illicit drugs of abuse. SERT activity is blocked by tricyclic and serotonin-selective reuptake inhibitors (SSRIs), drugs widely used in the treatment of depression, OCD, and other psychiatric diseases. Cocaine and amphetamines block SERT function, a process implicated in the psychoactive and addictive properties of the drugs (Eshleman et al. 1997; Verrico et al. 2005).

SERT can be regulated acutely by receptors, signal transducers, and kinase activity which rapidly modulate SERT expression (Anderson and Horne 1992; Miller and Hoffman 1994; Myers and Pitt 1988). In common with the DAT and NET, PKC activation leads to a rapid loss of surface-expressed SERT (Apparsundaram et al. 1998; Blakely et al. 1998; Miranda et al. 2005; Qian et al. 1997). In contrast to the DAT or NET, however, SERT substrates promote retention of the SERT on the membrane surface, whereas substrates for the DAT and NET might promote transporter internalization (Mao et al. 2004; Saunders et al. 2000).

The amino acid composition of the SERT is approximately 50% homologous to the DAT and NET, and its protein structure models similarly to the DAT and NET, with 12 transmembrane hydrophobic domains and intra- and extracellular loops of varying lengths. SERT messenger RNA (mRNA) expression is restricted to serotonin cell bodies in the raphe nuclei, whereas SERT protein is widely distributed in brain, following the pattern of extensive innervation of cortical and subcortical regions of brain, as noted above.

2.2
SERT Genomic Variants

2.2.1
Introduction

The SERT (or 5-HTT) genomic gene sequence is smaller (37.8 kb) than the DAT but highly polymorphic. More than 100 SNPs have been identified in this gene (from exons 1 to 15) and more than 35 SNPs were found in the 5′ region (12.6 kb from the upstream bleomycin hydrolase gene to exon 1) and deposited in dbSNP (Build 124). Less than 20 polymorphisms located in the 5′ promoter region, intron 2, and exons have been documented in the literature, with the majority also documented in dbSNP (Fig. 1B).

2.2.2
Coding Region Variants

Overall, the coding region of the SERT is highly conserved. From re-sequencing of more than 450 individuals, only 11 missense mutants have been reported and nine of them have a very low frequency (0.1%, Table 1, but see following section). This information suggests that wildtype function of SERT is required for human survival.

2.2.3
Non-coding Variants

The SERT-linked polymorphic region (5-HTTLPR or SERTLPR), located at approximately −1.4 kb upstream of exon 1, has 14 (*s*), 15, 16 (*l*), 19, 20 (*xl*), 22-repeats of 20–23 bp and both the 14- and 16-repeat alleles have minor variations (Delbruck et al. 1997; Heils et al. 1996; Nakamura et al. 2000). This 5-HTTLPR of the 5-HTT has been used as a marker in hundreds of association studies. Between 5-HTTLPR and exon 1 is a 381-bp indel, but there is little information available on this indel (Flattem and Blakely 2000). The second-most-used marker is the intron 2 VNTR (also termed STin2, Fig. 1B). VNTR contains 9, 10, and 12 repeats of 17 bp (STin2.9, STin2.10 and STin2.12) (Ogilvie et al. 1996).

The allelic frequency for 5-HTTLPR and VNTR varies significantly from population to population (Table 3). The *s* allele is associated with alcoholism, anxiety, suicide, and depression. The lowest frequency of the *s* allele is found in African populations and highest frequency in Chinese and Japanese cohorts. Nonetheless, the association of the *s*-allele with these disorders in the various populations has not been systematically investigated. In contrast, the *l* allele has the highest frequency in African populations. It would be interesting to

Table 3 Allele frequency of 5-HTTLPR and VNTR in different populations

5-HTTLPR[a]h	Allele	American	European[b]	East Asian[c]	African[d]
	l	57%–69%	54%–61%	10%–37%	79%–84%
	s	29%–44%	39%–46%	63%–79%	9%–21%
	xl	2%	0%	0%–1.6%	6%
VNTR[e]		European-American	African-American	Japanese	
	STin2.9	1.1%	1%	0%	
	STin2.10	46.6%	26.0%	2.3%	
	STin2.12	52.3%	73.1%	97.7%	

[a]Delbruck et al. 2001; Nakamura et al. 2000; Ogilvie et al. 1996, [b]French and German, [c]Chinese and Japanese, [d]Central African, [e]Gelernter et al. 1997

determine whether Africans have higher frequency of *l*-associated diseases such as pulmonary hypertension and ADHD (see Sect. 2.5). The 15-, 19-, and 22-repeat alleles of 5-HTTLPR are reported to have frequency of less than 0.8% in Japanese, but not in Caucasians (Nakamura et al. 2000). For the VNTR, STin2.10 and STin.12 are the major alleles in European Americans and African Americans, but in Japanese populations STin.12 dominates, with a frequency of 97.7% (Gelernter et al. 1997). The sequencing information indicates that the SERT is considerably variable in individuals and among different populations. Besides the 5-HTTLPR and VNTR regions, an SNP, 2892G/T, located the 3′-end of exon 15 has been used in association studies (Table 4). Eleven other

Table 4 *SLC6A4* association with human diseases

Marker/disease	Association (No. of studies)[a]		Reference(s)
5-HTTLPR			
Trait anxiety	Yes	(Meta of 26)	Schinka et al. 2004
Major depression	Yes	(2 meta-analyses)	Furlong et al. 1998; Lotrich and Pollock 2004
	No	(Meta of 15)	Lasky-Su et al. 2005
Alcohol dependence	Yes	(18)	Feinn et al. 2005; Gorwood et al. 2004; Hu et al. 2005
Suicide behavior	Yes	(Meta of >18)	Anguelova et al. 2003; Lin and Tsai 2004
Aggression	Yes	(7)	Cadoret et al. 2003; Courtet et al. 2001; Han et al. 2004; Klauck et al. 1997; Retz et al. 2004; Zalsman et al. 2001

Table 4 (continued)

Marker/disease	Association	(No. of studies)[a]	Reference(s)
	No	(2)	Baca-Garcia et al. 2004; Patkar et al. 2002
TCI/TPQ	Yes	(Meta of 16)	Munafo et al. 2005
Harm avoidance	No	(Meta of 8)	Sen et al. 2004
NEO neuroticism	Yes	(Meta of 10)	Sen et al. 2004
	No	(Meta of 13)	Munafo et al. 2005
Bipolar disorder	Yes	(4 meta-analyses)	Furlong et al. 1998; Lasky-Su et al. 2005; Levinson 2005; Lotrich and Pollock 2004
	No	(Meta of 43)	Cho et al. 2005
Autism	Yes	(2)	Cook et al. 1997; Klauck et al. 1997
Anorexia nervosa	Yes	(2)	Fumeron et al. 2001; Matsushita et al. 2004
	No	(1)	Sundaramurthy et al. 2000
Smoking (s)	Yes	(1)	Gerra et al. 2005
Alzheimer's disease	Yes	(1)	Li et al. 1997
dPIBS[c] in women	Yes	(1)	Cook et al. 1995
Pulmonary hypertension	Yes	(1)	Eddahibi et al. 2003
Intron 2 VNTR			
Schizophrenia	Yes	(Meta of 12)	Fan and Sklar 2005
Aggression	Yes	(1)	Davidge et al. 2004
5′ haplotypes[b]			
Autism	Yes	(1)	Conroy et al. 2004
Haplotype 1 ± STin2.12			
Smoking	Yes	(1)	Kremer et al. 2005
Exon 9 rs6353 (I425 V)			
OCD	Yes	(1)	Ozaki et al. 2003
3′ SNP 2892G/T			
Unipolar disorder	Suggestive	(1)	102 ($p = 0.034$)
5-HTT haplotypes			
ADHD	Yes	(Meta of > 8)	Bobb et al. 2005b; Curran et al. 2005

[a] Meta as meta-analysis; no single method is sufficient for assessing evidence of publication bias (Furlong et al. 1998) and this is why multiple meta-analyses are listed here
[b] Inferred from 5-HTTLPR, SNPs, and STin2 (VNTR)
[c] Diarrhea-predominant irritable bowel syndrome

polymorphisms listed in Fig. 1B cause missense mutations (indicated in the parentheses) in the SERT protein.

Based on information available in the HapMap database, the hSERT gene displays high LD for most populations. The D' value [for the region between rs2020933 located at 0.95 kb after exon 1 and rs1042173 located 0.1 kb to the 3'-end of exon 15 that covers almost the entire transcribed region (Fig. 1B)], is 1.0 for Caucasians ($n = 56$), Japanese ($n = 44$) and Africans ($n = 60$), but is 0.128 for Chinese ($n = 45$). The D' value between rs2066713 located 1.7 kb before exon 2 and rs1042173 in exon 15 is 0.78 for Caucasians, 0.70 for Chinese, 1.0 for Japanese, and 0.609 for Africans.

2.3
SERT Protein Variants

Ten human SERT protein variants have been discovered. Nine of them have been reported in the literature and another one (K201 N) is described in the NCBI dbSNP database, with a frequency of 15.7% or heterozygosity of 0.265 out of 51 individuals genotyped (Fig. 1B). I425V is the only one that has been characterized pharmacologically. I425 is located in the middle of hSERT TM8. It is conserved in all eight SERTs listed in Fig. 2 except ceSERT with Phe instead. At the same position, Ile and Val are present in DATs and NETs as well, suggesting—and experimentally confirmed—that mutation I425V would not influence plasma membrane expression of the transporter protein. In vitro assays demonstrated that I425V increases both substrate uptake affinity (K_m values $\downarrow$) by 1.2–1.6-fold and V_{max} by 1.7- to 2.2-fold, attributable to the mutation-mediated defect in regulation by nitric oxide and resulting constitutive activation of hSERT (Kilic et al. 2003). This gain-of-function mutant is found in two families with clustering of OCD and other serotonin-related disorders and represents a risk factor for these human diseases (Ozaki et al. 2003).

Of several variants characterized functionally, T4 is conserved in all four mammalian proteins but not in non-mammalian proteins. Since Ser8 with Lys10—conserved in the mammalian proteins as well—is a potential phosphorylation site for protein kinase C, T4 becomes a part of the consensus site "XTXXXSp" (Sp denotes a phosphor-serine) for kinase GSK3 (Fig. 2; Kennelly and Krebs 1991). Therefore, T4 might be important for hSERT regulation and the mutation T4A could result in a dysfunctional transporter protein. L255 in TM4 and L362 in TM7 are both conserved, not only in the SERT but also in the DAT and/or the NET, indicative of their importance in transport activity. Based on this, L255M and L362M conceivably would have altered transporter activity. Another variant, P339L in TM6, might have altered substrate affinity because P339 is highly conserved in all eight SERTs. The other SERT variants, with G56A (N-terminal), K201N (ECL2), E215K (ECL2), S293F (TM5), K605N (C-terminal), and P621S (C-terminal), could have normal transporter activity, either because of their benign locations in the protein or low conservation (Fig. 2).

Ten hSERT coding variants and the hSERT were recently transfected and investigated in parallel to define their total and surface protein expression, antagonist recognition, and transporter modulation by posttranslational, regulatory pathways. Two variants, Pro339Leu and Ile425Val, demonstrated significant changes in surface expression, supporting alterations in 5-HT transport capacity. Each of the SERT variants was capable of rapid, phorbol ester-triggered downregulation, but five variants (Thr4Ala, Gly56Ala, Glu215Lys, Lys605Asn, and Pro612Ser) demonstrated no capacity for 5-HT uptake stimulation after acute protein kinase G (PKG)/p38 mitogen-activated protein kinase (MAPK) activation. Epstein-Barr virus (EBV)-transformed lymphocytes natively expressing the most common of these variants (Gly56Ala) exhibited a similar loss of 5-HT uptake stimulation by PKG/p38 MAPK activators. HeLa cells transfected with the Gly56Ala variant demonstrated elevated basal phosphorylation and, unlike hSERT, could not be further phosphorylated after 8-bromo cyclic guanosine monophosphate (cGMP) (8BrcGMP) treatments. Taken together, spontaneously occurring human SERT coding variants displayed different regulatory patterns when transfected into cell lines. These findings suggest that spontaneous variants of the transporter structure can lead to a diversity of regulatory responses that may confer risks for disorders attributable to compromised 5-HT signaling (Prasad et al. 2005). Whether serotonin affects SERT localization and mobility similarly in various forms of SERT-containing coding region SNPs is a current and notable void in the literature.

2.4
Variations in SERT Expression Levels

hSERT expression levels can be detected by positron emission tomography (PET) imaging of SERT availability or in postmortem tissues by mRNA or protein levels. In different individuals, SERT availability (a measure of density) varied by fourfold as detected by PET imaging, or sixfold as measured by mRNA in postmortem tissue of selected brain regions (Frankle et al. 2005; Little et al. 1998). A number of lines of evidence have linked SERT polymorphisms to SERT expression levels. Several studies have shown that the *l* allele of 5-HTTLPR is associated with higher SERT expression levels. In vitro, a 1.4-kb fragment containing the *l* allele is associated with higher expression levels than one containing the *s* allele in transiently expressed cells (Heils et al. 1996) and in cultured human lymphoblast cell lines carrying the l/l alleles (Lesch et al. 1996). In vivo, the *l/l* carriers consistently display twofold higher uptake activity by the SERT in small muscle cells (Eddahibi et al. 2001) and higher mRNA levels in postmortem brain tissues (Little et al. 1998) than the s/s carriers. Imaging studies demonstrate that hSERT availability declines during aging, but there is no evidence for individual differences in the rate of declination (van Dyck et al. 2000).

Altered SERT expression is associated with human diseases, particularly depression and suicide. In healthy individuals ($n = 96$), SPECT imaging data revealed no correlation between 5-HTTLPR and SERT availability in the brainstem-diencephalon (van Dyck et al. 2004). In a preliminary PET imaging study, hSERT availability was decreased in the midbrain and correlated with severity of major depression (Newberg et al. 2005), possibly indicative of reduced serotonin availability in this brain region. Consistent with these findings was a postmortem study that investigated the association of SERT mRNA with suicide. The suicide group had 54% fewer dorsal raphe nucleus (DRN) neurons expressing SERT mRNA compared with controls, but in the serotonin neurons that expressed the SERT gene, the mRNA level per neuron was greater (Arango et al. 2001), again implying that overactive SERT could reduce serotonin availability. In a different postmortem study of schizophrenic brain, SERT mRNA levels in left superior frontal gyrus were higher but were lower in the left middle temporal gyrus (Hernandez and Sokolov 1997), indicative of a regional specificity of SERT expression levels. In platelets derived from alcoholics, [^{3}H]5-HT uptake and [^{3}H]paroxetine binding assays showed the *l/l* carriers displayed higher binding B_{max} values than s/s carriers, consistent with previous findings that *l/l* is associated with higher SERT expression. To add to the complexity of these findings, the higher SERT did not correlate with higher transport capacity in the same alcoholic individuals, suggesting that alcoholism may downregulate hSERT function (Javors et al. 2005). Accordingly, transporter density—as well as transporter function—needs to be considered in relating polymorphic regions of the SERT gene to function.

In addition to the 5-HTTLPR, the intron 2 VNTR influences reporter gene expression. Based on a luciferase (Luc) reporter system in vitro that reflects changes in gene expression, the 12-repeat allele of the intron 2 VNTR supported several-fold higher Luc expression than the 10-repeat allele in embryonic stem cells (Fiskerstrand et al. 1999). In order to compare the relative significance of the STin2.10 versus STin2.12 alleles for promoter regulation in vivo, two reporter plasmids were constructed by tagging the two alleles to the 5′-end of an hβg-LacZ hybrid (human β-globin minimal promoter sequence-βgal gene). These two constructs were then introduced into mice by transgenic technology and the βgal activity was examined in the transgenic mice at different developmental stages. At embryonic day (E)10.5, the STin2.12 mice expressed much higher βgal activity in the midbrain and rostral hindbrain floor plate than the STin2.10 mice (MacKenzie and Quinn 1999), indicating that the STin2.12 was a more influential modifier of promoter regulation. Further analysis demonstrated that the transcription factor YB-1 binds to this VNTR to enhance the 12-repeat-associated higher promoter activity in vitro (Klenova et al. 2004). Accordingly, differential allelic interactions with transcription factors may underlie allele-specific promoter activity.

2.5
SERT Association with Human Diseases

The SERT gene has been investigated in an astonishing 330-plus association studies with more than 30 different human diseases. Collectively, more than 10 diseases are positively associated with several markers on the SERT gene, including the promoter 5-HTTLPR and intron 2 VNTR (STin2), as seen in Table 4.

The association of 5-HTTLPR with psychiatric disorders (Table 4) is consistent with the principal involvement of hSERT in regulating serotonergic neuronal activity. The robustness of these association studies benefited from that fact that 5-HTTLPR is a functional polymorphism that significantly influences promoter activity (Heils et al. 1996). Meta-analysis of data from 26 different studies suggests that the SERT gene is a risk factor for anxiety-related traits (Lesch et al. 1996; Schinka et al. 2004). Individuals carrying s/s are more prone to acquire a conditioned fear response (Garpenstrand et al. 2001), which is reflected by the blood oxygen level-dependent (BOLD) functional magnetic resonance imaging (fMRI)-based clinical observations, suggesting that the right amygdala of *s/s* carriers displays greater response to fearful stimuli (Hariri et al. 2002). The association of SERT polymorphisms to major depression was anticipated because serotonin-selective reuptake inhibitors (SSRIs) are widely used in treating depression. Indeed, the association between depression and the 5-HTTLPR or the functional intron 2 VNTR has been confirmed in the majority of studies (Furlong et al. 1998; Lotrich and Pollock 2004; Ogilvie et al. 1996). Imaging-based morphometric and functional analyses have demonstrated that *s* allele carriers have reduced gray matter volume in the perigenual cingulate and amygdala, decreased coupling of the amygdala-cingulate feedback circuit (implicated in extinction of negative effect) and impaired emotion regulation (Pezawas et al. 2005). These findings provide biological evidence for SERT variants as a risk factor for major depression.

The *l* allele appears associated with a number of diseases as well. In a recent clinical study in France and Great Britain, *l/l* carriers, associated with higher levels of SERT expression in pulmonary artery smooth muscle cells than *l/s* and *s/s* carriers, displayed higher pulmonary artery pressure and were at higher risk for pulmonary hypertension in chronic obstructive pulmonary disease (Eddahibi et al. 2003). Another consistent finding is association of the *l* allele with ADHD (Bobb et al. 2005a, b; Curran et al. 2005).

5-HTTLPR, VNTR, and other SERT gene markers may or may not collectively show positive association with diseases. Both 5-HTTLPR and 5′ haplotypes are positively associated with autism and smoking and both 5-HTTLPR and VNTR are associated with aggression. In contrast, the 5-HTTLPR, but not VNTR, is positively associated with major depression (Lasky-Su et al. 2005). For schizophrenia, only VNTR, but not 5-HTTLPR, has shown a positive association, whereas for ADHD only the haplotypes, but not 5-HTTLPR, are positively associated. The missense mutant I425V and the exon 15 SNP 2892G/T are positively associated with OCD. Conceivably, there is low LD between 5-HTTLPR and other markers of the SERT in certain populations. A mechanism-based model of the relationship between SERT function and these neuropsychiatric disorders awaits insight into the functional consequences of each of these variants.

It is important to point out that association studies in different populations can be frequently inconsistent. For example, the "*s*" allele is associated with smoking (Gerra et al. 2005) and consistent with decreased SERT expression in platelets of African-American smokers (Patkar et al. 2003). However, haplotype analysis of 5-HTTLPR and VNTR indicates that that "*l*"+STin2.12, or each of these alleles, is associated with higher SERT expression in vitro and is associated with smoking in a Jewish cohort (Kremer et al. 2005). In another example, 5-HTTLPR, but not the VNTR, is associated very significantly ($p = 3.07 \times 10^{-5}$) with diarrhea-predominant irritable bowel syndrome [dPIBS] in Caucasian women from North America (Yeo et al. 2004), with the *s* allele over-represented in the patients with an odds ratio (OR) of 2.23. This is consistent with the findings that decreased SERT expression is associated with dPIBS in Italians (Bellini et al. 2003), but not in a Korean (Lee et al. 2004) or a Turkish population (Pata et al. 2002). A U.S. group reported the "*s*" allele association with autism (Cook et al. 1997) but the opposite association with autism ("*l*" allele) was reported by a German group (Klauck et al. 1997). Even for major depression, SERT may not be an essential risk factor (Anguelova et al. 2003). For OCD, the "*l*" allele of 5-HTTLPR was positively associated in two U.S. studies (McDougle et al. 1998) but not in South African, Israeli, German, and Canadian studies (Billett et al. 1997; Frisch et al. 2000; Kinnear et al. 2000; Walitza et al. 2004). In a final example, "*s*" association with anorexia nervosa was observed in both Japanese and French, but not in British Caucasian females (Fumeron et al. 2001; Matsushita et al. 2004; Sundaramurthy et al. 2000). These findings point to the complexity of polygenic diseases that are modulated by ethnic genotype, phenotype, and environment.

3
The Human Norepinephrine Transporter (*SLC6A2*)

3.1
Introduction

Central and peripheral norepinephrine is implicated in a range of physiological functions. Arousal, attention, memory, and mood, as well as autonomic regulation of heart rate and blood pressure, fall within the purview of norepinephrine function. Released norepinephrine in central and peripheral synapses is terminated by the NET, which regulates noradrenergic receptor-mediated neurotransmission. Dysfunction of the NET in cardiac sympathetic nerve terminals is implicated in hypertension, diabetes, cardiomyopathy, and congestive heart failure (Bohm et al. 1995; Esler et al. 1981; Garland et al. 2002; Mao et al. 2004; Merlet et al. 1992; Shannon et al. 2000). Compromised NET function has been suspected in mood disorders and in ADHD, although the latter association is controversial (Bobb et al. 2005a; Klimek et al. 1997; Xu et al. 2005; Yang et al. 2004). Centrally, the NET is a target of anti-hyperactivity and antidepressant medications as well as psychostimulant drugs of abuse (cocaine, amphetamines).

3.2
NET Genomic VARIANTS

The NET gene is smaller than the DAT gene (47.4 kb). Once again, the NET genomic DNA sequence displays a high density of polymorphisms (∼4/KB). More than 200 SNPs are located in this gene (from exons 1 to 15) and more than 40 SNPs are located in the 5′ 10-kb region (the upstream gene *FLJ20481* is 74 kb away from exon 1). Approximately 33 polymorphisms have been documented in the literature and most of them have been included in the dbSNP.

3.2.1
Non-coding Region Variants

One of the non-coding region variants, designated NETpPR, covers 363 bp between −3935 bp and −4297 upstream of exon 1 (Urwin et al. 2002). NETpPR contains six "AAGG"-repeat islands (AAGG1–6) and each island has 3–4 repeats. AAGG1 has 3/4 repeats (*s/l* alleles) and AAGG4 has 2/3 repeats (*s/l* alleles and *s* has a minor allele frequency (MAF) of 26% in Japanese). Both of the *s* alleles lose the consensus site for transcription factor Elk-1, but these two alleles have not yet been found on the same chromosomes. In Japanese populations, the AAGG4 *l* allele has been found to double the risk for anorexia nervosa (restrictive subtype) as seen in Table 5, but its frequency in other populations has not yet been reported. In some individuals, the 343 bp at the 3′ side of the 363 bp NETpPR can be deleted (Kim et al. 1999; Urwin et al. 2002). Accordingly, the NET gene has a highly polymorphic promoter region and is subject to extensive variations among individuals. Conceivably, some of these variations confer risk factors for NET-related diseases.

Certain polymorphisms either have been used in association studies or are known to cause missense mutations (indicated in the parentheses) in the hNET protein (Fig. 1C). rs3760019, rs2397771, and rs168924, located between NETpPR and exon 1, are T/C, C/G, and A/G SNPs and have been used in a single association study (Ono et al. 2003). Originally designated as T-182C, rs2242446 is located at 131 bp to the 5′-end of exon 2. With an MAF of 25.4% for the C allele, this stretch has been assessed in three association studies (Table 5; Inoue et al. 2004; Ryu et al. 2004; Zill et al. 2002). The intron 9C/A SNP was originally

Table 5 *SLC6A2* association with human diseases

Marker	Diseases	Association (No. of studies)		Reference(s)
5′ region				
NETpPR	Anorexia nervosa (restrictive subtype)	Yes	(1)	Urwin et al. 2002
rs168924	Hypertension	Yes	(1)	Ono et al. 2003
rs2242446	Major depression	Yes	(2)	Inoue et al. 2004; Ryu et al. 2004
		No	(1)	Zill et al. 2002
3′ region				
1429G/C (A457P)	Orthostatic intolerance	Yes	(1)	Shannon et al. 2000
Intron 8 SNP rs3785157	ADHD	Yes	(1)	Bobb et al. 2005
		No	(1)	McEvoy et al. 2002
Intron 10 SNP rs998424	ADHD	Yes	(1)	Bobb et al. 2005
		No	(2)	De Luca et al. 2004; McEvoy et al. 2002

an intron 8 SNP, without considering the discovery of the new exon 1 (Kim et al. 1999; Ono et al. 2003). Intron 9C/A was not associated with hypertension in Japanese (Ono et al. 2003) and has not yet been used for other populations. rs5569, located in exon 10 and originally reported as 1287G/A, has an MAF of 30%–35% for the A allele (Stober et al. 1996; and dbSNP Build 124) and has been used in four association studies (see Sect. 3.5).

LD between SNPs located in intron 1/intron 2 and the SNP rs5569 in exon 10 displays great diversity and varies significantly from subregion to subregion within the gene and from population to population as well. The D′ value is 0.259 in 60 Caucasians for LD between rs2242446 located 0.1 kb before exon 2 and rs5569 in exon 10. rs2242446 has not been genotyped, and no D′ information is available yet for other populations (HapMap). The D′ value for LD between an intron 2 SNP rs3785143 located 4 kb after exon 2 and the exon 10 SNP rs5569 is 0.2 for 60 Caucasians, but it is 1.0 for 45 Chinese, 44 Japanese, and 59 Africans. The D′ value for another region—between the same intron 2 SNP and an intron 11 SNP (rs36009, 0.1 kb 5′ to exon 12)—is 1.0 for Caucasians and Africans, 0.06 for Chinese, and 0.355 for Japanese. A recent systematic haplotype analysis using 26 SNPs across this gene for 384 individuals of Finnish-Caucasian, American-Caucasian, Plains American Indian, and African-American populations showed that there are three high-LD blocks of haplotypes that are conserved among these populations (Fig. 1C; Belfer et al. 2004). The D′ value for a 19.5-kb 5′ region—between the 5′ SNP rs4783899 (located at 3 kb upstream of exon 1) and the intron 4 SNP rs187714 (0.5 kb after exon 4)—is 0.95, 0.93, 0.99, and 0.81 for the four populations, respectively. However, the frequency of the most common haplotype in each block varies by up to 4- to 5-fold from population to population. The D' value for LD between the same 5' SNP and any of other 18 SNPs located downstream of rs187714 is less than 0.12, less than 0.45, less than 0.39, and less than 0.24, respectively. It seems that there are recombination hot spots within intron 4.

3.2.2
Coding Region Variants

Similarly, the coding region of *SLC6A2* displays great diversity as well, and the DNA sequences are more variable than *SLC6A3* and *SLC6A4*. More than 200 individuals have been re-sequenced and 20 missense mutants have been described; the frequency of the variants can be as high as 5.5% (Table 1). This information might suggest that human survivors are more tolerant to hNET-related diseases than to hDAT or hSERT-related diseases.

3.3
NET Protein Variants

hNET has the largest variability in amino acid sequence among the three transporters in question here. Of the 20 protein variants that have been discovered, 15 of them are reported in the literature and the other five (N7K, A31P, V247I, T283R, and I549T) described in dbSNP (Fig. 1C). Sixteen of them have been characterized pharmacologically.

The orthostatic intolerance-associated A457P mutation in TM9 disrupts plasma membrane expression not completely but dominantly to the wildtype and almost eliminates uptake activity based on in vitro analyses (Hahn et al. 2003; Shannon et al. 2000). A457 is highly conserved among the five NETs listed in Fig. 2 and is probably important for norepinephrine transport. The other residues at this position throughout DATs and SERTs are either Ser or Gly with small side-chains. Substitution with the bulky side-chained Pro for this Ala likely changes not only the substrate affinity but, more importantly, the protein conformation, leading to loss of uptake activity.

A similar mutant A369P located in ECL4 loses both plasma membrane expression and uptake activity completely (Hahn et al. 2005). A369 is conserved in neurotransmitter transporters except a Ser residue in ceDAT and msSERT. These findings are consistent with the notion that mutations of conserved residues in this ECL4 often disrupt transporter expression (Lin and Uhl 2003; Lin et al. 1999). R121Q, N292T, and Y548H all lose 22%–44% uptake activity with decreases accordingly in plasma membrane expression. R121 in intracellular loop 1 (ICL1) and Y548 in ECL6 are both highly conserved and might contribute to transporter configuration and/or expression. N292 is present only in the mammalian NETs and is probably important for NET-related transporter features.

Two of the characterized variants appear to be gain-of-function mutants. F528C displays a 31% increase in [^{3}H]norepinephrine uptake rate compared to the wildtype, apparently through an increase in plasma membrane expression. F528 is conserved throughout the transporters with the exception of an Ile residue in ceDAT. Removal of the phenyl ring from this Phe caused an almost tenfold decrease in uptake V_{max} without influencing expression in DAT (Lin et al. 1999), suggesting the importance of F528 for transport activity. G478S increased uptake affinity (K_m values ↓) by 2.5-fold and displayed an average V_{max}/K_m value of 65, comparing to 175 of the wildtype. G478S represents another gain-of-function mutant (Runkel et al. 2000). G478 in TM10 is conserved in 89% of the transporters or all of the mammalian transporters and, at the same position, an alternative is Ser in ceSERT or Thr in ceDAT (Fig. 2). G478 is apparently important for norepinephrine translocation.

Variants V69I, T99I, V2441I, V245I, V356L, N375S, V449I, and I549T appear to have normal transport activity (Hahn et al. 2005; Runkel et al. 2000). None of them influences uptake activity at statistically significant levels. V245I increased desipramine affinity by 2.8-fold. V449I increased the V_{max}/K_m value by 1.6-fold and desipramine affinity by 2.4-fold (Runkel et al. 2000). V69 is located in TM1 and highly conserved among the transporters (Fig. 2). It is little surprising that Ile substitution did not influence the uptake activity and increased desipramine affinity by 1.4-fold.

Among the four uncharacterized hNET variants (Fig. 1C), N7 and A31 are located in the N-terminus and not conserved among NETs. It would be surprising to see any functional alteration in N7K and A31P. At position V247 in TM4, Ile is an alternative, and V247I probably has normal transport activity as well. T283 in TM5 is conserved in all of the mammalian transporters. T283R has a substitution with a positively charged large side-chain and is thus expected to have altered transport activity. All of the speculations concerning these four variants, especially T283R, need to be verified by experimental studies.

3.4
Variations in hNET Expression Levels

There is some information available about genotypic association with hNET expression levels. Variations in hNET expression may be a significant factor in consideration of ADHD treatment strategies (Fischman and Madras 2005; Madras et al. 2005). It has been reported that expression levels in placenta vary widely from individual to individual and lower mRNA levels are associated with pre-eclampsia (Bottalico et al. 2004; Szot et al. 2000). Brain imaging data are not available currently due to lack of appropriate labeling agents (McConathy et al. 2004). Based on this summary, the human genetics of *SLC6A2* warrants further investigation.

3.5
NET Association with Human Diseases

There are approximately 20 association studies reported for the NET (Table 5). Findings from these studies suggest that this gene is associated with major depression, hypertension, orthostatic intolerance, and anorexia nervosa (restrictive type). The intron 8 and 10 SNPs rs3785157 and rs998424 were positively associated with ADHD in a U.S. study, but not in Canadian and Irish studies (Bobb et al. 2005a, b; De Luca et al. 2004; McEvoy et al. 2002).

The exon 10 SNP rs5569 (1287G/A) was not associated with major depression in Caucasian Germans and Canadians (Owen et al. 1999; Zill et al. 2002), personality disorders in Chinese (Tsai et al. 2002), nor bipolar disorder or schizophrenia in Poles (Leszczynska-Rodziewicz et al. 2002). This SNP—82 bp away from and probably in strong LD with the A457P mutant—has not yet been examined for potential association with orthostatic intolerance. None of the five SNPs rs1805064 (V69I), rs1805065 (T99I), rs1805066 (V245I), rs2234910 (V449I), and rs1805067 (G478S) was associated with panic disorder, Tourette's syndrome, bipolar disorder, or schizophrenia in Caucasian Germans (Sand et al. 2002; Stober et al. 1999; Stober et al. 1996). Because of the low LD between the 5'- and the 3'-ends, markers located in each of the three blocks of high LD should be used in future association studies (Belfer et al. 2004). Markers including NETpPR in the 5' promoter region merit investigation for ADHD because hNET is a therapeutic target for this disease (Madras et al. 2005).

4
Transporter Gene Knockout Mice: Implications

Studies in mice with full or partial deletions of transporter genes have provided important insights into the genes' contributions to neurochemistry, morphology, and behavior. Null mutations of the DAT gene are not lethal, suggesting that humans can sustain significant variability in hDAT expression levels. However, null mutant mice (DAT −/−) display profound adaptive phenotypic changes, including reduced D1 and D2 DA receptor expression levels in basal ganglia, decreased tissue DA concentrations, increased extracellular DA concentrations, and retarded pituitary development. Mice with null mutations (−/−)—but not heterozygous cohorts (+/−)—display higher locomotor activity, anterior pituitary hypoplasia and dwarfism, and unresponsiveness to the locomotor stimulant effects of cocaine, but they retain cocaine self-administration unless the serotonin transporter is also deleted (Bosse et al. 1997; Giros et al. 1996; Sora et al. 2001; Sora et al. 1998). Mice lacking SERT display normal development, reduced 5-HT concentration, and increased locomotion activity in response to (+)-3,4-methylenedioxymethamphetamine ("ecstasy") treatments (Bengel et al. 1998). The DAT and SERT double knockouts, but none of the single knockouts, lose cocaine-induced place preference, suggesting that DAT and SERT are important for cocaine reward (Sora et al. 2001). Mice lacking NET have increased extracellular norepinephrine concentration, reduced norepinephrine tissue concentrations, and reduced extracellular DA concentrations. The −/− mice display supersensitivity to sensitization by amphetamine and cocaine treatments (Xu et al. 2000). Interestingly, none of the heterozygote mice displays any significant behavioral alterations.

5
Summary

Neurotransmitter transporters are principal regulators of neurotransmission. Genomic variations in these transporter genes may cause missense mutations in the proteins and alterations in expression levels in a brain region-specific manner, which confer risk factors for many related human diseases. The involvement of multiple endogenous and exogenous factors in pathogenesis of human diseases adds a level of complexity to fundamental transporter genomics. Genetic manipulation of these transporter genes in rodents can facilitate our understanding of the biological and pharmacological consequences of changes in transporter gene function, but cannot fully replicate human environmental and other factors. Considerable progress has been made in identifying genetic variations and relating these variations to biological function, pharmacological response, and human diseases during the last decade. Our current knowledge will guide future strategies to identify population-specific and environmental risk factors. Because of the low LD between discrete regions in the genes, particularly *SLC6A3* and *SLC6A2*, and variable frequency of these variations among different populations, numerous challenges remain. In addition to the few described herein, of which hundreds of polymorphisms in each gene are functionally important, questions remain concerning which haplotypes carry functionally relevant alleles and confer risk factors for diseases. Future research will provide the infrastructure to develop a comprehensive model of the contribution of haplotypes, other involved genes, population differences, and cultural and environmental influences in engendering pathophysiological states of polygenic disease and drug response.

References

Anderson GM, Horne WC (1992) Activators of protein kinase C decrease serotonin transport in human platelets. Biochim Biophys Acta 1137:331–337

Anguelova M, Benkelfat C, Turecki G (2003) A systematic review of association studies investigating genes coding for serotonin receptors and the serotonin transporter: II. Suicidal behavior. Mol Psychiatry 8:646–653

Apparsundaram S, Galli A, DeFelice LJ, Hartzell HC, Blakely RD (1998) Acute regulation of norepinephrine transport: I. protein kinase C-linked muscarinic receptors influence transport capacity and transporter density in SK-N-SH cells. J Pharmacol Exp Ther 287:733–743

Arango V, Underwood MD, Boldrini M, Tamir H, Kassir SA, Hsiung S, Chen JJ, Mann JJ (2001) Serotonin 1A receptors, serotonin transporter binding and serotonin transporter mRNA expression in the brainstem of depressed suicide victims. Neuropsychopharmacology 25:892–903

Audrain-McGovern J, Lerman C, Wileyto EP, Rodriguez D, Shields PG (2004) Interacting effects of genetic predisposition and depression on adolescent smoking progression. Am J Psychiatry 161:1224–1230

Baca-Garcia E, Vaquero C, Diaz-Sastre C, Garcia-Resa E, Saiz-Ruiz J, Fernandez-Piqueras J, de Leon J (2004) Lack of association between the serotonin transporter promoter gene polymorphism and impulsivity or aggressive behavior among suicide attempters and healthy volunteers. Psychiatry Res 126:99–106

Barr CL, Xu C, Kroft J, Feng Y, Wigg K, Zai G, Tannock R, Schachar R, Malone M, Roberts W, Nothen MM, Grunhage F, Vandenbergh DJ, Uhl G, Sunohara G, King N, Kennedy JL (2001) Haplotype study of three polymorphisms at the dopamine transporter locus confirm linkage to attention-deficit/hyperactivity disorder. Biol Psychiatry 49:333–339

Bastani B, Arora RC, Meltzer HY (1991) Serotonin uptake and imipramine binding in the blood platelets of obsessive-compulsive disorder patients. Biol Psychiatry 30:131–139

Battersby S, Ogilvie AD, Smith CA, Blackwood DH, Muir WJ, Quinn JP, Fink G, Goodwin GM, Harmar AJ (1996) Structure of a variable number tandem repeat of the serotonin transporter gene and association with affective disorder. Psychiatr Genet 6:177–181

Battersby S, Ogilvie AD, Blackwood DH, Shen S, Muqit MM, Muir WJ, Teague P, Goodwin GM, Harmar AJ (1999) Presence of multiple functional polyadenylation signals and a single nucleotide polymorphism in the 3′ untranslated region of the human serotonin transporter gene. J Neurochem 72:1384–1388

Belfer I, Phillips G, Taubman J, Hipp H, Lipsky RH, Enoch MA, Max MB, Goldman D (2004) Haplotype architecture of the norepinephrine transporter gene SLC6A2 in four populations. J Hum Genet 49:232–245

Bellini M, Rappelli L, Blandizzi C, Costa F, Stasi C, Colucci R, Giannaccini G, Marazziti D, Betti L, Baroni S, Mumolo MG, Marchi S, Del Tacca M (2003) Platelet serotonin transporter in patients with diarrhea-predominant irritable bowel syndrome both before and after treatment with alosetron. Am J Gastroenterol 98:2705–2711

Bengel D, Murphy DL, Andrews AM, Wichems CH, Feltner D, Heils A, Mossner R, Westphal H, Lesch KP (1998) Altered brain serotonin homeostasis and locomotor insensitivity to 3,4-methylenedioxymethamphetamine ("Ecstasy") in serotonin transporter-deficient mice. Mol Pharmacol 53:649–655

Billett EA, Richter MA, King N, Heils A, Lesch KP, Kennedy JL (1997) Obsessive compulsive disorder, response to serotonin reuptake inhibitors and the serotonin transporter gene. Mol Psychiatry 2:403–406

Blakely RD, Ramamoorthy S, Schroeter S, Qian Y, Apparsundaram S, Galli A, DeFelice LJ (1998) Regulated phosphorylation and trafficking of antidepressant-sensitive serotonin transporter proteins. Biol Psychiatry 44:169–178

Blum K, Braverman ER, Wu S, Cull JG, Chen TJ, Gill J, Wood R, Eisenberg A, Sherman M, Davis KR, Matthews D, Fischer L, Schnautz N, Walsh W, Pontius AA, Zedar M, Kaats G, Comings DE (1997) Association of polymorphisms of dopamine D2 receptor (DRD2), and dopamine transporter (DAT1) genes with schizoid/avoidant behaviors (SAB). Mol Psychiatry 2:239–246

Bobb AJ, Addington AM, Sidransky E, Gornick MC, Lerch JP, Greenstein DK, Clasen LS, Sharp WS, Inoff-Germain G, Wavrant-De Vrieze F, Arcos-Burgos M, Straub RE, Hardy JA, Castellanos FX, Rapoport JL (2005a) Support for association between ADHD and two candidate genes: NET1 and DRD1. Am J Med Genet B Neuropsychiatr Genet 134:67–72

Bobb AJ, Castellanos FX, Addington AM, Rapoport JL (2005b) Molecular genetic studies of ADHD: 1991 to 2004. Am J Med Genet B Neuropsychiatr Genet 132:109–125

Bohm M, La Rosee K, Schwinger RH, Erdmann E (1995) Evidence for reduction of norepinephrine uptake sites in the failing human heart. J Am Coll Cardiol 25:146–153

Bosse R, Fumagalli F, Jaber M, Giros B, Gainetdinov RR, Wetsel WC, Missale C, Caron MG (1997) Anterior pituitary hypoplasia and dwarfism in mice lacking the dopamine transporter. Neuron 19:127–138

Bottalico B, Larsson I, Brodszki J, Hernandez-Andrade E, Casslen B, Marsal K, Hansson SR (2004) Norepinephrine transporter (NET), serotonin transporter (SERT), vesicular monoamine transporter (VMAT2) and organic cation transporters (OCT1, 2 and EMT) in human placenta from pre-eclamptic and normotensive pregnancies. Placenta 25:518–529

Cadoret RJ, Langbehn D, Caspers K, Troughton EP, Yucuis R, Sandhu HK, Philibert R (2003) Associations of the serotonin transporter promoter polymorphism with aggressivity, attention deficit, and conduct disorder in an adoptee population. Compr Psychiatry 44:88–101

Chen CK, Chen SL, Mill J, Huang YS, Lin SK, Curran S, Purcell S, Sham P, Asherson P (2003) The dopamine transporter gene is associated with attention deficit hyperactivity disorder in a Taiwanese sample. Mol Psychiatry 8:393–396

Chen WJ, Chen CH, Huang J, Hsu YP, Seow SV, Chen CC, Cheng AT (2001) Genetic polymorphisms of the promoter region of dopamine D2 receptor and dopamine transporter genes and alcoholism among four aboriginal groups and Han Chinese in Taiwan. Psychiatr Genet 11:187–195

Cho HJ, Meira-Lima I, Cordeiro Q, Michelon L, Sham P, Vallada H, Collier DA (2005) Population-based and family-based studies on the serotonin transporter gene polymorphisms and bipolar disorder: a systematic review and meta-analysis. Mol Psychiatry 10:771–781

Ciliax BJ, Heilman C, Demchyshyn LL, Pristupa ZB, Ince E, Hersch SM, Niznik HB, Levey AI (1995) The dopamine transporter: immunochemical characterization and localization in brain. J Neurosci 15:1714–1723

Comings DE, Wu S, Chiu C, Ring RH, Gade R, Ahn C, MacMurray JP, Dietz G, Muhleman D (1996) Polygenic inheritance of Tourette syndrome, stuttering, attention deficit hyperactivity, conduct, and oppositional defiant disorder: the additive and subtractive effect of the three dopaminergic genes—DRD2, D beta H, and DAT1. Am J Med Genet 67:264–288

Conroy J, Meally E, Kearney G, Fitzgerald M, Gill M, Gallagher L (2004) Serotonin transporter gene and autism: a haplotype analysis in an Irish autistic population. Mol Psychiatry 9:587–593

Contin M, Martinelli P, Mochi M, Albani F, Riva R, Scaglione C, Dondi M, Fanti S, Pettinato C, Baruzzi A (2004) Dopamine transporter gene polymorphism, spect imaging, and levodopa response in patients with Parkinson disease. Clin Neuropharmacol 27:111–115

Cook EH Jr, Stein MA, Krasowski MD, Cox NJ, Olkon DM, Kieffer JE, Leventhal BL (1995) Association of attention-deficit disorder and the dopamine transporter gene. Am J Hum Genet 56:993–998

Cook EH Jr, Courchesne R, Lord C, Cox NJ, Yan S, Lincoln A, Haas R, Courchesne E, Leventhal BL (1997) Evidence of linkage between the serotonin transporter and autistic disorder. Mol Psychiatry 2:247–250

Courtet P, Baud P, Abbar M, Boulenger JP, Castelnau D, Mouthon D, Malafosse A, Buresi C (2001) Association between violent suicidal behavior and the low activity allele of the serotonin transporter gene. Mol Psychiatry 6:338–341

Curran S, Mill J, Tahir E, Kent L, Richards S, Gould A, Huckett L, Sharp J, Batten C, Fernando S, Ozbay F, Yazgan Y, Simonoff E, Thompson M, Taylor E, Asherson P (2001) Association study of a dopamine transporter polymorphism and attention deficit hyperactivity disorder in UK and Turkish samples. Mol Psychiatry 6:425–428

Curran S, Purcell S, Craig I, Asherson P, Sham P (2005) The serotonin transporter gene as a QTL for ADHD. Am J Med Genet B Neuropsychiatr Genet 134:42–47

Daniels GM, Amara SG (1999) Regulated trafficking of the human dopamine transporter. Clathrin-mediated internalization and lysosomal degradation in response to phorbol esters. J Biol Chem 274:35794–35801

Davidge KM, Atkinson L, Douglas L, Lee V, Shapiro S, Kennedy JL, Beitchman JH (2004) Association of the serotonin transporter and 5HT1Dbeta receptor genes with extreme, persistent and pervasive aggressive behaviour in children. Psychiatr Genet 14:143–146

Daws LC, Callaghan PD, Moron JA, Kahlig KM, Shippenberg TS, Javitch JA, Galli A (2002) Cocaine increases dopamine uptake and cell surface expression of dopamine transporters. Biochem Biophys Res Commun 290:1545–1550

De La Garza R 2nd, Madras BK (2000) [(3)H]PNU-101958, a D(4) dopamine receptor probe, accumulates in prefrontal cortex and hippocampus of non-human primate brain. Synapse 37:232–244

De Luca V, Muglia P, Jain U, Kennedy JL (2004) No evidence of linkage or association between the norepinephrine transporter (NET) gene MnlI polymorphism and adult ADHD. Am J Med Genet B Neuropsychiatr Genet 124:38–40

Delbruck SJ, Wendel B, Grunewald I, Sander T, Morris-Rosendahl D, Crocq MA, Berrettini WH, Hoehe MR (1997) A novel allelic variant of the human serotonin transporter gene regulatory polymorphism. Cytogenet Cell Genet 79:214–220

Delbruck SJ, Kidd KU, Hoehe MR (2001) Identification of an additional allelic variant (XLS) of the human serotonin transporter gene (SLC6A4): -1201Cins66. Hum Mutat 17:524

Di Bella D, Catalano M, Balling U, Smeraldi E, Lesch KP (1996) Systematic screening for mutations in the coding region of the human serotonin transporter (5-HTT) gene using PCR and DGGE. Am J Med Genet 67:541–545

Dobashi I, Inada T, Hadano K (1997) Alcoholism and gene polymorphisms related to central dopaminergic transmission in the Japanese population. Psychiatr Genet 7:87–91

Doucette-Stamm LA, Blakely DJ, Tian J, Mockus S, Mao JI (1995) Population genetic study of the human dopamine transporter gene (DAT1). Genet Epidemiol 12:303–308

Eddahibi S, Humbert M, Fadel E, Raffestin B, Darmon M, Capron F, Simonneau G, Dartevelle P, Hamon M, Adnot S (2001) Serotonin transporter overexpression is responsible for pulmonary artery smooth muscle hyperplasia in primary pulmonary hypertension. J Clin Invest 108:1141–1150

Eddahibi S, Chaouat A, Morrell N, Fadel E, Fuhrman C, Bugnet AS, Dartevelle P, Housset B, Hamon M, Weitzenblum E, Adnot S (2003) Polymorphism of the serotonin transporter gene and pulmonary hypertension in chronic obstructive pulmonary disease. Circulation 108:1839–1844

Erblich J, Lerman C, Self DW, Diaz GA, Bovbjerg DH (2004) Stress-induced cigarette craving: effects of the DRD2 TaqI RFLP and SLC6A3 VNTR polymorphisms. Pharmacogenomics J 4:102–109

Eshleman AJ, Stewart E, Evenson AK, Mason JN, Blakely RD, Janowsky A, Neve KA (1997) Metabolism of catecholamines by catechol-O-methyltransferase in cells expressing recombinant catecholamine transporters. J Neurochem 69:1459–1466

Esler M, Jackman G, Bobik A, Leonard P, Kelleher D, Skews H, Jennings G, Korner P (1981) Norepinephrine kinetics in essential hypertension. Defective neuronal uptake of norepinephrine in some patients. Hypertension 3:149–156

Falkenburger BH, Barstow KL, Mintz IM (2001) Dendrodendritic inhibition through reversal of dopamine transport. Science 293:2465–2470

Fan JB, Sklar P (2005) Meta-analysis reveals association between serotonin transporter gene STin2 VNTR polymorphism and schizophrenia. Mol Psychiatry 10:928–938, 891

Feinn R, Nellissery M, Kranzler HR (2005) Meta-analysis of the association of a functional serotonin transporter promoter polymorphism with alcohol dependence. Am J Med Genet B Neuropsychiatr Genet 133:79–84

Fischman AJ, Madras BK (2005) The neurobiology of attention-deficit/hyperactivity disorder. Biol Psychiatry 57:1374–1376

Fiskerstrand CE, Lovejoy EA, Quinn JP (1999) An intronic polymorphic domain often associated with susceptibility to affective disorders has allele dependent differential enhancer activity in embryonic stem cells. FEBS Lett 458:171–174

Flattem NL, Blakely RD (2000) Modified structure of the human serotonin transporter promoter. Mol Psychiatry 5:110–115

Fleckenstein AE, Haughey HM, Metzger RR, Kokoshka JM, Riddle EL, Hanson JE, Gibb JW, Hanson GR (1999) Differential effects of psychostimulants and related agents on dopaminergic and serotonergic transporter function. Eur J Pharmacol 382:45–49

Franke P, Schwab SG, Knapp M, Gansicke M, Delmo C, Zill P, Trixler M, Lichtermann D, Hallmayer J, Wildenauer DB, Maier W (1999) DAT1 gene polymorphism in alcoholism: a family-based association study. Biol Psychiatry 45:652–654

Frankle WG, Narendran R, Huang Y, Hwang DR, Lombardo I, Cangiano C, Gil R, Laruelle M, Abi-Dargham A (2005) Serotonin transporter availability in patients with schizophrenia: a positron emission tomography imaging study with [11C]DASB. Biol Psychiatry 57:1510–1516

Frisch A, Michaelovsky E, Rockah R, Amir I, Hermesh H, Laor N, Fuchs C, Zohar J, Lerer B, Buniak SF, Landa S, Poyurovsky M, Shapira B, Weizman R (2000) Association between obsessive-compulsive disorder and polymorphisms of genes encoding components of the serotonergic and dopaminergic pathways. Eur Neuropsychopharmacol 10:205–209

Fumeron F, Betoulle D, Aubert R, Herbeth B, Siest G, Rigaud D (2001) Association of a functional 5-HT transporter gene polymorphism with anorexia nervosa and food intake. Mol Psychiatry 6:9–10

Furlong RA, Ho L, Walsh C, Rubinsztein JS, Jain S, Paykel ES, Easton DF, Rubinsztein DC (1998) Analysis and meta-analysis of two serotonin transporter gene polymorphisms in bipolar and unipolar affective disorders. Am J Med Genet 81:58–63

Gainetdinov RR, Caron MG (2003) Monoamine transporters: from genes to behavior. Annu Rev Pharmacol Toxicol 43:261–284

Galeeva AR, Iur'ev EB, Khusnutdinova EK (2001) [The evaluation of VNTR-polymorphism in the gene of dopamine transporter in men of different nationalities with acute alcoholic psychosis]. Zh Nevrol Psikhiatr Im S S Korsakova 101:43–45

Garland EM, Hahn MK, Ketch TP, Keller NR, Kim CH, Kim KS, Biaggioni I, Shannon JR, Blakely RD, Robertson D (2002) Genetic basis of clinical catecholamine disorders. Ann N Y Acad Sci 971:506–514

Garpenstrand H, Annas P, Ekblom J, Oreland L, Fredrikson M (2001) Human fear conditioning is related to dopaminergic and serotonergic biological markers. Behav Neurosci 115:358–364

Gelernter J, Kranzler HR, Satel SL, Rao PA (1994) Genetic association between dopamine transporter protein alleles and cocaine-induced paranoia. Neuropsychopharmacology 11:195–200

Gelernter J, Kranzler H, Cubells JF (1997) Serotonin transporter protein (SLC6A4) allele and haplotype frequencies and linkage disequilibria in African- and European-American and Japanese populations and in alcohol-dependent subjects. Hum Genet 101:243–246

Gerra G, Garofano L, Zaimovic A, Moi G, Branchi B, Bussandri M, Brambilla F, Donnini C (2005) Association of the serotonin transporter promoter polymorphism with smoking behavior among adolescents. Am J Med Genet B Neuropsychiatr Genet 135:73–78

Gill M, Daly G, Heron S, Hawi Z, Fitzgerald M (1997) Confirmation of association between attention deficit hyperactivity disorder and a dopamine transporter polymorphism. Mol Psychiatry 2:311–313

Giros B, Jaber M, Jones SR, Wightman RM, Caron MG (1996) Hyperlocomotion and indifference to cocaine and amphetamine in mice lacking the dopamine transporter. Nature 379:606–612

Glatt CE, DeYoung JA, Delgado S, Service SK, Giacomini KM, Edwards RH, Risch N, Freimer NB (2001) Screening a large reference sample to identify very low frequency sequence variants: comparisons between two genes. Nat Genet 27:435–438

Gorwood P, Limosin F, Batel P, Hamon M, Ades J, Boni C (2003) The A9 allele of the dopamine transporter gene is associated with delirium tremens and alcohol-withdrawal seizure. Biol Psychiatry 53:85–92

Gorwood P, Lanfumey L, Hamon M (2004) [Alcohol dependence and polymorphisms of serotonin-related genes]. Med Sci (Paris) 20:1132–1138

Greenwood TA, Kelsoe JR (2003) Promoter and intronic variants affect the transcriptional regulation of the human dopamine transporter gene. Genomics 82:511–520

Greenwood TA, Alexander M, Keck PE, McElroy S, Sadovnick AD, Remick RA, Kelsoe JR (2001) Evidence for linkage disequilibrium between the dopamine transporter and bipolar disorder. Am J Med Genet 105:145–151

Greenwood TA, Alexander M, Keck PE, McElroy S, Sadovnick AD, Remick RA, Shaw SH, Kelsoe JR (2002) Segmental linkage disequilibrium within the dopamine transporter gene. Mol Psychiatry 7:165–173

Grunhage F, Schulze TG, Muller DJ, Lanczik M, Franzek E, Albus M, Borrmann-Hassenbach M, Knapp M, Cichon S, Maier W, Rietschel M, Propping P, Nothen MM (2000) Systematic screening for DNA sequence variation in the coding region of the human dopamine transporter gene (DAT1). Mol Psychiatry 5:275–282

Gulley JM, Zahniser NR (2003) Rapid regulation of dopamine transporter function by substrates, blockers and presynaptic receptor ligands. Eur J Pharmacol 479:139–152

Hahn MK, Robertson D, Blakely RD (2003) A mutation in the human norepinephrine transporter gene (SLC6A2) associated with orthostatic intolerance disrupts surface expression of mutant and wild-type transporters. J Neurosci 23:4470–4478

Hahn MK, Mazei-Robison MS, Blakely RD (2005) Single nucleotide polymorphisms in the human norepinephrine transporter gene affect expression, trafficking, antidepressant interaction, and protein kinase C regulation. Mol Pharmacol 68:457–466

Halushka MK, Fan JB, Bentley K, Hsie L, Shen N, Weder A, Cooper R, Lipshutz R, Chakravarti A (1999) Patterns of single-nucleotide polymorphisms in candidate genes for blood-pressure homeostasis. Nat Genet 22:239–247

Han DH, Park DB, Na C, Kee BS, Lee YS (2004) Association of aggressive behavior in Korean male schizophrenic patients with polymorphisms in the serotonin transporter promoter and catecholamine-O-methyltransferase genes. Psychiatry Res 129:29–37

Hariri AR, Mattay VS, Tessitore A, Kolachana B, Fera F, Goldman D, Egan MF, Weinberger DR (2002) Serotonin transporter genetic variation and the response of the human amygdala. Science 297:400–403

Heils A, Teufel A, Petri S, Stober G, Riederer P, Bengel D, Lesch KP (1996) Allelic variation of human serotonin transporter gene expression. J Neurochem 66:2621–2624

Heinz A, Goldman D, Jones DW, Palmour R, Hommer D, Gorey JG, Lee KS, Linnoila M, Weinberger DR (2000) Genotype influences in vivo dopamine transporter availability in human striatum. Neuropsychopharmacology 22:133–139

Hernandez I, Sokolov BP (1997) Abnormal expression of serotonin transporter mRNA in the frontal and temporal cortex of schizophrenics. Mol Psychiatry 2:57–64

Hersch SM, Yi H, Heilman CJ, Edwards RH, Levey AI (1997) Subcellular localization and molecular topology of the dopamine transporter in the striatum and substantia nigra. J Comp Neurol 388:211–227

Holmes J, Payton A, Barrett JH, Hever T, Fitzpatrick H, Trumper AL, Harrington R, McGuffin P, Owen M, Ollier W, Worthington J, Thapar A (2000) A family-based and case-control association study of the dopamine D4 receptor gene and dopamine transporter gene in attention deficit hyperactivity disorder. Mol Psychiatry 5:523–530

Hong CJ, Cheng CY, Shu LR, Yang CY, Tsai SJ (2003) Association study of the dopamine and serotonin transporter genetic polymorphisms and methamphetamine abuse in Chinese males. J Neural Transm 110:345–351

Hu X, Oroszi G, Chun J, Smith TL, Goldman D, Schuckit MA (2005) An expanded evaluation of the relationship of four alleles to the level of response to alcohol and the alcoholism risk. Alcohol Clin Exp Res 29:8–16

Ingram SL, Prasad BM, Amara SG (2002) Dopamine transporter-mediated conductances increase excitability of midbrain dopamine neurons. Nat Neurosci 5:971–978

Inoue K, Itoh K, Yoshida K, Shimizu T, Suzuki T (2004) Positive association between T-182C polymorphism in the norepinephrine transporter gene and susceptibility to major depressive disorder in a Japanese population. Neuropsychobiology 50:301–304

Iwasa H, Kurabayashi M, Nagai R, Nakamura Y, Tanaka T (2001) Genetic variations in five genes involved in the excitement of cardiomyocytes. J Hum Genet 46:549–552

Jacobs BL, Azmitia EC (1992) Structure and function of the brain serotonin system. Physiol Rev 72:165–229

Jacobsen LK, Staley JK, Zoghbi SS, Seibyl JP, Kosten TR, Innis RB, Gelernter J (2000) Prediction of dopamine transporter binding availability by genotype: a preliminary report. Am J Psychiatry 157:1700–1703

Javors MA, Seneviratne C, Roache JD, Ait-Daoud N, Bergeson SE, Walss-Bass MC, Akhtar FZ, Johnson BA (2005) Platelet serotonin uptake and paroxetine binding among allelic genotypes of the serotonin transporter in alcoholics. Prog Neuropsychopharmacol Biol Psychiatry 29:7–13

Jorm AF, Henderson AS, Jacomb PA, Christensen H, Korten AE, Rodgers B, Tan X, Easteal S (2000) Association of smoking and personality with a polymorphism of the dopamine transporter gene: results from a community survey. Am J Med Genet 96:331–334

Jucaite A, Fernell E, Halldin C, Forssberg H, Farde L (2005) Reduced midbrain dopamine transporter binding in male adolescents with attention-deficit/hyperactivity disorder: association between striatal dopamine markers and motor hyperactivity. Biol Psychiatry 57:229–238

Keikhaee MR, Fadai F, Sargolzaee MR, Javanbakht A, Najmabadi H, Ohadi M (2005) Association analysis of the dopamine transporter (DAT1)-67A/T polymorphism in bipolar disorder. Am J Med Genet B Neuropsychiatr Genet 135:47–49

Kennelly PJ, Krebs EG (1991) Consensus sequences as substrate specificity determinants for protein kinases and protein phosphatases. J Biol Chem 266:15555–15558

Khodayari N, Garshasbi M, Fadai F, Rahimi A, Hafizi L, Ebrahimi A, Najmabadi H, Ohadi M (2004) Association of the dopamine transporter gene (DAT1) core promoter polymorphism −67T variant with schizophrenia. Am J Med Genet B Neuropsychiatr Genet 129:10–12

Kilic F, Murphy DL, Rudnick G (2003) A human serotonin transporter mutation causes constitutive activation of transport activity. Mol Pharmacol 64:440–446

Kim CH, Kim HS, Cubells JF, Kim KS (1999) A previously undescribed intron and extensive 5′ upstream sequence, but not Phox2a-mediated transactivation, are necessary for high level cell type-specific expression of the human norepinephrine transporter gene. J Biol Chem 274:6507–6518

Kimmel HL, Carroll FI, Kuhar MJ (2000) Dopamine transporter synthesis and degradation rate in rat striatum and nucleus accumbens using RTI-76. Neuropharmacology 39:578–585

Kinnear CJ, Niehaus DJ, Moolman-Smook JC, du Toit PL, van Kradenberg J, Weyers JB, Potgieter A, Marais V, Emsley RA, Knowles JA, Corfield VA, Brink PA, Stein DJ (2000) Obsessive-compulsive disorder and the promoter region polymorphism (5-HTTLPR) in the serotonin transporter gene (SLC6A4): a negative association study in the Afrikaner population. Int J Neuropsychopharmacol 3:327–331

Kirley A, Lowe N, Hawi Z, Mullins C, Daly G, Waldman I, McCarron M, O'Donnell D, Fitzgerald M, Gill M (2003) Association of the 480 bp DAT1 allele with methylphenidate response in a sample of Irish children with ADHD. Am J Med Genet B Neuropsychiatr Genet 121:50–54

Klauck SM, Poustka F, Benner A, Lesch KP, Poustka A (1997) Serotonin transporter (5-HTT) gene variants associated with autism? Hum Mol Genet 6:2233–2238

Klenova E, Scott AC, Roberts J, Shamsuddin S, Lovejoy EA, Bergmann S, Bubb VJ, Royer HD, Quinn JP (2004) YB-1 and CTCF differentially regulate the 5-HTT polymorphic intron 2 enhancer which predisposes to a variety of neurological disorders. J Neurosci 24:5966–5973

Klimek V, Stockmeier C, Overholser J, Meltzer HY, Kalka S, Dilley G, Ordway GA (1997) Reduced levels of norepinephrine transporters in the locus coeruleus in major depression. J Neurosci 17:8451–8458

Kremer I, Bachner-Melman R, Reshef A, Broude L, Nemanov L, Gritsenko I, Heresco-Levy U, Elizur Y, Ebstein RP (2005) Association of the serotonin transporter gene with smoking behavior. Am J Psychiatry 162:924–930

Lasky-Su JA, Faraone SV, Glatt SJ, Tsuang MT (2005) Meta-analysis of the association between two polymorphisms in the serotonin transporter gene and affective disorders. Am J Med Genet B Neuropsychiatr Genet 133:110–115

Lee DY, Park H, Kim WH, Lee SI, Seo YJ, Choi YC (2004) [Serotonin transporter gene polymorphism in healthy adults and patients with irritable bowel syndrome]. Korean J Gastroenterol 43:18–22

Lerman C, Caporaso NE, Audrain J, Main D, Bowman ED, Lockshin B, Boyd NR, Shields PG (1999) Evidence suggesting the role of specific genetic factors in cigarette smoking. Health Psychol 18:14–20

Lerman C, Shields PG, Wileyto EP, Audrain J, Hawk LH Jr, Pinto A, Kucharski S, Krishnan S, Niaura R, Epstein LH (2003) Effects of dopamine transporter and receptor polymorphisms on smoking cessation in a bupropion clinical trial. Health Psychol 22:541–548

Lesch KP, Bengel D, Heils A, Sabol SZ, Greenberg BD, Petri S, Benjamin J, Muller CR, Hamer DH, Murphy DL (1996) Association of anxiety-related traits with a polymorphism in the serotonin transporter gene regulatory region. Science 274:1527–1531

Leszczynska-Rodziewicz A, Czerski PM, Kapelski P, Godlewski S, Dmitrzak-Weglarz M, Rybakowski J, Hauser J (2002) A polymorphism of the norepinephrine transporter gene in bipolar disorder and schizophrenia: lack of association. Neuropsychobiology 45:182–185

Levinson DF (2005) Meta-analysis in psychiatric genetics. Curr Psychiatry Rep 7:143–151

Li LB, Chen N, Ramamoorthy S, Chi L, Cui XN, Wang LC, Reith ME (2004) The role of N-glycosylation in function and surface trafficking of the human dopamine transporter. J Biol Chem 279:21012–21020

Li T, Holmes C, Sham PC, Vallada H, Birkett J, Kirov G, Lesch KP, Powell J, Lovestone S, Collier D (1997) Allelic functional variation of serotonin transporter expression is a susceptibility factor for late onset Alzheimer's disease. Neuroreport 8:683–686

Limosin F, Loze JY, Boni C, Fedeli LP, Hamon M, Rouillon F, Ades J, Gorwood P (2004) The A9 allele of the dopamine transporter gene increases the risk of visual hallucinations during alcohol withdrawal in alcohol-dependent women. Neurosci Lett 362:91–94

Lin PY, Tsai G (2004) Association between serotonin transporter gene promoter polymorphism and suicide: results of a meta-analysis. Biol Psychiatry 55:1023–1030

Lin Z, Uhl GR (2003) Human dopamine transporter gene variation: effects of protein coding variants V55A and V382A on expression and uptake activities. Pharmacogenomics J 3:159–168

Lin Z, Wang W, Kopajtic T, Revay RS, Uhl GR (1999) Dopamine transporter: transmembrane phenylalanine mutations can selectively influence dopamine uptake and cocaine analog recognition. Mol Pharmacol 56:434–447

Ling D, Niu T, Feng Y, Xing H, Xu X (2004) Association between polymorphism of the dopamine transporter gene and early smoking onset: an interaction risk on nicotine dependence. J Hum Genet 49:35–39

Little KY, McLaughlin DP, Zhang L, Livermore CS, Dalack GW, McFinton PR, DelProposto ZS, Hill E, Cassin BJ, Watson SJ, Cook EH (1998) Cocaine, ethanol, and genotype effects on human midbrain serotonin transporter binding sites and mRNA levels. Am J Psychiatry 155:207–213

Little KY, Zhang L, Desmond T, Frey KA, Dalack GW, Cassin BJ (1999) Striatal dopaminergic abnormalities in human cocaine users. Am J Psychiatry 156:238–245

Little KY, Elmer LW, Zhong H, Scheys JO, Zhang L (2002) Cocaine induction of dopamine transporter trafficking to the plasma membrane. Mol Pharmacol 61:436–445

Loder MK, Melikian HE (2003) The dopamine transporter constitutively internalizes and recycles in a protein kinase C-regulated manner in stably transfected PC12 cell lines. J Biol Chem 278:22168–22174

Lotrich FE, Pollock BG (2004) Meta-analysis of serotonin transporter polymorphisms and affective disorders. Psychiatr Genet 14:121–129

MacKenzie A, Quinn J (1999) A serotonin transporter gene intron 2 polymorphic region, correlated with affective disorders, has allele-dependent differential enhancer-like properties in the mouse embryo. Proc Natl Acad Sci U S A 96:15251–15255

Madras BK, Miller GM, Fischman AJ (2002) The dopamine transporter: relevance to attention deficit hyperactivity disorder (ADHD). Behav Brain Res 130:57–63

Madras BK, Miller GM, Fischman AJ (2005) The dopamine transporter and attention-deficit/hyperactivity disorder. Biol Psychiatry 57:1397–1409

Maher BS, Marazita ML, Ferrell RE, Vanyukov MM (2002) Dopamine system genes and attention deficit hyperactivity disorder: a meta-analysis. Psychiatr Genet 12:207–215

Malison RT, Price LH, Berman R, van Dyck CH, Pelton GH, Carpenter L, Sanacora G, Owens MJ, Nemeroff CB, Rajeevan N, Baldwin RM, Seibyl JP, Innis RB, Charney DS (1998) Reduced brain serotonin transporter availability in major depression as measured by [123I]-2 beta-carbomethoxy-3 beta-(4-iodophenyl)tropane and single photon emission computed tomography. Biol Psychiatry 44:1090–1098

Mao W, Qin F, Iwai C, Vulapalli R, Keng PC, Liang CS (2004) Extracellular norepinephrine reduces neuronal uptake of norepinephrine by oxidative stress in PC12 cells. Am J Physiol Heart Circ Physiol 287:H29–39

Martinez D, Gelernter J, Abi-Dargham A, van Dyck CH, Kegeles L, Innis RB, Laruelle M (2001) The variable number of tandem repeats polymorphism of the dopamine transporter gene is not associated with significant change in dopamine transporter phenotype in humans. Neuropsychopharmacology 24:553–560

Matsushita S, Suzuki K, Murayama M, Nishiguchi N, Hishimoto A, Takeda A, Shirakawa O, Higuchi S (2004) Serotonin transporter regulatory region polymorphism is associated with anorexia nervosa. Am J Med Genet B Neuropsychiatr Genet 128:114–117

McConathy J, Owens MJ, Kilts CD, Malveaux EJ, Camp VM, Votaw JR, Nemeroff CB, Goodman MM (2004) Synthesis and biological evaluation of [11C]talopram and [11C]talsupram: candidate PET ligands for the norepinephrine transporter. Nucl Med Biol 31:705–718

McDougle CJ, Epperson CN, Price LH, Gelernter J (1998) Evidence for linkage disequilibrium between serotonin transporter protein gene (SLC6A4) and obsessive compulsive disorder. Mol Psychiatry 3:270–273

McEvoy B, Hawi Z, Fitzgerald M, Gill M (2002) No evidence of linkage or association between the norepinephrine transporter (NET) gene polymorphisms and ADHD in the Irish population. Am J Med Genet 114:665–666

Merlet P, Dubois-Rande JL, Adnot S, Bourguignon MH, Benvenuti C, Loisance D, Valette H, Castaigne A, Syrota A (1992) Myocardial beta-adrenergic desensitization and neuronal norepinephrine uptake function in idiopathic dilated cardiomyopathy. J Cardiovasc Pharmacol 19:10–16

Mill J, Asherson P, Browes C, D'Souza U, Craig I (2002) Expression of the dopamine transporter gene is regulated by the 3' UTR VNTR: Evidence from brain and lymphocytes using quantitative RT-PCR. Am J Med Genet 114:975–979

Miller GM, Madras BK (2002) Polymorphisms in the 3'-untranslated region of human and monkey dopamine transporter genes affect reporter gene expression. Mol Psychiatry 7:44–55

Miller KJ, Hoffman BJ (1994) Adenosine A3 receptors regulate serotonin transport via nitric oxide and cGMP. J Biol Chem 269:27351–27356

Miranda M, Wu CC, Sorkina T, Korstjens D, Sorkin A (2005) Enhanced ubiquitylation and accelerated degradation of the dopamine transporter mediated by protein kinase C. J Biol Chem 280:35617–35624

Mortensen OV, Amara SG (2003) Dynamic regulation of the dopamine transporter. Eur J Pharmacol 479:159–170

Muglia P, Jain U, Inkster B, Kennedy JL (2002) A quantitative trait locus analysis of the dopamine transporter gene in adults with ADHD. Neuropsychopharmacology 27:655–662

Muller-Vahl KR, Berding G, Brucke T, Kolbe H, Meyer GJ, Hundeshagen H, Dengler R, Knapp WH, Emrich HM (2000) Dopamine transporter binding in Gilles de la Tourette syndrome. J Neurol 247:514–520

Munafo MR, Clark T, Flint J (2005) Does measurement instrument moderate the association between the serotonin transporter gene and anxiety-related personality traits? A meta-analysis. Mol Psychiatry 10:415–419

Muramatsu T, Higuchi S (1995) Dopamine transporter gene polymorphism and alcoholism. Biochem Biophys Res Commun 211:28–32

Myers CL, Pitt BR (1988) Selective effect of phorbol ester on serotonin removal and ACE activity in rabbit lungs. J Appl Physiol 65:377–384

Nakamura M, Ueno S, Sano A, Tanabe H (2000) The human serotonin transporter gene linked polymorphism (5-HTTLPR) shows ten novel allelic variants. Mol Psychiatry 5:32–38

Nakatome M, Honda K, Islam MN, Terada M, Yamazaki M, Kuroki H, Ogura Y, Bai H, Wakasugi C (1995) Amplification of DAT1 (human dopamine transporter gene) 3′ variable region in the Japanese population. Hum Hered 45:262–265

Newberg AB, Amsterdam JD, Wintering N, Ploessl K, Swanson RL, Shults J, Alavi A (2005) 123I-ADAM binding to serotonin transporters in patients with major depression and healthy controls: a preliminary study. J Nucl Med 46:973–977

Newport DJ, Owens MJ, Knight DL, Ragan K, Morgan N, Nemeroff CB, Stowe ZN (2004) Alterations in platelet serotonin transporter binding in women with postpartum onset major depression. J Psychiatr Res 38:467–473

Nirenberg MJ, Vaughan RA, Uhl GR, Kuhar MJ, Pickel VM (1996) The dopamine transporter is localized to dendritic and axonal plasma membranes of nigrostriatal dopaminergic neurons. J Neurosci 16:436–447

Ogilvie AD, Battersby S, Bubb VJ, Fink G, Harmar AJ, Goodwim GM, Smith CA (1996) Polymorphism in serotonin transporter gene associated with susceptibility to major depression. Lancet 347:731–733

Oh KS, Shin DW, Oh GT, Noh KS (2003) Dopamine transporter genotype influences the attention deficit in Korean boys with ADHD. Yonsei Med J 44:787–792

Ono K, Iwanaga Y, Mannami T, Kokubo Y, Tomoike H, Komamura K, Shioji K, Yasui N, Tago N, Iwai N (2003) Epidemiological evidence of an association between SLC6A2 gene polymorphism and hypertension. Hypertens Res 26:685–689

Owen D, Du L, Bakish D, Lapierre YD, Hrdina PD (1999) Norepinephrine transporter gene polymorphism is not associated with susceptibility to major depression. Psychiatry Res 87:1–5

Owens MJ, Nemeroff CB (1998) The serotonin transporter and depression. Depress Anxiety 8 Suppl 1:5–12

Ozaki N, Goldman D, Kaye WH, Plotnicov K, Greenberg BD, Lappalainen J, Rudnick G, Murphy DL (2003) Serotonin transporter missense mutation associated with a complex neuropsychiatric phenotype. Mol Psychiatry 8:895, 933–936

Pata C, Erdal ME, Derici E, Yazar A, Kanik A, Ulu O (2002) Serotonin transporter gene polymorphism in irritable bowel syndrome. Am J Gastroenterol 97:1780–1784

Patkar AA, Berrettini WH, Hoehe M, Thornton CC, Gottheil E, Hill K, Weinstein SP (2002) Serotonin transporter polymorphisms and measures of impulsivity, aggression, and sensation seeking among African-American cocaine-dependent individuals. Psychiatry Res 110:103–115

Patkar AA, Gopalakrishnan R, Berrettini WH, Weinstein SP, Vergare MJ, Leone FT (2003) Differences in platelet serotonin transporter sites between African-American tobacco smokers and non-smokers. Psychopharmacology (Berl) 166:221–227

Paul SM, Rehavi M, Skolnick P, Ballenger JC, Goodwin FK (1981) Depressed patients have decreased binding of tritiated imipramine to platelet serotonin "transporter". Arch Gen Psychiatry 38:1315–1317

Payton A, Holmes J, Barrett JH, Hever T, Fitzpatrick H, Trumper AL, Harrington R, McGuffin P, O'Donovan M, Owen M, Ollier W, Worthington J, Thapar A (2001a) Examining for association between candidate gene polymorphisms in the dopamine pathway and attention-deficit hyperactivity disorder: a family-based study. Am J Med Genet 105:464–470

Payton A, Holmes J, Barrett JH, Sham P, Harrington R, McGuffin P, Owen M, Ollier W, Worthington J, Thapar A (2001b) Susceptibility genes for a trait measure of attention deficit hyperactivity disorder: a pilot study in a non-clinical sample of twins. Psychiatry Res 105:273–278

Pezawas L, Meyer-Lindenberg A, Drabant EM, Verchinski BA, Munoz KE, Kolachana BS, Egan MF, Mattay VS, Hariri AR, Weinberger DR (2005) 5-HTTLPR polymorphism impacts human cingulate-amygdala interactions: a genetic susceptibility mechanism for depression. Nat Neurosci 8:828–834

Prasad HC, Zhu CB, McCauley JL, Samuvel DJ, Ramamoorthy S, Shelton RC, Hewlett WA, Sutcliffe JS, Blakely RD (2005) Human serotonin transporter variants display altered sensitivity to protein kinase G and p38 mitogen-activated protein kinase. Proc Natl Acad Sci U S A 102:11545–11550

Qian Q, Wang Y, Li J, Yang L, Wang B, Zhou R (2003) [Association studies of dopamine D4 receptor gene and dopamine transporter gene polymorphisms in Han Chinese patients with attention deficit hyperactivity disorder]. Beijing Da Xue Xue Bao 35:412–418

Qian Y, Galli A, Ramamoorthy S, Risso S, DeFelice LJ, Blakely RD (1997) Protein kinase C activation regulates human serotonin transporters in HEK-293 cells via altered cell surface expression. J Neurosci 17:45–57

Retz W, Retz-Junginger P, Supprian T, Thome J, Rosler M (2004) Association of serotonin transporter promoter gene polymorphism with violence: relation with personality disorders, impulsivity, and childhood ADHD psychopathology. Behav Sci Law 22:415–425

Roman T, Schmitz M, Polanczyk G, Eizirik M, Rohde LA, Hutz MH (2001) Attention-deficit hyperactivity disorder: a study of association with both the dopamine transporter gene and the dopamine D4 receptor gene. Am J Med Genet 105:471–478

Roman T, Szobot C, Martins S, Biederman J, Rohde LA, Hutz MH (2002) Dopamine transporter gene and response to methylphenidate in attention-deficit/hyperactivity disorder. Pharmacogenetics 12:497–499

Rowe DC, Stever C, Gard JM, Cleveland HH, Sanders ML, Abramowitz A, Kozol ST, Mohr JH, Sherman SL, Waldman ID (1998) The relation of the dopamine transporter gene (DAT1) to symptoms of internalizing disorders in children. Behav Genet 28:215–225

Rowe DC, Stever C, Chase D, Sherman S, Abramowitz A, Waldman ID (2001) Two dopamine genes related to reports of childhood retrospective inattention and conduct disorder symptoms. Mol Psychiatry 6:429–433

Runkel F, Bruss M, Nothen MM, Stober G, Propping P, Bonisch H (2000) Pharmacological properties of naturally occurring variants of the human norepinephrine transporter. Pharmacogenetics 10:397–405

Ryu SH, Lee SH, Lee HJ, Cha JH, Ham BJ, Han CS, Choi MJ, Lee MS (2004) Association between norepinephrine transporter gene polymorphism and major depression. Neuropsychobiology 49:174–177

Sand PG, Mori T, Godau C, Stober G, Flachenecker P, Franke P, Nothen MM, Fritze J, Maier W, Lesch KP, Riederer P, Beckmann H, Deckert J (2002) Norepinephrine transporter gene (NET) variants in patients with panic disorder. Neurosci Lett 333:41–44

Sander T, Harms H, Podschus J, Finckh U, Nickel B, Rolfs A, Rommelspacher H, Schmidt LG (1997) Allelic association of a dopamine transporter gene polymorphism in alcohol dependence with withdrawal seizures or delirium. Biol Psychiatry 41:299–304

Saunders C, Ferrer JV, Shi L, Chen J, Merrill G, Lamb ME, Leeb-Lundberg LM, Carvelli L, Javitch JA, Galli A (2000) Amphetamine-induced loss of human dopamine transporter activity: an internalization-dependent and cocaine-sensitive mechanism. Proc Natl Acad Sci U S A 97:6850–6855

Schinka JA, Busch RM, Robichaux-Keene N (2004) A meta-analysis of the association between the serotonin transporter gene polymorphism (5-HTTLPR) and trait anxiety. Mol Psychiatry 9:197–202

Schmidt LG, Harms H, Kuhn S, Rommelspacher H, Sander T (1998) Modification of alcohol withdrawal by the A9 allele of the dopamine transporter gene. Am J Psychiatry 155:474–478

Sen S, Burmeister M, Ghosh D (2004) Meta-analysis of the association between a serotonin transporter promoter polymorphism (5-HTTLPR) and anxiety-related personality traits. Am J Med Genet B Neuropsychiatr Genet 127:85–89

Shannon JR, Flattem NL, Jordan J, Jacob G, Black BK, Biaggioni I, Blakely RD, Robertson D (2000) Orthostatic intolerance and tachycardia associated with norepinephrine-transporter deficiency. N Engl J Med 342:541–549

Sonders MS, Zhu SJ, Zahniser NR, Kavanaugh MP, Amara SG (1997) Multiple ionic conductances of the human dopamine transporter: the actions of dopamine and psychostimulants. J Neurosci 17:960–974

Sora I, Wichems C, Takahashi N, Li XF, Zeng Z, Revay R, Lesch KP, Murphy DL, Uhl GR (1998) Cocaine reward models: conditioned place preference can be established in dopamine- and in serotonin-transporter knockout mice. Proc Natl Acad Sci U S A 95:7699–7704

Sora I, Hall FS, Andrews AM, Itokawa M, Li XF, Wei HB, Wichems C, Lesch KP, Murphy DL, Uhl GR (2001) Molecular mechanisms of cocaine reward: combined dopamine and serotonin transporter knockouts eliminate cocaine place preference. Proc Natl Acad Sci U S A 98:5300–5305

Sorkina T, Doolen S, Galperin E, Zahniser NR, Sorkin A (2003) Oligomerization of dopamine transporters visualized in living cells by fluorescence resonance energy transfer microscopy. J Biol Chem 278:28274–28283

Stober G, Nothen MM, Porzgen P, Bruss M, Bonisch H, Knapp M, Beckmann H, Propping P (1996) Systematic search for variation in the human norepinephrine transporter gene: identification of five naturally occurring missense mutations and study of association with major psychiatric disorders. Am J Med Genet 67:523–532

Stober G, Hebebrand J, Cichon S, Bruss M, Bonisch H, Lehmkuhl G, Poustka F, Schmidt M, Remschmidt H, Propping P, Nothen MM (1999) Tourette syndrome and the norepinephrine transporter gene: results of a systematic mutation screening. Am J Med Genet 88:158–163

Sundaramurthy D, Pieri LF, Gape H, Markham AF, Campbell DA (2000) Analysis of the serotonin transporter gene linked polymorphism (5-HTTLPR) in anorexia nervosa. Am J Med Genet 96:53–55

Szot P, Leverenz JB, Peskind ER, Kiyasu E, Rohde K, Miller MA, Raskind MA (2000) Tyrosine hydroxylase and norepinephrine transporter mRNA expression in the locus coeruleus in Alzheimer's disease. Brain Res Mol Brain Res 84:135–140

Todd RD, Jong YJ, Lobos EA, Reich W, Heath AC, Neuman RJ (2001) No association of the dopamine transporter gene 3' VNTR polymorphism with ADHD subtypes in a population sample of twins. Am J Med Genet 105:745–748

Torres GE, Carneiro A, Seamans K, Fiorentini C, Sweeney A, Yao WD, Caron MG (2003) Oligomerization and trafficking of the human dopamine transporter. Mutational analysis identifies critical domains important for the functional expression of the transporter. J Biol Chem 278:2731–2739

Tsai SJ, Wang YC, Hong CJ (2002) Norepinephrine transporter and alpha(2c) adrenoceptor allelic variants and personality factors. Am J Med Genet 114:649–651

Ueno S, Nakamura M, Mikami M, Kondoh K, Ishiguro H, Arinami T, Komiyama T, Mitsushio H, Sano A, Tanabe H (1999) Identification of a novel polymorphism of the human dopamine transporter (DAT1) gene and the significant association with alcoholism. Mol Psychiatry 4:552–557

Ujike H, Harano M, Inada T, Yamada M, Komiyama T, Sekine Y, Sora I, Iyo M, Katsu T, Nomura A, Nakata K, Ozaki N (2003) Nine- or fewer repeat alleles in VNTR polymorphism of the dopamine transporter gene is a strong risk factor for prolonged methamphetamine psychosis. Pharmacogenomics J 3:242–247

Urwin RE, Bennetts B, Wilcken B, Lampropoulos B, Beumont P, Clarke S, Russell J, Tanner S, Nunn KP (2002) Anorexia nervosa (restrictive subtype) is associated with a polymorphism in the novel norepinephrine transporter gene promoter polymorphic region. Mol Psychiatry 7:652–657

van Dyck CH, Malison RT, Seibyl JP, Laruelle M, Klumpp H, Zoghbi SS, Baldwin RM, Innis RB (2000) Age-related decline in central serotonin transporter availability with [(123)I]beta-CIT SPECT. Neurobiol Aging 21:497–501

van Dyck CH, Malison RT, Staley JK, Jacobsen LK, Seibyl JP, Laruelle M, Baldwin RM, Innis RB, Gelernter J (2004) Central serotonin transporter availability measured with [123I]beta-CIT SPECT in relation to serotonin transporter genotype. Am J Psychiatry 161:525–531

van Dyck CH, Malison RT, Jacobsen LK, Seibyl JP, Staley JK, Laruelle M, Baldwin RM, Innis RB, Gelernter J (2005) Increased dopamine transporter availability associated with the 9-repeat allele of the SLC6A3 gene. J Nucl Med 46:745–751

Vandenbergh DJ, Persico AM, Hawkins AL, Griffin CA, Li X, Jabs EW, Uhl GR (1992) Human dopamine transporter gene (DAT1) maps to chromosome 5p15.3 and displays a VNTR. Genomics 14:1104–1106

Vandenbergh DJ, Thompson MD, Cook EH, Bendahhou E, Nguyen T, Krasowski MD, Zarrabian D, Comings D, Sellers EM, Tyndale RF, George SR, O'Dowd BF, Uhl GR (2000) Human dopamine transporter gene: coding region conservation among normal, Tourette's disorder, alcohol dependence and attention-deficit hyperactivity disorder populations. Mol Psychiatry 5:283–292

Vandenbergh DJ, Bennett CJ, Grant MD, Strasser AA, O'Connor R, Stauffer RL, Vogler GP, Kozlowski LT (2002) Smoking status and the human dopamine transporter variable number of tandem repeats (VNTR) polymorphism: failure to replicate and finding that never-smokers may be different. Nicotine Tob Res 4:333–340

Verrico C, Miller GM, Madras BK (2005) MDMA (ecstasy) and human dopamine, norepinephrine and serotonin transporters: implications for MDMA-Induced neurotoxicity and treatment. Psychopharmacology (Berl) 12:1–15

Waldman ID, Rowe DC, Abramowitz A, Kozel ST, Mohr JH, Sherman SL, Cleveland HH, Sanders ML, Gard JM, Stever C (1998) Association and linkage of the dopamine transporter gene and attention-deficit hyperactivity disorder in children: heterogeneity owing to diagnostic subtype and severity. Am J Hum Genet 63:1767–1776

Walitza S, Wewetzer C, Gerlach M, Klampfl K, Geller F, Barth N, Hahn F, Herpertz-Dahlmann B, Gossler M, Fleischhaker C, Schulz E, Hebebrand J, Warnke A, Hinney A (2004) Transmission disequilibrium studies in children and adolescents with obsessive-compulsive disorders pertaining to polymorphisms of genes of the serotonergic pathway. J Neural Transm 111:817–825

Wernicke C, Smolka M, Gallinat J, Winterer G, Schmidt LG, Rommelspacher H (2002) Evidence for the importance of the human dopamine transporter gene for withdrawal symptomatology of alcoholics in a German population. Neurosci Lett 333:45–48

Winsberg BG, Comings DE (1999) Association of the dopamine transporter gene (DAT1) with poor methylphenidate response. J Am Acad Child Adolesc Psychiatry 38:1474–1477

Xu F, Gainetdinov RR, Wetsel WC, Jones SR, Bohn LM, Miller GW, Wang YM, Caron MG (2000) Mice lacking the norepinephrine transporter are supersensitive to psychostimulants. Nat Neurosci 3:465–471

Xu X, Knight J, Brookes K, Mill J, Sham P, Craig I, Taylor E, Asherson P (2005) DNA pooling analysis of 21 norepinephrine transporter gene SNPs with attention deficit hyperactivity disorder: no evidence for association. Am J Med Genet B Neuropsychiatr Genet 134:115–118

Yang L, Wang YF, Li J, Faraone SV (2004) Association of norepinephrine transporter gene with methylphenidate response. J Am Acad Child Adolesc Psychiatry 43:1154–1158

Yeo A, Boyd P, Lumsden S, Saunders T, Handley A, Stubbins M, Knaggs A, Asquith S, Taylor I, Bahari B, Crocker N, Rallan R, Varsani S, Montgomery D, Alpers DH, Dukes GE, Purvis I, Hicks GA (2004) Association between a functional polymorphism in the serotonin transporter gene and diarrhoea predominant irritable bowel syndrome in women. Gut 53:1452–1458

Zalsman G, Frisch A, Bromberg M, Gelernter J, Michaelovsky E, Campino A, Erlich Z, Tyano S, Apter A, Weizman A (2001) Family-based association study of serotonin transporter promoter in suicidal adolescents: no association with suicidality but possible role in violence traits. Am J Med Genet 105:239–245

Zhuang X, Oosting RS, Jones SR, Gainetdinov RR, Miller GW, Caron MG, Hen R (2001) Hyperactivity and impaired response habituation in hyperdopaminergic mice. Proc Natl Acad Sci U S A 98:1982–1987

Zill P, Engel R, Baghai TC, Juckel G, Frodl T, Muller-Siecheneder F, Zwanzger P, Schule C, Minov C, Behrens S, Rupprecht R, Hegerl U, Moller HJ, Bondy B (2002) Identification of a naturally occurring polymorphism in the promoter region of the norepinephrine transporter and analysis in major depression. Neuropsychopharmacology 26:489–493

ADHD and the Dopamine Transporter: Are There Reasons to Pay Attention?

M. S. Mazei-Robison · R. D. Blakely (✉)

Vanderbilt School of Medicine, Suite 7140, MRB III, Nashville TN, 37232-8548, USA
randy.blakely@vanderbilt.edu

1	Overview of ADHD	374
1.1	Diagnosis and Prevalence of ADHD	374
1.2	Cognitive Deficits in ADHD, the Search for Endophenotypes	375
1.3	Treatment of ADHD	377
2	The Human Dopamine Transporter	380
2.1	Cloning the hDAT Gene	380
2.2	DAT Protein: Structure/Function	383
2.3	DAT Protein: Regulation	385
2.4	DAT Protein: Pharmacological Impact on Transporter Regulation	386
3	DAT Transgenics as Animal Models of ADHD	388
3.1	DAT Knockout Mice as a Model of ADHD	388
3.2	DAT Knockdown Mice as a Model of ADHD	389
4	Neuroimaging DAT in Human Subjects	390
4.1	DAT in Adult ADHD Subjects	390
4.2	DAT in Children with ADHD	392
4.3	Influence of DAT 3′VNTR on DAT Levels	393
5	Genetic Linkage in ADHD and the Impact of DAT Gene Variants	394
5.1	ADHD Linkage Studies	394
5.2	Association Studies of hDAT and ADHD	396
5.3	DAT 3′VNTR and Response to Methylphenidate	398
5.4	Investigation of hDAT Coding Variants in ADHD	399
	References	403

Abstract The catecholamine dopamine (DA) plays an important role as a neurotransmitter in the brain in circuits linked to motor function, reward, and cognition. The presynaptic DA transporter (DAT) inactivates DA following release and provides a route for non-exocytotic DA release (efflux) triggered by amphetamines. The synaptic role of DATs first established through antagonist studies and more recently validated through mouse gene-knockout experiments, raises questions as to whether altered DAT structure or regulation support clinical disorders linked to compromised DA signaling, including drug abuse, schizophrenia, and attention deficit hyperactivity disorder (ADHD). As ADHD appears to have highly heritable components and the most commonly prescribed therapeutics for ADHD target DAT, studies ranging from brain imaging to genomic and genetic analyses have begun to probe the DAT gene and its protein for possible contributions to the disorder and/or its treatment. In this review, after a brief overview of ADHD prevalence and diagnostic criteria, we examine the rationale and experimental findings surrounding a role for human DAT in ADHD. Based on the available evidence from our lab and labs of workers in the field, we suggest that although a common variant within the human DAT (hDAT) gene (*SLC6A3*) is unlikely to play a major role in the ADHD, contributions of hDAT to risk may be most evident in phenotypic subgroups. The in vitro and in vivo validation of functional variants, pursued for contributions to endophenotypes in a within family approach, may help elucidate DAT and DA contributions to ADHD and its treatment.

Keywords Dopamine · Transporter · Gene · Attention · ADHD · Methylphenidate

1
Overview of ADHD

1.1
Diagnosis and Prevalence of ADHD

Attention deficit hyperactivity disorder (ADHD) is one of the most commonly diagnosed disorders of childhood, affecting a reported 3%–5% of the school-age children in the U.S. (American Psychiatric Association 1994). Most investigators now make clinical diagnoses of ADHD using the diagnostic and statistical manual of the American Psychiatric Association (DSM-IV; American Psychiatric Association 1994). There are currently no biological tests for ADHD, and thus a diagnosis is based purely on a clinical assessment that also involves parent and teacher observations. The DSM-IV assessment allows for a diagnosis of three subtypes of ADHD: a predominately inattentive subtype, a predominately hyperactive-impulsive subtype, and a combined subtype that displays components of both primary subtypes. In order to meet DSM-IV diagnostic criteria for ADHD, a subject must be ascribed 6 out of 9 symptoms of hyperactivity/impulsivity and/or 6 out of 9 symptoms of inattention (Table 1). These symptoms must have been present for at least 6 months and they must be observable in at least two settings (most often at home and school). Additionally, symptoms must generate evidence of clinically significant impairment in functioning and that at least some of the symptoms should be present before 7 years of age. This is a stringent definition of ADHD and it has been suggested that some of these criteria may be too restrictive, especially the age of onset criteria, which may be sensitive to stressors exacerbating the phenotype as well as the availability of clinical referrals (Applegate et al. 1997; Barkley and Biederman 1997).

Although affecting 3%–5% of the school-age population in the U.S. overall, ADHD also exhibits a male:female bias of roughly 4:1 (American Psychiatric Association 1994). Reasons for this bias may include differences in gender-reinforced behavioral patterns as well as gender-influenced biological factors supporting the disorder itself. With respect to cultural differences, a simi-

Table 1 Example of DSM-IV symptom criteria for ADHD

Inattention (6 or more)	Hyperactivity-impulsivity (6 or more)
Fails to attend to details	Blurts out answers
Has difficulty sustaining attention	Difficulty awaiting turn
Does not seem to listen	Interrupts or intrudes
Fails to finish	Talks excessively
Has difficulty organizing tasks	Fidgets with hands or feet
Avoids sustained effort	Leaves seat in classroom
Loses things	Runs about or climbs
Is distracted by extraneous stimuli	Difficulty playing quietly
Is forgetful	Motor excess

larity of overall frequency and male bias was reported in a British mental health survey examining the general prevalence of DSM-IV disorders (including ADHD) in over 10,000 children. The ADHD prevalence was noted at 4.47% ±0.26% with a male:female bias of 4.2:1 (Ford et al. 2003). Interestingly, the strength of the gender bias depended on ADHD diagnostic subtype. For predominantly inattentive ADHD, the male:female bias was reported at 3:1, whereas for the combined subtype the bias was 4.9:1, and for the predominately hyperactive-impulsive 7:1. (Ford et al. 2003). In agreement with previous reports, the most common diagnosis of ADHD was the combined type (63.3%), followed by predominately inattentive (29.4%), and predominately hyperactive-impulsive (7.2%) (Ford et al. 2003). However, the 3%–5% prevalence estimate has been challenged by some investigators, since no large-scale epidemiological study of ADHD had been completed in the U.S. to date, only smaller clinic-based studies (Paule et al. 2000; Rowland et al. 2001). Preliminary data undertaken by National Institute of Environmental Health Sciences suggest that prevalence might actually be 2–3 times that estimated by DSM-IV, complicated both by significant under- (40%) and over-diagnosis (20%) (Paule et al. 2000).

1.2
Cognitive Deficits in ADHD, the Search for Endophenotypes

Although the implementation of DSM-IV-based diagnostic criteria for ADHD has allowed for greater consistency across studies, this development may not be particularly useful in the clarification of the etiology of ADHD, as little success has occurred when using symptom-based diagnostic criteria to map susceptibility genes in psychiatric disorders (Cornblatt and Malhotra 2001). Rather, a more quantitative measure of ADHD traits would be useful, and this has led to interest in identifying endophenotypes linked to the disorder.

An endophenotype is a quantitative measure that predicts the presence of the disorder, and may be closer to the causative agent(s) than diagnostic criteria (Almasy and Blangero 2001). Endophenotypes may also describe alterations in brain circuits shared with other disorders, providing opportunities to understand common mechanisms displayed in distinct ways due to phenotypic and genetic complexity. In reviewing the literature of neuroimaging, neuropsychological tests, and genetics in ADHD, Castellanos and Tannock (2002) recently proposed a few potential endophenotypes for ADHD.

One potential endophenotype for ADHD is shortened delay gradient (Castellanos and Tannock 2002). Shortened delay gradient is evident in a behavioral test of delay aversion, where subjects are allowed to choose between an immediate, smaller reward, or must wait for a larger reward (Sonuga-Barke 2002). Children with ADHD have been found to exhibit delay aversion more readily than controls, as they more often select the immediate reward (Barkley et al. 2001; Solanto et al. 2001; Sonuga-Barke 2002). Additionally, a shortened delay gradient has also been observed in a rat model of ADHD (Johansen et al. 2002). As will be discussed below, brain alterations observed in neuroimaging work in ADHD subjects could contribute to this phenotype including increased dopamine transporter (DAT) density in the striatum (Cheon et al. 2003; Dougherty et al. 1999; Dresel et al. 2000; Krause et al. 2000), striatal lesions (Herskovits et al. 1999; Max et al. 2002), and cerebellar vermis hypoplasia (Berquin et al. 1998; Castellanos et al. 2001; Mostofsky et al. 1998).

Another possible endophenotype for ADHD is a deficit in working memory (Castellanos and Tannock 2002). Working memory deficits have been observed in subjects with ADHD (Barnett et al. 2001; Kempton et al. 1999), and might be even a stronger predictor of a predominately inattentive form of ADHD (Chhabildas et al. 2001). Working memory deficits manifest in behaviors that require focused attention and executive functions. Diagnostic instruments that interrogate these processes include the Wisconsin card sorting test and the Stroop test. Interestingly, performance on tests such as the Stroop has been found to be heritable in both nonclinical (Stins et al. 2004) and ADHD (Slaats-Willemse et al. 2005) samples. Catecholamine modulation in the prefrontal cortex (PFC) is important for working memory (Sawaguchi 2001). Thus, in support of this phenotype, heritable disruption of dopamine (DA) or norepinephrine (NE) signaling in the PFC may exist, for instance by genetic variability in DAT and/or NE transporters (NET), catecholamine receptors, or metabolic pathways [e.g., catechol-O-methyl transferase (COMT); Egan et al. 2001].

Another potential endophenotype of ADHD is a deficit in response inhibition. Using a continuous performance task (CPT) as a measure of response inhibition, Strandburg and coworkers found that children with ADHD had a significant increase in reaction time, as well as a significantly greater number of false alarms than normal controls (Strandburg et al. 1996). A similar result was reported by Slaats-Willemse (2003) and coworkers, where they found an increase in accidental responses in a visual CPT in ADHD subjects compared to controls. Further, they found that non-affected siblings of ADHD subjects also had deficits in response inhibition, as measured by both the CPT and Go-NoGo tasks, reinforcing a genetic contribution to these tasks (Slaats-Willemse et al. 2003). One potential contribution to the genetic component of response inhibition is the variable number tandem repeat (VNTR) in the $3'$-untranslated region (UTR) of the DAT gene (DAT $3'$VNTR). The structure of this VNTR and its possible functional role will be described in greater detail later in this review. With respect to response inhibition, Loo and coworkers found that ADHD subjects that were homozygous for the 10-repeat $3'$VNTR allele performed more poorly on a CPT than ADHD subjects that had at least one 9-repeat $3'$VNTR allele (Loo et al. 2003). Additionally, differences in EEG measures were noted between the two groups, where 9-repeat $3'$VNTR allele carriers showed a decrease in CPT-β activity and CPT-θ/β ratio in response to methylphenidate (MPH) treatment, whereas those homozygous for the 10-repeat allele exhibited the opposite effect (Loo et al. 2003). Cornish and coworkers (2005) reported a similar result using a quantitative trait loci approach in a survey of British children in the general population. They found that homozygosity of the 10-repeat allele was present at a significantly higher frequency in the most affected ADHD subjects, and that subjects that were homozygous for the 10-repeat allele performed more poorly on a task of response inhibition (Cornish et al. 2005). Although there has been some discussion on the difficulty of using diminished response inhibition as an endophenotype of ADHD (Castellanos and Tannock 2002), work by multiple investigators suggests that response inhibition may be a tenable endophenotype, perhaps especially for the inattentive aspects of ADHD (Chhabildas et al. 2001). Together these studies point to heritable cognitive deficits in ADHD that may allow for more powerful investigations of underlying neurobiological determinants through the sharpening of phenotypic categories. Investigations have already commenced to determine if DAT is a candidate in support of altered performance on these tasks, and time will tell if initial indications of association are reliable.

1.3
Treatment of ADHD

The most common treatment for ADHD is pharmacological and typically involves the administration of psychostimulants such as MPH (Ritalin, Novartis Pharmaceuticals, East Hanover; Concerta, McNeil Consumer and Specialty Pharmaceuticals, Fort Washington) and amphetamine (AMPH) (Adderall, Shire Pharmaceuticals Group, Basingstoke; Dexedrine, GlaxoSmithKline, London). These drugs are known from animal (Dresel et al. 1998) and human (Volkow et al. 2001) studies to primarily target DAT, but also have action at NET and possibly the serotonin transporter (SERT). Orally delivered MPH blocks DAT in the striatum and increases extracellular DA (Volkow et al. 2001). Treatment of ADHD with psychostimulants has often been labeled paradoxical, since these drugs are known to increase motor activity in normal animals (Gainetdinov et al. 1999), yet effectively decrease these behaviors in children with ADHD and animal models of ADHD (Gainetdinov et al. 1999). Whether these differential actions reflect differences in age of subjects being administered medication or true endophenotypic traits of the disorder remains a matter of debate. Regardless, treatment with psychostimulants improves symptoms in most ADHD subjects (Swanson et al. 1993), but some subjects are refractive to treatment, or must discontinue use due to side effects such as insomnia and decreased appetite (Stein et al. 2003). These effects, along with a growing recognition of the role of NET in DA clearance in the PFC (Gresch et al. 1995; Mazei et al. 2002), have led to the development of NET antagonists as non-stimulant pharmacological treatments for ADHD, atomoxetine (Strattera, Eli Lilly and Company, Indianapolis) being the first approved agent of this class (Corman et al. 2004). Both atomoxetine and MPH raise extracellular DA and NE levels in the rat PFC, whereas only MPH elevates DA in the striatum (Bymaster et al. 2002), consistent with the high levels of DAT but not NET in the latter region. Aside from regulation of DA terminals in the PFC, ADHD treatments may also modulate the activity of the DA cell bodies in the ventral tegmental area (VTA) that project to the PFC. The activity of DA neurons in the VTA can be modulated by DA levels, as well as NE and NET blockers (reviewed in Adell and Artigas 2004). Consistent with this idea, Choong and Shen (2004) found that decreased activity of DA VTA neurons in a rat model of fetal alcohol syndrome was normalized by systemic administration of MPH. They observed that increases in extracellular DA, not NE, mediated this response, which could be clinically relevant in ADHD, as MPH treatment has been reported to improve attention deficits in children with fetal alcohol syndrome (Choong and Shen 2004).

Aside from pharmacological treatment, behavioral modification involving a variety of psychosocial interventions is also used in the treatment of ADHD (Pelham et al. 1998). Although both pharmacological and behavioral treatments have been documented to work well in the short term, less evidence exists to make definitive statements regarding long-term therapeutic outcomes. To address this issue, the National Institute of Mental Health (NIMH) sponsored a large, multi-site prospective study to examine the long-term effectiveness of pharmacological and psychosocial interventions in children with ADHD, which came to be known as the Multimodality Treatment Study of ADHD (MTA) (Arnold et al. 1997). This study compared four treatment strategies: medication management alone, behavioral modification alone, combined medication and behavioral treatment, and an active control where subjects were allowed to seek normal treatment options available in the community

(Arnold et al. 1997). In this study, 96 subjects that met DSM-IV criteria for combined-type ADHD were collected at six different project locations, and 144 subjects were randomly assigned to each of the four treatment groups. Treatment was provided over a period of 14 months. The medication treatment group was prescribed MPH using a double-blind, dose-response titration design. If subjects did not respond to increasing doses of MPH then other psychostimulants (AMPH) or non-stimulant treatments (imipramine) were administered. The behavioral treatment was multi-faceted and involved measures directed at the parents (35 parent-training sessions and school consultations with teachers), children (an 8-week summer treatment program), and the school system (placement of an extra part-time classroom assistant in the child's regular classroom). Researchers found significant treatment effects at 14 months, and the medication management and combined medication and behavioral treatment groups reported greater improvements than the behavioral treatment and active control treatment groups (Paule et al. 2000). This is not to say that behavioral modification was ineffective, as the behavioral treatment group also reported an improvement in symptoms. Rather, behavioral modification gave similar improvement scores as the active control treatment (which involved the administration of stimulant medication in ~2/3 families). Furthermore, at 24 months (10 months after completing the intensive MTA treatment strategy), the subjects that had either the combined medication and behavioral treatment or the medication treatment alone still reported greater improvements than the other two treatment groups, although the overall treatment effect was smaller than at 14 months (Group 2004).

Overall, these data indicate that psychostimulant treatment is safe and effective in the majority of cases of ADHD, whereas non-stimulant medications and behavioral modification can be effective alternatives for children who have adverse reactions or do not respond to psychostimulant treatment. Treatment of ADHD in some form is important, as untreated ADHD has been reported to contribute to academic failure, unemployment, and an increased risk for substance abuse with attendant economic impact for the individual and society (Biederman et al. 1999; Birnbaum et al. 2005; Harpin 2005). The most widely utilized therapeutic agents target DAT, although more recent studies indicate utility for augmenting cortical DA availability through NET blockade. Whether DAT and NET differentially support distinct subtypes of the disorder, or their pharmacological management, awaits comparative studies, hopefully supported by in vivo assessments of transporter occupancy and/or extracellular catecholamine levels. These studies also raise the question of whether genetic variability associated with DAT (or NET) impacts the risk for ADHD or utility of specific therapeutics. In the next section, we review the structure of the DAT gene and protein to set a foundation for discussion of studies examining genetic and drug modulation of DAT availability.

2
The Human Dopamine Transporter

2.1
Cloning the hDAT Gene

Human DAT (hDAT) complementary DNAs (cDNAs) were initially cloned from human substantia nigra libraries using probes derived from sequence of rat DAT (rDAT) (Giros et al. 1992; Vandenbergh et al. 1992b). The isolated clones bore open reading frames of 1,860 nucleotides, predicting a 620 amino acid protein (Fig. 1B). The hDAT protein exhibits 92% amino acid identity with rDAT and significant, but more reduced, identity with other members of the Na^+/Cl^--dependent neurotransmitter transporter family (SLC6) such as hNET, with which it is 66% identical. A number of DATs have been cloned from both mammalian (mouse, rat, bovine, primate) and non-mammalian species (zebrafish, *Caenorhabditis elegans*, *Drosophila*). Similar to hNET and the human SERT (hSERT) with which hDAT shares psychostimulant recognition, hydropathy analysis predicts 12 transmembrane domains (TMs) with intracellular N- and C-termini. Additionally, there are three *N*-glycosylation sites, and multiple consensus sequences for protein phosphorylation. The chromosomal localization of hDAT is on the short arm of chromosome 5, band 15.3 (5p15.3) (Giros et al. 1992; Vandenbergh et al. 1992a). The hDAT gene (*SLC6A3*) is large, spanning over 64 kb and consisting of 15 exons separated by 14 introns (Kawarai et al. 1997; Fig. 1A). Evidence of alternative splicing of hDAT has not been reported to date.

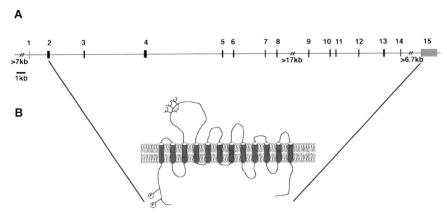

Fig. 1 A,B Gene and protein structure of the human dopamine transporter (hDAT). **A** The hDAT gene (*SLC6A3*) consists of 15 exons (*black boxes*) spanning over 60 kb with both 5'- and 3'-untranslated regions (*gray boxes*) (accession number AF11917 and Kawarai et al. 1997); scale is indicated in *lower left corner*. **B** The hDAT protein is 620 amino acids long with 12 transmembrane domains (TMs) and intracellular N- and C-termini

The 5′-flanking region of hDAT does not contain a canonical TATA nor CAT box, but does contain multiple consensus sequences for regulatory elements including two E box sites, an SP-1 site, and a conserved CCAGGAG motif that might contribute to neuron-specific expression (Kawarai et al. 1997; Kouzmenko et al. 1997). The 5′-region also contains a strong non-tissue specific promoter and a silencing element in intron 1 that limits cell specific expression (Kouzmenko et al. 1997; Sacchetti et al. 1999). Furthermore, single nucleotide polymorphisms (SNPs) in the 5′-sequence may also affect gene expression (Greenwood and Kelsoe 2003). Greenwood and Kelsoe (2003) found that specific combinations of 5′-polymorphisms, or haplotypes, influenced luciferase expression in SN4741 cells. While the haplotypes studied by Greenwood and Kelsoe (2003) were derived from pedigrees of bipolar disorder, polymorphisms in the 5′-region of DAT may also be relevant in ADHD.

hDAT messenger RNA (mRNA) also possesses a large 3′-UTR that contains a VNTR polymorphism (Vandenbergh et al. 1992a, b; Fig. 2A). The VNTR is an approximately 40-bp sequence that repeats 3 to 13 times, with 9- and 10-repeats being the most common in almost all human populations (Doucette-Stamm et al. 1995; Gelernter et al. 1998; Kang et al. 1999; Fig. 2B). The 3′VNTR is not conserved in rodent DATs, but nonhuman primates exhibit a variety of 3′VNTR sequences. Gorillas, orangutans, and chimpanzees have 1- or 2-repeat 3′VNTR sequences that are highly conserved between the three species, but neither the 1- nor 2-repeat VNTR is observed in human populations (Inoue-Murayama et al. 2002). The cynomolgus macaque has a 3′VNTR with 11- or 12-repeats, but the sequence of the 11-repeat is divergent from the hDAT 11-repeat 3′VNTR (Inoue-Murayama et al. 2002). Whereas the 3′VNTR repeat number varies in different human populations, the 10-repeat is almost universally the most common, with an allele frequency of approximately 70% (Doucette-Stamm et al. 1995; Gelernter et al. 1998; Kang et al. 1999; Mitchell et al. 2000; Vandenbergh et al. 1992a). Although there are differences in the population in 3′VNTR repeat number, the sequence of the repeats seems to be well conserved between individuals with the same 3′VNTR repeat number (Fuke et al. 2001; Mill et al. 2005; Fig. 2A).

Since the 3′VNTR repeat number varies in human subjects, it has been suggested that it might play a role in DAT mRNA stability or translation, with alterations possibly contributing to changes in DAT abundance in disorders such as ADHD. Multiple studies have been conducted but, as yet, there is not a clear consensus on what, if any, affect the 3′VNTR plays in vivo. Fuke and coworkers (2001) found that the 10-repeat allele increased luciferase expression in COS-7 cells compared to the expression of the 7- or 9-repeat alleles. However, Inoue-Murayama and coworkers (2002) found that the 9-repeat allele induced greater luciferase activity in the human neuroblastoma cell line SK-N-SH than the 10- and 11-repeat alleles. A similar result was reported by Miller and Madras (2002), where they found that the 9-repeat allele induced a greater luciferase expression than the 10-repeat allele in HEK-293 cells. A similar

Fig. 2 A,B 3′-UTR variable number tandem repeat (VNTR) in hDAT. **A** Sequence of the two most common alleles of the hDAT 3′VNTR, the 9-repeat and 10-repeat alleles (Sequence from Fuke et al. 2001 and Mill et al. 2005). **B** Frequency of the hDAT 3′VNTR in different human populations. [a]Gelernter et al. (1998); [b]Mitchell et al. (2000); [C]Doucette-Stamm et al. (1995)

luciferase expression study conducted by Mill and coworkers (2005) found no significant differences between 9- and 10-repeat alleles in HEK-293 and SH-SY5Y cells. The differing results between these studies may be due in part to the variety of cell lines used, as recent work has shown that the transcription factor Hesr1 downregulates reporter gene expression, and does so in a 3′VNTR repeat number-dependent and cell context-dependent manner (Fuke et al. 2005). Fuke and coworkers (2005) found that the 6- and 11-repeat alleles had significantly less expression than the 7–9- and 10-repeat alleles in HEK-293 cells, whereas the 6-, 7-, and 9-repeat alleles had greater expression than the 10- and 11-repeat alleles in the SH-SY5Y human neuroblastoma cell line. In COS-7 and Neuro2A cells, the 6-, 7-, 9-, and 10-repeat alleles had significantly higher activity than cells transfected with the 11-repeat allele. Although these studies in several cases utilize neuronal hosts, DA cell lines are not widely available, and thus it appears that to have a better understanding of the functional relevance of the 3′VNTR, a more physiologically relevant model needs to be developed. Moreover, results may depend on whether the entire 3′-UTR or just a portion surrounding the 3′VNTR region is used. Perhaps a better method to understand the physiological relevance of the 3′VNTR is to adopt a larger bacterial artificial chromosome (BAC) vector (Frengen et al. 2000) that could encompass genomic DNA comprising the native hDAT promoter as well as the coding and non-coding regions. Evaluation of promoter efficacy in transgenic mice could then be evaluated, with the clear caveat that both sites of integration and the murine nature of the host need to be considered.

2.2
DAT Protein: Structure/Function

Since hDAT protein has to date only been identified in brain, characterization of the transporter's basic properties has largely relied on heterologous expression systems studied in vitro. Data from studies expressing DAT and NET chimeras (Buck and Amara 1994; Giros et al. 1994; Syringas et al. 2000) first suggested that discrete domains within the transporter were involved in aspects of substrate recognition, translocation, and affinity. For example, TMs 1–3 and 10–11 were identified as important in the affinity for substrates, whereas TMs 5–8 were important in substrate translocation and affinity and selectivity of inhibitors (Buck and Amara 1994; Giros et al. 1994). Using similar approaches, TMs 1–3 and 9–12 were inferred as important for Na^+ and Cl^--dependence (Syringas et al. 2000).

Whereas chimera studies typically examine regions of roughly 50–200 amino acids, mutagenesis of specific residues can refine structure/function relationships and point to important motifs for posttranslational modification. One prominent feature of hDAT is the three canonical sites for N-linked glycosylation within extracellular loop (EL) 2 (Fig. 1B). Mutation of either a single site or a combination of the three sites revealed that glycosylation occurs at all three sites but is not necessary for hDAT expression or function in HEK-293 cells (Li et al. 2004a). Rather, the data suggest that N-glycosylation is important for optimal function of hDAT, as non-glycosylated DAT exhibits a decreased DA V_{max} as well as decreased WIN 35,428 binding, in addition to an increased endocytosis rate (Li et al. 2004a). Aside from the glycosylation status, EL2 may also play a role in, or be influenced by, inhibitor binding, as binding of inhibitors has been shown to protect DAT from trypsin digestion in this region (Gaffaney and Vaughan 2004). In contrast, DAT substrates do not protect against digestion (Gaffaney and Vaughan 2004). However, zinc immobilization of EL2 blocks DA transport, suggesting that EL2 movements are critical to the transport process, possibly to permit structural alterations in the adjoining TMs 3 and 4 (Norregaard et al. 1998). Zinc immobilization of EL2 and EL4 also enhances an uncoupled anion conductance whose presence negatively impacts transport and stimulates efflux (Meinild et al. 2004). These latter studies remind us of the importance of utilizing electrophysiological approaches to provide mechanistic insights underlying mutation-induced changes in DAT function, efforts that may be critical for defining the functional consequences of naturally occurring DAT gene variants (see Sect. 5.4).

In order to gain a more specific understanding of how substrates such as DA and antagonists such as cocaine interact with DAT, a number of mutational studies have been conducted to investigate specific residues that may be important for these interactions. In these studies, alanine substitution has been employed for residues in or near TMs including all of the tryptophan residues (Lin et al. 2000b), proline residues (Lin et al. 2000a), phenylalanine residues (Lin et al. 1999), as well as the substitution of 38 polar residues (including aspartic acid, tyrosine, asparagine, serine, threonine, and glutamic acid) (Itokawa et al. 2000). Additionally, the importance of D79 (Kitayama et al. 1992; Wang et al. 2003) and F105 (Wu and Gu 2003) in cocaine sensitivity has been studied in detail. These studies, as well as in silico modeling approaches, point to discontinuous segments of DATs as comprising ligand binding sites, with overlapping determinants evident for substrates and antagonists (for a review see Uhl and Lin 2003; Goldberg et al. 2003). Further understanding of the detailed structure of transporter–drug interactions may be useful for the design of improved therapeutics, as well as clarification of the impact of rare DAT coding variants (see Sect. 5.4).

While investigators initially assumed that DAT existed as a monomer, recent evidence suggests the DAT can form homomultimers. Hastrup and coworkers (2001) found that they could capture homodimers of hDAT in HEK-293 cells using crosslinking reagents, and that this interaction was dependent on the presence of Cys-306 at the top of TM6 (Hastrup et al. 2001). Further studies by this group also revealed an additional dimer interface at TM4 sensitive to cocaine binding (Hastrup et al. 2003). Confirmation of the formation of homodimers by mutational and biochemical analysis was provided by Torres and coworkers (2003), although they implicated a leucine-repeat in TM2 as important for DAT assembly and function (Torres et al. 2003). Further studies using fluorescence resonance energy transfer (FRET) imaging found that hDAT oligomers formed in the endoplasmic reticulum (ER) and were also present at the plasma membrane (Sorkina et al. 2003). In addition to interacting with itself, DAT has recently been found to interact with a number of synaptic proteins including PICK-1 (Torres et al. 2001), Hic-5 (Carneiro et al. 2002), PP2Ac (Bauman et al. 2000), and syntaxin (Lee et al. 2004). Although initial characterization of these interactions has been conducted and has suggested roles for interacting proteins in ER export and plasma membrane localization, the functional importance of these various interactions singly, and in combination, remains to be established in vivo.

2.3
DAT Protein: Regulation

Aside from structural determinants and interacting proteins, DAT activity is also regulated by cell signaling pathways. Many pathways have been implicated in the alteration of DAT activity (reviewed in Zahniser and Doolen 2001), but for the sake of brevity we focus attention on the four most well-studied pathways involving protein kinase C (PKC), protein kinase A (PKA), phosphoinositol 3-kinase (PI3-K), and mitogen-activated protein kinase (MAPK, p42,44 also known as ERK).

The most replicated regulation of DAT has been PKC modulation, where either direct PKC activation by phorbol esters (Chang et al. 2001; Granas et al. 2003; Huff et al. 1997; Kitayama et al. 1994; Loder and Melikian 2003; Melikian and Buckley 1999; Pristupa et al. 1998; Sorkina et al. 2005; Vaughan et al. 1997; Zhang et al. 1997) or indirect activation through the activation of a $G_{\alpha q}$-coupled G protein-coupled receptor (Granas et al. 2003) has been shown to decrease DAT activity. This modulation has been observed in multiple heterologous cell models (PC12, HEK-293, COS-7, LLC-PK1, N2A, PAE, C6 glioma, and Sf9) as well as in vivo (Vaughan et al. 1997). Current models suggest that this downregulation of activity is primarily due to a redistribution of DAT from the plasma membrane to intracellular compartments (Chang et al. 2001; Granas et al. 2003; Loder and Melikian 2003; Melikian and Buckley 1999; Pristupa et al. 1998). Whereas multiple studies have found that DAT is phosphorylated in response to PKC activation (Chang et al. 2001; Cowell et al. 2000; Foster et al. 2002; Granas et al. 2003; Huff et al. 1997; Vaughan et al. 1997), current studies suggest that this phosphorylation may not be necessary for PKC-mediated internalization (Chang et al. 2001; Granas et al. 2003). Recent evidence suggests that PKC phosphorylation may actually modulate a different activity of the transporter than is traditionally studied, reverse transport. Cowell and coworkers have previously shown that PKC activation via phorbol esters increases DA release from striatal slices and that this effect is blocked by DAT inhibitors (Cowell et al. 2000). Recent collaborative studies by the Galli, Javitch and Gnegy labs described alanine substitution of five N-terminal serine residues in hDAT, postulated to be sites for PKC phosphorylation, revealing that although much phosphorylation was lost, this change did not affect basal DA transport or PKC-dependent endocytosis, but instead greatly inhibited AMPH-induced DA efflux (Khoshbouei et al. 2004). When aspartates were substituted for the five serine residues, simulating phosphorylation, AMPH-induced DA efflux was restored (Khoshbouei et al. 2004). However, it should be mentioned that direct phosphorylation of these serine residues was not demonstrated in this work (Khoshbouei et al. 2004). These data suggest multiple roles for kinase modulation of DAT activity whereby PKC activation decreases DAT activity primarily through increased DAT internalization with direct phosphorylation of DAT by PKC or another kinase stabilizing an "efflux-willing" confirmation of DAT.

In addition to PKC, there is evidence that other second messenger systems also regulate DAT activity. Carvelli and coworkers (2002) found that insulin stimulated the activity of hDAT in HEK-293 cells and that this activation was dependent on the stimulation of PI3-K activity that induced a redistribution of hDAT to the plasma membrane (Carvelli et al. 2002). Inhibition of PI3-K, on the other hand, induced a decrease in DAT transport activity in both HEK-293 cells and striatal synaptosomes (Carvelli et al. 2002), an effect that has now been replicated by other groups in COS-7 (Lin et al. 2003) and SH-SY5Y cells (Mazei-Robison and Blakely 2005). Importantly, a dominant-negative form of Akt, a protein kinase effector downstream of PI3-K, could block insulin's effect on DAT activity (Garcia et al. 2005). The MAPK family has also been implicated in DAT regulation (Lin et al. 2003; Moron et al. 2003). These studies found that inhibitors of p42 and p44 MAPK decreased DAT transport activity and increased DAT internalization in HEK-293 and COS-7 cells, as well as in striatal synaptosomes (Lin et al. 2003; Moron et al. 2003). In contrast to PI3-K and MAPK, inhibition of PKA has been reported to increase DAT activity and DAT expression at the plasma membrane in heterologous cells (Pristupa et al. 1998). However, the opposite effect was observed in rat striatal synaptosomes where PKA activation via 8-bromo-cAMP increased DAT transport activity (Page et al. 2004).

Together, these data suggest that DAT activity can be modulated both in vitro and in vivo by multiple cellular signaling pathways, and that this modulation often involves trafficking of DAT to or away from the plasma membrane. The further elucidation of signaling pathways that alter DAT activity, as well as studies that elaborate whether these pathways are altered in disorders such as ADHD, may lead to the identification of a larger network of candidate genes supporting disease risk or therapeutic interventions.

2.4
DAT Protein: Pharmacological Impact on Transporter Regulation

It is not surprising, given the role of DAT in maintaining synaptic DA signaling, that drugs that target DAT have profound biochemical and physiological effects. These effects can be beneficial, for example the administration of MPH and bupropion for the treatment of ADHD and depression respectively, or have abuse liability, as is the case for many DAT-targeted psychostimulants. Owing to their roles in addiction and drug abuse, the most studied class of drugs that target the dopamine transporter includes the psychostimulants cocaine and AMPH. It is well understood that cocaine physically interacts with DAT and that this blockade is a critical component of the subjective "high" experienced by drug abusers (Ritz et al. 1987). However, in addition to its action as a DAT blocker, cocaine appears to alter DAT activity through disruption of the transporter's normal trafficking itinerary. Using N2A neuroblastoma cells stably transfected with hDAT, Little and coworkers (2002) found that incubating cells with 1 or 10 µM cocaine for 24 h increased WIN 35,428 binding as well as DAT transport capacity. Importantly, this modulation was reversible, as when cocaine was washed out for 24 h the binding and activity returned to pretreatment levels (Little et al. 2002). Cell surface biotinylation and immunofluorescence experiments revealed that the cocaine-induced increase in binding and transport were due to a redistribution of DAT to the plasma membrane. Consistent with this idea, neither DAT mRNA nor protein synthesis was altered by cocaine treatments (Little et al. 2002). A similar result was reported on a shorter time scale, where HEK-293 cells stably expressing hDAT exhibited a significant increase in DAT transport activity after only 5 or 10 min of 10 µM cocaine treatment (Daws et al. 2002). This effect also appeared to be due to a redistribution of hDAT to the plasma membrane. Furthermore, this increase in DAT activity could be measured in synaptosomes from rat nucleus accumbens 90 min after a single intraperitoneal injection of cocaine (Daws et al. 2002). Excitingly, this increase in DAT binding and transport also appears to occur in human cocaine abusers. Striatal synaptosomes obtained from postmortem brains of human cocaine abusers exhibit a significant increase in both the B_{max} of WIN 35,428 binding and the V_{max} for DA transport (Mash et al. 2002), relative to samples from age-matched drug-free controls or from victims of excited cocaine delirium (Mash et al. 2002). Mechanistically speaking, caution should be taken in relating the acute effects of cocaine in controlled settings (cell culture, animal studies) to the chronic effects described for human cocaine abusers. Regardless, work from both in vitro and in vivo studies suggests that cocaine acts both as an acute DAT blocker and as a modulator of DAT activity by impacting DAT trafficking to the plasma membrane. Studies are needed to determine whether another DAT blocker, MPH, has similar effects on DAT activity.

In contrast to cocaine, which primarily acts as a DAT blocker, the psychostimulant AMPH acts primarily as a DAT substrate and a DA releaser, facilitating the reverse operation of DAT (Jones et al. 1999b; Sulzer et al. 1993). Given the difference in how AMPH and cocaine interact with DAT, it is not surprising that the two agents differentially regulate DAT activity and surface expression. Whereas cocaine can increase DAT expression at the plasma membrane, treatment with AMPH triggers the opposite effect, a decrease in DAT surface levels supported by enhanced DAT endocytosis (Saunders et al. 2000). Using HEK-293 cells stably transfected with a FLAG-tagged hDAT, Saunders and coworkers (2000) found that a 1-h incubation with 2 µM AMPH significantly decreased DAT levels at the plasma membrane. Kahlig and coworkers combined patch clamp and amperometric recordings of DAT activity to establish that the accompanying decrease in DAT transport capacity triggered by AMPH arises from a redistribution of active transporters, as opposed to a functional inactivation prior to internalization (Kahlig et al. 2004). These in vitro studies have in vivo correlates. It has been shown that a single subcutaneous injection of AMPH results in a decrease in DAT transport in rat striatal synaptosomes isolated 1 h after injection (Fleckenstein et al. 1999). Whether or not this downregulation of DAT occurs in human AMPH abusers has not been reported, owing to an inability of current imaging approaches to define surface versus intracellular DAT proteins. Regardless, since treatments with blockers and AMPHs can both result in diminished DAT availability, by competitive occupancy of the DA binding site and via internalization of DAT proteins respectively, further pursuit of these regulatory changes as it relates to ADHD therapeutics is warranted.

3
DAT Transgenics as Animal Models of ADHD

3.1
DAT Knockout Mice as a Model of ADHD

Although pharmacological targeting of DAT first defined its importance in DA signaling, the critical role for DAT in maintaining normal presynaptic DA homeostasis has become more clearly evident from studies examining the biochemical and behavioral consequences of targeted DAT gene deletion in the mouse (Giros et al. 1996). Compared to freely moving wildtype littermates, DAT knockout mice have a fivefold increase in basal extracellular DA concentration (Jones et al. 1998). Additionally, when DA release is evoked in a striatal slice preparation, DAT knockout mice display a 100-fold increase in DA clearance time compared to wildtype mice, even though the amount of released DA is almost 75% less than that of wildtype littermates (Giros et al. 1996). The elevated extracellular levels and altered release of DA in DAT knockout animals suppresses the expression of dopamine receptor mRNA for both D1 and D2 receptors while increasing D3 receptor expression. These changes are accompanied by a near complete loss in the functional activity of the D2 autoreceptor (Fauchey et al. 2000; Giros et al. 1996; Jones et al. 1999b) as well as a disruption of D1 receptor trafficking (Dumartin et al. 2000).

The alterations in DA tone and signaling arising in the knockout disrupt multiple DA-dependent physiological processes including growth, lactation, colonic motility [DAT is expressed in the mouse small intestine (Chen et al. 2001; Li et al. 2004b)], immune response, sleep, cognitive performance, and motor activity (Bosse et al. 1997; Gainetdinov et al. 1999; Giros et al. 1996; Kavelaars et al. 2005; Rodriguiz et al. 2004; Spielewoy et al. 2000; Walker et al. 2000; Wisor et al. 2001). DAT knockout mice exhibit deficits in learning and memory, sensorimotor gating, and social interactions (Gainetdinov et al. 1999; Ralph et al. 2001; Rodriguiz et al. 2004). Additionally, Pogorelov and coworkers (2005) recently observed that when they tested DAT knockout mice in open field and zero maze tasks, the mice initially have an anxiety-type response when presented with the novel conditions that dissipates over time.

However, the most overt behavioral changes are an increase in spontaneous locomotor activity in a novel environment and a lack of locomotor habituation, leading to the hypothesis that DAT knockout mice might serve as a model for ADHD (Gainetdinov and Caron 2001; Giros et al. 1996; Spielewoy et al. 2000). In support of this idea, the most common pharmacological treatments for ADHD, psychostimulants that target DAT such as MPH and AMPH, decrease locomotor activity in these animals while increasing activity in wildtype mice (Gainetdinov et al. 1999). Given that these animals lack the most prominent target of psychostimulants, DAT, it may seem counterintuitive that these drugs are effective. However, as has been mentioned earlier, MPH and AMPH are not specific for DAT and are known to have activity at the transporters for NE and 5-hydroxytryptamine (5-HT) as well. Gainetdinov and coworkers (1999) addressed the issue of the target of these drugs in the DAT knockout mice and concluded that the alternative target was the serotonin system and SERT, since the SERT-specific blocker fluoxetine provided similar benefits, whereas the NET blocker nisoxetine did not. This conclusion was met with controversy, however, and the importance of SERT for the therapeutic efficacy of psychostimulants in humans is still a matter of debate (Marx 1999; Sarkis 2000; Volkow et al. 2000).

3.2
DAT Knockdown Mice as a Model of ADHD

In addition to DAT knockout mice, DAT "knockdown" mice have been generated that express approximately 10% of the normal amount of DAT protein (Zhuang et al. 2001). The targeting construct used to generate this line was intended to generate transgenic mice that could have their DAT level modulated by tetracycline; however, tetracycline-dependent transactivator tTA expression was not observed in the mutant animals. DAT knockdown mice do not display the dwarfism present in DAT knockout mice, but display similar biochemical alterations as DAT knockout mice, although to a lesser magnitude. For example, DAT knockdown mice have a 70% increase in extracellular DA levels, and in amperometry experiments conducted using striatal slices they exhibit a decreased evoked DA release and an increased DA clearance time compared to wildtype littermates (Zhuang et al. 2001). Importantly, these mice possess a hyperlocomotive phenotype that can be remedied by AMPH treatment (Zhuang et al. 2001). Additionally, these mice display an excessive sequential stereotypy of grooming behaviors that has been suggested as a mouse model of obsessive-compulsive disorder (Berridge et al. 2005). Together, studies from DAT knockout and knockdown mice paint a compelling picture that genetic disruption in DAT-supported DA signaling can establish behavioral phenotypes bearing the requisite pharmacological sensitivity to be bona fide ADHD models. Certainly these animals may merely phenocopy several of the more prominent facets of the disorder without true structural parallels at the level of the human DAT gene. Indeed, as we will discuss in Sect. 5.4, no severe loss-of-function DAT alleles have yet been described in human ADHD populations, although DAT gene evaluation is still ongoing, and weaker effects of existing DAT alleles could be amplified by the appropriate expression context or by the presence of modifier genes linked to regulatory pathways. The limited accessibility of investigators to native DAT expression in the living human brain is one obstacle to such correlations. In this regard, we now turn to efforts to examine DAT via non-invasive imaging approaches and the use of these techniques to examine DAT in ADHD subjects.

4
Neuroimaging DAT in Human Subjects

4.1
DAT in Adult ADHD Subjects

Given that most pharmacological treatments for ADHD target DAT, and that DAT levels and activity can be altered in response to psychostimulant treatment, researchers have been interested for several years in whether DAT levels are altered in human subjects with ADHD (Table 2). The first report of changes in DAT density in humans with ADHD was in 1999 (Dougherty et al. 1999). Using the single-photon emission computed tomography (SPECT) ligand ^{123}I-altropane to measure DAT density, Dougherty and coworkers (1999) observed that binding was elevated 70% in adult subjects with ADHD compared to age-matched controls. However, this study only examined six ADHD subjects, and four of the subjects had been previously treated with psychostimulants, although they were off medications for at least 1 month prior to the study. This issue of previous treatment led some to speculate that the differences observed could be due to psychostimulant treatment, rather than ADHD itself (Baughman 2000; Swanson 2000), similar to the results from postmortem brains from cocaine abusers that have elevated DAT levels (Mash et al. 2002). However, another group using a different SPECT agent specific for DAT, [Tc-99m]TRODAT-1, also found that DAT binding was elevated in a study examining ten previously untreated adults with ADHD and ten healthy

Table 2 Summary of neuroimaging studies of DAT density in the striatum of ADHD subjects

Number of subjects		Age	DAT ligand	DAT binding increased	MPH treatment effect	Study
Control	ADHD					
30	6	Adult	^{123}I-Altropane	Yes	–	Dougherty et al. 1999
10	10	Adult	[Tc-99m]TRODAT-1	Yes	Yes	Krause et al. 2000
14	17	Adult	[Tc-99m]TRODAT-1	Yes	Yes	Dresel et al. 2000
–	1[a]	Adult	[Tc-99m]TRODAT-1	Yes	Yes	Krause et al. 2002
9	9	Adult	^{123}β-CIT	No	Yes[b]	van Dyck et al. 2002
6	9	Child	^{123}I-IPT	Yes	–	Cheon et al. 2003
–	6	Child	^{123}I-Ioflupane	–	Yes	Vles et al. 2003

[a] A case study of a subject with ADHD and Tourette's syndrome, [b] Subjects responded behaviorally to methylphenidate, binding was not examined

age- and sex-matched controls (Krause et al. 2000). Although the effect was smaller than reported in the previous study, the DAT density in the striatum in the ADHD group was significantly higher (~15%) than controls (Krause et al. 2000). Furthermore, after the ADHD subjects were treated for 4 weeks with 5 mg MPH (3 times daily) their DAT availability decreased almost 30%, and their overall DAT levels were equal to or lower than the controls (Krause et al. 2000). In addition to decreasing DAT binding in ADHD subjects, this low dose of MPH was clinically effective, as all patients reported an improvement in ADHD symptoms (Krause et al. 2000). These results were extended in a larger group of 17 adult ADHD subjects, where a roughly 15% increase in DAT binding was observed in ADHD subjects that decreased 43% after MPH treatment (Dresel et al. 2000). A similar result was obtained in a case study of a drug-naïve adult with Tourette's syndrome and ADHD where a 24% elevation in [Tc-99m]TRODAT-1 binding was reduced 40% after 5 months of a low-dose MPH treatment (2.5 mg 3×daily; Krause et al. 2002) implicating the role of DA and DAT in not only ADHD, but other psychiatric disorders such as Tourette's syndrome. Additionally, it was observed that dopamine D2 receptor binding was unchanged, but that the symptoms of both disorders improved with MPH treatment (Krause et al. 2002).

There has been one report, however, that did not find an elevation of DAT binding in adults with ADHD (van Dyck et al. 2002). This study examined 9 adults with ADHD (8 of which were stimulant naïve) and employed a different SPECT ligand, [^{123}I]β-CIT, that, in contrast to the previously described ligands, binds SERT in addition to binding DAT. Another difference in this study is that more male subjects with ADHD ($n = 6$) were examined than female subjects ($n = 3$), whereas the previously mentioned studies all examined more female subjects than male [female:male—4:2 (Dougherty et al. 1999), 7:3 (Dresel et al. 2000), 10:7 (Krause et al. 2000), 1:0 (Krause et al. 2002), respectively]. With the exception of the Dougherty et al. study, all studies employed gender-matched control subjects; however, whether there is an intrinsic difference in the neurobiology of ADHD between females and males has not been examined. Given the male:female bias in ADHD overall, as well as the difference in this bias depending on ADHD diagnosis type, this may be an important avenue to explore. Although they did not exhibit an elevation of DAT binding, the subjects examined by van Dyck and coworkers (2002) did respond clinically to MPH, with a mean decrease in ADHD rating scale score from 35.5 at baseline, to 19.0 after treatment. Whether DAT binding also changed after treatment is unknown, as SPECT imaging was not completed after treatment (van Dyck et al. 2002).

4.2
DAT in Children with ADHD

Although the increase in DAT binding in adults with ADHD is intriguing, most work investigating ADHD targets children, as ADHD is typically thought to be a disorder that manifests during childhood. Cheon and coworkers (2003) were the first to examine DAT levels in the basal ganglia of children with ADHD using SPECT. They found an approximate 30% elevation in [^{123}I]IPT binding in the nine drug-naïve children with ADHD (mean age 9.67), compared to six healthy age-matched children; however, the DAT density levels did not correlate to ADHD severity scores (Cheon et al. 2003). Additionally, it has recently been shown that MPH treatment decreases DAT availability in children with ADHD similarly to that shown in adults (Vles et al. 2003). Six drug-naïve boys with ADHD (mean age 8.67) had baseline levels of striatal DAT measured using the SPECT ligand ^{123}I-Ioflupane, then were treated with MPH for 3–4 months and re-evaluated (Vles et al. 2003). DAT binding decreased in all six subjects after MPH treatment (28%–75%), and when treatment was withdrawn for 1 month from a single subject, the DAT binding increased to above pretreatment levels (Vles et al. 2003). D2 receptor binding was also evaluated in these subjects, and although the effect was more modest than DAT, MPH treatment decreased D2 binding as well (Vles et al. 2003).

Neuroimaging work has been fairly consistent in finding an elevation in DAT binding in the striatum; however, whether these changes translate to an increase in DAT activity is still unknown. For example, an increase in DAT binding potential could result from an alteration in DAT confirmation that increases DAT affinity, an increase in the synthesis of DAT molecules that are either intracellularly located or inserted into the membrane in an inactive state, or an increase in the number of active DAT molecules at the plasma membrane. The first hypothesis is unlikely, as multiple radioligands have been used to assess DAT levels, but whether the increased DAT binding is due to an increase in active DAT molecules has yet to be directly tested. It is known that oral MPH administration increases extracellular DA in vivo (measured via competition of D2 DA receptor binding; Volkow et al. 2001) and since MPH decreases DAT binding, a reasonable hypothesis would be that there is an excess number of active DAT molecules in ADHD subjects and that MPH decreases this overactivity. Additionally, all of these studies have only addressed DAT levels in the striatum, but both pharmacological and neuropsychological work suggests that DAT levels in the PFC may also be important in ADHD and in MPH response. Perhaps as neuroimaging methods improve, DAT levels in lower density regions such as the PFC can also be examined. Extension of studies to NET proteins that clear cortical DA in the PFC is also needed.

4.3
Influence of DAT 3′VNTR on DAT Levels

Given the presence of a 3′VNTR in hDAT, suggestions that 3′VNTR status may affect gene expression—and the availability of tools that can assess total DAT density in vivo—several groups have assessed the effect of the DAT 3′VNTR genotype on DAT availability in vivo using a number of different subject populations (Table 3). The reported effect of 3′VNTR genotype on DAT binding is varied, which may be due, at least in part, to intrinsic differences in the populations studied. For example, Heinz and coworkers (2000) in an examination of 14 abstinent alcoholics and 11 control subjects using the SPECT ligand [^{123}I]β-CIT observed that the 9/10 3′VNTR genotype had a mean reduction in DAT availability of 22% compared to 10/10 genotypes (Heinz et al. 2000). In contrast, when a group of 30 healthy control subjects was examined using the same radioligand, subjects with the 3′VNTR 10/10 genotype had significantly less DAT binding than 9/9 and 9/10 subjects (Jacobsen et al. 2000). Further complicating analysis, Martinez and coworkers (2001) did not find any difference in [^{123}I]β-CIT binding between subjects (31 healthy controls or 29 patients with schizophrenia) that were homozygous for the 3′VNTR 10-repeat allele, or those that carried a copy of the 9-repeat allele (Martinez et al. 2001). A similar lack of 3′VNTR effect on DAT availability using [Tc-99m]TRODAT-1 was reported in a study that examined 100 patients diagnosed with Parkinson's disease (PD) and 66 asymptomatic controls (Lynch et al. 2003), as well as another study that examined 36 PD subjects for levodopa response (Contin et al. 2004). Perhaps the most informative study for the purposes of this discussion is the recent report by Cheon and coworkers (2005) that examined DAT avail-

Table 3 Influence of DAT 3'VNTR on DAT binding in the striatum

VNTR 9/9, 9/10 repeat	10/10 repeat	Population studied	DAT ligand	Binding increased in	Study
10	15	Alcoholics and controls	$^{123}\beta$-CIT	10/10	Heinz et al. 2000
9	18	Controls	$^{123}\beta$-CIT	9/9,9/10	Jacobsen et al. 2000
23	36	Controls and schizophrenics	$^{123}\beta$-CIT	No diff.	Martinez et al. 2001
82	74	Parkinson's disease and controls	[Tc-99m] TRODAT-1	No diff.	Lynch et al. 2003
20	16	Parkinson's disease	^{123}FP-CIT	No diff.	Contin et al. 2004
4[a]	7	ADHD	^{123}I-IPT	10/10	Cheon et al. 2005

[a] This group contained 2–9/10 and 2–10/11 genotypes

ability via [^{123}I]IPT binding in 11 drug-naïve children with ADHD that were subsequently treated with MPH. They found that the children with 3'VNTR 10/10 genotype had significantly increased DAT binding (∼70%) compared to ADHD subjects not homozygous for the 10-repeat allele. While interpretation of this study is limited due to the small number of subjects ($n = 7$ for 10/10, $n = 4$ for non-10/10), the findings suggest that there might be a link between ADHD severity and treatment response and DAT availability and VNTR status. These studies also raise the question of whether other genetic modifiers exist to differentially regulate the impact of the DAT 3'VNTR. Clues to this notion may arise from further studies of DAT regulatory proteins or alternatively from genome-wide studies seeking risk genes for ADHD in a more unbiased fashion.

5
Genetic Linkage in ADHD and the Impact of DAT Gene Variants

5.1
ADHD Linkage Studies

Twin studies have shown that ADHD is a highly heritable disorder, with most estimates of heritability around 0.8 (Faraone and Biederman 1998; Thapar et al. 1999). Genome-wide linkage scans have been completed in the attempt to identify genetic loci that contribute to ADHD. The first study was completed by Fisher and coworkers (2002) using 126 affected sib pairs and a roughly 10-cM grid of microsatellite markers. They did not identify any significant linkage peaks, but found moderate logarithm of differences (lod) scores (>1.5) at 5p12, 10q26, 12q23, and 16p13 (Fisher et al. 2002).

Extending this work by adding additional affected sib pairs, Smalley and coworkers (2002) looked for linkage in a total of 277 affected sib pairs (ASPs). Using the combined sample and fine mapping on chromosome 16, a significant lod score was identified at 16p13 and narrowed the locus to a 12-cM region, which was described as the first susceptibility locus identified for ADHD (Smalley et al. 2002). When a genome-wide scan was completed on the larger sample, in addition to the peak at 16p13, suggestive linkage was found at 17p11 (Ogdie et al. 2003). However, a genome-wide linkage scan of 164 Dutch ASPs failed to replicate the finding at 16p13 or 17p11, instead suggestive evidence of linkage was observed at 15q and 7p (Bakker et al. 2003). Ogdie and coworkers (2004) attempted to sort these differences out by enlarging their original sample from 277 to 308 ASPs, and fine mapping nine regions, the five regions with the highest lod score in their previous genome-wide screen (Ogdie et al. 2003) (except 16p13 which was completed in 2002), and the four highest lod scores from the Dutch genome-wide screen (Bakker et al. 2003). Significant evidence of linkage was observed at 6q12 and 17p11 and suggestive linkage was founds at 5p13, the only region identified in both genome-wide scans (Ogdie et al. 2004). Neither significant nor suggestive linkage was observed at the three other regions identified in the Dutch sample.

Instead of using ASPs, 16 extended and multigenerational pedigrees from the genetic isolate of Paisa, Colombia, were used by Arcos-Burgos and coworkers (2004) in a genome-wide screen for ADHD. They identified four regions of significant linkage at 4q13.2, 5q33.3, 11q22, and 17p11 (Arcos-Burgos et al. 2004). The 17p11 region was also identified by Ogdie and coworkers (2004). So in the genome-wide screens conducted to date, 5p13 and 17p11 are the only two regions that have been significantly linked to ADHD in two independent samples, and no single region was identified in all three populations examined (Arcos-Burgos et al. 2004; Bakker et al. 2003; Ogdie et al. 2004). This lack of a consistent finding of linkage in ADHD could be due to a number of factors, including differences in ADHD diagnosis and the population of ADHD subjects examined. However, another possibility is that instead of being a complex disorder produced by common polymorphisms in a few different genes, that ADHD could result from a large number of rare variants, possibly localized to a limited pathway of functionally related genes. This type of model has recently been proposed for autism, where there is also a high heritability with a relative lack of consistent findings in genome-wide linkage studies (Veenstra-Vanderweele et al. 2004). In this model, multiple epigenetic and genetic and de novo and inherited factors have been postulated to contribute to the etiology of autism (Jiang et al. 2004), and this model may also have relevance for ADHD.

5.2
Association Studies of hDAT and ADHD

Unlike linkage studies, where only three groups have attempted to identify risk loci for ADHD, many groups sought evidence for a role of specific genes in ADHD via a candidate gene approach. Although many genes have been examined, the dopaminergic system has been most widely studied and VNTRs in the DA D4 receptor (DRD4) and hDAT have been particularly targeted (reviewed in DiMaio et al. 2003). To date, over 16 studies have examined whether the 3′VNTR of hDAT is associated/linked with childhood ADHD (Table 4). As shown in Table 4, the results have been mixed, with approximately half of the studies finding a positive linkage/association and the other half finding no evidence for linkage/association. This is not an uncommon occurrence in psychiatric genetics. Potential sources for this variance include differences between studies in diagnosis criteria, diagnosis type or severity, ethnicity, sex, and study size.

In an attempt to make sense of these disparate results, meta-analysis has been employed to determine whether there is an overall trend for DAT 3′VNTR association with ADHD. Neither a meta-analysis examining 11 (Maher et al. 2002) nor a meta-analysis examining 13 of the family-based association studies (Purper-Ouakil et al. 2005) found a significant association between DAT and ADHD (odds ratios were 1.27 and 1.13, respectively). However, Purper-Ouakil and coworkers (2004) noted that there was a significant between-samples heterogeneity and noted that this may be due in part to the wide range of ADHD subtype composition between studies (Purper-Ouakil et al. 2005).

Future analyses examining only specific subtypes of ADHD (i.e., the predominately hyperactive subtype) or only cohorts with a family history of ADHD may yield more consistent results. Subject stratification may be a particularly promising idea for DAT, as it has already been observed that the 10-repeat allele is more associated with hyperactive versus inattentive symptoms (Waldman et al. 1998), and that subjects with conduct disorder and ADHD may have a more heritable form of ADHD (Faraone et al. 2000). However, statistical power is also of concern, as predominately hyperactive subjects are typically in the minority, so larger studies are needed to determine if the DAT 3′VNTR contributes to a small but significant portion of ADHD. Furthermore, dimensional scaling of subjects along cognitive, impulsivity, and hyperactivity dimensions may be fruitful. In conjunction with DAT 3′VNTR association studies, it would be useful to examine in vivo DAT binding levels. This has been completed in a small set of samples (Cheon et al. 2005), but it would be much more informative in a larger ADHD sample. A combination of genetic and neuroimaging approaches may be able to define a DAT phenotype within ADHD more effectively, especially when genotype and diagnostic subtype are taken into account.

Table 4 Results from association studies using the hDAT 3′VNTR and ADHD

Study	N (probands)	Ethnicity (%)	Sex (% male)	Design	Association
Cook et al. 1995	57	82[a]/16[b]/2[c]	–	HRR	Yes
Gill et al. 1997	40	Irish	–	HRR	Yes
Waldman et al. 1998	117	68[a]/12[b]/4[c]/16[d]	74	TDT	Yes
Daly et al. 1999	103	Irish	86	HRR	Yes
				TDT	Yes
Palmer et al. 1999	209	80[a]	75	TDT	No
Holmes et al. 2000	133	UK Caucasian	92	Case/control	No
	108			TDT	No
Swanson[g] et al. 1998	105	–	–	HRR	No
Todd et al. 2001	219	Missouri twins	54	TDT	No
Roman et al. 2001	81	86 European Brazilian	86	HRR	No
				Case/control	No
Curran et al. 2001	66	UK Caucasian	–	TDT	Yes
	111	Turkish	86	TDT	No
Barr et al. 2001	102	96 White European, 4 African Canadian	–	TDT	No/yes
Chen et al. 2003	110	Taiwanese	84	HRR	Yes
				TDT	Yes
				TRANSMIT	Yes
Smith et al. 2003[f]	105	94[a], 5[b], 1[c]	91	Case/control	No
Kustanovich et al. 2004	535	79[a]/2[b]/4[c]/15[d]	74	TDT	No
Qian et al. 2004[e]	332	Han Chinese	87	Case/control	Yes
	188			TDT	No
Bakkar et al. 2005	238	Dutch	83	TDT	No

HRR, haplotype relative risk; TDT, transmission disequilibrium test, [a]White American, [b]African American, [c]Hispanic, [d]Mixed ethnicity, [e]Examined long (11–12 repeat allele) vs short VNTR (6–10 repeat allele), [f]Not actually clinically defined ADHD, [g]Subjects did not have any comorbid disorders and were all treated with MPH

5.3
DAT 3′VNTR and Response to Methylphenidate

Since most pharmacological treatments for ADHD target DAT, it is conceivable that a functional polymorphism in DAT could impact treatment efficacy. Recent studies have attempted to determine whether DAT 3′VNTR status specifically plays a role in response to MPH (Table 5). Initially, Winsberg and Comings (1999) observed that homozygosity of the 10-repeat DAT 3′VNTR allele was correlated with poor MPH response in 30 African American children with ADHD (Winsberg and Comings 1999). This result was replicated in a blind study of 50 Brazilian boys with ADHD, 30 homozygous for the 10-repeat allele and 20 with other genotypes, where this group found that 15/20 (75%) non-homozygotes demonstrated a greater than 50% improvement in basal scores after 30 days of MPH treatment, whereas only 14/30 (47%) 10/10 3′VNTR exhibited a similar positive response (Roman et al. 2002). A poor response to MPH was also observed in a small study of Korean children with ADHD, where only 2/7 (29%) of those with the 10/10 3′VNTR genotype had a favorable response after 8 weeks of treatment, whereas all four of the subjects without the 10/10 3′VNTR genotype responded favorably (Cheon et al. 2005). An opposite finding was reported by Kirley and coworkers (2003), who found that transmission of the 10-repeat allele by heterozygous parents was associated with a favorable response to MPH in a retrospective study of 119 Irish ADHD children (Kirley et al. 2003). The major caveat to this study is that it was a retrospective study, and response to MPH was determined as a parental assessment of "no response," "mediocre," and "very good" (Kirley et al. 2003). However,

Table 5 Influence of DAT 3′VNTR on methylphenidate response in ADHD

VNTR 9/9, 9/10 repeat	10/10 repeat	Ethnicity	Poor methylphenidate response in	Study
13	17	African American	10/10	Winsberg and Comings 1999
20	30	Brazilian	10/10	Roman et al. 2002
4[a]	7	Korean	10/10	Cheon et al. 2005
		Irish	10[b]	Kirley et al. 2003
6, 22	19	89% Caucasian, 4% African American, 2% Hispanic, 4% other	9/9	Stein et al. 2005

[a]This group contained 2–9/10 and 2–10/11 genotypes, [b]Transmission of the 10-repeat allele by parent, 109 ADHD subjects rated

a somewhat complementary finding was recently reported, where homozygosity of the less-frequent 9-repeat allele was associated with a poor response to MPH (Stein et al. 2005). Given the differences in findings between these studies, further work is needed, which preferably will utilize larger samples and take diagnostic subtype into account. Again, having neuroimaging data as a correlate to these studies might help to identify the underlying similarities and differences between these studies. For example, a study reporting that a response to MPH treatment is associated with the 10/10 genotype might have subjects with a higher DAT binding level than a similarly conducted study that finds the 10/10 genotype does not improve MPH response. Additionally, since the functional relevance of the hDAT 3'VNTR is still unknown, there might be other variants in hDAT that alter its levels as well as the response to treatments such as MPH. Thus, ADHD subjects at the extreme high or low end of DAT binding might be good candidates for the investigation of genetic variation in DAT that could contribute to ADHD.

5.4
Investigation of hDAT Coding Variants in ADHD

Despite efforts to explore DAT in ADHD-related phenotypes in animal studies, pharmacological treatment response, and the focus of neuroimaging and genetic studies on DAT 3'VNTR, little screening of the hDAT coding region for functional variants in subjects with ADHD has been entertained. Prior to work of our group, only three studies had sought evidence for variation in the hDAT coding region, resulting in the identification of five nonsynonymous coding SNPS: V55A, R237Q, V382A, A559V, and E602G (Cargill et al. 1999; Grunhage et al. 2000; Vandenbergh et al. 2000) (Table 6). The R237Q mutation was identified in a large screen for coding SNPs that examined 106 genes in an average of 57 subjects (Cargill et al. 1999). The study by Grunhage et al. (2000) specifically examined the hDAT gene, and the A559V and E602G mutations were identified in single individuals out of 45 subjects who suffered from bipolar disorder, whereas no nonsynonymous SNPs were identified in their screen of 46 control subjects (Grunhage et al. 2000). In the Vandenbergh study (2000), almost half of the subjects screened were controls, whereas 109 subjects had Tourette's syndrome, 64 subjects were alcohol-dependent, and only 15 subjects were diagnosed with ADHD (Vandenbergh et al. 2000). Unfortunately, the diagnostic groups in which V55A and V382A were identified are not available, but these two SNPs were also rare, likely occurring in single individuals.

Given functional evidence that DAT dysfunction could contribute to ADHD, we felt that it was important to screen a larger cohort of ADHD subjects for functional coding variants in hDAT. To this end, we screened the coding region and splice junctions of hDAT in 66 subjects with ADHD (Mazei-Robison et al. 2005). The mean age of our sample was 10.5 ± 0.3 years and was 79% male and 21% female, and the majority of subjects were Caucasian (82%), with the

Table 6 Single nucleotide polymorphisms (SNPs) identified in the coding region of hDAT

SNP type	Base pair change	Exon	Amino acid change	Frequency (%)
Synonymous	242 C/T	2	–	5–15[a], 3[b], 1[c], 12[d], 8[e]
Synonymous	278 G/T	2	–	< 5[a], 18[d], < 1[e]
Synonymous	290 C/T	2	–	< 5[a], 1.5[e]
Nonsynonymous	292 T/C	2	V55A	< 1[d]
Synonymous	299 C/T	2	–	5[d]
Synonymous	674 C/T	4	–	< 1[e]
Nonsynonymous	838 G/A	5	R237Q	NA[a]
Synonymous	938 C/T	6	–	< 5[a], 14[d], 4[e]
Synonymous	1106 C/T	7	–	6[d]
Synonymous	1178 C/A	8	–	5[d]
Synonymous	1196 G/A	8	–	34[d]
Nonsynonymous	1273 T/C	8	V382A	< 1[d]
Synonymous	1343 A/G	9	–	> 15–50[a], 27[b], 21[c], 33[d], 24[e]
Synonymous	1541 C/G	11	–	< 1[e]
Synonymous	1655 G/A	12	–	< 1[d]
Nonsynonymous	1804 C/T	13	A559 V	1[b], 1.5[e]
Synonymous	1820 C/T	13	–	< 1[e]
Synonymous	1859 C/T	13	–	< 5[a], 5[d], < 1[e]
Nonsynonymous	1933 A/G	14	E602G	1[b]

NA, frequency not available, [a]Cargill et al. 1999, [b]Grunhage et al. 2000, bipolar subjects, [c]Grunhage et al. 2000, control subjects, [d]Vandenbergh et al. 2000, [e]Mazei-Robison et al. 2005, ADHD subjects

number of African American subjects (18%) representative of the middle Tennessee area. As expected, the majority of our cohort had combined type ADHD (61%), with predominantly inattentive (30%) and predominantly hyperactive-impulsive (9%) diagnoses less prevalent. The hDAT 3′VNTR genotype was determined for all 66 subjects. The frequency of 9- and 10-repeat alleles in our sample was 77% for the 10-repeat allele and 22% for the 9-repeat allele, respectively. The frequency of the 9 and 10 alleles did not seem to differ by gender, but there was a difference depending on ADHD diagnosis type. In our sample, predominantly inattentive and combined type subjects had a similar 10/10 genotype frequency (50% and 55%, respectively), whereas the predominantly hyperactive-impulsive subjects were almost exclusively 10/10 (83%).

To search for novel hDAT variants, we used the Reveal Mutation Discovery System (Spectrumedix; Li et al. 2002) to screen the hDAT coding region and, in total, almost 250,000 bp were screened for the presence of polymorphisms, resulting in the identification of 21 variants (Fig. 3). Sequence variants were identified in or around 10 out of the 14 coding exons, and we found that 45 of the 66 subjects had at least one variant present. Of the 21 variants identified, 10 were located within exons (Table 6). Only one of these SNPs, 1804 C/T, produced a change in amino acid sequence, converting alanine 559 to valine (A559V). Five novel variants were identified in this cohort and all were rare. The DAT variants were not evenly distributed between subjects with different diagnosis types of ADHD. About half of the subjects with the combined phenotype had 0–1 or 2+ DAT variants per subject (52.5% 0–1 variant, 47.5% 2+ variant), whereas, in the predominantly inattentive group, 15 of the 20 subjects had 0–1 variants (75%). In the predominantly hyperactive group, 5 of the 6 subjects had 2+ variants (83%). This suggests that the predominantly hyperactive ADHD subjects may represent a subphenotype more enriched for hDAT variants.

In our efforts, we identified one nonsynonymous mutation, A559V, in two male siblings with ADHD. Both probands were heterozygous for the A559V allele, which was found by pedigree analysis to be carried by the mother and grandmother as well. One sibling was diagnosed with combined-type ADHD and the other was diagnosed with predominantly hyperactive-impulsive ADHD. Both siblings were being successfully treated with combined AMPH-dextroamphetamine treatment (Adderall). Although neither the mother nor grandmother had Conner's Adult ADHD Rating Scale scores indicative of clinically relevant ADHD symptoms, the grandmother did score above average on the hyperactivity/restlessness and impulsivity/emotional liability items and was above the 90th percentile in DSM-IV hyperactive-impulsive symptoms. Interestingly, this variant was previously identified by Grunhage et al. in a subject with bipolar disorder (Grunhage et al. 2000). Unfortunately, the pedigree analysis was not informative, as the mother that also suffered

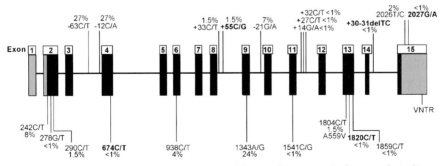

Fig. 3 Sequence variants identified in hDAT in children with ADHD. *Black rectangles* indicate hDAT coding sequence, *gray rectangles* indicate hDAT noncoding exon sequence, and exon numbers are listed *above* each rectangle. Coding sequence variants identified in the screen of ADHD subjects are listed *below*, with the corresponding allele frequency and noncoding sequence variants listed *above*. Variants in *bold* represent novel polymorphisms identified in our screen (Mazei-Robison and Blakely 2005)

from bipolar disorder did not carry the allele, and the phenotype of other carriers in the pedigree is unknown as the father was deceased and the only living member ascertained on the father's side also did not carry the allele (Grunhage et al. 2000). Without the ability to track the A559V allele within the family, association with bipolar disorder cannot be established, but there is evidence that central DA systems may also be involved bipolar disorder (Bunney and Garland 1982; Diehl and Gershon 1992; Gerner et al. 1976). Further, DAT dysfunction itself may contribute to bipolar disorder, as the clinically effective bipolar disorder treatment valproate significantly decreased locomotor activity and perseverative locomotor behavior in DAT knockdown mice, but not in wildtype littermates (Ralph-Williams et al. 2003). Consistent with the ability of ADHD and bipolar treatments to decrease locomotor activity in DAT knockdown mice, there seems to be emerging evidence of comorbidity of ADHD with mania and/or bipolar disorder, and this cohort may represent a unique clinical phenotype that could be useful in screens for functional variants of hDAT (Chang et al. 2000; Dienes et al. 2002; Faraone et al. 1997; Krishnan 2005; West et al. 1995; Wozniak et al. 1995a; Wozniak et al. 1995b).

Despite finding the A559V mutation in siblings with ADHD (Mazei-Robison et al. 2005) and in a subject with bipolar disorder (Grunhage et al. 2000), we did not identify any significant differences between hDAT and A559V using in vitro methodologies (Mazei-Robison and Blakely 2005). It is possible that A559V impacts an aspect of hDAT function that is not well replicated using our current in vitro methods or that a native cell context is essential to reveal phenotypic differences, which is also likely true for the hDAT 3′VNTR. Future studies expressing hDAT variants in DA neurons that lack endogenous DAT expression, such as we have established for characterization of *C. elegans* DAT mutants (Carvelli et al. 2004), may be of use. Additionally, tests of A559V under voltage-clamp may be able to reveal cellular phenotypes that are difficult to mimic in non-clamped cells.

In contrast to the A559V mutation, we found that the V382A mutation identified by Vandenbergh et al. (2000) altered the function of hDAT with similar properties evident in two different in vitro model systems. Similar to data reported by Lin and Uhl (2003), the V382A variant exhibited a decreased V_{max} for DA transport, a decreased surface expression, and a higher affinity for cocaine (Lin and Uhl 2003). Additionally, we found that the transport of NE was more compromised than that of DA, revealing that surface-trafficked transporters exhibit conformational alterations that can be sensed by the subtle chemical variation in catecholamine substrates. We also observed that the regulation of V382A by the phorbol ester PMA (phorbol myristate acetate) is significantly different from that exhibited by hDAT. When transiently transfected cells were incubated with PMA, the decrease in V382A transport was greater than that of hDAT. Whereas the downregulation of DA transport activity and protein internalization induced by PMA were reasonably well correlated for hDAT, this was not the case for V382A, where loss of DA transport was not matched by the extent of transporter internalization. Based on these studies, we suggest that the V382A mutation may stabilize a normally rapidly exited conformation that links functional inactivation of DAT proteins to internalization from the plasma membrane.

In summary, existing studies on the presence of DAT gene variants do not at present support a model whereby common variants with large effect support risk for the disorder. This conclusion does not rule out a role for DAT in ADHD but shifts the focus to a more complex contribution, one that may best be revealed through within-family analyses of key pedigrees identified as harboring DAT variants or variants of DAT regulators. Our studies have revealed a number of rare genomic variants at the DAT locus whose contributions within lineages have yet to be evaluated. Association studies using the hDAT 3'VNTR have not clearly established an association between hDAT and ADHD, and further effort is needed to clarify whether this polymorphism directly impacts DAT protein expression and in which type of subject. Future work that targets DAT and DA signaling in this disorder will likely need to consider both the collective impact of rare variants as well as contributions from DAT gene and protein regulators. It is amazing in some ways that as important as DAT is for DA signaling, no subjects have been identified where variation in DAT supports their disorder. Possibly, the A559V pedigrees that Grunhage et al. (2000) and we have identified represent steps in that direction, but stronger functional in vitro and in vivo data are needed for this variant to support such a possibility. The independent identification of this variant in ADHD and bipolar subjects is also of interest given evidence of comorbidity between the two disorders. In this regard, endophenotypes linked more closely to DAT and altered central DA signaling may allow for the elucidation of shared biological determinants despite what appear to be rather disparate clinical syndromes.

References

Adell A, Artigas F (2004) The somatodendritic release of dopamine in the ventral tegmental area and its regulation by afferent transmitter systems. Neurosci Biobehav Rev 28:415–431

Almasy L, Blangero J (2001) Endophenotypes as quantitative risk factors for psychiatric disease: rationale and study design. Am J Med Genet 105:42–44

American Psychiatric Association (1994) Attention-deficit and disruptive behavior disorders. Diagnostic and statistical manual of mental disorders: DSM-IV, 4th edn. American Psychiatric Association, Washington

Applegate B, Lahey BB, Hart EL, Biederman J, Hynd GW, Barkley RA, Ollendick T, Frick PJ, Greenhill L, McBurnett K, Newcorn JH, Kerdyk L, Garfinkel B, Waldman I, Shaffer D (1997) Validity of the age-of-onset criterion for ADHD: a report from the DSM-IV field trials. J Am Acad Child Adolesc Psychiatry 36:1211–1221

Arcos-Burgos M, Castellanos FX, Pineda D, Lopera F, Palacio JD, Palacio LG, Rapoport JL, Berg K, Bailey-Wilson JE, Muenke M (2004) Attention-deficit/hyperactivity disorder in a population isolate: linkage to loci at 4q13.2, 5q33.3, 11q22, and 17p11. Am J Hum Genet 75:998–1014

Arnold LE, Abikoff HB, Cantwell DP, Conners CK, Elliott G, Greenhill LL, Hechtman L, Hinshaw SP, Hoza B, Jensen PS, Kraemer HC, March JS, Newcorn JH, Pelham WE, Richters JE, Schiller E, Severe JB, Swanson JM, Vereen D, Wells KC (1997) National Institute of Mental Health Collaborative Multimodal Treatment Study of Children with ADHD (the MTA). Design challenges and choices. Arch Gen Psychiatry 54:865–870

Bakker SC, van der Meulen EM, Buitelaar JK, Sandkuijl LA, Pauls DL, Monsuur AJ, van 't Slot R, Minderaa RB, Gunning WB, Pearson PL, Sinke RJ (2003) A whole-genome scan in 164 Dutch sib pairs with attention-deficit/hyperactivity disorder: suggestive evidence for linkage on chromosomes 7p and 15q. Am J Hum Genet 72:1251–1260

Barkley RA, Biederman J (1997) Toward a broader definition of the age-of-onset criterion for attention-deficit hyperactivity disorder. J Am Acad Child Adolesc Psychiatry 36:1204–1210

Barkley RA, Edwards G, Laneri M, Fletcher K, Metevia L (2001) Executive functioning, temporal discounting, and sense of time in adolescents with attention deficit hyperactivity disorder (ADHD) and oppositional defiant disorder (ODD). J Abnorm Child Psychol 29:541–556

Barnett R, Maruff P, Vance A, Luk ES, Costin J, Wood C, Pantelis C (2001) Abnormal executive function in attention deficit hyperactivity disorder: the effect of stimulant medication and age on spatial working memory. Psychol Med 31:1107–1115

Baughman FA Jr (2000) Dopamine-transporter density in patients with ADHD. Lancet 355:1460–1461; author reply 1461–1462

Bauman AL, Apparsundaram S, Ramamoorthy S, Wadzinski BE, Vaughan RA, Blakely RD (2000) Cocaine and antidepressant-sensitive biogenic amine transporters exist in regulated complexes with protein phosphatase 2A. J Neurosci 20:7571–7578

Berquin PC, Giedd JN, Jacobsen LK, Hamburger SD, Krain AL, Rapoport JL, Castellanos FX (1998) Cerebellum in attention-deficit hyperactivity disorder: a morphometric MRI study. Neurology 50:1087–1093

Berridge KC, Aldridge JW, Houchard KR, Zhuang X (2005) Sequential super-stereotypy of an instinctive fixed action pattern in hyper-dopaminergic mutant mice: a model of obsessive compulsive disorder and Tourette's. BMC Biol 3:4

Biederman J, Wilens T, Mick E, Spencer T, Faraone SV (1999) Pharmacotherapy of attention-deficit/hyperactivity disorder reduces risk for substance use disorder. Pediatrics 104:e20

Birnbaum HG, Kessler RC, Lowe SW, Secnik K, Greenberg PE, Leong SA, Swensen AR (2005) Costs of attention deficit-hyperactivity disorder (ADHD) in the US: excess costs of persons with ADHD and their family members in 2000. Curr Med Res Opin 21:195–206

Bosse R, Fumagalli F, Jaber M, Giros B, Gainetdinov RR, Wetsel WC, Missale C, Caron MG (1997) Anterior pituitary hypoplasia and dwarfism in mice lacking the dopamine transporter. Neuron 19:127–138

Buck KJ, Amara SG (1994) Chimeric dopamine-norepinephrine transporters delineate structural domains influencing selectivity for catecholamines and 1-methyl-4-phenylpyridinium. Proc Natl Acad Sci U S A 91:12584–12588

Bunney WE Jr, Garland BL (1982) A second generation catecholamine hypothesis. Pharmacopsychiatria 15:111–115

Bymaster FP, Katner JS, Nelson DL, Hemrick-Luecke SK, Threlkeld PG, Heiligenstein JH, Morin SM, Gehlert DR, Perry KW (2002) Atomoxetine increases extracellular levels of norepinephrine and dopamine in prefrontal cortex of rat: a potential mechanism for efficacy in attention deficit/hyperactivity disorder. Neuropsychopharmacology 27:699–711

Cargill M, Altshuler D, Ireland J, Sklar P, Ardlie K, Patil N, Lane CR, Lim EP, Kalayanaraman N, Nemesh J, Ziaugra L, Friedland L, Rolfe A, Warrington J, Lipshutz R, Daley GQ, Lander ES (1999) Characterization of single-nucleotide polymorphisms in coding regions of human genes. Nat Genet 22:231–238

Carneiro A, Ingram SL, Beaulieu J-M, Sweeney A, Amara SG, Thomas SM, Caron MG, Torres GE (2002) The multiple LIM domain-containing adaptor protein Hic-5 synaptically colocalizes and interacts with the dopamine transporter. J Neurosci 22:7045–7054

Carvelli L, Moron JA, Kahlig KM, Ferrer JV, Sen N, Lechleiter JD, Leeb-Lundberg LM, Merrill G, Lafer EM, Ballou LM, Shippenberg TS, Javitch JA, Lin RZ, Galli A (2002) PI3-kinase regulation of dopamine uptake. J Neurochem 81:859–869

Carvelli L, McDonald PW, Blakely RD, Defelice LJ (2004) Dopamine transporters depolarize neurons by a channel mechanism. Proc Natl Acad Sci U S A 101:16046–16051

Castellanos FX, Tannock R (2002) Neuroscience of attention-deficit/hyperactivity disorder: the search for endophenotypes. Nat Rev Neurosci 3:617–628

Castellanos FX, Giedd JN, Berquin PC, Walter JM, Sharp W, Tran T, Vaituzis AC, Blumenthal JD, Nelson J, Bastain TM, Zijdenbos A, Evans AC, Rapoport JL (2001) Quantitative brain magnetic resonance imaging in girls with attention-deficit/hyperactivity disorder. Arch Gen Psychiatry 58:289–295

Chang KD, Steiner H, Ketter TA (2000) Psychiatric phenomenology of child and adolescent bipolar offspring. J Am Acad Child Adolesc Psychiatry 39:453–460

Chang M, Lee S-H, Kim J-H, Lee K-H, Kim Y-S, Son H, Lee Y-S (2001) Protein kinase C-mediated functional regulation of dopamine transporter is not achieved by direct phosphorylation of the dopamine transporter protein. J Neurochem 77:754–761

Chen JJ, Li Z, Pan H, Murphy DL, Tamir H, Koepsell H, Gershon MD (2001) Maintenance of serotonin in the intestinal mucosa and ganglia of mice that lack the high-affinity serotonin transporter: abnormal intestinal mobility and the expression of cation transporters. J Neurosci 21:6348–6361

Cheon KA, Ryu YH, Kim YK, Namkoong K, Kim CH, Lee JD (2003) Dopamine transporter density in the basal ganglia assessed with [123I]IPT SPET in children with attention deficit hyperactivity disorder. Eur J Nucl Med Mol Imaging 30:306–311

Cheon KA, Ryu YH, Kim JW, Cho DY (2005) The homozygosity for 10-repeat allele at dopamine transporter gene and dopamine transporter density in Korean children with attention deficit hyperactivity disorder: relating to treatment response to methylphenidate. Eur Neuropsychopharmacol 15:95–101

Chhabildas N, Pennington BF, Willcutt EG (2001) A comparison of the neuropsychological profiles of the DSM-IV subtypes of ADHD. J Abnorm Child Psychol 29:529–540

Choong K, Shen R (2004) Prenatal ethanol exposure alters the postnatal development of the spontaneous electrical activity of dopamine neurons in the ventral tegmental area. Neuroscience 126:1083–1091

Contin M, Martinelli P, Mochi M, Albani F, Riva R, Scaglione C, Dondi M, Fanti S, Pettinato C, Baruzzi A (2004) Dopamine transporter gene polymorphism, SPECT imaging, and levodopa response in patients with Parkinson disease. Clin Neuropharmacol 27:111–115

Corman SL, Fedutes BA, Culley CM (2004) Atomoxetine: the first nonstimulant for the management of attention-deficit/hyperactivity disorder. Am J Health Syst Pharm 61:2391–2399

Cornblatt BA, Malhotra AK (2001) Impaired attention as an endophenotype for molecular genetic studies of schizophrenia. Am J Med Genet 105:11–15

Cornish KM, Manly T, Savage R, Swanson J, Morisano D, Butler N, Grant C, Cross G, Bentley L, Hollis CP (2005) Association of the dopamine transporter (DAT1) 10/10-repeat genotype with ADHD symptoms and response inhibition in a general population sample. Mol Psychiatry (in press)

Cowell RM, Kantor L, Hewlett GH, Frey KA, Gnegy ME (2000) Dopamine transporter antagonists block phorbol ester-induced dopamine release and dopamine transporter phosphorylation in striatal synaptosomes. Eur J Pharmacol 389:59–65

Daws LC, Callaghan PD, Moron JA, Kahlig KM, Shippenberg TS, Javitch JA, Galli A (2002) Cocaine increases dopamine uptake and cell surface expression of dopamine transporters. Biochem Biophys Res Commun 290:1545–1550

Diehl DJ, Gershon S (1992) The role of dopamine in mood disorders. Compr Psychiatry 33:115–120

Dienes KA, Chang KD, Blasey CM, Adleman NE, Steiner H (2002) Characterization of children of bipolar parents by parent report CBCL. J Psychiatr Res 36:337–345

DiMaio S, Grizenko N, Joober R (2003) Dopamine genes and attention-deficit hyperactivity disorder: a review. J Psychiatry Neurosci 28:27–38

Doucette-Stamm LA, Blakely DJ, Tian J, Mockus S, Mao JI (1995) Population genetic study of the human dopamine transporter gene (DAT1). Genet Epidemiol 12:303–308

Dougherty DD, Bonab AA, Spencer TJ, Rauch SL, Madras BK, Fischman AJ (1999) Dopamine transporter density in patients with attention deficit hyperactivity disorder. Lancet 354:2132–2133

Dresel S, Krause J, Krause KH, LaFougere C, Brinkbaumer K, Kung HF, Hahn K, Tatsch K (2000) Attention deficit hyperactivity disorder: binding of [99mTc]TRODAT-1 to the dopamine transporter before and after methylphenidate treatment. Eur J Nucl Med 27:1518–1524

Dresel SH, Kung MP, Plossl K, Meegalla SK, Kung HF (1998) Pharmacological effects of dopaminergic drugs on in vivo binding of [99mTc]TRODAT-1 to the central dopamine transporters in rats. Eur J Nucl Med 25:31–39

Dumartin B, Jaber M, Gonon F, Caron MG, Giros B, Bloch B (2000) Dopamine tone regulates D1 receptor trafficking and delivery in striatal neurons in dopamine transporter-deficient mice. Proc Natl Acad Sci U S A 97:1879–1884

Egan MF, Goldberg TE, Kolachana BS, Callicott JH, Mazzanti CM, Straub RE, Goldman D, Weinberger DR (2001) Effect of COMT Val108/158 Met genotype on frontal lobe function and risk for schizophrenia. Proc Natl Acad Sci U S A 98:6917–6922

Faraone SV, Biederman J (1998) Neurobiology of attention-deficit hyperactivity disorder. Biol Psychiatry 44:951–958

Faraone SV, Biederman J, Wozniak J, Mundy E, Mennin D, O'Donnell D (1997) Is comorbidity with ADHD a marker for juvenile-onset mania? J Am Acad Child Adolesc Psychiatry 36:1046–1055

Faraone SV, Biederman J, Monuteaux MC (2000) Toward guidelines for pedigree selection in genetic studies of attention deficit hyperactivity disorder. Genet Epidemiol 18:1–16

Fauchey V, Jaber M, Caron MG, Bloch B, Le Moine C (2000) Differential regulation of the dopamine D1, D2 and D3 receptor gene expression and changes in the phenotype of the striatal neurons in mice lacking the dopamine transporter. Eur J Neurosci 12:19–26

Fisher SE, Francks C, McCracken JT, McGough JJ, Marlow AJ, MacPhie IL, Newbury DF, Crawford LR, Palmer CG, Woodward JA, Del'Homme M, Cantwell DP, Nelson SF, Monaco AP, Smalley SL (2002) A genomewide scan for loci involved in attention-deficit/hyperactivity disorder. Am J Hum Genet 70:1183–1196

Fleckenstein AE, Haughey HM, Metzger RR, Kokoshka JM, Riddle EL, Hanson JE, Gibb JW, Hanson GR (1999) Differential effects of psychostimulants and related agents on dopaminergic and serotonergic transporter function. Eur J Pharmacol 382:45–49

Ford T, Goodman R, Meltzer H (2003) The British Child and Adolescent Mental Health Survey 1999: the prevalence of DSM-IV disorders. J Am Acad Child Adolesc Psychiatry 42:1203–1211

Foster JD, Pananusorn B, Vaughan RA (2002) Dopamine transporters are phosphorylated on N-terminal serines in rat striatum. J Biol Chem 277:25178–25186

Frengen E, Zhao B, Howe S, Weichenhan D, Osoegawa K, Gjernes E, Jessee J, Prydz H, Huxley C, de Jong PJ (2000) Modular bacterial artificial chromosome vectors for transfer of large inserts into mammalian cells. Genomics 68:118–126

Fuke S, Suo S, Takahashi N, Koike H, Sasagawa N, Ishiura S (2001) The VNTR polymorphism of the human dopamine transporter (DAT1) gene affects gene expression. Pharmacogenomics J 1:152–156

Fuke S, Sasagawa N, Ishiura S (2005) Identification and characterization of the Hesr1/Hey1 as a candidate trans-acting factor on gene expression through the 3′ non-coding polymorphic region of the human dopamine transporter (DAT1) gene. J Biochem (Tokyo) 137:205–216

Gaffaney JD, Vaughan RA (2004) Uptake inhibitors but not substrates induce protease resistance in extracellular loop two of the dopamine transporter. Mol Pharmacol 65:692–701

Gainetdinov RR, Caron MG (2001) Genetics of childhood disorders: XXIV. ADHD, part 8: hyperdopaminergic mice as an animal model of ADHD. J Am Acad Child Adolesc Psychiatry 40:380–382

Gainetdinov RR, Wetsel WC, Jones SR, Levin ED, Jaber M, Caron MG (1999) Role of serotonin in the paradoxical calming effect of psychostimulants on hyperactivity. Science 283:397–401

Garcia B, Wei Y, Moron JA, Lin RZ, Javitch JA, Galli A (2005) Akt is essential for insulin modulation of amphetamine-induced human dopamine transporter cell surface redistribution. Mol Pharmacol 68:102–109

Gelernter J, Kranzler H, Lacobelle J (1998) Population studies of polymorphisms at loci of neuropsychiatric interest (tryptophan hydroxylase (TPH), dopamine transporter protein (SLC6A3), D3 dopamine receptor (DRD3), apolipoprotein E (APOE), mu opioid receptor (OPRM1), and ciliary neurotrophic factor (CNTF)). Genomics 52:289–297

Gerner RH, Post RM, Bunney WE Jr (1976) A dopaminergic mechanism in mania. Am J Psychiatry 133:1177–1180

Giros B, el Mestikawy S, Godinot N, Zheng K, Han H, Yang-Feng T, Caron MG (1992) Cloning, pharmacological characterization, and chromosome assignment of the human dopamine transporter. Mol Pharmacol 42:383–390

Giros B, Wang YM, Suter S, McLeskey SB, Pifl C, Caron MG (1994) Delineation of discrete domains for substrate, cocaine, and tricyclic antidepressant interactions using chimeric dopamine-norepinephrine transporters. J Biol Chem 269:15985–15988

Giros B, Jaber M, Jones SR, Wightman RM, Caron MG (1996) Hyperlocomotion and indifference to cocaine and amphetamine in mice lacking the dopamine transporter. Nature 379:606–612

Goldberg NR, Beuming T, Soyer OS, Goldstein RA, Weinstein H, Javitch JA (2003) Probing confirmational changes in neurotransmitter transporters: a structural context. Eur J Pharmacol 479:3–12

Granas C, Ferrer J, Loland CJ, Javitch JA, Gether U (2003) N-terminal truncation of the dopamine transporter abolishes phorbol ester- and substance P receptor-stimulated phosphorylation without impairing transporter internalization. J Biol Chem 278:4990–5000

Greenwood TA, Kelsoe JR (2003) Promoter and intronic variants affect the transcriptional regulation of the human dopamine transporter gene. Genomics 82:511–520

Gresch PJ, Sved AF, Zigmond MJ, Finlay JM (1995) Local influence of endogenous norepinephrine on extracellular dopamine in rat medial prefrontal cortex. J Neurochem 65:111–116

Group MC (2004) National Institute of Mental Health Multimodal Treatment Study of ADHD follow-up: 24-month outcomes of treatment strategies for attention-deficit/hyperactivity disorder. Pediatrics 113:754–761

Grunhage F, Schulze TG, Muller DJ, Lanczik M, Franzek E, Albus M, Borrmann-Hassenbach M, Knapp M, Cichon S, Maier W, Rietschel M, Propping P, Nothen MM (2000) Systematic screening for DNA sequence variation in the coding region of the human dopamine transporter gene (DAT1). Mol Psychiatry 5:275–282

Harpin VA (2005) The effect of ADHD on the life of an individual, their family, and community from preschool to adult life. Arch Dis Child 90 Suppl 1:i2–7

Hastrup H, Karlin A, Javitch JA (2001) Symmetrical dimer of the human dopamine transporter revealed by cross-linking Cys-306 at the extracellular end of the sixth transmembrane segment. Proc Natl Acad Sci U S A 98:10055–10060

Hastrup H, Sen N, Javitch JA (2003) The human dopamine transporter forms a tetramer in the plasma membrane: cross-linking of a cysteine in the fourth transmembrane segment is sensitive to cocaine analogs. J Biol Chem 278:45045–45048

Heinz A, Goldman D, Jones DW, Palmour R, Hommer D, Gorey JG, Lee KS, Linnoila M, Weinberger DR (2000) Genotype influences in vivo dopamine transporter availability in human striatum. Neuropsychopharmacology 22:133–139

Herskovits EH, Megalooikonomou V, Davatzikos C, Chen A, Bryan RN, Gerring JP (1999) Is the spatial distribution of brain lesions associated with closed-head injury predictive of subsequent development of attention-deficit/hyperactivity disorder? Analysis with brain-image database. Radiology 213:389–394

Huff RA, Vaughan RA, Kuhar MJ, Uhl GR (1997) Phorbol esters increase dopamine transporter phosphorylation and decrease transport Vmax. Mol Chem Neuropathol 68:225–232

Inoue-Murayama M, Adachi S, Mishima N, Mitani H, Takenaka O, Terao K, Hayasaka I, Ito S, Murayama Y (2002) Variation of variable number of tandem repeat sequences in the 3′-untranslated region of primate dopamine transporter genes that affects reporter gene expression. Neurosci Lett 334:206–210

Itokawa M, Lin Z, Cai NS, Wu C, Kitayama S, Wang JB, Uhl GR (2000) Dopamine transporter transmembrane domain polar mutants: ΔG and $\Delta\Delta G$ values implicate regions important for transporter functions. Mol Pharmacol 57:1093–1103

Jacobsen LK, Staley JK, Zoghbi SS, Seibyl JP, Kosten TR, Innis RB, Gelernter J (2000) Prediction of dopamine transporter binding availability by genotype: a preliminary report. Am J Psychiatry 157:1700–1703

Jiang YH, Sahoo T, Michaelis RC, Bercovich D, Bressler J, Kashork CD, Liu Q, Shaffer LG, Schroer RJ, Stockton DW, Spielman RS, Stevenson RE, Beaudet AL (2004) A mixed epigenetic/genetic model for oligogenic inheritance of autism with a limited role for UBE3A. Am J Med Genet 131:1–10

Johansen EB, Aase H, Meyer A, Sagvolden T (2002) Attention-deficit/hyperactivity disorder (ADHD) behaviour explained by dysfunctioning reinforcement and extinction processes. Behav Brain Res 130:37–45

Jones SR, Gainetdinov RR, Jaber M, Giros B, Wightman RM, Caron MG (1998) Profound neuronal plasticity in response to inactivation of the dopamine transporter. Proc Natl Acad Sci U S A 95:4029–4034

Jones SR, Gainetdinov RR, Hu XT, Cooper DC, Wightman RM, White FJ, Caron MG (1999a) Loss of autoreceptor functions in mice lacking the dopamine transporter. Nat Neurosci 2:649–655

Jones SR, Joseph JD, Barak LS, Caron MG, Wightman RM (1999b) Dopamine neuronal transport kinetics and effects of amphetamine. J Neurochem 73:2406–2414

Kahlig KM, Javitch JA, Galli A (2004) Amphetamine regulation of dopamine transport. Combined measurements of transporter currents and transporter imaging support the endocytosis of an active carrier. J Biol Chem 279:8966–8975

Kang AM, Palmatier MA, Kidd KK (1999) Global variation of a 40-bp VNTR in the 3′-untranslated region of the dopamine transporter gene (SLC6A3). Biol Psychiatry 46:151–160

Kavelaars A, Cobelens PM, Teunis MA, Heijnen CJ (2005) Changes in innate and acquired immune responses in mice with targeted deletion of the dopamine transporter gene. J Neuroimmunol 161:162–168

Kawarai T, Kawakami H, Yamamura Y, Nakamura S (1997) Structure and organization of the gene encoding human dopamine transporter. Gene 195:11–18

Kempton S, Vance A, Maruff P, Luk E, Costin J, Pantelis C (1999) Executive function and attention deficit hyperactivity disorder: stimulant medication and better executive function performance in children. Psychol Med 29:527–538

Khoshbouei H, Sen N, Guptaroy B, Johnson L, Lund D, Gnegy ME, Galli A, Javitch JA (2004) N-terminal phosphorylation of the dopamine transporter is required for amphetamine-induced efflux. PLoS Biol 2:387–393

Kirley A, Lowe N, Hawi Z, Mullins C, Daly G, Waldman I, McCarron M, O'Donnell D, Fitzgerald M, Gill M (2003) Association of the 480 bp DAT1 allele with methylphenidate response in a sample of Irish children with ADHD. Am J Med Genet B Neuropsychiatr Genet 121:50–54

Kitayama S, Shimada S, Xu H, Markham L, Donovan DM, Uhl GR (1992) Dopamine transporter site-directed mutations differentially alter substrate transport and cocaine binding. Proc Natl Acad Sci U S A 89:7782–7785

Kitayama S, Dohi T, Uhl G (1994) Phorbol esters alter functions of the expressed dopamine transporter. Eur J Pharmacol 268:115–119

Kouzmenko AP, Pereira AM, Singh BS (1997) Intronic sequences are involved in neural targeting of human dopamine transporter gene expression. Biochem Biophys Res Commun 240:807–811

Krause KH, Dresel SH, Krause J, Kung HF, Tatsch K (2000) Increased striatal dopamine transporter in adult patients with attention deficit hyperactivity disorder: effects of methylphenidate as measured by single photon emission computed tomography. Neurosci Lett 285:107–110

Krause KH, Dresel S, Krause J, Kung HF, Tatsch K, Lochmuller H (2002) Elevated striatal dopamine transporter in a drug naive patient with Tourette syndrome and attention deficit/hyperactivity disorder: positive effect of methylphenidate. J Neurol 249:1116–1118

Krishnan KR (2005) Psychiatric and medical comorbidities of bipolar disorder. Psychosom Med 67:1–8

Kustanovich V, Ishii J, Crawford L, Yang M, McGough JJ, McCracken JT, Smalley SL, Nelson SF (2004) Transmission disequilibrium testing of dopamine-related candidate gene polymorphisms in ADHD: confirmation of association of ADHD with DRD4 and DRD5. Mol Psychiatry 9:711–717

Lee KH, Kim MY, Kim DH, Lee YS (2004) Syntaxin 1A and receptor for activated C kinase interact with the N-terminal region of human dopamine transporter. Neurochem Res 29:1405–1409

Li LB, Chen N, Ramamoorthy S, Chi L, Cui XN, Wang LC, Reith ME (2004a) The role of N-glycosylation in function and surface trafficking of the human dopamine transporter. J Biol Chem 279:21012–21020

Li Q, Liu Z, Monroe H, Culiat CT (2002) Integrated platform for detection of DNA sequence variants using capillary array electrophoresis. Electrophoresis 23:1499–1511

Li ZS, Pham TD, Tamir H, Chen JJ, Gershon MD (2004b) Enteric dopaminergic neurons: definition, developmental lineage, and effects of extrinsic denervation. J Neurosci 24:1330–1339

Lin Z, Uhl GR (2003) Human dopamine transporter gene variation: effects of protein coding variants V55A and V382A on expression and uptake activities. Pharmacogenomics J 3:159–168

Lin Z, Wang W, Kopajtic T, Revay RS, Uhl GR (1999) Dopamine transporter: transmembrane phenylalanine mutations can selectively influence dopamine uptake and cocaine analog recognition. Mol Pharmacol 56:434–447

Lin Z, Itokawa M, Uhl GR (2000a) Dopamine transporter proline mutations influence dopamine uptake, cocaine analog recognition, and expression. Faseb J 14:715–728

Lin Z, Wang W, Uhl GR (2000b) Dopamine transporter tryptophan mutants highlight candidate dopamine- and cocaine-selective domains. Mol Pharmacol 58:1581–1592

Lin Z, Zhang PW, Zhu X, Melgari JM, Huff R, Spieldoch RL, Uhl GR (2003) Phosphatidylinositol 3-kinase, protein kinase C, and MEK1/2 kinase regulation of dopamine transporters (DAT) require N-terminal DAT phosphoacceptor sites. J Biol Chem 278:20162–20170

Little KY, Elmer LW, Zhong H, Scheys JO, Zhang L (2002) Cocaine induction of dopamine transporter trafficking to the plasma membrane. Mol Pharmacol 61:436–445

Loder MK, Melikian HE (2003) The dopamine transporter constitutively internalizes and recycles in a protein kinase C-regulated manner in stably transfected PC12 cell lines. J Biol Chem 278:22168–22174

Loo SK, Specter E, Smolen A, Hopfer C, Teale PD, Reite ML (2003) Functional effects of the DAT1 polymorphism on EEG measures in ADHD. J Am Acad Child Adolesc Psychiatry 42:986–993

Lynch DR, Mozley PD, Sokol S, Maas NM, Balcer LJ, Siderowf AD (2003) Lack of effect of polymorphisms in dopamine metabolism related genes on imaging of TRODAT-1 in striatum of asymptomatic volunteers and patients with Parkinson's disease. Mov Disord 18:804–812

Maher BS, Marazita ML, Ferrell RE, Vanyukov MM (2002) Dopamine system genes and attention deficit hyperactivity disorder: a meta-analysis. Psychiatr Genet 12:207–215

Martinez D, Gelernter J, Abi-Dargham A, van Dyck CH, Kegeles L, Innis RB, Laruelle M (2001) The variable number of tandem repeats polymorphism of the dopamine transporter gene is not associated with significant change in dopamine transporter phenotype in humans. Neuropsychopharmacology 24:553–560

Marx J (1999) How stimulant drugs may calm hyperactivity. Science 283:306

Mash DC, Pablo J, Ouyang Q, Hearn WL, Izenwasser S (2002) Dopamine transport function is elevated in cocaine users. J Neurochem 81:292–300

Max JE, Fox PT, Lancaster JL, Kochunov P, Mathews K, Manes FF, Robertson BA, Arndt S, Robin DA, Lansing AE (2002) Putamen lesions and the development of attention-deficit/hyperactivity symptomatology. J Am Acad Child Adolesc Psychiatry 41:563–571

Mazei MS, Pluto CP, Kirkbride B, Pehek EA (2002) Effects of catecholamine uptake blockers in the caudate-putamen and subregions of the medial prefrontal cortex of the rat. Brain Res 936:58–67

Mazei-Robison MS, Blakely RD (2005) Expression studies of naturally occurring human dopamine transporter variants identifies a novel state of transporter inactivation associated with val382ala. Neuropharmacology 49:737–749

Mazei-Robison MS, Couch RS, Shelton RC, Stein MA, Blakely RD (2005) Sequence variation in the human dopamine transporter gene in children with attention deficit hyperactivity disorder. Neuropharmacology 49:724–736

Meinild AK, Sitte HH, Gether U (2004) Zinc potentiates an uncoupled anion conductance associated with the dopamine transporter. J Biol Chem 279:49671–49679

Melikian H, Buckley K (1999) Membrane trafficking regulates the activity of the human dopamine transporter. J Neurosci 19:7699–7710

Mill J, Asherson P, Craig I, D'Souza UM (2005) Transient expression analysis of allelic variants of a VNTR in the dopamine transporter gene (DAT1). BMC Genet 6:3

Miller GM, Madras BK (2002) Polymorphisms in the 3'-untranslated region of human and monkey dopamine transporter genes affect reporter gene expression. Mol Psychiatry 7:44–55

Mitchell RJ, Howlett S, Earl L, White NG, McComb J, Schanfield MS, Briceno I, Papiha SS, Osipova L, Livshits G, Leonard WR, Crawford MH (2000) Distribution of the 3' VNTR polymorphism in the human dopamine transporter gene in world populations. Hum Biol 72:295–304

Moron JA, Zakharova I, Ferrer JV, Merrill GA, Hope B, Lafer EM, Lin ZC, Wang JB, Javitch JA, Galli A, Shippenberg TS (2003) Mitogen-activated protein kinase regulates dopamine transporter surface expression and dopamine transport capacity. J Neurosci 23:8480–8488

Mostofsky SH, Reiss AL, Lockhart P, Denckla MB (1998) Evaluation of cerebellar size in attention-deficit hyperactivity disorder. J Child Neurol 13:434–439

Norregaard L, Frederiksen D, Nielsen EO, Gether U (1998) Delineation of an endogenous zinc-binding site in the human dopamine transporter. EMBO J 17:4266–4273

Ogdie MN, Macphie IL, Minassian SL, Yang M, Fisher SE, Francks C, Cantor RM, McCracken JT, McGough JJ, Nelson SF, Monaco AP, Smalley SL (2003) A genomewide scan for attention-deficit/hyperactivity disorder in an extended sample: suggestive linkage on 17p11. Am J Hum Genet 72:1268–1279

Ogdie MN, Fisher SE, Yang M, Ishii J, Francks C, Loo SK, Cantor RM, McCracken JT, McGough JJ, Smalley SL, Nelson SF (2004) Attention deficit hyperactivity disorder: fine mapping supports linkage to 5p13, 6q12, 16p13, and 17p11. Am J Hum Genet 75:661–668

Page G, Barc-Pain S, Pontcharraud R, Cante A, Piriou A, Barrier L (2004) The up-regulation of the striatal dopamine transporter's activity by cAMP is PKA-, CaMK II- and phosphatase-dependent. Neurochem Int 45:627–632

Paule MG, Rowland AS, Ferguson SA, Chelonis JJ, Tannock R, Swanson JM, Castellanos FX (2000) Attention deficit/hyperactivity disorder: characteristics, interventions and models. Neurotoxicol Teratol 22:631–651

Pelham WE Jr, Wheeler T, Chronis A (1998) Empirically supported psychosocial treatments for attention deficit hyperactivity disorder. J Clin Child Psychol 27:190–205

Pogorelov VM, Rodriguiz RM, Insco ML, Caron MG, Wetsel WC (2005) Novelty seeking and stereotypic activation of behavior in mice with disruption of the dat1 gene. Neuropsychopharmacology 30:1818–1831

Pristupa ZB, McConkey F, Liu F, Man HY, Lee FJ, Wang YT, Niznik HB (1998) Protein kinase-mediated bidirectional trafficking and functional regulation of the human dopamine transporter. Synapse 30:79–87

Purper-Ouakil D, Wohl M, Mouren MC, Verpillat P, Ades J, Gorwood P (2005) Meta-analysis of family-based association studies between the dopamine transporter gene and attention deficit hyperactivity disorder. Psychiatr Genet 15:53–59

Ralph RJ, Paulus MP, Fumagalli F, Caron MG, Geyer MA (2001) Prepulse inhibition deficits and perseverative motor patterns in dopamine transporter knock-out mice: differential effects of D1 and D2 receptor antagonists. J Neurosci 21:305–313

Ralph-Williams RJ, Paulus MP, Zhuang X, Hen R, Geyer MA (2003) Valproate attenuates hyperactive and perseverative behaviors in mutant mice with a dysregulated dopamine system. Biol Psychiatry 53:352–359

Ritz MC, Lamb RJ, Goldberg SR, Kuhar MJ (1987) Cocaine receptors on dopamine transporters are related to self-administration of cocaine. Science 237:1219–1223

Rodriguiz RM, Chu R, Caron MG, Wetsel WC (2004) Aberrant responses in social interaction of dopamine transporter knockout mice. Behav Brain Res 148:185–198

Roman T, Szobot C, Martins S, Biederman J, Rohde LA, Hutz MH (2002) Dopamine transporter gene and response to methylphenidate in attention-deficit/hyperactivity disorder. Pharmacogenetics 12:497–499

Rowland AS, Umbach DM, Catoe KE, Stallone L, Long S, Rabiner D, Naftel AJ, Panke D, Faulk R, Sandler DP (2001) Studying the epidemiology of attention-deficit hyperactivity disorder: screening method and pilot results. Can J Psychiatry 46:931–940

Sacchetti P, Brownschidle LA, Granneman JG, Bannon MJ (1999) Characterization of the 5′-flanking region of the human dopamine transporter gene. Brain Res Mol Brain Res 74:167–174

Sarkis EH (2000) "Model" behavior. Science 287:2160–2162

Saunders C, Ferrer JV, Shi L, Chen J, Merrill G, Lamb ME, Leeb-Lundberg LM, Carvelli L, Javitch JA, Galli A (2000) Amphetamine-induced loss of human dopamine transporter activity: An internalization-dependent and cocaine-sensitive mechanism. Proc Natl Acad Sci U S A 97:6850–6855

Sawaguchi T (2001) The effects of dopamine and its antagonists on directional delay-period activity of prefrontal neurons in monkeys during an oculomotor delayed-response task. Neurosci Res 41:115–128

Slaats-Willemse D, Swaab-Barneveld H, de Sonneville L, van der Meulen E, Buitelaar J (2003) Deficient response inhibition as a cognitive endophenotype of ADHD. J Am Acad Child Adolesc Psychiatry 42:1242–1248

Slaats-Willemse D, Swaab-Barneveld H, De Sonneville L, Buitelaar J (2005) Familial clustering of executive functioning in affected sibling pair families with ADHD. J Am Acad Child Adolesc Psychiatry 44:385–391

Smalley SL, Kustanovich V, Minassian SL, Stone JL, Ogdie MN, McGough JJ, McCracken JT, MacPhie IL, Francks C, Fisher SE, Cantor RM, Monaco AP, Nelson SF (2002) Genetic linkage of attention-deficit/hyperactivity disorder on chromosome 16p13, in a region implicated in autism. Am J Hum Genet 71:959–963

Smith KM, Daly M, Fischer M, Yiannoutsos CT, Bauer L, Barkley R, Navia BA (2003) Association of dopamine beta hydroxylase gene with attention deficit hyperactivity disorder: genetic analysis of the Milwaukee longitudinal study. Am J Med Genet B Neuropsychiatr Genet 119:77–85

Solanto MV, Abikoff H, Sonuga-Barke E, Schachar R, Logan GD, Wigal T, Hechtman L, Hinshaw S, Turkel E (2001) The ecological validity of delay aversion and response inhibition as measures of impulsivity in AD/HD: a supplement to the NIMH multimodal treatment study of AD/HD. J Abnorm Child Psychol 29:215–228

Sonuga-Barke EJ (2002) Psychological heterogeneity in AD/HD—a dual pathway model of behaviour and cognition. Behav Brain Res 130:29–36

Sorkina T, Doolen S, Galperin E, Zahniser NR, Sorkin A (2003) Oligomerization of dopamine transporters visualized in living cells by fluorescence resonance energy transfer microscopy. J Biol Chem 278:28274–28283

Sorkina T, Hoover BR, Zahniser NR, Sorkin A (2005) Constitutive and protein kinase C-induced internalization of the dopamine transporter is mediated by a clathrin-dependent mechanism. Traffic 6:157–170

Spielewoy C, Roubert C, Hamon M, Nosten-Bertrand M, Betancur C, Giros B (2000) Behavioural disturbances associated with hyperdopaminergia in dopamine-transporter knockout mice. Behav Pharmacol 11:279–290

Stein MA, Sarampote CS, Waldman ID, Robb AS, Conlon C, Pearl PL, Black DO, Seymour KE, Newcorn JH (2003) A dose-response study of OROS methylphenidate in children with attention-deficit/hyperactivity disorder. Pediatrics 112:e404

Stein MA, Waldman ID, Sarampote CS, Seymour KE, Robb AS, Conlon C, Kim SJ, Cook EH (2005) Dopamine transporter genotype and methylphenidate dose response in children with ADHD. Neuropsychopharmacology 30:1374–1382

Stins JF, van Baal GC, Polderman TJ, Verhulst FC, Boomsma DI (2004) Heritability of Stroop and flanker performance in 12-year old children. BMC Neurosci 5:49

Strandburg RJ, Marsh JT, Brown WS, Asarnow RF, Higa J, Harper R, Guthrie D (1996) Continuous-processing—related event-related potentials in children with attention deficit hyperactivity disorder. Biol Psychiatry 40:964–980

Sulzer D, Maidment NT, Rayport S (1993) Amphetamine and other weak bases act to promote reverse transport of dopamine in ventral midbrain neurons. Mol Chem Neuropathol 60:527–535

Swanson JM (2000) Dopamine-transporter density in patients with ADHD. Lancet 355:1461; author reply 1461–1462

Swanson JM, McBurnett K, Wigal T, Pfiffner LJ, Lerner MA, Williams L, Christian D, Tamm L, WillCutt E, Crowley K, Clevenger W, Khouzam N, Woo C, Crinella FM, Fisher TD (1993) The effect of stimulant medication on ADD children: a 'review of reviews'. Except Child 60:154–162

Syringas M, Janin F, Mezghanni S, Giros B, Costentin J, Bonnet JJ (2000) Structural domains of chimeric dopamine-noradrenaline human transporters involved in the Na(+)- and Cl(−)-dependence of dopamine transport. Mol Pharmacol 58:1404–1411

Thapar A, Holmes J, Poulton K, Harrington R (1999) Genetic basis of attention deficit and hyperactivity. Br J Psychiatry 174:105–111

Torres GE, Yao WD, Mohn RR, Quan H, Kim K, Levey AI, Staudinger J, Caron MG (2001) Functional interaction between monoamine plasma membrane transporters and the synaptic PDZ domain-containing protein PICK1. Neuron 30:121–134

Torres GE, Carneiro A, Seamans K, Fiorentini C, Sweeney A, Yao WD, Caron MG (2003) Oligomerization and trafficking of the human dopamine transporter. Mutational analysis identifies critical domains important for the functional expression of the transporter. J Biol Chem 278:2731–2739

Uhl GR, Lin Z (2003) The top 20 dopamine transporter mutants: structure-function relationships and cocaine actions. Eur J Pharmacol 479:71–82 van Dyck CH, Quinlan DM, Cretella LM, Staley JK, Malison RT, Baldwin RM, Seibyl JP, Innis RB (2002) Unaltered dopamine transporter availability in adult attention deficit hyperactivity disorder. Am J Psychiatry 159:309–312

Vandenbergh DJ, Persico AM, Hawkins AL, Griffin CA, Li X, Jabs EW, Uhl GR (1992a) Human dopamine transporter gene (DAT1) maps to chromosome 5p15.3 and displays a VNTR. Genomics 14:1104–1106

Vandenbergh DJ, Persico AM, Uhl GR (1992b) A human dopamine transporter cDNA predicts reduced glycosylation, displays a novel repetitive element and provides racially-dimorphic TaqI RFLPs. Brain Res Mol Brain Res 15:161–166

Vandenbergh DJ, Thompson MD, Cook EH, Bendahhou E, Nguyen T, Krasowski MD, Zarrabian D, Comings D, Sellers EM, Tyndale RF, George SR, O'Dowd BF, Uhl GR (2000) Human dopamine transporter gene: coding region conservation among normal, Tourette's disorder, alcohol dependence and attention-deficit hyperactivity disorder populations. Mol Psychiatry 5:283–292

Vaughan RA, Huff RA, Uhl GR, Kuhar MJ (1997) Protein kinase C-mediated phosphorylation and functional regulation of dopamine transporters in striatal synaptosomes. J Biol Chem 272:15541–15546

Veenstra-Vanderweele J, Christian SL, Cook EH Jr (2004) Autism as a paradigmatic complex genetic disorder. Annu Rev Genomics Hum Genet 5:379–405

Vles JS, Feron FJ, Hendriksen JG, Jolles J, van Kroonenburgh MJ, Weber WE (2003) Methylphenidate down-regulates the dopamine receptor and transporter system in children with attention deficit hyperkinetic disorder (ADHD). Neuropediatrics 34:77–80

Volkow ND, Gatley SJ, Fowler JS, Wang GJ, Swanson J (2000) Serotonin and the therapeutic effects of ritalin. Science 288:11

Volkow ND, Wang G, Fowler JS, Logan J, Gerasimov M, Maynard L, Ding Y, Gatley SJ, Gifford A, Franceschi D (2001) Therapeutic doses of oral methylphenidate significantly increase extracellular dopamine in the human brain. J Neurosci 21:RC121

Waldman ID, Rowe DC, Abramowitz A, Kozel ST, Mohr JH, Sherman SL, Cleveland HH, Sanders ML, Gard JM, Stever C (1998) Association and linkage of the dopamine transporter gene and attention-deficit hyperactivity disorder in children: heterogeneity owing to diagnostic subtype and severity. Am J Hum Genet 63:1767–1776

Walker JK, Gainetdinov RR, Mangel AW, Caron MG, Shetzline MA (2000) Mice lacking the dopamine transporter display altered regulation of distal colonic motility. Am J Physiol Gastrointest Liver Physiol 279:G311–318

Wang W, Sonders MS, Ukairo OT, Scott H, Kloetzel MK, Surratt CK (2003) Dissociation of high-affinity cocaine analog binding and dopamine uptake inhibition at the dopamine transporter. Mol Pharmacol 64:430–439

West SA, Strakowski SM, Sax KW, Minnery KL, McElroy SL, Keck PE Jr (1995) The comorbidity of attention-deficit hyperactivity disorder in adolescent mania: potential diagnostic and treatment implications. Psychopharmacol Bull 31:347–351

Winsberg BG, Comings DE (1999) Association of the dopamine transporter gene (DAT1) with poor methylphenidate response. J Am Acad Child Adolesc Psychiatry 38:1474–1477

Wisor JP, Nishino S, Sora I, Uhl GH, Mignot E, Edgar DM (2001) Dopaminergic role in stimulant-induced wakefulness. J Neurosci 21:1787–1794

Wozniak J, Biederman J, Kiely K, Ablon JS, Faraone SV, Mundy E, Mennin D (1995a) Mania-like symptoms suggestive of childhood-onset bipolar disorder in clinically referred children. J Am Acad Child Adolesc Psychiatry 34:867–876

Wozniak J, Biederman J, Mundy E, Mennin D, Faraone SV (1995b) A pilot family study of childhood-onset mania. J Am Acad Child Adolesc Psychiatry 34:1577–1583

Wu X, Gu HH (2003) Cocaine affinity decreased by mutations of aromatic residue phenylalanine 105 in the transmembrane domain 2 of dopamine transporter. Mol Pharmacol 63:653–658

Zahniser NR, Doolen S (2001) Chronic and acute regulation of Na^+/Cl^--dependent neurotransmitter transporters: drugs, substrates, presynaptic receptors, and signaling systems. Pharmacol Ther 92:21–55

Zhang L, Coffey LL, Reith MEA (1997) Regulation of the functional activity of the human dopamine transporter by protein kinase C. Biochem Pharmacol 53:677–688

Zhuang X, Oosting RS, Jones SR, Gainetdinov RR, Miller GW, Caron MG, Hen R (2001) Hyperactivity and impaired response habituation in hyperdopaminergic mice. Proc Natl Acad Sci U S A 98:1982–1987

Inactivation of 5HT Transport in Mice: Modeling Altered 5HT Homeostasis Implicated in Emotional Dysfunction, Affective Disorders, and Somatic Syndromes

K. P. Lesch (✉) · R. Mössner

Molecular and Clinical Psychobiology, Department of Psychiatry and Psychotherapy, University of Würzburg, Füchsleinstr. 15, 97080 Würzburg, Germany
kplesch@mail.uni-wuerzburg.de

1	Introduction	418
2	Basic Features of 5HT Transporter Gene Inactivation	421
2.1	Neurochemistry	421
2.2	Receptor Expression and Function	422
2.3	Electrophysiology	424
2.4	Heterologous 5HT Clearance	425
2.5	Modeling Anxiety- and Depression-Like Behavior	426
2.6	$5HT_{1A}$ Receptor in Anxiety and Depression	428
3	Brain Development and Plasticity	430
3.1	Somatosensory Cortex and Visual System	430
3.2	Dentate Gyrus and Neurogenesis	431
4	Gene–Gene Interaction	432
4.1	5HT Transporter, Monoamine Oxidase A, and $5HT_{1B}$ Receptor	432
4.2	5HT Transporter and Brain-Derived Neurotrophic Factor	433
5	*5HTT* Inactivation as a Model for Serotonin-Related Somatic Disorders	434
5.1	Primary Pulmonary Hypertension	435
5.2	Irritable Bowel Syndrome	435
5.3	Multiple Sclerosis	436
5.4	Bone Growth	437
5.5	Neuropathic Pain	437
6	Alcohol and Other Substances of Abuse	437
7	Linking 5HTT to Affective Spectrum Disorders	439
7.1	Genetic Epidemiology	439
7.2	Molecular Genetics	440
7.3	Anxiety- and Depression-Related Traits	441
8	Gene–Environment Interaction	442
9	Molecular Imaging of Emotionality: A Risk Assessment Strategy for Depression?	444
References		447

Abstract Animal models have not only become an essential tool for investigating the neurobiological function of genes that are involved in the etiopathogenesis of human behavioral and psychiatric disorders but are also fundamental in the development novel therapeutic strategies. As an example, inactivation of the serotonin (5HT) transporter (*5Htt, Slc6a4*) gene in mice expanded our view of adaptive 5HT uptake regulation and maintenance of 5HT homeostasis in the developing human brain and molecular processes underlying anxiety-related traits, as well as affective spectrum disorders including depression. *5Htt*-deficient mice have been employed as a model complementary to direct studies of genetically complex traits and disorders, with important findings in biochemical, morphological, behavioral, and pharmacological areas. Based on growing evidence for a critical role of the 5HTT in the integration of synaptic connections in the rodent, nonhuman primate, and human brain during critical periods of development and adult life, more in-depth knowledge of the molecular mechanisms implicated in these fine-tuning processes is currently evolving. Moreover, demonstration of a joint influence of the *5HTT* variation and environmental sources during early brain development advanced our understanding of the mechanism of gene×gene and gene×environment interactions in the developmental neurobiology of anxiety and depression. Lastly, imaging techniques, which become increasingly elaborate in displaying the genomic influence on brain system activation in response to environmental cues, have provided the means to bridge the gap between small effects of *5HTT* variation and complex behavior, as well as psychopathological dimensions. The combination of elaborate genetic, epigenetic, imaging, and behavioral analyses will continue to generate new insight into *5HTT*'s role as a master control gene of emotion regulation.

Keywords Serotonin transporter · Gene knockout · Anxiety · Depression · Neurodevelopment

1
Introduction

Even though serotonin (5HT) controls a highly complex system of neurocircuits mediated by multiple pre- and postsynaptic 5HT receptor subtypes, high-affinity 5HT transport into the presynaptic neuron is mediated by no more than a single protein, the 5HT transporter (5HTT). The 5HTT is regarded as the initial site of action of antidepressant drugs and several neurotoxic compounds. Tricyclic antidepressants, such as prototypical imipramine, and the selective 5HT reuptake inhibitors (SSRIs)—paroxetine, citalopram, and sertraline—occupy several pharmacologically distinct sites overlapping at least partially the substrate binding site and are widely used in the treatment of depression, anxiety, and impulse control disorders, as well as substance abuse including alcoholism.

While 5HTT expression appears to be restricted to raphe neurons in adult brain, it has also been detected in the sensory areas of the cortex and thalamus during perinatal development. Cloning of the human 5HTT has identified a protein with 12 transmembrane domains (TMDs), and studies using site-directed mutagenesis and deletion mutants indicate that distinct amino acid residues participate in substrate translocation and competitive antagonist binding (Lesch et al. 1994; Penado et al. 1998). External loops do not appear to be the primary determinants of substrate and inhibitor binding sites but may represent active elements responsible for maintaining the stability and conformational flexibility of the transporter (Ravary et al. 2001). Evidence from studies with deletion mutants as well as computational analyses indicate that 5HTT function may be dependent on the formation of quaternary structures, such as dimers and tetramers (Chang et al. 1998). Several conducting states have been reported for the 5HTT: (1) during gating, a transport-associated current caused by a flux of ions which that are not involved in the transport but pass through the channel; (2) a transient current triggered by extreme negative potentials; and (3) a leakage current (Quick 2003). 5HTT function is acutely modulated by posttranslational modification including phosphorylation through calmodulin-dependent protein kinase A (PKA), protein kinase C (PKC), protein kinase G (PKG), and p38 mitogen-activated protein kinase as well as via nitric oxide and cyclic guanosine monophosphate (cGMP) (Jayanthi et al. 2005; Miller and Hoffman 1994; Prasad et al. 2005). Several intracellular signal transduction pathways converge on the transcriptional apparatus of the 5HTT gene and expression is regulated by cyclic AMP (cAMP)-, PKC-, and tyrosine kinase-dependent mechanisms (Ramamoorthy et al. 1993; Sakai et al. 1997).

The human 5HTT gene (*SERT, SLC6A4*) has been mapped to the chromosome 17q11.2, is composed of 14 exons spanning roughly 40 kb, and the sequence of the transcript predicts a protein of 630 amino acids with 12 TMDs (Lesch et al. 1994). Alternative promoters in combination with differential splicing involving exon 1A, B, and C, in brain versus gut and alternate polyadenylation site usage resulting in multiple mRNA species are likely to participate in the regulation of gene expression in humans (Heils et al. 1995; Ozsarac et al. 2002). In addition to several regulatory domains controlling selective expression in serotonergic neurons, transcriptional activity of the human *5HTT* is modulated by a length variation of a repetitive element, the *5HTT* gene-linked polymorphic region (*5HTTLPR*), located upstream of the transcription start site. Low *5HTTLPR*-dependent 5HTT function is associated with anxiety- and depression-related personality traits (also see Sect. 7.3; Lesch 2003; Lesch et al. 1996). Likewise, a role of regulatory and structural *5HTT* variation has been suggested in a variety of diseases such as depression, bipolar disorder, anxiety disorders, eating disorders, substance abuse, autism, schizophrenia, and neurodegenerative disorders (Lesch and Murphy 2003; Murphy et al. 2004). Additional variations at the *5HTT* locus have been described, although most single nucleotide polymorphisms (SNPs) that change the structure of 5HTT protein are rare. Their potential to alter 5HTT activity and their association with behavioral phenotypes or disorders remain to be determined (Di Bella et al. 1996; Glatt et al. 2001; Sutcliffe et al. 2005).

The orthologous murine *5Htt* gene (*Slc6a4*) located in a syntenic region on chromosome 11qB5 is composed of 14 exons extending approximately 34 kb. A TATA-like motif within 26–31 bp with respect to the transcription initiation site and several potential binding sites for transcription factors as well as cAMP-response element (CRE)- and GC response element (GRE)-like motifs are present in the GC-rich 5′-flanking region (Bengel et al. 1997). The core promoter contains a trinucleotide repeat (CTG/ATG), whereas a *5HTTLPR*-like sequence in humans or nonhuman primates was absent. Functional mapping of the transcriptional control region indicated constitutive promoter activity and several cell-specific enhancer/repressor elements within roughly 2 kb of *5Htt*"s 5′-flanking sequence (Heils et al. 1998).

Converging evidence that 5HTT deficiency plays a role in depression and related disorders, as well as the demand to model the neurobiological implications of allelic variation of 5HTT function, led to the generation of mice with a targeted inactivation of *5Htt*. Over the past decade, neurochemical, morphological, electrophysiological, behavioral, and pharmacological consequences of *5Htt* inactivation have been studied (Fig. 1). In the present review, fundamental features of *5Htt* knockout (KO) mice are described. An appraisal of morphological, physiological, and behavioral alterations in this mouse model

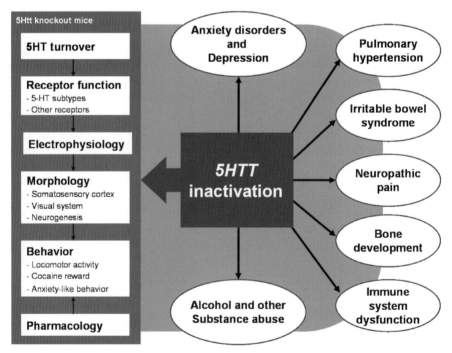

Fig. 1 Modeling traits of emotionality as well as affective spectrum and somatic disorders in mice with a targeted inactivation of the serotonin transporter (*5HTT*) gene

is also provided. Finally, conceptual issues of modeling traits of emotionality and affective spectrum and somatic disorders in *5Htt* KO mice are pointed out.

2
Basic Features of 5HT Transporter Gene Inactivation

2.1
Neurochemistry

For the generation of *5Htt* KO mice, a genomic segment containing exon 2 of the murine *5Htt* was deleted by homologous recombination (Bengel et al. 1998). This segment contains the start codon, a portion of the substrate binding domain, and several posttranslational modification sites. Deletion of exon 2 resulted in a shortened gene transcript and a truncated 5Htt protein that is retained in the endoplasmic reticulum of the somatodendritic compartment (Ravary et al. 2001). Since it is not transported to axons of the raphe neurons and does not reach the cell membrane, the truncated 5Htt protein remains nonfunctional.

Autoradiography of the 5Htt protein confirmed a roughly 50% reduction in uptake site density in $5Htt^{+/-}$ mice, and a complete absence of transporter binding across various regions in null mutants (Bengel et al. 1998). 5Htt-dependent function was further studied by examining 5HT uptake kinetics in brain synaptosomal preparations. The saturation isotherms of [^{3}H]5HT uptake were similar for the wildtype and $5Htt^{+/-}$ mice with no differences in the V_{max} and K_m in brain stem or cortex, while [^{3}H]5HT uptake was absent in $5Htt^{-/-}$ mutants. Lack of 5Htt-mediated 5HT clearance resulted in a persistent 7- to 13-fold increase of 5HT concentrations in the extracellular space as assessed by in vivo microdialysis in different brain regions including prefrontal cortex, striatum, nucleus accumbens, and substantia nigra (Fabre et al. 2000; Shen et al. 2004).

In contrast to synaptic and perineuronal extracellular space, $5Htt^{-/-}$ mutants exhibited a 60%–80% reduction in brain tissue 5HT concentrations and, to a lesser extent, in 5-hydroxyindoleacetic acid concentrations but nearly normal levels of 5HT in $5Htt^{+/-}$ mice in brain stem, hippocampus, hypothalamus, striatum, and frontal cortex (Bengel et al. 1998). Moreover, 5HT synthesis rates were increased 30%–60% as determined by the decarboxylase inhibition method (Kim et al. 2005). Brain regions with the capacity to increase 5HT synthesis and turnover, such as striatum and frontal cortex, demonstrated lesser reductions in 5HT concentrations. While female $5Htt^{+/-}$ mice had more than threefold greater increases in 5HT synthesis than males, no change in either Tph1 or Tph2 messenger RNA (mRNA) levels or in maximal in vitro Tph activity was detected. Brain tissue concentrations as well as synthesis and turnover of dopamine and norepinephrine were unchanged. Taken together, these findings indicate markedly altered homeostasis of the central 5HT but not catecholaminergic systems in *5Htt* KO mice, resulting in brain region- and gender-specific depleted 5HT tissue stores that are inadequately compensated for by increased 5HT synthesis.

2.2
Receptor Expression and Function

Adaptive changes in 5HT neurotransmission were investigated at the level of pre- and postsynaptic 5HT receptors. Pharmacological classification based on ligand binding to receptor subtypes, on signal transduction pathway responses to agonists/antagonists, and eventually on gene identification delineated four 5HT receptor subfamilies, $5HT_{1-4}$. The effects of 5HT on neuronal cells is conferred by at least 14 distinct 5HT receptors (Barnes and Sharp 1999). Due to multiple lines of evidence implicating $5HT_{1A}$ receptors in the pathophysiology and treatment response of affective spectrum disorders, their expression and function have systematically been studied in *5Htt* KO mice. $5HT_{1A}$ receptors operate both as somatodendritic autoreceptors and as postsynaptic receptors. Somatodendritic $5HT_{1A}$ autoreceptors are predominantly located on 5HT neurons and dendrites in the brainstem raphe complex, and their activation by 5HT or $5HT_{1A}$ agonists decreases the firing rate of serotonergic neurons and subsequently reduces the synthesis, turnover, and release of 5HT from nerve terminals in projection areas. Postsynaptic $5HT_{1A}$ receptors are widely distributed in forebrain regions that receive serotonergic input, notably in the cortex, hippocampus, septum, amygdala, and hypothalamus.

Specific labeling with radioligands, antibodies, and competitive RT-PCR, showed that $5Ht_{1a}$ receptor mRNA and protein are gene dose-dependently decreased in the dorsal raphe nucleus (DRN), slightly increased in the hippocampus, and unchanged in other forebrain areas of $5Htt^{-/-}$ mice compared to wildtype control mice (Fabre et al. 2000). In order to assess functionality of $5HT_{1A}$ receptors, hypothermic and neuroendocrine responses to the $5HT_{1A}$ agonist 8-hydroxy-2-(di-*n*-propylamino)tetraline (8-OH-DPAT) were examined. Like most $5HT_{1A}$ receptor agonists, 8-OH-DPAT dose-dependently produces hypothermia and increases the concentrations of plasma oxytocin, corticotropin, and corticosterone that is blocked by the $5HT_{1A}$ receptor antagonists such as WAY 100635 (Li et al. 1999). In $5Htt^{-/-}$ mice, the hypothermic response to 8-OH-DPAT was completely abolished, while plasma oxytocin and corticosterone responses were significantly attenuated. The functional desensitization of $5HT_{1A}$ receptor in $5Htt^{-/-}$ mice was reflected by a decrease in [^{3}H]8-OH-DPAT- and [^{125}I]MPPI [4-(29-methoxyphenyl)-1-[29-[*N*-(20-pyridinyl)-iodobenzamido]*ethyl*]piperazine]-binding sites in the hypothalamus and [^{125}I]MPPI-binding sites in the DRN. No adaptational changes in the hypothermic or hormonal responses to 8-OH-DPAT were observed in heterozygous $5Htt^{+/-}$ mice. Regarding gender-specific effects, the density of

5HT$_{1A}$ receptors in the DRN, but not in the hypothalamus or hippocampus, and 8-OH-DPAT-induced functional responses were reduced in both male and female *5Htt$^{-/-}$* mice, although the alterations were more extensive in females (Bouali et al. 2003; Li et al. 2000).

Similar regional differences were also detected for 5Ht$_{1b}$ receptors, with a decrease in the substantia nigra but not in the globus pallidus of mutant mice (Fabre et al. 2000). [^{35}S]GTPγS binding evoked by potent 5HT$_1$ receptor agonists confirmed a functional decrease in the DRN and the substantia nigra but not other brain areas in KO mice. Somatodendritic 5Ht$_{1a}$ autoreceptor-mediated decrease in brain 5HT turnover rate following administration of the 5HT$_{1A}$ partial agonist ipsapirone—as well as terminal 5Ht$_{1b}$ autoreceptor-elicited increase of 5HT outflow in the substantia nigra upon local application of the 5HT$_{1B/1D}$ antagonist GR 127935—is eliminated in *5Htt$^{-/-}$* mice.

Autoradiographic labeling by the selective antagonist MDL 100,907 and saturation experiments with cortical membranes revealed analogous region-specific adaptive changes in the density of 5Ht$_{2a}$ receptors in the lateral striatum, claustrum, and cortex of *5Htt$^{-/-}$* mutants (Rioux et al. 1999). In contrast, the density of 5Ht$_{2a}$ receptors was increased in the hypothalamus and septum, whereas 5Ht$_{2a}$ receptor mRNA was unchanged in all brain regions investigated (Li et al. 2003). Densities of 5Ht$_{2c}$ receptors were greater in the amygdala and choroid plexus of *5Htt$^{-/-}$* mice but reduced in the striatum, while 5Ht$_{2c}$ receptor mRNA was reduced in the choroid plexus and lateral habenula nucleus. Unlike other subtypes, the 5HT$_3$ receptor is an ionotropic receptor with ligand-gated cation channel function. Similar to G protein-coupled 5HT receptors, quantitative autoradiography also showed extensive adaptive changes of 5Ht$_3$ receptors. In *5Htt$^{-/-}$* mice, 5Ht$_3$ receptors were upregulated in frontal cortex, parietal cortex, and in stratum oriens of the hippocampal CA3 region (Mössner et al. 2004). Changes in 5Ht$_3$ receptor mRNA, as determined by in situ hybridization, were less pronounced.

Since 5HT release is regulated by adenosine, *5Htt* KO mice were also studied for adaptive changes of adenosine A$_1$ and A$_{2a}$ receptors by quantitative autoradiography with [^{3}H]8-cyclopentyl-1,3-dipropylxanthine and [^{3}H]CGS 21680, respectively (Mössner et al. 2000). Comparison of *5Htt$^{-/-}$* versus wildtype control mice revealed increased densities of A$_1$ receptors in the DRN—but not in any of the serotonergic projection areas—and downregulation of A$_{2a}$ receptors in basal ganglia. 5Htt×Maoa double KO mice showed adenosine receptor alterations similar to *5Htt$^{-/-}$* mice The adaptive changes of A$_1$ and A$_{2a}$ receptors in *5Htt$^{-/-}$* mice are likely to represent a compensatory neuroprotective effect mediated by the adenosinergic modulatory system.

Taken together, these brain region- and gender-specific adaptive changes in 5HT receptor expression and function constitute a part of the complex regulatory pattern of the central 5HT system likely to be related to the altered behaviors in *5Htt* KO mice (also see Sect. 2.5).

2.3
Electrophysiology

Compared with wildtype littermates, the mean spontaneous firing rate of DRN 5HT neurons was decreased gene dose-dependently in $5Htt^{+/-}$ and $5Htt^{-/-}$ mice (Gobbi et al. 2001). As a correlate of enhanced synaptic availability of 5HT at inhibitory $5HT_{1A}$ receptors, the selective $5HT_{1A}$ receptor antagonist WAY 100635 enhanced 5HT neuronal firing in $5Htt^{-/-}$ mice, whereas somatodendritic $5HT_{1A}$ autoreceptors are desensitized in 5Htt KO mice. At the postsynaptic level, the recovery time of the firing rate of hippocampus CA3 pyramidal neurons following iontophoretic applications of 5HT was significantly prolonged only in $5Htt^{-/-}$ mice. The SSRI paroxetine significantly prolonged the recovery time in wildtype and $5Htt^{+/-}$ mice, without altering the maximal inhibitory effect of 5HT. In $5Htt^{-/-}$ mice these neurons showed an attenuated response to 8-OH-DPAT, but not to 5HT itself.

Likewise, Mannoury la Cour and associates (2001, 2004) reported that application of the SSRIs paroxetine and citalopram onto brainstem slices resulted in a concentration-dependent inhibition of 5HT neuron firing in the DRN of wildtype mice, but not $5Htt^{-/-}$ mutants. Although the $5HT_{1A}$ receptor agonists ipsapirone and 5-carboxamidotryptamine inhibited the discharge, the potency of these agonists was markedly decreased in $5Htt^{-/-}$ mice by 55-fold and 6-fold, respectively. Similarly, intracellular recordings showed that the potency of 5-carboxamidotryptamine to hyperpolarize 5HT neurons in the DRN was significantly lower in $5Htt^{-/-}$ than in wildtype mice. These data contrasted with those obtained with hippocampal slices in which 5-carboxamidotryptamine was equipotent to hyperpolarize CA1 pyramidal neurons in both mutant and wildtype mice. As expected from their mediation through $5HT_{1A}$ receptors, the effects of ipsapirone and 5-carboxamidotryptamine were competitively inhibited by the selective $5HT_{1A}$ antagonist WAY 100635.

The effects of $5HT_{1A}$ receptor selective agents, such as the agonist 8-OH-DPAT, and the partial agonists ipsapirone, on anxiety-like behavior have previously been studied in rodents (De Vry 1995). Both agonists and partial agonists induce a dose-dependent anxiolytic effect that correlates with the inhibition of serotonergic neuron firing and decrease of 5HT release, as well as the reduction of 5HT signaling at postsynaptic target receptors. Blockade of the negative feedback by selective $5HT_{1A}$ receptor antagonists, such as WAY 100635, increases firing of the serotonergic neurons but exerts no effect on 5HT neurotransmission or behavior, while the combination with SSRIs augments increases in 5HT concentrations in terminal regions (Olivier and Miczek 1999).

These findings indicate that inactivation of the 5HTT induces a marked desensitization of $5HT_{1A}$ autoreceptors in the DRN without altering postsynaptic $5HT_{1A}$ receptor functioning in the hippocampus. Differential desensitization at pre- and postsynaptic levels is characteristic for $5Htt^{-/-}$ mice, whereas in $5Htt^{+/-}$ mice $5HT_{1A}$ receptor adaptation takes place only at the presynaptic level. Similarities between these changes and those evoked by chronic treatment with SSRIs emphasize the existence of regional differences in $5HT_{1A}$ receptor regulatory mechanisms and validate *5Htt* KO mice as a model to further investigate the molecular mechanisms underlying the differential regulation of $5HT_{1A}$ autoreceptors versus postsynaptic $5HT_{1A}$ receptors. The contrasting changes in $5HT_{1A}$ receptors in the DRN versus the hippocampus are also likely correlate with the anxiety-like behavioral alterations in *5Htt* KO mice (also see Sect. 2.5).

2.4
Heterologous 5HT Clearance

Excess extracellular 5HT concentrations potentially induce neurotoxic processes in *5Htt* KO mice that are likely be counteracted by passive diffusion, or more efficiently, by active uptake via heterologous transport mechanisms. Several studies using microdialysis in dopamine transporter ($Dat^{-/-}$)×$5Htt^{-/-}$ double KO mice and pharmacological inhibition of 5HT clearance in midbrain–hindbrain primary neuronal cultures confirmed that in *5Htt* KO mice 5HT is taken up into dopaminergic neurons of both striatum and substantia nigra, mediated via the DAT (Pan et al. 2001; Ravary et al. 2001; Shen et al. 2004; Zhou et al. 2002).

Although lacking the high affinity and selectivity of the 5HTT, the brain expresses a considerable number of other transporters, including the polyspecific organic cation transporters (OCTs). OCT1 and OCT3, two members of the potential-sensitive organic cation transporter gene family, physiologically transport a wide spectrum of organic cations. In addition, they mediate low-affinity 5HT transport and participate in the clearance of excessive 5HT. Investigation of *Oct1* and *Oct3* expression in the brain of *5Htt* KO mice by semi-quantitative competitive PCR and in situ hybridization revealed increased *Oct3* mRNA concentrations in the hippocampus, but not in other brain regions including cortex, striatum, cerebellum, and brainstem (Schmitt et al. 2003). No change in *Oct1* expression was detected between *5Htt* KO and control mice.

Adaptational changes within the glutathione detoxification system also reflect 5HT excess and its potential deleterious effects. While levels of reduced and oxidized glutathione were unchanged, glutathione metabolizing enzymes showed a differential pattern of modulation. A trend toward reduced glutathione peroxidase was detected in frontal cortex, brainstem, and cerebellum of *5Htt* KO mice, while the π-isotype of the detoxifying enzyme glutathione-S-transferase was decreased in a number of brain regions, especially in brainstem. At the level of the DNA, an increase of oxidative DNA adducts in the hippocampus of *5Htt* KO mice was found.

Cross-clearance of 5HT into dopaminergic neurons and alternative uptake of monoamines by different neurons appears to represent a compensatory route adopted under circumstances when their corresponding transporter function is no longer adequate. Given the importance of the hippocampus in learning and memory, upregulation of Oct3 expression and resulting heterologous low-affinity 5HT uptake may limit the adverse effects of elevated extracellular 5HT and may play a critical role in maintaining 5HT-dependent functions of the hippocampus in the absence of 5HTT.

2.5
Modeling Anxiety- and Depression-Like Behavior

Targeted gene inactivation approaches are remarkably expanding our understanding of the neurobiological basis of anxiety- and depression-related behavior in mice (Lesch 2001). However, the neural substrates that regulate emotional processes in humans or cause anxiety disorders and depression remain remarkably elusive. A neural circuit composed of several regions of the ventromedial prefrontal cortex, anterior cingulate, amygdala, hippocampus, ventral striatum, hypothalamus, and several other interconnected structures and involving multiple neurotransmitter systems have been implicated in emotion regulation (for review see Lesch et al. 2003). In humans, nonhuman primates, and other mammals, substantial evidence has accumulated that the serotonergic signaling pathway integrates elementary brain functions of cognition, sensory processing, and motor activity.

Serotonergic raphe neurons diffusely project to brain regions implicated in emotionality. During development, 5HT shapes various brain systems, above all the emotion-regulating limbic system circuitry. The diversity of the outcome is due to the capacity of 5HT to orchestrate the activity and interaction of several other neurotransmitter systems, particularly activity-, cognition-, and reward-related dopaminergic pathways. Given these links, the 5HT system constitutes the major modulator of emotional behavior and considerable evidence links serotonergic dysfunction to anxiety, depression, and comorbid conditions.

Among the reasons for the slow progress are conceptional deficiencies regarding the psychobiology of anxiety and depression, which make it difficult to develop and validate reliable models. The clinical presentation of affective spectrum disorders and the lack of consensus on clinical phenotypes or categories further complicates the development of mouse models for specific syndromes. In addition, affective disorders encompass not only the behavioral trait of inappropriate anxiety and depression but also the cognitive response towards this disposition. This response, however, is substantially modulated by environmental factors including cultural determinants such as rearing and education as well as sociocultural and -economic context. Investigations on the neurobiological basis of affective disorders therefore rely on the accurate dissection of behavioral dysfunction from other factors. The dilemma that no single paradigm of murine behavior mimics diagnostic entities or treatment response of anxiety disorders and depression reflects the fact that current classification systems are not based on the neurobiology of the respective disorder, rather than the failure to develop valid mouse models.

Various approaches have been employed to detect and quantify "anxiety-like" behaviors in mice, and most of them postulate that aversive stimuli, such as novelty or potentially harmful environments, induce a central state of fear and defensive reactions that can be assessed and quantified through physiologic and behavioral paradigms (Crawley 1999; Crawley and Paylor 1997). When rodents are introduced into a novel environment, they tend to move around the perimeter of the environment (open field). They stop occasionally and rear up, sniffing the walls and the floor. They initially spend very little time in the open center of the area. If they have a choice, they will spend more time in a dark than in a brightly lit area (light-dark box). Rodents will also spend more time in a small, elevated area enclosed by walls than in an elevated area without walls (elevated plus maze). When they move from one delimited area into another, they often engage in a type of stretching-out behavior. Anxiety-like behavior often appears to contrast with exploratory behavior, indicating that avoidance and curiosity/novelty seeking are related and share common mechanisms.

While substantial similarities between human and murine avoidance, defense, aggression, and escape responses exist, it remains obscure whether mice also experience associated cognitive processes similar to humans. In general, pathological anxiety may reflect an inappropriate activation of a normally adaptive, evolutionarily conserved defense reaction. It should therefore be practicable to elucidate both physiologic and pathologic anxiety by studying avoidant and defensive behavior in mice using a broad range of anxiety models to ensure comprehensive characterization of this behavioral phenotype.

Behavior of *5Htt* KO mice was tested in a variety of conditions evaluating fear, avoidance, conflict, stress responsiveness, and effects of various pharmacological agents. In particular, anxiety-like behaviors were characterized using a battery of tests including open field, elevated plus maze, and light-dark box. In these tests, both male and female *5Htt* KO mice show consistently increased anxiety-like behavior and inhibited exploratory locomotion (Holmes et al. 2003). The selective $5HT_{1A}$ receptor antagonist WAY 100635 produced an anxiolytic effect in the elevated plus maze in *5Htt* KO mice, suggesting that the abnormalities in anxiety-like and exploratory behavior is mediated by the $5HT_{1A}$ receptor. $5Htt^{-/-}$ mice, in which transporter binding sites are reduced by approximately 50%, were similar to controls on most measures of anxiety-like behavior. However, changes in exploratory behavior in $5Htt^{+/-}$ mice were limited to specific measures under baseline conditions, but extended to additional measures under more stressful test conditions. This observation is in accordance with reduced aggressive behavior in $5Htt^{+/-}$ mice that is limited to specific measures and test conditions. While male $5Htt^{-/-}$ mice are slower to attack an intruder and attacked with less frequency than control littermates, heterozygous $5Htt^{+/-}$ mice were as quick to attack, but made fewer overall attacks, as compared to wildtype controls. Aggression increased with repeated exposure to an intruder in $5Htt^{+/-}$ and control mice, but not in $5Htt^{-/-}$ mice.

As outlined above, these behavioral alterations in $5Htt^{+/-}$ mice are paralleled by subtle but consistent perturbations in 5HT homeostasis that are intermediate between $5Htt^{-/-}$ and wildtype mice in a gene dose-dependent manner, including elevated extracellular 5HT, decreased 5HT neuron firing in the DRN, and reduced $5HT_{1A}$ receptor expression and function. The evidence that serotonergic dysfunction in $5Htt^{+/-}$ mice may manifest and become noticeable as behavioral abnormalities only under challenging environmental conditions such as an inadequate mothering style strongly supports the disposition-stress model of anxiety disorders and depression (also see Sect. 8; Murphy et al. 2003; V. Carola, K.P. Lesch, and C. Gross, manuscript in preparation).

Excess extracellular 5HT that is expected to cause enhanced activation of pre- and postsynaptic 5HT receptors is accountable for the neural mechanisms underlying increased anxiety-related behavior in *5Htt* KO mice. High 5HT drives the negative autoinhibitory feedback and reduces cellular 5HT availability by stimulating $5HT_{1A}$ receptors, which results in their desensitization and downregulation in the midbrain raphe complex and, to a lesser extent, in hypothalamus, septum, and amygdala but not in the frontal cortex and hippocampus (also see Sects. 2.2 and 2.3; Li et al. 2000). Although postsynaptic $5HT_{1A}$ receptors appear to be unchanged in frontal cortex and hippocampus, indirect evidence for decreased presynaptic serotonergic activity but reduced 5HT clearance resulting in elevated synaptic 5HT is provided by compensatory alterations in 5HT synthesis and turnover as well as downregulation of terminal 5HT release-inhibiting $5HT_{1B}$ receptors (Fabre et al. 2000; Kim et al. 2005). Therefore, a partial downregulation of postsynaptic $5HT_{1A}$ receptors in some forebrain regions but a several-fold increase in extracellular concentrations of 5HT in *5Htt* KO mice could still cause excess net activation of postsynaptic $5HT_{1A}$ receptors, resulting in increased anxiety-like behavior and its reversal by WAY 100635 (Holmes et al. 2003). However, administration of WAY 100635 antagonizes not only postsynaptic $5HT_{1A}$ receptors in forebrain regions but also acts at somatodendritic autoreceptors in the raphe nuclei. Electrophysiological studies show that WAY 100635 causes a reversal of markedly reduced spontaneous firing rates of 5HT neurons in the DRN of $5Htt^{-/-}$ mice, indicating that the net effect of WAY 100635 on serotonergic neurotransmission in *5Htt* KO mice may be more complex than anticipated (Gobbi et al. 2001).

2.6
$5HT_{1A}$ Receptor in Anxiety and Depression

The $5HT_{1A}$ receptor has long been implicated in the pathophysiology of anxiety and depression; its role as a molecular target of anxiolytic and antidepressant drugs is well established (Griebel 1995; Griebel et al. 2000; Olivier et al. 1999). Patients with panic disorder and depression display an attenuation of $5HT_{1A}$ receptor-mediated hypothermic and neuroendocrine responses, reflecting a reduced responsivity of both pre- and postsynaptic $5HT_{1A}$ receptors (Lesch et al. 1990b, 1992). Likewise, a decrease in $5HT_{1A}$ ligand binding has been shown in postmortem brains of depressed suicide victims (Cheetham et al. 1990) as well as in forebrain areas such as the medial temporal lobe and in the raphe of depressed patients elicited by positron emission tomography (PET) (Drevets et al. 1999; Sargent et al. 2000). Both glucocorticoid administration and chronic stress, a pathogenetic factor in affective disorders, have also been demonstrated to result in downregulation of $5HT_{1A}$ receptors in the hippocampus in animals (Flugge 1995; Lopez et al. 1998; Wissink et al. 2000). While deficits in hippocampal $5HT_{1A}$ receptor function may contribute to the cognitive abnormalities associated with affective disorders, recent work suggests that activation of this receptor stimulates neurogenesis in the dentate gyrus of the hippocampus. By using both a mouse model with a targeted ablation of the $5HT_{1A}$ receptor and radiological methods, Santarelli and coworkers (2003) have provided persuasive evidence that $5HT_{1A}$-activated hippocampal neurogenesis is essentially required for the behavioral effects of long-term antidepressant treatment with SSRIs.

Intriguingly, downregulation and hyporesponsivity of $5HT_{1A}$ receptors in patients with major depression are not reversed by antidepressant drug treatment (Lesch et al. 1990a, 1991; Sargent et al. 2000), raising the possibility that low receptor function is a trait feature and therefore a pathogenetic mechanism of the disease. In line with this notion, evidence is accumulating that a polymorphism in the transcriptional control region of the $5HT_{1A}$ receptor gene (*HTR1A*)—resulting in allelic variation of $5HT_{1A}$ receptor expression—is associated with personality traits of negative emotionality. These include anxiety and depression (neuroticism and harm avoidance; Strobel et al. 2003), as well as major depressive disorder, suicidality, and panic disorder (Lemonde et al. 2003; Rothe et al. 2004). The converging lines of evidence that receptor deficiency or dysfunction is involved in mood and anxiety disorders encouraged investigators to genetically manipulate the $5HT_{1A}$ receptor in mice (Lesch and Mössner 1999; Ramboz et al. 1998). Mice with a targeted inactivation of the *Htr1a* show a complete lack of ligand binding to brain $5HT_{1A}$ receptors in $Htr1a^{-/-}$ mutants, with intermediate binding in the heterozygous $Htr1a^{+/-}$ mice.

Taken together, these findings add to an emerging picture of abnormalities in *5Htt* KO mice across a range of neurochemical, physiological, and behavioral parameters associated with affective spectrum disorders, including changes in $5HT_{1A}$ receptor function (Fabre et al. 2000; Li et al. 1999), altered sensitivity to alcohol and drugs of abuse such as cocaine (Kelai et al. 2003; Sora et al. 1998, 2001), dysregulated gastrointestinal motility (Chen et al. 2001), and disturbed REM sleep (Wisor et al. 2003). Finally, given the absence of the 5HTT throughout ontogeny, *5Htt* KO mice also provide a tool for studying the potential for neurodevelopment abnormalities affecting anxiety-like behavior.

3
Brain Development and Plasticity

3.1
Somatosensory Cortex and Visual System

Morphological analyses of brain structures where 5HT has been suggested to act as a differentiation signal in development revealed a destructive effect of *5Htt* inactivation on the formation and plasticity of cortical and subcortical structures. Investigations of 5HT participation in neocortical development and plasticity have been concentrated on the rodent somatosensory cortex (SSC), due to its one-to-one correspondence between each whisker and its cortical barrel-like projection area (Di Pino et al. 2004). The processes underlying patterning of projections in the SSC have been intensively studied with a widely held view that the formation of somatotopic maps does not depend on neural activity. The timing of serotonergic innervation coincides with pronounced growth of the cortex, the period when incoming axons begin to establish synaptic interactions with target neurons and to elaborate a profuse branching pattern. Additional evidence for a role of 5HT in the development of neonatal rodent SSC derives from the transient barrel-like distribution of 5HT, $5HT_{1B}$, and $5HT_{2A}$ receptors, and of the 5HTT (Mansour-Robaey et al. 1998). The transient barrel-like 5HT pattern visualized in layer IV of the SSC of neonatal rodents stems from 5HT uptake and vesicular storage in thalamocortical neurons, transiently expressing at this developmental stage both 5Htt and the vesicular monoamine transporter (VMAT)2 despite their ultimate glutamatergic phenotype (Lebrand et al. 1996).

5Htt inactivation profoundly disturbs formation of the SSC with altered cytoarchitecture of cortical layer IV, the layer that contains synapses between thalamocortical terminals and their postsynaptic target neurons (Persico et al. 2001). Brains of *5Htt* KO mice display a lack of characteristic barrel-like clustering of layer IV neurons in S1, despite relatively preserved trigeminal and thalamic patterns. Cell bodies as well as terminals, typically more dense in barrel septa, appear homogeneously distributed in layer IV of adult *Htt* KO brains. Injections of the 5HT synthesis inhibitor parachlorphenylalanine (pCPA) within a narrow time window of 2 days (postnatal) completely rescued formation of SSC barrel fields. Notably, heterozygous $5Htt^{+/-}$ KO mice develop all SSC barrel fields, but frequently present irregularly shaped barrels and fewer defined cell gradients between septa and barrel hollows.

These findings demonstrate that excessive concentrations of extracellular 5HT are deleterious to SSC development and suggest that transient 5HTT expression in thalamocortical neurons is responsible for barrel patterns in neonatal rodents, and its permissive action is required for normal barrel pattern formation, presumably by maintaining extracellular 5HT concentrations below a critical threshold. Because normal synaptic density in SSC layer IV of *5Htt* KO mice was shown, it is more likely that 5HT affects SSC cytoarchitecture by promoting dendritic growth toward the barrel hollows, as well as by modulating cytokinetic movements of cortical granule cells, similar to concentration-dependent 5HT modulation of cell migration described in other tissues.

Visual system abnormalities include irregular segregation of contralateral and ipsilateral retinogeniculate projections (Upton et al. 2002). 5HT overload seems also responsible for these alterations, since normal development of retinogeniculate projections is restored by 5HT synthesis inhibition. The effect of elevated extracellular 5HT concentration on the modulation of programmed cell death during neural development was also investigated in early postnatal brains of *5Htt* KO mice.

Lastly, *5Htt* inactivation also leads to a reduced number of apoptotic cells in striatum, thalamus, hypothalamus, cerebral cortex, and hippocampus on postnatal day 1 (P1) with differences displaying an increasing fronto-caudal gradient and regional specificity (Persico et al. 2003). These findings underscore the role of 5HT in the regulation of programmed cell death during brain development, and suggest that pharmacological enhancement of serotoninergic neurotransmission may minimize pathological apoptosis.

3.2
Dentate Gyrus and Neurogenesis

The subventricular zone (SVZ) of the lateral ventricles and the dentate gyrus of the hippocampus continue to generate new neurons throughout life in many mammalian species, and newly formed neurons may have the potential to integrate both structurally and functionally into pre-existing neuronal networks, although only 5%–10% of all newborn cells may eventually contribute to this set of connections (van Praag et al. 2002). While little is known about the regulation of adult neurogenesis, pharmacological compounds, stress, age, and repetitive physical exercise, as well as a variety of messenger molecules, transcription factors, and growth factors have been implicated (Duman 2002). An evolving concept assigns a central role to adult neurogenesis in both the pathogenesis of affective disorders and in the mechanisms of action of antidepressant drugs (Kempermann and Kronenberg 2003). Since the dentate gyrus is one of the few neurogenic foci within the brain and a target of a dense network of serotonergic fibers, 5HT has been implicated in the modulation of hippocampal neurogenesis (Goergen et al. 2002). Reflecting the considerable time required for processes of neuronal plasticity, 5HT-induced neurogenesis and resulting neuroplastic changes may be essential, or at least contribute critically, to the delayed onset of action of and effectiveness of antidepressive therapy (Santarelli et al. 2003). As inactivation of the *5Htt* reduces 5HT clearance resulting in persistently increased concentrations of synaptic 5HT, *5Htt* KO mice represent an effective model to investigate the effects of lifelong enhanced 5HT function on adult stem cell proliferation, survival, and differentiation in the hippocampus.

Using 5-bromo-2-deoxyuridine (BrdU) incorporation followed by quantitative immunohistochemistry and double-labeling techniques, aged *5Htt* KO mice (~14.5 months) showed an increase in proliferative capacity of adult neural stem cells (A. Schmitt, J. Benninghoff, C. Doenitz, S. Gross, M. Hermann, M. Rizzi, A. Gritti, S. Fritzen, A. Reif, D. Murphy, A. Vescovi, and K.P. Lesch, submitted). In contrast, in vivo analyses of young adult *5Htt* KO mice (~3.0 months) and cultures of neurospheres from their hippocampus did not reveal significant changes in proliferation of neural stem cells or survival of newborn cells. These analyses demonstrated that the cellular fate of newly generated cells in *5Htt* KO mice is not different with respect to the total number and percentage of neurons or glial cells from wildtype controls. The findings indicate that elevation of synaptic 5HT concentration throughout early development and later life of 5Htt-deficient mice does not induce adult neurogenesis but may influence stem cell proliferation in senescent mice.

Since the gene dose-dependent reduction in 5HTT availability in heterozygous *5Htt*$^{+/-}$ mice that leads to a modest delay in 5HT uptake but distinct irregularities in SSC barrel and septum shape as well as altered apoptotic and neurogenetic processes, is similar to those reported in humans carrying the low-activity *5HTTLPR* variant, it may be speculated that allelic variation in 5HTT function also affects the human brain during development with due consequences for disease liability and therapeutic response (Ansorge et al. 2004; Lesch and Gutknecht 2005). The evidence that changes in brain 5HT homeostasis exert long-term effects on cortical development and adult plasticity may be an important step forward in establishing the psychobiological groundwork for a neurodevelopmental hypothesis of negative emotionality (Lesch 2003). Although there is converging evidence that serotonergic dysfunction contributes to anxiety- and depression-related behavior, the precise mechanism that renders *5Htt* KO mice more anxious and stress responsive as well as less aggressive remains to be elucidated.

4
Gene–Gene Interaction

4.1
5HT Transporter, Monoamine Oxidase A, and 5HT$_{1B}$ Receptor

Two key players of serotonergic neurotransmission appear to mediate the deleterious effects of excess 5HT: the 5HTT and the 5HT$_{1B}$ receptor. Both molecules are expressed in primary sensory thalamic nuclei during the period when the segregation of thalamocortical projections occurs (Bennett-Clarke et al. 1996; Hansson et al. 1998). 5HT is internalized via 5HTT in thalamic neurons and is detectable in axon terminals (Cases et al. 1998; Lebrand et al.

1996). The presence of VMAT2 within the same neurons allows internalized 5HT to be stored in vesicles and used as a cotransmitter of glutamate. Lack of 5HT degradation in monoamine oxidase A (MAOA) KO mice as well as severe impairment of 5HT clearance in mice with an inactivation of the *5Htt* results in an accumulation of 5HT and overstimulation of 5HT receptors all along thalamic neurons (Cases et al. 1998).

Since $5HT_{1B}$ receptors are known to inhibit the release of glutamate in the thalamocortical somatosensory pathway, excessive activation of $5HT_{1B}$ receptors is likely to prevent activity-dependent processes involved in the patterning of afferents and barrel structures. This hypothesis is supported by a recent study using a strategy of combined KO of *Maoa*, *5Htt*, and *Htr1b*. While only partial disruption of the patterning of somatosensory thalamocortical projections was observed in *5Htt* KO, *Maoa*×*5Htt* double KO mice showed that 5HT accumulation in the extracellular space causes total disruption of the patterning of these projections (Salichon et al. 2001). Moreover, the targeted removal of $5HT_{1B}$ receptors in both *Maoa* and *5Htt* KO, yielding *Maoa*×*Htr₁ᵦ* and *5Htt*×*Htr₁ᵦ* double KOs, as well as in *Maoa*×*5Htt* double KO mice, allows a normal segregation of the somatosensory projections and retinal axons in the lateral geniculate nucleus (Salichon et al. 2001; Upton et al. 2002). These findings point to an essential role of the $5HT_{1B}$ receptor in mediating the deleterious effects of excess 5HT in the SSC.

4.2
5HT Transporter and Brain-Derived Neurotrophic Factor

Brain-derived neurotrophic factor (BDNF) is involved in a variety of trophic and modulatory effects that include a critical role in the development and plasticity of dopaminergic, serotonergic, and other neurons (Bonhoeffer 1996; Schuman 1999). Specifically, BDNF—in conjunction with $5HT_{1A}$ receptor activation—appears to regulate differentiation of 5HT neurons during embryonic development and prevents neurotoxin-induced serotonergic denervation in adult brain (Frechilla et al. 2000; Galter and Unsicker 2000a; 2000b). Furthermore, human fetal mesencephalic cultured cells treated with BDNF exhibit greater neuronal survival and increased tissue 5HT concentrations (Spenger et al. 1995). BDNF treatment of embryonic day (E)14 rat embryos induced a twofold increase in the number of raphe 5HT neurons and produced a marked extension and ramification of their neurites with greater expression of 5HTT, $5HT_{1A}$, and $5HT_{1B}$ receptors (Galter and Unsicker 2000a; Zhou and Iacovitti 2000). Reduced expression of BDNF modifies synaptic plasticity, resulting in specific alterations in spatial learning and memory processes, emotionality, and motor activity in *Bdnf* KO mice (Carter et al. 2002; Kernie et al. 2000; Minichiello et al. 1999), whereas targeted inactivation of the BDNF receptor, TrkB, leads to neuronal loss and cortical degenerative changes (Vitalis et al. 2002). In addition, BDNF mediates the effects of repeated stress exposure and long-term antidepressant treatment on neurogenesis and neuronal survival in the hippocampus (D'Sa and Duman 2002). These findings converge with reduced hippocampal plasticity reflected by a reduced hippocampal volume, and hippocampus-related memory deficiency plays a critical role in the pathophysiology of emotional and stress-related disorders (Duman 2002).

Bdnf KO mice have reduced sensory neuron survival, other neuronal deficits, and are viable for only a few weeks (Ernfors et al. 1994). Heterozygous $Bdnf^{+/-}$ mice exhibit gene dose-dependent reductions in BDNF expression in forebrain, hippocampus, and some hypothalamic nuclei (Kernie et al. 2000; MacQueen et al. 2001), as well as decreased striatal dopamine content, decreased potassium-elicited dopamine release, and some evidence for decreased forebrain 5HT concentrations and fiber densities at 18 months of age (Lyons et al. 1999). Furthermore, learning deficits and hyperactivity was revealed in $Bdnf^{+/-}$ mice. They also develop intermale aggressiveness in the resident-intruder test, but do not show increased anxiety, nor differences in an antidepressant-sensitive test. However, conditional inactivation of *Bdnf* in the postnatal brain leads to increased anxiety-like behavior, deficits in context-dependent learning in a fear conditioning paradigm, and hyperactivity (Rios et al. 2001).

For investigation of potential gene-interactive effects of altered BDNF expression on a brain 5HT system with diminished 5HT transport capability, a double-mutant mouse model was developed by interbreeding *5Htt* KO mice with heterozygous *Bdnf*-deficient mice ($Bdnf^{+/-}$), producing $5Htt^{-/-} \times Bdnf^{+/-}$ (sb) mice. Interestingly, male but not female sb mice showed further decreases in brain 5HT and 5-hydroxyindole acetic acid (5HIAA) concentrations, as well as further increases in anxiety-like behavior and stress reactivity compared with wildtype SB controls, *Bdnf*-deficient Sb mice, and *5Htt* KO sB mice (Ren-Patterson et al. 2005). Analysis of neuronal morphology showed that hypothalamic and hippocampal neurons exhibited 25%–30% reductions in dendrites in sb mice compared with SB controls. These findings further support the notion of a critical role of gene×gene interaction in brain plasticity, as well as underscoring the usefulness of this approach in evaluating epistatic consequences of *BDNF* and *5HTT* variations in anxiety and related disorders.

Of related interest, the *5Htt* KO mouse model also confirmed a role of 5HTT×BDNF interaction in the development and plasticity of serotonergic raphe neurons (Rumajogee et al. 2004).

5
5HTT Inactivation as a Model for Serotonin-Related Somatic Disorders

Beyond its role in the brain, the effect of *5Htt* inactivation has also been investigated in peripheral tissues of the mouse. The general principle underlying the implication of the 5HTT as a vulnerability factor for multiple peripheral disorders is that variants of *5HTT* may act both in the brain and in specific organs like heart, lung, gut, and adrenal plus other endocrine glands during development and adulthood (Yavarone et al. 1993). Since somatic disorders, such as primary pulmonary hypertension, autoimmune diseases like multiple sclerosis, disorders of bone development, irritable bowel syndrome, neuropathic pain, and myocardial infarction have been linked to dysregulation of 5HTT function, *5Htt* KO mice also represent a practical model system to elucidate the pathophysiology of these somatic syndromes.

5.1
Primary Pulmonary Hypertension

The progressive and frequently fatal primary pulmonary hypertension (PPH) is characterized by blood pressure increases associated with abnormal vascular proliferation in the pulmonary artery bed. 5HT is a potent inducer of smooth muscle cell proliferation in lung vessels, an effect that is attributable to 5HTT-mediated clearance of 5HT. Iatrogenic pulmonary hypertension has been identified as the result of treatment of obesity by the 5HT-releasing compound fenfluramine. Patients with spontaneous PPH more frequently have the high *5HTTLPR*-dependent 5HTT function in lung tissue and platelets (Eddahibi et al. 2001). While hypoxia induces smooth muscle hyperplasia and increases expression of 5HTT in lung vessels, a causative role for hypoxia in pulmonary hypertension has been inferred from evidence that *5Htt* KO mice are protected against hypoxia-induced pulmonary hypertension (Eddahibi et al. 2000). Unlike PPH patients, patients with chronic obstructive pulmonary disease (COPD) do not differ from controls in *5HTTLPR* genotype frequencies. However, COPD patients homozygous for the high-activity variant of *5HTTLPR* and who also display higher 5HTT expression in pulmonary artery cells, had significantly more severe pulmonary hypertension than patients with the low-activity variant, a finding which is in agreement with the rodent models (Eddahibi et al. 2003).

5.2
Irritable Bowel Syndrome

The gut is the only organ that displays reflexes and integrative neuronal activity even when isolated from the brain. Gut tissue contains by far the greatest amount of 5HT in the body (Gershon 2003). Epithelial enterochromaffin cells act as sensory transducers that activate the mucosal processes of both intrinsic and extrinsic primary afferent neurons through their release of 5HT. Acting at $5HT_4$ receptors, 5HT enhances the release of transmitters from their terminals and from other terminals in prokinetic reflex pathways. Signaling to the CNS, serotonergic transmission within the enteric nervous system, and the activation of myenteric intrinsic primary afferent neurons are mediated by the 5HT$_3$ receptor. Serotonergic signaling in the mucosa and the enteric nervous system is terminated by the 5HTT. Successive potentiation of 5HT and/or desensitization of its receptor could account for the symptoms seen in diarrhea-predominant and constipation-predominant irritable bowel syndrome (IBS), respectively. In one case-control study of IBS, homozygosity for the low-activity variant of *5HTTLPR* was significantly more frequent in a diarrhea-predominant subgroup, although no overall genotype difference from controls was found (Pata et al. 2002). IBS treatment responses (reductions in colonic transit time) to a 5HT$_3$ receptor antagonist were significantly more prevalent in diarrhea-predominant patients with the high-activity genotype (Camilleri et al. 2002).

Symptoms associated with the downregulation of the 5HTT in the human mucosa in IBS are similar to the symptoms observed in 5Htt KO mice. They have increased gut motility together with frequent diarrhea alternating with occasional constipation in an IBS-like fashion (Chen et al. 2001). Adaptive changes occur in the subunit composition of enteric 5HT$_3$ receptors of 5Htt KO mice that seem protective against the pathogenetic effects of excessive 5HT (Liu et al. 2002). Decreased expression of the 5HT$_{3B}$ subunit and reduced affinity of 5HT$_3$ receptors for 5HT coincide with a greater tendency of 5HT$_3$ receptors to desensitize.

5.3
Multiple Sclerosis

Between the nervous and the immune system, 5HT has been suggested to serve as a mediator of bidirectional interactions (Mössner et al. 2001; Mössner and Lesch 1998). In 5Htt KO mice, this interaction was investigated using experimental autoimmune encephalomyelitis (EAE), a well-defined animal model of autoimmune disease of the CNS mimicking features of the human disease multiple sclerosis. EAE was induced by immunization with the autoantigens myelin basic protein (MBP) or the immunodominant peptide of myelin oligodendrocyte glycoprotein (MOG). After immunization with either MBP or MOG, the disease course of the 5Htt KO mice was attenuated as compared to wildtype control mice, with a more pronounced difference in female animals (Hofstetter et al. 2005). Histological examination of the CNS and cytokine measurements in mononuclear cells from the spleens of *5Htt* KO mice revealed a reduction of the inflammatory infiltrate in the CNS and of the neuroantigen-specific production of interferon (IFN)-γ in splenocytes, again accompanied by a gender difference. These findings suggest a potential role of 5HT homeostasis in the fine-tuning of neuroantigen-specific immune responses.

5.4
Bone Growth

Evidence on the role of 5HT in bone development and preliminary clinical evidence demonstrating detrimental effects of SSRIs on bone growth have raised questions regarding the effects of these drugs on the growing skeleton (Bliziotes et al. 2001). In a recent study, the impact of 5HTT inhibition on the skeleton was investigated in growing mice treated with a SSRI and in *5Htt* KO mice. In both models, 5HTT inhibition had detrimental effects on bone mineral accrual (Warden et al. 2005). *5Htt* KO mice had a consistent skeletal phenotype of reduced mass, altered architecture, and inferior mechanical properties, whereas bone mineral accrual was impaired in growing mice treated with an SSRI. These phenotypes resulted from a reduction in bone formation without an increase in bone resorption and were not influenced by effects on skeletal mechanosensitivity or serum biochemistries. These findings implicate the 5HTT in the regulation of bone accrual in the growing skeleton.

5.5
Neuropathic Pain

Antidepressants are widely used in the treatment of neuropathic pain and are thought to exert their effect, at least partially, by 5HTT inhibition, thus resulting in activation of central antinociceptive pathways. Since *5Htt* inactivation is regarded as a model of lifelong treatment with antidepressants, the consequences of pain-related behavior was assessed in *5Htt* KO mice and wildtype littermates after a unilateral chronic constrictive sciatic nerve injury (CCI) (Vogel et al. 2003). Wildtype mice reproducibly developed ipsilateral thermal hyperalgesia and mechanical allodynia after CCI, whereas *5Htt* KO mice did not develop thermal hyperalgesia, but showed bilateral mechanical allodynia after the nerve injury. In *5Htt* KO mice, reduced 5HT levels in the injured peripheral nerves correlated with diminished behavioral signs of thermal hyperalgesia, a pain-related symptom caused by peripheral sensitization. In contrast, bilateral mechanical allodynia, a centrally mediated phenomenon, was associated with decreased spinal 5HT concentrations and may possibly be caused by a lack of spinal inhibition.

6
Alcohol and Other Substances of Abuse

Alcohol dependence is an etiologically and clinically heterogeneous syndrome caused by a complex interaction of genetic and environmental factors. Therefore, the differentiation of psychobiological traits of addictive behavior is of particular importance for the dissection of the complex genetic susceptibility of alcoholism. A central 5HT deficit reflected by lower concentrations of 5HT metabolites in cerebrospinal fluid (CSF) and low platelet 5HT content is thought to be involved in the pathogenesis of alcohol preference and dependence by modulating motivational behavior and adaptive processes. 5HT-related impulsive, aggressive, and suicidal behavior has been linked to a primordial personality that is susceptible to alcoholism, suggesting a link between alcohol-seeking behavior and low central 5HT. Although variations in several genes that encode receptors, enzymes, and transporters of the 5HT system have been tested as risk factors in patients with alcoholism, population and family-based association analyses between genetically influenced 5HT system function and alcohol dependence have mainly been focused on the *5HTT* due to its central role in the fine-tuning of serotonergic neurotransmission (Lesch 2005a). The influence of 5HT reuptake on alcohol consumption in *5Htt* KO mice was tested using a free-choice paradigm. Alcohol intake was lower in *5Htt*$^{-/-}$ mutant than in wildtype mice, and pharmacological blockade of 5HTT by the SSRI fluoxetine reduced alcohol use in wildtype mice only (Kelai et al. 2003). Although the efficacy of SSRIs in the treatment of alcohol dependence is controversial, these findings confirm the inhibitory effect of *5Htt* inactivation on alcohol consumption.

Behavioral consequences of psychostimulants and drugs of abuse were investigated by amphetamine-induced locomotor activity and cocaine-elicited place preference. While (+)-amphetamine-induced hyperactivity did not differ across *5Htt* genotypes, the locomotor enhancing effects of (+)-3,4-methylenedioxymethamphetamine (MDMA, "ecstasy"), a substituted amphetamine that releases 5HT via a transporter-dependent mechanism, was attenuated in *5Htt*$^{+/-}$ and completely absent in *5Htt*$^{-/-}$ mutants (Bengel et al. 1998). Lacking the molecular target of MDMA, *5Htt* KO mice also failed to show evidence for adaptive chances in synaptic plasticity elicited by MDMA as reflected by down- or upregulation of synaptotagmin IV and I, respectively (Peng et al. 2002). These data suggest that the presence of a functional 5HTT is essential for brain 5HT homeostasis and for MDMA-induced hyperactivity and effects on neuroplasticity.

Preferential dopamine uptake inhibitory and releasing action at the dopamine transporter (DAT) by cocaine and its analogs—in addition to their interaction with norepinephrine transporter (NET) and 5HTT—induces multiple neurochemical and behavioral effects that are thought to represent the molecular basis of euphoria/reward, sensitization, and addictive behavior. Quite unexpectedly, investigation of genetically modified mice indicated that cocaine reward/reinforcement, elicited by cocaine place preferences, is similar in *Dat* KO and wildtype mice (Sora et al. 2001; Sora et al. 1998). *5Htt* KO mice even displayed enhanced cocaine reward. However, deletions of both the *Dat* and *5Htt* in double KO mice completely eliminated cocaine reward, thus defining the minimal set of gene products necessary for the euphoric and addictive effects of cocaine. In contrast, *5Htt*×*Net* double KOs not only failed to reduce cocaine reward but strongly enhanced place preference, further supporting the notion of an involvement of multiple transporters in the rewarding and aversive actions of psychostimulants and drugs of abuse (Hall et al. 2002). Taken together, monoamine transporter KO mice represent a powerful tool for the investigation of drugs of abuse, including MDMA and cocaine, which in addition to having psychotropic and addictive effects, may also produce 5HT- and dopamine-induced neurotoxicity.

7
Linking 5HTT to Affective Spectrum Disorders

Among affective spectrum disorders, depression is the most prevalent and clinically relevant condition. Depression comprises an etiologically heterogeneous group of brain disorders that are characterized by a wide range of symptoms that reflect alterations in cognitive, psychomotor, and emotional processes. Affected individuals differ remarkably in the profile of clinical features, severity, and course of illness as well as response to drug treatment and reintegration efforts.

7.1
Genetic Epidemiology

Epidemiology has assembled convincing evidence that affective spectrum disorders including depression are substantially influenced by genetic factors and that the genetic component is highly complex, polygenic, and epistatic (for review see Lesch 2005b). Unipolar major depression maintains a population prevalence of 2%–19% and an age-adjusted risk for first-degree relatives of 5%–25%. In a meta-analysis of five large and rigorously selected family studies of major depression, familiality in this disease was demonstrated by a relative risk of 2.8 for affected subject versus first-degree relative status. Early age of onset and multiple episodes of depression seem to increase the familial aggregation, and different affective spectrum disorders are often present in the same family. Relatives of patients with bipolar disorder also have an increased risk of unipolar depression, and affective disorders tend to co-exist with anxiety in many families.

Twin and family-based studies have accrued considerable evidence that a complex genetic mechanism is involved in the vulnerability to depressive disorders (for review see Malhi et al. 2000). In general, twin studies of depressive adults suggest that genes and specific environmental factors are critical, and that shared environmental factors, although important in less severe subtypes of depression, are possibly of less significance. The heritability of unipolar depression appears to be remarkable, with estimates between 40% and 70%. Depression-associated genetic factors are largely shared with generalized anxiety disorder, while environmental determinants seem to be distinct. This notion is consistent with recent models of emotional disorders that view depression and anxiety as sharing common vulnerabilities but differing on dimensions including, for instance, focus of attention or psychosocial liability. Although life events may precipitate depression, examination of familial liability along with social adversity reveals that environmental effects tend to be contaminated by genetic influences. The predisposition to suffer life events is likely influenced by shared family environment and some events may be associated with genetic factors.

While genetic research has routinely focused either on depression-related traits or on depressive disorders, with few investigations evaluating the genetic and epigenetic relationship between the two, it is crucial to answer the question whether a certain quantitative trait etiopathogenetically either influences the disorder or represents a syndromal dimension of the disorder (or both). This more-apparent-than-real dichotomy also underscores the assumption that a genetic predisposition, coupled with early life stress, in critical stages of development may result in a phenotype that is neurobiologically vulnerable or resilient to stress and may influence an individual's threshold for developing depression on later life stress exposure. Because the mode of inheritance of depression is complex, it has been concluded that multiple genes of modest effect, in interaction with each other and in conjunction with environmental events, produce vulnerability to the disorder. Investigation of epigenetic mechanisms in humans, nonhuman primates, and rodents are currently stepping up the process of identification of genes that are critically involved in formation and plasticity of neurocircuitries basic to the pathogenesis of depression (also see Sect. 8).

7.2
Molecular Genetics

Linkage analysis in extended pedigrees is a practical approach for identifying disease genes for monogenic diseases displaying a Mendelian mode of inheritance. While past linkage studies in unipolar depression suffered from methodological deficiency, two state-of-the-art, genome-wide linkage scans have recently been published. Zubenko and associates (2003) reported several chromosomal loci that may influence the development of recurrent major depressive disorder in 81 families. The highest logarithm of differences (lod) score observed occurred on chromosome 2 (205 cM), located 121 kb proximal to the cyclic AMP response element binding protein 1 (CREB1), a ubiquitous transcription factor that is likely to participate in the regulation of *5HTT* expression. Abkevich and coworkers (2003) conducted a genome-wide scan in 1,890 individuals from 110 pedigrees with a strong family history of major depression and provide strong evidence for the existence of a sex-specific disposition locus for major depression on chromosome 12q22–12q23.2. Interestingly, a previous linkage analysis for quantitative trait loci (QTLs) influencing variation in the personality trait neuroticism also identified a gender-specific locus on chromosome 12q23.1 (Fullerton et al. 2003). The findings of these three linkage analyses, however, require rigorous replication.

Because the power of linkage analysis to detect small gene effects is quite limited, at least with realistic cohort sizes, molecular genetic research in depression has primarily relied on association analysis using DNA variants in or near candidate genes with etiological or pathophysiological relevance. Gene variants with a significant impact on the functionality of components of brain neurotransmission, such as the 5HT system, are a rational beginning. Based on converging lines of evidence that the 5HT and serotonergic gene expression are involved in a myriad of processes during brain development as well as synaptic plasticity in adulthood, depression-related temperamental predispositions and behavior is likely to be influenced by genetically driven variability of 5HT function. Consequently, the contribution of the *5HTT* variants to the risk of affective spectrum disorders including depression and bipolar disorder was explored in numerous independent population- and family-based studies (for review see Lesch 2003; Lesch et al. 2002).

The work on *5HTT* as a risk gene for emotional dysregulation and affective spectrum disorders was initiated by the discovery of a length variation of a repetitive sequence in the 5′-flanking transcriptional control region of the 5HTT gene (*5HTTLPR*) resulting in allelic variation of 5HTT expression and function and the demonstration of an influence of *5HTTLPR* genotype on personality traits of negative emotionality including anxiety, depression, and aggressiveness (neuroticism and agreeableness; Lesch et al. 1996). The short and long *5HTTLPR* variants differentially modulate: (1) transcriptional activity of the *5HTT* promoter, (2) 5HTT mRNA/protein concentration and 5HT uptake activity in lymphoblastoid cells, (3) *5HTT* mRNA concentrations in the raphe complex of human postmortem brain, (4) platelet 5HT uptake and content, (5) 5HT system responsivity elicited by pharmacological challenge tests, (6) mood changes following tryptophan depletion, and (7) in vivo SPECT/PET imaging of 5HTT in human brain with the short variant associated with lower 5HTT expression and function (for review see Lesch 2005a). Associated with allelic variation of 5HTT expression and function are adaptive changes of human brain $5HT_{1A}$ receptor revealed by pharmacological challenge tests and PET imaging (David et al. 2005; Lesch and Canli 2005).

7.3
Anxiety- and Depression-Related Traits

A growing body of evidence implicates personality traits, such as neuroticism or the anxiety-related cluster, in the comorbidity of mood disorders (Kendler 1996). The dimensional structure of neuroticism comprising fearfulness, depression, negative emotionality, and stress reactivity has been delineated by systematic research. As indexed by the personality scale of neuroticism, general vulnerability is likely to overlap genetically with both anxiety and depression. Separation of depression from depression-related personality disorders in current consensual diagnostic systems therefore enhanced interest in the link between temperament, personality, and mood disorders, as well as the impact of this interrelationship on the heterogeneity within diagnostic entities, prediction of long-term course, and treatment response. This concept may predict that when a QTL, such as the *5HTTLPR*, is found for neuroticism, the same QTL should be associated with symptoms of anxiety and depression (Lesch 2003, 2004). Anxiety and affective disorders are therefore likely to represent the extreme end of variation in negative emotionality. The genetic factor contributing to the extreme ends of dimensions of variation commonly recognized as a disorder may be quantitatively, not qualitatively, different from the rest of the distribution. This vista has important implications for identifying genes for complex traits related to distinct disorders.

However, the effect sizes for the *5HTTLPR*-personality associations indicate that this polymorphism has only a moderate influence on behavioral predispositions and disease risk that corresponds to less than 5% of the total variance, based on estimates from twin studies using these and related measures that have consistently demonstrated that genetic factors contribute 40%–60% of the variance in neuroticism and other related personality traits (Lesch et al. 1996). The associations therefore represent only a modest share of the genetic contribution to depression- and anxiety-related traits. Although additive contributions of comparable size or epistatic interaction have, in fact, been found in studies of other quantitative traits, a reasonably conservative conclusion at the current stage of research is expected to comprise the view that the influence of a single, common polymorphism on continuously distributed traits is likely to be modest, if not minimal.

8
Gene–Environment Interaction

Since the genetic basis of present-day temperamental and behavioral traits is already laid out in many mammalian species including mice and may reflect selective forces among our remote ancestors, research efforts have recently been focused on nonhuman primates, especially rhesus macaques. In this primate model, environmental influences are probably less complex, can be more easily controlled for, and are thus less likely to confound associations between behavior and genes. All forms of emotionality in rhesus monkeys—major categories are anxiety, aggression, and depression—appear to be modulated by environmental factors, and marked disruptions to the mother-infant relationship likely confer increased risk. In rhesus monkeys, maternal separation and replacement of the mother by an inanimate surrogate mother during the first months of life results in long-term consequences for the functioning of the central 5HT system and defects in peer interaction and social adaptation, and it is associated with increases in anxiety- and depression-related behaviors like rocking and grooming (Higley et al. 1991). This suggests that early environmental trauma can directly induce long-term plastic changes in the brain that alter anxiety-related responses in adulthood.

One of the most replicated findings in psychobiology is the observation of lower 5HIAA, the major metabolite of 5HT, in the brain and CSF in impulsive aggression and suicidal behavior. In rhesus monkeys, brain 5HT turnover, as measured by cisternal CSF 5HIAA concentrations, shows a strong heritable component and is trait-like, with demonstrated stability over an individual's lifespan (Higley et al. 1991). Early experiences have long-term consequences for the function of the central 5HT system, as indicated by robustly altered CSF 5HIAA levels, as well as anxiety, depression, and aggression-related behavior in rhesus monkeys deprived of their mother at birth and raised only with peers. This model of maternal separation was therefore used to study gene×environment interaction by testing for associations between central 5HT turnover and an orthologous polymorphism in the transcriptional control region of the rhesus *rh5HTT* gene *(rhHTTLPR)* (Lesch et al. 1997). The findings suggest that the *rh5HTTLPR* genotype is predictive of CSF 5HIAA concentrations, but that early experiences make unique contributions to variation in later 5HT system function and thus provide evidence of an environment-dependent association between *5HTT* and a direct measure of brain 5HT function (Bennett et al. 2002). The interactive effect of the *rh5HTTLPR* genotype and early rearing environment on infant traits, social play, aggressive behavior, and stress reactivity, as well alcohol preference and dependence, was also explored (Barr et al. 2003, 2004a, b, c; Champoux et al. 2002). The consequences of the deleterious early experience of maternal separation all seem consistent with the notion that the *5HTTLPR* may influence the risk for affective disorders, possibly in a sexually dimorphic pattern.

All in all, these findings provide evidence of an environment-dependent association between allelic variation of 5HTT expression and central 5HT function, and illustrate the possibility that specific genetic factors play a role in 5HT-mediated behavior in primates. Because rhesus monkeys exhibit temperamental and behavioral traits that parallel anxiety, depression, and aggression-related personality dimensions associated in humans with the low-actitvity *5HTTLPR* variant, it may be possible to search for evolutionary continuity in this genetic mechanism for individual differences. Nonhuman primate studies may also be useful to help identify environmental factors that either compound the vulnerability conferred by a particular genetic makeup or, conversely, act to improve the behavioral outcome associated with a distinct genetic makeup.

Consequently, it is increasingly accepted that much of the impact of genes on emotionality, including anxiety and depression, depends on interactions between genes and the environment. Such interactions would lead to the expression of environmental effects only in the presence of a permissive genetic background. Not unexpected, a recent study by Caspi and coworkers (2003) confirmed that individuals with one or two low-activity versions of the *5HT-TLPR* are up to twofold more likely to get depressed after stressful events such as bereavement, romantic disasters, illnesses, or losing their job. Moreover, childhood maltreatment significantly increased probability to develop depressive syndromes in later life in individuals with the low-activity variant of the *5HTTLPR*. These results further support the conception of how a combination of genetic disposition and specific life events may interact to facilitate the development of mental illness. What went largely unnoticed, though, was its implications for the relevance of studying the genetics of personality. Depression is strongly associated with anxiety- and depression-related traits (e.g., neuroticisms, harm avoidance), the factual personality dimensions that have been linked to allelic variation of 5HTT function. Given the high comorbidity between anxiety and depression and the evidence for their modulation by common genetic factors (Kendler 1996), it is likely that predisposition to mood disorders will also be determined by environmental influences whose impact on the brain is under genetic control. The substantiation that interaction between early trauma inflicted by childhood maltreatment and allelic variation of 5HTT function increases the vulnerability to develop mood disorders is particularly interesting. A remarkable body of evidence suggests that emotionality and stress reactivity can be influenced by experiences early in life, and it has long been supposed that severe early life trauma may increase the risk for anxiety and affective disorders. The notion that during early developmental stages individuals are particularly susceptible to adverse environmental influences receives further confirmation through investigation in *5Htt* KO mice that are designed to demonstrated epigenetic effects of the quality of maternal care in interaction with altered gene expression on life-long emotional behavior and brain functioning (V. Carola, K.P. Lesch, and C. Gross, manuscript in preparation).

9
Molecular Imaging of Emotionality: A Risk Assessment Strategy for Depression?

Imaging techniques become increasingly elaborate in displaying the genomic influence on brain system activation in response to environmental cues, thus representing a tool to bridge the gap between multiple alleles with small effects and complex behavior as well as psychopathological dimensions (Fig. 2). Evidence for a modulatory effect of the *5HTTLPR* on prefrontal cortex activity suggests that genotype–phenotype correlations may be accessible to structural and functional imaging of the brain. In two subsequent studies, Fallgatter and coworkers (1999, 2004) were the first to report an association between *5HT-*

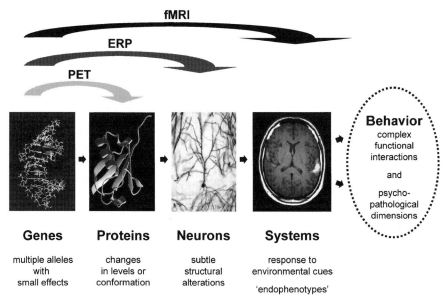

Fig. 2 Molecular functional imaging of genotype–endophenotype correlations as a tool to bridge the gap between alleles with small effects and complex behavior, as well as psychopathological dimensions. *ERP*, event-related potentials; *fMRI*, functional magnetic resonance imaging; *PET*, positron emission tomography

TLPR genotype and prefrontal cortex-limbic excitability detected by event-related potentials (ERP) with two different tasks of cognitive response control (the Go-NoGo and error-processing tasks). Individuals with one or two short alleles of the *5HTTPLR* showed higher prefrontal brain activity compared to subjects homozygous for the long variant, thus indicating that the low-activity short variant is associated with enhanced responsiveness of the prefrontal cortex, particularly the anterior cingulate cortex (ACC). These findings strongly suggest a relationship between cognitive brain function and allelic variation of 5HTT function.

Using functional magnetic resonance imaging (fMRI), Hariri and associates (2005) likewise observed that individuals with one or two copies of the low-activity short variant of the *5HTTLPR* exhibit greater neuronal activity of the human amygdala, a brain system central to emotionality and social behavior, as assessed by fMRI, in response to fearful stimuli compared with individuals homozygous for the long allele. The *5HTTLPR*-related effects on the response bias of amygdala reactivity to environmental threat were subsequently confirmed in a large cohort of both healthy men and women. They indicated that the allelic variation of 5HTT function might represent a classic susceptibility factor for affective disorders by biasing the functional reactivity of the amygdala in the context of stressful life experiences and/or deficient cortical regulatory input.

Subsequent investigations were focused on the modifying impact of the limbic cortex in the context of *5HTTLPR*'s role in depression risk. Although patients with depression display a decreased volume in the subgenual division of the ACC together with altered activity of the limbic circuit components involving the anterior cingulate and amygdala, an enduring disagreement remained whether these abnormalities predispose for the development of depression and comorbid conditions or are a consequence of the depressed state. To address this question, Pezawas et al. (2005) used a complementary fMRI approach to confirm that the low-expressing short allele of the *5HTTLPR* is associated with depression-typical structural and functional alterations. Carriers of the "s" variant showed reduced gray matter volume of both the perigenual anterior cingulate and the amygdala. In addition, the findings revealed a *5HTT* genotype-dependent correlation of amygdala activity with the activity of the rostral and caudal segment of the anterior cingulate, indicating a genetically regulated dynamic coupling that renders the amygdala more responsive to emotional stimuli. Heinz and colleagues (2005) also observed increased amygdala activation in carriers of the *5HTTLPR* short variant that was paralleled by enhanced functional coupling with the ventromedial prefrontal cortex. Moreover, it was shown that positively valenced emotional scenes also evoked amygdala activation, consistent with the general task for the amygdala in both positive and negative emotion regulation; but only the response to negative stimuli was associated with the *5HTT* genotype.

Despite its obvious relevance beyond traits of emotionality and affective disorders, the assumption that the neutral baseline as the control condition does not itself produce changes in activation as a function of the *5HTT* genotype was not investigated in these studies. Recently, Canli and coworkers (2005) showed that allelic variation in 5HTT function is associated with differential activation to negative, positive, and neutral stimuli in limbic, striatal, and cortical regions using the Stroop, the attentional interference task. This task is sensitive to individual differences in personality and mood, and activates either the cognitive division or affective division of the ACC and amygdala, depending on whether the stimuli are neutral or affectively valenced. While increased amygdala activation to negative, relative to neutral, stimuli, in *5HTTLPR* short allele carriers was confirmed, the differences were determined by decreased activation to neutral stimuli, rather than increased activation to negative stimuli. High-resolution structural images and automated processes also revealed *5HTT* genotype-related volumetric and gray matter density differences in frontal cortical regions, anterior cingulate, and cerebellum. These findings are consistent with the notion that the ACC plays an integral role in both affective and (non-affective) cognitive processes and is heavily interconnected with a number of cortical and subcortical regions and therefore implicate 5HT transport efficiency in a wide-ranging spectrum of brain processes that affect neural systems controlling affective, cognitive, and motor processes.

The database of functional 5HTT imaging was further extended and its relevance is supported by several flanking investigations. Bertolino and coworkers (2005) studied two groups of volunteers categorized by contrasting cognitive/personality styles characteristic of variable salience to fearful stimuli: phobic-prone versus eating disorder-prone subjects. The fMRI results showed that phobic-prone individuals selectively recruited the amygdala to a greater extent than eating disorder-prone subjects. Interestingly, amygdala activation was independently predicted by cognitive style and *5HTT* genotype, suggesting that responsivity of the amygdala may represent an emergent property that is based on the association between genetic and psychological factors. It is concluded that certain aspects of the cognitive/personality style are rooted in physiological responses of the fear circuitry that interact with processing of environmental stimuli. Furmark and colleagues (2004) scanned social phobic patients using PET of cerebral blood flow to report greater amygdala activation during a public, compared to a private, speaking task as a function of *5HTT* genotype.

The combination of elaborate genetic, imaging, and behavioral analyses will continue to generate new and exciting insight into the role of the 5HTT in modulating neural function and behavior. Altogether, these numerous lines of evidence impressively support the notion that allelic variation of 5HTT function—and consequently serotonergic signaling—contributes to the response of brain regions underlying human emotionality and its resulting behavior. They also indicate that differential excitability of limbic circuits to emotional stimuli might contribute to exaggerated anxiety-related responses, as well as increased risk for affective spectrum disorders.

Acknowledgements The author is supported by the European Commission (NEWMOOD LSHM-CT-2003-503474), Bundesministerium für Bildung und Forschung (IZKF 01 KS 9603), and the Deutsche Forschungsgemeinschaft (SFB 581, KFO 125/1–1).

References

Abkevich V, Camp NJ, Hensel CH, Neff CD, Russell DL, Hughes DC, Plenk AM, Lowry MR, Richards RL, Carter C, Frech GC, Stone S, Rowe K, Chau CA, Cortado K, Hunt A, Luce K, O'Neil G, Poarch J, Potter J, Poulsen GH, Saxton H, Bernat-Sestak M, Thompson V, Gutin A, Skolnick MH, Shattuck D, Cannon-Albright L (2003) Predisposition locus for major depression at chromosome 12q22–12q23.2. Am J Hum Genet 73:1271–1281

Ansorge MS, Zhou M, Lira A, Hen R, Gingrich JA (2004) Early-life blockade of the 5-HT transporter alters emotional behavior in adult mice. Science 306:879–881

Barnes NM, Sharp T (1999) A review of central 5-HT receptors and their function. Neuropharmacology 38:1083–1152

Barr CS, Newman TK, Becker ML, Parker CC, Champoux M, Lesch KP, Goldman D, Suomi SJ, Higley JD (2003) The utility of the non-human primate; model for studying gene by environment interactions in behavioral research. Genes Brain Behav 2:336–340

Barr CS, Newman TK, Lindell S, Shannon C, Champoux M, Lesch KP, Suomi SJ, Goldman D, Higley JD (2004a) Interaction between serotonin transporter gene variation and rearing condition in alcohol preference and consumption in female primates. Arch Gen Psychiatry 61:1146–1152

Barr CS, Newman TK, Schwandt M, Shannon C, Dvoskin RL, Lindell SG, Taubman J, Thompson B, Champoux M, Lesch KP, Goldman D, Suomi SJ, Higley JD (2004b) Sexual dichotomy of an interaction between early adversity and the serotonin transporter gene promoter variant in rhesus macaques. Proc Natl Acad Sci U S A 101:12358–12363

Barr CS, Newman TK, Shannon C, Parker C, Dvoskin RL, Becker ML, Schwandt M, Champoux M, Lesch KP, Goldman D, Suomi SJ, Higley JD (2004c) Rearing condition and rh5-HTTLPR interact to influence limbic-hypothalamic-pituitary-adrenal axis response to stress in infant macaques. Biol Psychiatry 55:733–738

Bengel D, Heils A, Petri S, Seemann M, Glatz K, Andrews A, Murphy DL, Lesch KP (1997) Gene structure and 5′-flanking regulatory region of the murine serotonin transporter. Brain Res Mol Brain Res 44:286–292

Bengel D, Murphy DL, Andrews AM, Wichems CH, Feltner D, Heils A, Mössner R, Westphal H, Lesch KP (1998) Altered brain serotonin homeostasis and locomotor insensitivity to 3,4-methylenedioxymethamphetamine ("Ecstasy") in serotonin transporter-deficient mice. Mol Pharmacol 53:649–655

Bennett AJ, Lesch KP, Heils A, Long JC, Lorenz JG, Shoaf SE, Champoux M, Suomi SJ, Linnoila MV, Higley JD (2002) Early experience and serotonin transporter gene variation interact to influence primate CNS function. Mol Psychiatry 7:118–122

Bennett-Clarke CA, Chiaia NL, Rhoades RW (1996) Thalamocortical afferents in rat transiently express high-affinity serotonin uptake sites. Brain Res 733:301–306

Bertolino A, Arciero G, Rubino V, Latorre V, De Candia M, Mazzola V, Blasi G, Caforio G, Hariri A, Kolachana B, Nardini M, Weinberger DR, Scarabino T (2005) Variation of human amygdala response during threatening stimuli as a function of 5′HTTLPR genotype and personality style. Biol Psychiatry 57:1517–1525

Bliziotes MM, Eshleman AJ, Zhang XW, Wiren KM (2001) Neurotransmitter action in osteoblasts: expression of a functional system for serotonin receptor activation and reuptake. Bone 29:477–486

Bonhoeffer T (1996) Neurotrophins and activity-dependent development of the neocortex. Curr Opin Neurobiol 6:119–126

Bouali S, Evrard A, Chastanet M, Lesch KP, Hamon M, Adrien J (2003) Sex hormone-dependent desensitization of 5-HT1A autoreceptors in knockout mice deficient in the 5-HT transporter. Eur J Neurosci 18:2203–2212

Camilleri M, Atanasova E, Carlson PJ, Ahmad U, Kim HJ, Viramontes BE, McKinzie S, Urrutia R (2002) Serotonin-transporter polymorphism pharmacogenetics in diarrhea-predominant irritable bowel syndrome. Gastroenterology 123:425–432

Canli T, Omura K, Haas BW, Fallgatter A, Constable RT, Lesch KP (2005) Beyond affect: a role for genetic variation of the serotonin transporter in neural activation during a cognitive attention task. Proc Natl Acad Sci U S A 102:12224–12229

Carter AR, Chen C, Schwartz PM, Segal RA (2002) Brain-derived neurotrophic factor modulates cerebellar plasticity and synaptic ultrastructure. J Neurosci 22:1316–1327

Cases O, Lebrand C, Giros B, Vitalis T, De Maeyer E, Caron MG, Price DJ, Gaspar P, Seif I (1998) Plasma membrane transporters of serotonin, dopamine, and norepinephrine mediate serotonin accumulation in atypical locations in the developing brain of monoamine oxidase A knock-outs. J Neurosci 18:6914–6927

Caspi A, Sugden K, Moffitt TE, Taylor A, Craig IW, Harrington H, McClay J, Mill J, Martin J, Braithwaite A, Poulton R (2003) Influence of life stress on depression: moderation by a polymorphism in the 5-HTT gene. Science 301:386–389

Champoux M, Bennett A, Shannon C, Higley JD, Lesch KP, Suomi SJ (2002) Serotonin transporter gene polymorphism, differential early rearing, and behavior in rhesus monkey neonates. Mol Psychiatry 7:1058–1063

Chang AS, Starnes DM, Chang SM (1998) Possible existence of quaternary structure in the high-affinity serotonin transport complex. Biochem Biophys Res Commun 249:416–421

Cheetham SC, Crompton MR, Katona CL, Horton RW (1990) Brain 5-HT1 binding sites in depressed suicides. Psychopharmacology (Berl) 102:544–548

Chen JJ, Li Z, Pan H, Murphy DL, Tamir H, Koepsell H, Gershon MD (2001) Maintenance of serotonin in the intestinal mucosa and ganglia of mice that lack the high-affinity serotonin transporter: Abnormal intestinal motility and the expression of cation transporters. J Neurosci 21:6348–6361

Crawley JN (1999) Behavioral phenotyping of transgenic and knockout mice: experimental design and evaluation of general health, sensory functions, motor abilities, and specific behavioral tests. Brain Res 835:18–26

Crawley JN, Paylor R (1997) A proposed test battery and constellations of specific behavioral paradigms to investigate the behavioral phenotypes of transgenic and knockout mice. Horm Behav 31:197–211

D'Sa C, Duman RS (2002) Antidepressants and neuroplasticity. Bipolar Disord 4:183–194

David SP, Murthy NV, Rabiner EA, Munafo MR, Johnstone EC, Jacob R, Walton RT, Grasby PM (2005) A functional genetic variation of the serotonin (5-HT) transporter affects 5-HT1A receptor binding in humans. J Neurosci 25:2586–2590

De Vry J (1995) 5-HT1A receptor agonists: recent developments and controversial issues. Psychopharmacology (Berl) 121:1–26

Di Bella D, Catalano M, Balling U, Smeraldi E, Lesch KP (1996) Systematic screening for mutations in the coding region of the 5-HTT gene using PCR and DGGE. Am J Med Genet 67:541–545

Di Pino G, Mössner R, Lesch KP, Lauder JM, Persico AM (2004) Serotonin roles in neurodevelopment: more than just neural transmission. Curr Neuropharmacol 2:403–417

Drevets WC, Frank E, Price JC, Kupfer DJ, Holt D, Greer PJ, Huang Y, Gautier C, Mathis C (1999) PET imaging of serotonin 1A receptor binding in depression. Biol Psychiatry 46:1375–1387

Duman RS (2002) Synaptic plasticity and mood disorders. Mol Psychiatry 7 Suppl 1:S29–34

Eddahibi S, Hanoun N, Lanfumey L, Lesch KP, Raffestin B, Hamon M, Adnot S (2000) Attenuated hypoxic pulmonary hypertension in mice lacking the 5-hydroxytryptamine transporter gene. J Clin Invest 105:1555–1562

Eddahibi S, Humbert M, Fadel E, Raffestin B, Darmon M, Capron F, Simonneau G, Dartevelle P, Hamon M, Adnot S (2001) Serotonin transporter overexpression is responsible for pulmonary artery smooth muscle hyperplasia in primary pulmonary hypertension. J Clin Invest 108:1141–1150

Eddahibi S, Chaouat A, Morrell N, Fadel E, Fuhrman C, Bugnet AS, Dartevelle P, Housset B, Hamon M, Weitzenblum E, Adnot S (2003) Polymorphism of the serotonin transporter gene and pulmonary hypertension in chronic obstructive pulmonary disease. Circulation 108:1839–1844

Ernfors P, Lee KF, Jaenisch R (1994) Mice lacking brain-derived neurotrophic factor develop with sensory deficits. Nature 368:147–150

Fabre V, Beaufour C, Evrard A, Rioux A, Hanoun N, Lesch KP, Murphy DL, Lanfumey L, Hamon M, Martres MP (2000) Altered expression and functions of serotonin 5-HT1A and 5-HT1B receptors in knock-out mice lacking the 5-HT transporter. Eur J Neurosci 12:2299–2310

Fallgatter A, Jatzke S, Bartsch A, Hamelbeck B, Lesch K (1999) Serotonin transporter promoter polymorphism influences topography of inhibitory motor control. Int J Neuropsychopharmacol 2:115–120

Fallgatter AJ, Herrmann MJ, Roemmler J, Ehlis AC, Wagener A, Heidrich A, Ortega G, Zeng Y, Lesch KP (2004) Allelic variation of serotonin transporter function modulates the brain electrical response for error processing. Neuropsychopharmacology 29:1506–1511

Flugge G (1995) Dynamics of central nervous 5-HT1A-receptors under psychosocial stress. J Neurosci 15:7132–7140

Frechilla D, Insausti R, Ruiz-Golvano P, Garcia-Osta A, Rubio MP, Almendral JM, Del Rio J (2000) Implanted BDNF-producing fibroblasts prevent neurotoxin-induced serotonergic denervation in the rat striatum. Brain Res Mol Brain Res 76:306–314

Fullerton J, Cubin M, Tiwari H, Wang C, Bomhra A, Davidson S, Miller S, Fairburn C, Goodwin G, Neale MC, Fiddy S, Mott R, Allison DB, Flint J (2003) Linkage analysis of extremely discordant and concordant sibling pairs identifies quantitative-trait loci that influence variation in the human personality trait neuroticism. Am J Hum Genet 72:879–890

Furmark T, Tillfors M, Garpenstrand H, Marteinsdottir I, Langstrom B, Oreland L, Fredrikson M (2004) Serotonin transporter polymorphism related to amygdala excitability and symptom severity in patients with social phobia. Neurosci Lett 362:189–192

Galter D, Unsicker K (2000a) Brain-derived neurotrophic factor and trkB are essential for cAMP-mediated induction of the serotonergic neuronal phenotype. J Neurosci Res 61:295–301

Galter D, Unsicker K (2000b) Sequential activation of the 5-HT1(A) serotonin receptor and TrkB induces the serotonergic neuronal phenotype. Mol Cell Neurosci 15:446–455

Gershon MD (2003) Plasticity in serotonin control mechanisms in the gut. Curr Opin Pharmacol 3:600–607

Glatt CE, DeYoung JA, Delgado S, Service SK, Giacomini KM, Edwards RH, Risch N, Freimer NB (2001) Screening a large reference sample to identify very low frequency sequence variants: comparisons between two genes. Nat Genet 27:435–438

Gobbi G, Murphy DL, Lesch K, Blier P (2001) Modifications of the serotonergic system in mice lacking serotonin transporters: an in vivo electrophysiological study. J Pharmacol Exp Ther 296:987–995

Goergen EM, Bagay LA, Rehm K, Benton JL, Beltz BS (2002) Circadian control of neurogenesis. J Neurobiol 53:90–95

Griebel G (1995) 5-Hydroxytryptamine-interacting drugs in animal models of anxiety disorders: more than 30 years of research. Pharmacol Ther 65:319–395

Griebel G, Belzung C, Perrault G, Sanger DJ (2000) Differences in anxiety-related behaviours and in sensitivity to diazepam in inbred and outbred strains of mice. Psychopharmacology (Berl) 148:164–170

Hall FS, Li XF, Sora I, Xu F, Caron M, Lesch KP, Murphy DL, Uhl GR (2002) Cocaine mechanisms: enhanced cocaine, fluoxetine and nisoxetine place preferences following monoamine transporter deletions. Neuroscience 115:153–161

Hansson SR, Mezey E, Hoffman BJ (1998) Serotonin transporter messenger RNA in the developing rat brain: early expression in serotonergic neurons and transient expression in non-serotonergic neurons. Neuroscience 83:1185–1201

Hariri AR, Drabant EM, Munoz KE, Kolachana BS, Mattay VS, Egan MF, Weinberger DR (2005) A susceptibility gene for affective disorders and the response of the human amygdala. Arch Gen Psychiatry 62:146–152

Heils A, Teufel A, Petri S, Seemann M, Bengel D, Balling U, Riederer P, Lesch KP (1995) Functional promoter and polyadenylation site mapping of the human serotonin (5-HT) transporter gene. J Neural Transm 102:247–254

Heils A, Wichems C, Mössner R, Petri S, Glatz K, Bengel D, Murphy DL, Lesch KP (1998) Functional characterization of the murine serotonin transporter gene promoter in serotonergic raphe neurons. J Neurochem 70:932–939

Heinz A, Braus DF, Smolka MN, Wrase J, Puls I, Hermann D, Klein S, Grusser SM, Flor H, Schumann G, Mann K, Buchel C (2005) Amygdala-prefrontal coupling depends on a genetic variation of the serotonin transporter. Nat Neurosci 8:20–21

Higley JD, Suomi SJ, Linnoila M (1991) CSF monoamine metabolite concentrations vary according to age, rearing, and sex, and are influenced by the stressor of social separation in rhesus monkeys. Psychopharmacology (Berl) 103:551–556

Hofstetter HH, Mössner R, Lesch KP, Linker RA, Toyka KV, Gold R (2005) Absence of reuptake of serotonin influences susceptibility to clinical autoimmune disease and neuroantigen-specific interferon-gamma production in mouse EAE. Clin Exp Immunol 142:39–44

Holmes A, Lit Q, Murphy DL, Gold E, Crawley JN (2003) Abnormal anxiety-related behavior in serotonin transporter null mutant mice: the influence of genetic background. Genes Brain Behav 2:365–380

Jayanthi LD, Samuvel DJ, Blakely RD, Ramamoorthy S (2005) Evidence for biphasic effects of protein kinase C on serotonin transporter function, endocytosis, and phosphorylation. Mol Pharmacol 67:2077–2087

Kelai S, Aissi F, Lesch KP, Cohen-Salmon C, Hamon M, Lanfumey L (2003) Alcohol intake after serotonin transporter inactivation in mice. Alcohol Alcohol 38:386–389

Kempermann G, Kronenberg G (2003) Depressed new neurons—adult hippocampal neurogenesis and a cellular plasticity hypothesis of major depression. Biol Psychiatry 54:499–503

Kendler KS (1996) Major depression and generalised anxiety disorder. Same genes, (partly) different environments—revisited. Br J Psychiatry Suppl 68–75

Kernie SG, Liebl DJ, Parada LF (2000) BDNF regulates eating behavior and locomotor activity in mice. EMBO J 19:1290–1300

Kim DK, Tolliver TJ, Huang SJ, Martin BJ, Andrews AM, Wichems C, Holmes A, Lesch KP, Murphy DL (2005) Altered serotonin synthesis, turnover and dynamic regulation in multiple brain regions of mice lacking the serotonin transporter. Neuropharmacology 49:798–810

Lebrand C, Cases O, Adelbrecht C, Doye A, Alvarez C, Elmestikawy S, Seif I, Gaspar P (1996) Transient uptake and storage of serotonin in developing thalamic neurons. Neuron 17:823–835

Lemonde S, Turecki G, Bakish D, Du L, Hrdina PD, Bown CD, Sequeira A, Kushwaha N, Morris SJ, Basak A, Ou XM, Albert PR (2003) Impaired repression at a 5-hydroxytryptamine 1A receptor gene polymorphism associated with major depression and suicide. J Neurosci 23:8788–8799

Lesch KP (2001) Mouse anxiety: the power of knockout. Pharmacogenomics J 1:187–192

Lesch KP (2003) Neuroticism and serotonin: a developmental genetic perspective. In: Plomin R, DeFries J, Craig I, McGuffin P (eds) Behavioral genetics in the postgenomic era. American Psychiatric Press, Washington, pp 389–423

Lesch KP (2004) Gene-environment interaction and the genetics of depression. J Psychiatry Neurosci 29:174–184

Lesch KP (2005a) Alcohol dependence and emotion regulation: is serotonin the link. Eur J Pharmacol 526:113–124

Lesch KP (2005b) Neurogenomics of depression. In: Licinio J, Wong ML (eds) From novel insights to therapeutic strategies. Wiley-VCH, Weinheim, pp 713–732

Lesch KP, Canli T (2005) 5-HT1A receptor and anxiety-related traits: pharmacology, genetics, and imaging. In: Canli T (ed) Biology of personality and individual differences. Guilford, New York, pp 713–732

Lesch KP, Gutknecht L (2005) Pharmacogenetics of the serotonin transporter. Prog Neuropsychopharmacol Biol Psychiatry 29:1062–1073

Lesch KP, Mössner R (1999) Knockout Corner: 5-HT(1A) receptor inactivation: anxiety or depression as a murine experience. Int J Neuropsychopharmacol 2:327–331

Lesch KP, Murphy DL (2003) Molecular genetics of transporters for norepinephrine, dopamine, and serotonin in behavioral traits and complex diseases. In: Broeer S, Wagner CA (eds) Membrane transport diseases: molecular basis of inherited transport defects. Kluwer Academic/Plenum, New York, pp 349–364

Lesch KP, Disselkamp-Tietze J, Schmidtke A (1990a) 5-HT1A receptor function in depression: effect of chronic amitriptyline treatment. J Neural Transm Gen Sect 80:157–161

Lesch KP, Mayer S, Disselkamp-Tietze J, Hoh A, Wiesmann M, Osterheider M, Schulte HM (1990b) 5-HT1A receptor responsivity in unipolar depression. Evaluation of ipsapirone-induced ACTH and cortisol secretion in patients and controls. Biol Psychiatry 28:620–628

Lesch KP, Hoh A, Schulte HM, Osterheider M, Muller T (1991) Long-term fluoxetine treatment decreases 5-HT1A receptor responsivity in obsessive-compulsive disorder. Psychopharmacology (Berl) 105:415–420

Lesch KP, Wiesmann M, Hoh A, Muller T, Disselkamp-Tietze J, Osterheider M, Schulte HM (1992) 5-HT1A receptor-effector system responsivity in panic disorder. Psychopharmacology (Berl) 106:111–117

Lesch KP, Balling U, Gross J, Strauss K, Wolozin BL, Murphy DL, Riederer P (1994) Organization of the human serotonin transporter gene. J Neural Transm Gen Sect 95:157–162

Lesch KP, Bengel D, Heils A, Sabol SZ, Greenberg BD, Petri S, Benjamin J, Muller CR, Hamer DH, Murphy DL (1996) Association of anxiety-related traits with a polymorphism in the serotonin transporter gene regulatory region. Science 274:1527–1531

Lesch KP, Meyer J, Glatz K, Flugge G, Hinney A, Hebebrand J, Klauck SM, Poustka A, Poustka F, Bengel D, Mössner R, Riederer P, Heils A (1997) The 5-HT transporter gene-linked polymorphic region (5-HTTLPR) in evolutionary perspective: alternative biallelic variation in rhesus monkeys. Rapid communication. J Neural Transm 104:1259–1266

Lesch KP, Greenberg BD, Higley JD, Murphy DL (2002) Serotonin transporter, personality, and behavior: toward dissection of gene-gene and gene-environment interaction. In: Benjamin J, Ebstein R, Belmaker RH (eds) Molecular genetics and the human personality. American Psychiatric Press, Washington, pp 109–135

Lesch KP, Zeng Y, Reif A, Gutknecht L (2003) Anxiety-related traits in mice with modified genes of the serotonergic pathway. Eur J Pharmacol 480:185–204

Li Q, Wichems C, Heils A, Van De Kar LD, Lesch KP, Murphy DL (1999) Reduction of 5-hydroxytryptamine (5-HT)(1A)-mediated temperature and neuroendocrine responses and 5-HT(1A) binding sites in 5-HT transporter knockout mice. J Pharmacol Exp Ther 291:999–1007

Li Q, Wichems C, Heils A, Lesch KP, Murphy DL (2000) Reduction in the density and expression, but not G-protein coupling, of serotonin receptors (5-HT1A) in 5-HT transporter knock-out mice: gender and brain region differences. J Neurosci 20:7888–7895

Li Q, Wichems CH, Ma L, Van de Kar LD, Garcia F, Murphy DL (2003) Brain region-specific alterations of 5-HT2A and 5-HT2C receptors in serotonin transporter knockout mice. J Neurochem 84:1256–1265

Liu MT, Rayport S, Jiang Y, Murphy DL, Gershon MD (2002) Expression and function of 5-HT3 receptors in the enteric neurons of mice lacking the serotonin transporter. Am J Physiol Gastrointest Liver Physiol 283:G1398–1411

Lopez JF, Chalmers DT, Little KY, Watson SJ (1998) A.E. Bennett Research Award. Regulation of serotonin1A, glucocorticoid, and mineralocorticoid receptor in rat and human hippocampus: implications for the neurobiology of depression. Biol Psychiatry 43:547–573

Lyons WE, Mamounas LA, Ricaurte GA, Coppola V, Reid SW, Bora SH, Wihler C, Koliatsos VE, Tessarollo L (1999) Brain-derived neurotrophic factor-deficient mice develop aggressiveness and hyperphagia in conjunction with brain serotonergic abnormalities. Proc Natl Acad Sci U S A 96:15239–15244

MacQueen GM, Ramakrishnan K, Croll SD, Siuciak JA, Yu G, Young LT, Fahnestock M (2001) Performance of heterozygous brain-derived neurotrophic factor knockout mice on behavioral analogues of anxiety, nociception, and depression. Behav Neurosci 115:1145–1153

Malhi GS, Moore J, McGuffin P (2000) The genetics of major depressive disorder. Curr Psychiatry Rep 2:165–169

Mannoury la Cour C, Boni C, Hanoun N, Lesch KP, Hamon M, Lanfumey L (2001) Functional consequences of 5-HT transporter gene disruption on 5-HT(1a) receptor-mediated regulation of dorsal raphe and hippocampal cell activity. J Neurosci 21:2178–2185

Mannoury la Cour C, Hanoun N, Melfort M, Hen R, Lesch KP, Hamon M, Lanfumey L (2004) GABA(B) receptors in 5-HT transporter- and 5-HT1A receptor-knock-out mice: further evidence of a transduction pathway shared with 5-HT1A receptors. J Neurochem 89:886–896

Mansour-Robaey S, Mechawar N, Radja F, Beaulieu C, Descarries L (1998) Quantified distribution of serotonin transporter and receptors during the postnatal development of the rat barrel field cortex. Brain Res Dev Brain Res 107:159–163

Miller KJ, Hoffman BJ (1994) Adenosine A3 receptors regulate serotonin transport via nitric oxide and cGMP. J Biol Chem 269:27351–27356

Minichiello L, Korte M, Wolfer D, Kuhn R, Unsicker K, Cestari V, Rossi-Arnaud C, Lipp HP, Bonhoeffer T, Klein R (1999) Essential role for TrkB receptors in hippocampus-mediated learning. Neuron 24:401–414

Mössner R, Lesch KP (1998) Role of serotonin in the immune system and in neuroimmune interactions. Brain Behav Immun 12:249–271

Mössner R, Albert D, Persico AM, Hennig T, Bengel D, Holtmann B, Schmitt A, Keller F, Simantov R, Murphy D, Seif I, Deckert J, Lesch KP (2000) Differential regulation of adenosine A(1) and A(2A) receptors in serotonin transporter and monoamine oxidase A-deficient mice. Eur Neuropsychopharmacol 10:489–493

Mössner R, Daniel S, Schmitt A, Albert D, Lesch KP (2001) Modulation of serotonin transporter function by interleukin-4. Life Sci 68:873–880

Mössner R, Dringen R, Persico AM, Janetzky B, Okladnova O, Albert D, Götz M, Benninghoff J, Schmitt A, Henneberg A, Gerlach P, Riederer P, Lesch KP (2002) Increased hippocampal DNA oxidation in serotonin transporter deficient mice. J Neural Transm 109:557–565

Mössner R, Schmitt A, Hennig T, Benninghoff J, Gerlach M, Riederer P, Deckert J, Lesch KP (2004) Quantitation of 5HT3 receptors in forebrain of serotonin transporter deficient mice. J Neural Transm 111:27–35

Murphy DL, Uhl GR, Holmes A, Ren-Patterson R, Hall FS, Sora I, Detera-Wadleigh S, Lesch KP (2003) Experimental gene interaction studies with SERT mutant mice as models for human polygenic and epistatic traits and disorders. Genes Brain Behav 2:350–364

Murphy DL, Lerner A, Rudnick G, Lesch KP (2004) Serotonin transporter: gene, genetic disorders, and pharmacogenetics. Mol Interv 4:109–123

Olivier B, Miczek KA (1999) Fear and anxiety: mechanisms, models and molecules. In: Dodman N, Shuster I (eds) Psychopharmacology of animal behavior disorders. Blackwell, London, pp 105–121

Olivier B, Soudijn W, van Wijngaarden I (1999) The 5-HT1A receptor and its ligands: structure and function. Prog Drug Res 52:103–165

Ozsarac N, Santha E, Hoffman BJ (2002) Alternative non-coding exons support serotonin transporter mRNA expression in the brain and gut. J Neurochem 82:336–344

Pan Y, Gembom E, Peng W, Lesch KP, Mössner R, Simantov R (2001) Plasticity in serotonin uptake in primary neuronal cultures of serotonin transporter knockout mice. Brain Res Dev Brain Res 126:125–129

Pata C, Erdal ME, Derici E, Yazar A, Kanik A, Ulu O (2002) Serotonin transporter gene polymorphism in irritable bowel syndrome. Am J Gastroenterol 97:1780–1784

Penado KM, Rudnick G, Stephan MM (1998) Critical amino acid residues in transmembrane span 7 of the serotonin transporter identified by random mutagenesis. J Biol Chem 273:28098–28106

Peng W, Premkumar A, Mössner R, Fukuda M, Lesch KP, Simantov R (2002) Synaptotagmin I and IV are differentially regulated in the brain by the recreational drug 3,4-methylenedioxymethamphetamine (MDMA). Brain Res Mol Brain Res 108:94–101

Persico AM, Revay RS, Mössner R, Conciatori M, Marino R, Baldi A, Cabib S, Pascucci T, Sora I, Uhl GR, Murphy DL, Lesch KP, Keller F (2001) Barrel pattern formation in somatosensory cortical layer IV requires serotonin uptake by thalamocortical endings, while vesicular monoamine release is necessary for development of supragranular layers. J Neurosci 21:6862–6873

Persico AM, Baldi A, Dell'Acqua ML, Moessner R, Murphy DL, Lesch KP, Keller F (2003) Reduced programmed cell death in brains of serotonin transporter knockout mice. Neuroreport 14:341–344

Pezawas L, Meyer-Lindenberg A, Drabant EM, Verchinski BA, Munoz KE, Kolachana BS, Egan MF, Mattay VS, Hariri AR, Weinberger DR (2005) 5-HTTLPR polymorphism impacts human cingulate-amygdala interactions: a genetic susceptibility mechanism for depression. Nat Neurosci 8:828–834

Prasad HC, Zhu CB, McCauley JL, Samuvel DJ, Ramamoorthy S, Shelton RC, Hewlett WA, Sutcliffe JS, Blakely RD (2005) Human serotonin transporter variants display altered sensitivity to protein kinase G and p38 mitogen-activated protein kinase. Proc Natl Acad Sci U S A 102:11545–11550

Quick MW (2003) Regulating the conducting states of a mammalian serotonin transporter. Neuron 40:537–549

Ramamoorthy S, Cool DR, Mahesh VB, Leibach FH, Melikian HE, Blakely RD, Ganapathy V (1993) Regulation of the human serotonin transporter: cholera toxin-induced stimulation of serotonin uptake in human placental choriocarcinoma cells is accompanied by increased serotonin transporter mRNA levels and serotonin transporter-specific ligand binding. J Biol Chem 268:21626–21631

Ramboz S, Oosting R, Amara DA, Kung HF, Blier P, Mendelsohn M, Mann JJ, Brunner D, Hen R (1998) Serotonin receptor 1A knockout: an animal model of anxiety-related disorder. Proc Natl Acad Sci U S A 95:14476–14481

Ravary A, Muzerelle A, Darmon M, Murphy DL, Moessner R, Lesch KP, Gaspar P (2001) Abnormal trafficking and subcellular localization of an N-terminally truncated serotonin transporter protein. Eur J Neurosci 13:1349–1362

Ren-Patterson RF, Cochran LW, Holmes A, Sherrill S, Huang SJ, Tolliver T, Lesch KP, Lu B, Murphy DL (2005) Loss of brain-derived neurotrophic factor gene allele exacerbates brain monoamine deficiencies and increases stress abnormalities of serotonin transporter knockout mice. J Neurosci Res 79:756–771

Rios M, Fan G, Fekete C, Kelly J, Bates B, Kuehn R, Lechan RM, Jaenisch R (2001) Conditional deletion of brain-derived neurotrophic factor in the postnatal brain leads to obesity and hyperactivity. Mol Endocrinol 15:1748–1757

Rioux A, Fabre V, Lesch KP, Moessner R, Murphy DL, Lanfumey L, Hamon M, Martres MP (1999) Adaptive changes of serotonin 5-HT2A receptors in mice lacking the serotonin transporter. Neurosci Lett 262:113–116

Rothe C, Gutknecht L, Freitag CM, Tauber R, Franke P, Fritze J, Wagner G, Peikert G, Wenda B, Sand P, Jacob C, Rietschel M, Nöthen MM, Garritsen H, Fimmers R, Deckert J, Lesch KP (2004) Association of a functional −1019C>G 5-HT1A receptor gene polymorphism with panic disorder with agoraphobia. Int J Neuropsychopharmacol 7:189–192

Rumajogee P, Verge D, Hanoun N, Brisorgueil MJ, Hen R, Lesch KP, Hamon M, Miquel MC (2004) Adaption of the serotoninergic neuronal phenotype in the absence of 5-HT autoreceptors or the 5-HT transporter: involvement of BDNF and cAMP. Eur J Neurosci 19:937–944

Sakai N, Sasaki K, Nakashita M, Honda S, Ikegaki N, Saito N (1997) Modulation of serotonin transporter activity by a protein kinase C activator and an inhibitor of type 1 and 2A serine/threonine phosphatases. J Neurochem 68:2618–2624

Salichon N, Gaspar P, Upton AL, Picaud S, Hanoun N, Hamon M, De Maeyer EE, Murphy DL, Mössner R, Lesch KP, Hen R, Seif I (2001) Excessive activation of serotonin (5-HT) 1B receptors disrupts the formation of sensory maps in monoamine oxidase A and 5-HT transporter knock-out mice. J Neurosci 21:884–896

Santarelli L, Saxe M, Gross C, Surget A, Battaglia F, Dulawa S, Weisstaub N, Lee J, Duman R, Arancio O, Belzung C, Hen R (2003) Requirement of hippocampal neurogenesis for the behavioral effects of antidepressants. Science 301:805–809

Sargent PA, Kjaer KH, Bench CJ, Rabiner EA, Messa C, Meyer J, Gunn RN, Grasby PM, Cowen PJ (2000) Brain serotonin 1A receptor binding measured by positron emission tomography with [11C]WAY-100635: effects of depression and antidepressant treatment. Arch Gen Psychiatry 57:174–180

Schmitt A, Mössner R, Gossmann A, Fischer IG, Gorboulev V, Murphy DL, Koepsell H, Lesch KP (2003) Organic cation transporter capable of transporting serotonin is upregulated in serotonin transporter-deficient mice. J Neurosci Res 71:701–709

Schuman EM (1999) Neurotrophin regulation of synaptic transmission. Curr Opin Neurobiol 9:105–109

Shen HW, Hagino Y, Kobayashi H, Shinohara-Tanaka K, Ikeda K, Yamamoto H, Yamamoto T, Lesch KP, Murphy DL, Hall FS, Uhl GR, Sora I (2004) Regional differences in extracellular dopamine and serotonin assessed by in vivo microdialysis in mice lacking dopamine and/or serotonin transporters. Neuropsychopharmacology 29:1790–1799

Sora I, Wichems C, Takahashi N, Li XF, Zeng Z, Revay R, Lesch KP, Murphy DL, Uhl GR (1998) Cocaine reward models: conditioned place preference can be established in dopamine- and in serotonin-transporter knockout mice. Proc Natl Acad Sci U S A 95:7699–7704

Sora I, Hall FS, Andrews AM, Itokawa M, Li XF, Wei HB, Wichems C, Lesch KP, Murphy DL, Uhl GR (2001) Molecular mechanisms of cocaine reward: combined dopamine and serotonin transporter knockouts eliminate cocaine place preference. Proc Natl Acad Sci U S A 98:5300–5305

Spenger C, Hyman C, Studer L, Egli M, Evtouchenko L, Jackson C, Dahl-Jorgensen A, Lindsay RM, Seiler RW (1995) Effects of BDNF on dopaminergic, serotonergic, and GABAergic neurons in cultures of human fetal ventral mesencephalon. Exp Neurol 133:50–63

Strobel A, Gutknecht L, Zheng Y, Reif A, Brocke B, Lesch KP (2003) Allelic variation of serotonin receptor 1A function is associated with anxiety- and depression-related traits. J Neural Transm 110:1445–1453

Sutcliffe JS, Delahanty RJ, Prasad HC, McCauley JL, Han Q, Jiang L, Li C, Folstein SE, Blakely RD (2005) Allelic heterogeneity at the serotonin transporter locus (SLC6A4) confers susceptibility to autism and rigid-compulsive behaviors. Am J Hum Genet 77:265–279

Upton AL, Ravary A, Salichon N, Moessner R, Lesch KP, Hen R, Seif I, Gaspar P (2002) Lack of 5-HT(1B) receptor and of serotonin transporter have different effects on the segregation of retinal axons in the lateral geniculate nucleus compared to the superior colliculus. Neuroscience 111:597–610

van Praag H, Schinder AF, Christie BR, Toni N, Palmer TD, Gage FH (2002) Functional neurogenesis in the adult hippocampus. Nature 415:1030–1034

Vitalis T, Cases O, Gillies K, Hanoun N, Hamon M, Seif I, Gaspar P, Kind P, Price DJ (2002) Interactions between TrkB signaling and serotonin excess in the developing murine somatosensory cortex: a role in tangential and radial organization of thalamocortical axons. J Neurosci 22:4987–5000

Vogel C, Mössner R, Gerlach M, Heinemann T, Murphy DL, Riederer P, Lesch KP, Sommer C (2003) Absence of thermal hyperalgesia in serotonin transporter-deficient mice. J Neurosci 23:708–715

Warden SJ, Robling AG, Sanders MS, Bliziotes MM, Turner CH (2005) Inhibition of the serotonin (5-hydroxytryptamine) transporter reduces bone accrual during growth. Endocrinology 146:685–693

Wisor JP, Wurts SW, Hall FS, Lesch KP, Murphy DL, Uhl GR, Edgar DM (2003) Altered rapid eye movement sleep timing in serotonin transporter knockout mice. Neuroreport 14:233–238

Wissink S, Meijer O, Pearce D, van Der Burg B, van Der Saag PT (2000) Regulation of the rat serotonin-1A receptor gene by corticosteroids. J Biol Chem 275:1321–1326

Yavarone MS, Shuey DL, Tamir H, Sadler TW, Lauder JM (1993) Serotonin and cardiac morphogenesis in the mouse embryo. Teratology 47:573–584

Zhou FC, Lesch KP, Murphy DL (2002) Serotonin uptake into dopamine neurons via dopamine transporters: a compensatory alternative. Brain Res 942:109–119

Zhou J, Iacovitti L (2000) Mechanisms governing the differentiation of a serotonergic phenotype in culture. Brain Res 877:37–46

Zubenko GS, Maher B, Hughes HB 3rd, Zubenko WN, Stiffler JS, Kaplan BB, Marazita ML (2003) Genome-wide linkage survey for genetic loci that influence the development of depressive disorders in families with recurrent, early-onset, major depression. Am J Med Genet 123B:1–18

Lessons from the Knocked-Out Glycine Transporters

J. Gomeza[1] (✉) · W. Armsen[2] · H. Betz[2] · V. Eulenburg[2]

[1]Department of Pharmacology, The Panum Institute, University of Copenhagen, Blegdamsvej 3, 2200 Copenhagen, Denmark
jesus.gomeza@farmakol.ku.dk

[2]Department of Neurochemistry, Max-Planck-Institute for Brain Research, Deutschordenstrasse 46, 60528 Frankfurt, Germany

1	Neurotransmitter Functions of Glycine in the CNS	458
2	GlyTs Are Members of the Na^+/Cl^--Dependent Transporter Family	459
2.1	GlyT Gene Structures	459
2.2	GlyT Protein Structure	460
2.3	Glycine Uptake: Electrogenic Properties and Transport Mechanism	462
2.4	Plasma Membrane Localization and Modulation of GlyTs	463
3	Distribution of GlyTs in the CNS	464
3.1	Tissue and Cellular Distribution	464
3.2	Transcriptional Control	466
4	Functional Roles of GlyTs	467
4.1	Generation of GlyT-Deficient Mice	467
4.2	GlyTs Are Essential for Vital Postnatal Functions	468
4.3	Essential Functions of GlyTs at Glycinergic Synapses	469
4.4	Pharmacology of GlyTs	471
4.5	Function of GlyTs at Glutamatergic Synapses	473
4.5.1	Pharmacological Studies In Vitro	473
4.5.2	Pharmacology and Genetics In Vivo	474
5	GlyTs and Human Diseases	475
5.1	GlyT Genes: Candidate Disease Loci?	475
5.2	GlyTs as Potential Drug Targets	476
6	Conclusions and Perspectives	477
	References	478

Abstract Glycine has multiple neurotransmitter functions in the central nervous system (CNS). In the spinal cord and brainstem of vertebrates, it serves as a major inhibitory neurotransmitter. In addition, it participates in excitatory neurotransmission by modulating the activity of the *N*-methyl-D-aspartate (NMDA) subtype of glutamate receptors. The extracellular concentrations of glycine are regulated by Na^+/Cl^--dependent glycine transporters (GlyTs), which are expressed in neurons and adjacent glial cells. Considerable progress has been made recently towards elucidating the in vivo roles of GlyTs in the CNS. The generation and analysis of animals carrying targeted disruptions of GlyT genes (GlyT knockout mice) have allowed investigators to examine the different contributions of individual GlyT subtypes to synaptic transmission. In addition, they have provided animal models for two hereditary human diseases, glycine encephalopathy and hyperekplexia. Selective GlyT inhibitors have been shown to modulate neurotransmission and might constitute promising therapeutic tools for the treatment of psychiatric and neurological disorders such as schizophrenia and pain. Therefore, pharmacological and genetic studies indicate that GlyTs are key regulators of both glycinergic inhibitory and glutamatergic excitatory neurotransmission. This chapter describes our present understanding of the functions of GlyTs and their involvement in the fine-tuning of neuronal communication.

Keywords Glycine · Glycine uptake · Glycine transporter · Inhibitory glycine receptor · N-Methyl-D-aspartate receptor · Transport inhibitors · Knockout mice · Schizophrenia · Pain · Hyperekplexia · Glycine encephalopathy

1
Neurotransmitter Functions of Glycine in the CNS

More than 40 years ago, the amino acid glycine was recognized to have neurotransmitter functions in the mammalian central nervous system (CNS). Initially, glycine was found to be highly enriched in spinal cord and brainstem and to inhibit the firing of spinal cord neurons. This effect was blocked by strychnine, a selective antagonist of inhibitory glycine receptors (GlyRs) (Laube et al. 2002; Lynch 2004). Today it is well established that glycine serves, along with γ-aminobutyric acid (GABA), as a principal inhibitory neurotransmitter in the adult mammalian CNS (Betz et al. 2000). At inhibitory synapses, glycine is released from the presynaptic nerve terminal upon depolarization and Ca^{2+}-influx via the exocytotic fusion of glycine-filled synaptic vesicles with the presynaptic plasma membrane (Legendre 2001). The released glycine binds to postsynaptic GlyRs, thereby causing an increase in the chloride conductance of the plasma membrane (Betz et al. 2000). GlyRs are pentameric membrane proteins composed of α- and β-subunits which, upon glycine binding, undergo a conformational change that opens an intrinsic anion channel. In the adult CNS, GlyR activation results in an influx of chloride ions into the cytoplasm of the postsynaptic cell, thus leading to hyperpolarization and thereby an increase in the threshold for the initiation of action potentials (Lynch 2004). In contrast, embryonic neurons are depolarized, i.e. excited upon GlyR activation. This excitatory action of glycine is due to high intracellular chloride concentrations at early stages of development (Singer et al. 1998), which result in chloride efflux from the cytoplasm upon opening of the GlyR anion channels. Around birth, the intracellular chloride concentration is reduced due to the expression of a very effective chloride export system, the K^+/Cl^--cotransporter KCC2 (Hubner et al. 2001; Rivera et al. 1999). This leads to a conversion of glycine responses from excitatory to inhibitory.

In addition to its role in GlyR-mediated inhibitory neurotransmission, glycine also has essential functions at excitatory glutamatergic synapses. Here, glycine acts, together with glia-derived D-serine, as a co-agonist of glutamate at ionotropic glutamate receptors of the N-methyl-D-aspartate receptor (NMDAR) subtype (Johnson and Ascher 1987; Verdoorn et al. 1987). NMDARs are heterotetrameric proteins composed of glycine binding (NR1, NR3) and glutamate binding (NR2) subunits (Hollmann 1999; Laube et al. 1998). The most widely expressed NMDARs are heterotetramers of NR1 and NR2, which display a high binding affinity ($K_d \sim 500$ nM) for glycine. Since the glycine concentration in the cerebrospinal fluid has been estimated to be in the low micromolar range, the high-affinity glycine binding site at the NMDAR was thought to be permanently saturated. However, recent studies have shown effects of exogenously applied glycine on NMDAR-mediated currents, demonstrating that alterations in the ambient glycine concentration have modulatory effects on NMDAR activity. First, the application of up to 20 µM glycine causes a potentiation of the NMDAR component of glutamatergic neurotransmission (Berger et al. 1998), supporting the idea that the glycine binding site of NMDARs is not permanently saturated at synapses. In contrast, application of a high concentration of glycine (≥ 100 µM) results in a decrease of NMDAR-mediated neurotransmission due to enhanced receptor internalization (Nong et al. 2003).

The neurotransmitter functions of glycine imply that extracellular glycine concentrations have to be tightly controlled. At inhibitory synapses, the released glycine must be removed efficiently after GlyR binding, thereby allowing for high-frequency firing and thus precise temporal control of neurotransmission. Furthermore, cytosolic glycine pools have to be replenished for the reloading of synaptic vesicles after endocytosis. A stringent regulation of extracellular glycine levels is also essential for NMDAR modulation at excitatory synapses. All these tasks are accomplished by highly selective transport systems, the glycine transporters (GlyTs). This chapter describes our present understanding of GlyTs and their involvement in the fine-tuning of neuronal transmission.

2
GlyTs Are Members of the Na$^+$/Cl$^-$-Dependent Transporter Family

2.1
GlyT Gene Structures

Following the demonstration that glycine acts as an inhibitory neurotransmitter, high-affinity [^{3}H]glycine uptake into neurons and glial cells was found in rat spinal cord and other regions of the rodent CNS (Lopez-Corcuera and Aragon 1989). This glycine uptake was dependent on Na$^+$ and Cl$^-$ ion gradients and resembled the glycine transport system previously described for human erythrocytes, the so-called 'system Gly' (Ellory et al. 1981). A hallmark of 'system Gly' is its inhibition by sarcosine (*N*-methyl-glycine). In the early 1990s, the first purification of a functional GlyT protein from pig brainstem with a molecular weight of about 100 kDa was reported (Lopez-Corcuera et al. 1991). Interestingly, although [^{3}H]glycine uptake driven by the reconstituted transporter was Na^+- and Cl^--dependent, sarcosine was unable to inhibit it. This was the first convincing evidence that more than one GlyT exists in the mammalian CNS. Subsequently, two GlyT isoforms, GlyT1 and GlyT2, were identified by homology screening with previously cloned Na^+/Cl^--dependent neurotransmitter transporter complementary (c)DNAs (Guastella et al. 1992; Liu et al. 1993; Smith et al. 1992). Both, GlyT1 and GlyT2 exist in different isoforms which result from alternative splicing and/or usage of different promoters (Guastella et al. 1992; Liu et al. 1992).

The human GlyT1 gene (*SLC6A9*) encompasses 14 exons distributed over 44.1 Mb and is found on human chromosome 1 (p31.3–p32) (Jones et al. 1995; Kim et al. 1994). To date, in humans and other mammals three variants, GlyT1 a–c, have been identified which differ only in their extreme N-terminal regions (Adams et al. 1995; Borowsky et al. 1993; Kim et al. 1994). GlyT1a and GlyT1b are generated by differential promoter usage. GlyT1c is an alternative splice product of GlyT1b (Adams et al. 1995; Borowsky and Hoffman 1998). Comparison of the human N-terminal sequences revealed that GlyT1c has an initial 15-residue sequence equivalent to that of GlyT1b, which is followed by a unique 54-residue sequence not found in the other two isoforms (Kim et al. 1994). Additionally, for bovine GlyT1 two C-terminal splice variants (d, e) have been described (Hanley et al. 2000) that have not been found in other species so far.

The GlyT2 gene (*SLC6A5*) is located on human chromosome 11 (p15.1–15.2) (Morrow et al. 1998). Its 16 exons span a region of 20.6 Mb. In mouse, three alternative N-terminal splice variants (GlyT2a–c) have been described and analysed (Ebihara et al. 2004). Mouse GlyT2a contains eight additional amino acids at the extreme N-terminus, whereas GlyT2b and c, which are identical at the amino acid level, have different 5′-untranslated regions (UTRs).

2.2
GlyT Protein Structure

GlyTs share common structural features with other members of the Na^+/Cl^--dependent transporter family (*SLC6*) that utilize neurotransmitters and metabolites as substrates (Nelson 1998). *N*-Glycosylation and cysteine scanning mutagenesis, in addition to in vitro glycosylation reporter fusion assays, have shown that GlyTs are polytopic membrane proteins with 12 transmembrane domains (TMDs), which are connected by six extracellular (EL) and five intracellular (IL) loops (Olivares et al. 1997) (Fig. 1a). Site-directed mutagenesis indicates that GlyT1 and GlyT2 are extensively glycosylated at four asparagine residues within their large EL2 connecting the hydrophobic TMDs 3 and 4

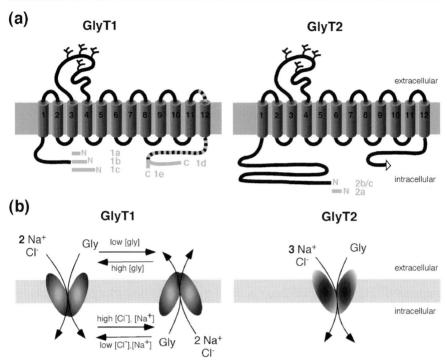

Fig. 1 a,b Membrane topology and transport properties of GlyTs. **a** GlyTs are characterized by 12 putative TMDs with intracellular N- and C-termini. Different splice variants are indicated in grey. For GlyT1, three N-terminal splice variants (a–c) and two C-terminal splice variants (d, e) have been identified. The *dashed line* indicates the shorter GlyT1e variant, identified only for bovine GlyT1, whose TMD12 may cross the membrane not as an α-helix. Alternate promoter usage generates three N-terminal GlyT2 isoforms (a–c) with eight additional amino acids for GlyT2a and shorter identical protein sequences for GlyT2b and c. *N*-Linked carbohydrates are attached to the large EL2. The C-terminal PDZ domain-binding motif of GlyT2 is drawn as a *triangle*. **b** Both GlyTs catalyse the symport of glycine in the presence of Na^+ and Cl^- but differ in their transport stoichiometries, with two Na^+ being co-transported by GlyT1 (*left*) and three Na^+ for GlyT2 (*right*). It is therefore suggested that GlyT1 is working close to equilibrium, and that it may either import or export glycine depending on ion gradients and membrane potential. For GlyT2, glycine transport is unlikely to be reversed by changes in the intracellular ionic concentrations due to its higher Na^+ transport stoichiometry. (Adapted with permission from Eulenburg et al. 2005)

(Olivares et al. 1995), thereby establishing the extracellular location of this hydrophilic loop region. Immunofluorescence and electron microscopy studies employing sequence-specific antibodies have confirmed that the N- and C-terminal ends of GlyTs are located intracellularly (Olivares et al. 1994). Almost all members of the *SLC6* family show C- and N-termini of comparable length, e.g. approximately 30–90 amino acids for the N-terminus and about 30–45 residues for the C-terminus. However, GlyT2 has a unique N-terminal region encompassing about 200 amino acids. Moreover, the C-termini of both GlyTs, with 60–75 amino acids, are relatively long compared with the other family members. Up to now, however, no unique functions of these large intracellular domains have been identified.

2.3
Glycine Uptake: Electrogenic Properties and Transport Mechanism

The uptake of glycine mediated by GlyTs is strictly Na^+-dependent and only partially Cl^--dependent (Lopez-Corcuera et al. 1998). Extracellular binding of both ions together with glycine is supposed to induce a structural change in GlyTs, from an 'outwardly open' to an 'inwardly open' conformation, thus enabling intracellular release of substrates (Rudnick 1998). In transient transfection studies, kinetic analyses of GlyT1 and GlyT2 showed similar K_m values for uptake activity. However, a new combination of electrophysiology and radio-tracing techniques demonstrated that GlyTs differ in their ionic stoichiometries (Lopez-Corcuera et al. 1998; Roux and Supplisson 2000). The stoichiometry of substrate/ion co-transport has been determined to be 3 Na^+/Cl^- per glycine for GlyT2, whereas GlyT1 has a transport stoichiometry of 2 Na^+/Cl^- per glycine (Fig. 1b). Therefore, the uptake process is electrogenic, resulting in the intracellular accumulation of two positive charges per transport cycle for GlyT2 and only one positive charge for GlyT1. This difference of one Na^+ in ionic coupling between GlyTs has important consequences for their function. First, it implies that, under physiological conditions, the driving force available for uphill glycine transport by GlyT2 is much higher than for that by GlyT1. Therefore, GlyT2 has a higher capacity in maintaining millimolar intracellular versus submicromolar extracellular glycine levels than GlyT1. Second, it suggests that GlyT1 might function in a reverse uptake mode, i.e. releasing glycine from the cytosol into the extracellular space, in case of alterations in substrate concentration gradients or membrane potential (Sakata et al. 1997). In agreement with this interpretation, GlyT1 expressed in *Xenopus* oocytes generated an efflux of glycine upon intracellular glycine application or depolarization, whereas GlyT2 showed no efflux unless high concentrations of Na^+, Cl^- and glycine were co-injected into the oocyte cytoplasm (Supplisson and Roux 2002). Therefore, it has been proposed that GlyT1 might mediate Ca^{2+}-independent glycine release from cells and thereby contributes to the regulation of the extracellular glycine concentration.

The amino acid side chains and protein domains that participate in the function of GlyTs are not yet well defined. Mutational analysis has identified residues within transmembrane regions which are thought to have key roles in the transport cycle. Conservative substitutions of Tyr289, located in TMD3 of GlyT2, decreased the apparent K_m of glycine and severely altered Na^+ and Cl^- dependence (Ponce et al. 2000). Thus, TMD3 may form part of a common permeation pathway for glycine and co-transported ions. In contrast, deletion of the N- and C-terminal regions did not affect glycine transport upon reconstitution into liposomes (Olivares et al. 1994), demonstrating that these intracellular domains are not required for transporter function. Likewise, N-glycosylation of EL2 of GlyT1 is not important for transport activity because treatment of the purified and liposome-reconstituted transporter with N-glycosidase resulted in the loss of its carbohydrate moiety but did not alter transport characteristics (Olivares et al. 1995). Mutation of the four N-glycosylation sites of GlyT2, however, induced a substantial decrease in glycine uptake activity, demonstrating a significant role of the carbohydrate moiety in stabilizing the active conformation of this transporter (Martinez-Maza et al. 2001; Nunez and Aragon 1994). The extracellular loop regions are also thought to contribute to substrate binding and uptake activity. Cysteine scanning mutagenesis has demonstrated that the EL1 of GlyT2, although it is not involved in substrate binding, acts as a fluctuating hinge that upon binding of glycine undergoes sequential conformational changes, which are thought to be essential for substrate translocation (Lopez-Corcuera et al. 2001). In summary, all available data indicate that extensive structural rearrangements, involving many different domains of GlyTs, take place during the transport cycle, thereby enabling the transporters to switch in an alternate access mode (Rudnick 1998) from an 'outwardly' to an 'inwardly' facing conformation.

2.4
Plasma Membrane Localization and Modulation of GlyTs

Considerable information exists regarding structural domains that are required for the insertion of GlyTs into the plasma membrane. In transfection studies, the intracellular N- and C-terminal domains have been found to be important for proper targeting of GlyT1 to the cell surface. Deletion of the C-terminal domain of GlyT1, which contains a putative PDZ-domain binding motif, results in impaired trafficking to the plasma membrane (Olivares et al. 1994). In contrast, deletion of the GlyT1 N-terminal region does not affect membrane insertion of the transporter. Interestingly, these N-terminally truncated mutants display an apical distribution in polarized Madin–Darby canine kidney (MDCK) cells as compared to a basolateral localization observed with full-length GlyT1b (Poyatos et al. 2000). Likewise, replacement of two dileucine motifs by alanines in the cytoplasmic tail of GlyT1b induces a relocation of the protein to the apical surface (Poyatos et al. 2000).

Extensively glycosylated asparagine residues within EL2 are also essential for insertion of GlyTs into the plasma membrane. Progressive mutation of the four N-glycosylation consensus sequences of GlyT1 results in missorting of the unglycosylated mutants in transfected cells (Olivares et al. 1995). Similar mutations in GlyT2 are responsible for a nonpolarized versus an apical localization in MDCK cells (Martinez-Maza et al. 2001). Furthermore, a C-terminal PDZ-domain binding motif in GlyT2 might be important for its targeting to the synapse. Biochemical experiments and yeast two-hybrid screening have identified interaction partners of GlyT2, such as the PDZ domain protein syntenin-1, which binds to the GlyT2 C-terminal tail (Ohno et al. 2004). In contrast, the presynaptic SNARE (soluble *N*-ethylmaleimide-sensitive factor attachment protein receptor) protein syntaxin 1A (Geerlings et al. 2000) and Ulip6 (Horiuchi et al. 2005), a member of the collapsin response mediator protein family which has been implicated in endocytosis (Nishimura et al. 2003), bind to the N-terminal domain of GlyT2. Co-expression experiments of syntaxin 1A with GlyT1 or GlyT2 in transfected cells modified the number of GlyTs on the cell surface (Geerlings et al. 2000). This might have been due to changes in the rates of exocytosis or internalization of the GlyTs upon their interaction with syntaxin 1A. In neurons, constitutive delivery of GlyT2 to the plasma membrane has been found to be mediated by syntaxin 1A, whereas GlyT2 internalization is independent of this interaction (Geerlings et al. 2001).

Other intracellular regions of GlyTs also contribute to plasma membrane localization. Mutations of charged amino acids in the IL2 of GlyT2 have been shown to abolish its internalization upon phorbol ester treatment (Fornes et al. 2004), suggesting that protein kinase C (PKC)-regulated proteins interact with this GlyT2 intracellular domain and modulate its post-plasma membrane trafficking.

In summary, distinct domains of GlyTs are required for proper insertion into and removal from the plasma membrane mainly through interactions with intracellular regulatory binding proteins.

3
Distribution of GlyTs in the CNS

3.1
Tissue and Cellular Distribution

GlyTs share overlapping expression patterns in caudal regions of the CNS (spinal cord, brainstem and cerebellum) (Jursky and Nelson 1996; Luque et al. 1995; Zafra et al. 1995) in which glycine acts as inhibitory neurotransmitter (Fig. 2), thus supporting an important role of both transporters in the regulation of the extracellular glycine levels at inhibitory glycinergic synapses. In addition, GlyT1 transcripts are also detectable in forebrain regions and retina, where only minor GlyT2 expression is found.

Throughout the CNS, GlyT1 is almost exclusively localized on glial cells, in particular astrocytes (Zafra et al. 1995). However, GlyT1 in mammals is absent from Müller cells, the major glial cell type in the retina. Here, GlyT1 is expressed by a subset of neuronal cells, i.e. amacrine and some ganglion cells (Pow and Hendrickson 1999). Recently, it has been shown that, in forebrain regions, GlyT1 is expressed not only in astrocytes but is found presynaptically

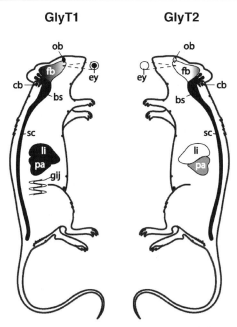

Fig. 2 Schematic representation of the expression patterns of the GlyTs. GlyT1 (*left*) and GlyT2 (*right*) expression patterns overlap in several regions of the CNS, particularly in the spinal cord (*sc*), brainstem (*bs*) and cerebellum (*cb*). In forebrain regions (*fb*), GlyT1 is transcribed, whereas for GlyT2 only very weak immunoreactivity has been reported in the hippocampus. GlyT1 protein can be additionally found in the retina of the eye (*ey*) and predominantly in the olfactory bulb (*ob*). Outside of the CNS, both GlyTs are expressed in the pancreas (*pa*). Furthermore, GlyT1 immunoreactivity is found in the liver (*li*) and the gastrointestinal junctions (*gij*)

in a subset of most likely glutamatergic neurons which synapse onto NMDARs (Cubelos et al. 2005). This is consistent with a role of GlyT1 in the regulation of glycine levels at glutamatergic synapses.

GlyT2 is the main neuronal transporter isoform found in caudal regions of the CNS (Zafra et al. 1995). Ultrastructural analysis of GlyT2 localization using immunogold electron microscopy techniques has shown that GlyT2 is enriched in close proximity to, although excluded from, the presynaptic active zone of glycinergic nerve terminals (Mahendrasingam et al. 2003; Zafra et al. 1995). The overall distribution of GlyT2 parallels that of GlyRs. Therefore, GlyT2 is considered a reliable marker of glycinergic terminals (Jursky and Nelson 1995; Poyatos et al. 1997; Zeilhofer et al. 2005). Interestingly, GlyT2 has recently been detected in the hippocampus, a brain region previously thought to be devoid of glycinergic neurotransmission (Danglot et al. 2004). Although glycine-induced currents have not been observed in adult hippocampal neurons, GlyT2 co-localizes with GlyRs and the vesicular inhibitory amino acid transporter (VIAAT), an additional marker of inhibitory synapses (Danglot et al. 2004). Also, expression of GlyT2 outside of the CNS has been demonstrated in A-cells of the islets of Langerhans in the pancreas (Gammelsaeter et al. 2004).

During development, both GlyT genes are expressed already prenatally, with GlyT1 transcripts appearing earlier than those of GlyT2. GlyT1 activity has been observed already in the pre-implantation embryo, where glycine is used as an organic osmolyte which regulates the cell volume at the cleavage stage (Steeves and Baltz 2005). After implantation, GlyT1 immunoreactivity (IR) can be identified at embryonic day 10 (E10) in the midbrain floor plate, with its expression spreading to median parts of the spinal cord at E12 (Jursky and Nelson 1996). The expression of GlyT1 extends during E13–14 to radial glial processes in forebrain regions. An exception is the cerebellum, where GlyT1 is detectable only postnatally, correlating with the late onset of synaptogenesis in this brain region (Jursky and Nelson 1996). At E13, GlyT1 expression is already detectable in tissues outside the CNS, i.e. in liver, pancreas and the gastroduodenal junction (Jursky and Nelson 1996).

Compared with GlyT1, GlyT2 immunoreactivity appears at later stages of development. At E12, parallel to the onset of synaptogenesis, GlyT2 can be first detected immunohistochemically in the white matter of the ventral spinal cord (Jursky and Nelson 1996). This might reflect the enrichment of GlyT2 in growth cones of outgrowing axons during this period (Poyatos et al. 1997). The expression of GlyT2 increases and extends upon subsequent development until GlyT2 is also found in the thalamus and the dorsal spinal cord at the time when all circuits are established in the spinal cord (E16–E17). At birth, the overall pattern of GlyT2 expression is basically established, except for the cerebellum, where the onset of synaptogenesis occurs later. During early postnatal development, the expression of GlyT2 increases further, and GlyT2 immunoreactivity changes to punctate structures in grey matter (Jursky and Nelson 1996). Maximal expression of GlyT2 is reached at postnatal day 14 (P14). Thereafter, expression decreases until P21 when the adult expression levels are established (Friauf et al. 1999).

3.2
Transcriptional Control

Although their expression patterns have been extensively studied and their gene structure is established, little is currently known about the transcriptional regulation of GlyT genes. Analyses of their promoter regions have revealed the presence of sequence elements important for transcriptional control. However, the precise mechanisms that govern the expression of GlyTs during development and in the mature CNS are currently unknown.

It has been shown that GlyT2 expression is controlled by neuronal activity. In neurons from the rat dorsal cochlear nucleus, a controlled increase in synaptic activity by acoustic stimulation evoked a local increase in GlyT2 messenger (m)RNA levels (Barmack et al. 1999). The expression of GlyT1 is also dependent on the surrounding tissue. In primary cultures, glial GlyT1 expression is only induced in mixed cultures consisting of neurons and glia but not in pure glial cultures (Zafra et al. 1997). Upon cytotoxic killing of the neurons, a down-regulation of GlyT1 mRNA has been observed. This transcriptional down-regulation of the GlyT1 gene could be mediated by members of the HMGN nuclear factor family. The HMGN family comprises small, basic proteins that bind specifically to nucleosomes, thereby promoting chromatin unfolding and modulation of transcription. Two splice variants of one member of this family, HMGN3a and HMGN3b, bind to the GlyT1 gene in vivo, up-regulating its expression (West et al. 2004). Moreover, the HMGN3 genes are highly expressed in glia cells and retina, co-localizing with GlyT1. These results suggest that HMGN3s regulate GlyT1 expression. However, it is currently not known whether HMGN3a and HMGN3b are indiscriminate transcriptional regulators, designed to stimulate overall gene expression during a particularly active period such as differentiation, or whether they specifically target selected genes. Another transcriptional regulator known to increase GlyT1 expression is Trb-1, which is a neuron-specific T-box transcription factor. In cultured hippocampal neurons, overexpression of Trb-1 led to an increase in GlyT1 promotor-dependent luciferase expression (Wang et al. 2004). A Trb-1 binding motif, a non-palindromic so-called T element, is found in the 5′-flanking region of the GlyT1 gene. The in vivo role of this element remains to be elucidated.

4
Functional Roles of GlyTs

4.1
Generation of GlyT-Deficient Mice

GlyTs are supposed to have specialized functions at glycinergic inhibitory synapses. Since the localization of GlyT2 parallels the distribution of GlyRs (Jursky and Nelson 1995; Zafra et al. 1995), it was previously thought that GlyT2 was the main isoform mediating the clearance of presynaptically released glycine at glycinergic synapses. However, GlyT1 also might play a significant role in controlling the extracellular concentration of glycine, as suggested by its overlapping expression patterns with GlyT2 in caudal regions of the CNS (Zafra et al. 1995). Consistent with their locations, glial and neuronal glycine transporters might compete in the rapid re-uptake of synaptically released glycine at glycinergic synapses. Alternatively, the GlyTs could have different but complementary roles in controlling the extracellular levels of glycine. To address this issue, we have generated mouse lines lacking functional GlyT1 and GlyT2 and analysed the effects of GlyT inactivation on glycinergic inhibitory neurotransmission (Gomeza et al. 2003a; Gomeza et al. 2003b).

To generate mice lacking functional GlyT1 and GlyT2, we targeted both GlyT genes in mouse E14 (129/OLA) embryonic stem (ES) cells via homologous recombination (Gomeza et al. 2003a, b). The GlyT1 gene was inactivated by replacing its exon 3, which encodes the second transmembrane region of the GlyT1 protein (Adams et al. 1995), with a neomycin-resistance cassette. Likewise, the GlyT2 gene was inactivated by removing the exon encoding the fourth transmembrane domain of the transporter (Liu et al. 1993; Ebihara et al. 2004). Through standard transgenic and mouse breeding techniques, chimeric mice were obtained and used to generate F1 offspring heterozygous for both GlyTs. These animals appeared phenotypically normal and showed undisturbed development and fertility. Intercrossing of the heterozygous mice generated wild-type (+/+) as well as heterozygous (+/−) and homozygous (−/−) knockout (KO) mutant mice at the expected Mendelian ratios, excluding gross defects during embryonic development. The absence of the wild-type GlyT transcripts and proteins in the homozygous KO animals was confirmed by RT-PCR and Western blot analysis, respectively. Membrane fractions prepared from both forebrain and brainstem regions of GlyT1$^{-/-}$ mice exhibited significantly reduced [^{3}H]glycine uptake (Gomeza et al. 2003a). Similarly, GlyT2$^{-/-}$ mice displayed low glycine transport activity in brainstem and spinal cord, whereas no change was observed with membrane fractions prepared from frontal cortex (Gomeza et al. 2003b), where GlyT1 is the major transporter isoform expressed (Jursky and Nelson 1996). Hence, as expected from the normal distribution of both GlyT isoforms, glycine uptake was selectively impaired, consistent with a complete loss of the respective GlyT isoform in the different homozygous mutant mice.

4.2
GlyTs Are Essential for Vital Postnatal Functions

Both homozygous GlyT1$^{-/-}$ and GlyT2$^{-/-}$ mice appear grossly normal at birth. However, loss of either transporter severely shortens the lifespan of the animals. GlyT1$^{-/-}$ mice die on the day of birth, demonstrating that GlyT1 is dispensable for embryonic development but essential for postnatal survival. They fail to suckle, as is obvious from a lack of milk in their stomachs, and show an abnormal body posture with dropping forelimbs (Gomeza et al. 2003a). In addition, they display weak spontaneous motor activity in response to mild tactile stimuli and react poorly to intense pain stimuli. Plethysmographic recordings (Chatonnet et al. 2002) demonstrated that dysfunction of motor activity extends to the respiratory system. Breathing patterns and respiratory frequencies are severely depressed in GlyT1$^{-/-}$ newborn animals, resulting in long periods of apnoea interrupted by gasp-like inspirations instead of the regular breathing characteristic of wild-type animals (Gomeza et al. 2003a). In summary, GlyT1$^{-/-}$ newborn mice display motosensory deficits characterized by lethargy, hypotonia, hypo-responsivity and a severe respiratory deficiency.

GlyT2-deficient mice also show a lethal phenotype, although death occurs only during the second postnatal week and after developing an acute neuromotor disorder (Gomeza et al. 2003b). GlyT2 KO mice display a readily observable tremor in their limbs. Also, when suspended by the tail, they show an abnormal behaviour, clasping their hind feet and holding the front paws together. These are typical symptoms of muscular rigidity (Becker et al. 2000, Hartenstein et al. 1996). Finally, GlyT2$^{-/-}$ mice are unable to right themselves onto all four legs after being turned on their backs, indicating reduced motor co-ordination (Hartenstein et al. 1996). In conclusion, GlyT2-deficient mice display a lethal neuromotor disorder whose symptoms are entirely different from those seen upon GlyT1 deletion, and include tremor, muscular spasticity and impaired motor co-ordination.

The severe phenotypes observed in GlyT1 and GlyT2 KO mice cannot be attributed to anatomical malformation or insufficient synaptic differentiation of the CNS. GlyT KO mice show no histological defects in skeletal muscle or visceral organs. Similarly, no differences were found when examining specific regions of the CNS where GlyTs are expressed, such as the brainstem, spinal cord and forebrain. Western blot and immunohistochemistry analyses revealed no major changes in the expression level, distribution and density of proteins localized in glycinergic and glutamatergic synapses. Further analysis showed that absence of GlyT2 caused no change in GlyT1 expression in GlyT2 KO animals. Likewise, GlyT2 remained unaffected in homozygous GlyT1$^{-/-}$ animals. This demonstrates that ablation of GlyT genes does not cause major adaptive alterations in synapse biochemistry and excludes the possibility that loss of one specific transporter might lead to compensatory changes in the expression of the other GlyT isoform.

4.3
Essential Functions of GlyTs at Glycinergic Synapses

The neuromotor disorders observed in the homozygous KO mice suggested wide-ranging deficits of glycinergic neurotransmission in those animals. To examine the effects of GlyT gene deletion on glycinergic neurotransmission, electrophysiological recordings (Hulsmann et al. 2000) were performed from hypoglossal motoneurons in brainstem slices of GlyT mutant mice. Loss of either transporter induced abnormal glycinergic inhibitory currents in the KO animals. Neurons from GlyT1-deficient mice generated glycinergic spontaneous inhibitory postsynaptic currents (IPSCs) with increased frequencies and longer decay time constants than those in wild-type mice (Gomeza et al. 2003a). In addition, augmented tonic chloride conductances mediated by GlyRs were also recorded. These findings demonstrate hyperactive glycinergic signalling in GlyT1 KO mice and indicate that deletion of GlyT1 causes elevated extracellular glycine concentrations which result in a sustained activation of postsynaptic GlyRs. Furthermore, they suggest a fundamental role of GlyT1 in lowering extracellular glycine levels at inhibitory glycinergic synapses. In contrast, GlyT2-deficient hypoglossal neurons displayed glycinergic IPSCs with markedly reduced amplitudes compared to those from wild-type cells, a finding that revealed reduced glycinergic activity in the KO animals (Gomeza et al. 2003b). As mentioned above, no postsynaptic changes in the distribution and localization of GlyRs were found in the mutant animals. This excludes the possibility that postsynaptic receptor currents are reduced due to malfunction or down-regulation of GlyRs and indicates that glycine release from glycinergic terminals is impaired in the GlyT2 KO mice. Apparently, loss of GlyT2 decreases the cytosolic glycine concentration in the presynaptic terminal, leading to inefficient synaptic vesicle refilling with glycine, and hence reduced transmitter

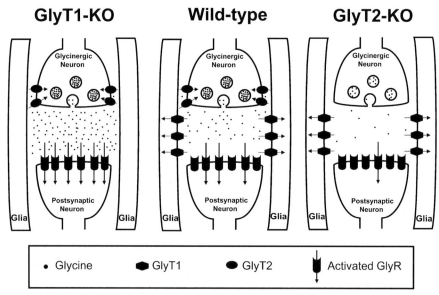

Fig. 3 Models of glycinergic synapse structure in GlyT knockout mice. The scheme depicts glycinergic synapses from GlyT1 knockout (*left*), wild-type (*centre*) and GlyT2 knockout (*right*) mice, respectively. The neuronal and glial location of GlyT2 and GlyT1 are indicated. In the wild-type (*centre*), GlyT2 is essential for glycine uptake into the nerve terminal cytosol, thereby replenishing the supply of presynaptic transmitter. Glycine released from the presynaptic terminal is removed from the synaptic cleft by GlyT1, which terminates glycinergic postsynaptic currents and maintains a low extracellular glycine concentration. In the GlyT1 KO mouse (*left*), the absence of GlyT1 leads to an increased glycine concentration in the synaptic cleft, which results in a sustained activation of postsynaptic GlyRs, and thus a potentiation of glycinergic inhibitory neurotransmission. In the GlyT2-KO mouse (*right*), deletion of GlyT2 results in an impairment of presynaptic vesicle loading with glycine. This leads to a low content of the neurotransmitter within synaptic vesicles. Hence, the amount of glycine released into the synaptic cleft upon stimulation is reduced compared with the wild-type synapse, which in turn results in diminished activation of GlyRs and reduced glycine-mediated neurotransmission

release upon presynaptic stimulation, which results in smaller postsynaptic currents. Together, these data show that GlyT2 does not play an important role in clearing glycine from the synaptic cleft of glycinergic synapses but is essential for glycine uptake into the presynaptic cytosol, and hence glycine recycling.

The perturbation of glycinergic neurotransmission found in GlyT-deficient mice explains the strong phenotype displayed by these animals. Slice recordings of neuronal activity in the GlyT1 KO brainstem circuitry responsible for generating the respiratory rhythm revealed a slowed and irregular pattern that was normalized upon application of the GlyR inhibitor strychnine. This indicates that the motosensory deficits caused by the deletion of GlyT1 are a result of glycinergic over-inhibition induced by the sustained activation of GlyRs in the presence of high levels of extracellular glycine. In contrast, loss of GlyT2 causes severe glycinergic under-inhibition that produces the 'spastic' phenotype observed in GlyT2 KO mice.

In summary, GlyT1 and GlyT2 have different roles at glycinergic synapses (Fig. 3): GlyT1 eliminates glycine from the synaptic cleft, thus terminating glycine neurotransmission and maintaining a low extracellular glycine concentration throughout the caudal regions of the CNS. This prevents excessive tonic activation of GlyRs. On the other hand, GlyT2 is required for replenishing the pool of cytosolic glycine from which vesicles are filled for release. Thus, both GlyTs have complementary functions at glycinergic synapses, which fine-tune the efficacy of glycinergic inhibition in the mammalian CNS.

4.4
Pharmacology of GlyTs

Important functions of different neurotransmitter transporters have been disclosed by pharmacological approaches, and the identification of potent transport inhibitors has contributed essentially to our recent understanding of their in vivo functions. Furthermore, transporters are known to constitute important drug targets. Widely used drugs, which act on neurotransmitter transporters, include the tricyclic antidepressants, which block transporters specific for serotonin and norepinephrine (Blakely et al. 1994), and anticonvulsants such as tiagabine, which selectively inhibits the neuronal GABA transporter GAT1 (Ashton and Young 2003).

Also in case of the GlyTs, the initial hypothesis that two different transport systems exist emerged from differences in susceptibility to N-methyl-glycine (sarcosine). Sarcosine competitively inhibits GlyT1, but not GlyT2 (Lopez-Corcuera et al. 1998). Electrophysiological and radioactive tracer experiments revealed that this glycine analogue acts as a substrate, thereby competing with glycine for GlyT1 binding. Further substitutions at the amino group of sarcosine with lipophilic heterocycles has led to several novel high-affinity blockers of GlyT1, like ALX-5407, its stereoisomer (R)-N-[3-(4'-fluorophenyl)-3-(4'-

phenylphenoxy)propyl]sarcosine ((*R*)-NFPS) (Aubrey and Vandenberg 2001), (3-(4-chloro-phenyl)-3[4-(thiazole-2-carbonyl)-phenoxy]-propyl-methyl-amino)-acetic acid) (CP 802,079) (Martina et al. 2004) and (*R,S*)-(±)-*N*-methyl-*N*-[(4-trifluoromethyl)phenoxy]-3-phenyl-propyl-glycine (Org-24461) (Harsing et al. 2003; Fig. 4). Although the mode of action of these compounds is not fully understood, these inhibitors have been shown to act as noncompetitive inhibitors, which do not change the accessibility of the Na$^+$- and Cl$^-$-binding sites (Mallorga et al. 2003). Moreover, these substances appear to be very potent, as glycine uptake by GlyT1 expressed in heterologous systems or in synaptosomal preparations was irreversibly inhibited for up to 1 h.

Initially reported inhibitors of GlyT2 comprise the antidepressant amoxapine and several alkanols, including ethanol (Nunez et al. 2000a; Nunez et al. 2000b). Due to their low affinity and selectivity for GlyT2, these compound are not suitable for pharmacological studies. The first inhibitors of high-affinity and specificity for GlyT2 were generated recently. To date, several organic compounds, including 4-benzyloxy-3,5-dimethoxy-*N*-[(1-di-methylaminoacyclopentyl)methyl]-benzamide (Ho et al. 2004), a series of 5,5-diaryl-2-amino-4-pentenoates (Isaac et al. 2001) and 4-benzyloxy-3,5-dimethoxy-*N*[(1-dimethylaminocyclopentyl)methyl]benzamide (Org-25543) (Caulfield et al. 2001), have been found to potently inhibit GlyT2 without affecting the activity of GlyT1, GlyR or NMDAR (Fig. 4). All these inhibitors constitute new tools for the more detailed in vivo and in vitro analysis of GlyT functions.

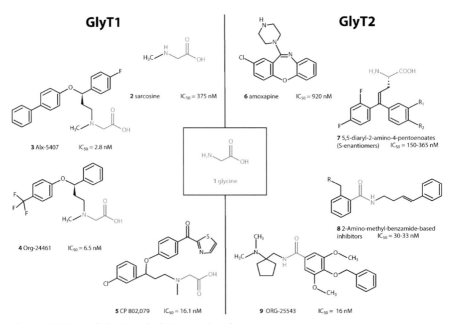

Fig. 4 Inhibitors of GlyT1 and GlyT2. For details, see text

4.5
Function of GlyTs at Glutamatergic Synapses

4.5.1
Pharmacological Studies In Vitro

As described above, glycine not only constitutes a major inhibitory neurotransmitter but is also essential for glutamatergic neurotransmission by serving as a co-agonist of the NMDAR. It is now widely accepted that the glycine concentration at glutamatergic synapses is close to the K_m of the glycine binding site of the NMDARs, thereby allowing for modulation of NMDAR activity by small changes in the ambient glycine concentration. Since glycine is not co-released with glutamate from synaptic vesicles, the source of glycine at excitatory synapses is at present unclear. In brain areas which are rich in glycinergic synapses, i.e. brainstem and spinal cord, glycine might reach NMDARs through spill-over from highly active neighbouring glycinergic synapses (Ahmadi et al. 2003). In these locations, insufficient clearance of glycine from the synaptic cleft (or small distances between excitatory and inhibitory synapses) might allow for diffusion of the neurotransmitter to adjacent glutamatergic synapses. However, such mechanisms do not seem to play a role in higher brain regions lacking strong glycinergic neurotransmission, where GlyTs might then have a pivotal role as regulators of glutamatergic neurotransmission by modulating synaptic glycine concentration.

Supporting evidence for this hypothesis came from studies using the GlyT-specific inhibitors Org-24958 and Org-25543, which specifically block GlyT1 and GlyT2, respectively (Bradaia et al. 2004). Application of these compounds led to a facilitation of NMDAR activity in spinal cord slice preparations. Comparable results were obtained in recordings performed on forebrain neurons (i.e. cortex or hippocampus), where the application of low doses of the GlyT1 inhibitor CP 802,079 caused a strong facilitation of the NMDAR component of glutamatergic neurotransmission (Martina et al. 2004). This included some forms of synaptic plasticity such as long-term potentiation (LTP), which has been shown to be NMDAR-dependent. Strikingly, the application of high concentrations of GlyT1 inhibitors or glycine only produces a transient facilitation of the NMDAR currents. At later time points, however, this effect is inversed (Martina et al. 2004). This has been attributed to priming of NMDARs for internalization and their subsequent removal from the cell membrane by clathrin-mediated endocytosis (Nong et al. 2003). In summary, these findings support the assumption that GlyT1 controls extracellular glycine levels at glutamatergic synapses, resulting in modulatory effects on NMDAR function.

4.5.2
Pharmacology and Genetics In Vivo

Acute in vivo application by reverse dialysis or systemic application of GlyT1 inhibitors like sarcosine, NFPS or ALX5407 has been shown to cause a significant increase of the glycine concentration in the cerebrospinal fluid, whereas the concentration of other amino acids was not changed (Martina et al. 2004; Whitehead et al. 2004). Electrophysiological recordings showed, comparable to the in vitro experiments, increased NMDAR-mediated glutamatergic currents (Lim et al. 2004). Interestingly, the concentration of citrulline was also remarkably increased upon the application of GlyT1 inhibitors (Whitehead et al. 2004). At synapses, citrulline can be generated as a stoichiometric byproduct of NO synthesis by neuronal NO synthase. The activity of this enzyme has been shown to be dependent on NMDAR-mediated Ca^{2+}-influx (Brenman and Bredt 1997); the elevated levels of citrulline thus provide further evidence for an increased NMDAR activity, resulting from GlyT1 inhibition.

Subsequent studies of heterozygous GlyT1-deficient animals also indicate a role of GlyT1 in NMDAR regulation. Membrane preparations from GlyT1$^{+/-}$ animals display approximately 50% reduced [^{3}H]glycine uptake activity compared to wild-type samples (Gabernet et al. 2005; Gomeza et al. 2003a; Tsai et al. 2004). Thus, the loss of one functional GlyT1 allele is not compensated by up-regulation of the second allele. In contrast to what is seen in wild-type animals, the glycine binding site of NMDARs seems to be saturated in GlyT1$^{+/-}$ mice, because application of exogenous glycine or D-serine did not lead to a facilitation of NMDAR-mediated currents (Gabernet et al. 2005; Tsai et al. 2004). The constitutive saturation of the NMDAR glycine binding site results in an increase in the ratio of NMDAR/AMPAR (α-amino-3-hydroxy-5-methyl-4-isoxazole propionic acid receptor)-mediated currents. However, whether this effect is due to a stronger NMDAR response or a reduction of the AMPAR component of glutamatergic neurotransmission is an unsolved issue.

Kinetic analysis of spontaneous and evoked synaptic events in GlyT1$^{+/-}$ animals revealed faster deactivation time constants of the NMDAR currents compared with their wild-type litters (Martina et al. 2005), suggesting a different subunit composition of NMDARs. This hypothesis received support from an altered susceptibility to the NMDAR modulators ifenprodil and zinc, which inhibit NMDARs containing NR2B or NR2A subunits, respectively. Together, these results demonstrate that constitutive saturation of the glycine binding site of the NMDAR leads to compensatory changes at glutamatergic synapses.

Although heterozygous GlyT1-deficient animals do not display any obvious phenotype (Gabernet et al. 2005; Gomeza et al. 2003a; Tsai et al. 2004), the behavioural analysis of GlyT1$^{+/-}$ animals revealed additional differences when compared with wild-type litters. Assessment of spatial orientation in the Morris water maze revealed a better performance of the mutant mice (Tsai et al. 2004). Additionally prepulse inhibition (PPI), the reduction of the sound-induced startle response by pre-exposition to a milder acoustic stimulus, was tested. Although PPI was not changed in GlyT$^{+/-}$ animals, Tsai et al. demonstrated changes in the pharmacology of this behavioural paradigm (Tsai et al. 2004). In GlyT1$^{+/-}$ animals, PPI was partially resistant against disruption with amphetamines, but displayed an increased susceptibility for the NMDAR open channel blocker MK801. In addition, a recent study has reported that in mice the application of the GlyT1 inhibitor ALX 5407 caused different effects depending on dosage (Lipina et al. 2005). A low concentration of the inhibitor was able to restore the disruption of PPI induced by subsaturating concentrations of MK801, thus mimicking the results obtained with mice displaying reduced GlyT1 activity. The application of high concentrations, however, resulted in a reduction of PPI. Comparable results were obtained in electrophysiological records in which high concentrations of the GlyT1 inhibitors reduced the NMDAR current amplitude (Martina et al. 2004). This is consistent with a removal of NMDARs from synaptic sites. Taken together, these data further underline the importance of GlyT1 in glutamatergic neurotransmission. Here, lowering of the extracellular glycine concentration by GlyT1 does not only allow a glycine-mediated potentiation of NMDAR activity, but additionally prevents NMDARs from internalization.

5
GlyTs and Human Diseases

5.1
GlyT Genes: Candidate Disease Loci?

The symptoms observed in GlyT1- and GlyT2-deficient mice are similar to those associated with human hereditary diseases which develop in early postnatal life or during adolescence. Hyperglycinergic GlyT1 KO mice present symptoms similar to glycine encephalopathy or non-ketotic hyperglycinaemia (NKH) (Applegarth and Toone 2001). This disorder is characterized by muscular hypotonia, lethargy and poor feeding with hiccups (Boneh et al. 1996; Tada et al. 1992). NKH can rapidly progress to a lethal symptomatology associated with respiratory insufficiency, apnoea, coma and death. Often patients with NKH have a primary defect in the glycine cleavage system (GCS) (Applegarth and Toone 2001). The GCS consists of four protein components (named as P-, H-, T- and L-proteins) (Sakata et al. 2001), which are located in the inner mitochondrial membrane and catalyse the degradation of intracellular glycine. As a consequence, the affected individuals accumulate elevated levels of glycine in the blood and, particularly, in cerebrospinal fluid. The high glycine concentration is supposed to cause potent activation of GlyRs, thus leading to symptoms of glycinergic over-inhibition. On the other hand, hypoglycinergic GlyT2 KO animals resemble patients suffering from hyperekplexia (Becker et al. 2000;

Zhou et al. 2002). This neuromotor disorder is characterized by exaggerated startle responses and, in severe cases, a 'stiff baby syndrome'. Some dominant and recessive forms of hyperekplexia are known to be associated with mutations in the GlyR α1- and β-subunit genes that impair postsynaptic GlyR function (Laube et al. 2002; Lynch 2004) and cause a phenotype characterized by spasticity, muscular rigidity and tremor.

Although neither of the human GlyT genes has been linked to a disease phenotype, it is noteworthy that about 50% of the patients diagnosed with NKH or hyperekplexia carry no mutations in GCS or GlyR genes (Applegarth and Toone 2001; Vergouwe et al. 1997). This indicates that glycine encephalopathy- and hyperekplexia-like syndromes may be caused through other genetic mechanisms, and it is consistent with the idea that mutations in the human GlyT gene might cause such hereditary neurological disorders. If so, GlyT-deficient mice should constitute valuable animal models to analyse the underlying pathomechanisms.

5.2
GlyTs as Potential Drug Targets

The facilitation of the NMDAR component of glutamatergic neurotransmission upon acute application of GlyT1 inhibitors identified this transporter as a prime pharmacological target for diseases thought to be associated with NMDAR hypofunction, such as schizophrenia (Jentsch and Roth 1999). This widespread psychiatric disorder is characterized by hyperactivity, cognitive deficits and stereotyped behaviour. Partial NMDAR inhibitors like ketamine or phenylcyclidine (PCP) are known to induce some of the symptomatology characteristic for schizophrenia in healthy humans (Jentsch and Roth 1999). In rodents, these compounds cause hyperactivity and stereotyped behaviour, as well as impaired spatial orientation. Therefore, ketamine- and PCP-treated animals are thought to constitute useful pharmacological models of schizophrenia. Notably, in these animals both the application of glycine and the (partial) inhibition of GlyT1 by its antagonists have positive effects on most symptoms (Javitt 2002; Javitt et al. 2003; Kinney et al. 2003; Sur and Kinney 2004). These animal studies have recently been extended to clinical trials by using a diet containing a high glycine content or the GlyT1 substrate analogue sarcosine (Javitt 2002). Both strategies, in combination with classical medication, seem to have a beneficial effect on the symptoms in human schizophrenic patients.

The use of high-affinity GlyT1 inhibitors like NFPS (N-[3-(4'-fluorophenyl)-3-(4'-phenylphenoxy)propyl]sarcosine) as therapeutic compounds is controversial. Up to now, possible side-effects have prevented clinical trials. First, a strong increase in the extracellular glycine concentration might lead to a facilitation of glycinergic neurotransmission, which could result in severe side-effects such as respiratory suppression. Second, increased glycine concentrations at glutamatergic synapses might result in a persistent saturation of the glycine binding site of NMDAR. In heterozygous GlyT1-deficient mice, which seem to be a good genetic model for the long-term application of GlyT1 inhibitors, the chronic saturation of the glycine binding site of the NMDAR apparently led to adaptive changes in the electrophysiological properties of glutamatergic synapses (Martina et al. 2005). Although GlyT1$^{+/-}$ animals do not display any obvious phenotypic changes (Gabernet et al. 2005; Gomeza et al. 2003a; Tsai et al. 2004), this does not exclude negative effects in humans. Furthermore, in rodents the application of high concentrations of GlyT1 inhibitors induces hyperactivity and strong stereotyped behaviour in vivo, symptoms that are also seen after the administration of partial NMDAR inhibitors (Lipina et al. 2005). Together with studies showing a reduction in NMDAR amplitude upon GlyT1 inhibition (Martina et al. 2004), these data suggest that a strong inhibition of GlyT1 could result in a down-regulation of NMDARs. Thus, overmedication with GlyT1 inhibitors might boost the disease phenotype.

Data obtained in GlyT1-deficient mice, as well as pharmacological studies, indicate that GlyT1 inhibitors also have the potential to increase the efficacy of inhibitory neurotransmission (Bradaia et al. 2004; Gomeza et al. 2003a). This increased inhibitory tone in spinal circuitries should reduce motoneuron activity and decrease pain perception. Therefore, the local application of GlyT1 inhibitors in spinal cord might be useful for muscle relaxation and analgesia during narcosis. Whether a partial inhibition of GlyT2, which transiently may also increase the activity of postsynaptic GlyRs (Bradaia et al. 2004), could also be beneficial is not yet clear.

6
Conclusions and Perspectives

In summary, the use of newly developed antagonists and the generation of mouse models have greatly improved our understanding of the precise functions of GlyTs in the CNS. At inhibitory glycinergic synapses, GlyT1 is essential for the rapid removal of the neurotransmitter from the synaptic cleft, thereby terminating neurotransmission via GlyRs. Inhibition or the loss of GlyT1 leads to an accumulation of glycine in the synaptic cleft, resulting in overinhibition. Furthermore, GlyT1 prevents the saturation of the glycine binding site of NMDARs, thereby enabling glycine potentiation of glutamatergic neurotransmission. Inactivation of a single GlyT1 allele leads to full saturation of the NMDAR glycine binding site, which then results in major changes in the physiology of glutamatergic synapses. This modulatory function of GlyT1 makes it a highly promising drug target in psychiatric diseases such as schizophrenia, for which no effective medication is currently available. However, the possible side-effects, which may arise from the dual functions of GlyT1 at inhibitory and excitatory synapses, as well as its functions in non-neuronal tissues, will have to be examined carefully. In contrast, the major function of GlyT2 seems the reuptake of neurotransmitter into the inhibitory presynaptic terminals to allow efficient refilling of synaptic vesicles with glycine. The loss of GlyT2 activity results in a strong reduction of glycinergic inhibition and causes a severe hyperekplexia phenotype.

An area of research that will have to be extended in future studies concerns the potential role of GlyT1-mediated glycine release from astrocytes or even neurons. Such studies are essential for further understanding of the functional role of these transporters. Furthermore, although the analysis of GlyT-deficient mice has already provided important details on GlyT functions, more sophisticated genetic models will be required to dissect the variant- and region-specific tasks of GlyTs. These approaches need to overcome the early lethality of GlyT deficiency, a prerequisite for studying GlyT functions in the adult nervous system.

Acknowledgements Work in the authors' laboratories has been supported by grants from the Deutsche Forschungsgemeinschaft (SFB 269 and SPP 1172), European Community (TMR ERBFMRXCT9), Fonds der chemischen Industrie, Lundbeck Foundation, Novo Nordisk Foundation, Carlsberg Foundation and the Danish Medical Research Council.

References

Adams RH, Sato K, Shimada S, Tohyama M, Puschel AW, Betz H (1995) Gene structure and glial expression of the glycine transporter GlyT1 in embryonic and adult rodents. J Neurosci 15:2524–2532

Ahmadi S, Muth-Selbach U, Lauterbach A, Lipfert P, Neuhuber WL, Zeilhofer HU (2003) Facilitation of spinal NMDA receptor currents by spillover of synaptically released glycine. Science 300:2094–2097

Applegarth DA, Toone JR (2001) Nonketotic hyperglycinemia (glycine encephalopathy): laboratory diagnosis. Mol Genet Metab 74:139–146

Ashton H, Young AH (2003) GABA-ergic drugs: exit stage left, enter stage right. J Psychopharmacol 17:174–178

Aubrey KR, Vandenberg RJ (2001) N[3-(4′-fluorophenyl)-3-(4′-phenylphenoxy)propyl]sarcosine (NFPS) is a selective persistent inhibitor of glycine transport. Br J Pharmacol 134:1429–1436

Barmack NH, Guo H, Kim HJ, Qian H, Qian Z (1999) Neuronally modulated transcription of a glycine transporter in rat dorsal cochlear nucleus and nucleus of the medial trapezoid body. J Comp Neurol 415:175–188

Becker L, Hartenstein B, Schenkel J, Kuhse J, Betz H, Weiher H (2000) Transient neuromotor phenotype in transgenic spastic mice expressing low levels of glycine receptor beta-subunit: an animal model of startle disease. Eur J Neurosci 12:27–32

Berger AJ, Dieudonne S, Ascher P (1998) Glycine uptake governs glycine site occupancy at NMDA receptors of excitatory synapses. J Neurophysiol 80:3336–3340

Betz H, Harvey RJ, Schloss P (2000) Structures, diversity and pharmacology of glycine receptors and transporters. In: Möhler H (ed) Pharmacology of GABA and glycine neurotransmission. Springer-Verlag, Heidelberg Berlin New York, pp 375–401

Blakely RD, De Felice LJ, Hartzell HC (1994) Molecular physiology of norepinephrine and serotonin transporters. J Exp Biol 196:263–281

Boneh A, Degani Y, Harari M (1996) Prognostic clues and outcome of early treatment of nonketotic hyperglycinemia. Pediatr Neurol 15:137–141

Borowsky B, Hoffman BJ (1998) Analysis of a gene encoding two glycine transporter variants reveals alternative promoter usage and a novel gene structure. J Biol Chem 273:29077–29085

Borowsky B, Mezey E, Hoffman BJ (1993) Two glycine transporter variants with distinct localization in the CNS and peripheral tissues are encoded by a common gene. Neuron 10:851–863

Bradaia A, Schlichter R, Trousard J (2004) Role of glial and neuronal glycine transporters in the control of glycinergic and glutamatergic synaptic transmission in lamina X of the rat spinal cord. J Physiol 559:169–186

Brenman JE, Bredt DS (1997) Synaptic signaling by nitric oxide. Curr Opin Neurobiol 7:374–378

Caulfield WL, Collie IT, Dickins RS, Epemolu O, McGuire R, Hill DR, McVey G, Morphy JR, Rankovic Z, Sundaram H (2001) The first potent and selective inhibitors of the glycine transporter type 2. J Med Chem 44:2679–2682

Chatonnet F, del Toro ED, Voiculescu O, Charnay P, Champagnat J (2002) Different respiratory control systems are affected in homozygous and heterozygous kreisler mutant mice. Eur J Neurosci 15:684–692

Cubelos B, Gimenez C, Zafra F (2005) Localization of the GLYT1 Glycine Transporter at Glutamatergic Synapses in the Rat Brain. Cereb Cortex 15:448–459

Danglot L, Rostaing P, Triller A, Bessis A (2004) Morphologically identified glycinergic synapses in the hippocampus. Mol Cell Neurosci 27:394–403

Ebihara S, Yamamoto T, Obata K, Yanagawa Y (2004) Gene structure and alternative splicing of the mouse glycine transporter type-2. Biochem Biophys Res Commun 317:857–864

Ellory JC, Jones SE, Young JD (1981) Glycine transport in human erythrocytes. J Physiol 320:403–422

Eulenburg V, Armsen W, Betz H, Gomeza J (2005) Glycine transporters: essential regulators of neurotransmission. Trends Biochem Sci 30:325–333

Fornes A, Nunez E, Aragon C, Lopez-Corcuera B (2004) The second intracellular loop of the glycine transporter 2 contains crucial residues for glycine transport and phorbol ester-induced regulation. J Biol Chem 279:22934–22943

Friauf E, Aragon C, Lohrke S, Westenfelder B, Zafra F (1999) Developmental expression of the glycine transporter GLYT2 in the auditory system of rats suggests involvement in synapse maturation. J Comp Neurol 412:17–37

Gabernet L, Pauly-Evers M, Schwerdel C, Lentz M, Bluethmann H, Vogt K, Alberati D, Mohler H, Boison D (2005) Enhancement of the NMDA receptor function by reduction of glycine transporter-1 expression. Neurosci Lett 373:79–84

Gammelsaeter R, Froyland M, Aragon C, Danbolt NC, Fortin D, Storm-Mathisen J, Davanger S, Gundersen V (2004) Glycine, GABA and their transporters in pancreatic islets of Langerhans: evidence for a paracrine transmitter interplay. J Cell Sci 117:3749–3758

Geerlings A, Lopez-Corcuera B, Aragon C (2000) Characterization of the interactions between the glycine transporters GLYT1 and GLYT2 and the SNARE protein syntaxin 1A. FEBS Lett 470:51–54

Geerlings A, Nunez E, Lopez-Corcuera B, Aragon C (2001) Calcium- and syntaxin 1-mediated trafficking of the neuronal glycine transporter GLYT2. J Biol Chem 276:17584–17590

Gomeza J, Hulsmann S, Ohno K, Eulenburg V, Szoke K, Richter D, Betz H (2003a) Inactivation of the glycine transporter 1 gene discloses vital role of glial glycine uptake in glycinergic inhibition. Neuron 40:785–796

Gomeza J, Ohno K, Hulsmann S, Armsen W, Eulenburg V, Richter DW, Laube B, Betz H (2003b) Deletion of the mouse glycine transporter 2 results in a hyperekplexia phenotype and postnatal lethality. Neuron 40:797–806

Guastella J, Brecha N, Weigmann C, Lester HA, Davidson N (1992) Cloning, expression, and localization of a rat brain high-affinity glycine transporter. Proc Natl Acad Sci U S A 89:7189–7193

Hanley JG, Jones EM, Moss SJ (2000) GABA receptor rho1 subunit interacts with a novel splice variant of the glycine transporter, GLYT-1. J Biol Chem 275:840–846

Harsing LG Jr, Gacsalyi I, Szabo G, Schmidt E, Sziray N, Sebban C, Tesolin-Decros B, Matyus P, Egyed A, Spedding M, Levay G (2003) The glycine transporter-1 inhibitors NFPS and Org 24461: a pharmacological study. Pharmacol Biochem Behav 74:811–825

Hartenstein B, Schenkel J, Kuhse J, Besenbeck B, Kling C, Becker CM, Betz H, Weiher H (1996) Low level expression of glycine receptor beta subunit transgene is sufficient for phenotype correction in spastic mice. EMBO J 15:1275–1282

Ho KK, Appell KC, Baldwin JJ, Bohnstedt AC, Dong G, Guo T, Horlick R, Islam KR, Kultgen SG, Masterson CM, McDonald E, McMillan K, Morphy JR, Rankovic Z, Sundaram H, Webb M (2004) 2-(Aminomethyl)-benzamide-based glycine transporter type-2 inhibitors. Bioorg Med Chem Lett 14:545–548

Hollmann M (1999) Structure of ionotropic glutamate receptors. In: Jonas P, Monyer H (eds) Ionotropic glutamate receptors in the CNS (Handbook of experimental pharmacology). Springer-Verlag, Heidelberg Berlin New York, pp 1–98

Horiuchi M, Loebrich S, Brandstaetter JH, Kneussel M, Betz H (2005) Cellular localization and subcellular distribution of Unc-33-like protein 6, a brain-specific protein of the collapsin response mediator protein family that interacts with the neuronal glycine transporter 2. J Neurochem 94:307–315

Hubner CA, Stein V, Hermans-Borgmeyer I, Meyer T, Ballanyi K, Jentsch TJ (2001) Disruption of KCC2 reveals an essential role of K-Cl cotransport already in early synaptic inhibition. Neuron 30:515–524

Hulsmann S, Oku Y, Zhang W, Richter DW (2000) Metabolic coupling between glia and neurons is necessary for maintaining respiratory activity in transverse medullary slices of neonatal mouse. Eur J Neurosci 12:856–862

Isaac M, Slassi A, Silva KD, Arora J, MacLean N, Hung B, McCallum K (2001) 5,5-Diaryl-2-amino-4-pentenoates as novel, potent, and selective glycine transporter type-2 reuptake inhibitors. Bioorg Med Chem Lett 11:1371–1373

Javitt DC (2002) Glycine modulators in schizophrenia. Curr Opin Investig Drugs 3:1067–1072

Javitt DC, Balla A, Burch S, Suckow R, Xie S, Sershen H (2003) Reversal of phencyclidine-induced dopaminergic dysregulation by N-methyl-D-aspartate receptor/glycine-site agonists. Neuropsychopharmacology 29:300–307

Jentsch JD, Roth RH (1999) The neuropsychopharmacology of phencyclidine: from NMDA receptor hypofunction to the dopamine hypothesis of schizophrenia. Neuropsychopharmacology 20:201–225

Johnson JW, Ascher P (1987) Glycine potentiates the NMDA response in cultured mouse brain neurons. Nature 325:529–531

Jones EM, Fernald A, Bell GI, Le Beau MM (1995) Assignment of SLC6A9 to human chromosome band 1p33 by in situ hybridization. Cytogenet Cell Genet 71:211

Jursky F, Nelson N (1995) Localization of glycine neurotransmitter transporter (GLYT2) reveals correlation with the distribution of glycine receptor. J Neurochem 64:1026–1033

Jursky F, Nelson N (1996) Developmental expression of the glycine transporters GLYT1 and GLYT2 in mouse brain. J Neurochem 67:336–344

Kim KM, Kingsmore SF, Han H, Yang-Feng TL, Godinot N, Seldin MF, Caron MG, Giros B (1994) Cloning of the human glycine transporter type 1: molecular and pharmacological characterization of novel isoform variants and chromosomal localization of the gene in the human and mouse genomes. Mol Pharmacol 45:608–617

Kinney GG, Sur C, Burno M, Mallorga PJ, Williams JB, Figueroa DJ, Wittmann M, Lemaire W, Conn PJ (2003) The glycine transporter type 1 inhibitor N-[3-(4'-fluorophenyl)-3-(4'-phenylphenoxy)propyl]sarcosine potentiates NMDA receptor-mediated responses in vivo and produces an antipsychotic profile in rodent behavior. J Neurosci 23:7586–7591

Laube B, Kuhse J, Betz H (1998) Evidence for a tetrameric structure of recombinant NMDA receptors. J Neurosci 18:2954–2961

Laube B, Maksay G, Schemm R, Betz H (2002) Modulation of glycine receptor function: a novel approach for therapeutic intervention at inhibitory synapses? Trends Pharmacol Sci 23:519–527

Legendre P (2001) The glycinergic inhibitory synapse. Cell Mol Life Sci 58:760–793

Lim R, Hoang PD, Berger AJ (2004) Blockade of glycine transporter-1 (GLYT-1) potentiates NMDA-receptor mediated synaptic transmission in hypoglossal motoneurons. J Neurophysiol 92:2530–2537

Lipina T, Labrie V, Weiner I, Roder J (2005) Modulators of the glycine site on NMDA receptors, D: -serine and ALX 5407, display similar beneficial effects to clozapine in mouse models of schizophrenia. Psychopharmacology (Berl) 179:54–67

Liu QR, Nelson H, Mandiyan S, Lopez-Corcuera B, Nelson N (1992) Cloning and expression of a glycine transporter from mouse brain. FEBS Lett 305:110–114

Liu QR, Lopez-Corcuera B, Mandiyan S, Nelson H, Nelson N (1993) Cloning and expression of a spinal cord- and brain-specific glycine transporter with novel structural features. J Biol Chem 268:22802–22808

Lopez-Corcuera B, Aragon C (1989) Solubilization and reconstitution of the sodium-and-chloride-coupled glycine transporter from rat spinal cord. Eur J Biochem 181:519–524

Lopez-Corcuera B, Vazquez J, Aragon C (1991) Purification of the sodium- and chloride-coupled glycine transporter from central nervous system. J Biol Chem 266:24809–24814

Lopez-Corcuera B, Martinez-Maza R, Nunez E, Roux M, Supplisson S, Aragon C (1998) Differential properties of two stably expressed brain-specific glycine transporters. J Neurochem 71:2211–2219

Lopez-Corcuera B, Nunez E, Martinez-Maza R, Geerlings A, Aragon C (2001) Substrate-induced conformational changes of extracellular loop 1 in the glycine transporter GLYT2. J Biol Chem 276:43463–43470

Luque JM, Nelson N, Richards JG (1995) Cellular expression of glycine transporter 2 messenger RNA exclusively in rat hindbrain and spinal cord. Neuroscience 64:525–535

Lynch JW (2004) Molecular structure and function of the glycine receptor chloride channel. Physiol Rev 84:1051–1095

Mahendrasingam S, Wallam CA, Hackney CM (2003) Two approaches to double post-embedding immunogold labeling of freeze-substituted tissue embedded in low temperature Lowicryl HM20 resin. Brain Res Brain Res Protoc 11:134–141

Mallorga PJ, Williams JB, Jacobson M, Marques R, Chaudhary A, Conn PJ, Pettibone DJ, Sur C (2003) Pharmacology and expression analysis of glycine transporter GlyT1 with [3H]-(N-[3-(4'-fluorophenyl)-3-(4'phenylphenoxy)propyl])sarcosine. Neuropharmacology 45:585–593

Martina M, Gorfinkel Y, Halman S, Lowe JA, Periyalwar P, Schmidt CJ, Bergeron R (2004) Glycine transporter type 1 blockade changes NMDA receptor-mediated responses and LTP in hippocampal CA1 pyramidal cells by altering extracellular glycine levels. J Physiol 557:489–500

Martina M, ME BT, Halman S, Tsai G, Tiberi M, Coyle JT, Bergeron R (2005) Reduced glycine transporter type 1 expression leads to major changes in glutamatergic neurotransmission of CA1 hippocampal neurones in mice. J Physiol 563:777–793

Martinez-Maza R, Poyatos I, Lopez-Corcuera B, E Nu, Gimenez C, Zafra F, Aragon C (2001) The role of N-glycosylation in transport to the plasma membrane and sorting of the neuronal glycine transporter GLYT2. J Biol Chem 276:2168–2173

Morrow JA, Collie IT, Dunbar DR, Walker GB, Shahid M, Hill DR (1998) Molecular cloning and functional expression of the human glycine transporter GlyT2 and chromosomal localisation of the gene in the human genome. FEBS Lett 439:334–340

Nelson N (1998) The family of Na^+/Cl^- neurotransmitter transporters. J Neurochem 71:1785–1803

Nishimura T, Fukata Y, Kato K, Yamaguchi T, Matsuura Y, Kamiguchi H, Kaibuchi K (2003) CRMP-2 regulates polarized Numb-mediated endocytosis for axon growth. Nat Cell Biol 5:819–826

Nong Y, Huang YQ, Ju W, Kalia LV, Ahmadian G, Wang YT, Salter MW (2003) Glycine binding primes NMDA receptor internalization. Nature 422:302–307

Nunez E, Aragon C (1994) Structural analysis and functional role of the carbohydrate component of glycine transporter. J Biol Chem 269:16920–16924

Nunez E, Lopez-Corcuera B, Martinez-Maza R, Aragon C (2000a) Differential effects of ethanol on glycine uptake mediated by the recombinant GLYT1 and GLYT2 glycine transporters. Br J Pharmacol 129:802–810

Nunez E, Lopez-Corcuera B, Vazquez J, Gimenez C, Aragon C (2000b) Differential effects of the tricyclic antidepressant amoxapine on glycine uptake mediated by the recombinant GLYT1 and GLYT2 glycine transporters. Br J Pharmacol 129:200–206

Ohno K, Koroll M, El Far O, Scholze P, Gomeza J, Betz H (2004) The neuronal glycine transporter 2 interacts with the PDZ domain protein syntenin-1. Mol Cell Neurosci 26:518–529

Olivares L, Aragon C, Gimenez C, Zafra F (1994) Carboxyl terminus of the glycine transporter GLYT1 is necessary for correct processing of the protein. J Biol Chem 269:28400–28404

Olivares L, Aragon C, Gimenez C, Zafra F (1995) The role of N-glycosylation in the targeting and activity of the GLYT1 glycine transporter. J Biol Chem 270:9437–9442

Olivares L, Aragon C, Gimenez C, Zafra F (1997) Analysis of the transmembrane topology of the glycine transporter GLYT1. J Biol Chem 272:1211–1217

Ponce J, Biton B, Benavides J, Avenet P, Aragon C (2000) Transmembrane domain III plays an important role in ion binding and permeation in the glycine transporter GLYT2. J Biol Chem 275:13856–13862

Pow DV, Hendrickson AE (1999) Distribution of the glycine transporter glyt-1 in mammalian and nonmammalian retinae. Vis Neurosci 16:231–239

Poyatos I, Ponce J, Aragon C, Gimenez C, Zafra F (1997) The glycine transporter GLYT2 is a reliable marker for glycine-immunoreactive neurons. Brain Res Mol Brain Res 49:63–70

Poyatos I, Ruberti F, Martinez-Maza R, Gimenez C, Dotti CG, Zafra F (2000) Polarized distribution of glycine transporter isoforms in epithelial and neuronal cells. Mol Cell Neurosci 15:99–111

Rivera C, Voipio J, Payne JA, Ruusuvuori E, Lahtinen H, Lamsa K, Pirvola U, Saarma M, Kaila K (1999) The K^+/Cl^- co-transporter KCC2 renders GABA hyperpolarizing during neuronal maturation. Nature 397:251–255

Roux MJ, Supplisson S (2000) Neuronal and glial glycine transporters have different stoichiometries. Neuron 25:373–383

Rudnick G (1998) Bioenergetics of neurotransmitter transport. J Bioenerg Biomembr 30:173–185

Sakata K, Sato K, Schloss P, Betz H, Shimada S, Tohyama M (1997) Characterization of glycine release mediated by glycine transporter 1 stably expressed in HEK-293 cells. Brain Res Mol Brain Res 49:89–94

Sakata Y, Owada Y, Sato K, Kojima K, Hisanaga K, Shinka T, Suzuki Y, Aoki Y, Satoh J, Kondo H, Matsubara Y, Kure S (2001) Structure and expression of the glycine cleavage system in rat central nervous system. Brain Res Mol Brain Res 94:119–130

Singer JH, Talley EM, Bayliss DA, Berger AJ (1998) Development of glycinergic synaptic transmission to rat brain stem motoneurons. J Neurophysiol 80:2608–2620

Smith KE, Borden LA, Hartig PR, Branchek T, Weinshank RL (1992) Cloning and expression of a glycine transporter reveal colocalization with NMDA receptors. Neuron 8:927–935

Steeves CL, Baltz JM (2005) Regulation of intracellular glycine as an organic osmolyte in early preimplantation mouse embryos. J Cell Physiol 204:273–279

Supplisson S, Roux MJ (2002) Why glycine transporters have different stoichiometries. FEBS Lett 529:93–101

Sur C, Kinney GG (2004) The therapeutic potential of glycine transporter-1 inhibitors. Expert Opin Investig Drugs 13:515–521

Tada K, Kure S, Takayanagi M, Kume A, Narisawa K (1992) Non-ketotic hyperglycinemia: a life-threatening disorder in the neonate. Early Hum Dev 29:75–81

Tsai G, Ralph-Williams RJ, Martina M, Bergeron R, Berger-Sweeney J, Dunham KS, Jiang Z, Caine SB, Coyle JT (2004) Gene knockout of glycine transporter 1: characterization of the behavioral phenotype. Proc Natl Acad Sci U S A 101:8485–8490

Verdoorn TA, Kleckner NW, Dingledine R (1987) Rat brain N-methyl-D-aspartate receptors expressed in Xenopus oocytes. Science 238:1114–1116

Vergouwe MN, Tijssen MA, Shiang R, van Dijk JG, al Shahwan S, Ophoff RA, Frants RR (1997) Hyperekplexia-like syndromes without mutations in the GLRA1 gene. Clin Neurol Neurosurg 99:172–178

Wang TF, Ding CN, Wang GS, Luo SC, Lin YL, Ruan Y, Hevner R, Rubenstein JL, Hsueh YP (2004) Identification of Tbr-1/CASK complex target genes in neurons. J Neurochem 91:1483–1492

West KL, Castellini MA, Duncan MK, Bustin M (2004) Chromosomal proteins HMGN3a and HMGN3b regulate the expression of glycine transporter 1. Mol Cell Biol 24:3747–3756

Whitehead KJ, Pearce SM, Walker G, Sundaram H, Hill D, Bowery NG (2004) Positive N-methyl-D-aspartate receptor modulation by selective glycine transporter-1 inhibition in the rat dorsal spinal cord in vivo. Neuroscience 126:381–390

Zafra F, Aragon C, Olivares L, Danbolt NC, Gimenez C, Storm-Mathisen J (1995) Glycine transporters are differentially expressed among CNS cells. J Neurosci 15:3952–3969

Zafra F, Poyatos I, Gimenez C (1997) Neuronal dependency of the glycine transporter GLYT1 expression in glial cells. Glia 20:155–162

Zeilhofer HU, Studler B, Arabadzisz D, Schweizer C, Ahmadi S, Layh B, Bosl MR, Fritschy JM (2005) Glycinergic neurons expressing enhanced green fluorescent protein in bacterial artificial chromosome transgenic mice. J Comp Neurol 482:123–141

Zhou L, Chillag KL, Nigro MA (2002) Hyperekplexia: a treatable neurogenetic disease. Brain Dev 24:669–674

The Norepinephrine Transporter in Physiology and Disease

H. Bönisch (✉) · M. Brüss†

Department of Pharmacology and Toxicology, University of Bonn, Reuterstr. 2b,
53115 Bonn, Germany
boenisch@uni-bonn.de

1	Introduction	486
2	Properties, Physiology and Pharmacology of the NET	487
2.1	Basic Properties and Mechanisms of Transport	487
2.1.1	Co-substrates and Direction of Transport	489
2.1.2	NET Inhibitors	491
2.2	Physiological Importance and Knockout of the NET	492
3	Tissue Expression	493
4	Regulation of NET Function and Expression	494
5	Structure–Function Relationship	497
6	Gene Structure, Promoter and Alternative Splicing	501
7	hNET: Significance in Disease, Therapy and Diagnosis	502
7.1	Genetic Variations	502
7.2	NET and Dysautonomia	504
7.3	NET and Hypertension	505
7.4	NET and Myocardial Ischemia	505
7.5	NET and Obesity	506
7.6	NET and Anorexia Nervosa	506
7.7	NET and ADHD	507
7.8	NET and Depression	508
7.9	NET and Addiction	509
7.10	NET and Pain	510
7.11	NET Ligands in Diagnosis and Therapy	511
References		512

Abstract The norepinephrine transporter (NET) terminates noradrenergic signalling by rapid re-uptake of neuronally released norepinephrine (NE) into presynaptic terminals. NET exerts a fine regulated control over NE-mediated behavioural and physiological effects including mood, depression, feeding behaviour, cognition, regulation of blood pressure and heart rate. NET is a target of several drugs which are therapeutically used in the treatment or diagnosis of disorders among which depression, attention-deficit hyperactivity disorder and feeding disturbances are the most common. Individual genetic variations in the gene encoding the human NET (hNET), located at chromosome 16q12.2, may contribute to the pathogenesis of those diseases. An increasing number of studies concerning the identification of single nucleotide polymorphisms in the hNET gene and their potential association with disease as well as the functional investigation of naturally occurring or induced amino acid variations in hNET have contributed to a better understanding of NET function, regulation and genetic contribution to disorders. This review will reflect the current knowledge in the field of NET from its initial discovery until now.

Keywords Norepinephrine transporter · Re-uptake · Storage · Synthesis · Disease · Genetic variation

1
Introduction

The neurotransmitter norepinephrine (NE), released from noradrenergic neurons of the peripheral (PNS) or central nervous system (CNS), is rapidly removed from the synaptic cleft by means of the NE transporter (NET) located in the plasma membrane of noradrenergic neurons. After re-uptake, NE is then either deaminated by mitochondrial monoamine oxidase (MAO) or taken up into the storage vesicles by means of the vesicular monoamine transporter 2 (VMAT2). The re-uptake of NE ensures fast termination of synaptic transmission; this fast inactivation enables NET to exert a fine control over its effector system.

The neuronal NE uptake system was first detected in the periphery. The availability of radiolabelled catecholamines enabled Axelrod and co-workers (Axelrod et al. 1959, 1961) to examine the fate of these amines after intravenous injection in laboratory animals. They observed a selective accumulation of radiolabelled epinephrine (EPI) and NE in sympathetically innervated organs (e.g. spleen and heart) which was dependent on intact sympathetic nerve terminals and which was inhibited by cocaine.

In 1963 Iversen described in more detail some pharmacological properties of the neuronal uptake of tritiated NE by the isolated perfused rat heart (Iversen 1963). After the discovery of a further and completely different extraneuronal NE uptake process, the neuronal NE uptake system was designated as "uptake1" and the extraneuronal transport system as "uptake2" (Iversen et al. 1965). The latter has come to be known as EMT (extraneuronal monoamine transporter) or OCT3, a member of the organic cation transporter (OCT) family. Since catecholamines are largely protonated at physiological pH (Mack and Bönisch 1979), they do not readily cross the blood-brain barrier; thus, it was not surprising that only small amounts of radiolabelled catecholamines were detected in the brain in the early experiments by Axelrod and co-workers.

In 1969 uptake of EPI and NE was described for the first time in brain tissues, namely in homogenates (Coyle and Snyder 1969) and in synaptosomes (Snyder and Coyle 1969), i.e. in brain nerve terminals isolated with a newly available method. By means of (1) radiolabelled NE, (2) the usage of synaptosomes, (3) slices of sympathetically innervated tissues, and (4) isolated perfused organs (e.g. hearts), a great deal of our present knowledge about the physiological and pharmacological properties of the neuronal uptake of NE—such as the dependence on sodium and chloride ions, reversal of transport by indirectly acting sympathomimetic amines or its inhibition by a variety of drugs—was established (for reviews see e.g. Iversen et al. 1967; Paton 1979; Bönisch and Trendelenburg 1988; Trendelenburg 1991).

With the availability of tritiated desipramine (DMI), a tricyclic antidepressant which specifically inhibits the NET, and subsequent binding studies using this radioligand (Langer et al. 1981; Raisman et al. 1982) or the selective NET blocker nisoxetine (Tejani-Butt 1992)—as well as the identification of clonal cell lines natively expressing the NET, such as PC12 and SKN-SH cells—it was possible to examine the NE transport system at a more molecular level (for review, see Bönisch and Brüss 1994).

A milestone in the field of NET research was the expression cloning of the human NET (hNET) from complementary DNA (cDNA) originating from the human neuroblastoma line SKN-SH by Amara and co-workers (Pacholczyk et al. 1991). This was possible due to the availability of new techniques in molecular biology. The knowledge of the cDNA sequence of the hNET and that of the related transporter for γ-aminobutyric acid (GABA) (Guastella et al. 1990) enabled cDNA cloning of the related transporters for dopamine (DAT) and serotonin (SERT) and of species homologues of the NET (Lingen et al. 1994; Brüss et al. 1997).

A further step forward was the construction of NET/DAT transporter chimeras (Buck and Amara 1994; Giros et al. 1994), chromosomal mapping of the hNET gene (Brüss et al. 1993) and the characterization of the genomic structure (Pörzgen et al. 1995) and promoter of the hNET gene (Meyer et al. 1998) and its naturally occurring hNET variants (Runkel et al. 2000), as well as the establishment of NET knockout (NET-KO) mice (Xu et al. 2000). These and further recent findings—such as the elucidation of rapid changes in transporter surface expression and new techniques to measure transport (Blakely et al. 2005) or the recent description of the crystal structure of a bacterial homologue of Na^+/Cl^--dependent neurotransmitter transporters (Yamashita et al. 2005)—resulted in new insights in the structure and function of the NET and in its role in physiology and disease summarized in this review.

2
Properties, Physiology and Pharmacology of the NET

2.1
Basic Properties and Mechanisms of Transport

Transport of the naturally occurring substrate NE by the NET is saturable and characterized by a half-saturation constant (K_m) of about 0.5 μM (Table 1).

Table 1 Substrates and inhibitors of the NET[a]

Substrates			
Catecholamines	(K_m, µM)	Other substances	(K_m, µM)
Dopamine (DA)	(~0.1)	Tranylcypromine	(~2.0)
Norepinephrine (NE)	(~0.8)	Selegiline	(~4.0)
Epinephrine (EPI)	(~3.0)	Amezinium	(~0.5)
Other amines		Bretylium	(~10)
Metaraminol	(~0.3)	Guanethidine	(~4.0)
Tyramine	(~0.5)	MIBG	(~3.0)
Phenylethylamine	(~0.5)	DSP-4	(~3.0)
d-Amphetamine	(~0.5)	Xylamine	(~3.0)
Ephedrine	(~3.0)	MPP$^+$	(~2.0)
Serotonin (5-HT)	(~20)	ASP$^+$	(~2.0)
Inhibitors			
Antidepressants	(K_i, nM)	Other substances	(K_i, nM)
Desipramine	(~4)	Nisoxetine	(~4)
Nortriptyline	(~8)	Sibutramine	(~180)
Reboxetine	(~8)	Atomoxetine	(~1)
Maprotiline	(~10)	RTI-55 (β-CIT)	(~1)
Nomifensine	(~10)	Cocaine	(~900)

ASP$^+$, 4-(4-(dimethylamino)styryl)-N-methyl-pyridinium; MIBG, m-Iodobenzyl-guanidine; MPP$^+$, N-methyl-4-phenylpyridinium [a]K_m and K_i values are means of published values taken from: Iversen 1965; Paton 1976; Bönisch and Harder 1986; Schömig et al. 1988; Graefe and Bönisch 1988; Pacholczyk et al. 1991; Cheetham et al. 1996; Apparsundaram et al. 1997; Eshleman et al. 1999; Rothman and Baumann 2003; Schwartz et al. 2003; Mason et al. 2005; Owens et al. 1997; Tatsumi et al. 1997; Olivier et al. 2000

In hNET cDNA-transfected cells, Apparsundaram et al. (1997) demonstrated that dopamine (DA) was transported by the NET with an eightfold lower K_m and an about two-fold lower V_{max}, resulting in a fivefold higher V_{max}/K_m value for DA than for NE. The V_{max}/K_m value is a measure of effectiveness of transport (Graefe and Bönisch 1988) or an indicator of the catalytic efficiency of translocation. That DA is a better substrate of the NET than NE was also seen by Giros et al. (1994) and Buck and Amara (1994); however, no difference in the kinetic constants of uptake of NE and DA by the NET was observed by (Burnette et al. 1996). EPI, the third endogenous catecholamine, is transported by the NET with an about fourfold higher K_m (Table 1), an about twofold lower V_{max} and an about ninefold lower effectiveness than NE (Apparsundaram et al. 1997). Thus, the presence of a β-hydroxyl group is not essential for transport by the NET, whereas alkylation of the primary amino group reduces transport effectiveness, and isopropyl-NE (isoprenaline) seems not to be transported by the NET but is an excellent substrate of the extraneuronal monoamine transporter (OCT3).

Introduction of a methyl group at the α-carbon of the side chain of catecholamines or phenylethylamines does not affect effectiveness of transport (Graefe and Bönisch 1988), since α-methyl-NE (the metabolite of the antihypertensive drug α-methyl-dopa) and metaraminol are well-transported NET substrates. Introduction of a methyl group at one of the phenolic hydroxy groups of catecholamines, as it occurs by O-methylation through catechol-O-methyl transferase (COMT) results in amines such as normetanephrine which are not transported by the NET. As shown in Table 1, the NET also transports many other amines such as serotonin (5-hydroxytryptamine, 5-HT) and the indirectly acting sympathomimetic amines tyramine, ephedrine and d-amphetamine. It has been questioned whether the lipophilic d-amphetamine is transported or whether it only competitively inhibits the NET; however, using tritiated d-amphetamine, specific uptake by the NET (in PC12 cells) has been clearly demonstrated (Bönisch 1984).

The list of "other transported compounds" includes drugs such as MAO inhibitors (tranylcypromine, selegiline, amezinium), adrenergic neuron-blocking agents (guanethidine, bretylium), meta-iodobenzylguanidine (MIBG), the covalently binding NET suicide substrates and noradrenergic neurotoxins DSP-4 and xylamine, the dopaminergic neurotoxin MPP^+ (which is also a substrate for the DAT and SERT), and the fluorescent model substrate ASP^+ (Table 1).

2.1.1
Co-substrates and Direction of Transport

Substrate transport by the NET is dependent on Na^+ and Cl^- and the Na^+-gradient (Na^+ outside high) across the plasma membrane is the main driving force, dictating the transport direction (normally from outside to inside; for review, see e.g. Masson et al. 1999; Rudnick 1997; Sonders et al. 2005). The inside negative membrane potential created by the K^+ gradient also contributes a driving force. Both ion gradients are maintained by the Na^+/K^+-ATPase, and inhibition of this enzyme (by e.g. ouabain) or a lack of ATP causes a suppression of NE uptake and a reversal of the transport direction (see below), a phenomenon observed e.g. in ischemia (Schömig et al. 1991). Outward transport can experimentally also be induced by an increase of intracellular Na^+ evoked by e.g. the sodium channel opener veratridine (Graefe and Bönisch 1988, Chen et al. 1998) or by a reduction of extracellular Na^+ (Graefe and Bönisch 1988; Pifl et al. 1997). In the latter situation, monovalent cations other than Na^+ are not able to take over the unique role of Na^+ (Graefe and Bönisch 1988); this seems to hold true also for the DAT and SERT (Shank et al. 1987; Bryan-Lluka and Bönisch 1997).

Since both Na^+ and Cl^- change K_m and V_{max} of NE transport, and since both ions are needed for binding of competitive NET inhibitors such as desipramine, we had proposed a transport model based on a single centre-gated pore mechanism with alternating access of the solute to the binding site. In this model, the empty and mobile carrier loses mobility by first binding (from the extracellular site with high Na^+) the co-substrate Na^+. In the next step, the substrate NE and the co-substrate Cl^- are bound, resulting in a regain of mobility. After translocation of substrate and co-substrates and dissociation at the inner face of substrate and co-substrates, the carrier returns in its unloaded state (to become "fixed" again by binding of extracellular Na^+) and thus catalyses net transport, or it returns in a substrate-loaded state catalysing exchange, i.e. carrier-mediated outward transport (Harder and Bönisch 1985; Bönisch and Trendelenburg 1988; Graefe and Bönisch 1988; Bönisch 1998). The model considers that NE is translocated as positively charged NE^+ and that coupling of NE^+ transport to co-transported ions occurs at a stoichiometry of 1 NE^+:1 Na^+:1 Cl^-. The turnover rate has been estimated to be between 1 and 2.5 transport cycles per second (Bönisch and Harder 1986; Gu et al. 1996). In our model, which is an extension of the "facilitated exchange diffusion model" proposed by Fischer and Cho (1979), the co-transported Na^+ facilitates substrate binding from the internal site, and thereby also facilitates carrier-mediated NE efflux by inward transport of substrates (together with Na^+) such as indirectly acting sympathomimetic amines (Bönisch 1986; Langeloh et al. 1987). In fact, substances which are NET (or DAT or SERT) substrates can be identified (and distinguished from transporter inhibitors) by their ability to induce outward transport (Bönisch and Trendelenburg 1988; Wölfel and Graefe 1992; Burnette et al. 1996; Chen et al. 1998).

The extended model of facilitated exchange diffusion is also supported by the fact that (1) a preload of noradrenergic neurons (Graefe and Bönisch 1988) or of NET-expressing cells (Chen et al. 1997) with a NET substrate causes saturation of carrier-mediated efflux and (2) IC_{50} or K_i values for competitive inhibition of NE uptake by NET substrates are identical (or very similar) to EC_{50} values for induction of carrier-mediated efflux (Bönisch and Trendelenburg 1988; Chen et al. 1997).

Although our model is compatible with the crystal structure and features of the recently described bacterial Na^+/Cl^--dependent leucine transporter related to Na^+/Cl^--dependent neurotransmitter transporters (Yamashita et al. 2005), questions still remain concerning e.g. the role of Na^+ for substrate binding and for its coupling to substrate transport as well as reversal of transport induced by amphetamine. Using real-time, spatially resolved analysis and the fluorescent substrate ASP^+, Schwartz et al. (2003, 2005) showed that substrate binding is in a narrow pore deep within the transporter and is independent of Na^+, and that binding rates by far exceed transport rates. The NET has also a channel mode (Galli et al. 1998), suggesting the appearance of an aqueous pore through the transporter protein during the transport cycle.

Amphetamine-induced reverse transport of DA by the DAT has recently been shown to occur by a slow carrier-like and a rapid channel-like mode involving protein kinase C (PKC)-mediated phosphorylation of the DAT (Kahlig et al. 2005). Furthermore, Seidel et al. (2005) provided evidence at a concatamer of the GABA transporter and SERT that substrate-induced efflux relies on sequential rather than concomitant counter-transport, and that the switch from concomitant to sequential mode is evoked by amphetamine-induced activation of PKC. It remains to be shown whether this holds true also for NET oligomers.

2.1.2
NET Inhibitors

As shown in Table 1, the NET is inhibited with potencies in the nanomolar range by various antidepressants (ADs) such as the tricyclic ADs desipramine and nortriptyline, the tetracyclic AD maprotiline, and the ADs nomifensine and reboxetine. Reboxetine is the most selective NET inhibitor among all of these drugs, while desipramine is the most selective among the tricyclics. Other high-affinity NET inhibitors listed in Table 1 are nisoxetine, atomoxetine [the (−)isomer of tomoxetine], sibutramine, cocaine and the cocaine analogue RTI-55 (also known as β-CIT). Among these compounds, nisoxetine is a very selective NET inhibitor (Tejani-Butt 1992); this holds true also for atomoxetine (Tatsumi et al. 1997; Bymaster et al. 2002), whereas sibutramine also inhibits the SERT (Luque and Rey 1999), while cocaine and RTI-55 additionally block the DAT and the SERT (Eshleman et al. 1999).

The high-affinity NET inhibitors desipramine and nisoxetine are experimentally used as radioligands to label the NET (e.g. in autoradiography studies) and to examine the density of transporter sites in tissues and cells as well as the binding properties of the NET, and inhibition of radioligand binding can be utilized to examine affinities of unlabelled compounds to the NET (Bönisch and Harder 1986; Michael-Hepp et al. 1992; Tejani-Butt 1992; Graham and Langer 1992, and see references cited in the legend to Table 1). These studies showed that binding to the NET of desipramine and nisoxetine is dependent on Na^+ and Cl^- and is displaced by NET substrates; this was interpreted to indicate a common binding site for substrates and inhibitors at the NET. However, differences in sodium dependence, in K_i values for inhibition of NE uptake and for inhibition of radioligand binding by NET substrates, as well as differences in sensitivity to exchanges of certain amino acids of the NET, indicate that these NET inhibitors bind to a site at the NET which is not identical with the substrate recognition site but may overlap with that site. It should be noted that, in the micromolar range, NET inhibitors also interact with ligand-gated ion channels such as the nicotinic acetylcholine (nACh) receptor (Hennings et al. 1999) or the 5-HT_3 receptor (Eisensamer et al. 2003).

2.2
Physiological Importance and Knockout of the NET

NE is an important neurotransmitter in the PNS and CNS. The major noradrenergic nucleus in the brain is the locus coeruleus (LC) in the brain stem. Projections from the LC innervate virtually all areas of the brain including the spinal cord. Projections to the prefrontal cortex (an area involved in drive and motion) and the hippocampus (involved in learning and memory) may play an important role in depression. LC activity is sensitive to environmental stimuli and changes in internal homeostasis. LC activity is involved in flight-or-fight responses, arousal, sleep–wake cycle and modulation of the sympathetic nervous system, including pulse rate and blood pressure. LC firing with NE release potentiates the firing of dopaminergic cells in the ventral tegmental area that project to the limbic forebrain, and projections to raphe nuclei influence the firing rate of 5-HT neurons. Thus, because of these interactions between the brain stem modulatory systems, any (pharmacological) intervention directed toward one system will most likely lead to changes within the others (for review, see Ressler and Nemeroff 1999; Anand and Charney 2000).

The physiological importance of the NET located on noradrenergic nerve terminals is the re-uptake of released NE; thus, the NET is important for the fine-tuning of the noradrenergic neurotransmission. The NET, which also transports DA (as mentioned already in Sect. 2.1), has been shown to be also implicated in the clearance of DA in brain regions with low levels of the DAT (Burnette et al. 1996; Carboni and Silvagni 2004; Moron et al. 2002).

The targeted disruption of the NET gene in NET-KO mice (Xu et al. 2000) has provided an opportunity to examine a NET defect in vivo. NET-KO mice show profound alterations in NE homeostasis. In prefrontal cortex, hippocampus and cerebellum the NE level was up to 70% lower than in wild-type mice, and the clearance rate of released NE was at least sixfold slower, indicating that diffusion becomes the main mechanism for NE clearance in NET-KO mice (Xu et al. 2000). The depletion of intraneuronal stores, in spite of increased activity of tyrosine hydroxylase (Xu et al. 2000) underlines the importance of the NET in the maintenance of a physiologically high intraneuronal NE content and in the prevention of volume transmission by NE escaping (by diffusion) from the synaptic cleft to reach more distant brain areas.

In NET-KO mice, α_1-adrenoceptor expression was decreased in the hippocampus (Xu et al. 2000) as well as in other brain regions (Dziedzicka-Wasylewska et al. 2006). On the other hand, in the spinal cord, expression of α_2-adrenoceptor tended to increase and morphine-induced analgesia was increased (Bohn et al. 2000), whereas in several other brain regions α_2-adrenoceptor expression was clearly elevated in NET-KO mice (Gilsbach et al. 2005). This is in accordance with an increased modulation of NE release by presynaptic α_2-adrenoceptor observed by Vizi et al. (2004) in NET-KO mice. In antidepressant tests, NET-KO mice behaved like antidepressant-treated wildtype mice. Furthermore, in NET-KO mice no additional effects of paroxetine and bupropion, antidepressants interacting primarily with the SERT and DAT, respectively, were observed (Xu et al. 2000). NET-KO mice showed enhanced responses to the psychostimulant cocaine, which correlated with suppression of presynaptic DA function and supersensitivity of postsynaptic D2 and D3 receptors (Xu et al. 2000). A reduced effectiveness of cocaine for inhibition of DA uptake in the nucleus accumbens of NET-KO mice has been interpreted as an indication that DA uptake in this brain region might primarily depend on the NET (Gainetdinov et al. 2002). On the other hand, after selective inhibition of the NET by reboxetine, results of microdialysis experiments indicate that NE is taken up and stored in striatal dopaminergic neurons (Gobert et al. 2004). That the DAT accepts NE as substrate had already been shown previously (Buck and Amara 1994; Giros et al. 1994). In NET-KO mice it was furthermore shown that NE is also taken up and stored in serotonergic neurons (Vizi et al. 2004). Uptake and storage of monoamines by heterologous monoamine transporters may contribute to effects at other monoaminergic targets of antidepressants which selectively inhibit a defined monoamine transporter.

NET-KO mice show reduced locomotor responses to novelty; they also show reduced temperature and reduced body weight (Xu et al. 2000), indicating that NE and the NET may be involved in their regulation. In addition, NET-deficient mice exhibit decreased vulnerability to seizure (Kaminski et al. 2005), and they show excessive tachycardia and elevated blood pressure with wakefulness and activity (Keller et al 2004).

In the periphery, the NET is not only expressed at noradrenergic nerve endings but also in the placenta. In the human placenta, the NET, together with further monoamine transporters (SERT, VMAT2 and OCTs) keeps low the concentration of circulating vasoactive monoamines as a protective mechanism preventing vasoconstriction in the placental vascular bed and thereby securing stable blood flow to the fetus (Bottalico et al. 2004).

3
Tissue Expression

The expression of the mRNA of the NET is, like the mRNAs for other monoamine transporters such as the DAT or the SERT, localized to monoaminergic cell bodies rather than to nerve terminals (Povlok and Amara 1997; Backs et al. 2001). Furthermore, mRNA expression is generally restricted to cells that synthesize the corresponding monoamine. Thus, in the brain, NET mRNA expression is an indicator for noradrenergic pathways with cell bodies primarily located in the brain stem, and there in the locus coeruleus complex in the dorsal pons, especially in the nucleus locus coeruleus proper (Lorang et al. 1994; Eymin et al. 1995). NET mRNA is additionally expressed (together with dopamine-β-hydroxylase mRNA) in the lateral tegmentum of the medulla and pons. All these regions encompass most of the noradrenergic cell bodies in the CNS.

Since noradrenergic neurons originating from the locus coeruleus project to many different brain regions, the NET can influence many neural pathways involved in e.g. autonomic and neuroendocrine regulation, arousal, attention, and complex behaviours that are associated with affect, emotion and depression. In the periphery, NET mRNA is expressed in sympathetic ganglia, in the adrenal medulla and the placenta. No mRNA could be detected in peripheral noradrenergic nerve endings (Ungerer et al. 1996; Backs et al. 2001; Li et al. 2001), indicating that NET mRNA is expressed only in noradrenergic cell bodies in which the synthesis of the transporter protein also takes place. The NET protein is expressed at noradrenergic cell bodies, dendrites, axons and nerve endings, and obviously not directly in the active zones of synapses but more laterally; this has at least been demonstrated for the closely related DAT (Nirenberg et al. 1996; Schroeter et al. 2000).

NET protein can be detected either functionally by its transport activity or by specific labelling of the NET by means of specific anti-NET antibodies (Brüss et al. 1995; Savchenko et al. 2003), or by radioligand binding using specific and high-affinity NET inhibitors such as desipramine or nisoxetine (Raisman et al. 1982; Michael-Hepp et al. 1992; Tejani-Butt 1992; Sucic et al. 2002; Distelmaier et al. 2004). NET protein expression was demonstrated in the brain not only in noradrenergic somata but also at dendrites and axons of noradrenergic fibres, e.g. within the cortex and hippocampus (Tejani-Butt 1992; Hoffman et al. 1998; Schroeter et al. 2000), and in the periphery in the plasma membrane of sympathetic ganglia and noradrenergic nerve terminals within sympathetically innervated tissues such as heart and blood vessels (Eisenhofer 2001).

Examination of ontogeny of NET expression showed that the NET appears early in the young embryo (Sieber-Blum and Ren 2000; Ren et al. 2003), and its postnatal expression in the brain stem changes during maturation (Sanders et al. 2005). The NET protein is also expressed in neuroendocrine tumour cell lines such as rat PC12 pheochromocytoma cells or the human neuroblastoma cell line SKN-SH and in some non-neuronal cells, e.g. in endothelial cells of small vessels of the lung, fibroblast-like cells of the dental pulp, myometrial cells, syncytiotrophoblasts and in cultured glial cells (Bönisch and Brüss 1994). However, NET does not seem to be expressed in situ in glia (Hoffman et al. 1998).

4
Regulation of NET Function and Expression

The NET (or its function) can be regulated acutely or chronically, and regulation may be due to changes in gene transcription, mRNA translation or stability, posttranslational modifications such as phosphorylation, protein trafficking, cytoskeleton interaction and oligomerization (Zahniser and Doolen 2001).

The membrane turnover rate of the NET is not known, but it may be relatively long if one assumes that it is similar to that of the closely related DAT whose turnover rate, determined by intraventricular administration of an irreversible DAT inhibitor, was characterized by a half-life of about 2 days (Kimmel et al. 2000). Long-term NET regulation by de novo synthesis may be relevant during long-term blockade of the NET by e.g. NET-inhibiting antidepressants. Thus, repeated administration of desipramine was shown to produce up-regulation of NET mRNA in the rat locus coeruleus (Szot et al. 1993), and up-regulation of the NET protein in placenta was observed during treatment of rats with cocaine (Shearman and Meyer 1999). However, in an autoradiographic study using tritiated nisoxetine, repeated administration of desipramine was shown to cause a reduction in the number of nisoxetine binding sites in rat hippocampus, amygdala and thalamus but not in other regions such as frontal cortex, hypothalamus or locus coeruleus (Bauer and Tejani-Butt 1992). In addition, in a recent and similarly designed study, no regulation of brain nisoxetine binding sites was observed (Hebert et al. 2001).

In in vitro studies on PC12 cells constitutively expressing the rat NET (rNET), prolonged exposure to desipramine or nisoxetine reduced the number of nisoxetine binding sites and produced a parallel reduction in NE uptake (Zhu and Ordway 1997). A similar antidepressant-induced down-regulation of the NET was also observed in HEK293 cells stably expressing the hNET (Zhu et al. 1998); this down-regulation was not accompanied with changes in NET mRNA levels. Similar results were also obtained in several other studies (for review, see Zahniser and Doolen 2001). The mechanism underlying this antidepressant-induced NET regulation remains unclear. However, Zhu and co-workers recently showed that it could also represent an experimental artefact, at least when expression is measured by radioligand binding, since antidepressants such as desipramine cause long-term NET occupancy by persistent membrane retention (Zhu et al. 2004; Ordway et al. 2005). Long-term regulation of the NET in cultured cells induced by extremely high concentration of NET substrates reported in the literature may also largely be due other effects such as toxic side effects (for details see: Zahniser and Doolen 2001). Administration to rats of reserpine, an inhibitor of VMAT2, was shown not only to increase tyrosine hydroxylase expression but also to decrease NET mRNA expression in locus coeruleus and adrenal medulla (Cubells et al. 1995), indicating a certain role of NET substrates in the regulation of the NET.

In cultured neurons from superior cervical ganglia of newborn rats, NET expression is reduced (concomitantly with tyrosine hydroxylase expression) by long-term cell treatment with the cholinergic differentiation factor/leukaemia inhibitory factor and ciliary neurotrophic factor, neurokines known to induce a switch from adrenergic to cholinergic phenotype, whereas retinoic acid increased NET expression (Matsuoka et al. 1997). In cultured neurons of the rat adrenal medulla, NET expression was decreased by long-term treatment with the glucocorticoid dexamethasone but increased by nerve growth factor (NGF) (Wakade et al. 1996). However, in PC12 cells, long-term exposure to NGF reduced NE uptake and mouse NET expression (Ikeda et al. 2001). Long-term exposure to insulin, another growth factor, has been reported to cause a reduction in NET mRNA expression in the locus coeruleus (Figlewicz et al. 1993). In cultured quail neural crest cells, fibroblast growth factor, neurotrophin-3 and transforming growth factor-β1 caused an increase in NET mRNA expression and function (Ren et al. 2001). In cultured sympathetic neurons, the inflammatory cytokine cardiotrophin-1 was shown to decrease NET mRNA expression and NET function (Li et al. 2003).

Acute and rapid changes in NET function were observed by B- and C-type natriuretic peptides which both caused enhanced NE uptake in rat hypothalamus and adrenal medulla (Vatta et al. 1996, 1997). These peptides activate specific membrane-bound guanylyl cyclase receptors, thus modulating cellular functions via the intracellular second messenger cyclic guanosine monophosphate (cGMP) and activation of protein kinase G. However, neither short-term nor long-term cGMP elevation had an effect on NE transport in rat PC12 and human SKN-SH cells constitutively expressing the NET or in cells transfected with human or rat NET cDNA (Bryan-Lluka et al. 2001). In the same study, elevation of cyclic AMP (cAMP), which activates protein kinase A, also had no effect in these cell systems, except at PC12 cells, where short-term cAMP elevation caused a decrease in NE uptake and long-term elevation additionally evoked a decrease in mRNA expression, indicating a cell type-specific regulation (Bryan-Lluka et al. 2001).

We were the first who showed that phorbol ester-induced activation of PKC causes a relatively fast reduction in NE uptake and membrane expression (measured by nisoxetine binding) in cells constitutively expressing the NET (SKN-SH) and in hNET cDNA-transfected cells. We also showed that this rapid down-regulation was observed in transfected cells expressing an hNET Ser^{259}Ala mutant in which the canonical PKC phosphorylation site had been destroyed (Bönisch 1998). PKC activation-induced down-regulation of the NET (in SKN-SH cells) by activation of muscarinic receptors was shown by Blakely and co-workers, who also demonstrated that this effect was due to NET internalization, a new type of trafficking-dependent regulation of monoamine transporters (Apparsundaram et al. 1998a; 1998b; Blakely et al. 2005).

PKC-triggered reduction in NET surface expression was recently shown not to involve dynamin- or clathrin-mediated internalization but to occur via lipid rafts, and thus to be similar to the regulation of the glucose transporter GLUT4 (Jayanthi et al. 2004). Interestingly, insulin had been shown to cause in SKN-SH cells a rapid stimulation of NE uptake with an increase in V_{max} but without an alteration of K_m or NET surface expression (Apparsundaram et al. 2001). The insulin-mediated stimulation of intrinsic NET activity was dependent on extracellular calcium and on p38 mitogen-activated protein kinase (p38

MAPK) and phosphatidylinositol 3-OH kinase (PI3K); this study also provided evidence that NET surface expression under basal conditions is dependent on a PI3K-linked pathway (Apparsundaram et al. 2001). In hypothalamus-brainstem neuronal cultures from rat brains, angiotensin II (Ang-II) causes both acute and chronic stimulation of NE uptake.

Long-term Ang-II-mediated activation of AT1 receptors induces an upregulation of NET mRNA and enhanced NE uptake through activation of the Ras-Raf-MAP kinase, including nuclear signalling via SRE (serum-response enhancer element) and AP-1 (activator protein-1), whereas in neurons from SHR (spontaneous hypertensive rats), signalling through PI3K (including protein kinase B and AP-1) contributes to the stimulation of NET gene transcription (Yang and Raizada 1999). Acute Ang-II-mediated increase in NE uptake is a posttranscriptional event (Lu et al. 1996) leading to a rapid increase in NET surface expression, an effect also observed after brief potassium-induced depolarization (Savchenko et al. 2003; Blakely et al. 2005). Thus, the movement of the NET [and of the other monoamine transporters (Torres et al. 2003)] to and from the plasma membrane provides a new and interesting regulation of the transport capacity. This process seems to be supported by a close localization of surface NETs to synaptic vesicle pools (Kippenberger et al. 1999; Blakely et al. 2005). In this rapid regulation of NET surface expression, several proteins associated with plasma membranes and storage vesicles have been shown to be involved, including: syntaxin 1A, a presynaptic soluble *N*-ethylmaleimide-sensitive factor attachment protein receptor (SNARE) protein, which colocalizes with the NET (Sung et al. 2003) and protein phosphatase 2A anchoring subunit, as well as 14-3-3 proteins (Sung et al. 2005).

Finally, acute regulation of the NET (as for all other Na^+/Cl^--dependent transporters) occurs also through the alteration of the electrical and chemical gradients that drive substrate transport.

5
Structure–Function Relationship

The NET, DAT and SERT belong to the family of Na^+/Cl^--dependent monoamine transporters. Hydropathy analysis of the deduced amino acid sequences suggests the presence of 12 putative transmembrane domains (TMs), intracellular localization of the amino and carboxyl termini and a large extracellular loop positioned between TM3 and TM4 with several potential *N*-glycosylation sites (Fig. 1). A comparison of the primary sequences of the human (h) monoamine transporters demonstrates 80% and 69% homology between hNET and hDAT and hDAT and hSERT, respectively. Figure 1 indicates amino acids conserved between hNET and hDAT as well as those conserved between all three transporters. This topological model has been confirmed for the NET (Brüss et al. 1995) and also for the DAT and SERT (see Torres

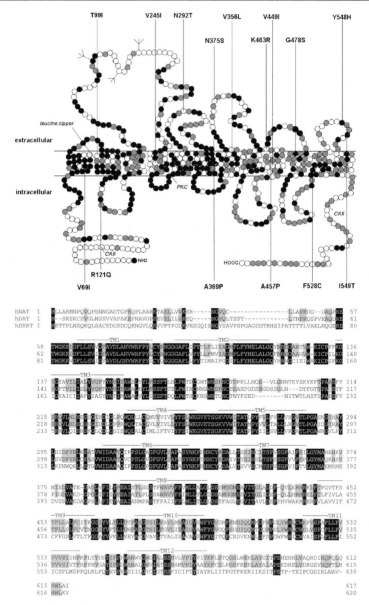

Fig. 1 Proposed topology of the human NET depicting 12 transmembrane domains (TMs) (*upper part*), and alignment of the human NET with the human DAT and human SERT (*lower part*). The *upper part* also indicates naturally occurring hNET variants, a leucine zipper in TM2, *N*-glycosylation sites within the second extracellular loop and intracellular residues which are potential sites for phosphorylation by protein kinase C (*PKC*) or casein kinase II (*CKII*). Amino acids conserved between hNET and hDAT are indicated in both parts in *grey* and those conserved between all three transporters in *black*

et al. 2003). Two cysteine residues located in the large extracellular loop are conserved among all members of the family; they may form a disulfide bond thus maintaining a functional transporter conformation (Povlok and Amara 1997). TM2 contains several conserved leucine residues resembling a leucine zipper, which may be implicated in mediating protein–protein interactions.

Residues and domains important for substrate binding and translocation have been identified by means of transporter chimeras, accessibility of substituted cysteines, and site-directed mutagenesis. Using chimeric constructs between NET and DAT, Giros et al. (1994) reported that regions from the amino-terminal through the first five TMs are likely to be involved in the uptake mechanism and ionic dependence. They postulated that regions within TMs 6–8 determine tricyclic antidepressant binding and cocaine interactions, and they said that the carboxyl-terminal region encompassing TM9 to the C-terminal tail appears to be responsible for the stereoselectivity and high affinity for substrates. However, Buck and Amara (1994), using similar chimeras to examine domains influencing substrate selectivity and translocation, proposed that the region from TM4 to TM8 is involved in substrate translocation.

The region between TM5 and TM8 of the NET has been examined in more detail for residues important for desipramine binding, using hNET mutants in which amino acids not conserved in the desipramine-insensitive hDAT were exchanged against those of the hDAT. Replacement in TM8 of serine (in position 399) and glycine (in position 400) against proline and leucine, respectively, resulted in an about 3,000-fold reduction in desipramine affinity without a change in the affinities for NE, DA and cocaine (Roubert et al. 2001). In a search for residues involved in binding and translocation of substrates and co-substrates, we exchanged residues in TM1 (Trp80 and Arg81) and TM2 (Glu113) of the hNET, which are absolutely conserved in all Na^+/Cl^--dependent neurotransmitter transporters including the glycine transporter (GLYT) and GABA transporter (GAT). In the rat GAT, replacement of these residues had been shown to cause an almost complete loss of transport without a change in membrane expression (Pantanowitz et al. 1993; Kleinberger-Doron and Kanner 1994; Keshet et al. 1995). None of the three hNET mutants (Trp80Ser, Arg81His and Glu113Asp) exhibited NE transport (Bönisch et al. 1999), indicating that the positive charged arginine may be involved in Cl^- binding, while tryptophan and the negatively charged glutamate are involved in binding of Na^+. In addition, several other glutamate residues conserved in monoamine transporters (beside Glu113) have been identified as being of functional importance in the hNET (Sucic et al. 2002). Mutation of an aspartate residue in TM1 (Asp79 in hNET), which is highly conserved in all monoamine transporters, abolished transport activity of hNET and hSERT (Bönisch et al. 1999; Barker et al. 1999), indicating that its carboxyl group may interact with the positive charge of the amine group of monoamines. For the DAT, further aspartate residues (Asp13, Asp435 and Asp476) have been shown to play a potential role in substrate recognition (Chen et al. 2001).

N-Glycosylation of asparagine residues in the large extracellular loop of the NET has been shown to be important for hNET protein stability, surface trafficking, and transport activity but not for ligand recognition (Melikian et al. 1996). There are still further conserved amino acids which may play a potential role in substrate and inhibitor binding (see Torres et al. 2003b).

Monoamine transporters might exist as oligomers (Sitte and Freissmuth 2003). In the DAT, mutation of the leucine zipper in TM2 was shown to abolish transporter delivery to the plasma membrane and interaction with wild-type

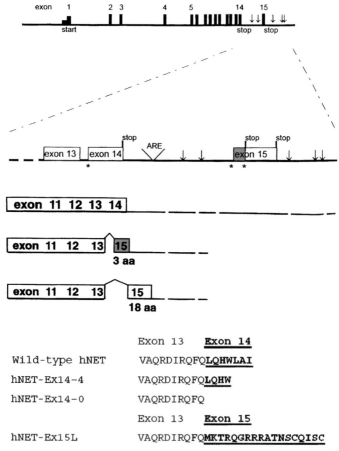

Fig. 2 Schematic representation of the hNET gene organization and of alternative splicing in the 3′ region of hNET transcripts. *Upper part* of Fig. 2, hNET gene structure and alternative splicing; *black vertical bars*, exons 1–15; *start*, translational start codon; *stop*, translational stop codon; *ARE*, adenylate-rich region; *arrows*, polyadenylation signals; *aa*, amino acid. *Lower part* of Fig. 2, deduced C-terminal amino acid sequences of wild-type hNET and two artificial (*Ex14-4* and *Ex14-0*) C-terminal hNET variants as well as the long (*Ex15L*) hNET splice variant (see Distelmaier et al. 2004 for further information)

DAT; therefore, it has been proposed that TM2 (of the DAT) is important for transporter assembly and oligomerization is essential for cell surface trafficking (Torres et al. 2003a).

An interaction between the PDZ domain-containing protein PICK1 and the amino acid motif Leu-Ala-Ile in the carboxyl termini of DAT (Torres et al. 2001) and NET (Distelmaier et al. 2004) has recently been shown to play an important role in transporter trafficking. A lack of this motif (in an hNET splice variant, see Fig. 2) or a mutation of these amino acids caused a strong reduction in hNET surface expression. In addition, the expression of the mutant hNET exerted a dominant-negative effect on plasma membrane expression of the wild-type hNET, indicating a physiological role of PICK1 interaction in the regulation of hNET surface expression. Since PICK1 has been shown to interact with PKC, it may also be involved in PKC-mediated trafficking of monoamine transporters.

6
Gene Structure, Promoter and Alternative Splicing

The human NET is encoded by a single gene which has been localized to chromosome 16q12.2 by somatic panel hybridization of fluorescence in situ hybridization (Brüss et al. 1993; Gelernter et al. 1993). The hNET gene consists of 14 coding exons and spans about 45 kb (Pörzgen et al. 1995). Downstream of exon 14, five consensus polyadenylation sites and a new coding exon (exon 15) which can be alternatively (by skipping exon 14) spliced to exon 13 have been identified. Due to alternative usage of two splice acceptor sites in exon 15, either 3 or 18 amino acids will be expressed at the C-terminus of hNAT following exon 13 (Pörzgen et al. 1998) (Fig. 2).

Functional expression of the long exon 15 splice variant (hNET-Ex15L) and two artificial hNET variants in which either three or all seven amino acids of exon 14 were deleted showed that all variants exhibited no difference in comparison to the wild-type hNET concerning the K_m of NE uptake and K_d of nisoxetine binding, but all variants were affected by reduced V_{max} of NE transport and a diminished B_{max} of nisoxetine binding (Distelmaier et al. 2004). The short hNET variant containing only three amino acids encoded by exon 15, and thus lacking the PICK1 recognition-motif, has been shown to be functionally inactive (Kitayama et al. 2001).

In addition to the coding exons, a noncoding exon 0 has been identified in the 5′-region of the hNET gene which is preceded by the promoter region (Meyer et al. 1998; Wiedemann et al. 1998; Kim et al. 1999). By primer extension and 5′ rapid amplification of cDNA ends (RACE) experiments, it was found that exon 0 is located from −343 to −765 upstream from the translation start codon. Through alternative splicing in exon 0, two transcripts which differ by 183 base pairs may be generated. The promoter of the hNET gene is located upstream of the transcription start site at −765 with respect to the translation start site. Functional analysis by means of luciferase assays of a 4-kb promoter fragment revealed a robust promoter-driven expression of the reporter gene only in hNET-expressing cells (SKN-SH) but not in hNET-negative JAR or COS-7 cells. The basal promoter is located in a short stretch (123 bp) upstream of the transcription start site. A cAMP response element (CRE) located upstream of the core promoter confers elevated hNET expression in a cAMP-dependent manner (Meyer et al. 1998). Kim et al. (1999) reported that when intron 1 (located between exon 0 and exon 1) is included in reporter gene constructs of hNET gene 5' constructs, this intron confers elevated expression of reporter constructs. The highest promoter activity was found when a 9-kb upstream fragment including intron 1 was used to transfect SK-N-BE(2)C cells. Later it was shown that an E-box motif within intron 1 is a least partly responsible for the promoter-enhancing activity of hNET intron 1 in hNET-positive and hNET-negative cell lines (Kim et al. 2001).

7
hNET: Significance in Disease, Therapy and Diagnosis

7.1
Genetic Variations

In the CNS, NE is involved in the regulation of mood, sleep, behaviour and the central control over the endocrine and sympathetic system (Young and Landsberg 1998).

Since the hNET is responsible for NE clearance in the periphery and in CNS, variations in the coding sequence affecting functional properties of the transporter or noncoding variations leading to elevated or decreased expression levels of hNET may be associated with a variety of physiological/pathophysiological consequences. The CNS hNET is a target of tricyclic antidepressants, and selective NE re-uptake inhibitors (SNRIs) are being used as antidepressants and in the therapy of attention deficit hyperactivity disorder (ADHD)—as well as in eating disorders. It is also notable that central hNETs are affected by addicting drugs such as cocaine and amphetamines (see Sect. 1). It is therefore not surprising that more studies than ever are concerned with the identification of single nucleotide polymorphisms (SNPs) or other genetic variations in the hNET gene of varying psychiatric-patient collectives and with corresponding association studies (see Table 2 and also Fig. 1).

The same is true for the periphery with respect to several autonomic dysfunctions such as orthostatic intolerance syndromes, essential hypertension and congestive heart failure, which are related to disturbances of the homeostasis of the sympathetic nervous system. Sympathetic homeostasis mainly depends on a fine-regulated concerted action of NE synthesis, release and re-uptake into sympathetic nerve endings or other tissues endowed with hNET as

Table 2 Naturally occurring amino acid variations of hNET as deduced from hNET gene SNPs

Variant	Position (aa)	Functional effect	Known association to disease	rs-number	Reference(s)
Asn → Lys	7	nd	None	rs11568323	1
Ala → Pro	31	nd	None	rs13306039	2
Val → Ile	69	None	None	rs1805064	3, 4
Thr → Ile	99	None	None	rs1805065	3, 4
Arg → Gln	121	TR ↓; SF ↓	None	rs13306041	5, 6
Asn → Lys	146	nd	None	na	1
Val → Ile	160	nd	None	na	1
Thr → Arg	193	nd	None	na	1
Val → Ile	244	None	None	na	6
Val → Ile	245	None	None	rs1805066	3, 4
Val → Ile	247	nd	None	rs11568341	7
Thr → Arg	283	nd	None	rs11568325	7
Asn → Thr	292	TR ↓; SF ↓	None	rs5563	6, 8
Val → Leu	356	None	None	rs5565	6, 8
Ala → Pro	369	TR ↓↓; SF nd	None	rs5566	6, 8
Asn → Ser	375	None	None	rs5567	6, 8
Val → Ile	449	None	None	rs2234910	3, 4
Ala → Pro	457	TR ↓↓; SF ↓↓	OI	Swiss:Var 010022	9
Lys → Arg	463	None	None	rs5570	6, 8
Gly → Ser	478	NE-K_m ↑	None	rs1805067	3, 4
Phe → Cys	528	TR ↑; SF ↑	None	rs5558	6, 8
Tyr → His	548	TR ↓; SF ↓	None	rs5559	6, 8
Ile → Thr	549	SF ↓	None	rs3743788	6, 10

Abbreviations: aa, amino acid position in hNET protein; na, not available; nd, not determined; NE-K_m, Michaelis–Menten constant for NE-uptake; OI, orthostatic intolerance; SF, surface expression; TR, NE transport rate; References: 1, http://www.pharmgkb.org/; 2, the Japan metabolic disease database (JMDBase); 3, Stöber et al. 1996; 4, Runkel et al. 2000; 5, Iwasa et al. 2001; 6, Hahn et al. 2005; 7, http://www.mutdb.org/; 8, Halushka et al. 9; Shannon et al. 2000; 10, NCBI-SNP-database (dbSNP)

the adrenal medullary (for review, see Blakely 2001). Of course, not only hNET variations but also other variations in the genes encoding adrenergic receptors or NE synthesizing as well as metabolizing enzymes and related genes encoding proteins interfering with the signalling pathways of adrenergic receptors may play a role in the often complex and multigenic genesis of such diseases.

Table 2 gives an overview of the currently identified hNET-gene nonsynonymous coding variations and their proven or putative functional consequences and possible associations to diseases, of which some are discussed in more detail elsewhere in this volume.

7.2
NET and Dysautonomia

The autonomic nervous system, which is divided into a sympathetic and a parasympathetic part, is responsible for the maintenance of body homeostasis. While the main transmitter of the parasympathetic system is ACh, postganglionic sympathetic neurons use NE as their main neurotransmitter, along with noradrenergic neurons in the CNS brainstem which are involved in the regulation of sleep-wake rhythm, food ingestion, blood pressure, learning and mood (Foote et al. 1983). Dysautonomia is a pathophysiological condition caused by abnormal function of the autonomic nervous system which may be transiently induced by drugs or may be permanent due to genetical disturbances of the autonomic system. The concentration of catecholamines in the sympathoadrenal system normally is relatively constant as a result of a balanced concerted action of catecholamine biosynthesis, storage, release, re-uptake and degradation. NE released at central and peripheral synapses is effectively inactivated by re-uptake via hNET, which recaptures 70%–90% of released NE, leaving the rest to spill over to into the circulation or uptake in other tissues expressing uptake2 (Iversen 1961; Esler et al. 1990). Re-uptake of NE by hNET in the heart is most noticeable, since in the heart noradrenergic synaptic clefts are narrower compared to other tissues (Novi 1968) and uptake2 is mostly absent (Eisenhofer et al. 1996).

Impaired function of NET through NET blockade by drugs (e.g. antidepressants) or through genetic alterations in the NET gene should affect the sympathoadrenal balance. Indeed, a missense mutation in hNET has been identified recently in a patient suffering from orthostatic intolerance (Shannon et al. 2000). This is the only currently identified variation in hNET which is definitely disease-related (Table 2). Orthostatic intolerance is an autonomous dysfunction characterized by elevated heart rate, light-headedness, weakness, sweating and other symptoms which generally occur after trying to get a standing position (for review, see Goldstein et al. 2002; Lu et al. 2004). The heterozygous missense variation found in hNET of this patient was exchange of alanine 457 to proline in TM9 of hNET protein. Functional analysis of the mutant hNET showed that this variant exhibited a 98% loss of function (Shannon et al. 2000). The functional impairment of this variant hNET has been shown to be the result of disturbed trafficking of hNET A457P to the plasma membrane and a pronounced increase in the K_m of NE (Paczkowski et al. 2002). Combined expression of this variant with wild-type hNET resulted in a dominant-negative effect on wild-type hNET expression, an explanation for the dramatic in vivo effect of this heterozygous mutation (Hahn et al. 2003).

Despite this genetic evidence of hNET involvement in dysautonomia, several other reports show more or less controversial effects of NET blockade on cardiovascular regulation. When NET was blocked by desipramine, female rats demonstrated elevated supine heart rate, reduced tyramine response and a diminished plasma ratio of the NE-metabolite dihydroxyphenylglycol relative to NE. DMI-treated rats exhibited an attenuated tachycardia after stimulation of the baroreflex by nitroprusside, indicating that a reduction of baroreflex function and sympathetic outflow are the main effects of NET inactivation in orthostatic intolerance (Carson et al. 2002).

7.3
NET and Hypertension

Several reports indicate that hNET may be involved in essential hypertension. A substantial fraction of patients suffering from essential hypertension show a neurogenic origin of disease as indicated by high NE spillover from heart and kidney (Esler et al. 1988) or from increased muscle sympathetic nerve activity measured at the tibial nerve (Yamada et al. 1989). These findings may be attributable to an elevated neuronal firing rate or increased density of sympathetic innervation but may also partly be related to impaired re-uptake of NE by the NET. This view is supported by the finding that selective blockade of hNET by reboxetine induces a slight but significant rise in blood pressure (Schroeder et al. 2002). In a study with overweight normotensive, obesity hypertensive and lean patients with essential hypertension, only the group of lean hypertensive patients displayed an elevated NE spillover from the heart. In addition, only in this group a reduced cardiac release of the NE metabolite dihydroxyphenylglycol was found, indicating impaired NET function (Rumantir et al. 2000). In a large study in which the nature and frequency of SNPs in 75 candidate genes for blood-pressure homeostasis and hypertension had been investigated by chip hybridization, seven nonsynonymous coding SNPs were detected in the hNET gene (Halushka et al. 1999). Functional investigation of these variants revealed varying effects on transporter function, expression and regulation (Hahn et al. 2005; see Table 2). Furthermore, a genetic association study including 1,950 Japanese subjects indicated an association between a variation in the hNET promoter and essential hypertension (Ono et al. 2003).

7.4
NET and Myocardial Ischemia

In protracted myocardial ischemia, the induced hypoxia, ATP depletion and pH decrease affect NE storage and elevate the activity of the Na^+/H^+-exchanger (NHE), which leads to increased intracellular NE and Na^+ concentrations. This condition leads to the reversal of NET, resulting in a massive carrier-mediated

NE release (Schömig et al. 1991; Kübler and Strasser 1994). In guinea-pig heart, it has been demonstrated that drugs which inhibit or stimulate NE overflow lead to reduced or prolonged time of ventricular fibrillation (Imamura et al. 1996; Hatta et al. 1999).

In a deduced cell-culture model stably expressing hNET, it was shown that e.g. desipramine inhibited NET-mediated NE efflux and that inhibition of NHE abolished this efflux, whereas inhibition of Na^+/K^+ ATPase by ouabain potentiated NE efflux (Smith and Levi 1999). In another cell culture model (SKN-SH cells stably transfected with angiotensin II AT1 receptor), it has been shown that Ang-II, which is known to be elevated in myocardial ischemia, activates NHE through AT1 receptors and thus should stimulate NET-mediated NE release during myocardial ischemia (Reid et al. 2004). It may be concluded from these findings that polymorphisms in the hNET gene may influence the risk and outcome of myocardial ischemia.

7.5
NET and Obesity

Several lines of evidence suggest that NE and thus NET may be implicated in feeding behaviour and eating disorders (see also the following section). Hypothalamically administered NE reduces food intake in animals (Grossman 1960). LY368975 [(R)-thionisoxetine], a selective inhibitor of NET, has been shown to decrease food intake of food-deprived rats by up to 44% (Gehlert et al. 1998). While selective 5-HT re-uptake inhibitors are able to reduce the frequency of binge eating, they were unable to reduce long-term body weight. Sibutramine, an inhibitor of 5-HT and NE re-uptake, is efficient in the promotion and maintenance of weight reduction and is well tolerated (Milano et al. 2005). In a recent study, 15 healthy probands were treated with 8 mg reboxetine or placebo. In the reboxetine group, carbohydrate oxidation and adipose tissue blood flow was higher compared to placebo. The isoproterenol-induced increase in interstitial NE concentration was higher in the placebo group compared to reboxetine-treated probands. It was concluded that NET blockade by reboxetine may sensitize adipose tissue to β-adrenergic stimulation (Boschmann et al. 2002). Thus, one may be tempted to speculate that genetic disturbances leading to elevated hNET expression or function also may be associated with an obese phenotype. This assumption would be consistent with the finding that only in lean hypertensive patients an elevated NE spillover from the heart was observed (see Sect. 7.3).

7.6
NET and Anorexia Nervosa

The fact that long-term weight-restored patients suffering from anorexia nervosa (AN) have reduced NE plasma levels compared to controls (Kaye et al.

1985; Pirke et al. 1992) and the finding that NET-KO mice exhibit a reduced body weight (Xu et al. 2000; see also Sect. 2.2) indicate that NET may be involved in the pathogenesis of AN. Recently a novel 343-bp GA-rich repeat containing six islands of consensus AAGG tetranucleotide repeats has been detected in the upstream region of hNET promoter (Urwin et al. 2002).This region was shown to be highly polymorphic in that a long and short variant of repeat 4 (L4 and S4) were found with frequencies of 0.74 and 0.26, respectively, in 50 DNA samples. Another deletion of an AAGG (S1) was found in repeat island 1 of few AN patients. Both S1 and S4 lead to the abolishment of a putative Elk-1 transcription factor binding site. Investigation of 87 trios (patients and parents) with DSM-IV AN restrictive type exhibited a significant association of the L4 genotype, or another variation in linkage disequilibrium with L4, with AN restrictive type (Urwin et al. 2002). Later the same group reported that the risk for the development of AN restrictive type more than doubles when individuals are homozygous for the L4 allele and additionally carry a functional polymorphism in the promoter of the MAO-A gene, which is X-chromosomally located (Urwin et al. 2003). We have analysed the polymorphic hNET region (about 500 bp containing the repeat islands) with reporter gene constructs and dual-luciferase assays in hNET-negative HEK293 cells and in hNET-expressing SKN-SH cells. None of the constructs containing the different repeat alleles exhibited a significant promoter activity in HEK293 cells and in SKN-SH cells, whereas a silencing activity was found when the constructs were tested in 5′ to the SV40 promoter (M. Brüss, A.K. Wübken, H. Bönisch unpublished). Since there was no striking difference between the particular alleles, it seems more likely that the polymorphism is in linkage disequilibrium to another variation in this chromosomal region.

7.7
NET and ADHD

ADHD is one of the most frequent (8%–12% worldwide) psychiatric disturbance in childhood. By means of family, twin and adoption studies, the predisposition to ADHD seems to be genetically determined. By genome-wide genetic scanning and association studies of candidate genes, several groups have tried to find significant association of genetic variation with ADHD (for review, see Faraone et al. 2005). Among the candidate genes—which include nearly all catecholaminergic receptors and transporters—hNET was included in some studies, not least because NET inhibitors are effective in the treatment of ADHD. In most studies, no association between hNET polymorphisms and ADHD was found (De Luca et al. 2004; Xu et al. 2005) but a significant association of two NET SNPs was reported recently (Bobb et al. 2005). Furthermore, an association between a NET polymorphism (G1287A) and responsiveness to methylphenidate therapy has been reported (Yang et al. 2004). Methylphenidate (Ritalin) and other psychostimulant drugs such as amphetamine are among the most commonly used drugs in the treatment of ADHD. This drug is thought to achieve its therapeutic benefit mainly by inhibition of the dopamine transporter and facilitation of dopamine release (Spencer et al. 2000). Nevertheless, there are strong indications that the noradrenergic and serotonergic systems are involved in ADHD. Neonatal rats in which dopaminergic neurons were destroyed by 6-hydroxydopamine show increased motor activity and learning deficits. Enhanced motor activity and learning deficiency respond to treatment with *d*-amphetamine, as well as to methylphenidate, in these dopamine-depleted rats (Shaywitz et al. 1978). In this animal model of ADHD, it has been shown that selective inhibitors of NET (desipramine and nisoxetine)—as well as of SERT (citalopram and fluvoxamine) but not DAT inhibitors—were able to reduce motor activity (Davids et al. 2002). The results of these animal studies are supported by the effective therapy of ADHD with SERT and/or NET inhibitors. Recently a large multicentre study has been conducted in which a combination of fluoxetine (a selective SERT inhibitor) and atomoxetine (a selective NET inhibitor) was compared with atomoxetine monotherapy in the effectiveness of therapy of ADHD with concurrent symptoms of depression and anxiety. The results of the study indicate that atomoxetine monotherapy was almost as effective as a combination therapy of both drugs (Kratochvil et al. 2005). These and other data indicate that disturbances of the noradrenergic system are involved in the aetiology of ADHD (for review, see Biederman and Spencer 1999). Regarding the fact that NET-inhibitors are effective in ADHD therapy, it seems likely that genetic variations which would either lead to enhanced NET function or expression (such as in the promoter region) or to diminished NE efficacy at target receptors may be involved in the pathogenesis of this disorder.

7.8
NET and Depression

The human NET is one of the main targets of commonly used antidepressants including non-selective and selective NET inhibitors. Elevation of central extracellular NE and/or 5-HT is thought to be the primary antidepressive action of such compounds. So it is self-evident that variations in the hNET gene may contribute to the pathogenesis of depression. A number of studies have reported changes in the level of NE and its metabolites in cerebrospinal fluid, plasma and urine, indicating the involvement of the NE system in depression (Lake et al. 1982; Yehuda et al. 1998). With respect to this view, a number of studies have been performed in which SNPs in the hNET gene were identified and the possible association to depression and related psychiatric disorders have been examined. In patients affected by bipolar disorder or schizophrenia, 13 NET SNPs have been identified but no associations to the disorders were found (Stöber et al. 1996). Investigation of a silent G1287A polymorphism in exon 9 of hNET in a collective suffering from major depression did not reveal an association to this disease (Owen et al. 1999). No genetic linkage of hNET

SNPs was found in patients suffering from Tourette's syndrome (Stöber et al. 1999) and manic disorder (Hadley et al. 1995). However, a recent report indicated a positive association of the T182C polymorphism in the hNET promoter to major depression in a Japanese and Korean population (Inoue et al. 2004; Ryu et al. 2004), a finding that previously could not be established in a German collective with major depression (Zill et al. 2002). Taken together, these data indicate that psychiatric disorders are complex and that population-specific multigenic contributions in addition to environmental factors determine the pathogenesis of these malignancies. Whole genome scanning of large patient collectives and controls of defined populations could provide further insight into the genetic factors which elevate the risk for the development of psychiatric disorders such as major depression.

7.9
NET and Addiction

Cocaine is a potent blocker of DAT, SERT and NET, and elevation of dopamine is thought to be the main mechanism of cocaine's psychotropic effects (Kuhar et al. 1991). Other drugs of abuse such as nicotine, opioids, amphetamine and ethanol may possess particular effects at various receptor systems, but the dopaminergic system seems to be involved in any rewarding effect and thus in the addictive state of drug abuse. Nevertheless, there is growing evidence that SERT and NET may play a complex role in the altered brain physiology in chronic drug consumption.

It is known that depressed patients are more often drug abusers than healthy controls and that antidepressants including NET inhibitors may reduce this addictive state. Rats treated for 1 month with cocaine exhibited elevated SERT density in the infralimbic cortex, nucleus accumbens and some other brain regions, and these changes were abolished after 4 days of cocaine withdrawal. In contrast, NET density was reduced in cocaine-treated rats in the bed nucleus of stria terminalis, lateral parabrachial area and inferior olive, but upon drug cessation a strong upregulation of NET has been observed in the paraventricular nucleus of the hypothalamus. In these experiments, no change in dopamine transporter level was seen (Belej et al. 1996).

In contrast to these data—which were determined by radioligand binding—another study investigated mRNA levels in chronically cocaine-treated rats by in situ hybridization. Interestingly, in this study NET and SERT mRNA levels were found to be unchanged during cocaine administration and after withdrawal, whereas DAT mRNA was upregulated in the ventral tegmental area upon cocaine cessation (Arroyo et al. 2000). In rhesus monkeys, chronic cocaine treatment led to strong upregulation of NET protein in the bed nucleus of the stria terminalis (Macey et al. 2003); and in postmortem human brains of cocaine abusers, upregulated NET protein was found in the insular cortex (Mash et al. 2005). In the conditioned place-preference test, DAT-KO mice and

SERT-KO mice still exhibited cocaine-conditioned place preference (Sora et al. 1998); but in double-knockout mice without DAT and SERT, this conditioned place preference was abolished (Sora et al. 2001). This finding indicates the complex nature of addiction and reward and implicates a special role of 5-HT in cocaine's reward action.

A known chronic cocaine effect is the down-regulation of prodynorphin (PDYN) expression in hypothalamus and an increase in caudate putamen. In rats, the selective NET inhibitor nisoxetine induced an increase in PDYN expression in hypothalamus and other brain regions but a decrease of PDYN in caudate putamen (Di Benedetto et al. 2004); thus, it may counteract some cocaine effects. Reboxetine, another selective NET inhibitor, has been shown to reduce nicotine self-administration in rats; but this effect may partly be a consequence of reboxetine's non-competitive inhibition of nACh receptors (Rauhut et al. 2002).

7.10
NET and Pain

It is generally accepted that, in addition to standard analgesic therapy with e.g. opioids in the management of chronic pain, the addition of antidepressants such as tricyclic compounds in antinociceptive therapy may improve the benefit for patients with chronic or neuropathic pain (Sindrup and Jensen 1999). 5-HT and NE are known to be mediators of endogenous analgesic mechanisms in the descending pain pathways which project to the spinal dorsal horn (Jones 1991; Willis and Westlund 1997). NET-KO mice have been shown to have increased extracellular NE levels (Xu et al. 2000), and these mice exhibit a potentiated analgesic effect of opioids compared to wild-type mice (Bohn et al. 2000). Recently it has been shown that intra-peritoneally administered duloxetine, a dual 5-HT and NE re-uptake inhibitor, is a potent analgesic to reduce late phase paw-licking behaviour in a rat formalin model of persistent pain (Iyengar et al. 2004). In addition, it was shown that orally applied duloxetine was an efficacious analgesic in the L5/L6 spinal nerve ligation model in contrast to a very low efficacy in the tail-flick model of acute nociceptive pain (Iyengar et al. 2004). In contrast, intrathecal NE has been shown to have antinociceptive effects in acute pain in tail-flick and hot-plate tests, and these effects are thought to be mediated through activation of α_2-adrenoceptors in spinal cord neurons (Reddy et al. 1980; Howe et al. 1983). In rats, the selective competitive NET inhibitor reboxetine and Xen2174—a structural analogue of the cone snail-derived peptide Mr1A (Sharpe et al. 2001), a non-competitive NET inhibitor—reduced tactile hypersensitivity after surgery when they were intrathecally administered. This effect could completely be blocked by the α_2-adrenoceptor antagonist idazoxan but only party by atropine (Obata et al. 2005). In contrast to this, an analgesic clonidine effect could completely be blocked by atropine in a rat model for neuropathic pain (Pan et al. 1999), indicating that NET-inhibition suppresses post-surgery hypersensitivity by a mechanism different from that in neuropathic pain. Taken together, these data demonstrate that variations in the hNET gene as well as in the pain-modulating and transmitting receptor systems may influence the interindividual pain sensitivity and the efficacy of applied analgesics.

7.11
NET Ligands in Diagnosis and Therapy

Substrates and inhibitors of the NET are used as drugs with various indications (Iversen 2000). Indirectly acting sympathomimetic amines which are not able to penetrate the blood–brain barrier (e.g. tyramine and related amines) are used as vasoconstrictors, while lipophilic amines (e.g. amphetamine and derivatives) possess not only misused psychostimulatory and anorectic properties, they are also utilized to treat ADHD. With the exception of cocaine—which possesses local anaesthetic and pronounced psychostimulatory properties—and RTI-55, all NET inhibitors are used, or have at least been developed, as antidepressants. However, NET inhibitors have also been shown to be useful drugs in the treatment of other disorders. Thus, atomoxetine is used to treat ADHD (Simpson and Plosker 2004), and sibutramine as anti-obesity drug (Luque and Rey 1999). Furthermore, high-affinity transporter ligands may also be utilized as diagnostic tools for single photon emission computed tomography (SPECT) or positron emission tomography (PET) (for review, see Laakso and Hietala 2000; McConathy et al. 2004; de Win et al. 2005).

An increasing number of studies underline the usefulness of hNET substrates for diagnosis and therapy of several tumour types. Cancer tissues overexpressing hNET such as pheochromocytoma and several neuroendocrine gastrointestinal tumours are sensitive to therapy with radiolabelled [^{131}I]-meta-iodobenzylguanidine (^{131}I-MIBG) (Hoefnagel et al. 1987; Hadrich et al. 1999; Zuetenhorst et al. 1999; Höpfner et al. 2002, 2004). In a bladder cancer cell line (EJ138) transfected with NET-cDNA under control of a tumour-specific telomerase promoter, ^{131}I-MIBG dose-dependently led to cell death (Fullerton et al. 2005). Thus, NET expression under tissue-specific promoters is a new tool for suicide gene therapy. Recently it has been shown that the combined application of interferon-γ and unlabelled MIBG was more effective than interferon-γ alone in its antiproliferative and apoptotic effects on neuroendocrine gastrointestinal and pancreatic carcinoid tumour cell lines. MIBG which is taken up by NET led to additive cytotoxic effects and increased S-phase arrest in these cells, indicating that a combined application of interferon-γ and cold MIBG may be a suitable treatment option in neuroendocrine gastrointestinal cancer therapy (Höpfner et al. 2004). In addition to these therapeutic implications of hNET, radiolabelled hNET substrates are also useful in diagnosis, e.g. in the detection of metastases of the above-mentioned tumour types. Myocardial uptake of ^{123}I-labelled MIBG reflects the relative distribution of sympathetic neurodensity and function in the myocardium (Bourachot et al. 1993; Hattori and Schwaiger 2000). This approach has also been shown to be useful in staging of diabetic peripheral nerve degeneration (Kiyono et al. 2005).

PET is a valuable tool for in vivo imaging of a growing number of malignancies. Since NET is involved in a variety of diseases including cardiovascular and CNS-related mood disorders, the usage of PET probes for in vivo monitoring of hNET density would be a valuable method for diagnosis and staging of several disorders, as well as for the understanding of disease pathology. So far, the measurement of NET levels by PET has been hampered by the lack of suitable and specific PET radioprobes for NET, due to the short half-life or defluorination of the ligand (Wilson et al. 2003; Schou et al. 2004). A recent study in monkeys showing successful PET imaging of NET by the usage of (S,S)-[18F]FMeNER-D2, a fluorinated reboxetine derivative, has suggested the usefulness of this compound in humans which, however, may be restricted due to radiation dose limitations by law (Seneca et al. 2005).

Acknowledgements We thank Birger Wenge and Ralf Gilsbach for their help with the figures.

References

Anand A, Charney DS (2000) Norepinephrine dysfunction in depression. J Clin Psychiatry 61:16–24

Apparsundaram S, Moore KR, Malone MD, Hartzell HC, Blakely RD (1997) Molecular cloning and characterization of an L-epinephrine transporter from sympathetic ganglia of the bullfrog, Rana catesbiana. J Neurosci 17:2691–2702

Apparsundaram S, Galli A, DeFelice LJ, Hartzell HC, Blakely RD (1998a) Acute regulation of norepinephrine transport. I. Protein kinase C-linked muscarinic receptors influence transport capacity and transporter density in SK-N-SH cells. J Pharmacol Exp Ther 287:733–743

Apparsundaram S, Schroeter S, Giovanetti E, Blakely RD (1998b) Acute regulation of norepinephrine transport: II. PKC-modulated surface expression of human norepinephrine transporter proteins. J Pharmacol Exp Ther 287:744–751

Apparsundaram S, Sung U, Price RD, Blakely RD (2001) Trafficking-dependent and -independent pathways of neurotransmitter transporter regulation differentially involving p38 mitogen-activated protein kinase revealed in studies of insulin modulation of norepinephrine transport in SK-N-SH cells. J Pharmacol Exp Ther 299:666–677

Arroyo M, Baker WA, Everitt BJ (2000) Cocaine self-administration in rats differentially alters mRNA levels of the monoamine transporters and striatal neuropeptides. Mol Brain Res 83:107–120

Axelrod J, Weil-Malherbe H, Tomchick R (1959) The physiological disposition of H3-epinephrine and its metabolite metanephrine. J Pharmacol Exp Ther 127:251–256

Axelrod J, Whitby LG, Herting G (1961) Effect of psychotropic drugs on the uptake of H3-norepinephrine by tissues. Science 133:383–384

Backs J, Haunstetter A, Gerber SH, Metz J, Borst MM, Strasser RH, Kubler W, Haass M (2001) The neuronal norepinephrine transporter in experimental heart failure: Evidence for a posttranscriptional downregulation. J Mol Cell Cardiol 33:461–472

Barker EL, Moore KR, Rakhshan F, Blakely RD (1999) Transmembrane domain I contributes to the permeation pathway for serotonin and ions in the serotonin transporter. J Neurosci 19:4705–4717

Bauer ME, Tejani-Butt SM (1992) Effects of repeated administration of desipramine or electroconvulsive shock on norepinephrine uptake sites measured by [3H]nisoxetine autoradiography. Brain Res 582:208–214

Belej T, Manji D, Sioutis S, Barros HMT, Nobrega JN (1996) Changes in serotonin and norepinephrine uptake sites after chronic cocaine: Pre- vs post-withdrawal effects. Brain Res 736:287–296

Biederman J, Spencer T (1999) Attention-deficit/hyperactivity disorder (ADHD) as a noradrenergic disorder. Biol Psychiatry 46:1234–1242

Blakely RD (2001) Physiological genomics of antidepressant targets: keeping the periphery in mind. J Neurosci 21:8319–8323

Blakely RD, DeFelice LJ, Galli A (2005) Biogenic amine neurotransmitter transporters: just when you thought you knew them. J Appl Physiol 20:225–231

Bobb AJ, Addington AM, Sidransky E, Gornick MC, Lerch JP, Greenstein DK, Clasen LS, Sharp WS, Inoff-Germain G, Vrieze FWD, Arcos-Burgos M, Straub RE, Hardy JA, Castellanos FX, Rapoport JL (2005) Support for association between ADHD and two candidate genes: NET1 and DRD1. Am J Med Genet B Neuropsychiatr Genet 134:67–72

Bohn LM, Xu F, Gainetdinov RR, Caron MG (2000) Potentiated opioid analgesia in norepinephrine transporter knock-out mice. J Neurosci 20:9040–9045

Bönisch H (1984) The transport of (+)-amphetamine by the neuronal noradrenaline carrier. Naunyn Schmiedebergs Arch Pharmacol 327:267–272

Bönisch H (1986) The role of co-transported sodium in the effect of indirectly acting sympathomimetic amines. Naunyn Schmiedebergs Arch Pharmacol 332:135–141

Bönisch H (1998) Transport and drug binding kinetics in membrane vesicle preparation. Methods Enzymol 296:259–278

Bönisch H, Brüss M (1994) The noradrenaline transporter of the neuronal plasma-membrane. Ann N Y Acad Sci 733:193–202

Bönisch H, Harder R (1986) Binding of 3H-desipramine to the neuronal noradrenaline carrier of rat phaeochromocytoma cells (PC-12 cells). Naunyn Schmiedebergs Arch Pharmacol 334:403–411

Bönisch H, Trendelenburg U (1988) The mechanism of action of indirectly acting sympathomimetic amines. In: Trendelenburg U, Weiner N (eds) Catecholamines. Handbook of experimental pharmacology, vol 90/I. Springer, Heidelberg, Berlin, New York, pp 247–277

Bönisch H, Runkel F, Roubert C, Giros B, Brüss M (1999) The human desipramine-sensitive noradrenaline transporter and the importance of defined amino acids for its function. J Auton Pharmacol 19:327–333

Boschmann M, Schroeder C, Christensen NJ, Tank J, Krupp G, Biaggioni I, Klaus S, Sharma AM, Luft FC, Jordan J (2002) Norepinephrine transporter function and autonomic control of metabolism. J Clin Endocrinol Metab 87:5130–5137

Bottalico B, Larsson I, Brodszki J, Hernandez-Andrade E, Casslen B, Marsal K, Hansson SR (2004) Norepinephrine transporter (NET), serotonin transporter (SERT), vesicular monoamine transporter (VMAT2) and organic cation transporters (OCT1, 2 and EMT) in human placenta from pre-eclamptic and normotensive pregnancies. Placenta 25:518–529

Bourachot ML, Merlet P, Pouillart F, Valette H, Bourguignon M, Scherrer M, Castaigne A, Syrota A (1993) I-123 metaiodobenzylguanidine scintigraphy as an index of severity in primary hypertrophic cardiomyopathy. J Nucl Med 34:14

Brüss M, Kunz J, Lingen B, Bönisch H (1993) Chromosomal mapping of the human gene for the tricyclic antidepressant-sensitive noradrenaline transporter. Hum Genet 91:278–280

Brüss M, Hammermann R, Brimijoin S, Bönisch H (1995) Antipeptide antibodies confirm the topology of the human norepinephrine transporter. J Biol Chem 270:9197–9201

Brüss M, Pörzgen P, Bryan-Lluka LJ, Bönisch H (1997) The rat norepinephrine transporter: molecular cloning from PC12 cells and functional expression. Mol Brain Res 52:257–262

Bryan-Lluka LJ, Paczkowski FA, Bönisch H (2001) Effects of short- and long-term exposure to c-AMP and c-GMP on the noradrenaline transporter. Neuropharmacology 40:607–617

Buck KJ, Amara SG (1994) Chimeric dopamine-norepinephrine transporters delineate structural domains influencing selectivity for catecholamines and 1-methyl-4-phenyl-pyridinium. Proc Natl Acad Sci U S A 91:12584–12588

Burnette WB, Bailey MD, Kukoyi S, Blakely RD, Trowbridge CG, Justice JB (1996) Human norepinephrine transporter kinetics using rotating disk electrode voltammetry. Anal Chem 68:2932–2938

Bymaster FP, Zhang W, Carter PA, Shaw J, Chernet E, Phebus L, Wong DT, Perry KW (2002) Fluoxetine, but not other selective serotonin uptake inhibitors, increases norepinephrine and dopamine extracellular levels in prefrontal cortex. Psychopharmacology (Berl) 160:353–361

Carboni E, Silvagni A (2004) Dopamine reuptake by norepinephrine neurons: exception or rule? Crit Rev Neurobiol 16:121–128

Carson RP, Diedrich A, Robertson D (2002) Autonomic control after blockade of the norepinephrine transporter: a model of orthostatic intolerance. J Appl Physiol 93:2192–2198

Chen N, Vaughan RA, Reith ME (2001) The role of conserved tryptophan and acidic residues in the human dopamine transporter as characterized by site-directed mutagenesis. J Neurochem 77:1116–1127

Chen NH, Trowbridge CG, Justice JB (1998) Voltammetric studies on mechanisms of dopamine efflux in the presence of substrates and cocaine from cells expressing human norepinephrine transporter. J Neurochem 71:653–665

Coyle JT, Snyder SH (1969) Catecholamine uptake by synaptosomes in homogenates of rat brain: stereospecificity in different areas. J Pharmacol Exp Ther 170:221–231

Cubells JF, Baker H, Volpe BT, Smith GP, Das SS, Joh TH (1995) Innervation-independent changes in the mRNAs encoding tyrosine hydroxylase and the norepinephrine transporter in rat adrenal medulla after high-dose reserpine. Neurosci Lett 193:189–192

Davids E, Zhang KH, Kula NS, Tarazi FI, Baldessarini RJ (2002) Effects of norepinephrine and serotonin transporter inhibitors on hyperactivity induced by neonatal 6-hydroxydopamine lesioning in rats. J Pharmacol Exp Ther 301:1097–1102

De Luca V, Muglia P, Jain U, Kennedy JL (2004) No evidence of linkage or association between the norepinephrine transporter (NET) gene MnlI polymorphism and adult ADHD. Am J Med Genet B Neuropsychiatr Genet 124B:38–40

De Win MM, Habraken JB, Reneman L, van den BW, den Heeten GJ, Booij J (2005) Validation of [(123)I]beta-CIT SPECT to assess serotonin transporters in vivo in humans: a double-blind, placebo-controlled, crossover study with the selective serotonin reuptake inhibitor citalopram. Neuropsychopharmacology 30:996–1005

Di Benedetto M, Feliciani D, D'Addario C, Izenwasser S, Candeletti S, Romualdi P (2004) Effects of the selective norepinephrine uptake inhibitor nisoxetine on prodynorphin gene expression in rat CNS. Mol Brain Res 127:115–120

Distelmaier F, Wiedemann P, Brüss M, Bönisch H (2004) Functional importance of the C-terminus of the human norepinephrine transporter. J Neurochem 91:537–546

Diziedzicka-Wasylewska M, Faron-Gorecka A, Kusmider M, Drozdowska E, Rogoz Z, Siwanowicz J, Caron MG, Bönisch H (2006) Effect of antidepressant drugs in mice lacking the norepinephrine transporter. Neuropsychopharmacology (in press)

Eisenhofer G (2001) The role of neuronal and extraneuronal plasma membrane transporters in the inactivation of peripheral catecholamines. Pharmacol Ther 91:35–62

Eisenhofer G, Friberg P, Rundqvist B, Quyyumi AA, Lambert G, Kaye DM, Kopin IJ, Goldstein DS, Esler MD (1996) Cardiac sympathetic nerve function in congestive heart failure. Circulation 93:1667–1676

Eisensamer B, Rammes G, Gimpl G, Shapa M, Ferrari U, Hapfelmeier G, Bondy B, Parsons C, Gilling K, Zieglgänsberger W, Holsboer F, Rupprecht R (2003) Antidepressants are functional antagonists at the serotonin type 3 (5-HT3) receptor. Mol Psychiatry 8:994–1007

Eshleman AJ, Carmolli M, Cumbay M, Martens CR, Neve KA, Janowsky A (1999) Characteristics of drug interactions with recombinant biogenic amine transporters expressed in the same cell type. J Pharmacol Exp Ther 289:877–885

Esler M, Jennings G, Korner P, Willett I, Dudley F, Hasking G, Anderson W, Lambert G (1988) Assessment of human sympathetic nervous-system activity from measurements of norepinephrine turnover. Hypertension 11:3–20

Esler M, Jennings G, Lambert G, Meredith I, Horne M, Eisenhofer G (1990) Overflow of catecholamine neurotransmitters to the circulation—source, fate, and functions. Physiol Rev 70:963–985

Eymin C, Charnay Y, Greggio B, Bouras C (1995) Localization of noradrenaline transporter messenger-RNA expression in the human locus-coeruleus. Neurosci Lett 193:41–44

Faraone SV, Perlis RH, Doyle AE, Smoller JW, Goralnick JJ, Holmgren MA, Sklar P (2005) Molecular genetics of attention-deficit/hyperactivity disorder. Biol Psychiatry 57:1313–1323

Figlewicz DP, Szot P, Israel PA, Payne C, Dorsa DM (1993) Insulin reduces norepinephrine transporter mRNA in vivo in rat locus coeruleus. Brain Res 602:161–164

Fischer JF, Cho AK (1979) Chemical release of dopamine from striatal homogenates: evidence for an exchange diffusion model. J Pharmacol Exp Ther 208:203–209

Foote SL, Bloom FE, Astonjones G (1983) Nucleus locus coeruleus—new evidence of anatomical and physiological specificity. Physiol Rev 63:844–914

Fullerton NE, Mairs RJ, Kirk D, Keith WN, Carruthers R, McCluskey AG, Brown M, Wilson L, Boyd M (2005) Application of targeted radiotherapy/gene therapy to bladder cancer cell lines. Eur Urol 47:250–256

Gainetdinov RR, Sotnikova TD, Caron MG (2002) Monoamine transporter pharmacology and mutant mice. Trends Pharmacol Sci 23:367–373

Galli A, Blakely RD, DeFelice LJ (1998) Patch-clamp and amperometric recordings from norepinephrine transporters: channel activity and voltage-dependent uptake. Proc Natl Acad Sci U S A 95:13260–13265

Gehlert DR, Dreshfield L, Tinsley F, Benvenga MJ, Gleason S, Fuller RW, Wong DT, Hemrick-Luecke SK (1998) The selective norepinephrine reuptake inhibitor, LY368975, reduces food consumption in animal models of feeding. J Pharmacol Exp Ther 287:122–127

Gelernter J, Kruger S, Pakstis AJ, Pacholczyk T, Sparkes RS, Kidd KK, Amara S (1993) Assignment of the norepinephrine transporter protein (Net1) locus to chromosome-16. Genomics 18:690–692

Gilsbach R, Faron-Gorecka A, Rogoz Z, Brüss M, Caron MG, Dziedzicka-Wasylewska M, Bönisch H (2005) Norepinephrine transporter knockout-induced upregulation of brain alpha2A/C-adrenergic receptors. J Neurochem 96:1111–1120

Giros B, Wang YM, Suter S, McLeskey SB, Pifl C, Caron MG (1994) Delineation of discrete domains for substrate, cocaine, and tricyclic antidepressant interactions using chimeric dopamine-norepinephrine transporters. J Biol Chem 269:15985–15988

Gobert A, Billiras R, Cistarelli L, Millan MJ (2004) Quantification and pharmacological characterization of dialysate levels of noradrenaline in the striatum of freely-moving rats: release from adrenergic terminals and modulation by alpha2-autoreceptors. J Neurosci Methods 140:141–152

Goldstein DS, Holmes C, Frank SM, Dendi R, Cannon RO, Sharabi Y, Esler MD, Eisenhofer G (2002) Cardiac sympathetic dysantonomia in chronic orthostatic intolerance syndromes. Circulation 106:2358–2365

Graefe KH, Bönisch H (1988) The transport of amines across the axonal membranes of noradrenergic and dopaminergic neurones. In: Trendelenburg U, Weiner N (eds) Catecholamines. Handbook of experimental pharmacology, vol 90/I. Springer, Heidelberg, Berlin, New York, pp 193–245

Graham D, Langer SZ (1992) Advances in sodium-ion coupled biogenic amine transporters. Life Sci 51:631–645

Grossman SP (1960) Eating or drinking elicited by direct adrenergic or cholinergic stimulation of hypothalamus. Science 132:301–302

Gu HH, Wall S, Rudnick G (1996) Ion coupling stoichiometry for the norepinephrine transporter in membrane vesicles from stably transfected cells. J Biol Chem 271:6911–6916

Guastella J, Nelson N, Nelson H, Czyzyk L, Keynan S, Miedel MC, Davidson N, Lester HA, Kanner BI (1990) Cloning and expression of a rat brain GABA transporter. Science 249:1303–1306

Hadley D, Hoff M, Holik J, Reimherr F, Wender P, Coon H, Byerley W (1995) Manic-depression and the norepinephrine transporter gene. Hum Hered 45:165–168

Hadrich D, Berthold F, Steckhan E, Bönisch H (1999) Synthesis and characterization of fluorescent ligands for the norepinephrine transporter: Potential neuroblastoma imaging agents. J Med Chem 42:3101–3108

Hahn MK, Robertson D, Blakely RD (2003) A mutation in the human norepinephrine transporter gene (SLC6A2) associated with orthostatic intolerance disrupts surface expression of mutant and wild-type transporters. J Neurosci 23:4470–4478

Hahn MK, Mazei-Robison MC, Blakely RD (2005) Single nucleotide polymorphisms in the human norepinephrine transporter gene affect expression, trafficking, antidepressant interaction, and protein kinase C regulation. Mol Pharmacol 68:457–466

Halushka MK, Fan JB, Bentley K, Hsie L, Shen NP, Weder A, Cooper R, Lipshutz R, Chakravarti A (1999) Patterns of single-nucleotide polymorphisms in candidate genes for blood-pressure homeostasis. Nat Genet 22:239–247

Harder R, Bönisch H (1985) Effects of monovalent ions on the transport of noradrenaline across the plasma membrane of neuronal cells (PC-12 cells). J Neurochem 45:1154–1162

Hatta E, Maruyama R, Marshall SJ, Imamura M, Levi R (1999) Bradykinin promotes ischemic norepinephrine release in guinea pig and human hearts. J Pharmacol Exp Ther 288:919–927

Hattori N, Schwaiger M (2000) Metalodobenzylguanidine scintigraphy of the heart: what have we learnt clinically? Eur J Nucl Med 27:1–6

Hebert C, Habimana A, Elie R, Reader TA (2001) Effects of chronic antidepressant treatments on 5-HT and NA transporters in rat brain: an autoradiographic study. Neurochem Int 38:63–74

Hennings EC, Kiss JP, De Oliveira K, Toth PT, Vizi ES (1999) Nicotinic acetylcholine receptor antagonistic activity of monoamine uptake blockers in rat hippocampal slices. J Neurochem 73:1043–1050

Hoefnagel CA, Voute PA, Dekraker J, Marcuse HR (1987) Radionuclide diagnosis and therapy of neural crest tumors using I-131 metaiodobenzylguanidine. J Nucl Med 28:308–314

Hoffman BJ, Hansson SR, Mezey E, Palkovits M (1998) Localization and dynamic regulation of biogenic amine transporters in the mammalian central nervous system. Front Neuroendocrinol 19:187–231

Höpfner M, Sutter AP, Beck NI, Barthel B, Maaser K, Jockers-Scherubl MC, Zeitz M, Scherubl H (2002) Meta-iodobenzylguanidine induces growth inhibition and apoptosis of neuroendocrine gastrointestinal tumor cells. Int J Cancer 101:210–216

Höpfner M, Sutter AP, Huether A, Ahnert-Hilger G, Scherubl H (2004) A novel approach in the treatment of neuroendocrine gastrointestinal tumors: additive antiproliferative effects of interferon-gamma and meta-iodobenzylguanidine. Bmc Cancer 4:23

Howe JR, Wang JY, Yaksh TL (1983) Selective antagonism of the anti-nociceptive effect of intrathecally applied alpha-adrenergic agonists by intrathecal prazosin and intrathecal yohimbine. J Pharmacol Exp Ther 224:552–558

Ikeda T, Kitayama S, Morita K, Dohi T (2001) Nerve growth factor down-regulates the expression of norepinephrine transporter in rat pheochromocytoma (PC12) cells. Mol Brain Res 86:90–100

Imamura M, Lander HM, Levi R (1996) Activation of histamine H-3-receptors inhibits carrier-mediated norepinephrine release during protracted myocardial ischemia—comparison with adenosine A(1)-receptors and alpha(2)-adrenoceptors. Circ Res 78:475–481

Inoue K, Itoh K, Yoshida K, Shimizu T, Suzuki T (2004) Positive association between T-182C polymorphism in the norepinephrine transporter gene and susceptibility to major depressive disorder in a Japanese population. Neuropsychobiology 50:301–304

Iversen LL (1963) The uptake of noradrenaline by the isolated perfused rat heart. Br J Pharmacol Chemother 21:523–537

Iversen LL (2000) Neurotransmitter transporters: fruitful targets for CNS drug discovery. Mol Psychiatry 5:357–362

Iversen LL, Glowinski J, Axelrod J (1965) The uptake and storage of H3-norepinephrine in the reserpine-pretreated rat heart. J Pharmacol Exp Ther 150:173–183

Iversen LL, De Champlain J, Glowinski J, Axelrod J (1967) Uptake, storage and metabolism of norepinephrine in tissues of the developing rat. J Pharmacol Exp Ther 157:509–516

Iwasa H, Kurabayashi M, Nagai R, Nakamura Y, Tanaka T (2001) Genetic variations in five genes involved in the excitement of cardiomyocytes. J Hum Genet 46:549–552

Iyengar S, Webster AA, Hemrick-Luecke SK, Xu JY, Simmons RMA (2004) Efficacy of duloxetine, a potent and balanced serotonin-norepinephrine reuptake inhibitor in persistent pain models in rats. J Pharmacol Exp Ther 311:576–584

Jones SL (1991) Descending noradrenergic influences on pain. Prog Brain Res 88:381–394

Kaminski RM, Shippenberg TS, Witkin JM, Rocha BA (2005) Genetic deletion of the norepinephrine transporter decreases vulnerability to seizures. Neurosci Lett 382:51–55

Kaye WH, Jimerson DC, Lake CR, Ebert MH (1985) Altered norepinephrine metabolism following long-term weight recovery in patients with anorexia-nervosa. Psychiatry Res 14:333–342

Keller NR, Diedrich A, Appalsamy M, Tuntrakool S, Lonce S, Finney C, Caron MG, Robertson D (2004) Norepinephrine transporter-deficient mice exhibit excessive tachycardia and elevated blood pressure with wakefulness and activity. Circulation 110:1191–1196

Keshet GI, Bendahan A, Su H, Mager S, Lester HA, Kanner BI (1995) Glutamate-101 is critical for the function of the sodium and chloride-coupled GABA transporter GAT-1. FEBS Lett 371:39–42

Kim CH, Kim HS, Cubells JF, Kim KS (1999) A previously undescribed intron and extensive 5' upstream sequence, but not Phox2a-mediated transactivation, are necessary for high level cell type-specific expression of the human norepinephrine transporter gene. J Biol Chem 274:6507–6518

Kimmel HL, Carroll FI, Kuhar MJ (2000) Dopamine transporter synthesis and degradation rate in rat striatum and nucleus accumbens using RTI-76. Neuropharmacology 39:578–585

Kippenberger AG, Palmer DJ, Comer AM, Lipski J, Burton LD, Christie DL (1999) Localization of the noradrenaline transporter in rat adrenal medulla and PC12 cells: evidence for its association with secretory granules in PC12 cells. J Neurochem 73:1024–1032

Kiyono Y, Kajiyama S, Fujiwara H, Kanegawa N, Saji H (2005) Influence of the polyol pathway on norepinephrine transporter reduction in diabetic cardiac sympathetic nerves: implications for heterogeneous accumulation of MIBG. Eur J Nucl Med Mol Imaging 32:438–442

Kleinberger-Doron N, Kanner BI (1994) Identification of tryptophan residues critical for the function and targeting of the gamma-aminobutyric acid transporter (subtype A). J Biol Chem 269:3063–3067

Kratochvil CJ, Newcorn JH, Arnold LE, Duesenberg D, Emslie GJ, Quintana H, Sarkis EH, Wagner KD, Gao HT, Michelson D, Biederman J (2005) Atomoxetine alone or combined with fluoxetine for treating ADHD with comorbid depressive or anxiety symptoms. J Am Acad Child Adolesc Psychiatry 44:915–924

Kübler W, Strasser RH (1994) Signal-transduction in myocardial-ischemia. Eur Heart J 15:437–445

Kuhar MJ, Ritz MC, Boja JW (1991) The dopamine hypothesis of the reinforcing properties of cocaine. Trends Neurosci 14:299–302

Laakso A, Hietala J (2000) PET studies of brain monoamine transporters. Curr Pharm Des 6:1611–1623

Lake CR, Pickar D, Ziegler MG, Lipper S, Slater S, Murphy DL (1982) High plasma norepinephrine levels in patients with major affective-disorder. Am J Psychiatry 139:1315–1318

Langeloh A, Bönisch H, Trendelenburg U (1987) The mechanism of the 3H-noradrenaline releasing effect of various substrates of uptake1: multifactorial induction of outward transport. Naunyn Schmiedebergs Arch Pharmacol 336:602–610

Langer SZ, Raisman R, Briley M (1981) High-affinity [3H] DMI binding is associated with neuronal noradrenaline uptake in the periphery and the central nervous system. Eur J Pharmacol 72:423–424

Li H, Ma SK, Hu XP, Zhang GY, Fei J (2001) Norepinephrine transporter (NET) is expressed in cardiac sympathetic ganglia of adult rat. Cell Res 11:317–320

Li W, Knowlton D, Woodward WR, Habecker BA (2003) Regulation of noradrenergic function by inflammatory cytokines and depolarization. J Neurochem 86:774–783

Lingen B, Brüss M, Bönisch H (1994) Cloning and expression of the bovine sodium- and chloride-dependent noradrenaline transporter. FEBS Lett 342:235–238

Lorang D, Amara SG, Simerly RB (1994) Cell-type-specific expression of catecholamine transporters in the rat brain. J Neurosci 14:4903–4914

Lu CC, Tseng CJ, Tang HS, Tung CS (2004) Orthostatic intolerance: potential pathophysiology and therapy. Chin J Physiol 47:101–109

Lu D, Yu K, Paddy MR, Rowland NE, Raizada MK (1996) Regulation of norepinephrine transport system by angiotensin II in neuronal cultures of normotensive and spontaneously hypertensive rat brains. Endocrinology 137:763–772

Luque CA, Rey JA (1999) Sibutramine: a serotonin-norepinephrine reuptake-inhibitor for the treatment of obesity. Ann Pharmacother 33:968–978

Macey DJ, Smith HR, Nader MA, Porrino LJ (2003) Chronic cocaine self-administration upregulates the norepinephrine transporter and alters functional activity in the bed nucleus of the stria terminalis of the rhesus monkey. J Neurosci 23:12–16

Mack F, Bönisch H (1979) Dissociation constants and lipophilicity of catecholamines and related compounds. Naunyn Schmiedebergs Arch Pharmacol 310:1–9

Mash DC, Ouyang QJ, Qin YJ, Pablo J (2005) Norepinephrine transporter immunoblotting and radioligand binding in cocaine abusers. J Neurosci Methods 143:79–85

Masson J, Sagne C, Hamon M, El Mestikawy S (1999) Neurotransmitter transporters in the central nervous system. Pharmacol Rev 51:439–464

Matsuoka I, Kumagai M, Kurihara K (1997) Differential and coordinated regulation of expression of norepinephrine transporter in catecholaminergic cells in culture. Brain Res 776:181–188

McConathy J, Owens MJ, Kilts CD, Malveaux EJ, Camp VM, Votaw JR, Nemeroff CB, Goodman MM (2004) Synthesis and biological evaluation of [C-11]talopram and [C-11]talsupram: candidate PET ligands for the norepinephrine transporter. Nucl Med Biol 31:705–718

Melikian HE, Ramamoorthy S, Tate CG, Blakely RD (1996) Inability to N-glycosylate the human norepinephrine transporter reduces protein stability, surface trafficking, and transport activity but not ligand recognition. Mol Pharmacol 50:266–276

Meyer J, Wiedemann P, Okladnova O, Brüss M, Staab T, Stober G, Riederer P, Bönisch H, Lesch KP (1998) Cloning and functional characterization of the human norepinephrine transporter gene promoter—rapid communication. J Neural Transm 105:1341–1350

Michael-Hepp J, Blum B, Bönisch H (1992) Characterization of the [H-3] desipramine binding-site of the bovine adrenomedullary plasma-membrane. Naunyn Schmiedebergs Arch Pharmacol 346:203–207

Milano W, Petrella C, Casella A, Capasso A, Carrino S, Milano L (2005) Use of sibutramine, an inhibitor of the reuptake of serotonin and noradrenaline, in the treatment of binge eating disorder: a placebo-controlled study. Adv Ther 22:25–31

Moron JA, Brockington A, Wise RA, Rocha BA, Hope BT (2002) Dopamine uptake through the norepinephrine transporter in brain regions with low levels of the dopamine transporter: evidence from knock-out mouse lines. J Neurosci 22:389–395

Nirenberg MJ, Vaughan RA, Uhl GR, Kuhar MJ, Pickel VM (1996) The dopamine transporter is localized to dendritic and axonal plasma membranes of nigrostriatal dopaminergic neurons. J Neurosci 16:436–447

Novi AM (1968) An electron microscopic study of innervation of papillary muscles in rat. Anat Rec 160:123–141

Obata H, Conklin D, Eisenach JC (2005) Spinal noradrenaline transporter inhibition by reboxetine and Xen2174 reduces tactile hypersensitivity after surgery in rats. Pain 113:271–276

Olivier B, Soudijn W, von Wijngaarden I (2000) Serotonin, dopamine and norepinephrine transporters in the central nervous system and their inhibitors. Prog Drug Res 54:61–119

Ono K, Iwanaga Y, Mannami T, Kokubo Y, Tomoike H, Komamura K, Shioji K, Yasui N, Tago N, Iwai N (2003) Epidemiological evidence of an association between SLC6A2 gene polymorphism and hypertension. Hypertens Res 26:685–689

Ordway GA, Jia WH, Li J, Zhu MY, Mandela P, Pan J (2005) Norepinephrine transporter function and desipramine: residual drug effects versus short-term regulation. J Neurosci Methods 143:217–225

Owen D, Du LS, Bakish D, Lapierre YD, Hrdina PD (1999) Norepinephrine transporter gene polymorphism is not associated with susceptibility to major depression. Psychiatry Res 87:1–5

Owens MJ, Morgan WN, Plott SJ, Nemeroff CB (1997) Neurotransmitter receptor and transporter binding profile of antidepressants and their metabolites. J Pharmacol Exp Ther 283:1305–1322

Pacholczyk T, Blakely RD, Amara SG (1991) Expression cloning of A cocaine-sensitive and antidepressant-sensitive human noradrenaline transporter. Nature 350:350–354

Paczkowski F, Bönisch H, Bryan-Lluka LJ (2002) Pharmacological properties of the naturally occurring Ala457Pro variant of the human norepinephrine transporter. Pharmacogenetics 12:165–173

Pan HL, Chen SR, Eisenach JC (1999) Intrathecal clonidine alleviates allodynia in neuropathic rats—interaction with spinal muscarinic and nicotinic receptors. Anesthesiology 90:509–514

Pantanowitz S, Bendahan A, Kanner BI (1993) Only one of the charged amino acids located in the transmembrane alpha-helices of the gamma-aminobutyric acid transporter (subtype A) is essential for its activity. J Biol Chem 268:3222–3225

Paton DM (1979) The mechanism of neuronal and extraneuronal transport of catecholamines. Raven Press, New York, pp 1–370

Pirke KM, Kellner M, Philipp E, Laessle R, Krieg JC, Fichter MM (1992) Plasma norepinephrine after a standardized test meal in acute and remitted patients with anorexianervosa and in healthy controls. Biol Psychiatry 31:1074–1077

Pörzgen P, Bönisch H, Brüss M (1995) Molecular-cloning and organization of the coding region of the human norepinephrine transporter gene. Biochem Biophys Res Commun 215:1145–1150

Povlok SL, Amara S (1997) The structure and function of norepinephrine, dopamine, and serotonin transporters. In: Reith MEA (ed) Neurotransmitter transporters. Humana Press, Totowa, New Jersey, pp 1–28

Raisman R, Sette M, Pimoule C, Briley M, Langer SZ (1982) High-affinity [3H]desipramine binding in the peripheral and central nervous system: a specific site associated with the neuronal uptake of noradrenaline. Eur J Pharmacol 78:345–351

Rauhut AS, Mullins SN, Dwoskin LP, Bardo MT (2002) Reboxetine: attenuation of intravenous nicotine self-administration in rats. J Pharmacol Exp Ther 303:664–672

Reddy SVR, Maderdrut JL, Yaksh TL (1980) Spinal-cord pharmacology of adrenergic agonist-mediated antinociception. J Pharmacol Exp Ther 213:525–533

Reid AC, Mackins CJ, Seyedi N, Levi R, Silver RB (2004) Coupling of angiotensin II AT(1) receptors to neuronal NHE activity and carrier-mediated norepinephrine release in myocardial ischemia. Am J Physiol Heart Circ Physiol 286:H1448–H1454

Ren ZG, Pörzgen P, Zhang JM, Chen XR, Amara SG, Blakely RD, Sieber-Blum M (2001) Autocrine regulation of norepinephrine transporter expression. Mol Cell Neurosci 17:539–550

Ren ZG, Pörgzen PP, Youn YH, Sieber-Blum M (2003) Ubiquitous embryonic expression of the norepinephrine transporter. Dev Neurosci 25:1–13

Ressler KJ, Nemeroff CB (1999) Role of norepinephrine in the pathophysiology and treatment of mood disorders. Biol Psychiatry 46:1219–1233

Roubert C, Cox P, Brüss M, Hamon M, Bönisch H, Giros B (2001) Determination of residues in the norepinephrine transporter that are critical for tricyclic antidepressant affinity. J Biol Chem 276:8254–8260

Rudnick G (1997) Mechanisms of biogenic amine neurotransmitter transporters. In: Reith MEA (ed) Neurotransmitter transporters. Humana Press, Totowa, pp 73–100

Rumantir MS, Kaye DM, Jennings GL, Vaz M, Hastings JA, Esler MD (2000) Phenotypic evidence of faulty neuronal norepinephrine reuptake in essential hypertension. Hypertension 36:824–829

Runkel F, Brüss M, Nöthen MM, Stöber G, Propping P, Bönisch H (2000) Pharmacological properties of naturally occurring variants of the human norepinephrine transporter. Pharmacogenetics 10:397–405

Ryu SH, Lee SH, Lee HJ, Cha JH, Ham BJ, Han CS, Choi MJ, Lee MS (2004) Association between norepinephrine transporter gene polymorphism and major depression. Neuropsychobiology 49:174–177

Sanders JD, Happe HK, Bylund DB, Murrin LC (2005) Development of the norepinephrine transporter in the rat CNS. Neuroscience 130:107–117

Savchenko V, Sung U, Blakely RD (2003) Cell surface trafficking of the antidepressant-sensitive norepinephrine transporter revealed with an ectodomain antibody. Mol Cell Neurosci 24:1131–1150

Schömig E, Haass M, Richardt G (1991) Catecholamine release and arrhythmias in acute myocardial ischaemia. Eur Heart J 12 Suppl F:38–47

Schou M, Halldin C, Sovago J, Pike VW, Hall H, Gulyas B, Mozley PD, Dobson D, Shchukin E, Innis RB, Farde L (2004) PET evaluation of novel radiofluorinated reboxetine analogs as norepinephrine transporter probes in the monkey brain. Synapse 53:57–67

Schroeder C, Tank J, Boschmann M, Diedrich A, Sharma AM, Biaggioni I, Luft FC, Jordan J (2002) Selective norepinephrine reuptake inhibition as a human model of orthostatic intolerance. Circulation 105:347–353

Schroeter S, Apparsundaram S, Wiley RG, Miner LH, Sesack SR, Blakely RD (2000) Immunolocalization of the cocaine- and antidepressant-sensitive l-norepinephrine transporter. J Comp Neurol 420:211–232

Schwartz JW, Blakely RD, DeFelice LJ (2003) Binding and transport in norepinephrine transporters—real-time, spatially resolved analysis in single cells using a fluorescent substrate. J Biol Chem 278:9768–9777

Schwartz JW, Novarino G, Piston DW, DeFelice LJ (2005) Substrate binding stoichiometry and kinetics of the norepinephrine transporter. J Biol Chem 280:19177–19184

Seidel S, Singer EA, Just H, Farhan H, Scholze P, Kudlacek O, Holy M, Koppatz K, Krivanek P, Freissmuth M, Sitte HH (2005) Amphetamines take two to tango: an oligomer-based counter-transport model of neurotransmitter transport explores the amphetamine action. Mol Pharmacol 67:140–151

Seneca N, Andree B, Sjoholm N, Schou M, Pauli S, Mozley PD, Stubbs JB, Liow JS, Sovago J, Gulyas B, Innis R, Halldin C (2005) Whole-body biodistribution, radiation dosimetry estimates for the PET norepinephrine transporter probe (S,S)-[F-18]FMeNER-D-2 in non-human primates. Nucl Med Commun 26:695–700

Sharpe IA, Gehrmann J, Loughnan ML, Thomas L, Adams DA, Atkins A, Palant E, Craik DJ, Adams DJ, Alewood PF, Lewis RJ (2001) Two new classes of conopeptides inhibit the alpha 1-adrenoceptor and noradrenaline transporter. Nat Neurosci 4:902–907

Shaywitz BA, Klopper JH, Gordon JW (1978) Methylphenidate in 6-hydroxydopamine-treated developing rat pups—effects on activity and maze performance. Arch Neurol 35:463–469

Shearman LP, Meyer JS (1999) Cocaine up-regulates norepinephrine transporter binding in the rat placenta. Eur J Pharmacol 386:1–6

Sieber-Blum M, Ren ZG (2000) Norepinephrine transporter expression and function in noradrenergic cell differentiations. Mol Cell Biochem 212:61–70

Simpson D, Plosker GL (2004) Atomoxetine: a review of its use in adults with attention deficit hyperactivity disorder. Drugs 64:205–222

Sindrup SH, Jensen TS (1999) Efficacy of pharmacological treatments of neuropathic pain: an update and effect related to mechanism of drug action. Pain 83:389–400

Sitte HH, Freissmuth M (2003) Oligomer formation by Na^+-Cl^--coupled neurotransmitter transporters. Eur J Pharmacol 479:229–236

Smith NCE, Levi R (1999) LLC-PK1 cells stably expressing the human norepinephrine transporter: A functional model of carrier-mediated norepinephrine release in protracted myocardial ischemia. J Pharmacol Exp Ther 291:456–463

Snyder SH, Coyle JT (1969) Regional differences in H3-norepinephrine and H3-dopamine uptake into rat brain homogenates. J Pharmacol Exp Ther 165:78–86

Sonders MS, Quick M, Javitch JA (2005) How did the neurotransmitter cross the bilayer? A closer view. Curr Opin Neurobiol 15:1–9

Sora I, Wichems C, Takahashi N, Li XF, Zeng ZZ, Revay R, Lesch KP, Murphy DL, Uhl GR (1998) Cocaine reward models: conditioned place preference can be established in dopamine- and in serotonin-transporter knockout mice. Proc Natl Acad Sci U S A 95:7699–7704

Sora I, Hall FS, Andrews AM, Itokawa M, Li XF, Wei HB, Wichems C, Lesch KP, Murphy DL, Uhl GR (2001) Molecular mechanisms of cocaine reward: combined dopamine and serotonin transporter knockouts eliminate cocaine place preference. Proc Natl Acad Sci U S A 98:5300–5305

Spencer T, Biederman J, Wilens T (2000) Pharmacotherapy of attention deficit hyperactivity disorder. Child Adolesc Psychiatr Clin N Am 9:77–97

Stöber G, Nöthen MM, Pörzgen P, Brüss M, Bönisch H, Knapp M, Beckman H, Propping P (1996) Systematic search for variation in the human norepinephrine transporter gene: Identification of five naturally occurring missense mutations and study of association with major psychiatric disorders. Am J Med Genet 67:523–532

Stöber G, Hebebrand J, Cichon S, Brüss M, Bönisch H, Lehmkuhl G, Poustka F, Schmidt M, Remschmidt H, Propping P, Nöthen MM (1999) Tourette syndrome and the norepinephrine transporter gene: results of a systematic mutation screening. Am J Med Genet 88:158–163

Sucic S, Paczkowski FA, Runkel F, Bönisch H, Bryan-Lluka LJ (2002) Functional significance of a highly conserved glutamate residue of the human noradrenaline transporter. J Neurochem 81:344–354

Sung U, Apparsundaram S, Galli A, Kahlig KM, Savchenko V, Schroeter S, Quick MW, Blakely RD (2003) A regulated interaction of syntaxin 1A with the antidepressant-sensitive norepinephrine transporter establishes catecholamine clearance capacity. J Neurosci 23:1697–1709

Sung U, Jennings JL, Link AJ, Blakely RD (2005) Proteomic analysis of human norepinephrine transporter complexes reveals associations with protein phosphatase 2A anchoring subunit and 14-3-3 proteins. Biochem Biophys Res Commun 333:671–678

Szot P, Ashliegh EA, Kohen R, Petrie E, Dorsa DM, Veith R (1993) Norepinephrine transporter mRNA is elevated in the locus coeruleus following short- and long-term desipramine treatment. Brain Res 618:308–312

Tatsumi M, Groshan K, Blakely RD, Richelson E (1997) Pharmacological profile of antidepressants and related compounds at human monoamine transporters. Eur J Pharmacol 340:249–258

Tejani-Butt SM (1992) [3H]Nisoxetine: a radioligand for quantitation of norepinephrine uptake sites by autoradiography or by homogenate binding. J Pharmacol Exp Ther 260:427–436

Torres GE, Yao WD, Mohn AR, Quan H, Kim KM, Levey AI, Staudinger J, Caron MG (2001) Functional interaction between monoamine plasma membrane transporters and the synaptic PDZ domain-containing protein PICK1. Neuron 30:121–134

Torres GE, Carneiro A, Seamans K, Fiorentini C, Sweeney A, Yao WD, Caron MG (2003a) Oligomerization and trafficking of the human dopamine transporter. Mutational analysis identifies critical domains important for the functional expression of the transporter. J Biol Chem 278:2731–2739

Torres GE, Gainetdinov RG, Caron MG (2003b) Plasma membrane monoamine transporters: structure, regulation and function. Nat Rev Neurosci 4:13–25

Trendelenburg U (1991) The TiPS lecture: functional aspects of the neuronal uptake of noradrenaline. Trends Pharmacol Sci 12:334–337

Ungerer M, Chlistalla A, Richardt G (1996) Upregulation of cardiac uptake 1 carrier in ischemic and nonischemic rat heart. Circ Res 78:1037–1043

Urwin RE, Bennetts B, Wilcken B, Lampropoulos B, Beumont P, Clarke S, Russell J, Tanner S, Nunn KP (2002) Anorexia nervosa (restrictive subtype) is associated with a polymorphism in the novel norepinephrine transporter gene promoter polymorphic region. Mol Psychiatry 7:652–657

Urwin RE, Bennetts BH, Wilcken B, Lampropoulos B, Beumont PJV, Russell JD, Tanner SL, Nunn KP (2003) Gene-gene interaction between the monoamine oxidase A gene and solute carrier family 6 (neurotransmitter transporter, noradrenalin) member 2 gene in anorexia nervosa (restrictive subtype). Eur J Hum Genet 11:945–950

Vatta MS, Presas M, Bianciotti LG, Zarrabeitia V, Fernandez BE (1996) B and C types natriuretic peptides modulate norepinephrine uptake and release in the rat hypothalamus. Regul Pept 65:175–184

Vatta MS, Presas MF, Bianciotti LG, Rodriguez-Fermepin M, Ambros R, Fernandez BE (1997) B and C types natriuretic peptides modify norepinephrine uptake and release in the rat adrenal medulla. Peptides 18:1483–1489

Vizi ES, Zsilla G, Caron MG, Kiss JP (2004) Uptake and release of norepinephrine by serotonergic terminals in norepinephrine transporter knock-out mice: implications for the action of selective serotonin reuptake inhibitors. J Neurosci 24:7888–7894

Wakade AR, Wakade TD, Poosch M, Bannon MJ (1996) Noradrenaline transport and transporter mRNA of rat chromaffin cells are controlled by dexamethasone and nerve growth factor. J Physiol 494:67–75

Wiedemann P, Pörzgen P, Bönisch H, Brüss M (1998) The human noradrenaline transporter gene: a new and alternatively spliced noncoding 5′ exon. Naunyn Schmiedebergs Arch Pharmacol 358:R409

Willis WD, Westlund KN (1997) Neuroanatomy of the pain system and of the pathways that modulate pain. J Clin Neurophysiol 14:2–31

Wilson AA, Johnson DP, Mozley D, Hussey D, Ginovart N, Nobrega J, Garcia A, Meyer J, Houle S (2003) Synthesis and in vivo evaluation of novel radiotracers for the in vivo imaging of the norepinephrine transporter. Nucl Med Biol 30:85–92

Wölfel R, Graefe KH (1992) Evidence for various tryptamines and related compounds acting as substrates of the platelet 5-hydroxytryptamine transporter. Naunyn Schmiedebergs Arch Pharmacol 345:129–136

Xu F, Gainetdinov RR, Wetsel WC, Jones SR, Bohn LM, Miller GW, Wang YM, Caron MG (2000) Mice lacking the norepinephrine transporter are supersensitive to psychostimulants. Nat Neurosci 3:465–471

Xu XH, Knight J, Brookes K, Mill J, Sham P, Craig I, Taylor E, Asherson P (2005) DNA pooling analysis of 21 norepinephrine transporter gene SNPs with attention deficit hyperactivity disorder: No evidence for association. Am J Med Genet B Neuropsychiatr Genet 134B:115–118

Yamada Y, Miyajima E, Tochikubo O, Matsukawa T, Ishii M (1989) Age-related-changes in muscle sympathetic-nerve activity in essential-hypertension. Hypertension 13:870–877

Yamashita A, Singh SK, Kawate T, Jin Y, Gouaux E (2005) Crystal structure of a bacterial homologue of Na^+/Cl^--dependent neurotransmitter transporters. Nature 437:215–223

Yang H, Raizada MK (1999) Role of phosphatidylinositol 3-kinase in angiotensin II regulation of norepinephrine neuromodulation in brain neurons of the spontaneously hypertensive rat. J Neurosci 19:2413–2423

Yang L, Wang YF, Li J, Faraone SV (2004) Association of norepinephrine transporter gene with methylphenidate response. J Am Acad Child Adolesc Psychiatry 43:1154–1158

Yehuda R, Siever LJ, Teicher MH, Levengood RA, Gerber DK, Schmeidler J, Yang RK (1998) Plasma norepinephrine and 3-methoxy-4-hydroxyphenylglycol concentrations and severity of depression in combat posttraumatic stress disorder and major depressive disorder. Biol Psychiatry 44:56–63

Young JB, Landsberg L (1998) Catecholamines and the adrenal medulla. In: Wilson JD, Foster DW, Kroenberg HM, Larsen PR (eds) Williams textbook of endocrinology, 9th edn. WB Saunders, Philadelphia, pp 665–728

Zahniser NR, Doolen S (2001) Chronic and acute regulation of Na^+/Cl^--dependent neurotransmitter transporters: drugs, substrates, presynaptic receptors, and signaling systems. Pharmacol Ther 92:21–55

Zhu MY, Ordway GA (1997) Down-regulation of norepinephrine transporters on PC12 cells by transporter inhibitors. J Neurochem 68:134–141

Zhu MY, Blakely RD, Apparsundaram S, Ordway GA (1998) Down-regulation of the human norepinephrine transporter in intact 293-hNET cells exposed to desipramine. J Neurochem 70:1547–1555

Zhu MY, Kyle PB, Hume AS, Ordway GA (2004) The persistent membrane retention of desipramine causes lasting inhibition of norepinephrine transporter function. Neurochem Res 29:419–427

Zill P, Engel R, Baghai TC, Juckel G, Frodl T, Muller-Siecheneder F, Zwanzger P, Schule C, Minov C, Behrens S, Rupprecht R, Hegerl W, Moller HJ, Bondy B (2002) Identification of a naturally occurring polymorphism in the promoter region of the norepinephrine transporter and analysis in major depression. Neuropsychopharmacology 26:489–493

Zuetenhorst H, Taal BG, Boot H, Olmos RV, Hoefnagel G (1999) Long-term palliation in metastatic carcinoid tumours with various applications of meta-iodobenzylguanidin (MIBC): pharmacological MIBG, I-131-labelled MIBG and the combination. Eur J Gastroenterol Hepatol 11:1157–1164

The High-Affinity Choline Transporter: A Critical Protein for Sustaining Cholinergic Signaling as Revealed in Studies of Genetically Altered Mice

M. H. Bazalakova · R. D. Blakely (✉)

Vanderbilt School of Medicine, Suite 7140, MRB III, Nashville TN, 37232-8548, USA
randy.blakely@vanderbilt.edu

1	CHT Function and Regulation	526
1.1	HACU and Cholinergic Neurotransmission	526
1.2	Pharmacological and Behavioral Modulation of HACU	528
1.3	Cloning the CHT Gene	529
1.4	CHT Provides the Molecular Basis of HACU Regulation	530
2	CHT and Genetic Mouse Models of Cholinergic Dysfunction	531
2.1	CHT Is Required for Survival: CHT Homozygosity Is Lethal	531
2.2	CHT Is Upregulated in the $AChE^{-/-}$ Mouse	534
2.3	CHT Is Also Upregulated in the AChE Transgenic Mouse	535
2.4	CHT Upregulation Provides Mechanism of Compensation in $ChAT^{+/-}$ Mice	536
2.5	CHT Downregulation in $\alpha 3^{-/-}$ Mice	537
3	$CHT^{+/-}$ Mice as Models of Cholinergic Dysfunction	537
3.1	Functional Compensation and Normal Basal Behaviors in $CHT^{+/-}$ Mice	538
3.2	Physical and Pharmacological Challenges Reveal Motor Phenotypes in $CHT^{+/-}$ Mice	538
3.3	Muscarinic Expression Is Altered in $CHT^{+/-}$ Mice	539
	References	540

Abstract In cholinergic neurons, the presynaptic choline transporter (CHT) mediates high-affinity choline uptake (HACU) as the rate-limiting step in acetylcholine (ACh) synthesis. It has previously been shown that HACU is increased by behaviorally and pharmacologically-induced activity of cholinergic neurons in vivo, but the molecular mechanisms of this change in CHT function and regulation have only recently begun to be elucidated. The recent cloning of CHT has led to the generation of new valuable tools, including specific anti-CHT antibodies and a CHT knockout mouse. These new reagents have allowed researchers to investigate the possibility of a presynaptic, CHT-mediated, molecular plasticity mechanism, regulated by and necessary for sustained in vivo cholinergic activity. Studies in various mouse models of cholinergic dysfunction, including acetylcholinesterase (AChE) transgenic and knock-out mice, choline acetyltransferase (ChAT) heterozygote mice, muscarinic (mAChR) and nicotinic (nAChR) receptor knockout mice, as well as CHT knockout and heterozygote mice, have revealed new information about the role of CHT expression and regulation in response to long-term alterations in cholinergic neurotransmission. These mouse models highlight the capacity of CHT to provide for functional compensation in states of cholinergic dysfunction. A better understanding of modes of CHT regulation should allow for experimental manipulation of cholinergic signaling in vivo with potential utility in human disorders of known cholinergic dysfunction such as Alzheimer's disease, Parkinson's disease, schizophrenia, Huntington's disease, and dysautonomia.

Keywords Choline · Transporter · Knockout · Transgenic · Heterozygote · Hemicholinium-3

1
CHT Function and Regulation

1.1
HACU and Cholinergic Neurotransmission

The presynaptic high-affinity choline transporter (CHT) derives its physiological importance from its support of the synthesis of the neurotransmitter acetylcholine (ACh). ACh plays an important role in both central and peripheral nervous systems, where it maintains vital biological functions such as motor function, attention, memory, reward, and autonomic regulation (Winkler et al. 1995; Sarter and Bruno 1997; Perry et al. 1999; Dani 2001; Misgeld et al. 2002; Fink-Jensen et al. 2003; Kitabatake et al. 2003; Wess 2004). Selective blockade or genetic ablation of the rate-limiting step in ACh synthesis—CHT-mediated high-affinity choline uptake (HACU)—reduces ACh synthesis and release in vitro, and impairs cholinergic transmission at central and peripheral synapses in vivo (Macintosh et al. 1956; Guyenet et al. 1973; Maire and Wurtman 1985; Ferguson et al. 2004). Breakdown of cholinergic signaling at the neuromuscular junction (NMJ) due to CHT dysfunction leads to lethal respiratory paralysis and demonstrates the essential physiological role of CHT (Schueler 1955; Ferguson et al. 2004).

The turnover of ACh in cholinergic neurons involves a well-studied cycle of synthesis, release, hydrolysis, and reuptake (Fig. 1). The enzyme choline acetyltransferase (ChAT) synthesizes ACh in the axoplasm from the precursors choline and acetyl coenzyme A (acetyl-CoA). ChAT is not believed to be saturated with the substrate choline, as presynaptic levels of choline are significantly lower than ChAT K_m. ACh is packaged into synaptic vesicles by the vesicular ACh transporter (VAChT), and released into the synaptic cleft upon depolarization of the neuron by an action potential. Once released in the synaptic cleft, ACh interacts with pre- and postsynaptic nicotinic (nAChR) and muscarinic (mAChR) receptors and is inactivated enzymatically, through hydrolysis, into acetate and choline by the enzyme acetylcholinesterase (AChE). The choline is recycled at the presynaptic terminal via a distinct, carrier-mediated HACU mechanism. Low-affinity choline uptake (LACU) transporters are found in all cell types, are not enriched in nerve terminals, and take up choline to be used in phosphatidylcholine synthesis in order to support cellular membrane maintenance and repair needs (Bussiere et al. 2001). Cholinergic neurons have an added requirement, elaborated in the presynaptic terminal,

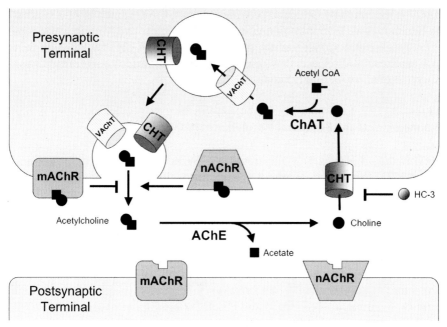

Fig. 1 Acetylcholine (ACh) turnover at the cholinergic synapse. The enzyme choline acetyltransferase (*ChAT*) synthesizes ACh in the axoplasm from the precursors choline and acetyl coenzyme A (*acetyl-CoA*). ACh is then packaged into synaptic vesicles by the vesicular ACh transporter (*VAChT*), and released into the synaptic cleft upon neuronal depolarization. In the synaptic cleft, ACh interacts with pre- and postsynaptic nicotinic (*nAChR*) and muscarinic (*mAChR*) receptors. Presynaptic mAChRs exert an autoinhibitory effect on ACh release, whereas presynaptic nAChR activation increases ACh release. Synaptic ACh is hydrolyzed into acetate and choline by the enzyme acetylcholine esterase (*AChE*). Choline is then transported into the presynaptic terminal by the hemicholinium-3 (*HC-3*) sensitive, high-affinity choline transporter (*CHT*), in the rate-limiting step of subsequent ACh synthesis

for choline, which they cannot synthesize de novo, but need as a substrate for ACh synthesis. This need is supported by the HACU mechanism.

The HACU process was described as early as 1961 (Birks and McIntosh 1961). Subsequent work showed it is present primarily in cholinergic nerve terminals (Kuhar et al. 1973; Kuhar and Murrin 1978), where it takes up choline specifically for the biosynthesis and replenishment of ACh pools (Yamamura and Snyder 1972; Haga and Noda 1973; Kuhar et al. 1973; Blusztajn 1998; Bussiere et al. 2001). HACU displays affinity for choline in the lower micromolar range ($K_m^{HACU} = 1\text{–}5$ μM, versus $K_m^{LACU} = 50$ μM) is a Na^+- and Cl^--dependent process (versus Na^+/Cl^--independent LACU), and is inhibited by the selective and competitive antagonist hemicholinium-3 (HC-3) ($K_i^{HACU} = 10\text{–}100$ nM, versus $K_i^{LACU} = 50$ μM) (Yamamura and Snyder 1972; Yamamura and Snyder 1973; Simon and Kuhar 1976; Kuhar and Murrin 1978). We now know that

HACU is mediated by CHT, and provides the rate-limiting step in ACh synthesis (Holden et al. 1975; Simon and Kuhar 1975; Kuhar and Murrin 1978; Jope 1979; Tucek 1985). Consequently, specific inhibition of CHT by HC-3 impairs ACh synthesis and release, thus compromising cholinergic neurotransmission (Guyenet et al. 1973; Murrin et al. 1977).

1.2
Pharmacological and Behavioral Modulation of HACU

CHT-mediated HACU is highly regulated by neuronal activity (Simon and Kuhar 1975), but the molecular mechanisms of this regulation remained unclear until recently. Traditionally, [^{3}H]HC-3 ligand binding assays have been used to quantitatively determine CHT levels in sealed nerve terminal preparations (synaptosomes) or tissue sections (Vickroy et al. 1984; Sandberg and Coyle 1985; Vickroy et al. 1985). Using uptake and binding techniques, studies have shown that CHT-mediated HACU is regulated by neuronal activity in vitro and in vivo (Murrin and Kuhar 1976). Accordingly, in vivo administration of drugs that affect the turnover and release of ACh (both indices of cholinergic activity) results in altered HACU and [^{3}H]HC-3 binding measured in vitro. For example, intraperitoneal (i.p.) administration of the muscarinic antagonist atropine, which induces cholinergic firing and ACh release, leads to increases in HACU V_{max} and [^{3}H]HC-3 binding in synaptosomes from rat striatum and hippocampus (Goldman and Erickson 1983; Lowenstein and Coyle 1986). In contrast, treatment with agents that inhibit cholinergic signaling, such as pentobarbital, results in lower ACh turnover, HACU, and [^{3}H]HC-3 binding (Lowenstein and Coyle 1986).

Additional studies have employed behavioral stimuli, such as training to criterion in a T maze or an 8-arm radial maze, to activate cholinergic neurons and assess CHT response in the whole animal (Burgel and Rommelspacher 1978; Wenk et al. 1984). The data demonstrate increases in synaptosomal HACU V_{max} from behaviorally trained versus non-trained animals, remarkably lasting more than 20 days beyond the final training or testing session (Wenk et al. 1984). In a modern refinement of these studies, Apparsundaram and colleagues have examined CHT-mediated HACU in a behavioral paradigm that specifically taxes attentional effort and therefore depends on intact cortical cholinergic circuits (McGaughy et al. 1996; Apparsundaram et al. 2005). The authors trained rats in a cognitive vigilance task (CVT), which required the animals to detect the presence or absence of a light stimulus and respond by making the correct choice between two levers. Control animals were either not subjected to the behavioral paradigm [non-performing (NP) rats] or were trained in a simple reaction time task (SRTT), where only the correct lever was present, thus decreasing demands on attentional processing. HACU V_{max} in synaptosomes from the right medial prefrontal cortex (mPFC), but not striatum, was accelerated in CVT compared to SRTT and NP control rats

(Apparsundaram et al. 2005). These results support a coupling of cholinergic signaling—demonstrated by increased extracellular ACh release in the mPFC of CVT-performing rats—and CHT capacity (Arnold et al. 2002).

Two explanations for a mechanism underlying activity-dependent changes in HACU V_{max} have been proposed: a change in the intrinsic rate of CHT activity (catalytic activation) or a change in the number of CHTs at the plasma membrane (altered trafficking). Although changes in [^{3}H]HC-3 binding have commonly been interpreted in the literature as an index of changing CHT protein levels, several studies suggest that alterations in [^{3}H]HC-3 binding may reflect not only changes in the number of binding sites (B_{max}) but also affinity for the ligand (K_D) (Lowenstein and Coyle 1986). Studies have shown Ca^{2+}-dependent, ATP-induced changes in [^{3}H]HC-3 binding in membrane preparations, which appear to be primarily due to changes in K_D (Chatterjee and Bhatnagar 1990; Saltarelli et al. 1990). This interpretation is supported by the use of membrane preparations which, unlike synaptosomal sealed nerve terminals, do not contain intracellular reserve pools of CHT that can be trafficked to the plasma membrane (Chatterjee and Bhatnagar 1990; Saltarelli et al. 1990). Therefore, any differences in [^{3}H]HC-3 binding in membrane preparations are likely the result of conformational changes of the CHT protein already present in the plasma membrane, due to phosphorylation or other protein modifications. The cloning and characterization of CHT have provided new tools, such as CHT-specific antibodies, which can be used to affect a more direct and reliable quantification of CHT protein levels at the plasma membrane.

1.3
Cloning the CHT Gene

CHT genes have now been identified in *Caenorhabditis elegans* (*Cho-1*) (Okuda et al. 2000), torpedo (*CTL1*) (O'Regan et al. 2000), rat (*CHT1*) (Okuda et al. 2000), human (*hCHT*) (Apparsundaram et al. 2000; Okuda and Haga 2000), and mouse (*mCHT*) (Apparsundaram et al. 2001a). The mCHT complementary DNA (cDNA) encodes a protein of 580 amino acids, with a molecular mass of approximately 63 kDa. Sequence analysis of mCHT demonstrates a 93% and 98% amino acid identity with the human and rat orthologues respectively (Apparsundaram et al. 2001a), and identifies the gene as a member of the Na^+/glucose solute carrier superfamily 5 (SLC5). Analysis of the amino acid sequence predicts a topology of 13 transmembrane domains, an extracellular consensus site for N-linked glycosylation, and several cytoplasmic protein kinase A (PKA) and protein kinase C (PKC) phosphorylation sites (Apparsundaram et al. 2001a; for review see Ferguson and Blakely 2004). Electron microscopy (EM) immunogold labeling studies, using antibodies directed towards C-terminal epitopes, support a cytoplasmic localization of the CHT carboxy-terminus (Ferguson et al. 2003), and epitope-tagging efforts confirm an extracellular localization of the N-terminus (Ribeiro et al. 2005). In addition to the generation of polyclonal and monoclonal CHT-specific antibodies, the cloning of mCHT has led to the generation of a CHT knockout mouse (Ferguson et al. 2004)—both useful new tools for the study of in vivo CHT function and regulation.

1.4
CHT Provides the Molecular Basis of HACU Regulation

The generation of CHT-specific antibodies allowed for the first detailed characterization of CHT expression and subcellular localization. Several studies utilizing immunoblotting and immunofluorescence have demonstrated that CHT is expressed predominantly in cholinergic terminals (Misawa et al. 2001; Lips et al. 2002; Ferguson et al. 2003; Kus et al. 2003), although expression has also been detected in human keratinocytes, rat tracheal epithelia, lung bronchial epithelial cells, and human T lymphocytes (Fujii et al. 2003; Kawashima and Fujii 2003; Pfeil et al. 2003a; Pfeil et al. 2003b; Proskocil et al. 2004).

Within cholinergic terminals, where HACU activity predominates and the transported choline is specifically targeted for ACh synthesis (Bussiere et al. 2001), CHT is localized in at least two distinct pools: an intracellular vesicular pool and a plasma membrane pool (Ferguson et al. 2003; Ferguson and Blakely 2004). Immunogold labeling and EM analysis in striatal cholinergic terminals and the NMJ reveal that the majority of CHT is present on intracellular small, clear synaptic vesicles (Ferguson et al. 2003; Nakata et al. 2004). At the NMJ, less than 5% of CHT immunoreactivity is associated with the plasma membrane (Nakata et al. 2004). Parallel biochemical approaches utilizing subcellular fractionation and velocity gradient purification confirm an enrichment of CHT in synaptic vesicles that are also enriched for synaptic vesicle proteins such as synaptophysin and VAChT, and can be demonstrated to contain ACh (Ferguson et al. 2003).

The pattern of synaptic vesicle CHT enrichment lends strong support to the hypothesis that CHT activity may be regulated, at least in part, by trafficking of CHT intracellular reserves to the plasma membrane under conditions of demand on cholinergic neurotransmission. Indeed, high potassium in vitro depolarization of striatal synaptosomal preparations (which leads to ACh release and hence mimics in vivo neuronal activity) results in increased HACU V_{max}, mediated largely, but not entirely, by redistribution of CHT from the intracellular reserve pool (Ferguson et al. 2003). Interestingly, while all CHT-positive synaptic vesicles also contain VAChT, only CHT appears to be captured at the plasma membrane for a period of time following synaptic terminal depolarization (Ferguson et al. 2003). The presence of dileucine-like internalization motifs in the cytoplasmic C-terminus of CHT and the transporter's largely intracellular localization suggest that specific mechanisms must exist to allow stabilization of CHT at the plasma membrane for prolonged enhancements of HACU. Indeed, experimental evidence from site-directed mutagenesis followed by in vitro biochemical studies in transfected cell lines, as well as primary neuronal cultures, points to clathrin-mediated endocytosis as a highly efficient mechanism for CHT trafficking (A.M. Ruggiero, S.M. Ferguson, H. Iwemoto, L. DeFelice, and R.D. Blakely, in preparation; Ribeiro et al. 2003, 2005). Gates and colleagues have used immunoprecipitation, immunoblotting, and synaptosomal biotinylation techniques to reveal that PKC and protein phosphatase 2A (PP2A)-induced alterations in CHT trafficking are paralleled by changes in CHT phosphorylation, although specific mechanisms linking phosphorylation with trafficking capacity are as yet unclear (Gates et al. 2004).

Finally, the Sarter and Apparsundaram labs elegantly combined in vivo behavioral paradigms (specifically taxing attentional cholinergic circuits) with in vitro synaptosomal subcellular fractionation (allowing quantification of CHT in the plasma membrane) in order to provide an explanation for the changes in HACU V_{max} observed following behavioral modulation of cholinergic activity (Apparsundaram et al. 2005). The authors observed significantly increased density of CHT, but not VAChT or other synaptic protein, at the plasma membrane in response to cholinergic activity. Such increases in density were not observed in a control behavioral paradigm that did not challenge cholinergically supported attentional processes specifically. The above studies provide strong evidence that one mechanism of CHT regulation, currently understood best, is CHT trafficking from intracellular reserves to the plasma membrane. Additional studies will be needed to describe the molecular detail of possible alternative modes of CHT regulation, including catalytic activation of plasma membrane resident CHT proteins. Of note, activation of biogenic amine transporters by p38 mitogen-activated protein kinase has been described recently, and may provide a model for non-trafficking-based activation of CHT (Apparsundaram et al. 2001b; Zhu et al. 2004).

2
CHT and Genetic Mouse Models of Cholinergic Dysfunction

Genetic approaches have been used to generate mice with overexpression or knockouts of various components of the system that supports ACh neurotransmitter turnover, including CHT itself, the ChAT and AChE enzymes, and various muscarinic and nicotinic receptor subtypes. These mouse models of cholinergic dysfunction have provided new evidence for a dynamic role of CHT as a critical modulator of cholinergic neurotransmission (Table 1).

2.1
CHT Is Required for Survival: CHT Homozygosity Is Lethal

The generation of CHT knockout mice via homologous recombination allowed the direct study of CHT's physiological impact in vivo (Ferguson et al. 2004).

Table 1 Compensatory changes in mouse models of cholinergic dysfunction

Mouse model	ACh metabolism	Total CHT expression	Synaptic HACU	mAChR expression	nAChR expression	Altered behavior	Reference(s)
CHT+/−	Normal bulk tissue ACh content (LC/MS, HPLC)	Reduced (Whole brain; specific regions include striatum, cortex, hippocampus, midbrain, hindbrain; 50%; Western)	Not changed	Reduced M1 (striatum, 25%); M2 (striatum, 50%; cortex, 35%; Western)	N/A	Hyposensitive to scopolamine Hypersensitive to oxotremorine, HC-3 Impaired treadmill performance	Ferguson et al. 2004; Bazalakova et al. 2004, submitted[a]
AChE−/−	Hypo hypercholinergic symptoms, i.e. tremor	Increased (Striatum; 50%; Western)	N/A	Reduced M1, M2, M4 (cortex and hippocampus, 50%–80%; Western)	N/A	Hypersensitive to scopolamine Hyposensitive to oxotremorine	Li et al. 2000, 2003; Volpicelli-Dalley et al. 2003a, b
AChE-Tg	Hyper Normal basal ACh extracellular levels (hippocampus, microdyalisis) ACh release hypersensitive to halothane	Increased (Striatum, 100%; HC-3 binding)	Increased (Striatal and hippocampal, but not cortical) synaptosomes	Increased M2 (striatum; AF-DX-284 binding)	Increased α4 (striatum; cytosine binding)	Hyposensitive to oxotremorine Hyposensitive to nicotine Hyperactivity in novel environment (open field)	Beeri et al. 1995, 1997; Erb et al. 2001; Svedberg et al. 2003

Table 1 (continued)

Mouse model	ACh metabolism	Total CHT expression	Synaptic HACU	mAChR expression	nAChR expression	Altered behavior	Reference(s)
ChAT[+/−]	Normal bulk content, in vitro ACh release (hippocampal slices)	Increased (Septum, 70% mRNA; RT-PCR) (Striatum, cortex, hippocampus, spinal cord; 150% protein; Western)	Increased (Hippocampal slices, 110%)	N/A	N/A	Normal baseline Morris water maze performance	Brandon et al. 2004
α3[−/−]	Reduced ACh release at sympathetic ganglia (electrophysiology)	Possibly reduced at preganglionic SGC terminals	N/A	N/A	N/A	ACh release is hyposensitive to HC-3	Rassadi et al. 2005;

I. ACh cycle: acetylcholine (ACh); acetylcholinesterase (AChE); choline acetyltransferase (ChAT); high-affinity choline transporter (CHT); high-affinity choline uptake (HACU); nicotinic acetylcholine receptor (nAChR); muscarinic acetylcholine receptor (mAChR); superior cervical ganglia (SCG). II. Pharmacological agents: hemicholinium-3 (HC-3); muscarinic agonist (oxotremorine); muscarinic antagonist (scopolamine); nicotinic agonist (nicotine). III. Genotypes: heterozygous (+/−); homozygous knockout (−/−); transgenic (Tg); wildtype (+/+). IV. Various: high-performance liquid chromatography (HPLC); liquid chromatography/mass spectrometry (LC/MS); reverse-transcriptase polymerase reaction (RT-PCR); data not available (N/A). [a]M.H. Bazalakova, J. Wright, E.J. Schneble, M.P. McDonald, C.J. Heilman, A.I. Levey, and R.D. Blakely

Homozygous ($CHT^{-/-}$) mice are born at expected Mendelian ratios and display normal gross anatomy. However, $CHT^{-/-}$ pups appear apneic and hypoxic, are largely immobile, and die within an hour after birth. Electrophysiological recordings at the sternomastoid NMJ show that the $CHT^{-/-}$ animals are born with wildtype stores of ACh, but are unable to sustain ACh synthesis and release under high demand for ACh turnover at the NMJ synapse, leading to motor and respiratory paralysis (Ferguson et al. 2004). Detailed immunofluorescence studies of the $CHT^{-/-}$ NMJ reveal a more subtle phenotype of immature NMJ formation, similar to the phenotype observed in another knockout mouse, $CHAT^{-/-}$, which is incapable of ACh synthesis and is also not viable (Misgeld et al. 2002; Brandon et al. 2003). However, $CHT^{+/-}$ and $CHAT^{+/-}$ mice survive, appear grossly indistinguishable from $CHT^{+/+}$ mice, and can be used as valuable tools to investigate CHT function and regulation.

2.2
CHT Is Upregulated in the $AChE^{-/-}$ Mouse

AChE hydrolysis of ACh in the synaptic cleft terminates cholinergic neurotransmission. Therefore, a genetically induced loss of AChE may be viewed as a hypercholinergic state, where diminished ACh metabolism leads to sustained cholinergic receptor stimulation. While such prolonged receptor activation can be lethal, $AChE^{-/-}$ mice survive to adulthood, although they are unable to eat solid food and their lifespan is reduced due to fatal seizures (Xie et al. 2000; Duysen et al. 2002). The survival of these animals raises two questions: What parameters change to compensate for a state of ACh hypometabolism and which genes show the most plastic response?

Whereas butyrylcholinesterase activity and messenger RNA (mRNA) levels of muscarinic receptors in these mice are not changed, [^{3}H]quinuclinyl benzilate ([^{3}H]QNB) binding studies, immunofluorescence, and immunoblotting with specific antibodies demonstrate 50%–80% reduced expression of M1, M2, and M4 receptors in $AChE^{-/-}$ cortex and hippocampus homogenates, as well as decreased cell surface density and increased internalization of the muscarinic receptors (Li et al. 2000; Mesulam et al. 2002; Li et al. 2003; Volpicelli-Daley et al. 2003a). Consequently, $AChE^{-/-}$ mice display reduced sensitivity to mAChR stimulation such as oxotremorine-induced hypothermia, tremor, salivation, and analgesia, but dose-dependent heightened sensitivity to mAChR inhibition, namely scopolamine-induced increases in locomotor activity (Volpicelli-Daley et al. 2003b).

Intriguingly, immunoblotting for CHT reveals a 60% increase in CHT expression in striatal homogenates, in the absence of changes in ChAT activity or VAChT expression (Volpicelli-Daley et al. 2003b). This compensatory change may reflect an attempt by the presynaptic terminal to increase substrate availability in a cholinergic system deprived of choline derived from ACh hydrolysis, which thus appears critical for the sustained re-synthesis of releasable neurotransmitter. Micromolar choline levels are measured in blood and CSF, but levels at cholinergic terminals may be more limited due to efficient clearance mechanisms (including both HACU and LACU). Regardless, $AChE^{-/-}$ phenotypes complement findings in $CHT^{+/-}$ mice that, in contrast to $AChE^{-/-}$ animals, have approximately 50% less CHT protein compared to CHT wildtype ($CHT^{+/+}$) siblings, are hyposensitive to scopolamine challenge, and are hyperresponsive to oxotremorine-induced tremor (Bazalakova et al. 2004; M.H. Bazalakova, J. Wright, E.J. Schneble, M.P. McDonald, C.J. Heilman, A.I. Levey, and R.D. Blakely, submitted). The above studies suggest that CHT upregulation and muscarinic receptor downregulation and internalization act in concert to provide a mechanism of compensation for decreased ACh metabolism in $AChE^{-/-}$ mice. These data also indicate that a mechanism must be in place to transduce changes in extracellular levels of ACh to CHT. This feedback mechanism could be mediated by presynaptic receptors or a retrograde signal that has yet to be identified.

2.3
CHT Is Also Upregulated in the AChE Transgenic Mouse

Interestingly, increased CHT protein expression is present in another model of cholinergic dysfunction—AChE transgenic (AChE-Tg) mice. Autoradiography in brain sections reveals a higher than twofold increase in [^{3}H]HC-3 binding in the striatum of AChE-Tg mice compared to wildtype controls (Beeri et al. 1997). These findings correlate with observations of approximately 70% higher HACU in striatal and hippocampal, but not cortical synaptosomal preparations from AChE transgenics (Erb et al. 2001). Whereas microdialysis revealed normal basal levels of extracellular ACh in the hippocampi of freely moving AChE-Tg mice, halothane anesthesia reduced both ACh levels and synaptosomal HACU to significantly lower levels in AChE transgenics compared to wildtype controls. In addition, administration of the AChE inhibitor physostigmine (i.p.) normalized the higher HACU observed in AChE transgenics to control levels (Erb et al. 2001). Although it may appear paradoxical that CHT is upregulated in both AChE-Tg and $AChE^{-/-}$ mice, CHT upregulation in the AChE-Tg brain may serve as an attempt to compensate for overall cholinergic hypofunction caused by hypermetabolism of ACh.

Whereas initial studies reported unchanged tritiated ligand binding for nicotinic ACh receptors (nAChRs) and muscarinic ACh receptors (mAChRs) in brain sections (Beeri et al. 1997), subsequent studies detected increased [^{3}H]cytosine (α4 nAChR subunit) and [^{3}H]AF-DX-384 (M2 mAChR subtype) binding in homogenates from AChE transgenic striata. No changes were observed in [^{125}I]a-bungarotoxin (a7 nicotinic receptor subunit) or [^{3}H]pirenzepine (M1 muscarinic receptor subtype) binding (Svedberg et al. 2003). The specific changes in CHT, mAChR, and nAChR expression suggest that these proteins work concurrently to overcome insults on cholinergic signaling in

AChE transgenics, a strategy for compensation comparable to that observed in AChE$^{-/-}$ mice.

Altered expression of CHT and cholinergic receptors may underlie differences in behavior and sensitivity to pharmacological challenges in the AChE-Tg mice. Similarly to AChE$^{-/-}$ animals, AChE transgenics are resistant to muscarinic (oxotremorine) agonist-induced hypothermia, as well as nicotinic (nicotine) agonist-induced hypothermia, but retain their ability to thermoregulate, as evidenced by normal responses to noradrenergic agents and exposure to cold (Beeri et al. 1995). AChE transgenics display normal motor behavior in familiar environments (repeated exposure to open field paradigm), but increased motor activity in novel environments (initial exposure to open field chambers) (Erb et al. 2001). The latter finding contrasts with observations of CHT$^{+/-}$ mice, which show a pattern of reduced running-wheel activity upon initial introduction to a novel environment, with activity gradually increasing to wildtype levels as the animals habituate to the surroundings (M.H. Bazalakova, preliminary findings). AChE transgenics also differ from CHT$^{+/-}$ mice in their performance in at least one cognitive behavioral task—the water maze spatial learning and memory task. While AChE transgenic mice are impaired in an age-dependent manner in both visible and hidden versions of the water maze, CHT$^{+/-}$ mice performance is not different from controls and is not affected by age (Beeri et al. 1995; Bazalakova et al. 2004; M.H. Bazalakova, J. Wright, E.J. Schneble, M.P. McDonald, C.J. Heilman, A.I. Levey, and R.D. Blakely, submitted).

In summary, the findings of increased [^{3}H]HC-3 binding and HACU in AChE transgenics suggest that CHT upregulation provides one compensatory mechanism in a hypocholinergic state due to excess ACh metabolism, possibly in conjunction with altered muscarinic and/or nicotinic receptor expression.

2.4
CHT Upregulation Provides Mechanism of Compensation in ChAT$^{+/-}$ Mice

Upregulation of CHT protein expression and function have also been observed in ChAT$^{+/-}$ mice. While ChAT activity is reduced by almost 50% in the ChAT$^{+/-}$ brain, both bulk ACh levels in the striatum, frontal cortex, and hippocampus, as well as high potassium depolarization-induced ACh release from ChAT$^{+/-}$ hippocampal slices, are normal. Not surprisingly, ChAT$^{+/-}$ performance in the water maze paradigm is indistinguishable from that of ChAT$^{+/+}$ controls (Brandon et al. 2004).

How then is functional compensation achieved in the ChAT$^{+/-}$ animals? The authors report that AChE activity is comparable between ChAT$^{+/-}$ and ChAT$^{+/+}$ mice, and while they do not investigate muscarinic or nicotinic receptor expression, reverse-transcriptase polymerase chain reaction and immunoblotting reveal 70% and 150% increases in CHT mRNA (septum) and protein (striatum, cortex, hippocampus, spinal cord), respectively. In addition, in vitro [^{3}H]ACh synthesis in hippocampal slices pre-loaded with [^{3}H]choline substrate is approximately 60% higher in ChAT$^{+/-}$ tissues compared to ChAT$^{+/+}$ controls (Brandon et al. 2004). The above studies point to CHT upregulation as an essential, compensatory response to the reduction in ChAT activity, sustaining wildtype levels of AChE activity, as well as ACh content and depolarization-evoked ACh release. Clearly, cholinergic neurons place a significant responsibility on CHT to balance deficits elsewhere in cholinergic signaling.

2.5
CHT Downregulation in α3$^{-/-}$ Mice

The α3 nicotinic receptor knockout (α3$^{-/-}$) mice provide the first illustration of CHT regulation as a compensatory response to altered cholinergic receptor function (Rassadi et al. 2005). Electrophysiological recordings from cholinergic sympathetic superior cervical ganglia in α3$^{-/-}$ mice demonstrate that absence of postsynaptic α3 receptors completely disrupts fast excitatory synaptic transmission. Immunofluorescence, confocal microscopy, and EM analyses demonstrate unaltered terminal localization of various synaptic proteins, including ChAT, VAChT, synaptobrevin, syntaxin 1a, and synaptotagmin, as well as the morphologic integrity of the cholinergic synapses under investigation, including normal levels of small clear synaptic vesicles, large dense core vesicles, docked vesicles, synaptic cleft width, etc. Instead, the authors identify the basis of the electrophysiological phenotype as a deficit in presynaptic ACh release (Rassadi et al. 2005).

Electrophysiological recordings reveal that repeated stimulation results in faster depletion of neurotransmitter in α3$^{-/-}$ terminals compared to α3$^{+/+}$ controls, reminiscent of the inability of the CHT$^{-/-}$ NMJ to sustain ACh turnover (Ferguson et al. 2004). Further emphasizing the parallel, HC-3 blockade has no effect on α3$^{-/-}$ ACh release, whereas it significantly reduces α3$^{+/+}$ ACh output, providing further support for the hypothesis that CHT downregulation provides a compensatory mechanism at α3$^{-/-}$ sympathetic ganglia synapses (Rassadi et al. 2005). It will be informative to use choline uptake, immunoblotting, and immunofluorescence approaches to assess potential loss of CHT function and expression as a conclusive mechanism by which ACh release capacity is lost in α3$^{-/-}$ mice. The findings in α3$^{-/-}$ synapses again point to the likely existence of a retrograde signal, whose ultimate target is regulation of CHT function and expression.

3
CHT$^{+/-}$ Mice as Models of Cholinergic Dysfunction

As described above, changes in CHT function and expression have been documented in various mouse models of cholinergic dysfunction, suggesting an active role for CHT regulation in response to insults on ACh neurotransmission (Table 1). These observations led to the hypothesis that reduced CHT reserves may result in impairment of cholinergically mediated behaviors under conditions of physical or pharmacological challenge to cholinergic circuits. In support of this idea, $CHT^{+/-}$ mice display an increase in sensitivity to the lethal effects of HC-3 compared to wildtype littermates (Ferguson et al. 2004).

3.1
Functional Compensation and Normal Basal Behaviors in $CHT^{+/-}$ Mice

Under normal conditions, $CHT^{+/-}$ mice survive to adulthood, are capable of sustaining basal sensory-motor behaviors, reproduce, and generally appear indistinguishable from their wildtype siblings. These observations are supported by the functional compensation evident in biochemical studies: Synaptosomal uptake assays show that choline transport in $CHT^{+/-}$ mice is comparable to uptake in $CHT^{+/+}$ controls (Ferguson et al. 2004). In addition, ChAT and AChE activities are unchanged in the $CHT^{+/-}$ brain (Ferguson et al. 2004; M.H. Bazalakova, J. Wright, E.J. Schneble, M.P. McDonald, C.J. Heilman, A.I. Levey, and R.D. Blakely, submitted). In our initial behavioral investigation of this model, $CHT^{+/-}$ performance is indistinguishable from wildtype in a wide variety of behavioral tasks, including sensory-motor (Irwin screen, pre-pulse inhibition), motor coordination (rotarod and wire hang), overall locomotor activity (open fields), anxiety (elevated plus maze, light-dark paradigm) and spatial learning and memory tests (Morris water maze, spontaneous Y maze alternation, rewarded T-maze alternation) (Bazalakova et al. 2004; M.H. Bazalakova, J. Wright, E.J. Schneble, M.P. McDonald, C.J. Heilman, A.I. Levey, and R.D. Blakely, submitted).

3.2
Physical and Pharmacological Challenges Reveal Motor Phenotypes in $CHT^{+/-}$ Mice

As opposed to relatively normal spontaneous behavior, focused challenges reveal that $CHT^{+/-}$ mice are vulnerable to sustained demands on cholinergically supported behaviors. For example, $CHT^{+/-}$ mice are capable of sustaining running on a treadmill at low speeds or for short periods of time, but they are unable to reach treadmill speeds as high as those attained by $CHT^{+/+}$ littermates. At a constant high speed, $CHT^{+/-}$ mice fatigue sooner than $CHT^{+/+}$ controls. Pharmacological challenges also reveal motor phenotypes in $CHT^{+/-}$ mice. Specifically, $CHT^{+/-}$ mice are hyposensitive to muscarinic antagonist (scopolamine)-induced increases in locomotion, and display a trend towards hypersensitivity to muscarinic agonist (oxotremorine)-induced tremor (Bazalakova et al. 2004; M.H. Bazalakova, J. Wright, E.J. Schneble, M.P. McDonald, C.J. Heilman, A.I. Levey, and R.D. Blakely, submitted).

3.3
Muscarinic Expression Is Altered in CHT$^{+/-}$ Mice

The altered sensitivity of CHT$^{+/-}$ mice to muscarinic agents may be explained by region-specific changes in mAChR expression. Whereas binding studies with the tritiated nonspecific muscarinic antagonist [^{3}H]QNB do not show dramatic overall changes in mAChR protein levels in the cortex, striatum, hippocampus and midbrain, immunoblot studies using mAChR-specific antibodies demonstrate decreased M1 (striatum) and M2 (cortex and striatum) protein expression (Bazalakova et al. 2004; M.H. Bazalakova, J. Wright, E.J. Schneble, M.P. McDonald, C.J. Heilman, A.I. Levey, and R.D. Blakely, submitted). These findings are particularly interesting in light of the alterations of muscarinic receptor expression found in conjunction with changes in CHT protein levels in other mouse models of cholinergic dysfunction, including AChE$^{-/-}$ and AChE-Tg mice (Erb et al. 2001; Svedberg et al. 2003; Volpicelli-Daley et al. 2003b).

It will be of interest to quantify M3, M4, and M5 expression in the CHT$^{+/-}$ brain, as well as evaluate CHT$^{+/-}$ mice performance in behaviors that are known to be altered in various mAChR knockouts. For example, cocaine self-administration is diminished in M5 knockouts, while amphetamine-stimulated locomotion is enhanced in M1 knockout mice (Gerber et al. 2001; Thomsen et al. 2005). Future studies will address these and similar questions, leading to a better understanding of the interplay between CHT and mAChRs in the maintenance of ACh turnover, physiological cholinergic signaling, and psychostimulant action.

In summary, CHT plays an important role in the maintenance of ACh turnover and cholinergic neurotransmission. Studies in various mouse models of cholinergic dysfunction demonstrate that CHT actively participates in compensatory pathways aimed at maintaining cholinergic signaling. Initial basic findings suggest that relocalization of vesicular pools to the plasma membrane surface constitutes a likely mechanism of CHT regulation. Studies utilizing genetically modified mouse models may reveal interesting alternatives. For example, if synaptosomal biotinylation or subcellular fractionation techniques demonstrate CHT protein presence on the plasma membrane of $\alpha 3^{-/-}$ brain terminals that lack HC-3 sensitive choline uptake, a mechanism of regulation other than vesicular translocation will be indicated. Future investigations are needed to provide detailed insight into the molecular mechanisms of CHT regulation and its interactions with additional neurotransmitter pathways, with the eventual goal of manipulating CHT function in disorders of cholinergic origin.

Acknowledgements We appreciate the generous support of the Alzheimer's Association (R.D.B.) and the NIH (MH073159, R.D.B.). M.H.B. is a member of the Medical Scientist Training Program (NIH award 5T32GM007347) at Vanderbilt University School of Medicine. We would like to thank Dr. Alicia M. Ruggiero for helpful discussion and critical review of the manuscript.

References

Apparsundaram S, Ferguson SM, George AL Jr, Blakely RD (2000) Molecular cloning of a human, hemicholinium-3-sensitive choline transporter. Biochem Biophys Res Commun 276:862–867

Apparsundaram S, Ferguson SM, Blakely RD (2001a) Molecular cloning and characterization of a murine hemicholinium-3-sensitive choline transporter. Biochem Soc Trans 29:711–716

Apparsundaram S, Sung U, Price RD, Blakely RD (2001b) Trafficking-dependent and -independent pathways of neurotransmitter transporter regulation differentially involving p38 mitogen-activated protein kinase revealed in studies of insulin modulation of norepinephrine transport in SK-N-SH cells. J Pharmacol Exp Ther 299:666–677

Apparsundaram S, Martinez V, Parikh V, Kozak R, Sarter M (2005) Increased capacity and density of choline transporters situated in synaptic membranes of the right medial prefrontal cortex of attentional task-performing rats. J Neurosci 25:3851–3856

Arnold HM, Burk JA, Hodgson EM, Sarter M, Bruno JP (2002) Differential cortical acetylcholine release in rats performing a sustained attention task versus behavioral control tasks that do not explicitly tax attention. Neuroscience 114:451–460

Bazalakova MH, Ferguson SM, Wright J, McDonald MP, Blakely RD (2004) Modeling presynaptic cholinergic hypofunction: pharmacological and behavioral studies in choline transporter heterozygous mice. Soc Neurosci Abstr 142.1

Beeri R, Andres C, Lev-Lehman E, Timberg R, Huberman T, Shani M, Soreq H (1995) Transgenic expression of human acetylcholinesterase induces progressive cognitive deterioration in mice. Curr Biol 5:1063–1071

Beeri R, Le Novere N, Mervis R, Huberman T, Grauer E, Changeux JP, Soreq H (1997) Enhanced hemicholinium binding and attenuated dendrite branching in cognitively impaired acetylcholinesterase-transgenic mice. J Neurochem 69:2441–2451

Birks R, McIntosh FC (1961) Acetylcholine metabolism of a sympathetic ganglion. Can J Biochem Physiol 39:787–825

Blusztajn JK (1998) Choline, a vital amine. Science 281:794–795

Brandon EP, Lin W, D'Amour KA, Pizzo DP, Dominguez B, Sugiura Y, Thode S, Ko CP, Thal LJ, Gage FH, Lee KF (2003) Aberrant patterning of neuromuscular synapses in choline acetyltransferase-deficient mice. J Neurosci 23:539–549

Brandon EP, Mellott T, Pizzo DP, Coufal N, D'Amour KA, Gobeske K, Lortie M, Lopez-Coviella I, Berse B, Thal LJ, Gage FH, Blusztajn JK (2004) Choline transporter 1 maintains cholinergic function in choline acetyltransferase haploinsufficiency. J Neurosci 24:5459–5466

Burgel P, Rommelspacher H (1978) Changes in high affinity choline uptake in behavioral experiments. Life Sci 23:2423–2427

Bussiere M, Vance JE, Campenot RB, Vance DE (2001) Compartmentalization of choline and acetylcholine metabolism in cultured sympathetic neurons. J Biochem (Tokyo) 130:561–568

Chatterjee TK, Bhatnagar RK (1990) Ca2(+)-dependent, ATP-induced conversion of the [3H]hemicholinium-3 binding sites from high- to low-affinity states in rat striatum: effect of protein kinase inhibitors on this affinity conversion and synaptosomal choline transport. J Neurochem 54:1500–1508

Dani JA (2001) Overview of nicotinic receptors and their roles in the central nervous system. Biol Psychiatry 49:166–174

Duysen EG, Stribley JA, Fry DL, Hinrichs SH, Lockridge O (2002) Rescue of the acetylcholinesterase knockout mouse by feeding a liquid diet; phenotype of the adult acetylcholinesterase deficient mouse. Brain Res Dev Brain Res 137:43–54

Erb C, Troost J, Kopf S, Schmitt U, Loffelholz K, Soreq H, Klein J (2001) Compensatory mechanisms enhance hippocampal acetylcholine release in transgenic mice expressing human acetylcholinesterase. J Neurochem 77:638–646

Ferguson SM, Blakely RD (2004) The choline transporter resurfaces: new roles for synaptic vesicles? Mol Interv 4:22–37

Ferguson SM, Savchenko V, Apparsundaram S, Zwick M, Wright J, Heilman CJ, Yi H, Levey AI, Blakely RD (2003) Vesicular localization and activity-dependent trafficking of presynaptic choline transporters. J Neurosci 23:9697–9709

Ferguson SM, Bazalakova M, Savchenko V, Tapia JC, Wright J, Blakely RD (2004) Lethal impairment of cholinergic neurotransmission in hemicholinium-3-sensitive choline transporter knockout mice. Proc Natl Acad Sci U S A 101:8762–8767

Fink-Jensen A, Fedorova I, Wortwein G, Woldbye DP, Rasmussen T, Thomsen M, Bolwig TG, Knitowski KM, McKinzie DL, Yamada M, Wess J, Basile A (2003) Role for M5 muscarinic acetylcholine receptors in cocaine addiction. J Neurosci Res 74:91–96

Fujii T, Okuda T, Haga T, Kawashima K (2003) Detection of the high-affinity choline transporter in the MOLT-3 human leukemic T-cell line. Life Sci 72:2131–2134

Gates J Jr, Ferguson SM, Blakely RD, Apparsundaram S (2004) Regulation of choline transporter surface expression and phosphorylation by protein kinase C and protein phosphatase 1/2A. J Pharmacol Exp Ther 310:536–545

Gerber DJ, Sotnikova TD, Gainetdinov RR, Huang SY, Caron MG, Tonegawa S (2001) Hyperactivity, elevated dopaminergic transmission, and response to amphetamine in M1 muscarinic acetylcholine receptor-deficient mice. Proc Natl Acad Sci U S A 98:15312–15317

Goldman ME, Erickson CK (1983) Effects of acute and chronic administration of antidepressant drugs on the central cholinergic nervous system. Comparison with anticholinergic drugs. Neuropharmacology 22:1215–1222

Guyenet P, Lefresne P, Rossier J, Beaujouan JC, Glowinski J (1973) Inhibition by hemicholinium-3 of (14C)acetylcholine synthesis and (3H)choline high-affinity uptake in rat striatal synaptosomes. Mol Pharmacol 9:630–639

Haga T, Noda H (1973) Choline uptake systems of rat brain synaptosomes. Biochim Biophys Acta 291:564–575

Holden JT, Rossier J, Beaujouan JC, Guyenet P, Glowinski J (1975) Inhibition of high-affinity choline transport in rat striatal synaptosomes by alkyl bisquaternary ammonium compounds. Mol Pharmacol 11:19–27

Jope RS (1979) High affinity choline transport and acetylCoA production in brain and their roles in the regulation of acetylcholine synthesis. Brain Res 180:313–344

Kawashima K, Fujii T (2003) The lymphocytic cholinergic system and its biological function. Life Sci 72:2101–2109

Kitabatake Y, Hikida T, Watanabe D, Pastan I, Nakanishi S (2003) Impairment of reward-related learning by cholinergic cell ablation in the striatum. Proc Natl Acad Sci U S A 100:7965–7970

Kuhar MJ, Murrin LC (1978) Sodium-dependent, high affinity choline uptake. J Neurochem 30:15–21

Kuhar MJ, Sethy VH, Roth RH, Aghajanian GK (1973) Choline: selective accumulation by central cholinergic neurons. J Neurochem 20:581–593

Kus L, Borys E, Ping Chu Y, Ferguson SM, Blakely RD, Emborg ME, Kordower JH, Levey AI, Mufson EJ (2003) Distribution of high affinity choline transporter immunoreactivity in the primate central nervous system. J Comp Neurol 463:341–357

Li B, Stribley JA, Ticu A, Xie W, Schopfer LM, Hammond P, Brimijoin S, Hinrichs SH, Lockridge O (2000) Abundant tissue butyrylcholinesterase and its possible function in the acetylcholinesterase knockout mouse. J Neurochem 75:1320–1331

Li B, Duysen EG, Volpicelli-Daley LA, Levey AI, Lockridge O (2003) Regulation of muscarinic acetylcholine receptor function in acetylcholinesterase knockout mice. Pharmacol Biochem Behav 74:977–986

Lips KS, Pfeil U, Haberberger RV, Kummer W (2002) Localisation of the high-affinity choline transporter-1 in the rat skeletal motor unit. Cell Tissue Res 307:275–280

Lowenstein PR, Coyle JT (1986) Rapid regulation of [3H]hemicholinium-3 binding sites in the rat brain. Brain Res 381:191–194

Macintosh FC, Birks RI, Sastry PB (1956) Pharmacological inhibition of acetylcholine synthesis. Nature 4543:1181

Maire JC, Wurtman RJ (1985) Effects of electrical stimulation and choline availability on the release and contents of acetylcholine and choline in superfused slices from rat striatum. J Physiol (Paris) 80:189–195

McGaughy J, Kaiser T, Sarter M (1996) Behavioral vigilance following infusions of 192 IgG-saporin into the basal forebrain: selectivity of the behavioral impairment and relation to cortical AChE-positive fiber density. Behav Neurosci 110:247–265

Mesulam MM, Guillozet A, Shaw P, Levey A, Duysen EG, Lockridge O (2002) Acetylcholinesterase knockouts establish central cholinergic pathways and can use butyrylcholinesterase to hydrolyze acetylcholine. Neuroscience 110:627–639

Misawa H, Nakata K, Matsuura J, Nagao M, Okuda T, Haga T (2001) Distribution of the high-affinity choline transporter in the central nervous system of the rat. Neuroscience 105:87–98

Misgeld T, Burgess RW, Lewis RM, Cunningham JM, Lichtman JW, Sanes JR (2002) Roles of neurotransmitter in synapse formation: development of neuromuscular junctions lacking choline acetyltransferase. Neuron 36:635–648

Murrin LC, Kuhar MJ (1976) Activation of high-affinity choline uptake in vitro by depolarizing agents. Mol Pharmacol 12:1082–1090

Murrin LC, DeHaven RN, Kuhar MJ (1977) On the relationship between (3H)choline uptake activation and (3H)acetylcholine release. J Neurochem 29:681–687

Nakata K, Okuda T, Misawa H (2004) Ultrastructural localization of high-affinity choline transporter in the rat neuromuscular junction: enrichment on synaptic vesicles. Synapse 53:53–56

O'Regan S, Traiffort E, Ruat M, Cha N, Compaore D, Meunier FM (2000) An electric lobe suppressor for a yeast choline transport mutation belongs to a new family of transporter-like proteins. Proc Natl Acad Sci U S A 97:1835–1840

Okuda T, Haga T (2000) Functional characterization of the human high-affinity choline transporter. FEBS Lett 484:92–97

Okuda T, Haga T, Kanai Y, Endou H, Ishihara T, Katsura I (2000) Identification and characterization of the high-affinity choline transporter. Nat Neurosci 3:120–125

Perry E, Walker M, Grace J, Perry R (1999) Acetylcholine in mind: a neurotransmitter correlate of consciousness? Trends Neurosci 22:273–280

Pfeil U, Haberberger RV, Lips KS, Eberling L, Grau V, Kummer W (2003a) Expression of the high-affinity choline transporter CHT1 in epithelia. Life Sci 72:2087–2090

Pfeil U, Lips KS, Eberling L, Grau V, Haberberger RV, Kummer W (2003b) Expression of the high-affinity choline transporter, CHT1, in the rat trachea. Am J Respir Cell Mol Biol 28:473–477

Proskocil BJ, Sekhon HS, Jia Y, Savchenko V, Blakely RD, Lindstrom J, Spindel ER (2004) Acetylcholine is an autocrine or paracrine hormone synthesized and secreted by airway bronchial epithelial cells. Endocrinology 145:2498–2506

Rassadi S, Krishnaswamy A, Pie B, McConnell R, Jacob MH, Cooper E (2005) A null mutation for the alpha3 nicotinic acetylcholine (ACh) receptor gene abolishes fast synaptic activity in sympathetic ganglia and reveals that ACh output from developing preganglionic terminals is regulated in an activity-dependent retrograde manner. J Neurosci 25:8555–8566

Ribeiro FM, Alves-Silva J, Volknandt W, Martins-Silva C, Mahmud H, Wilhelm A, Gomez MV, Rylett RJ, Ferguson SS, Prado VF, Prado MA (2003) The hemicholinium-3 sensitive high affinity choline transporter is internalized by clathrin-mediated endocytosis and is present in endosomes and synaptic vesicles. J Neurochem 87:136–146

Ribeiro FM, Black SA, Cregan SP, Prado VF, Prado MA, Rylett RJ, Ferguson SS (2005) Constitutive high-affinity choline transporter endocytosis is determined by a carboxyl-terminal tail dileucine motif. J Neurochem 94:86–96

Saltarelli MD, Yamada K, Coyle JT (1990) Phospholipase A2 and 3H-hemicholinium-3 binding sites in rat brain: a potential second-messenger role for fatty acids in the regulation of high-affinity choline uptake. J Neurosci 10:62–72

Sandberg K, Coyle JT (1985) Characterization of [3H]hemicholinium-3 binding associated with neuronal choline uptake sites in rat brain membranes. Brain Res 348:321–330

Sarter M, Bruno JP (1997) Cognitive functions of cortical acetylcholine: toward a unifying hypothesis. Brain Res Brain Res Rev 23:28–46

Schueler FW (1955) A new group of respiratory paralyzants. I. The "hemicholiniums". J Pharmacol Exp Ther 115:127–143

Simon JR, Kuhar MG (1975) Impulse-flow regulation of high affinity choline uptake in brain cholinergic nerve terminals. Nature 255:162–163

Simon JR, Kuhar MJ (1976) High affinity choline uptake: ionic and energy requirements. J Neurochem 27:93–99

Svedberg MM, Svensson AL, Bednar I, Nordberg A (2003) Neuronal nicotinic and muscarinic receptor subtypes at different ages of transgenic mice overexpressing human acetylcholinesterase. Neurosci Lett 340:148–152

Thomsen M, Woldbye DP, Wortwein G, Fink-Jensen A, Wess J, Caine SB (2005) Reduced cocaine self-administration in muscarinic M5 acetylcholine receptor-deficient mice. J Neurosci 25:8141–8149

Tucek S (1985) Regulation of acetylcholine synthesis in the brain. J Neurochem 44:11–24

Vickroy TW, Roeske WR, Yamamura HI (1984) Sodium-dependent high-affinity binding of [3H]hemicholinium-3 in the rat brain: a potentially selective marker for presynaptic cholinergic sites. Life Sci 35:2335–2343

Vickroy TW, Roeske WR, Gehlert DR, Wamsley JK, Yamamura HI (1985) Quantitative light microscopic autoradiography of [3H]hemicholinium-3 binding sites in the rat central nervous system: a novel biochemical marker for mapping the distribution of cholinergic nerve terminals. Brain Res 329:368–373

Volpicelli-Daley LA, Duysen EG, Lockridge O, Levey AI (2003a) Altered hippocampal muscarinic receptors in acetylcholinesterase-deficient mice. Ann Neurol 53:788–796

Volpicelli-Daley LA, Hrabovska A, Duysen EG, Ferguson SM, Blakely RD, Lockridge O, Levey AI (2003b) Altered striatal function and muscarinic cholinergic receptors in acetylcholinesterase knockout mice. Mol Pharmacol 64:1309–1316

Wenk G, Hepler D, Olton D (1984) Behavior alters the uptake of [3H]choline into acetylcholinergic neurons of the nucleus basalis magnocellularis and medial septal area. Behav Brain Res 13:129–138

Wess J (2004) Muscarinic acetylcholine receptor knockout mice: novel phenotypes and clinical implications. Annu Rev Pharmacol Toxicol 44:423–450

Winkler J, Suhr ST, Gage FH, Thal LJ, Fisher LJ (1995) Essential role of neocortical acetylcholine in spatial memory. Nature 375:484–487

Xie W, Stribley JA, Chatonnet A, Wilder PJ, Rizzino A, McComb RD, Taylor P, Hinrichs SH, Lockridge O (2000) Postnatal developmental delay and supersensitivity to organophosphate in gene-targeted mice lacking acetylcholinesterase. J Pharmacol Exp Ther 293:896–902

Yamamura HI, Snyder SH (1972) Choline: high-affinity uptake by rat brain synaptosomes. Science 178:626–628

Yamamura HI, Snyder SH (1973) High affinity transport of choline into synaptosomes of rat brain. J Neurochem 21:1355–1374

Zhu CB, Hewlett WA, Feoktistov I, Biaggioni I, Blakely RD (2004) Adenosine receptor, protein kinase G, and p38 mitogen-activated protein kinase-dependent up-regulation of serotonin transporters involves both transporter trafficking and activation. Mol Pharmacol 65:1462–1474

Subject Index

ρ 30
1-methyl-4-phenylpyridinium 200
10-repeat 330, 331, 340–342, 349
3-kinase 265
5-HT 60–63, 66–70, 158
5-HTTLPR 332, 344–346, 348–351
5HTT gene-linked polymorphic region 419
$5HT_{1A}$ receptor in anxiety and depression 428
$5Ht_{1a}$ receptor 422
$5Ht_{1b}$ receptor 423
$5Ht_{2a}$ receptor 423
$5Ht_{2c}$ receptor 423
$5Ht_3$ receptor 423
5Htt KO mice 421
6-hydroxydopamine 200, 216
9-repeat 340, 342

abuse 328, 341–343, 352
acetylcholine 174, 526, 528
– biological functions 526
– in $\alpha^{-/-}$ mice 537
– in $ChAT^{+/-}$ mice 536
– in $CHT^{-/-}$ mice 534
– turnover 528
acetylcholinesterase 526
– in AChE-Tg mice 535
– knockout ($AChE^{-/-}$) mice 534
– transgenic (AChE-Tg) mice 535
acetyltransferase
– ACh turnover 526
adaptive changes in 5HT neurotransmission 422
addiction 328
ADHD 328, 330, 340–342, 345, 346, 350, 352, 353, 355, 356
adrenaline 158
affective spectrum disorders 439

affinity 339, 347, 354, 355
aggression 343, 345, 346, 350
agmatine 166, 173
Akt 265
alcoholism 341, 342, 344, 349
alternate access model 96
alternating access 190
Alzheimer's disease 346
amacrine cells 145
amperometry 219
amphetamine 60, 61, 67, 105, 199, 200, 206–208, 215, 328, 330, 339, 357
amphetamine (ADHD) 401
amphetamine (AMPH) 377, 386–390
amphetamines 343, 352
amyotrophic lateral sclerosis 278
analgesia 477
aNET substrates
– mphetamine 489
angiotensin-converting enzyme 87
anorexia nervosa 346, 351–353, 356
antidepressants 67, 352, 488, 491–493, 495, 502, 504, 509–511
– Celexa 60
– imipramine 60, 67
– Paxil 60
– Prozac 60
– Zoloft 60
anxiety 343–345, 350
anxiety- and depression-like behavior 426
arachidonic acid 293
arbitrary fluorescent units (AFUs) 32
ASP^+ 32
astrocytes 142, 277
autism 343, 346, 350, 351
autoreceptors 225

basal ganglia 225
biotin 260
biotinylated 261
bipolar disorder 341, 342, 346, 356, 399, 401–403
bone growth 437
botulinum toxin 185, 186, 188, 189, 192
brain and NET 492
brain region 329, 342, 343, 348, 349, 357
– caudate 328, 329
– frontal cortex 329
– nucleus accumbens 328, 329
– prefrontal cortex 329
– putamen 328, 329
– substantia nigra 328, 329
– ventral tegmental area 328, 429
brain-derived neurotrophic factor 433
brown adipose tissue (BAT) 143

C6 glioma 261
calcium 221
cancer therapy 511
carboxypeptidase A 87
carnitine transporter 156
caudate putamen 216
caveolae 288
CB1 cannabinoid receptor 144
cell line 381, 383
– C6 glioma 385
– COS-7 381, 383, 385, 386
– HEK-293 381, 383–388
– LLC-PK1 385
– N2A 385, 387
– Neuro2A 383
– PAE 385
– PC12 385
– Sf9 385
– SH-SY5Y 383, 386
– SK-N-SH 381
CFTR 87
chaotropes 86
chloride 139, 219
chloride channel 119, 122, 127–130
– anion conductance 116
– chloride leak conductance 116
– substrate-activated chloride conductance 117
– uncoupled anion conductances 117
cholecystokinin$^+$ 144

choline 162, 173
choline acetyltransferase 526
– ACh turnover 526
– heterozygous $CHT^{+/-}$ mice 536
– homozygous ($CHAT^{+/-}$) mice 534
– homozygous ($CHAT^{-/-}$) mice 534
choline acetyltransferase (ChAT) 144
choline transporter 526, 528, 529
– catalytic activation 529, 531
– endocytosis 531
– expression 530
– heterozygous ($CHT^{+/-}$) mice 535, 536, 538, 539
– high-affinity choline uptake 526–528, 530
– homozygous ($CHT^{-/-}$) mice 534
– in $\alpha^{-/-}$ mice 537
– in AChE-Tg mice 535
– in $AChE^{-/-}$ mice 534
– in $ChAT^{+/-}$ mice 536
– phosphorylation 529, 531
– rate-limiting step in ACh synthesis 526
– subcellular localization 530
– trafficking 529–531
choline transporters
– ACh synthesis, rate-limiting step of 528
cimetidine 158
clearance 156
cocaine 60, 61, 66–69, 199, 207–209, 215, 328, 330, 339, 342, 343, 352, 356, 357, 387, 390, 402, 486, 488, 491, 493, 495, 499, 502, 509–511
conducting states 191
confocal microscopy 33
conformational change 61–64, 68–70
contextual fear conditioning 267
cotransport 61, 67
countertransport 62
creatinine 166
cysteine 63–65, 67–70
cysteine-scanning mutagenesis 63
cystic fibrosis 185

DAT 60, 61, 66, 67
delivery 263
dentate gyrus 431
depression 182, 328, 343–345, 349–351, 353, 356

Subject Index

desensitization 182
dicarboxylate amino acid cation symporter family (DAACS) 2
dicarboxylate amino acid cation transporter (DAACS) 5, 11
diffusion 221
disease 277
dopamine 60, 158, 307, 309, 313
– clearance 198
– efflux 199, 207–209
– transport 199
dopamine transporter 200, 205
– basal phosphorylation 201, 203
– binding partners 209
– dephosphorylation 202, 203
– downregulation 201, 205–208
– endocytosis 205, 206
– phosphorylation sites 200, 204
dopamine transporter (DAT) 187
dopamine transporters (DAT) 76–78, 87
drug abuse 341
DSM-IV 374, 375

EAAC1 188, 252, 280
EAAT 114, 116, 252
– ASCT1 114, 116, 124
– ASCT2 114, 116
– EAAC1 114, 124
– EAAT1 120, 122, 127
– EAAT1–5 114
– EAAT2 122
– EAAT4 127
– GLAST1 114
– GLT1 114, 120, 124
– Gltph 117
EAAT1 280
EAAT2 253, 280
EAAT2b 280
EAAT3 253, 280
EAAT4 253, 254, 280
EAAT5 253, 280
ecstasy 61, 356
efflux 182, 384, 386
Electrophysiological Background 98
– steady-state current 98
electrophysiology 424
EMT 152, 173
– mutation 154

– transporter activation 173
endocytosis 263, 292
endoplasmic reticulum 237
ergothioneine transporter 156
ETT 156
evans blue 145
excitatory amino acid transporter (EAAT) 2–4, 6, 7
– excitatory amino acid transporter 1 (EAAT1) 5–7, 9, 10, 17
– excitatory amino acid transporter 2 (EAAT2) 5, 9, 10
– excitatory amino acid transporter 4 (EAAT4) 5–7, 10, 17
– Glt$_{Ph}$ 4
– GLT-1 (EAAT2) 2
excitotoxicity 182, 253, 277
extracellular loop 63
extraneuronal monoamine transporter (EMT) 152

facilitated exchange diffusion hypothesis 105
fluorescence correlation spectroscopy (FCS) 45
fluorescence lifetime imaging microscopy (FLIM) 42
fluorescence recovery after photobleaching (FRAP) 48
functional magnetic resonance imaging 445

GABA 60, 70
GABA transporter 185
– regulation 185
GABA transporter (GAT1) 76–78
GLAST 252, 280
GLT-1 252
GLT1 280
GLT1b 280
glucocorticoid-inducible kinase 267
glucose transporter (GLUT1) 89
glucose transporters 185
glutamate 251, 306–308, 313, 314, 321
– glutamate uptake 317
– glutamatergic synapse 309
– glutamatergic terminals 312
– glutamatergic transmission 308
– glutamatergic vesicles 309, 314, 321

- homeostasis 281
- spillover 288
- uptake 279
glutamate binding site 122, 124, 126, 131
- l-serine-O-sulfate 122
- 3-methylglutamate 122
- 4-methylglutamate 122
- kainate 122
glutamate transporter (GLT-1) 81, 82
glutamate transporter homolog (Glt$_{Ph}$) 82, 83
glutamate transporters 187
glutamate–glutamine cycle 279
glutamatergic synapses 459, 469, 473, 474
glutathione detoxification 425
glycine encephalopathy 475
glycine receptor 458
glycine transporters 186
- GLYT1 186
- GLYT2 186
glycine uptake 459, 462, 468, 471, 474
- inhibitors 472, 476
- kinetic analyses 462
- stoichiometry 462
glycinergic synapse 470
glycinergic synapses 464, 467, 469, 471, 473
glycophorin A 234
glycosylation 286, 380, 383, 384
GlyT 460
- development 466
- interaction partners 464
- membrane insertion 463
- N-glycosylation 463
- permeation pathway 462
- phorbol ester 464
- polytopic 460
GlyT1 gene 460
- GlyT1 mRNA 467
GlyT2 gene 460
- transcriptional regulation 466
growth factor 282
guanidine 166

H89 267
haloperidol 225
halorhodopsin 87
haplotype 381

head trauma 253
hemicholinium-3 527
- binding 528, 529, 535
- cholinergic neurotransmission 528
- hypersensitive to CHT$_{+/-}$ mice 538
heterologous 5HT clearance 425
histamine 166, 173
HMGN 467
hNET and disease 502
- addiction 509
- ADHD 502, 507
- anorexia nervosa 506
- depression 508
- dysautonomia 504
- hypertension 505
- myocardial ischemia 505
- obesity 506
- orthostatic intolerance 503–505
- pain 510
hNET and PET 511, 512
hNET and SPECT 511
hNET gene 487, 501
hNET mutants 499
hNET promoter 487, 501, 502, 505, 507, 509
hNET SNPs 502, 503, 505, 508, 509
hNET splicing 500, 501
hNET variant 487, 498, 500, 501, 504
Hofmeister series 86
Hsc70 ATPase 85, 87
human 5HTT gene 419
hyperekplexia 475
hypertension 345, 346, 350, 352, 353, 356
hypoglossal motoneurons 469
hypoglycemia 253

immunofluorescence 223
inhibitory G protein 226
intermediolateral cell column (IML) 143
internalization 288, 496
intracellular loop 65
ion binding site 122, 131
- H$^+$ 116, 124
- K$^+$ 116, 124, 131
- Na$^+$ 116, 124, 127, 131
- Zn^{2+} 127
ion channels 184
- calcium channels 184

– CFTR   184
– K⁺ channels   185
– Na⁺ channels   185
ion gradients   60
IPSCs   469, 470
irritable bowel syndrome   435

KCC2   458
knockout   139
knockout mice   217
kosmotropes   86

LacS transporter   89
lactose permease (LacY)   85, 87
lateral superior olive (LSO)   146
leak current   102, 103
leucine heptad repeat   234
leucine transporter   76, 89
LeuT   62–70
lipid rafts   287
Long-Evans rat   225
long-term potentiation   267

MAP kinase   187
MAPK   386
mazindol   223
MDMA   328
medial nucleus of the trapezoid body (MNTB)   146
melibiose carrier (melB)   84, 85
membrane potential   60
messenger RNA   224
meta-analysis   396
metformin   160
methamphetamine   199, 200, 206, 207, 222, 328, 342
methylphenidate   218
MIBG   488, 489, 511
mitogen-activated protein kinase (MAPK)   385
Morris water maze   474
motor neurons   140
MPP⁺   151
multiple sclerosis   436
Munc-18   185, 186, 189
muscarinic   526
muscarinic receptors   528
– drug challenge   528, 536, 538
– drug challenges   534
– in AChE-Tg mice   535

– in AChE⁻/⁻ mice   534
– in CHT⁺/⁻ mice   539
muscle relaxation   477
mutation   401–403

N-isothiocyanatophenethyl spiperone   226
Na⁺/Cl⁻-dependent transporter   2–4, 6, 7
– dopamine transporter (DAT)   3, 5–8, 11–15, 17, 18
– GABA transporter 1 (GAT1)   5, 7, 8
– GABA transporter 2 (GAT2)   6, 7
– GABA transporter 3 (GAT3)   5, 7, 8
– GABA transporter 4 (GAT4)   5–8, 11, 17
– glycine transporter 1b (GLYT1b)   5–8, 11
– glycine transporter 2a (GLYT2a)   5, 7
– LeuT$_{Aa}$ 3, 4, 8, 9, 13, 14, 16
– norepinephrine transporter (NET)   5, 8, 14
– serotonin transporter (SERT)   5, 8, 14
Na⁺-dependent transport systems   251
NE transport   487, 488, 496, 497, 499, 503
NE transporters (NET)   376
neocortical development and plasticity   430
NET   60, 61, 67, 70
NET chimeras   499
NET expression   493
NET inhibitor
– nisoxetine   510
– reboxetine   510
NET inhibitors   488, 491, 494, 507–509, 511
– atomoxetine   488, 491, 508, 511
– desipramine   487, 488, 491, 494, 495, 499, 505, 506, 508
– DMI   487, 505
– duloxetine   510
– nisoxetine   487, 488, 491, 494–496, 501, 508
– reboxetine   488, 491, 493, 505, 506, 512
– sibutramine   488, 491, 506, 511
NET mRNA   493–497, 509
NET regulation   494, 495
NET SNPs   507

NET structure–function 497
NET substrate 490
– co-substrate 490
– NE 490, 495
NET substrates 488, 511
– amphetamine 488, 490, 491, 507–509, 511
– amphetamines 502
– DA 488, 492, 493, 499
– dopamine 488, 508, 509
– ED 493
– EPI 486, 488
– epinephrine 486, 488
– NE 486–490, 492, 496, 497, 499, 501–506, 508, 510
– norepinephrine 486, 488
NET topology 498
NET transport 489
– co-substrates 489
– driving force 489
– outward transport 489, 490
NET transport model 490
NET-KO 487, 492, 493, 507, 510
– adrenoceptor 492
– blood pressure 493
– locomotor responses 493
neurodegeneration 281
neurogenesis 431
neuropathic pain 437
neuroprotection 288
neurotoxins 200
neurotransmitter 59–62, 137
neurotransmitter sodium symporter family (NSS) 2, 5, 11
neurotransmitter transporter 2, 4, 6, 7, 11, 16
BNF-κB 282
nicotinic 526
nicotinic receptors
– in α$^{-/-}$ mice 537
– in AChE-Tg mice 535
NMDA receptor 186
NMDAR 459, 473, 474
– compensatory changes 474
– glycine binding site 459
– priming 473
– regulators 473
– role of GlyT1 474
NO synthesis 474
nomifensine 223

non-neuronal monoamine transporter 152
non-specific uptake 155
noradrenaline 158, 307, 317, 318
norepinephrine 60, 70
norepinephrine transporter (NET) 187, 378–380, 383, 389, 393
NSF 183
NSS 60, 62–64, 67, 70
nucleus accumbens 216

OCD 343, 347, 351
OCT1 152
– mutation 154
OCT1h 154
OCT2 152
OCT2h 154
OCT3 152
oligomer formation 234
oligomer-based counter-transport model 244
oligomerization hypothesis 240
oocyte expression system 157
– artefacts 157
organic cation transporter 152
organic cation transporters 425
orthostatic intolerance 353, 354, 356

pain 477
Parkinson's disease 328, 342
patch clamp 219
PC12 221
PDGF 266
PDZ domain 186
peptide mapping 204
permeation 66
permeation pathway 65, 66, 69
pertussis toxin 226
pheochromocytoma 511
phorbol ester 385, 402
phosphatidylinositol 265
phosphatidylinositol 3-kinase 266
phosphoamino acid 204, 205
phosphoinositol 3-kinase (PI3-K) 385, 386
phospholamban 234
phosphorylation 289, 380, 385, 386
photomultiplier tube (PMT) 33
physiological importance 492
PI3K 282

pimozide to enhance   225
PKA   283
PKC   200, 202, 264, 386, 491, 496, 501
PKCα   265
platelet-derived growth factor   265
plethysmographic recordings   468
point spread function (PSF)   34
polymorphism   381, 395, 398, 400, 403
prefrontal cortex   216
prefrontal cortex (PFC)   376, 378, 393
prepulse inhibition   474
primary pulmonary hypertension   435
promoter   333, 340, 342, 344, 349, 350, 352, 356
protein kinase A   185, 267
protein kinase A (PKA)   385, 386
protein kinase C   185, 187, 188, 222, 263
protein kinase C (PKC)   385
psychostimulants   377–380, 387, 389, 390
Purkinje cells   146, 254

quality control   237
quantitative fluorescence microscopy   32
quantum yield   42
quinpirole   225

rab3a   188
raphe magnus (RMg)   143
receptor   59, 385, 388
– D1 dopamine   388
– D2 dopamine   388, 391–393
– D3 dopamine   388
– D4 dopamine   396
receptors   526
recycling   263
response inhibition   376, 377
reuptake   343
reverse transport   385, 491
rhodanese   85
rostral medullary raphe pallidus (RPa)   143
rostral ventral lateral medulla (RVLM)   140

sarcosine   460, 471, 476
schizophrenia   328, 341, 342, 346, 350, 356, 476

serotonergic raphe neurons   426
serotonin   187, 307, 310, 314, 315, 317, 318
– monoaminergic terminals   309
– serotonergic terminals   312
– serotonin uptake   317
serotonin transporter   187, 191
serotonin transporter (SERT)   87–89
serotonin transporters (SERT)   76, 77
SERT   60–70
serum   267
shortened delay gradient   376
sialin   139
single nucleotide polymorphism   399
single nucleotide polymorphisms   381
single nucleotide polymorphisms (SNPs)   419
single-file model   101
SKN-SH cells   487, 496, 502, 506, 507
SLC22A1   153
SLC22A2   153
SLC22A3   153
SLC22A4   156
SLC22A5   156
smoking   341, 346, 350, 351
SNAP   183
SNAP 25   183
SNARE hypothesis   183
sodium   219
sodium glucose cotransporter (SGLT1)   84
sodium/proline transporter (PutP)   83, 84
somatosensory cortex   430
specific uptake   155
spontaneous firing rate of DRN 5HT neurons   424
STin2   332, 344–346, 349–351
stoichiometry   61, 62, 67, 192
striatal synaptosome   386
striatal synaptosomes   387, 388
striatum   376, 378, 385, 388, 389, 392, 393
stroke   253
substances of abuse   437
substantia nigra   216
substrate specificity   175
suicide   343–345, 349
sulfhydryl reagents   120, 125, 129
– MTSEA   120

– MTSES 120, 129
– MTSET 120, 129
– N-ethylmaleimide 126
superior cervical ganglia neurons 40
SV2 188
synaptic strength 306
synaptic vesicle 60, 306, 309
synaptic vesicles 218
synaptophysin 188
synaptosomes 224
synprint 184
syntaxin 62, 70, 183
syntaxin 1A 187, 189
– antisense 185, 187
– helical domain 189
– helical domains 183
system 425

TCI/TPQ 346
TEA 173
tetraethylammonium (TEA) 152
thalamocortical neurons 192
thermogenesis 143
thermolysin 86
TIRF microscopy 40
TM 65–67
TNF-α 282
trafficking 217, 262, 495, 496, 500, 501, 504
tranporter stucture
– crystal structure 120
– topology 120
transcription factor 282
transient currents 101
translocation 219
transmembrane domain 62, 70
– TM 63, 64
transport efficiency 159
transport efficiency (TE) 155
transporter structure
– bacterial glutamate transporter 117

– crystal structure 117, 126
– hairpin loops 117
– topology 119
transporters 96
turnover rates 189
two-photon excitation 33
tyramine 158, 173

uptake 334, 339, 347–350, 354, 355

VAMP 183, 188, 189
ventral tegmental area 216
vesicle depletion 222
vesicular GABA transporter (VGAT) 145
VGAT 188
VGLUT1 139

weak base *see* vesicle depletion model
weak base hypothesis 106
WIN35428 224
Wistar rat 225
working memory 376

Xenopus laevis oocyte 157, 184, 185, 187, 189, 219

Zn^{2+} 4–18
– neuromodulator 4
– synaptic Zn^{2+} 16–18
– vesicular Zn^{2+}-transporter 16, 17
– vesicular Zn^{2+}-transporter (ZnT-3) 16, 17
– Zn^{2+} inhibition 7
Zn^{2+}-binding site 4, 5, 7, 8, 11
– artificial Zn^{2+}-binding site 4
– endogenous Zn^{2+}-binding site 4, 7, 8, 11, 16
– Zn^{2+} coordinating residues (coordinating histidines) 7–10
– Zn^{2+}-binding residues 8